Criminological Theory

Titles of Related Interest

Criminological Theory

A Text/Reader

Stephen G. Tibbetts
California State University, San Bernardino

Craig Hemmens
Boise State University

SAGE

Los Angeles | London | New Delhi
Singapore | Washington DC

For information:

SAGE Publications, Inc.
2455 Teller Road
Thousand Oaks, California 91320
E-mail: order@sagepub.com

SAGE Publications Ltd.
1 Oliver's Yard
55 City Road
London EC1Y 1SP
United Kingdom

SAGE Publications India Pvt. Ltd.
B 1/I 1 Mohan Cooperative Industrial Area
Mathura Road, New Delhi 110 044
India

SAGE Publications Asia-Pacific Pte. Ltd.
33 Pekin Street #02-01
Far East Square
Singapore 048763
Printed in the United States of America

Library of Congress Cataloging-in-Publication Data

Tibbetts, Stephen G.
Criminological theory: a text/reader / Stephen G. Tibbetts, Craig Hemmens.
    p. cm.
Includes both collection of key articles on criminological theory that have previously appeared in a number of leading criminology/criminal justice scholarly journals, along with authored textual material that serves to explain and synthesize the readings.
Includes bibliographical references and index.
ISBN 978-1-4129-5037-4 (pbk.)
    1. Criminology. 2. Crime. I. Hemmens, Craig. II. Title.

HV6018.T53 2010
364.01—dc22   2009026432

This book is printed on acid-free paper.

12   13   10   9   8   7   6   5   4   3

Acquisitions Editor:	Jerry Westby
Associate Editor	Lindsay Dutro
Editorial Assistant:	Eve Oettinger
Production Editor:	Catherine M. Chilton
Copy Editor:	Jacqueline Tasch
Typesetter:	C&M Digitals (P) Ltd.
Proofreader:	Doris Hus
Indexer:	Hyde Park Publishing Services LLC
Cover Designer:	Edgar Abarca
Marketing Manager:	Christy Guilbault

Brief Contents

Detailed Contents

READINGS   267

Agnew revises and updates his general strain theory in light of empirical evidence that is both supporting and not supporting of the theory. Acknowledging the scientific research regarding a given theory and revising a theory accordingly promotes good theoretical development, and this piece provides a good example of this process.

Section VII. Social Process/Learning Theories of Crime   437

INTRODUCTION   438

READINGS   471

> Examines modern policy implications and the frameworks
> on which such policies are based, as well as evaluations
> of their effectiveness.

Foreword

The Criminal Justice System

Craig Hemmens

You hold in your hands a book that we think is something new. It is billed a "text/reader." What that means is we have attempted to take the two most commonly used types of books, the textbook and the reader, and blend the two in a way that will appeal to both students and faculty.

Our experience as teachers and scholars has been that textbooks for the core classes in criminal justice (or any other social science discipline) leave many students and professors cold. The textbooks are huge, crammed with photographs, charts, highlighted material, and all sorts of pedagogical devices intended to increase student interest. Too often, however, these books end up creating a sort of sensory overload for students and suffer from a focus on "bells and whistles" such as fancy graphics at the expense of coverage of the most current research on the subject matter.

Readers, on the other hand, typically include recent and classic research articles on the subject matter. They generally suffer, however, from an absence of meaningful explanatory material. Articles are simply lined up and presented to the students, with little or no context or explanation. Students, particularly undergraduate students, are often confused and overwhelmed.

This text/reader represents our attempt to take the best of both the textbook and reader approaches to criminological theory. It can serve either as a supplement to a core textbook or as a stand-alone text. The book includes a combination of previously published articles and textual material introducing these articles and providing some structure and context for the selected readings. The book is broken up into a number of sections. The sections of the book track the typical content and structure of a textbook on the subject. Each section of the book has an introduction that serves to explain and provide context for the readings that follow. The readings are a selection of the best recent research that has appeared in academic journals, as well as some classic readings. The articles are edited as necessary to make them accessible to students. This variety of research and perspectives will provide the student with a grasp of the development of research, as well as an understanding of the current status of research in the subject area. This approach gives the student the opportunity to learn the basics (in the text portion of each section) and to read some of the most interesting research on the subject.

An introductory chapter explains the organization and content of the book and provides a context and framework for the text and articles that follow, as well as introducing relevant themes, issues, and concepts. This will assist the student in understanding the articles.

Each section will include a summary of the material covered and some discussion questions. These summaries and discussion questions should facilitate student thought and class discussion of the material.

It is our belief that this method of presenting the material will be more interesting for both students and faculty. We acknowledge that this approach may be viewed by some as more challenging than the traditional textbook. To that we say: Yes! It is! But we believe that if we raise the bar, our students will rise to the challenge. Research shows that students and faculty often find textbooks boring to read. We believe that many criminal justice instructors would welcome the opportunity to teach without having to rely on a "standard" textbook that covers only the most basic information and that lacks both depth of coverage and an attention to current research. This book provides an alternative for instructors who want to get more out of the basic criminal justice courses/curriculum than one can get from a basic textbook that is aimed at the lowest common denominator and filled with flashy but often useless features that merely serve to drive up the cost of the textbook. This book is intended for instructors who want to go beyond the ordinary, basic coverage provided in textbooks.

We also believe students will find this approach more interesting. They are given the opportunity to read current, cutting-edge research on the subject, while also being provided with background and context for this research. In addition to including the most topical and relevant research, we have included a short entry, "How to Read a Research Article." The purpose of this piece, which is placed at the beginning of the book, is to provide students with an overview of the components of a research article. It also serves to help walk them through the process of reading a research article, lessening their trepidation and increasing their ability to comprehend the material presented therein. Many students will be unfamiliar with reading and deciphering research articles; we hope this feature will help them to do so. In addition, we provide a student study site on the Internet. This site has additional research articles, study questions, practice quizzes, and other pedagogical material that will assist students in the process of learning the material. We chose to put these pedagogical tools on a companion study site rather than in the text to allow instructors to focus on the material, while still giving students the opportunity to learn more.

We hope that this unconventional approach will be more interesting, and thus make learning and teaching more fun. Criminal justice is a fascinating subject, and the topic deserves to be presented in an interesting manner. We hope you will agree.

Preface

A number of excellent criminology theory textbooks and readers are available to students and professors, so why this one? The reason is that stand-alone textbooks and readers (often assigned as an expensive addition to a textbook) have a pedagogical fault that we seek to rectify with the present book. Textbooks focus on providing a broad overview of the topic area and lack depth and a focus on current research, whereas readers often feature in-depth articles about a single topic with little or no text to describe the history and context in which each model was proposed, or text to unify the readings, and little in the way of pedagogy. This book provides more in the way of text and pedagogy and uses recent research-based articles to help students understand criminological theory. This book is unique in that it is a hybrid text/reader offering the best of both worlds. It includes a collection of articles on criminological theory that have previously appeared in a number of leading criminology journals along with original text that explains and synthesizes the readings. We have selected some of the best recent research and literature reviews and assembled them into this text/reader for an undergraduate or graduate theory class.

Journal articles and book chapters selected for inclusion have been selected based primarily on how they add to and complement the text and on how interesting we perceive them to be for students. Although some key seminal theoretical pieces are included among the previously published works, an emphasis was placed on selecting some of the best contemporary empirical studies that actually test the theories in each respective section. We believe this focus on including the most recent major empirical studies and scholarly works, most from leading criminology journals, regarding the various major theoretical models in criminology is one of the key distinguishing features of this text as compared to other books on criminological theory. In our opinion, these articles are the best contemporary work on the issues they address.

However, journal articles are written for professional audiences, not for students, and thus often contain quantitative material students are not expected to understand. They also often contain concepts and hair-splitting arguments over minutia that tend to turn students glassy-eyed. Mindful of this, the articles contained in this text/reader have been edited to make them as student friendly as possible. We have done this without doing injustice to the core points raised by the authors or detracting from the authors' key findings and conclusions. Those wishing to read these articles (and others) in their entirety are able to do so by accessing the Sage website provided for users of this book. Research-based quantitative articles are balanced by review/overview essay-type articles and qualitative articles providing subjective insight into criminal behavior from the points of view of offenders.

This book can serve as a supplemental reader or the primary text for an undergraduate course in criminological theory or as the primary text for a graduate course. In a graduate course, it would serve as both an introduction to the extant literature and a sourcebook for additional reading, as well as a springboard for enhanced class discussion. When used in an undergraduate course, this book can serve to provide greater depth than the standard textbook. It is important to note that the readings and the introductory texts provide a comprehensive survey of the current state of the existing scientific literature in virtually all areas of criminological theory, as well as giving a history of how we got to this point regarding each theoretical model and each topic area.

⊠ Structure of the Book

We use a rather typical outline for criminological theory textbook topics/sections, beginning with an introduction of the definitions of crime and criminology and measuring crime, as well as what such measures of crime reveal regarding the various characteristics that are most associated with higher offending rates. This is a very important aspect of the book because each theory or model must be judged by how well it explains the distribution of crime rates among these various characteristics. In the Introduction, we also discuss the criteria that are required for determining causality, including a discussion of how extremely difficult (often impossible) this is to do in criminological research, because we can't randomly assign individuals to bad parenting, unemployment, low IQ, and so on.

After the Introduction, we present chronologically the history and development of criminological theory, with an emphasis on when such perspectives became popular among theorists and mainstream society. Thus, we start with the earliest models (pre-classical and Classical School) of criminal theorizing of the 18th century. We then discuss the evolution of the Positive School perspective of the 19th century, which began with biological theories of crime. Then we present the various other positive theories that were proposed in the early 20th century, which include social structure models and social process theories that were presented in the early or mid-1900s. Then we examine theoretical models that were presented in the latter 20th century, such as social conflict, Marxist, and feminist models of criminality. We then present the more contemporary theoretical explanations of criminality, which include developmental and life-course models and integrated theories of crime. Finally, although policy implications are discussed at the end of each section introduction, we finish with a summary section that specifies the types of policies suggested by the various paradigms and recent scientific findings regarding each of the major theoretical models presented throughout the book.

This text/reader is divided into 11 sections that mirror the sections in a typical criminology textbook, each dealing with a particular type of subject matter in criminology. Thus, each of the section introductions concludes with an evaluation of the empirical support for the theories and the policy implications derivable from them. These sections are as follows:

Introduction

We first provide an introductory section dealing with what criminological theory is, as well as examining the concepts of crime and the criteria used to determine whether a theory is adequate for explaining behavior. This section introduces the facts and criteria by which all of the theoretical models presented in the following sections will be evaluated. We also include a discussion of the criteria involved in determining whether a given factor or variable actually causes criminal behavior.

I. Pre-Classical and Classical Theories of Crime

In this section, we examine the types of theories that were dominant before logical theories of crime were presented, namely supernatural or demonic theories of crime. Then we examine how the Age of Enlightenment led to more rational approaches for explaining criminal behavior, such as that of the Classical School and neoclassical theory. We also discuss the at length the major model that evolved from the Classical School, deterrence theory. We provide studies in this section that have put deterrence theory to empirical testing.

II. Modern Applications of the Classical Perspective: Deterrence, Rational Choice, and Routine Activities/Lifestyle Theories of Crime

In this section, we will review more contemporary theoretical models and empirical findings regarding explanations of crime that focus on deterrence, and other recent perspectives—such as rational choice theory, routine activities theory, and the lifestyle perspective—that are based on the assumption of individuals choosing their behavior or targets based on rational decisions. Some of these perspectives focus more on the individuals' choice of the perceived costs or benefits of a given act, whereas other models focus on the type of location that they choose to commit crime or their daily activities or lifestyles that predispose them to certain criminal behavior.

III. Early Positive School Perspectives of Criminality

This section will examine the early development of theoretical models that proposed that certain individuals or groups were predisposed to criminal offending. The earliest theories in the 19th century proposed that certain physical traits were associated with criminal behavior, whereas perspectives in the early 20th century proposed that such criminality was due to intelligence. This section also examines body-type theory, which proposes that the physical body type of an individual has an effect on criminality. We will also examine modern applications of this perspective and review the empirical support such theoretical models have received in modern times.

IV. Modern Biosocial Perspectives of Criminal Behavior

In this section, we will review the various forms of modern studies regarding the link between physiology and criminality. Such theoretical models include family studies, twin and adoption studies, cytogenetic studies, and studies on hormones and neurotransmitters. We will examine some of the primary methods used to examine this link as well as more rational and recent empirical studies, which show a relatively consistent link between physiological factors and criminal behavior.

V. Early Social Structure and Strain Theories of Crime

This section reviews the development of the social structure perspective, starting in the 19th century, and culminating with Merton's theory of strain in the early 20th century. A variety of perspectives based on Merton's strain theory will be examined, but all of these models have a primary emphasis on how the social structure produces criminal behavior. We will examine the many empirical studies that have tested the validity of these early social structure theories, as well as discussing policy implications that these models suggested.

VI. The Chicago School and Cultural/Subcultural Theories of Crime

In this section, we examine the evolution and propositions of the scholars at the University of Chicago, the most advanced form of criminological theorizing of the early 20th century. In addition to the evolution of the Chicago School and its application of ecological theory to criminal behavior, we will also examine the more modern applications of this theoretical framework for explaining criminal behavior among residents of certain neighborhoods. Finally, we will discuss several theoretical models that examine cultural or subcultural groups that differ drastically from conventional norms.

VII. Social Process/Learning Theories of Crime

This section examines the many perspectives that have proposed that criminal behavior is the result of being taught by significant others to commit crime. When it was first presented, it was considered quite novel. We will examine the evolution of various theories of social learning, starting with the earliest, which were based on somewhat outdated forms of learning theory, and then we will progress to more modern theories that incorporate contemporary learning models. We will also examine the most recent versions of this type of theoretical perspective, which incorporates all forms of social learning in explaining criminal behavior.

VIII. Social Reaction, Critical, and Feminist Models of Crime

In this section, we examine a large range of theories, with the common assumption that the reason for criminal behavior is factors outside of the traditional criminal justice system. Many social reaction theories, for example, are based on labeling theory, which proposes that it is not the individual offender who is to blame, but rather the societal reaction to such early antisocial behavior. Furthermore, this section will examine the critical perspective, which blames the existing legal and economical structure for most of the assigned "criminal" label that is used against most offenders. Also, we will discuss the major perspectives of criminal offending regarding females as compared to males, as well as their differential treatment by the formal criminal justice system. Finally, we look at how explaining low levels of female offending might be important for policies regarding males.

IX. Life-Course Perspectives of Criminality

This section will examine the various theoretical perspectives that emphasize the predisposition and influences that are present among individuals who begin committing crime at early ages versus later ages. We also examine the various stages of life that tend to have a high influence on an individual's state of criminality (e.g., marriage), as well as the empirical studies that have examined these types of transitions in life. Finally, we examine the various types of offenders and the types of transitions/trajectories that tend to influence their future behavior, along with various policy implications that can be suggested by such models of criminality.

X. Integrated Theoretical Models and New Perspectives of Crime

In this section, we present the general theoretical framework for integrated models. Then we introduce some criticisms of such integration of traditional theoretical models. In addition, we present several integrated models of criminality, some of which are based on micro-level factors and others that are based on macro-level factors. Finally, we examine the weaknesses and strengths of these various models, based on empirical studies that have tested their validity.

XI. Applying Criminological Theory to Policy

This final section will review the most recent empirical evidence regarding how the theoretical models and findings reviewed in this book can be used to inform policies to reduce criminal behavior among offenders. These studies show that many theoretical perspectives suggest some effective policy recommendations, whereas other theoretical frameworks have been shown to be less effective in terms of policy applications.

▧ Ancillaries

To enhance the use of this text/reader and to assist those using this book as a core text, we have developed high-quality ancillaries for instructors and students.

Instructor's Resource CD. A variety of instructor's materials are available. For each chapter, this includes summaries, PowerPoint slides, chapter activities, Web resources, and a complete set of test questions.

Student Study Site. This comprehensive student study site features chapter outlines students can print for class, flashcards, interactive quizzes, Web exercises, links to additional journal articles, links to Frontline videos, NPR and PBS radio shows, and more.

✎ Acknowledgments

We would first of all like to thank Executive Editor Jerry Westby. Jerry's faith in and commitment to the project are greatly appreciated, as are those of his very able developmental editor, Lindsay Dutro. They kept up a most useful three-way dialogue among authors, publisher, and a parade of excellent reviewers, making this text the best that it could possibly be. We would also like to thank Erin Conley, who wrote the guide, "How to Read a Research Article." Our copy editor, Jacqueline Tasch, spotted every errant comma, dangling participle, and missing reference in the manuscript, for which we are truly thankful. Thank you one and all.

Stephen Tibbetts also would like to thank the various individuals who helped him complete this book. First, he would like to thank the professors he had as an undergraduate at University of Florida, who first exposed him to criminological theory. These professors include Ronald Akers, and Lonn Lanza-Kaduce, with a special thanks to Donna Bishop, who was the instructor in his first criminological theory course. He would also like to thank the influential professors he had at the University of Maryland, including Denise Gottfredson, Colin Loftin, David McDowell, Lawrence Sherman, and Charles Wellford. He gives a very special acknowledgement to Raymond Paternoster, who was primary mentor and adviser and exerted an influence words can't describe. Ray and his wife, Ronet Bachman, provided rare support as surrogate parents when Tibbetts was in graduate school. Paternoster also introduced him to the passion for criminological theory that he hopes is reflected in this book.

Also at University of Maryland, he would like to thank Alex Piquero, with whom he had the great luck of sharing a graduate student office in the mid-1990s, and who is now there as a professor. Without the many collaborations and discussions about theory that he had with Alex, this book would be quite different and would likely not exist. Alex and his wife, Nicole Leeper Piquero, have been consistent key influences on Tibbetts's perspective and understanding of theories of crimes, especially contemporary perspectives (and both of the Piqueros' works are represented in this book).

In addition, Tibbetts would like to thank several colleagues who have helped him subsequent to his education. First, he would like to thank John Paul Wright at the University of Cincinnati and Chris Gibson at the University of Florida for inspiring him to further explore more biosocial and developmental areas of criminality. Also, he would like to thank Mary Schmidt and Joseph Schwartz at California State University, San Bernardino (CSUSB), who provided much help in the compilation of materials for this book. In addition, he would like to especially thank Pamela Schram and Larry Gaines, fellow professors at CSUSB, who have provided the highest possible level of support and guidance during his career. Furthermore, it should be noted that Pamela Schram provided key insights and materials that aided in the writing of several sections of this book.

Tibbetts owes the most gratitude to his wife, Kim, who patiently put up with him typing away for the past few years while working on this book. Her constant support and companionship are what keeps him going.

Both authors are grateful to the many reviewers who spent considerable time reading early drafts of their work and who provided helpful suggestions for improving both the textual material and the edited readings. Trying to please so many individuals is a challenge, but one that is ultimately satisfying and one that undoubtedly made the book better than it would otherwise have been. Heartfelt thanks to the following experts: Shannon Barton-Bellessa, Indiana State University; Michael L. Benson, University of Cincinnati; Robert Brame, University of South Carolina; Tammy Castle, University of West Florida; James Chriss, Cleveland State University; Toni DuPont-Morales, Pennsylvania State University; Joshua D. Freilich, John Jay College; Randy Gainey, Old Dominion University; Robert Hanser, Kaplan University; Heath Hoffman, College of Charleston; Thomas Holt, University of North Carolina, Charlotte; Rebecca Katz, Morehead State University; Dennis Longmire, Sam Houston State University; Gina Luby, DePaul University; Michael J. Lynch, University of South Florida; Michelle Hughes Miller, Southern Illinois University, Carbondale; J. Mitchell Miller, University of Texas, San Antonio; Travis Pratt, Arizona State University; Lois Presser, University of Tennessee; Robert Sarver, University of Texas, Arlington; Joseph Scimecca, George Mason University; Martin S. Schwartz, Ohio University; Ira Sommers, California State University, Los Angeles; Amy Thistlethwaite, Northern Kentucky University; Kimberly Tobin, Westfield State College; Michael Turner, University of North Carolina, Charlotte; Scott Vollum, James Madison University; Courtney Waid, North Dakota State University; and Barbara Warner, Georgia State University.

Dedication

Stephen G. Tibbetts dedicates this book to his daughter, Rian Sage, who has been really cool to hang out with watching Sponge-Bob over the last few years; Rian has been the best daughter anyone could ask for.

Craig Hemmens dedicates this book, all the books in this series, and everything of value he has ever done to his father, George Hemmens, who showed him the way; James Marquart and Rolando del Carmen, who taught him how; and Mary and Emily, for giving him something he loves even more than his work.

INTRODUCTION TO THE BOOK

An Overview of Issues in Criminological Theory

◪ Introduction

Welcome to the world of criminological theory! It is an exciting and complex endeavor that explains why certain individuals and groups commit crimes and why other people do not. This book will explore the conceptual history of this endeavor as well as current theories. Most of us can relate directly to many of these theories; we may know friends or family members who fit dominant models of criminal behavior.

This introduction begins by describing what criminology is; what distinguishes it from other perspectives of crime, such as religion, journalism, or philosophy; and how definitions of crime vary across time and place. Then, it examines some of the major issues used to classify different theories of criminology. After exploring the various paradigms and categories of criminological theory, we discuss what characteristics help to make a theory a good one—in criminology or in any scientific field. In addition, we review the specific criteria for proving causality—for showing what predictors or variables actually cause criminal behavior. We also explain why—for logistic and ethical reasons—few theories in criminology will ever meet the strict criteria required to prove that key factors actually cause criminal behavior. Finally, we look at the strengths and weaknesses of the various measures of crime, which are used to test the validity of all criminological theories, and what those measures reveal about how crime is distributed across various individuals and groups. Although the discussion of crime distribution, as shown by various measures of criminality, may seem removed from our primary discussion regarding theories of why certain individuals and groups commit more crime than others, nothing could be further from the truth. Ultimately, all theories of criminal behavior will be judged based on how much each theory can explain the observed rates of crime shown by the measures of criminality among individuals and groups.

What Is Criminology, and How Does It Differ From Other Examinations of Crime?

Criminology is the scientific study of crime, especially of why people commit crime. Although many textbooks have more complex definitions of crime, the word *scientific* separates our definition from other perspectives and examinations of crime.[1] Philosophical and legal examinations of crime are based on logic and deductive reasoning, for example, by developing propositions for what makes logical sense. Journalists play a vital role in examinations of crime: exploring what is happening in criminal justice and revealing injustices and new forms of crime; however, journalists tend to examine anecdotes or examples of crime, as opposed to examining objective measures of criminality.

Taken together, philosophical, legal, and journalistic perspectives of crime are not scientific because they do not involve the use of the **scientific method**. Specifically, they do not develop specific predictions, known scientifically as **hypotheses,** which are based on prior knowledge and studies, and then go out and test such predictions through observation. Criminology is based on this scientific method, whereas other examinations of crime are not.

Instead, philosophers and journalists tend to examine a specific case, make conclusions based on that one example of a crime incident, and then leave it at that. Experts in these nonscientific disciplines do not typically examine a multitude of stories that are similar to the one they are considering, nor do they apply the elements of their story to an existing theoretical framework that offers specific predictions or hypotheses and then test those predictions by further observation. The method of testing predictions through observation and then applying the findings to a larger body of knowledge, as established by theoretical models, is solely the domain of criminologists, and it separates criminology from other fields. The use of the scientific method is a distinguishing criterion for many studies of human behavior, such as psychology, economics, sociology, and anthropology, which is why these disciplines are generally classified as **social sciences;** criminology is one.

To look at another perspective on crime, religious accounts are almost entirely based on dogmatic, authoritarian, or reasoning principles, meaning that they are typically based on what some authority (e.g., the pope, the Bible, the Torah, or the Koran) had to say about the primary causes of crime and the best ways to deal with such violations. These

---

[1]Stephen Brown, Finn Esbensen, and Gilbert Geis, *Criminology,* 6th ed. (Cincinnati, OH: LexisNexis, 2007).

ideas are not based on observations. A science like criminology is based not on authority or anecdotes but on empirical research, even if that research is conducted by a 15-year-old who performs a methodologically sound study. In other words, the authority of the scientist performing the study does not matter; rather, the observed evidence is of utmost importance, not who is performing the research. Criminology is based on science, and its work is accomplished through direct observation and testing of hypotheses, even if those findings do not fit neatly into logical principles or the general feelings of the public.

What Is Theory?

Theory can be defined as a set of concepts linked together by a series of statements to explain why an event or phenomenon occurs. A simple way of thinking about theories is that they provide explanations of why the world works the way it does. In other words, a theory is a model of the phenomenon that is being discussed, which in this case is criminal behavior. Sometimes, perhaps quite often, theories are simply wrong, even if the predictions they give are highly accurate.

For example, in the early Middle Ages, most people, including expert scientists, believed the Earth was the center of the universe because everything seemed to rotate and revolve around our home planet. If we wake up day after day and see the sun (or moon) rise and set in close to the same place, it appears that these celestial bodies are revolving around the Earth, especially considering the fact that we don't feel the world around us moving. Furthermore, calendars predicting the change of seasons, as well as the location and phases of these celestial bodies (such as the moon's phases), were quite accurate. However, although the experts were able to predict the movements of celestial objects quite well and they developed extremely accurate calendars, they had absolutely no understanding of what was actually happening. Later, when some individuals tried to convince the majority that they were wrong, specifically that the Earth was not the center of the universe, they were condemned as heretics and persecuted, even though their theoretical models were correct.

The same type of argument could be made about the Earth being flat; at one time, observations and all existing models seemed to claim it as proven and true. Some disagreed and decided to test their own predictions, which is how America was discovered by European explorers. Still, many who believed the Earth was round were persecuted or outcast from the mainstream society in Europe at the time.

▲ **Photo I.1**   Theories of the Earth as the center of the universe were dominant for many centuries, and scientists who proposed that the Earth was not the center of the universe were often persecuted. Over time, the theory was proved false.

Two things should be clear, then: theories can be erroneous, and accurate predictions can be made (e.g., early calendars and moon/star charts) using them, even though there is no true understanding of what is actually happening. One way to address both of these issues is to base knowledge and theories on scientific observation and testing. All respected theories of crime in the modern era are based on science; thus, we try to avoid buying into and applying theories that are inaccurate, and we continuously refine and improve our theories (based on findings from scientific testing) to gain a better understanding of what causes people to commit crime. Criminology, as a science, always allows and even welcomes criticism to its existing theoretical models. There is no emphasis on authority but rather on the scientific method and the quality of the observations that take place in testing the predictions. All scientific theories can be improved, and they are improved only through observation and empirical testing.

What Is Crime?

Definitions of crime vary drastically. For example, some take a **legalistic approach** toward defining crime, including only acts that are specifically prohibited in the legal codes of a given jurisdiction. The problem with such a definition is that what is a crime in one jurisdiction is not necessarily a crime in other jurisdictions. To clarify, some acts, such as murder and armed robbery, are against the law in virtually all countries and all regions of the United States, across time and culture. These are known as acts of *mala in se*, literally meaning "evil in itself."[2] Typically, these crimes involve serious violence and shock the society in which they occur.

Other crimes are known as acts of *mala prohibita*, which has the literal meaning of "evil because prohibited." This acknowledges that these are not inherently evil acts; they are bad only because the law says so.[3] A good example is prostitution, which is illegal in most of the United States but is quite legal and even licensed in most counties of Nevada. The same can be said about gambling and drug possession or use. These are just some examples of acts that are criminal in certain places or at certain times and thus are not agreed upon by most members of a given community.

This book examines both *mala in se* and *mala prohibita* types of offenses, as well as other acts of **deviance**, which are not against the law in many places but are statistically atypical and may be considered more immoral than illegal. For example, in Nevada in the 1990s, a young man watched his friend (who was later criminally prosecuted) kill a young girl in the bathroom at a casino, but he told no one. Although most people would claim that this was highly immoral, at that time, the Nevada state laws did not require people who witnessed a killing to report it to authorities (Note: As a result of this event, Nevada made withholding such information a criminal act). Therefore, this act was *deviant* because most people would find it immoral, but it was not criminal because it was not technically against the laws in the jurisdiction at that time.

Other acts of deviance are not necessarily immoral but are certainly statistically unusual and violate social norms, such as purposely "farting" at a formal dinner. Such activities are relevant for our discussion, even if they are not defined as criminal by the

---

[2]Ibid.

[3]Ibid.

law, because they show a disposition toward antisocial behavior often found in individuals who are likely to become criminal offenders. Furthermore, some acts are moving from deviant to illegal all the time, such as the use of cell phones while driving or smoking cigarettes in public; many jurisdictions are moving to have these behaviors made illegal and have been quite successful to date, especially in New York and California.

Most *mala in se* activities (e.g., murder) are highly deviant, too, meaning they are not typically found in society, but many, if not most, *mala prohibita* acts—say, speeding on a highway—are not deviant because they are committed by most people at some point. This is a good example of a *mala prohibita* act that is illegal but not deviant. This book will examine theories for all of these types of activities, even those that do not violate the law at the present time in a given jurisdiction.

How Are Criminological Theories Classified?
The Major Theoretical Paradigms

Scientific theories of crime can be categorized based on several important concepts, assumptions, and characteristics. To begin, most criminological theories are classified by the **paradigm** they emphasize. Paradigms are distinctive theoretical models or perspectives; in the case of crime, they vary based largely on opposing assumptions of human behavior. There are four major paradigms.[4]

The first of these are deterrence/rational choice theories, commonly referred to as the **Classical School** perspective, which we will discuss at length later in this book. It assumes that individuals have free will and choose to commit crimes based on rational, hedonistic decisions; they weigh out the potential costs and benefits of offending and then choose what will maximize their pleasure and minimize their pain. The distinguishing characteristic of these theories is that they emphasize the free choice that individuals have in committing crime. The other paradigms are based on the influence of factors other than free will or rational decision making—for example, biology, culture, parenting, and economics.

Another category of theories is positivism, which is somewhat the opposite of the rational choice theories. These theories argue that individuals do *not* have free will or rationality in making decisions to commit crime. Rather, the **Positive School** perspective assumes that individuals are passive subjects of determinism, which means that people do not freely choose their behavior. Instead, their behavior is determined by factors outside of their free will, such as genetics, IQ, education, employment, peer influences, parenting, and economics's.[5] Most of the highly respected and scientifically validated criminological theories of the modern era fall into this category.[6]

Another group of criminological theories belong to the conflict/critical perspective, which emphasizes the use of law as a reaction or tool to enforce restraint on others by those in power or authority; it also involves how society reacts when a person (often a juvenile) is caught doing something wrong. These theories emphasize group behavior

---

[4]George Vold, Thomas Bernard, and Jeffrey Snipes. *Theoretical Criminology*, 5th ed. (Oxford, UK: Oxford University Press, 2002).

[5]Ibid.

[6]Lee Ellis and Anthony Walsh, "Criminologists' Opinions about Causes and Theories of Crime and Delinquency," *The Criminologist* 24 (1999):1–4.

over individual behavior: Groups that are in power use the criminal codes as a tool in keeping people who have limited power restrained or confined.

Finally, over the last few decades, a new category has emerged, namely the integrated theoretical models, which attempt to combine the best aspects of explanatory models into a single, better theoretical framework for understanding crime. These models tend to suffer from the logical inconsistencies of integrating theoretical models that have opposing assumptions or propositions. All of these categories will become clearer as we progress through this book.

Additional Ways to Classify Criminological Theories

Although the major paradigms are the primary way that criminological theories are classified, there are several other ways that they can be categorized. Specifically, theoretical models can be classified based on whether they focus on individuals or groups as their primary units of examination. For instance, some theories emphasize why certain individuals do or do not commit crime. This level of investigation, in which the focus is on the individual, is often referred to as the **micro-level of analysis**, much as microeconomics is the study of economics on the individual (person) level of analysis. When your instructors score each student on an exam, this is a microlevel analysis.

On the other hand, many theories emphasize primarily the group or **macro-level of analysis**, much as macroeconomics is the study of economic principles at the aggregate or group level of analysis. In this book, some sections are separated by whether the individual or group level of analysis is emphasized. For example, social process theories tend to be more microlevel oriented, whereas social structure theories are more macrolevel oriented). Here's a good example. If instructors compare the mean score (i.e., average) of one class to the mean score in another course they are teaching, this is a comparison of group rates, regardless of the performance on any individual in either class. Ultimately, a great theory would explain both the micro- and macrolevels of analysis, but we will see that very few attempt to explain or account for both levels.

Criminological theories can also be classified by the way they view the general perspective of how laws are made. Some theories assume that laws are made to define acts as criminal to the extent that they violate rights of individuals, and thus, virtually everyone agrees that such acts are immoral. This type of perspective is considered a **consensual perspective** (or nonconflict model). On the other hand, many modern forms of criminological theories fall into an opposite type of theory, commonly known as the **conflict perspective**, which assumes that different groups disagree about the fairness of laws and that laws are used as a tool by those in power to keep down other, lower-power groups. There are many forms of both of these types of consensual and conflict theoretical models, and both will be specifically noted as we progress through the book.

A final, but perhaps most important way to classify theories is in terms of their assumptions regarding human nature. Some theories assume that people are born good (e.g., giving, benevolent, etc.) and are corrupted by social or other developmental influences that lead them to crime. A good example is strain theory, which claims that people are born innocent and with good intentions but that society causes them to commit crime. On the other hand, many of the most popular current theories claim that virtually all individuals are born with a disposition toward being bad (e.g., selfish, greedy, etc.) and must be socialized or restrained from following their inherent propensities for engaging in crime.[7] A good example of this is control theory, which assumes that all individuals have a predisposition to be greedy, selfish, violent, and so on. (i.e., they are criminally disposed).

Another variation on this issue involves theories that are often referred to as *tabula rasa*, literally translated as "blank slate." This assumes that people are born with no leaning toward good or bad but are simply influenced by the balance of positive or negative influences that are introduced socially during their development. A good example is differential association/reinforcement theory, which assumes that all individuals are born with a "blank slate" and that they learn whether to be good or bad based on what they experience. Although the dominant assumption tends to vary across these three models from time to time, the most popular theories today (which are self- and social-control theories) seem to imply the second option, specifically that people are born selfish and greedy and must be socialized and trained to be good and conforming.[8] There are other ways that criminological theories can be classified, but the various characteristics that we have discussed in this section summarize the most important factors.

Characteristics of Good Theories

Respected scientific theories in all fields, whether it be chemistry, physics, or criminology, tend to have the same characteristics. After all, the same scientific review process (i.e., blind peer review by experts) is used in all sciences to determine which studies and theoretical works are of high quality. The criteria that characterize a good theory in chemistry are the same ones used to judge what makes a good criminological theory. Such characteristics include: parsimony, scope, logical consistency, testability, empirical validity, and policy implications.[9] Each of these characteristics is examined here (it should be noted that our discussion and many of the examples provided for the characteristics are taken from Akers & Sellers, 2004).[10]

Parsimony is achieved by explaining a given phenomenon, in our case criminal activity, in the simplest way as possible. Other characteristics being equal, the simpler a theory is the better. The problem with criminal behavior is that it is highly complex. However, that has not stopped some criminologists from attempting to explain this convoluted phenomenon in highly simple ways. For example, one of the most recent and popular theories (at least regarding the amount of related research and what theories the experts believe are the most important) is the theory of low self-control (which we discuss later in this book). This very simple model holds that one personality factor—low self-control—is responsible for all criminal activity. The originators of this theory, Michael Gottfredson and Travis Hirschi, assert that every single act of crime and deviance is caused by this same factor: low self-control[11]—everything from speeding, smoking tobacco, not wearing a seat

---

[7]Ibid.

[8]Ibid.

[9]Ronald Akers and Christine Sellers, *Criminological Theories*, 4th ed. (Los Angeles: Roxbury, 2004).

[10]Ibid, Chapter 1.

[11]Michael Gottfredson and Travis Hirschi, *A General Theory of Crime* (Palo Alto, CA: Stanford University Press, 1990).

belt while driving, and having numerous sex partners to serious crimes such as murder and armed robbery are caused by low self-control.

Although this theory has been disputed by much of the subsequent research on this model, it remains one of the most popular and accepted models of the modern era.[12] Furthermore, despite the criticisms of this theory, many notable criminologists still believe that this is the best single model of offending that has been presented to date. In addition, there is little doubt that this model has become the most researched theoretical model over the last two decades.[13]

Perhaps the most important reason why so much attention has been given to this theory is its simplicity, putting all of the focus on a single factor. Virtually all other theoretical models have specified multiple factors that are proposed to play a major part in determining processes that explain why individuals commit crime. After all, how can low self-control explain white-collar crime? Some self-control is required to obtain a white-collar position of employment. Regardless, it is true that a simple theory is better than a more complex one, as long as other characteristics are equivalent. However, given a complex behavior like criminal behavior, it is likely that a simple explanation, such as naming one factor to explain everything, is unlikely to be adequate.

Scope is the characteristic that indicates how much of a given phenomenon the theory seeks to explain. Other characteristics being equal, the larger the scope the better the theory. This is somewhat related to parsimony in the sense that some theories, like the theory of low self-control, seek to explain all crimes and all deviant acts as well. So the theory of low self-control has a very wide scope. Other theories of crime may seek to explain only property crime, such as some versions of strain theory, or drug usage. However, the wider the scope of what a theory can explain, the better the theory, assuming other characteristics are equal.

Logical consistency is the extent to which a theory makes sense in terms of its concepts and propositions. It is easier to see what is meant by logical consistency by showing examples of what does *not* fit this criterion. Some theories simply don't make sense because of the face value of its propositions. For example, Cesare Lombroso, called the father of criminology, claimed that the most serious offenders are "born criminals," biological throwbacks to an earlier stage of evolutionary development who can be identified by their physical features.[14] Lombroso, who is discussed at more length later in this book, claimed that tattoos were one of the physical features that identified these born criminals. This doesn't make sense, however, because tattoos are not biological physical features— no baby has ever been born with a tattoo. This criticism will make even more sense when we discuss the criteria for determining causality later in this chapter.

Another prominent example of theories that lack logical consistency is the work of early feminist theorists, such as Freda Adler, who argued that as females gain educational and employment opportunities, they will be more likely to converge with males on crime rates.[15] Such hypotheses were logically inconsistent with the data available at the time

[12]Ellis and Walsh, "Criminologists' Opinions."

[13]Anthony Walsh and Lee Ellis, "Political Ideology and American Criminologists' Explanations for Criminal Behavior," *The Criminologist* 24 (1999):1, 14.

[14]Cesare Lombroso, *The Criminal Man* (Milan: Hoepli, 1876).

[15]Freda Adler, *Sisters in Crime* (New York: McGraw Hill, 1975).

they were presented and even more today; the facts show that females who are given the most opportunities commit the fewest crimes. On the contrary, females who have not been given these benefits commit the most crime. These are just two examples of how past theories have not been logically consistent with the data at the time they were created, not to mention future research findings, which have completely dismissed their hypotheses.

Testability is the extent to which a theory can be put to empirical, scientific testing. Some theories simply cannot be tested. A good example is Freud's theory of the psyche. Freud described three domains of the psyche: the conscious ego, the subconscious id, and the superego, but none of these domains can be observed or tested.[16] Although some theories can be quite influential without being testable (as was Freud's theory), other things being equal, it is a considerable disadvantage for a theoretical model to be untestable and unobservable. Fortunately, most established criminological theories can be examined through empirical testing.

Empirical validity is the extent to which a theoretical model is supported by scientific research. Obviously, this is highly related to the previous characteristic of being testable. Virtually all accepted modern criminological theories are testable, but that does not mean they are equal in terms of empirical validity. Although some integrated models (meaning two or more traditional theories being merged together, which will be examined later in this book) have gained a large amount of empirical validity, these models sort of cheat because they merge the best of two or more models, even when the assumptions of these models are not compatible. Therefore, the best empirical validity from an independent theoretical model, by itself, has been found for differential reinforcement theory, which has been strongly supported for various crime types (ranging from tobacco usage to violence) among a wide variety of populations (ranging from young children to elderly subjects).[17]

Ultimately, assuming other characteristics being equal, empirical validity is perhaps one of the most important characteristics used in determining how good a theory is at explaining a given phenomenon or behavior. If a theory has good empirical validity, it is an accurate explanation of behavior; if it does not have good empirical validity, it should be revised or dismissed because it is simply not true.

Policy implications is the extent to which a theory can create realistic and useful guidance for changing the way that society deals with a given phenomena. In our case, this means providing a useful model for informing authorities of how to deal with crime. An example is the broken windows theory, which says that to reduce serious crimes, authorities should focus on the minor incivilities that occur in a given area. This theory has been

---

[16]Much of this discussion is taken from George Vold et al., *Theoretical Criminology.*

[17]See studies including Ronald Akers and Gang Lee, "A Longitudinal Test of Social Learning Theory: Adolescent Smoking," *Journal of Drug Issues* 26 (1996): 317–43; Ronald Akers and Gang Lee, "Age, Social Learning, and Social Bonding in Adolescent Substance Abuse," *Deviant Behavior* 19 (1999): 1–25; Ronald Akers and Anthony J. La Greca, "Alcohol Use Among the Elderly: Social Learning, Community Context, and Life Events," in *Society, Culture, and Drinking Patterns Re-examined*, ed. David J. Pittman and Helene Raskin White (New Brunswick, NJ: Rutgers Center of Alcohol Studies, 1991), 242–62; Sunghyun Hwang, *Substance Use in a Sample of South Korean Adolescents: A Test of Alternative Theories* (Ann Arbor, MI: University Microfilms, 2000); Sunghyun Hwang and Ronald Akers, "Adolescent Substance Use in South Korea: A Cross-Cultural Test of Three Theories," in *Social Learning Theory and the Explanation of Crime: A Guide for the New Century*, ed. Ronald Akers and Gary F. Jensen (New Brunswick, NJ: Transaction Publishers, 2003).

used successfully by many police agencies (most notably by New York City police, who reduced their homicide rate by more than 75% in the last decade). Other theories may not be as useful in terms of reducing crime because they are too abstract or propose changes that are far too costly or impossible to implement, such as theories that emphasize changing family structure or the chromosomal makeup of individuals. So other things being equal, a theory that has readily available policy implications would be advantageous as compared to theories that do not.

Criteria for Determining Causality

There are several criteria for determining whether a certain variable causes another variable to change—in other words, causality. For this discussion, we will be using the commonly used scientific notation of a predictor variable—called X—as causing an explanatory variable—called Y; such variables are also commonly referred to as independent or predictor variable (X), and a dependent or explanatory (Y) variable. Such criteria are used for all scientific disciplines, whether chemistry, physics, biology, or criminology. In this book, we are discussing crime, so we will concentrate on examples that relate to this goal, but some examples will be given that are not crime related. Unfortunately, we will also see that given the nature of our field, there are some important problems with determining causality, largely because we are dealing with human beings as opposed to a chemical element or biological molecule.

The three criteria that are needed to show causality are: temporal ordering, covariation/correlation, and accounting for spuriousness.

Temporal ordering requires that the predictor variable (X) must precede the explanatory variable (Y) if we are to determine that X causes Y. Although this seems like a "no-brainer," it is sometimes violated in criminological theories. For example, you'll remember that Lombroso claimed "born criminals" could be identified by tattoos, which obviously goes against this principle.

A more recent scientific debate has focused on whether delinquency is an outcome variable (Y) due to associations with delinquent peers/associates (X), or whether delinquency (X) causes associations with delinquent peers/associates (Y), which then leads to even more delinquency. This can be seen as the argument of which came first, the

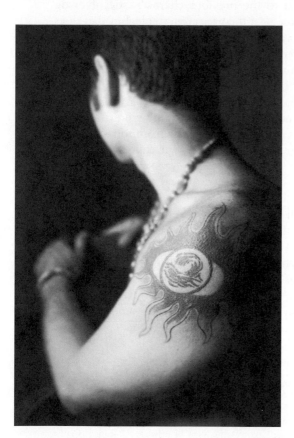

▲ **Photo I.2**   Early theories identified criminals by whether they had tattoos; at that time, this might have been true. In contemporary times, many individuals have tattoos, so this would not apply.

chicken or the egg. Studies show that both processes are often taking place, meaning that delinquency and associations with delinquent peers are likely to be both predictor and explanatory variables in most cases, and this forms a reciprocal/feedback loop that encourages both causal paths.[18] Thus, temporal ordering is an important question, and often it is quite complex and must be examined to fully understand causal order.

Correlation or covariation is the extent to which a change in the predictor (X) is associated with a change (either higher or lower) in the explanatory variable (Y). In other words, a change in X leads to a change in Y. For example, a rise in unemployment (X) someplace is likely to lead to a rise in crime rates (Y) in the same area; this would be a positive association because both increased. Similarly, an increase in employment (X) is likely to lead to a decrease in crime rates (Y) in that area; this would be a negative, or inverse, association, because one decreased and the other increased. The criterion of covariance is not met when a change in X does *not* produce any change in Y. Thus, if a significant change in X does not lead to a significant change in Y, then this criterion is not met.

However, correlation alone does not mean that X causes Y. For example, ice cream sales (X) tend to be highly associated with crime rates (Y). However, this does not mean that ice cream sales cause higher crime rates. Rather, other factors, in this case, warm weather, lead to increases in both sales of ice cream and the number of people who are outdoors in public areas and interacting, which naturally leads to greater opportunities and tendencies to engage in criminal activity. This leads to the final criterion for determining causality.

Accounting for **spuriousness** is a complicated way of saying that to determine that X causes Y, other factors (typically called Z factors) that could be causing the observed association must be accounted for before one is sure that it is actually X that is causing Y. In other words, these other Z factors may account for the observed association between X and Y. What often happens is that a third factor (Z) causes two events to occur together in time and place.

A good example is the observation that a greater number of fire officers showing up at a fire is correlated with more damage at a fire. If only the first two criteria of causality were followed, this would lead to the conclusion that the increased number of fire officers (X) causes the heavier fire damage (Y). This conclusion meets the temporal ordering and covariance criteria. However, a third Z variable or factor is causing both X and Y to appear together. This Z variable is the size of the fire, which is causing more officers to show up and also causing more damage. Once this Z factor is accounted for, the effect of X on Y becomes nonexistent.

Using the Lombroso example, tattoos may have predicted criminality at the time he wrote (although criminals weren't born with them). However, Lombroso did not account for an important Z factor, namely associates or friends who also had tattoos. This Z factor caused the simultaneous occurrence of both other factors. To clarify, individuals who had friends or associates with tattoos tended to get tattoos, and (especially at that time in the 1800s), friends or associates who had tattoos also tended to commit more crime. In that era, pirates and incarcerated individuals were most likely to get tattoos. Therefore,

---

[18]Terence Thornberry, "Toward an Interactional Theory of Delinquency," *Criminology* 25 (1987): 863–87.

had Lombroso controlled for the number of tattooed associates of the criminals he studied, he likely would have found no causal effect on crime from body art.

Ultimately, researchers in criminology are fairly good at determining the first two criteria of causality, temporal ordering and covariance/correlation. Most scientists can perform classical experiments that randomly assign subjects either to receive or not to receive the experimental manipulation to examine the effect on outcomes. However, the dilemma for criminologists is that the factors that appear to be important (according to police officers, parole agents, or corrections officers) are family variables, personality traits, employment variables, intelligence, and other similar characteristics that cannot be experimentally manipulated to control for possible Z factors. After all, how can we randomly assign certain people or groups to bad parents or bad educations, no jobs, low IQ, bad genetics, or delinquent peers? Even if we could manage such manipulations, ethical constraints would prohibit them.

Thus, as criminologists, we may never be able to meet all the criteria of causality, so we are essentially stuck with building a case for the factors we think are causing crime by amassing as much support as we can regarding temporal ordering and covariance/correlation, and perhaps accounting for other factors in advanced statistical models. Ultimately, social science, and especially criminology, is a difficult field in terms of establishing causality, and we shall see that the empirical validity of various criminological theories is hindered by such issues.

Measures of Crime

Crime can be measured in an infinite number of ways. To some extent, readers have measured crime to some extent by observing what they have seen happening in their own neighborhood or reading/watching the news every day. However, some measures of crime go beyond these anecdotal or personal experiences, and these more exacting measures are what criminologists commonly use to gauge rates of crime across time and place.

Specifically, three major categories of crime measures are used by social scientists to examine crime. The first and most used measure is the **Uniform Crime Report** (UCR). Police send reports about certain crimes and arrests for committing them to the Federal Bureau of Investigation (FBI), which combines the reports from many thousands of police agencies across the nation and publishes the UCR annually.

The second measure is the **National Crime Victimization Survey** (NCVS; prior to the early 1990s, it was known as the National Crime Survey [NCS]). Like the UCR, the report is issued in the U.S. Department of Justice (DOJ), but the data are collected in an entirely different way. Specifically, interviews are conducted with a large, random sample of U.S. households asking how much crime they have experienced in half-year intervals. The NCVS is collected by the research branch of the DOJ called the Bureau of Justice Statistics (BJS), in conjunction with the U.S. Bureau of the Census, which is one of the earliest agencies to collect information about citizens and thus the most experienced at such endeavors.

The third measure, which is perhaps the most important for purposes of this book, is **self-report data** (SRD), which are primarily collected by independent academic scientists or "think tank" agencies such as RAND Corporation. Participating in surveys or interviews, individuals report crimes against themselves or crimes they have committed. This measure is the most important for the purposes of this book because the UCR and NCVS do not provide in-depth information on the offenders or the victims, such as personality, biology/physiology, family life, and economic information. These factors are of the utmost importance for our purposes because there is a broad consensus that they cause people to commit crime, and yet, they are missing from the most commonly used measures of crime. Self-report data are the best, and in most cases the only, measure that can be used to figure out why some people offend and others do not. However, like the other measures, self-reports have numerous weaknesses (and strengths).

Each of these three measures is briefly examined here. Although the measures are not the primary emphasis of this book, it is important to understand their strengths and weaknesses.

The Uniform Crime Report

The UCR is the oldest and most used measure of crime rates in the United States for purposes of examining trends and distribution of crime. It began in the early 1930s, and although changes have been made along the way, it is relatively stable in terms of comparing various years and decades. As mentioned before, it is collected by many thousands of independent police agencies at federal, state, and local levels. These thousands of agencies send their reports of crimes and arrests to their respective state capitals, which then forward their synthesized reports to the FBI headquarters, where all reports are combined to provide an overview of crime in the nation.

The FBI definitions of crimes often differ from state categorizations, and the way that they differentiate crimes are important in terms of future discussions in this section. The FBI concentrates on eight (four violent and four property) **index offenses**, or Part I offenses. The four violent crimes are murder/non-negligent manslaughter, forcible rape (not statutory), robbery, and aggravated assault (which involves intentions of serious injury on the victim). The four property offenses are burglary (which includes a breaking/entering or trespass), motor

▲ **Photo I.3** The annual Uniform Crime Reports are produced by the FBI. Local, county, and state criminal justice agencies send their annual crime data to the J. Edgar Hoover Building in Washington, D.C. UCR data are, by their nature, incomplete, as many crimes are never reported to the police at all. This "dark figure of crime" might be as high as 50% of all crime incidents.

vehicle theft, larceny (which does not involve a trespass; an example is shoplifting), and arson (which was added to the crime index count in the late 1970s). All reports to police for these eight offenses are included in the crime index, whether or not they resulted in an arrest. This information is often referred to as crimes known to police (CKP).

The UCR also includes about two dozen other offenses known as **non-index offenses** (or Part II offenses), which are reported only if an arrest is made in a case. These offenses range from violent crimes (such as simple assault), to embezzlement and fraud, to offenses that are considered violations of the law only if an individual is under 18 years of age (such as running away from home). The major problem with the estimates of these non-index offenses is that the likelihood of arresting someone for such crimes is less than 10% of the actual occurrence of such offenses, so the data regarding the non-index offenses is highly inaccurate. The official count from the FBI is missing at least 90% of the actual offenses that take place in the United States. Therefore, we will primarily concentrate on the index offenses for the purposes of our discussion.

Even the count of index offenses has numerous problems. The most important and chronic problem with using the UCR as a measure of crime is that most of the time victims fail to report crimes—yes, even aggravated assault, forcible rape, robbery, burglary, and larceny. Recent studies estimate that about 70% to 80% of these serious crimes are not reported to police. Criminologists call this "missing" amount of crime the **dark figure** because it never shows up in the police reports that get sent to the FBI.

There are many reasons why victims do not report these serious crimes to the police. One of the most important is that they consider it a personal matter. Many times, the offense is committed by a family member, a close friend, or an acquaintance. For instance, police are rarely informed about aggravated assaults among siblings. Rape victims are often assaulted on a date or by someone they know; they may feel that they contributed to their attack by choosing to go out with the offender or they may believe that police won't take such a claim seriously. Regardless of the case, many crime victims prefer to handle it informally and not involve the police.

Another major reason why police are not called is that victims don't feel the crime is important enough to report. For example, thieves may steal a small item that the victim won't miss, so they don't see the need to report what the police or FBI would consider a serious crime. This is likely related to another major reason why people do not report crime to the police: They have no confidence that reporting the case to law enforcement will do any good. Many people, often residents of the neighborhoods that are most crime-ridden, are likely to feel that police are not competent or will not seriously investigate their charges.

There are many other reasons why people do not report their victimizations to police. Some may fear retaliation, for example, in cases involving gang activity; many cities, especially those with many gangs, have seen this occur even more in recent years. The victims may also fail to report a crime for fear that their own illegal activities will be exposed; an example is a prostitute who has been brutally beaten by her pimp. In U.S. society, much crime is committed against businesses, but they are very reluctant to report crimes because they do not want a reputation for being a hot spot for criminal activity. Sometimes, victims call the police or 911, but they leave the scene if the police fail to show up in a reasonable amount of time. This has become a chronic problem, despite efforts by police departments to prioritize calls.

Perhaps the most chronic, most important, and often ignored failure to report crimes can be traced to U.S. school systems. Most studies of crime and victimization in schools show that much and maybe even most juvenile crimes occur in schools, but these offenses almost never get reported to police, even when school resource officers (SROs) are assigned to the school. Schools—and especially private schools—have a vested interest in not reporting crimes that occur on their premises to the police. After all, no school (or school system) wants to become known as a crime-ridden place/system.

Schools are predisposed to being crime ridden because the most likely offenders and the most likely victims—young people—interact there in close quarters for most of the day. The school is much happier, however, if teachers and administrators deal informally with the parties involved in an on-campus fight; the school does not want these activities reported to and by the media. In addition, the student parties involved in the fight do not want to be formally arrested and charged with the offense, so they are also happy with the case being handled informally. Finally, the parents of the students are also generally pleased with the informal process because they do not want their children involved with a formal legal case.

Even universities and colleges follow this model of not reporting crimes when they occur on campus. A good example can be seen on most of the Websites of virtually all major colleges, which are required by federal law to report crimes on campus. Official reports of crimes, ranging from rapes to liquor law violations, are often in the single digits each year for campuses housing many thousands of students. Of course, some crimes may not be reported to the school, and others may be dealt with administratively rather than calling police. The absence of school data is a big weakness of the UCR.

Besides the Dark Figure, there are many other criticisms of the UCR as a measure of crime. For example, the way that crimes are counted can be misleading. Specifically, the UCR counts only the most serious crime that is committed in a given incident. For example, if a person or persons rob, rape, and murder a victim, only the murder would show up in the UCR; the robbery and rape would not be recorded. Furthermore, there are inconsistencies in how the UCR counts the incidents; for example, if a person walks into a bar and assaults eight people there, it would be counted as eight assaults, but if the same person walks into the bar and robs every person of wallets and purses, it is counted as one robbery. This makes little sense, but it is the official count policy traditionally used by the UCR.

Some other more important criticisms of the UCR involve political considerations, such as the fact that many police departments (such as Philadelphia's) have systematically altered the way that crimes are defined, for example, by manipulating the way that classifications and counts of crimes are recorded (e.g., the difference between aggravated assault [an index crime] and simple assault [a non-index crime]), so that official estimates make it seem that major crimes have decreased in the city when in fact they may have actually increased.

It is important also to note the strengths of the UCR measure. First, it was started in 1930, making it the longest, systematic measure of crime in the United States. This is a very important advantage if one wants to examine crime over most of the 20th century. We will see later that there have been extremely high crime rates at certain times (such as during the 1930s and the 1970s/1980) and very low crime rates at other points (such as the early 1940s and recent years [late 1990s-present]). Other measures, such as the NCVS and national self-report data, did not come into use until much later, so the UCR is important for the fact that it started so early.

Another important strength of this measure is that two of the offenses that the UCR concentrates on are almost always reported and therefore overcome the weakness of the Dark Figure or lack of reporting to police. These two offenses are murder/non-negligent manslaughter and motor vehicle theft. Murder is almost always reported because a dead body is found; very few murders go unreported to authorities. Although a few may elude official recording, for example, if the body is transported elsewhere or carefully hidden, almost all murders are recorded. Similarly, motor vehicle thefts, a property crime, are almost always reported because any insurance claims must provide a police report. Most cars are worth thousands (or at least many hundreds) of dollars, so victims tend to report when their vehicle(s) have been stolen; this provides a rather valid estimate of property crime in areas. The rest of the offenses (yes, even the other index crimes) counted by the UCR are far less reliable. If someone is doing a study on homicide or motor vehicle theft, the UCR is likely the best source of data, but for any other crime, researchers should probably look elsewhere.

This is even further advised for studies examining non-index offenses, which the UCR counts only when someone is arrested for a given offense. The vast majority of non-index offenses do not result in an arrest. To shed some light on how much actual non-index crime is not reported to police, it is useful to examine the **clearance rate** of the Index offenses in the UCR, our best indicator of solving crimes. Even for the crimes that the FBI considers most serious, the clearance rate is about 21% of crimes reported to police, meaning that they took a report for a crime. Of course, the more violent offenses have higher clearance rates because (outside of murder), they inherently have a witness, namely the victim, and because police place a higher priority on solving violent crimes. However, for some of the index crimes, especially serious property offenses, the clearance rates are very low. Furthermore, it should be noted that the clearance rate of serious index crimes has not improved at all over the last few decades, despite much more advanced resources and technology, such as DNA, fingerprints, and faster cars. This data on the clearance rates is only for the most serious, or index, crimes; thus, the reporting of the UCR regarding non-index crimes is even more inaccurate because there is even less reporting (i.e., the Dark Figure) and less clearance of these less serious offenses. In other words, the data provided by the UCR regarding the non-index offenses is totally invalid and for the most part completely worthless.

Ultimately, the UCR is a good measure for (1) measuring the overall crime rate in the United States over time, (2) for examining what crime was like prior to the 1970s, and (3) for investigating murder and motor vehicle theft because they are almost always reported. Outside of these offenses, the UCR has serious problems, and fortunately, we have better measures to use for examining crime rates in the United States.

The National Crime Victimization Survey

Another commonly used measure of crime is the NCVS (the NCS until the early 1990s), which is distinguished from other key measures of crime because it concentrates on the victims of crime, whereas other measures tend to emphasize the offenders. In fact, that is the key reason why this measure was started in 1973, after several years of preparation and pretesting. To clarify, one of the key recommendations of Lyndon Johnson's President's Commission on Law Enforcement and Administration of Justice in the late 1960s was to learn more about the characteristics of victims of crime; at that time, virtually no studies had been done on the subject, whereas much research had been done on criminal offenders. So the efforts of this commission set into motion the creation of the NCVS.

Since it began, the NCVS has been designed and collected by two agencies: the U.S. Bureau of the Census and the Bureau of Justice Statistics, which is one of the key research branches of the U.S. Department of Justice. The NCVS is collected in a completely different way from the other commonly used measures of crimes; the researchers select tens of thousands of U.S. households, and each member over 12 years of age in these households is interviewed every six months about crime that occurred in that previous six-month period (each selected household remains in the survey for three years, resulting in seven collection periods, which includes the initial interview). Although the selection of households is to some extent random, the way this sampling is designed guarantees that a certain proportion of houses selected have certain characteristics. For example, before the households are selected, they are first categorized according to factors such as region of the country, type of area (urban, suburban, rural), income level, and racial/ethnic composition. This type of sampling, called a multistage, stratified cluster sampling design, ensures that enough households are sampled in the survey to make conclusions regarding these important characteristics. As you will see later in this section, some of the most victimized groups (by rate) in the United States do not make up a large portion of the population or households. So if the sampling design was not set up to select a certain number of people from certain groups, it is quite likely the researchers would not obtain enough cases to draw conclusions about them.

The data gathered from this sample are then adjusted, and statistical estimates are made about crime across the United States, with the NCVS estimates showing about three times more crime than the UCR rates. Some may doubt the ability of this selected sample to represent crime in the nation, but most studies find that its estimates are far more accurate than those provided by the UCR (with the exception of homicide and maybe motor vehicle theft). This is largely due to the expertise and professionalism of the agencies that collect and analyze the data, as well as the carefully thought out and well-administered survey design, which is indicated by the interview completion rates (typically more than 90%, higher than virtually all other crime/victimization surveys).

One of the biggest strengths of the NCVS is that it directly addresses the worst problem with the previously discussed measure, the *Uniform Crime Report* (UCR). Specifically, the greatest weakness of the UCR is the Dark Figure, or the crimes that victims fail to report, which happens most of the time (except in cases of homicide or motor vehicle theft). The NCVS interviews victims about crimes that happened to them, even those that were not reported to police. Thus, the NCVS captures far more crime events than the UCR, especially crimes that are highly personal (such as rape). So the extent to which the NCVS captures much of this Dark Figure of crime is its greatest strength.

Despite this important strength, the NCVS, like the other measures of crime, has numerous weaknesses. Probably the biggest problem is that two of the most victimized groups in U.S. society are systematically not included in the NCVS. Specifically, homeless people are completely left out because they do not have a home and the participants are contacted through households; yet, they remain one of the most victimized groups per capita in our society.

Another highly victimized group in our society that is systematically not included in the NCVS is young children. Studies consistently show that the younger a person is, the more likely he or she is to be victimized. Infants, in particular, especially in their first hours or days of life, face great risk of death or other sort of victimization, typically from parents or caregivers. This is not very surprising, especially in light of the fact that very young children cannot defend themselves or run away. They can't even tell anyone until they are old enough to speak, and then most are too afraid to do so or are not given an opportunity. Although, to some extent, it is understandable to exempt young children from such sensitive questions, the loss of this group is huge in terms of estimating victimization in the United States.

The NCVS also misses the crimes suffered by American businesses, which cumulatively is an enormous amount. In the early years of the NCVS, businesses were also sampled, but that was discontinued in the late 1970s. Had it continued, it would be invaluable information for social scientists and policymakers, not to mention the businesses that are losing billions of dollars each year as a result of crimes committed against them.

Many find it surprising that the NCVS does not collect data on homicide, which most people and agencies consider the most serious and important crime. Researchers studying murder cannot get information from the NCVS but must rely on the UCR, which is most accurate in its reporting for this crime type.

The NCVS also has issues with people accurately reporting the victimization that has occurred to them in the previous six months. However, studies show that their reports are surprisingly accurate most of the time. Often when participants report the incidents inaccurately, they make unknowing mistakes rather than intentionally lying. Obviously, victims sometimes forget incidents that occur to them, probably because most of the time they know or are related to the person offending against them, so they never think of it as a crime per se but rather as a personal disagreement. When asked if they were a victim of theft, they may not think to report the time that a brother or uncle borrowed a tool without asking and never returned it.

Although the NCVS researchers go to great lengths to prevent it, a common phenomenon known as **telescoping** tends to occur, which leads to overreporting of incidents. Telescoping is the human tendency to perceive events as having occurred more recently than they actually did. This is one of the key reasons why the NCVS researchers interview the household subjects every six months, but it still happens. For instance, a larceny may have occurred eight months ago, but it seems like it happened just a few months ago to the participant, so it is reported to the researchers as occurring in the last six months when it really didn't. Thus, this inflates the estimates for that crime interval for the nation.

As mentioned before, an additional weakness is that the NCVS did not start until 1973, so it cannot provide any estimates of victimization prior to that time. A study of national crime rates prior to the 1970s has little choice but to use the UCR. Still, for most crimes, the NCVS has provided a more accurate estimate over the past three decades. Since the NCVS was created, the crime trends it reveals tend to be highly consistent with those shown by the UCR. For example, both measures show violent crime rates peaking at the same time (about 1980), and both agree on when the rates increased (the 1970s) and decreased (the last decade [late-1990s to present]) the most. This is very good because if they did not agree, that would mean at least one (or both) were wrong. Before we discuss the national trends in crime rates, however, we will examine the strengths and weaknesses of the third measure of crime.

Self-Report Studies of Crime

The final measure of crime consists of various self-report studies of crime, in which individuals report (either in a written survey or interview) the extent of their own past criminal offending or victimization and other information. There is no one systematic study providing a yearly estimate of crime in the United States; rather, self-report studies tend to be conducted by independent researchers or institutes. Even when they do involve a national sample (such as the National Youth Survey), they almost never use such data to make estimates of the extent of crime or victimization across the nation.

This lack of a long-term, systematic study that can be used to estimate national crime rates may be the greatest weakness of self-report studies, however, this very weakness—not having a universal consistency in collection—is also its greatest strength. To clarify, researchers can develop their questionnaires to best fit the exact purposes of their study. For example, if researchers are doing a study on the relationship between a given person-ality trait (e.g., narcissism) and criminal offending, they can simply give participants a questionnaire that contains a narcissism scale and items that ask about past criminal behavior. Of course, these scales and items must be checked for their reliability and valid-ity, but it is a relatively easy way to directly measure and test the hypotheses with which the researcher is most concerned.

Some question the accuracy of the self-report data because they believe participants typically lie, but most studies have concluded that participants generally tell the truth. Specifically, researchers have compared self-reported offenses to lie detector machine results, given the same survey to the same individuals to see if they answer the same each time (called "test-retest" reliability), cross-checked their self-reported arrests to police arrest data. All of these methods showed that most people tend to be quite truthful when answering surveys.[19]

The most important aspect of self-report surveys is that they are the only available source of data for determining the social psychological reasons for why people commit crime. The UCR and NCVS have virtually no data on the personality, family life, biolog-ical development, or other characteristics of criminal offenders, which are generally con-sidered key factors in the development of criminality. Therefore, although we examine the findings of all three measures in the next section, the vast majority of the content we cover in this book will be based on findings from self-report studies.

What Do the Measures of Crime Show Regarding the Distribution of Crime?

It is important to examine the most aggregated trends of crime, namely the ups and downs of overall crime rates in the United States across different decades. We will start with crime in the early 1900s, largely because the best data started being collected during this era and also because the 20th century (especially the most recent decades) is most rel-evant to our understanding of the reasons for our current crime rates. However, most experts believe that the U.S. crime rate, whether in terms of violent or property offend-ing, used to be far higher prior to the 20th century, historians have concluded based on sporadic, poorly recorded documentation in the 18th and 19th centuries. By virtually all accounts, crime per capita (especially homicide) was far higher in the 1700s and 1800s than at any point after 1900, which is likely due to many factors, but perhaps most important

---

[19]Michael Hindelang, Travis Hirschi, and J. Weis. *Measuring Delinquency* (Beverly Hills, CA: Sage, 1981).

because formal agencies of justice, such as police and corrections (i.e., prisons, parole, etc.), did not exist in most of the United States until the middle or end of the 1800s. Up to that time, it was up to individual communities or vigilantes to deal with offenders.

Therefore, there was little ability to investigate or apprehend criminals, or a means to imprison them. But as industrialization increased, the need to establish formal police agencies and correctional facilities evolved as a way to deal with people who offended in modern cities. By 1900, most existing states had formed police and prison systems, which is where our discussion will begin.

The level of crime in the United States, particularly homicide, was relatively low at the beginning of the 20th century, perhaps because of the formal justice agencies that had been created during the 19th century. For example, the first metropolitan U.S. police departments were formed in Boston and then New York during the 1830s; in the same decade but a bit earlier, the first state police department, Texas Rangers, was established, and the federal marshals were founded still earlier. Although prisons started in the late 1790s, they did not begin to resemble their modern form or proliferate rapidly until the mid-1800s. The first juvenile court was formed in the Chicago area in 1899. The development of these formal law enforcement and justice agencies may have contributed to the low level of crime and homicide in the very early 1900s. (Note: our discussion of the crime rate in the early 1900s will primarily deal with homicide because it is the only valid record of crime at that time; UCR did not start until 1930s. Most people consider this the most serious crime, and its frequency typically reflects the overall crime rate).

The effect of the creation of these formal agencies did not persist long into the 20th century. Looking at the level of homicides that occurred in the United States, it is obvious that large increases took place between 1910 and 1920, which is likely due to the extremely high increases in industrialization, as the U.S. economy moved from agriculture to an industrial emphasis. More important, population growth was rapid as a result of urbanization. Whenever high numbers of people move into an area (in this case, cities) and form a far more dense population (think of New York City at that time, or Las Vegas in current times), it creates a crime problem. This is likely due to more opportunity for crime against others; after all, when people are crammed together, it creates a situation in which there are far more potential offenders in close proximity with far more potential victims. A good modern example of this is high schools, which studies show have higher crime rates than city subways or other crime-ridden areas, largely because they densely pack people together, and in such conditions, opportunities for crime are readily available. Thus, the rapid industrialization and urbanization of the early 1900s probably are the most important reasons for the increase in homicide, and crime in general, at that time in the United States.

The largest increases in U.S. homicide in the early 1900s occurred during the 1920s and early 1930s, with the peak level of homicide coming in the early 1930s. Criminologists and historians have concluded that this huge increase in the homicide rate was primarily due to two factors, beyond the industrialization and urbanization that explained the increase prior to the 1920s. First, the U.S. Congress passed a constitutional amendment that banned distribution and consumption of alcohol beginning in 1920. The period that followed is known as Prohibition. This legal action proved to be a disaster, at least in terms of crime, and Congress later agreed—but not until the 1930s—by passing another amendment to do away with the previous amendment that banned

▲ **Photo I.4**   Group portrait of a police department liquor squad posing with cases of confiscated alcohol and distilling equipment during Prohibition.

alcohol. For about 14 years, which notably recorded the highest U.S. rates of homicide and crime before 1950, the government attempted to stop people from drinking. Prior to Prohibition, gangsters were relatively passive and did not hold much power.

This political mandate gave the black market a lot of potential in terms of monetary profit and reasons for killing off competition. Some of the greatest massacres of rival gangs of organized crime syndicates (e.g., Italian Mafia) occurred during the Prohibition era. The impact on crime was likely only one of the many problems with Prohibition, but it was a very important, and deadly, one for our purposes. Once Prohibition ended in the early 1930s, the homicide and crime rates decreased significantly, which may have implications regarding modern drug policies. According to studies, many banned substances today are less violence-producing than alcohol, which studies show is the most violence-causing substance. For example, most criminologists believe that the current War on Drugs may actually be causing far more crime than it seeks to prevent (even if it may be lowering the number of drug addicts), due to the black market it creates for drugs that are in demand, much like the case with alcohol during Prohibition.

Another major reason why the homicide rate and overall crime levels increased so much during the early 1930s was the Great Depression, which sent the United States into an unprecedented state of economic upheaval. Most historians and criminologists agree that the stock market crash of the late 1920s was a primary contributor to the large levels of homicide in the early 1930s. We will return to this subject later, when we examine the strain theory of crime, which places an emphasis of economic structure and poverty as the primary causes of crime.

| Figure I.1 | Homicide Rates in the United States in the 20th Century |

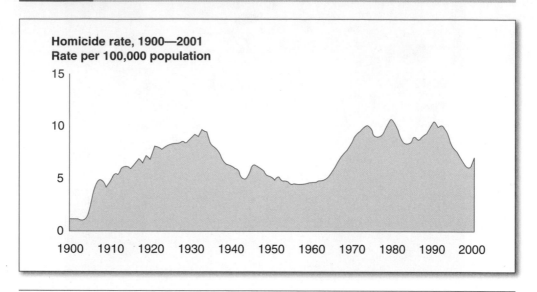

Homicide rate, 1900—2001
Rate per 100,000 population

NOTE: The 2001 rate includes deaths attributed to the 9/11 terrorism attacks.

Although the homicide/crime rate experienced a significant drop after Prohibition was eliminated, another reason for this decrease likely was due to social policies of the New Deal, which was implemented by President Franklin D. Roosevelt. Such policies included creating new jobs for the people hit hardest by the Depression through such programs as Job Corps and the Tennessee Valley Authority, both of which still exist today. Although such programs likely aided the economic (and thus crime) recovery for the United States, world events of the early 1940s provided the greatest reasons for the huge decreases that were seen at that time.

The entry of the United States into World War II was probably the biggest contributor to decreasing U.S. crime in the early 20th century. As you will notice, the homicides decreased dramatically in the four years (1941–1945) that hundreds of thousands of young men (the most likely offenders and victims) were sent overseas to fight on two fronts, Europe and the South Pacific. Any time a society loses a huge portion of the most common offenders, namely young (teenage to twenties) males, they can expect a drop in crime like the one the United States experienced while these giant portions of men were elsewhere. However, at the end of 1945, most of these men returned and began making babies, which triggered the greatest increase in babies that U.S. society had ever seen. This generation of babies started what historians call the Baby Boom, which would have the greatest impact on crime levels that has ever been recorded in U.S. history.

Although crime rates increased after soldiers returned home from overseas in the late 1940s, they did not rise to anywhere near the levels of Prohibition and the Great Depression. Alcohol was legal, and the economy was doing relatively well after World War II. During the 1950s, the crime level remained relatively low and stable until the early 1960s, when the impact of the Baby Boom emerged in terms of the crime rate. If a very large share of the population is made up of young people, particular teenage or early-20s males, the crime rate will almost inevitably go up. This is exactly what occurred in the United States, starting in the early 1960s, and it led to the largest 10-year increase in crime that the country has ever seen.

The Baby-Boom Effect

The UCR shows that the greatest single-decade increase in the crime rate occurred between 1965 and 1975. In that time, the overall crime rate more than doubled, an increase that was unprecedented. Notably, this increase occurred during the War on Poverty, which was set into motion by President Lyndon B. Johnson in a program he termed the Great Society; it turned many people and policymakers against having the government address economic issues for bettering society. However, the demographic factor that they did not take into account was that this was the era in which most people in society were in young age groups, which predisposed the nation for the high crime rates experienced at this time. In contrast, the following generation, called Generation X, which is those individuals born between 1965 and 1978, had a very low birthrate, which may have contributed to the low crime rates observed in recent years.

The high level of young people in society was not the only societal trend going on during the late 1960s and early 1970s that would lead to higher crime rates. For example, a large number of people were being arrested as a result of their participation in the civil rights movement, the women's rights movement, and anti-Vietnam War activities. Perhaps most important, the 1970s had the highest levels of drug usage and favorable attitudes toward drugs since accurate national studies have been conducted. Virtually all measures of drug usage peaked during the 1970s or early 1980s.

So ultimately, many things came together between the mid-1960s and the late 1970s that predisposed this generation of Baby Boomers to lead to the greatest increase in the crime rate that the United States has ever seen, culminating in the peak of crime and homicide in 1980. All of our measures of crime agree that crime, especially homicide, reached their highest level about that year. Although other periods, such as the early 1990s, showed similar increases in crime, largely due to juvenile offending, no other period was higher than the 1980 period, and it was most likely due to the many baby boomers coming of age and to the high drug usage.

Crime levels declined somewhat in the early 1980s and then rose again in the late 1980s and early 1990s, but the crime and homicide rate never exceeded the 1980 peak. Furthermore, after 1994, the crime rate dropped drastically every year for about a decade, to the point where the crime rate was as low as it had been in the early 1960s. Thus, the U.S. crime rate is currently about where it was about 40 years ago.

There are many reasons for this huge decrease over the last decade and a half. One of the biggest reasons is that the population has a relatively smaller proportion of young people than during the 1960s and 1970s, but obviously, there is more to the picture. Drug usage, as well as attitudes favorable toward drugs, has dropped a lot in recent years. Furthermore, the incarceration rate of prisoners is about 400% to 500% what it was in the early 1970s. Although many experts have claimed that locking up criminals does not reduce crime levels in a society, it is hard to deny that imprisoning five times more people will not result in catching many, if not most of the habitual offenders in our society. It makes sense: By rate, the United States currently locks up more citizens in prison than virtually any other developed country.

Almost all crime tends to be non-random. Consistent with this, the crime measures show a number of trends in which crime occurs among certain types of people, in certain places, at certain times, and so on. We turn to an examination of such concentrations of crime, starting with large, macro differences in crime rates across regions of the United States.

Regional and City Differences in Rates of Crime

Crime tends to be higher in certain regions of the country. According to the UCR, the United States is separated into four regions: Northeast, Midwest, South, and West. For the last few decades, crime rates (based on crime per capita) have been significantly higher in two of these regions: the South and the West. These two regions consistently have higher rates than the other regions, with one or the other having the highest rates for violence or property offenses or both each year. Some studies have found that when poverty levels are accounted for, much of the regional differences are explained away. Although this is a very simple conclusion, the studies seem to be consistent in tying regional differences to variations in social factors, notably socioeconomic levels.

Regardless of the region, there seems to be extreme variation from high to low crime rates across states and cities within each of these large regions. For example, crime measures consistently show that certain U.S. states and jurisdictions have consistently high crime rates. The two standouts are Louisiana and the District of Columbia, with the latter having an extremely high rate, typically more than eight times the national average for homicide). It is quite ironic that arguably the most powerful city in the world has an extremely serious crime problem, certainly one of the worst in our nation.

Another question is how does variation in states or jurisdictions, or in cities and counties, vary drastically from one region to the next. For instance, Camden, New Jersey, one of the cities in the lower-rate Northeast region according to the UCR, had the highest rate of crime among all U.S. cities in the years 2004 and 2005. Detroit, Michigan, was second worst for both of these years; it used to be No. 1 before Camden "outdid" Detroit. At the same time, however, it is important to note that New Jersey also had some of the safest cities in the nation for these years–two of the 10 safest cities are in New Jersey, which shows how much crime can vary from place to place, even those in relatively close proximity. Notably, in the most recent estimates from the FBI, the cities of St. Louis and New Orleans were the highest rates of serious violent crimes in 2006 and 2007, respectively. An important factor regarding New Orleans was the devastation of the city's infrastructure after Hurricane Katrina, which essentially wiped out the city's criminal justice system and resources.

Crime has also been found to cluster within a given city, whether the overall city is relatively low-crime or high-crime. Virtually every area (whether urban, suburban, or rural) has what is known as **hot spots** or places that have high levels of crime activity. Such places are often bars or liquor stores or other types of businesses, such as bus stops/depots, convenience stores, shopping malls, motels/hotels, fast-food restaurants, check-cashing businesses, tattoo parlors, discount stores (such as one-dollar stores), and so on. However, hot spots can also be residential places, such as a home that police are constantly called to for domestic violence or certain apartment complexes that are crime ridden, something often seen in subsidized housing areas.

Even the nicest of cities or areas have hot spots, and even the worst cities or areas also tend to have most of their police calls coming from specific addresses or businesses. This is one of the best examples of how crime does not tend to be random. Many police agencies have begun using spatial software on computer systems to analyze where the hot spots in a given city are and to predict where certain crimes are likely to occur in the future. This allows more preventive patrols and proactive strategies in such zones. One criminological theory, routine activities theory, is largely based on explaining why hot spots exist.

Another way that crime is known to cluster is time of day. This varies greatly depending on the type of group one is examining. For example, juvenile delinquency and victimization, especially for violence, tends to peak sharply at 3 p.m. on school days (about half of the days of the year for children), which is the time that youths are most likely to lack supervision (i.e., children are let out of school and are often not supervised by adults until they get home). On the other hand, adult crime and victimization, especially for violence, tends to peak much later on almost all days at about 11 p.m., which is a sharp contrast from the juvenile peak in midafternoon. These estimates are primarily based on FBI/UCR data.

To some extent, the peak hour for juveniles is misleading; other non–police-based estimates show that just as much crime is going on during school but schools tend to not report it. This widespread lack of reporting by schools occurs because none of the parties involved want a formal police report taken. Typically, the youth doesn't want to be arrested, the child's parents do not want their son or daughter to be formally processed by police, and most important, no school wants to become known as a dangerous place. So the police are typically called only in extreme cases, for example, if a student pulls a gun on another student or actually uses a weapon.

This also occurs in colleges/universities because such institutions largely depend on tuition for funding, and this goes down if enrollment levels decline. After all, no parents want to send their teenager to a college that is high in crime. Federal law now requires virtually all colleges to report to the public their crime levels, so there is a lot at stake if police take a formal report on crime events. Thus, most colleges, like the K–12 school systems, have an informal process in place in which even violent crimes are often handled informally within the school systems.

Crimes tend to peak significantly during the summer. Studies show that criminals tend to be highly opportunistic, meaning that they happen to be going about their normal activities when they see an opportunity to commit a crime, as compared to a more hydraulic model, in which an offender actually goes out looking to commit a crime. Because criminals are like everyone else in most ways, they tend to be out and about more in the summer, so they are more likely to see opportunities at that time. Furthermore, youths are typically out of school during the summer, so they are often bored and not supervised by adults as much as during the traditional school year. Burglary tends to rise 300% during the summer, an increase that may be linked to the fact that people go on vacation and leave their home vacant for weeks or months at a time. All of the factors come together to produce much higher rates in the summer than in any other season.

A couple of crimes, such as murder and robbery, tend to have a second peak in the winter, which most experts believe is due to high emotions during the holidays, additional social interaction (and often drinking of alcohol) during the holidays, and an increase in wanting money or goods for gift-giving, which would explain robbery increases at that time. These offenses are the exception, however, not the rule. Most offenses, including murder and robbery, tend to have their highest peaks during summer/warmer months.

Age is perhaps the most important way that crime and victimization tend to cluster in certain groups. Almost no individuals get arrested before the age of 10; if they do, it is a huge predictor that the child is likely going to become a habitual, chronic offender. However, from the age of 10 to 17, the offending rate for all serious street crimes (i.e., FBI index crimes) goes up drastically, peaking at an average age of 17. Then, the offending rates begin decreasing significantly, such that by the time people reach the age of 20, the likelihood of being arrested has fallen in half as compared to the mid-teenage years. This offending level continues to decline throughout life for all of the serious index crimes, although other crimes are likely to be committed more often at later ages, such as white-collar crimes, tax evasion, and gambling.

The extraordinarily high levels of offending in the teenage years have implications for how we prevent and deal with juvenile delinquents and for why we are so bad at preventing habitual offenders from committing so many crimes. We are often good at predicting who the most chronic, serious offenders are by the time they are in their 20s, but that does little good because most offenders have committed most of their crimes before they hit 20 years old.

Another characteristic important in the way crime clusters is gender. In every society, at every time in recorded history, males commit far more serious street crimes (both violent and property) than do females. It appears there is almost no closing of this gap in offending, at least for FBI index crimes. Even in the most recent years, males are the offenders in 80% to 98% of all serious violent crimes (murder, robbery, aggravated assault, forcible rape), and make up the vast majority of offenders in the property index crimes (burglary, motor vehicle theft, arson, and larceny). The last offense is often surprising to people because the most common type of larceny is shoplifting, which people often perceive as being done mostly by women. All studies show that men commit most of the larcenies in the United States. It is important to realize that males in *all* societies throughout the world commit the vast majority of offenses, and the more violent and serious the crimes, the more men are represented in such offenses.

However, there are a few non-index crimes that females commit as much as, or more than males. Specifically, in terms of property crimes, embezzlement and fraud are two offenses that females commit at comparable rates to men, which likely has to do with enhanced opportunities to commit such crimes. Most of the workforce is now female, which wasn't true in past decades, and many women work in banking and other businesses that tempt employees by having large amounts of money available for embezzling. In terms of public disorder, prostitution arrests tend to be mostly female, which is not too surprising.

The only other offense in which females are well-represented is running away from home, which is a status offense (illegal for juveniles only). However, virtually all other sources and studies of offending rates (e.g., self-report studies) show that male juveniles run away far more than females, but because of societal norms and values, females get arrested far more than males. Feminist theories of the patriarchal model (in short, men are in charge and dominate/control females) and the chivalry model (simply, females are treated differently because they are seen as more innocent) argue that females are protected as a type of property. This may be important in light of the contrasting findings regarding female and male rates of running away, as compared to the opposite trends regarding female and male rates of being arrested for running away. The bottom line is that families are more likely to report the missing child and to press law enforcement agencies to pursue girls rather than boys who have run away. We will explore explanations for such differences in later sections, particularly in the section in which we cover conflict/feminist theories of crime.

Victimization and offending rates are also clustered according to the density of a given area. All sources of crime data show that rates of offending and victimization are far higher per capita in urban areas than in suburban and rural regions. Furthermore, this trend is linear: The more rural an area is, the lower the rates for crime and victimization. To clarify, urban areas have, by far, the highest rates for offending, followed by suburban areas; the least amount of crime and victimization is found in rural (e.g., farming) areas. This trend has been shown for many decades and is undisputed. Keep in mind that such rates are based on population in such areas, so even per capita of citizens in a given region, this trend holds true. This is likely due to enhanced opportunities to commit crime in urban and even suburban areas, as well as the fact that rural communities tend to have stronger **informal controls**, such as family, church, and community ties. Studies consistently show that informal controls are far more effective in preventing and solving crimes than are the **formal controls** of justice, which include all official aspects of law enforcement and justice, such as police, courts, and corrections (i.e., prisons, probation, parole).

A good example of the effectiveness of informal sanctions can be seen in the early formation of the U.S. colonies, such as the Massachusetts Bay Colony. In the early 1600s, crime per capita was at an all-time low in what would become the United States. It may surprise many that police and prisons did not exist then; rather, the low crime rate was due to high levels of informal controls: When people committed crimes, they were banished from the society or were shunned.

As in Nathaniel Hawthorne's novel *The Scarlet Letter*, even for what would now be considered a relatively minor offense (adultery), a person was forced to wear a large letter (such as an A for adultery or a T for theft), and they were shunned by all others in their social world. Such punishments were (and still are) highly effective punishments and deterrents for offenders, but they work only in communities with high levels of informal controls. Studies have shown that in such societies, crime tends to be extremely low, in fact, so low that such communities may "invent" serious crimes so that they can have an identifiable group on which to blame societal problems. We see this occur in the Massachusetts Bay Colony, with the creation of a new offense, witchcraft, for which hundreds of people were put on trial and many were executed.

Such issues will be raised again later when we discuss the sociological theories of Emile Durkheim. Nevertheless, it is interesting to note that some judges and communities have gone back to such public shaming punishments, such as making people carry signs of their offenses and putting the names of offenders in the local newspapers. But the ultimate conclusion at this point is that rural communities tend to have far higher informal controls, which keeps their crime rates low. On the other hand, urban (especially inner-city areas) tend to have very low levels of informal controls, which leads to little attempt by the community to police itself.

Crime also tends to cluster according to social class, with the lower classes experiencing far more violent offending and victimization. This is now undisputed and consistently shown across all sources of data regarding criminal offending. Interestingly, the characteristics thus far examined for offending rates tend to be a mirror image for victimization rates. To clarify, young urban males tend to have the highest rates of criminal offending—and this group also has the highest rates of victimization as a result of violent offending. This mirror image phenomenon is often referred to as the **equivalency hypothesis.** However, this equivalency hypothesis is not true regarding the relationship of social class and property crimes.

Specifically, members of middle- to upper-class households tend to experience more victimization for property crimes than do lower-class households, but the most likely offenders in most property crimes are from the lower class. This makes sense; offenders will tend to steal from the people/places that have the most property or money to steal. This tendency has been found since criminological data has been collected, even back to the early 1800s, and it is often found in present times, although not consistently shown each year (such as in the NCVS data). Nevertheless, the equivalency hypothesis still holds true for violent crimes: lower-class individuals commit more violent crimes, and they are victimized more as a result of such violent crimes.

Race/Ethnicity Rates of Crime

Another important characteristic for how crimes are clustered in U.S. society is race/ethnicity. Regarding most violent crimes, the most victimized group by far in the United States is Native Americans/American Indians. According to NCVS data, Native Americans are victimized at almost twice the rate of any other racial/ethnic group. This is likely due to the extreme levels of poverty, unemployment, and so on that exists on virtually all reservations for American Indians. Although some Indian tribes have recently gained profits from operating gaming casinos on their lands, the vast majority of tribes in most states are not involved in such endeavors, so deprivation and poverty are still the norm.

Although there is little offending data regarding this group (the UCR does not adequately measure this group in their arrest data), it is generally assumed that this group has the highest rates of offending as well. This is a fairly safe assumption because research has clearly shown that the vast majority of criminal offending or victimization is **intraracial.** This means that crime tends to occur within a race/ethnicity (e.g., Whites offending against Whites), as opposed to being **interracial,** or across race/ethnicity (e.g., Whites offending against Blacks).

Another major group that experiences an extremely high rate of victimization, particularly for homicide, is Blacks. (The term *Black* is used here, as opposed to African American, because that is what most measures [UCR/NCVS] use and because many African Americans are not Black, for example, many citizens from Egypt or South Africa). According to UCR data for homicide, which the NCVS does not report, Blacks have by far the highest rates of victimization and offending. Again, this is likely due to the extreme levels of poverty experienced by this group, as well as the high levels of single-headed households among this population (which likely explains much of the poverty).

A good example of the extent of the high levels of homicide rates among Blacks, in terms of both offending and victimization, can be seen in previously examined rates of certain U.S. cities. Washington, D.C., New Orleans, St. Louis, and Detroit are some U.S. cities with the highest murder rates. They also have some of the highest proportions of Black residents, as compared to other U.S. cities. Notably, studies have shown that when researchers control for poverty rates and single-headed households, the racial/ethnicity effect seems to be explained away.

For example, data shows that Washington, D.C., and the state of Louisiana are the top two jurisdictions for high rates of poverty, with most of the poor being children or teenagers, who are most prone to commit crimes, especially violent offenses (such as murder). Thus, the two highest racial/ethnic groups for violent crime—Native Americans and Blacks—both as victims and offenders, also tend to have the highest rates in the nation for poverty and broken families. It should also be noted that Hispanics, a common ethnic group in the United States, also has relatively high offending and victimization rates as compared to Whites and other minorities, such as Asians, who tend to be quite low in terms of crime and victimization rates.

Although there are numerous other ways that crime tends to be clustered in our society, we have covered most of the major variables in this section. The rest of this book will now deal with examining the various theories for why crime tends to occur in such groups and individuals, and the way that we determine how accurate these theories are will directly relate to how well they explain the high rates of the groups that we have discussed.

Policy Implications

One of the key elements of a good theory is that it can help inform policymakers in making decisions about how to reduce crime. This is a theme that will be reviewed at the end of each chapter in this book. After all, a criminological theory is only truly useful in the "real world" if it helps to reduce criminal offending there. Many theories have been used as the basis of such changes in policy, and we will present in each chapter of the book examples of how theories of crime discussed in each chapter have guided policy making. We will also present empirical evidence for such policies, specifically, whether or not such policies were successful (and many are not).

A good overview of how criminological theory has impacted policy making is provided in the reading selection that was chosen for this section. In this article, Lawrence Sherman, who is one of the most highly respected scholars on policing, discusses how theoretical models have informed criminal justice practitioners in reducing crime. Furthermore, we have also provided a four-page instructional guide about how to read a research article, using examples from the Sherman article. Many articles advise policy changes, usually at the end. Regardless of whether or not the scientific studies presented in this book specifically point to certain policies, readers should be constantly thinking about the types of policy implications that can be deduced from the findings of each study/article.

◪ Section Summary

- ◆ Criminology is the scientific study of crime, which involves the use of the scientific method and testing hypotheses and distinguishes it from other perspectives of crime (e.g., journalism, religious, legal) because these fields are not based on science.

- Criminological theory seeks to do more than simply predict criminal behavior; rather, the goal is to more fully understand the reasons why certain individuals offend or do not.
- Definitions of crime vary drastically across time and place; most criminologists tend to focus on deviant behaviors, whether or not they violate the criminal codes.
- There are various paradigms or unique perspectives of crime, which have contrasting assumptions and propositions. The Classical School of criminological theory assumes free will and choice among individuals, whereas the Positive School rejects the notion of free will and focuses on biological, social, and psychological factors that negate the notion of choice among criminal offenders. Other theoretical paradigms, such as conflict/critical offending and integrated models, also exist.
- Criminological theories can be classified by their fundamental assumptions as well as other factors, such as the unit of measure on which they focus—the individual or the group, micro or macro—or the assumptions they make regarding basic human nature: good, bad, or blank slate.
- There are more than a handful of characteristics that good criminological theories should have and that bad criminological theories do not have.
- Three criteria are required for determining causality, which is essential in determining whether or not an independent (predictive) variable (X) actually affects a dependent (consequent) variable (Y)
- This introduction also discussed the primary measures of crime that are used for estimating the crime committed in the United States; these primary measures are police reports (e.g., the UCR), victimization surveys (e.g., NCVS), and offenders' self-reports of the crimes they have committed.
- We also discussed in this section what the measures of crime have shown regarding the distribution of criminal offending in terms of region, race/ethnicity, time of day, gender, and socioeconomic status. It is important to note that various theories presented in this book will be tested on the extent to which each theory can explain the clustering of crime among certain individuals and groups.

KEY TERMS

Classical School	Equivalency hypothesis	Logical consistency
Clearance rate	Formal controls	Macro-level of analysis
Conflict perspective	Hot spots	Mala in Se
Consensual perspective	Hypotheses	Mala Prohibita
Correlation or covariation	Index offenses	Micro-level of analysis
Criminology	Informal controls	National Crime Victimization Survey
Dark figure	Interracial	
Deviance	Intraracial	Non-index offenses
Empirical validity	Legalistic approach	Paradigm

Parsimony

Policy implications

Positive School

Scientific method

Scope

Self-report data

Social sciences

Spuriousness

Telescoping

Temporal Ordering

Testability

Theory

Uniform Crime Report (UCR)

DISCUSSION QUESTIONS

1. How does criminology differ from other perspectives of crime?

2. How does a good theoretical explanation of crime go beyond simply predicting crime?

3. Should criminologists emphasize only crimes, made illegal by law, or should they also study acts that are deviant but not illegal? Explain why you feel this way.

4. Even though you haven't been exposed to all the theories in this book, do you favor the classical or positive assumptions regarding criminal behavior? Explain why.

5. Which types of theories do you think are most important, those that explain individual behavior (micro) or those that explain criminality among groups (macro)? Explain your decision.

6. Do you tend to favor the idea that human beings are born bad, good, or blank slate? Discuss your decision and provide an example.

7. Which characteristics of a good theory do you find most important? Which are least important? Make sure to explain why you feel that way.

8. Of the three criteria for determining causality between a predictor variable and a consequent variable, which do you think is easiest to show, and which is hardest to show? Why?

9. Which of the measures of crime do you think is the best estimate of crime in the United States? Why? Also, which measure is the best to use for determining associations between crime and personality traits? Why?

10. Looking at what the measures of crime show, which finding surprises you the most? Explain why.

11. What types of policies do you feel reduce crime the most? Which do you think have little or no effect in reducing crime?

WEB RESOURCES

Measures of Crime:

Overview of Criminological Theory (by Dr. Cecil Greek at Florida State University): http://www.criminology.fsu.edu/crimtheory/

FBI's Uniform Crime Report (UCR): http://www.fbi.gov/ucr/ucr.htm

Department of Justice, National Crime Victimization Survey (NCVS) FAQs: http://www.ojp.usdoj.gov/bjs/abstract/ncsrqa.htm

NCVS Results: http://www.ojp.usdoj.gov/bjs/cvict.htm

How To Read A Research Article

As you travel through your criminal justice and criminology studies, you will soon learn that some of the best-known and emerging explanations of crime and criminal behavior come from research articles in academic journals. This book is full of research articles, and you may be asking yourself, "How do I read a research article?" It is my hope to answer this question with a quick summary of the key elements of any research article, followed by the questions you should be answering as you read through the assigned sections.

Every research article published in a social science journal will have the following elements: (1) introduction, (2) literature review, (3) methodology, (4) results, and (5) discussion/conclusion.

In the introduction, you will find an overview of the purpose of the research. Within the introduction, you will also find the hypothesis or hypotheses. A hypothesis is most easily defined as an educated statement or guess. In most hypotheses, you will find that the format usually followed is: If X, Y will occur. For example, a simple hypothesis may be: "If the price of gas increases, more people will ride bikes." This is a testable statement that the researcher wants to address in his or her study. Usually, authors will state the hypothesis directly, but not always. Therefore, you must be aware of what the author is actually testing in the research project. If you are unable to find the hypothesis, ask yourself what is being tested or manipulated and what are the expected results.

The next section of the research article is the literature review. At times, the literature review will be separated from the text in its own section, and at other times, it will be found within the introduction. In any case, the literature review is an examination of what other researchers have already produced in terms of the research question or hypothesis. For example, returning to my hypothesis on the relationship between gas prices and bike riding, we may find that five researchers have previously conducted studies on the increase of gas prices. In the literature review, the author will discuss their findings and then discuss what his or her study will add to the existing research. The literature review may also be used as a platform of support for the hypothesis. For example, one researcher may have already determined that an increase in gas prices causes more people to roller-skate to work. The author can use this study as evidence to support his or her hypothesis that increased gas prices will lead to more bike riding.

The methods used in the research design are found in the next section of the research article. In the methodology section, you will find the following: who/what was studied, how many subjects were studied, the research tool (e.g., interview, survey, observation), how long the subjects were studied, and how the data that was collected was processed. The methods section is usually very concise, with every step of the research project recorded. This is important because a major goal of the researcher is reliability; describing exactly how the research was done allows it to be repeated. Reliability is determined by whether the results are the same.

The results section is an analysis of the researcher's findings. If the researcher conducted a quantitative study, using numbers or statistics to explain the research, you will find statistical tables and analyses that explain whether or not the researcher's hypothesis is supported. If the researcher conducted a qualitative study, non-numerical research for the purpose of theory construction, the results will usually be displayed as a theoretical analysis or interpretation of the research question.

The research article will conclude with a discussion and summary of the study. In the discussion, you will find that the hypothesis is usually restated, and there may be a small discussion of why this was the hypothesis. You will also find a brief overview of the methodology and results. Finally, the discussion section looks at the implications of the research and what future research is still needed.

Now that you know the key elements of a research article, let us examine a sample article from your text.

⬚ The Use and Usefulness of Criminology, 1751–2005: Enlightened Justice and Its Failures

By Lawrence W. Sherman

1. What is the thesis or main idea from this article?

   - The thesis or main idea is found in the introductory paragraph of this article. Although Sherman does not point out the main idea directly, you may read the introduction and summarize the main idea in your own words. For example: The thesis or main idea is that criminology should move away from strict analysis and toward scientific experimentation to improve the criminal justice system and crime control practices.

2. What is the hypothesis?

   - The hypothesis is found in the introduction of this article. It is first stated in the beginning paragraph: "As experimental criminology provides more comprehensive evidence about responses to crime, the prospects for better basic science—and better policy—will improve accordingly." The hypothesis is also restated in the middle of the second section of the article. Here, Sherman actually distinguishes the hypothesis by stating: "The history of criminology . . . provides an experimental test of this hypothesis about analytic versus experimental social science: *that social science has been most useful, if not most used, when it has been most experimental, with visibly demonstrable benefits (or harm avoidance) from new inventions.*"

3. Is there any prior literature related to the hypothesis?

   - As you may have noticed, this article does not have a separate section for a literature review. However, you will see that Sherman devotes attention to prior literature under the heading Enlightenment, Criminology, and Justice. Here, he offers literature regarding the analytical and experimental history of criminology. This brief overview helps the reader understand the prior research, which explains why social science became primarily analytic.

4. What methods are used to support the hypothesis?

    ◆ Sherman's methodology is known as a historical analysis. In other words, rather than conducting his own experiment, Sherman is using evidence from history to support his hypothesis regarding analytic and experimental criminology. When conducting a historical analysis, most researchers use archival material from books, newspapers, journals, and so on. Although Sherman does not directly state his source of information, we can see that he is basing his argument on historical essays and books, beginning with Henry Fielding's *An Enquiry Into the Causes of the Late Increase of Robbers* (1751) and continuing through the social experiments of the 1980s by the National Institute of Justice. Throughout his methodology, Sherman continues to emphasize his hypothesis about the usefulness of experimental criminology, along with how experiments have also been hidden in the shadows of analytic criminology throughout history.

5. Is this a qualitative study or quantitative study?

    ◆ To determine whether a study is qualitative or quantitative, you must look at the results. Is Sherman using numbers to support his hypothesis (quantitative), or is he developing a non-numerical theoretical argument (qualitative)? Because Sherman does not use statistics in this study, we can safely conclude that this is a qualitative study.

6. What are the results, and how does the author present the results?

    ◆ Because this is a qualitative study, as we earlier determined, Sherman offers the results as a discussion of his findings from the historical analysis. The results may be found in the section titled Criminology: Analytic, Useful, and Used. Here, Sherman explains that "*the vast majority of published criminology remains analytic and nonexperimental.*" He goes on to say that although experimental criminology has been shown to be useful, it has not always been used or has not been used correctly. Because of the misuse of experimental criminology, criminologists have steered toward the safety of analysis rather than experimentation. Therefore, Sherman concludes that "analytic social science still dominates field experiments by 100 to 1 or better in criminology. . . . Future success of the field may depend upon a growing public image based on experimental results."

7. Do you believe that the author/s provided a persuasive argument? Why or why not?

    ◆ This answer is ultimately up to the reader, but looking at this article, I believe that it is safe to assume that readers will agree that Sherman offered a persuasive argument. Let us return to his major premise: The advancement of theory may depend on better experimental evidence, but as history has illustrated, the vast majority of criminology remains analytical. Sherman supports this proposition with a historical analysis of the great thinkers of criminology and the absence of experimental research throughout a major portion of history.

8.   Who is the intended audience of this article?

   ◆ A final question that will be useful for the reader deals with the intended audience. As you read the article, ask yourself, to whom is the author wanting to speak. After you read this article, you will see that Sherman is writing for students, professors, criminologists, historians, and criminal justice personnel. The target audience may most easily be identified if you ask yourself, "Who will benefit from reading this article?"

9.   What does the article add to your knowledge of the subject?

   ◆ This answer is best left up to the reader because the question is asking how the article improved your knowledge. However, one way to answer the question is as follows: This article helps the reader to understand that criminology is not just about theoretical construction. Criminology is both an analytical and experimental social science, and to improve the criminal justice system as well as criminal justice policies, more attention needs to be given to the usefulness of experimental criminology.

10.   What are the implications for criminal justice policy that can be derived from this article?

   ◆ Implications for criminal justice policy are most likely to be found in the conclusion or the discussion sections of the article. This article, however, emphasizes the implications throughout the article. From this article, we are able to derive that crime prevention programs will improve greatly if they are embedded in well-funded experiment-driven data rather than strictly analytical data. Therefore, it is in the hands of policymakers to fund criminological research and apply the findings in a productive manner to criminal justice policy.

Now that we have gone through the elements of a research article, it is your turn to continue through your text, reading the various articles and answering the same questions. You may find that some articles are easier to follow than others but do not be dissuaded. Remember that each article will follow the same format: introduction, literature review, methods, results, and discussion. If you have any problems, refer to this introduction for guidance.

READING

In this selection, Lawrence Sherman provides an excellent review of the policies that have resulted from the beginning of the very early stages of classical theories, through the early positivist era, and into modern times. Sherman's primary point is that experimental research is highly important in determining the policies that should be used with offenders and potential offenders. Although many important factors can never be experimentally manipulated—bad parents, poor schooling, negative peer influences—there are, as Sherman asserts, numerous types of variables that can be experimentally manipulated by criminological researchers. The resulting findings can help guide policymakers to push forward more efficient and effective policies regarding the prevention of and reaction to various forms of criminal offending. There is no better scholar to present such an argument and support for it; Sherman is perhaps the best-known scholar who has applied the experimental method to criminological research, given his experience with studies regarding domestic violence and other criminal offenses.

Readers are encouraged to consider other variables or aspects of crime that can be examined via experimental forms of research. Furthermore, readers should consider the vast number of variables that are important causes of crime or delinquency but could never be experimentally manipulated for logistic or ethical reasons.

The Use and Usefulness of Criminology, 1751–2005

Enlightened Justice and Its Failures

Lawrence W. Sherman

Criminology was born in a crime wave, raised on a crusade against torture and execution, and then hibernated for two centuries of speculation. Awakened by the rising crime rates of the latter twentieth century, most of its scholars chose to pursue analysis over experiment. The twenty-first century now offers more policy-relevant science than ever, even if basic science still occupies center stage. Its prospects for integrating basic and "clinical" science are growing, with more scholars using multiple tools rather than pursuing single-method work. Criminology contributes only a few drops of science in an ocean of decision making, but the number of drops is growing steadily. As experimental criminology provides

Source: Lawrence Sherman, "The Use and Usefulness of Criminology, 1751–2005," *The ANNALS of the American Academy of Political and Social Science* 600, no. 1 (2005): 115–35. The selection here is taken from A. Walsh and C. Hemmens, *Introduction to Criminology: A Text/Reader* (Thousand Oaks, CA: Sage, 2008), 24–32. Copyright © 2008 Sage Publications, Inc. Reprinted by permission of Sage Publications, Inc.

more comprehensive evidence about responses to crime, the prospects for better basic science — and better policy—will improve accordingly.

⬛ Enlightenment, Criminology, and Justice

The entire history of social science has been shaped by key choices scholars made in that transformative era, choices that are still made today. For criminology more than most disciplines, those Enlightenment choices have had enormous consequences for the use and usefulness of its social science. The most important of these consequences is that justice still remains largely un-Enlightened by empirical evidence about the effects of its actions on public safety and public trust.

Historians may despair at defining a coherent intellectual or philosophical content in the Age of Enlightenment, but one idea seems paramount: "that we understand nature and man best through the use of our natural faculties" (May 1976, xiv) by systematic empirical methods, rather than through ideology, abstract reasoning, common sense, or claims of divine principles made by competing religious authorities. Kant, in contrast, stressed the receiving end of empirical science in his definition of Enlightenment: the time when human beings regained the courage to "use one's own mind without another's guidance" (Gay 1969, 384).

Rather than becoming *experimental* in method, social science became primarily *analytic*. This distinction between experimental manipulation of some aspect of social behavior versus detached (if systematic) observation of behavioral patterns is crucial to all social science (even though not all questions for social science offer a realistic potential for experiment). The decision to cast social science primarily in the role of critic, rather than of inventor, has had lasting consequences for the enterprise, especially for the credibility of its conclusions. There may be nothing so practical as a good theory, but it is hard to visibly— or convincingly—demonstrate the benefits of social analysis for the reduction of human misery. The absence of "show-and-tell" benefits of analytic social science blurred its boundaries with ideology, philosophy, and even emotion. This problem has plagued analytic social science ever since, with the possible exception of times (like the Progressive Era and the 1960s) when the social order itself was in crisis. As sociologist E. Digby Baltzell (1979) suggested about cities and other social institutions, "as the twig is bent, so grows the tree." Social science may have been forged in the same kind of salon discussions as natural science, but without some kind of empirical reports from factories, clinics, or farm fields. Social science has thus famously "smelled too much of the lamp" of the library (Gay 1969). Even when analytic social science has been most often used, it is rarely praised as useful.

That is not to say that theories (with or without evidence) have lacked influence in criminology, or in any social science. The theory of deterrent effects of sanctions was widely used to reduce the severity of punishment long before the theory could be tested with any evidence. The theories of "anomie" and "differential association" were used to plan the 1960s "War on Poverty" without any clear evidence that opportunity structures could be changed. Psychological theories of personality transformation were used to develop rehabilitation programs in prisons long before any of them were subject to empirical evaluation. Similarly, evidence (without theory) of a high concentration of crime among a small proportion of criminal offenders was used to justify more severe punishment for repeat offenders, also without empirical testing of those policies.

The criminologists' general preference for analysis over experiment has not been universal in social science. Enlightenment political science was, in an important—if revolutionary— sense, experimental, developing and testing new forms of government soon after they were suggested in print. The Federalist Papers, for example, led directly to the "experiment" of the Bill of Rights.

Perhaps the clearest exception to the dominance of analytic social science was within criminology itself in its very first work during the Enlightenment. The fact that criminologists do not remember it this way says more about its subsequent dominance by analytic methods than about the true history of the field. Criminology was born twice in the eighteenth century, first (and forgotten) as an experimental science and then (remembered) as an analytic one. And though experimental criminology in the Enlightenment had an enormous impact on institutions of justice, it was analytic criminology that was preserved by law professors and twentieth-century scholars as the foundation of the field.

The history of criminology thus provides an experimental test of this hypothesis about analytic versus experimental social science: *that social science has been most useful, if not most used, when it has been most experimental, with visibly demonstrable benefits (or harm avoidance) from new inventions.* The evidence for this claim in eighteenth-century criminology is echoed by the facts of criminology in the twentieth century. In both centuries, the fraternal twins of analysis and experiment pursued different pathways through life, while communicating closely with each other. One twin was critical, the other imaginative; one systematically observational, the other actively experimental; one detached with its integrity intact, the other engaged with its integrity under threat. Both twins needed each other to advance their mutual field of inquiry. But it has been experiments in every age that made criminology most useful, as measured by unbiased estimates of the effects of various responses to crime.

The greatest disappointment across these centuries has been the limited usefulness of experimental criminology in achieving "geometric precision" (Beccaria 1764/1964) in

the pursuit of "Enlightened Justice," defined as "the administration of sanctions under criminal law guided by (1) inviolate principles protecting human rights of suspects and convicts while seeking (2) consequences reducing human misery, through means known from (3) unbiased empirical evidence of what works best" (Sherman 2005). While some progress has been made, most justice remains unencumbered by empirical evidence on its effects. To understand why this disappointment persists amid great success, we must begin with the Enlightenment itself.

⊠ Inventing Criminology: Fielding, Beccaria, and Bentham

The standard account of the origin of criminology locates it as a branch of moral philosophy: part of an aristocratic crusade against torture, the death penalty, and arbitrary punishment, fought with reason, rhetoric, and analysis. This account is true but incomplete. Criminology's forgotten beginnings preceded Cesare Beccaria's famous 1764 essay in the form of Henry Fielding's 1753 experiments with justice in London. Inventing the modern institutions of a salaried police force and prosecutors, of crime reporting, crime records, employee background investigations, liquor licensing, and social welfare policies as crime prevention strategies, Fielding provided the viable preventive alternatives to the cruel excesses of retribution that Beccaria denounced—before Beccaria ever published a word.

The standard account hails a treatise on "the science of justice" (Gay 1969, 440) that was based on Beccaria's occasional visits to courts and prisons, followed by many discussions in a salon. The present alternative account cites a far less famous treatise based on more than a thousand days of Fielding conducting trials and sentencing convicts in the world's (then) largest city, supplemented by his on-site inspections of tenements, gin joints, brothels, and public hangings. The standard account thus chooses a criminology of analytic detachment over a criminology of clinical engagement.

The standard account in twentieth-century criminology textbooks traced the origin of the field to this "classical school" of criminal law and criminology, with Cesare Beccaria's (1738–1794) treatise *On Crimes and Punishments* (1764) as the first treatise in scientific criminology. (Beccaria is also given credit [incorrectly], even by Enlightenment scholars, for first proposing that utility be measured by "the greatest happiness divided among the greatest number"— which Frances Hutcheson, a mentor to Adam Smith, had published in Glasgow in 1725 before Beccaria was born [Buchan 2003, 68–71]). Beccaria, and later Bentham, contributed the central claims of the deterrence hypothesis on which almost all systems of criminal law now rely: that punishment is more likely to prevent future crime to the extent that it is certain, swift, and proportionate to the offense (Beccaria) or more costly than the benefit derived from the offense (Bentham).

Fielding

This standard account of Beccaria as the *first* criminologist is, on the evidence, simply wrong. Criminology did not begin in a Milanese salon among the group of aristocrats who helped Beccaria formulate and publish his epigrams but more than a decade earlier in a London magistrate's courtroom full of gin-soaked robbery defendants. The first social scientist of crime to publish in the English—and perhaps any— language was Henry Fielding, Esq. (1707–1754). Fielding was appointed by the government as magistrate at the Bow Street Court in London. His years on that bench, supplemented by his visits to the homes of London labor and London poor, provided him with ample qualitative data for his 1751 treatise titled *An Enquiry Into the Causes of the Late Increase of Robbers*.

Fielding's treatise is a remarkable analysis of what would today be called the "environmental criminology" of robbery. Focused on the reasons for a crime wave and the policy alternatives to hanging as the only means of combating crime, Fielding singles out the wave of "that poison called gin" that hit mid-century London like crack hit New York in the 1980s. He theorizes that a drastic price increase (or tax) would make gin too expensive for most people to consume, thereby reducing violent crime. He also proposes more regulation of gambling, based on his interviews with arrested robbers who said they had to rob to pay their gambling debts. Observing the large numbers of poor and homeless people committing crime, he suggests a wider "safety net" of free housing and food. His emphasis is clearly on prevention without punishment as the best policy approach to crime reduction.

Fielding then goes on to document the failures of punishment in three ways. First, the system of compulsory "voluntary policing" by each citizen imposed after the Norman Conquest had become useless: "what is the business of every man is the business of no man." Second, the contemporary system of requiring crime victims to prosecute their own cases (or hire a lawyer at their own expense) was failing to bring many identified offenders to justice. Third, witnesses were intimidated and often unwilling to provide evidence needed for conviction. All this leads him to hint at, but not spell out, a modern system of "socialized" justice in which the state, rather than crime victims, pays for police to investigate and catch criminals, prosecutors to bring evidence to court, and even support for witnesses and crime victims.

His chance to present his new "invention" to the government came two years after he published his treatise on robbery. In August, 1753, five different robbery-murders were committed in London in one week. An impatient cabinet secretary summoned Fielding twice from his sickbed and asked him to propose a plan for stopping the murders. In four days, Fielding submitted a "grant proposal" for an experiment in policing that would cost £600 (about £70,000 or $140,000 in current value). The purpose of the money was to retain, on salary, the band of detectives Fielding worked with, and to pay a reward to informants who would provide evidence against the murderers.

Within two weeks, the robberies stopped, and for two months not one murder or robbery was reported in Westminster (Fielding 1755/1964, 191–193). Fielding managed to obtain a "no-cost extension" to the grant, which kept the detectives on salary for several years. After Henry's death, his brother John obtained new funding, so that the small team of "Bow Street Runners" stayed in operation until the foundation of the much larger—and uniformed—Metropolitan Police in 1829.

The birth of the Bow Street Runners was a turning point in the English paradigm of justice. The crime wave accompanying the penny-a-quart gin epidemic of the mid-eighteenth century had demonstrated the failure of relying solely on the *severity* of punishment, so excessive that many juries refused to convict people who were clearly guilty of offenses punishable by death—such as shoplifting. As Bentham would later write, there was good reason to think that the *certainty* of punishment was too low for crime to be deterrable. As Fielding said in his treatise on robbery, "The utmost severity to offenders [will not] be justifiable unless we take every possible method of preventing the offence." Fielding was not the only inventor to propose the idea of a salaried police force to patrol and arrest criminals, but he was the first to conduct an *experiment* testing that invention. While Fielding's police experiment would take decades to be judged successful (seventy-six years for the "Bobbies" to be founded at Scotland Yard in 1829), the role of experimental evidence proved central to changing the paradigm of practice.

Beccaria

In sharp contrast, Beccaria had no clinical practice with offenders, nor was he ever asked to stop a crime wave. Instead, he took aim at a wave of torture and execution that characterized European justice. Arguing the same ideology of prevention as Fielding (whose treatise he did not cite), Beccaria urged abolition of torture, the death penalty, and secret trials. Within two centuries, almost all Europe had adopted his proposals. While many other causes of that result can be cited, there is clear evidence of Beccaria's 1764 treatise creating a "tipping point" of public opinion on justice.

What Beccaria did not do, however, was to supply a shred of scientific evidence in support of his theories of the deterrent effects of non-capital penalties proportionate to the severity of the offense. Nor did he state his theories in a clearly falsifiable way, as Fielding had done. In his method, Beccaria varies little from law professors or judges (then and now) who argue a blend of opinion and factual assumptions they find reasonable, deeming it enlightened truth *ipse dixit* ("because I say so myself"). What he lacked by the light of systematic analysis of data, he made up for by eloquence and "stickiness" of his aphorisms. Criminology by slogan may be more readily communicated than criminology by experiment in terms of fame. But it is worth noting that the founding of the British police appears much more directly linked to Fielding's experiments than the steady abolition of the death penalty was linked to Beccaria's book.

Bentham

Beccaria the moral-empirical theorist stands in sharp contrast to his fellow Utilitarian Jeremy Bentham, who devoted twelve years of his life (and some £10,000) to an invention in prison administration. Working from a book he wrote on a "Panopticon" design for punishment by incarceration (rather than hanging), Bentham successfully lobbied for a 1794 law authorizing such a prison to be built. He was later promised a contract to build and manage such a prison, but landed interests opposed his use of the site he had selected. We can classify Bentham as an experimentalist on the grounds that he invested much of his life in "trying" as well as thinking. Even though he did not build the prison he designed, similar prisons (for better or worse) were built in the United States and elsewhere. Prison design may justifiably be classified as a form of invention and experimental criminology, as distinct from the analytic social science approach Bentham used in his writings—thereby making him as "integrated" as Fielding in terms of theory and practice. The demise of Bentham's plans during the Napoleonic Wars marked the end of an era in criminology, just as the Enlightenment itself went into retreat after the French Revolution and the rise of Napoleon. By 1815, experimentalism in criminology was in hibernation, along with most of criminology itself, not to stir until the 1920s or spring fully to life until the 1960s.

▧ Two Torpid Centuries— With Exceptions

Analytic criminology continued to develop slowly even while experimental criminology slumbered deeply, but neither had any demonstrable utility to the societies that fostered them. One major development was the idea of involuntary causes of crime "determined" by either social (Quetelet 1835/2004) or biological (Lombroso 1876/1918) factors that called into question the legal doctrines of criminal responsibility. The empirical evidence for these claims, however, was weak (and in Lombroso's case, wrong), leaving the theoretical approach to criminology largely

unused until President Johnson's War on Poverty in the 1960s.

Cambridge-Somerville

The first fully randomized controlled trial in American criminology appears to have been the Cambridge-Somerville experiment, launched in Massachusetts in the 1930s by Dr. Richard Clark Cabot. This project offered high-risk young males "friendly guidance and social support, healthful activities after school, tutoring when necessary, and medical assistance as needed" (McCord 2001). It also included a long-term "big brother" mentoring relationship that was abruptly terminated in most cases during World War II. While the long-term effects of the program would not be known until the 1970s, the critical importance of the experimental design was recognized at the outset. It was for that reason that the outcomes test could reach its startling conclusion: "The results showed that as compared with members of the control group, those who had been in the treatment program were more likely to have been convicted for crimes indexed by the Federal Bureau of Investigation as serious street crimes; they had died an average of five years younger; and they were more likely to have received a medical diagnosis as alcoholic, schizophrenic, or manic-depressive" (McCord 2001, 188). In short, the boys offered the program would have been far better off if they had been "deprived" of the program services in the randomly assigned control group.

No study in the history of criminology has ever demonstrated such clear, unintended, criminogenic effects of a program intended to prevent crime. To this day, it is "exhibit A" in discussions with legislators, students, and others skeptical of the value of evaluating government programs of any sort, let alone crime prevention programs. Its early reports in the 1950s also set the stage for a renaissance in experimental criminology, independently of the growth of analytic criminology.

⊠ Renaissance: 1950–1982

Amidst growing concern about juvenile delinquency, the Eisenhower administration provided the first federal funding for research on delinquency prevention. Many of the studies funded in that era, with both federal and non-federal support, adopted experimental designs. What follows is merely a highlighting of the renaissance of experimental criminology in the long twilight of the FDR coalition prior to the advent of the Reagan revolution.

Martinson and Wilson

While experimental evidence was on the rise in policing, it was on the decline in corrections. The comprehensive review of rehabilitation strategies undertaken by Lipton, Martinson, and Wilks (1975) initially focused on the internal validity of the research designs in rehabilitation experiments within prisons. Concluding that these designs were too weak to offer unbiased estimates of treatment effects, the authors essentially said "we don't know" what works to rehabilitate criminals. In a series of less scientific and more popular publications, the summary of the study was transformed into saying that there is no evidence that criminals can be rehabilitated. Even the title "What Works" was widely repeated in 1975 by word of mouth as "nothing works."

The Martinson review soon became the basis for a major change in correctional policies. While the per capita rates of incarceration had been dropping throughout the 1960s and early 1970s, the trend was rapidly reversed after 1975 (Ruth and Reitz 2003). Coinciding with the publication of Wilson's (1975) first edition of *Thinking About Crime*, the Martinson review arguably helped fuel a sea change from treating criminals as victims of society to treating

society as the victim of criminals. That, in turn, may have helped to feed a three-decade increase in prisoners (Laub 2004) to more than 2.2 million, the highest incarceration rate in the world.

⊠ Warp Speed: 1982–2005

Stewart

In September, 1982, a former Oakland Police captain named James K. Stewart was appointed director of the National Institute of Justice (NIJ). Formerly a White House Fellow who had attended a National Academy of Sciences discussion of the work of NIJ, Stewart had been convinced by James Q. Wilson and others that NIJ needed to invest more of its budget in experimental criminology. He acted immediately by canceling existing plans to award many research grants for analytic criminology, transferring the funds to support experimental work. This work included experiments in policing, probation, drug market disruption, drunk-driving sentences, investigative practices, and shoplifting arrests.

Schools

The 1980s also witnessed the expansion of experimental criminology into the many school-based prevention programs. Extensive experimental and quasi-experimental evidence on their effects—good and bad—has now been published. In one test, for example, a popular peer guidance group that was found effective as an alternative to incarceration was found to increase crime in a high school setting. Gottfredson (1987) found that high-risk students who were not grouped with other high-risk students in high school group discussions did better than those who were.

Drug Courts

The advent of (diversion from prosecution to medically supervised treatments administered by) "drug courts" during the rapid increase in experimental criminology has led to a large and growing volume of tests of drug court effects on recidivism. Perhaps no other innovation in criminal justice has had so many controlled field tests conducted by so many different independent researchers. The compilations of these findings into meta-analyses will shed increasing light on the questions of when, and how, to divert drug-abusing offenders from prison.

Boot Camps

Much the same can be said about boot camps. The major difference is that boot camp evaluations started off as primarily quasi-experimental in their designs (with matched comparisons or worse), but increasing numbers of fully randomized tests have been conducted in recent years (Mitchell, MacKenzie, and Perez 2005). Many states persist in using boot camps for thousands of offenders, despite fairly consistent evidence that they are no more effective than regular correctional programs.

Child Raising

Criminology has also claimed a major experiment in child raising as one of its own. Beginning at the start of the "warp speed" era, the program of nurse home visits to at-risk first mothers designed by Dr. David Olds and his colleagues (1986) has now been found to have long-term crime prevention effects. Both mothers and children show these effects, which may be linked to lower levels of child abuse or better anger management practices in child raising.

⬙ Criminology: Analytic, Useful, and Used

This recitation of a selected list of experiments in criminology must be labeled with a consumer warning: *the vast majority of published criminology remains analytic and nonexperimental*. While criminology was attracting funding and students during the period of rising crime of the 1960s to 1990s, criminologists put most of their efforts into the basic science of crime patterns and theories of criminality. Studies of the natural life course of crime among cohorts of males became the central focus of the field, as measured by citation patterns (Wolfgang, Figlio, and Thornberry 1978). Despite standing concerns that criminology would be "captured" by governments to become a tool for developing repressive policies, the evidence suggests that the greatest (or largest) generation of criminologists in history captured the field away from policymakers.

The renaissance in experimental criminology therefore addressed very intense debates over many key issues in crime and justice, providing the first unbiased empirical guidance available to inform those debates. That much made criminology increasingly useful, at least potentially. Usefulness alone, of course, does not guarantee that the information will be *used*. Police agencies today do make extensive use of the research on concentrating patrols in crime hot spots, yet they have few repeat offender units, despite two successful tests of the "invention." Correctional agencies make increasing use of the "what works" literature in the United States and United Kingdom, yet prison populations are still fed by people returned to prison on the unevaluated policy of incarcerating "technical" violators of the conditions of their release (who have not committed new crimes). Good evidence alone is not enough to change policy in any context. Yet absent good evidence, there is a far greater danger that bad policies will win out. Analytic criminology—well or badly done—poses fewer risks for society than badly done experimental criminology. It is not clear that another descriptive test of differential association theory will have any effect on policy making, unless it is embedded in a program evaluation. But misleading or biased evidence from poor-quality research designs—or even unreplicated experiments—may well cause the adoption of policies that ultimately prove harmful.

This danger is, in turn, reduced by the lack of influence criminology usually has on policy making or operational decisions. That, in turn, is linked to the absence of clear conclusions about the vast majority of criminal justice policies and decisions. Until experimental criminology can develop a more comprehensive basis of evidence for guiding operations, practitioners are unlikely to develop the habit of checking the literature before making a decision. The possibility of improving the quality of both primary evidence and systematic reviews offers hope for a future in which criminology itself may entail less risk of causing harm.

This is by no means a suggestion that analytic criminology be abandoned; the strength of experimental criminology may depend heavily on the continued advancement of basic (analytic) criminology. Yet the full partnership between the two has yet to be realized. Analytic social science still dominates field experiments by 100 to 1 or better in criminology, just as in any other field of research on human behavior. Future success of the field may depend upon a growing public image based on experimental results, just as advances in treatment attract funding for basic science in medicine.

⊠ Conclusion

Theoretical criminology will hold center stage for many years to come. But as Farrington (2000) has argued, the advancement of theory may depend on better experimental evidence. And that, in turn, may depend on a revival in the federal funding that has recently dropped to its lowest level in four decades. Such a revival may well depend on exciting public interest in the practical value of research, as perhaps only experiments can do.

"Show and tell" is hard to do while it is happening. Yet it is not impossible. Whether anyone ever sees a crime prevention program delivered, it is at least possible to embed an experimental design into every long-term analytic study of crime in the life course. As Joan McCord (2003) said in her final words to the American Society of Criminology, the era of purely observational criminology should come to an end. Given what we now know about the basic life-course patterns, McCord suggested, "all longitudinal studies should now have experiments embedded within them."

Doing what McCord proposed would become an experiment *in* social science as well as *of* social science. That experiment is already under way, in a larger sense. Criminology is rapidly becoming more multi-method, as well as multi-level and multi-theoretical. Criminology may soon resemble medicine more than economics, with analysts closely integrated with clinical researchers to develop basic science as well as treatment. The integration of diverse forms and levels of knowledge in "consilience" with each other, rather than a hegemony of any one approach, is within our grasp. It awaits only a generation of broadly educated criminologists prepared to do many things, or at least prepared to work in collaboration with other people who bring diverse talents to science.

⊠ References

Baltzell, D. (1979). *Puritan Boston and Quaker Philadelphia: Two Protestant ethics and the spirit of class authority and leadership.* New York: Free Press.

Beccaria, C. (1964). *On crimes and punishments* (J. Grigson, Trans.). Milan, Italy: Oxford University Press. (Original work published 1764)

Buchan, J. (2003). *Crowded with genius: The Scottish Enlightenment: Edinburgh's moment of the mind.* New York: HarperCollins.

Farrington, D. (2000). Explaining and preventing crime: The globalization of knowledge. The American Society of Criminology 1999 Presidential Address. *Criminology, 38,* 1–24.

Fielding, H. (1964). *The journal of a voyage to Lisbon.* London: Dent. (Original work published 1755)

Gay, P. (1969). *The Enlightenment: An interpretation. Vol. 2, The science of freedom.* New York: Knopf.

Gottfredson, G. (1987). Peer group interventions to reduce the risk of delinquent behavior: A selective review and a new evaluation. *Criminology, 25,* 671–714.

Laub, J. (2004). The life course of criminology in the United States: The American Society of Criminology 2003 Presidential Address. *Criminology, 42,* 1–26.

Lipton, D., Martinson, R., & Wilks, J. (1975). *The effectiveness of correctional treatment: A survey of treatment evaluation studies.* New York: Praeger.

Lombroso, C. (1918). *Crime, its causes and remedies.* Boston: Little, Brown. (Original work published 1876)

May, H. (1976). *The Enlightenment in America.* New York: Oxford University Press.

McCord, J. (2001). *Crime prevention: A cautionary tale.* Proceedings of the Third International, Inter-Disciplinary Evidence-Based Policies and Indicator Systems Conference, University of Durham. Retrieved April 22, 2005, from http://cem.dur.ac.uk

McCord, J. (2003). *Discussing age, crime, and human development. The future of life-course criminology.* Denver, CO: American Society of Criminology.

Mitchell, O., MacKenzie, D., & Perez, D. (2005). A randomized evaluation of the Maryland correctional boot camp for adults: Effects on offender anti-social attitudes and cognitions. *Journal of Offender Rehabilitation, 40*(4).

Olds, D., Henderson, C., Chamberlin, R., & Tatelbaum, R. (1986). *Pediatrics, 78,* 65–78.

Quetelet, A. (2004). A treatise on man. As cited in F. Adler, G. O. W. Mueller, and W. S. Laufer, *Criminology and the criminal justice system* (5th ed., p. N-6). New York: McGraw-Hill. (Original work published 1835)

Ruth, H., & Reitz, K. (2003). *The challenge of crime: Rethinking our response.* Cambridge, MA: Harvard University Press.

Sherman, L. (2005). *Enlightened justice: Consequentialism and empiricism from Beccaria to Braithwaite.* Address to the 14th World Congress of Criminology, International Society of Criminology, Philadelphia, August 8.

Wilson, J. (1975). *Thinking about crime.* New York: Basic Books.

Wolfgang, M., Figlio, R., & Thornberry, T. (1978). *Evaluating criminology.* New York: Elsevier.

REVIEW QUESTIONS

1. According to Sherman, what part did Henry Fielding's research play in the early stages of research on crime and the influence it had on stopping robberies in London? Explain in detail. What implications can be drawn about using resources (i.e., money) to stop a given crime in a certain location?

2. According to Sherman, what impact did the "Bow Street Runners" have?

3. What does Sherman have to say about what Beccaria and Bentham contributed to policies regarding crime? Do you agree with Sherman's assessment?

4. What does Sherman have to say about Martinson's review of rehabilitation programs and its impact on policy?

5. What does Sherman have to say about criminological research regarding schools, drug courts, boot camps, and child raising? Which recent programs does he claim had success? Which recent programs or designs does he suggest do not work?

❖

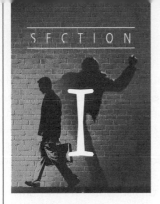

PRE-CLASSICAL AND CLASSICAL THEORIES OF CRIME

Introduction

This section will examine the earliest logical theories of rule-breaking, namely explanations of criminal conduct that emphasize the free will and ability of individuals to make rational decisions regarding the consequences of their behavior. The natural capabilities of human beings to make decisions based on expected costs and benefits were acknowledged during the **Age of Enlightenment** in the 17th and 18th centuries. This understanding of human capabilities led to what is considered the first rational theory of criminal activity, deterrence theory. The importance and impact of this theory had a more profound impact on justice systems in the United States than any other perspective to date. Furthermore, virtually all criminal justice systems (e.g., policing, courts, corrections) are based on this theoretical model, even today.

Such theories of human rationality were in stark contrast to the theories focused on religious or supernatural causes of crime, which had prevailed through most of human civilization up to the Age of Enlightenment. In addition, the **Classical School** theories of crime are distinguished from theories in subsequent sections of this book by their emphasis on the free will and rational decision making of individuals, which modern theories of crime tend to ignore. The theoretical perspectives discussed in this section all focus on the ability of human beings to choose their own behavior and destiny, whereas paradigms that existed before and after this period tend to emphasize the inability of individuals to control their behavior due to external factors. Therefore, the Classical School is perhaps the paradigm best suited for analysis of what types of calculations are going on in someone's head before they commit a crime.

The different Classical School theories presented in this section vary in many ways, most notably in what they propose as the primary constructs and processes individuals use to determine whether or not they are going to commit the crime. For example, some Classical School theories place an emphasis on the potential negative consequences of their actions, whereas others focus on the possible benefits of such activity. Still others concentrate on the opportunities and existing situations that predispose people to engage in criminal activity. Regardless of their differences, all of the theories examined in this section emphasize a common theme: Individuals commit crime because they identify certain situations and acts as beneficial due to the perceived lack of punishment and the perceived likelihood of profits, such as money or peer status. In other words, the potential offender weighs out the possible costs and pleasures of committing a given act and then behaves in a rational way based on this analysis.

The most important distinction of these Classical School theories, as opposed to those discussed in future sections, is that they emphasize individuals making their own decisions regardless of extraneous influences, such as the economy or bonding with society. Although many extraneous factors may influence the ability of an individual to rationally consider offending situations, the Classical School assumes that the individual takes all these influences into account when making the decision about whether or not to engage in criminal behavior. Given the focus placed on individual responsibility, it is not surprising that Classical School theories are used as the basis for the U.S. policies on punishment for criminal activity. The Classical School theories are highly compatible and consistent with the conservative "get tough" movement that has existed since the mid-1970s. Thus, the Classical School still retains the highest importance in terms of policy and pragmatic punishment in the United States, as well as throughout all countries in the Western world.

As you will see, the Classical School theoretical paradigm was presented as early as the mid-1700s, and its prominence as a model of offending behavior in criminal justice systems is still dominant. Scientific and academic circles, however, have dismissed many of the claims of this perspective. For reasons we shall explore in this section, the assumptions and primary propositions of the Classical School theories have been neglected by most recent theoretical models of criminologists. This dismissal is likely premature, given the impact that this perspective has had on understanding human nature, as well as the profound influence it has had on most criminal justice systems, especially that in the United States.

⊠ Pre-Classical Perspectives of Crime and Punishment

Over the long course of human civilization, people have mostly believed that criminal activity was caused by supernatural causes or religious factors. It has been documented that some primitive societies believed that crime increased during major thunderstorms or major droughts. Most primitive cultures believed that when a person engaged in behavior that violated the tribe's or clan's rules, the devil or evil spirits were making them do it.[1] For example, in many societies at that time, if someone had committed criminal activity, it was common to perform exorcisms or primitive surgeries, such as breaking open the skull of the perpetrator to allow the demons to leave his or her head. Of course, this almost always resulted in the death of the accused person, but it was seen as a liberating experience for the offender.

This was just one form of dealing with criminal behavior, but it epitomizes how primitive cultures understood the causes of crime. As the movie *The Exorcist* revealed, exorcisms were still being performed by representatives of a number of religions, including Catholicism, in the 21st century "to get the devil out of them." In June 2005, a Romanian monk and four nuns acknowledged engaging in an exorcism that led to the death of the victim, who was crucified, a towel stuffed in her mouth, and left without food for many days.[2] When this monk and the nuns were asked to explain why they did this, they defiantly said they were trying to take the devils out of the 23-year-old woman. Although they were prosecuted by Romanian authorities, many governments might not have done so because many societies around the world still believe in and condone such practices.

Readers may be surprised to learn that the Roman Catholic Church still authorizes college-level courses on how to perform exorcisms. Specifically, recent news reports revealed that a Vatican-recognized university was offering a course in exorcism and demonic possession for a second year because of their concern about the "devil's lure."[3] In fact, Pope Benedict XVI welcomed a large group of Italian exorcists who visited the Vatican in September 2005 and encouraged them to carry on their work for the Catholic Church.[4] Furthermore, the Roman Catholic Church in 1999 issued their revised guidelines for conducting exorcisms, which recommend consulting physicians when exorcisms are performed; it also provides an 84-page description of the language (in Latin) to be used in such rituals. It should be noted that the use of such exorcisms is quite rare, especially in more developed nations.

One of the most common supernatural beliefs in primitive cultures was that the full moon caused criminal activity. Then as now, there was much truth to the full-moon theory. In primitive times, people believed the crime was related to the influence of higher powers, including the "destructive influence" of the moon itself. Modern studies have shown, however, that the increase in crime is primarily due to a Classical School theoretical model: There are simply more opportunities to commit crime when the moon is full because there is more light at night, which results in more people out on the streets. In any case, nighttime is well-established as a high-risk period for adult crime, such as sexual assault.

Although some primitive theories had some validity in determining when crime was more common, virtually none of them accurately predicted who would commit the

---

[1]Stephen E. Brown, Finn-Aage Esbensen, and Gilbert Geis, *Criminology: Explaining Crime and Its Context,* 5th ed. (Cincinnati, OH: Anderson, 2004).

[2]Associated Press, "Five Charged in Deadly Exorcism," June 24, 2005.

[3]Associated Press, "University Offering Course on Exorcism," July 9, 2005.

[4]Reuters, "Pope Benedict XVI Warmly Greets Exorcist Convention," September 17, 2005.

offenses. During the Middle Ages, just about everyone was from the lower classes, and only a minority of that group engaged in offending against the normative society. So, for most of human civilization, there was virtually no rational theoretical understanding for why individuals violate the laws of society; instead people believed crime was caused by supernatural or religious factors of "the devil made me do it" variety.

Consistent with these views, the punishments related to offending during this period were quite harsh by modern standards. Given the assumption that evil spirits drove the motivations for criminal activity, the punishments of criminal acts, especially those deemed particularly offensive to the norms of the given society, were often quite inhumane. Common punishments at that time were beheading, torture, being burned alive at the stake, and being drowned, stoned, or quartered. Good discussions of such harsh examples of punishment, such as quartering, can be found in Brown et al.'s discussion of punishment.[5]

Although many would find the primitive forms of punishment and execution to be quite barbaric, many societies still practice them. For example, Islamic court systems as well as other religious/ethnic cultures are often allowed to carry out executions and other forms of corporal punishment. Fifteen individuals were whipped with a cane for gambling in Aceh, Indonesia, a highly conservative Muslim region. The caning was held in public and outside a mosque.[6] In the United States, gambling is a relatively minor crime—when it is not totally legal, as in many places in the United States. It is interesting to note, however, that a relatively recent Gallup Poll regarding the use of caning (i.e., public whipping) of convicted individuals was supported by most of the American public.[7]

Compared to U.S. standards, the more extreme forms of corporal punishment, particularly public executions carried out by many religious courts and countries, are drawn out and very painful. An example is stoning, in which people are buried up to the waist and local citizens throw small stones at them until they die (large stones are not allowed because they would lead to death too quickly). In most of the Western world, such brutal forms of punishment and execution were done away with in the 1700s due to the impact of the Age of Enlightenment.

⬙ The Age of Enlightenment

In the midst of the extremely draconian times of the mid-1600s, Thomas Hobbes, in his book *Leviathan* (1651), proposed a rational theory of why people are motivated to form democratic states of governance.[8] Hobbes started with a basic framework that all individuals are in a constant state of warfare with all other individuals. Hobbes used the extreme examples of primitive tribes and sects. He argued that the primitive state of man is selfish and greedy, so people live in a constant state of fear of everyone else. However,

---

[5]Brown et al., *Criminology.*

[6]Irwan Firdaus, "Gamblers Whipped in Muslim Indonesia," Associated Press, June 25, 2005.

[7]Philip Shenon, "U.S. Youth in Singapore Loses Appeal on Flogging," *New York Times,* April 1, 1994.

[8]Thomas Hobbes, *Leviathan* (New York: Library of Liberal Arts, 1958; first published in 1651).

Hobbes also proclaimed that people are also rational, so they will rationally organize and form a sound form of governance, which can create rules to avoid this constant state of fear. Interestingly, once a government is created, the state of warfare mutates from one waged among individuals or families to one between nations. This can be seen in modern times; after all, it is quite rare that a person or family declares war against another (although gangs may be an exception), but we often hear of governments declaring war.

Hobbes stated that the primitive state of fear—of constant warfare of everyone against everyone else—was the motivation for entering into the contract with others in creating a common authority. At the same time, Hobbes also specified that it was this exact emotion—fear—that was needed in making citizens conform to the given rules or laws in society. Strangely, it appears that the very emotion that inspires people to enter into an agreement to form a government is the same emotion that inspires these same individuals to follow the rules of the government that is created. It is ironic, but it is quite true.

Given the social conditions during the 1600s, this model appears somewhat accurate; there was little sense of community regarding working as a group toward progress. It had not been that long since most societies had experienced the Middle Ages—also known as the "Dark Ages"—when a third of the world's population had died from sickness, and most others were severely deprived or in extreme states of poverty. Just prior to the 1600s, the feudal system was the dominant model of governance in most of the Western world. During this feudal era, a very small group of aristocrats (less than 1% of the population) owned and operated the largely agricultural economy that existed. Virtually no rights were afforded to individuals in the Middle Ages or at the time Hobbes wrote his book. Most people had no individual rights regarding criminal justice, let alone a say in the existing governments of the time. Hobbes's book clearly took issue with this lack of say in the government, which has profound implications regarding the justice systems of that time.

Hobbes clearly stated that until the citizens were entitled to a certain degree of respect from their governing bodies, as well as their justice systems, the citizens would never fully buy into the authority of government or the system of justice. Hobbes proposed a number of extraordinary ideas that came to define the Age of Enlightenment. He presented a drastic paradigm shift to this new social structure, which had extreme implications for justice systems throughout the world.

Hobbes explicitly declared that human beings are rational beings who choose their destiny by creating a society. Hobbes further proposed that individuals in such societies democratically create rules of conduct that all members of that society must follow. These rules that all citizens decide upon become laws, and the result of not following them is punishment determined by the democratically instituted government. It is clear from Hobbes's statements that the government, as instructed by the citizens, not only has the authority to punish individuals who violate the rules of the society but, more important, has the duty to punish them. When such an authority fails to fulfill this duty, it can quickly result in a breakdown in the social order.

The arrangement of citizens promising to abide by the rules or laws set forth by a given society in return for protection is commonly referred to as the **social contract**. Hobbes introduced this idea, but it was also emphasized by all other Enlightenment theorists after him, such as Rousseau, Locke, Voltaire, Montesquieu, and others. The idea of the social contract is an extraordinarily important part of Enlightenment philosophy. Although the Enlightenment philosophers had significant differences in what they believed, the one thing they had in common was the belief in the social contract: the idea that people invest in the laws of their society, with the guarantee that they will be protected from others who violate such rules.

Another shared belief among Enlightenment philosophers was giving people a say in the government, especially the justice system. All of them emphasized fairness in determining who was guilty, as well as appropriate punishments or sentences. During the time in which they wrote, individuals who stole a loaf of bread to feed their family were sentenced to death, whereas upper-class individuals who stole large sums of money or committed murder were pardoned. This goes against common sense; moreover, it violates the social contract. If citizens observe people being excused for violating the law, then their belief in the social contract breaks down. This same feeling can be applied to modern times. When the Los Angeles police officers who were filmed beating a suspect were acquitted of criminal charges, a massive riot erupted among the citizens of the community. This was a good example of the social contract breaking down when people realize that the government failed to punish members of the community (ironically police officers) who had violated its rules.

The concept of the social contract was likely the most important contribution of the Enlightenment philosophers, but there were others. Another key concept of Enlightenment philosophers focused on democracy, emphasizing that every person in society should have a say via the government; specifically, they promoted the ideal of "one person, one vote." Granted, at the time they wrote, this meant one vote for each White, land-owning male and not for women, minorities, or the poor, it was a step in the right direction. Until then, no individuals outside of the aristocracy had any say in government or the justice system.

The Enlightenment philosophers also talked about every individual's right to the pursuit of life, liberty, and happiness. This probably sounds familiar because it is contained in the U.S. Constitution. Until the Enlightenment, individuals were not considered to have these rights; rather, they were seen as instruments for serving totalitarian governments. Although most citizens of the Western world take these rights for granted, they did not exist prior to the Age of Enlightenment.

Perhaps the most relevant concept that the Enlightenment philosophers emphasized, as mentioned above, was the idea that human beings are rational and therefore have free will. The philosophers of this age focused on the ability of individuals to consider the consequences of their actions, and they assumed that people freely choose their behavior (or lack thereof), especially in regard to criminal activity. The father of criminal justice made this assumption in his formulation of what is considered to be the first bona fide theory of why people commit crime.

The Classical School of Criminology

The foundation of the Classical School of criminological theorizing is typically traced to the Enlightenment philosophers, but the specific origin of the Classical School is considered to be the 1764 publication of *On Crimes and Punishments* by Italian scholar Cesare Bonesana, Marchese de Beccaria (1738–1794), commonly known as Cesare Beccaria. Amazingly, he wrote this book at the age of 26 and published it anonymously, but its almost instant popularity persuaded him to come forward as the author. Due to this significant work, most experts consider Beccaria the father of criminal justice and the father of the classical school of criminology, but perhaps most important, the father of deterrence theory. This section provides a comprehensive survey of the ideas and impact of Cesare Beccaria and the Classical School.

Influences on Beccaria and His Writings

The Enlightenment philosophers had a profound impact on the social and political climate of the late 1600s and 1700s. Growing up in this time period, Beccaria was a child of the Enlightenment, and as such, he was highly influenced by the concepts and propositions that these great thinkers proposed. The Enlightenment philosophy is readily evident in Beccaria's essay, and he incorporates much of its assumptions into his work. As a student of law, Beccaria had a good background for determining what was

▲ **Photo 1.1**   Cesare Beccaria (1738–1794)

and was not rational in legal policy. But his loyalty to the Enlightenment ideal was ever present throughout his work.

Beccaria places an emphasis on the idea of social contract and incorporates the idea that citizens give up certain rights in exchange for the state's or government's protection. He also asserts that acts or punishments by the government that violate the overall sense of unity will not be accepted by the populace, largely due to the need for the social contract to be a fair deal. Beccaria explicitly stated that laws are compacts of free individuals in a society. In addition, he specifically notes his appeal to the ideal of the greatest happiness shared by the greatest number, which is otherwise known as **utilitarianism**. This, too, was a focus of the Enlightenment philosophers. Finally, the emphasis on free will and individual choice by individuals is key to his propositions and theorizing. Indeed, as we shall see, Enlightenment philosophy is present in virtually all of his propositions; he directly cites Hobbes, Montesquieu, and other Enlightenment thinkers in his work.[9]

Beccaria's Proposed Reforms and Ideas of Justice

When Beccaria wrote, authoritarian governments ruled the justice systems, which were actually quite unjust during that time. For example, it was not uncommon that a person who stole a loaf of bread to feed his/her family would be imprisoned for life.

---

[9]Hobbes, *Leviathan;* Charles Louis de Secondat, Baron de la Brede et de Montesquieu, *The Spirit of the Laws* (New York: Library of Liberal Arts, 1949; first published in 1748).

A good example of this is seen in the story of *Les Miserables:* The protagonist, Jean Valjean, gets a lengthy prison sentence for stealing food for his starving loved ones. On the other hand, a judge might excuse a person who had committed several murders because the confessed killer was from a prominent family.

Beccaria sought to rid the justice system of such arbitrary acts on the part of individual judges. Specifically, Beccaria claimed in his essay that "only laws can decree punishments for crimes . . . judges in criminal cases cannot have the authority to interpret laws" (p. 14).[10] Rather, he believed that legislatures, elected by the citizens, must define crimes and the specific punishment for each criminal offense. One of his main goals was to prevent a single person from assigning an overly harsh sentence on a defendant and allowing another defendant in a similar case to walk free for the same criminal act, which was quite common at that time. Thus, Beccaria's writing calls for a set punishment for a given offense, without consideration of the presiding judge's personal attitudes or the defendant's background.

Beccaria believed that "the true measure of crimes is namely the harm done to society."[11] Thus, anyone at all who committed a given act against the society should face the same consequence. He was very clear that the law should impose a specific punishment for a given act, regardless of the contextual circumstances. One aspect of this principle was that it ignored the intent the offender had in committing the crime. Obviously, this is not true of most modern justice systems; intent often plays a key role in the charges and sentencing of defendants in many types of crimes. Most notably, the different degrees of homicide in most U.S. jurisdictions include first-degree murder, which requires proof of planning or "malice aforethought"; second-degree murder, which typically involves no evidence of planning, but rather a spontaneous act of killing; and various degrees of manslaughter, which generally include some level of provocation on the part of the victim. This is just one example of how important intent, legally known as ***mens rea*** (literally, "guilty mind"), is in most modern justice systems. Many types of offending are graded by degree of intent, as opposed to the focus on only the act itself, known legally as ***actus reus*** (literally, "guilty act"). Beccaria's propositions focus on only the actus reus because he claimed that an act against society was just as harmful, regardless of the intent, or mens rea. Despite his recommendations, most societies factor in the intent of the offender in criminal activity. Still, this proposal of "a given act should be given equal punishment" certainly seemed to represent a significant improvement over the arbitrary punishments in the regimes and justice systems of the 1700s.

Another important reform that Beccaria proposed was to do away with practices that were common in "justice" systems of the time—the word *justice* is in quotes because they were largely systems of injustice. Specifically, Beccaria claimed that secret accusations should not be permitted; rather, defendants should be able to confront and cross-examine witnesses. Writing about secret accusations, he says "their customary use makes men false and deceptive" (p. 25); he asks, "Who can defend himself against calumny when it comes armed with tyranny's strongest shield, *secrecy?*"[12] Although some modern countries still

---

[10]Cesare Beccaria, *On Crimes and Punishments*, trans. Henry Paolucci (New York: MacMillan,1963), 14.

[11]Ibid., 64.

[12]Ibid., 25–26.

accept and use secret accusations and disallow the cross-examination of witnesses, Beccaria set the standard in guaranteeing such rights to defendants in the United States and most Western societies.

In addition, Beccaria argued that torture should not be used against defendants:

> A cruelty consecrated by the practice of most nations is torture of the accused . . . either to make him confess the crime or to clear up contradictory statements, or to discover accomplices . . . to discover other crimes of which he might be guilty but of which he is not accused.[13] (p. 30)

Although some countries, such as Israel and Mexico, currently allow the use of torture for eliciting information or confessions, most countries abstain from the practice. There has been wide discussion about a memo, written by former U.S. Attorney General Alberto Gonzales when he was President George W. Bush's lead counsel at the White House, claiming that the U.S. military could use torture against terrorist suspects. However, the United States has traditionally agreed with Beccaria, who believed that any information or oaths obtained under torture were relatively worthless; our country apparently agrees, at least in terms of domestic criminal defendants. Beccaria's belief in the worthlessness of torture is further seen in his statement that "it is useless to reveal the author of a crime that lies deeply buried in darkness."[14]

It is likely that Beccaria believed the use of torture was one of the worst aspects of the criminal justice systems of his time and a horrible manifestation of the truly barbaric acts common in feudal times in the Middle Ages. This is seen in his further elaboration of torture:[15]

> This infamous crucible of truth is a still-standing memorial of the ancient and barbarous legislation of a time when trials by fire and by boiling water, as well as the uncertain outcomes of duels, were called "judgments of God."

Beccaria also expressed his doubt of the relevance of any information received via method of torture:[16]

> Thus the impression of pain may become so great that, filling the entire sensory capacity of the tortured person, it leaves him free only to choose what for the moment is the shortest way of escape from pain.

As Beccaria sees it, the policy implications from such use of torture are that "of two men, equally innocent or equally guilty, the strong and courageous will be acquitted, the weak and timid condemned."[17]

---

[13]Ibid., 30.

[14]Ibid., 30–31.

[15]Ibid., 31.

[16]Ibid., 32.

[17]Ibid., 32.

Beccaria also claimed that defendants should be tried by fellow citizens or peers, not by judges: "I consider an excellent law that which assigns popular jurors, taken by lot, to assist the chief judge . . . that each man ought to be judged by his peers."[18] It is clear that Beccaria felt that the responsibility of determining the facts of a case should be placed in the hands of more than one person, a belief driven by his Enlightenment beliefs about democratic philosophy, in which citizens of the society should have a voice and serve in judging the facts and deciding the verdicts of criminal cases. This proposition is representative of Beccaria's overall philosophy toward fairness and democratic process, which Enlightenment philosophers shared.

Today, U.S. citizens often take for granted the right to have a trial by a jury of their peers. It may surprise some readers to know that some modern developed countries have not provided this right. For example, Russia just recently held its first jury trials in 85 years. When Vladimir Lenin was in charge of Russia, he banished jury trials. Over the course of several decades, the bench-trials in Russia produced a 99.6% rate of convictions. This means that virtually every person in Russia who was accused of a crime was found guilty. Given the relatively high percentage of defendants found to be innocent of crimes in the United States, not to mention the numerous people who have been released from death row after DNA analysis showed they were not guilty, it is rather frightening to think of how many falsely accused individuals were convicted and unjustly sentenced in Russia over the last century.

Another important area of Beccaria's reforms involves the emphasis on making the justice system, particularly its laws and decisions, more public and better understood. This fits the Enlightenment assumption that individuals are rational: If people know the consequences of their actions, they will act accordingly. Beccaria stated that "when the number of those who can understand the sacred code of laws and hold it in their hands increases, the frequency of crimes will be found to decrease."[19] At the time, the laws were often unknown to the populace, in part because of widespread illiteracy but perhaps more as a result of the failure to publicly declare what the laws were. Even when laws were posted, they were often in languages (e.g., Latin) that the citizens did not read or speak. So Beccaria stressed the need for society to ensure that the citizens were educated about what the laws were; he believed that this alone would lead to a significant decrease in law violation.

Furthermore, Beccaria believed that the important stages and decision-making processes of any justice system should be public knowledge, rather than being held in secret or decided behind closed doors. As Beccaria stated, "punishment . . . must be essentially public."[20] This has a highly democratic and Enlightenment feel to it, in the sense that citizens of a society have the right to know what vital judgments are being made. After all, in a democratic society, citizens give the government the deeply profound responsibility of distributing punishment for crimes against the society. The citizens are entitled to know what decisions their government officials are making, particularly regarding justice. Besides providing knowledge and an understanding of what is going on, it also ensures a form of checks and balances on what is happening. Furthermore, the public nature of

---

[18]Ibid., 21.

[19]Ibid., 17.

[20]Ibid., 99.

trials and punishments inherently produces a form of deterrence for those individuals who may be considering such criminal activity.

One of Beccaria's most profound and important proposed reforms is one of the least noted. Beccaria said, "The surest but most difficult way to prevent crimes is by perfecting education."[21] We know of no other review of his work that notes this hypothesis, which is quite amazing because most of the reviews are done for an educational audience. Furthermore, this emphasis on education makes sense, given Beccaria's emphasis on knowledge of laws and consequences of criminal activity, as well as his focus on deterrence.

Beccaria's Ideas of Death Penalty

Another primary area of Beccaria's reforms dealt with the use—and, in his day, the abuse—of the death penalty. First, let it be said that Beccaria was against the use of capital punishment. (Interestingly, he was not against corporal punishment, which he explicitly stated was appropriate for violent offenders.) Perhaps this was due to the times in which he wrote, in which a large number of people were put to death, often by harsh methods. Still, Beccaria had several rational reasons for why he felt the death penalty was not an efficient and effective punishment.

First, Beccaria argued that the use of capital punishment inherently violated the social contract:[22]

> Is it conceivable that the least sacrifice of each person's liberty should include sacrifice of the greatest of all goods, life? . . . The punishment of death, therefore, is not a right, for I have demonstrated that it cannot be such; but it is the war of a nation against a citizen whose destruction it judges to be necessary or useful.

In a related theme, the second reason that Beccaria felt that the death penalty was an inappropriate form of punishment was that if the government endorsed the death of a citizen, it would provide a negative example to the rest of society. He said, "the death penalty cannot be useful, because of the example of barbarity it gives men."[23] Although some studies show some evidence that use of death penalty in the United States deters crime,[24] most

---

[21]Ibid., 98.

[22]Ibid., 45.

[23]Ibid., 50.

[24]For a review and analysis showing a deterrent effect of capital punishment, see S. Stack, "The Effect of Publicized Executions on Homicides in California," *Journal of Crime and Justice* 21 (1998): 1–16. Also see I. Ehrlich, "The Deterrent Effect of Capital Punishment: A Question of Life and Death," *American Economic Review* 65 (1975): 397–417; I. Ehrlich, "Capital Punishment and Deterrence," *Journal of Political Economy* 85 (1977): 741–88; S. K. Layson, "Homicide and Deterrence: A Reexamination of United States Time-Series Evidence," *Southern Economic Journal* 52 (1985): 68–89; D. P. Phillips, "The Deterrent Effect of Capital Punishment: Evidence on an Old Controversy," *American Journal of Sociology* 86(1980): 139–48; S. Stack, "Publicized Executions and Homicide, 1950–1980," *American Sociological Review* 52 (1987): 532–40; S. Stack, "Execution Publicity and Homicide in South Carolina," *Sociological Quarterly* 31 (1990): 599–611; S. Stack, "The Impact of Publicized Executions on Homicide," *Criminal Justice and Behavior* 22 (1995): 172–86.

studies show no effect or even a positive effect on homicides.[25] Researchers have called this increase of homicides after executions the **brutalization effect**, and a similar phenomenon can be seen at numerous sporting events when violence breaks out among spectators at boxing matches, hockey games, soccer/football games, and so on. There have even been incidents in recent years at youth sporting events.

To further complicate the possible contradictory effects of capital punishment, some analyses show that both deterrence and brutalization occur at the same time for different types of murder or crime, depending on the level of planning or spontaneity of a given act. For example, a sophisticated analysis of homicide data from California examined the effects of a high-profile execution in 1992, largely because it was the first one in the state in 25 years.[26] As predicted, the authors found that nonstranger felony murders, which typically involve some planning, significantly decreased after the high-profile execution, whereas the level of argument-based stranger murders, which are typically more spontaneous, significantly increased during the same time period. Thus, both the effects of deterrence and brutalization were observed at the same time and location following a given execution.

Another primary reason that Beccaria was against the use of capital punishment was that he believed it was an ineffective deterrent. Specifically, he thought that a punishment that was quick, such as death penalty, could not be an effective deterrent as compared to a drawn-out penalty. As Beccaria stated, "It is not the intensity of punishment that has the greatest effect on the human spirit, but its duration."[27] It is likely that many readers can relate to this type of argument, not that they necessarily agree with it; the idea of spending the rest of one's life in a cell is a very scary concept to most people. To many people, such a concept is more frightening than death, which supports Beccaria's idea that the duration of the punishment may be more of a deterrent than the short, albeit extremely intense, punishment, of execution.

Beccaria's Concept of Deterrence and the Three Key Elements of Punishment

Beccaria is generally considered the father of **deterrence theory** for good reason. Beccaria was the first known scholar to write a work that summarized such extravagant

---

[25]W. C. Bailey, "The Deterrent Effect of the Death Penalty for Murder in California," *Southern California Law Review* 52 (1979): 743–64; D. Lester, "The Deterrent Effect of Execution on Homicide," *Psychological Reports* 64 (1989): 306–14; W. J. Bowers, "The Effect of Execution is Brutalization, Not Deterrence," *Capital Punishment: Legal and Social Science Approaches*, ed. K. C. Haas and J. A. Inciardi (Newbury Park, CA: Sage, 1988), 49–89; W. C. Bailey and R. D. Peterson, "Murder and Capital Punishment: A Monthly Time Series Analysis of Execution Publicity," *American Sociological Review* 54 (1989): 722–43; J. A. Fox and Michael L. Radelet, "Persistent Flaws in Econometric Studies of the Deterrent Effect of the Death Penalty," *Loyola of Los Angeles Law Review* 23 (1990):.29–44; John K. Cochran, Mitchell Chamlin, and M. Seth, "Deterrence or Brutalization? An Impact Assessment of Oklahoma's Return to Capital Punishment," *Criminology* 32 (1994): 107–34; for a review, see W. C. Bailey and R. D. Peterson, "Capital Punishment, Homicide, and Deterrence: An Assessment of the Evidence," in *Studying and Preventing Homicide*, ed. M. D. Smith and M. A. Zahn (Thousand Oaks, CA: Sage, 1999), 223–45.

[26]John K. Cochran and Mitchell B. Chamlin, "Deterrence and Brutalization: The Dual Effects of Executions," *Justice Quarterly* (2000): 685–706.

[27]Beccaria, *On Crimes*, 46–47.

ideas regarding the direction of human behavior toward choice, as opposed to fate or destiny. Prior to his work, the common wisdom on the issue of human destiny was that it was chosen by the gods or God. At that time, governments and society generally believed that people were born either good or bad. Beccaria, as a child of the Enlightenment, defied this belief in proclaiming that people freely choose their destiny and, thus, their decisions to commit or not commit criminal behavior.

Beccaria suggested three characteristics of punishment that would make a significant difference in whether the individual decides to commit a criminal act. These vital deterrent characteristics of punishment included celerity (swiftness), certainty, and severity.

Swiftness. The first of these characteristics was celerity, which we will refer to as **swiftness of punishment.** Beccaria saw two reasons why swiftness of punishment was important. At the time he wrote, some defendants were spending many years awaiting trial. Often, this was a longer time than they would have been locked up as punishment for their alleged offenses, even if the maximum penalty was imposed. As Beccaria stated, "the more promptly and the more closely punishment follows upon the commission of a crime, the more just and useful will it be."[28] Thus, the first reason that Beccaria recommended swiftness of punishment was to reform a system that was severely lacking.

The second reason that Beccaria emphasized a swift sentencing was related to the deterrence aspect of punishment. A swift trial and punishment was important, Beccaria said,[29] "because of privation of liberty, being itself a punishment, should not precede the sentence." This not only was unjust, in the sense that some of these defendants would not have been incarcerated for such a long period even if they were convicted and sentenced to the maximum for the charges they were accused of committing, but it was detrimental because the individual would not link the sanction with the violation(s) they committed. Specifically, Beccaria believed that people build an association between the pains of punishment and their criminal acts. He asserted:[30]

> Promptness of punishments is more useful because when the length of time that passes between the punishment and the misdeed is less, so much the stronger and more lasting the human mind is the association of these two ideas, crime and punishment; they then come insensibly to be considered, one as the cause, the other as the necessary inevitable effect. It has been demonstrated that the association of ideas is the cement that forms the entire fabric of the human intellect. (p. 56)

An analogy can be made with training animals or children; you have to catch them in the act or soon after or the punishment does not matter because the offender does not know why he or she is being punished. Ultimately, Beccaria argued that for both reform and deterrent reasons, punishment should occur quickly after the act. Despite the common sense aspects of making punishments swift, this has not been examined by

---

[28]Ibid., 55.

[29]Ibid., 55.

[30]Ibid., 56.

modern empirical research and therefore is the most neglected of the three elements of punishment Beccaria emphasized.

Certainty. The second characteristic that Beccaria felt was vital to the effectiveness of deterrence was **certainty of punishment.** Beccaria considered this the most important quality of punishment: "even the least of evils, when they are certain, always terrify men's minds."[31] He also said,[32] "The certainty of punishment, even if it be moderate, will always make a stronger impression than the fear of another which is more terrible but combined with the hope of impunity." As scientific studies later showed, Beccaria was accurate in his assumption that perceived certainty or risk of punishment was the most important aspect of deterrence.

It is interesting to note that certainty is the least likely characteristic of punishment to be enhanced in modern criminal justice policy. Over the last few decades, the risk of criminals being caught and arrested has not increased. Law enforcement officials have been able to clear only about 21% of known felonies. Such clearance rates are based on the rate at which known suspects are apprehended for crimes that are reported to police. Law enforcement officials are no better at solving serious crimes known to police than they were in past decades, despite increased knowledge and resources put toward solving such crimes.

Severity. The third characteristic that Beccaria emphasized was **severity of punishment.** Specifically, Beccaria claimed that for a punishment to be effective, the possible penalty must outweigh the potential benefits (e.g., financial payoff) of the given crime. However, this came with a caveat. This aspect of punishment was perhaps the most complicated part of Beccaria's philosophy, primarily because he thought that too much severity would lead to more crime, but the punishment must exceed any benefits expected from the crime. Beccaria said:[33]

> For a punishment to attain its end, the evil which it inflicts has only to exceed the advantage derivable from the crime; in this excess of evil one should include the . . . loss of the good which the crime might have produced. All beyond this is superfluous and for that reason tyrannical.

Beccaria makes clear in this statement that punishments should equal or outweigh any benefits of a crime to deter individuals from engaging in such acts. However, he also explicitly states that any punishments that largely exceed the reasonable punishment for a given crime are inhumane and may lead to further criminality.

A modern example of how punishments can be taken to an extreme and thereby cause more crime rather than deter it is the current "three-strikes-you're-out" laws. Such laws have become common in many states, such as California. In such jurisdictions, individuals who have committed two prior felonies can be sentenced to life imprisonment for

---

[31]Ibid., 58.

[32]Ibid., 58.

[33]Ibid., 43.

committing a crime, even a nonviolent crime, that the state statutes consider a "serious felony." Such laws have been known to drive some relatively minor property offenders to become violent when they know that they will be incarcerated for life when they are caught. A number of offenders have even wounded or killed people to avoid apprehension, knowing that they would face life imprisonment even for a relatively minor property offense. In a recent study, the authors analyzed the impact of three-strikes laws in 188 large cities in the 25 states that have such laws and concluded that there was no significant reduction in crime rates as a result. Furthermore, the areas with three-strikes laws typically had higher rates of homicide.[34]

Ultimately, Beccaria's philosophy on the three characteristics of good punishment in terms of deterrence—swiftness, certainty, and severity—is still highly respected and followed in most Western criminal justice systems. Despite its contemporary flaws and caveats, perhaps no other traditional framework is so widely adopted. With only one exception, Beccaria's concepts and propositions are still considered the ideal in virtually all Western criminal justice systems.

Beccaria's Conceptualization of Specific and General Deterrence

Beccaria also defined two identifiable forms of deterrence: specific and general. Although these two forms of deterrence tend to overlap in most sentences by judges, they can be distinguished in terms of the intended target of the punishment. Sometimes, the emphasis is clearly on one or the other, as Beccaria noted in his work.

Although Beccaria did not coin the terms specific and general deterrence, he clearly makes the case that both are important. Regarding punishment, he said, "the purpose can only be to prevent the criminal from inflicting new injuries on its citizens and to deter others from similar acts" (p. 42). The first portion of this statement—preventing the criminal from reoffending—focuses on the defendant and the defendant alone, regardless of any possible offending by others. Punishments that focus primarily on the individual are considered **specific deterrence**, also referred to as special or individual deterrence. This concept is appropriately labeled because the emphasis is on the specific individual who offended. On the other hand, the latter portion of Beccaria's quotation emphasizes the deterrence of others, regardless of whether the individual criminal is deterred. Punishments that focus primarily on other potential criminals and not on the actual criminal are referred to as **general deterrence**.

Readers may wonder how a punishment would not be inherently both a specific and general deterrent. After all, in today's society, virtually all criminal punishments given to individuals (i.e., specific deterrence) in our society are done in court, a public venue, so people are somewhat aware of the sanctions (i.e., general deterrence). However, when Beccaria wrote in the 18th century, many if not most of sentencing was done behind closed doors and was not known to the public and therefore had no way to deter other potential offenders. Therefore, Beccaria saw much utility in letting the public know what punishments were handed out for given crimes. This fulfilled the goal of general deterrence, which was essentially scaring others into not committing such criminal acts, while

---

[34]Tomislav V. Kovandzic, John J. Sloan, III, and Lynne M. Vieraitis, "'Striking Out' as Crime Reduction Policy: The Impact of 'Three Strikes' Laws on Crime Rates in the U.S. Cities," *Justice Quarterly* 21 (2004): 207–40.

it also furthered his reforms of letting the public know if fair and balanced justice was being administered.

Despite the obvious overlap, there are some identifiable distinctions between specific and general deterrence seen in modern sentencing strategy. For example, some judges have chosen to hand out punishments to defendants in which they are obligated, as a condition of their probation or parole, to walk along their town's main streets while wearing a sign that says "Convicted Child Molester" or "Convicted Shoplifter." Other cities have implemented policies in which pictures and identifying information of those individuals who are arrested, such as prostitutes or men who solicit them, are put in the newspaper or placed on billboards.

These punishment strategies are not likely to be much of a specific deterrent. Having now been labeled, these individuals may actually be psychologically encouraged to engage in doing what the public expects them to do. The specific deterrent effect may not be particularly strong. However, authorities are hoping for a strong general deterrent effect in most of these cases. They expect that many of the people who see these sign-laden individuals on the street or in public pictures are going to be frightened from engaging in similar activity.

There are also numerous diversion programs, particularly for juvenile, first-time, and minor offenders, which seek to punish offenders without engaging them in public hearings or trials. The goal of such programs is to hold the individuals accountable and have them fulfill certain obligations without having them dragged through the system, which is often public. Thus, the goal is obviously to instill specific deterrence without using the person as a "poster child" for the public, which obviously negates any aspects of general deterrence.

Although most judges invoke both specific and general deterrence in many of the criminal sentences that they hand out, there are notable cases in which either specific or general deterrence is emphasized, sometime exclusively. Ultimately, Beccaria seemed to emphasize general deterrence, and overall crime prevention, which is suggested by his statement, "It is better to prevent crimes than to punish them. This is the ultimate end of every good legislation."[35] This claim implies that it is better to deter potential offenders before they offend, rather than imposing sanctions on already convicted criminals. Beccaria's emphasis on prevention (over reaction) and general deterrence is also evident in his claim that education is likely the best way to reduce crime. After all, the more educated an individual is regarding the law and potential punishments, as well as public cases in which offenders have been punished, the less likely he or she will be to engage in such activity. Beccaria's identification of the differential emphases in terms of punishment was a key element in his work that continues to be important in modern times.

Summary of Beccaria's Ideas and His Influence on Policy

Ultimately, Beccaria summarized his ideas on reforms and deterrence with the statement:[36]

---

[35]Ibid., 93.

[36]Beccaria, *On Crimes*, 99.

In order for punishment not to be, in every instance, an act of violence of one or of many against a private citizen, it must be essentially public, prompt, necessary, the least possible in the given circumstances, proportionate to the crimes, dictated by the laws.

In this statement, Beccaria is saying that the processing and punishment administered by justice systems must be known to the public, which delegates to the state the authority to make such decisions. Furthermore, he asserts that the punishment must be appropriately swift, certain (i.e., necessary), and appropriately severe, which fits his concept of deterrence. Finally, he reiterates the need to have equal punishment for a given criminal act, as opposed to having arbitrary punishments imposed by one judge. These are just some of many ideas that Beccaria proposed, but he apparently saw these points as being most important.

Although we, as U.S. citizens, take for granted the rights proposed by Beccaria, they were quite unique concepts during the 18th century. In fact, the ideas proposed by Beccaria were so unusual and revolutionary then that he published his book anonymously. It is obvious that Beccaria was considerably worried about being accused of blasphemy by the church and of being persecuted by governments for his views.

Regarding the first claim, Beccaria was right; the Roman Catholic Church excommunicated Beccaria when it became known that he wrote the book. In fact, his book remained on the list of condemned works until relatively recently (the 1960s). On the other hand, government officials of the time surprisingly embraced his work. The Italian government and most of European and other world officials, particularly dictators, embraced his work as well. Beccaria was invited to visit many other country capitals, even those of the most authoritarian states at that time, to help reform their criminal justice systems. For example, Beccaria was invited to meet with Catherine the Great, the czarina of Russia, during the late 1700s to help revise and improve their justice system. Most historical records suggest that Beccaria was not a great diplomat or representative of his ideas, largely because he was not physically or socially adequate for such endeavors. However, his ideas were strong and stood on their own merit.

Dictators and authoritarian governments may have liked his reform framework so much because it explicitly stated that treason was the most serious crime. As Beccaria said, "The first class of crime, which are the gravest because most injurious, are those known as crimes of *lese majesty* [high treason]. . . . Every crime . . . injures society, but it is not every crime that aims at its immediate destruction."[37] According to Enlightenment philosophy, violations of law are criminal acts not only against the direct victims, but also against the entire society because they break the social contract. As Beccaria stated, the most heinous criminal acts are those that directly violate the social contract, which would be treason and espionage.

In his reform proposals, dictators may have seen a chance to pacify revolutionary citizens, who might be aiming to overthrow their governments. In many cases, reforms were only a temporary solution. After all, the American Revolution occurred in the 1770s, the French Revolution occurred in the 1780s, and other revolutions occurred soon after this period.

---

[37]Ibid., 68.

Governments that tried to apply his ideas to the letter experienced problems, but generally, most European (and American) societies that incorporated Beccaria's ideas had fairer and more democratic justice systems than they had before Beccaria. This is why he is to this day considered the father of criminal justice.

Impact of Beccaria's Work on Other Theorists

Beccaria's work had an immediate impact on the political and philosophical state of affairs in the late 18th century. He was invited to many other countries to reform their justice systems, and his propositions and theoretical model of deterrence were incorporated into many of the newly formed constitutions of countries, most formed after major revolutions. The most notable of these was the Constitution and Bill of Rights of the United States.

It is quite obvious that the many founding documents constructed before and during the American Revolution in the late 1700s were heavily influenced by Beccaria and other Enlightenment philosophers. Specifically, the concept that the U.S. government is "of the people, by the people, and for the people" makes it clear that the Enlightenment idea of democracy and voice in the government is of utmost importance. Another clear example is the emphasis on due process and individual rights in the U.S. Bill of Rights. Among the important concepts derived from Beccaria's work are the right to trial by jury, the right to confront and cross-examine witnesses, the right to a speedy trial, and the right to be informed about decisions of the justice system (such as charges, pleas, trials, verdicts, sentences, etc.).

The impact of Beccaria's work on the working ideology of our system of justice cannot be overstated. The public nature of our justice system comes from Beccaria, as does the emphasis on deterrence. The United States, as well as virtually all Western countries, incorporates in its justice system the certainty and severity of punishment to reduce crime. This system of deterrence remains the dominant model in criminal justice: The goal is to deter potential and previous offenders from committing crime by enforcing a punishment that will make them reconsider the next time they think about engaging in such activity. This model assumes a rational thinking human being, as described by Enlightenment philosophy, who can learn from past experiences or from seeing others punished for offenses that he or she is rationally thinking about committing. Thus, Beccaria's work has had a profound impact on the existing philosophy and working of most justice systems throughout the world.

Beyond this, Beccaria also had a large impact on further theorizing about human decision making related to committing criminal behavior. One of the more notable theorists inspired by Beccaria's ideas was Jeremy Bentham (1748–1832) of England, who has become a well-known classical theorist in his own right, perhaps because he helped spread the Enlightenment/Beccarian philosophy to Britain. His influence in the development of classical theorizing is debated, with a number of major texts not covering his writings at all.[38] Although he did not add a significant amount of theorizing beyond Beccaria's propositions regarding reform and deterrence, Bentham did further refine the ideas presented by previous theorists, and his legacy is well known.

---

[38]George B. Vold, Thomas J. Bernard, and Jeffrey B. Snipes, *Theoretical Criminology*, 5th ed. (New York: Oxford University Press, 2002).

One of the more important contributions of Bentham was the concept of the "hedonistic calculus," which was essentially the weighing of pleasure versus pain. This, of course, is strongly based on the Enlightenment/Beccarian concept of rational choice and utility. After all, if the expected pain outweighs the expected benefit of doing a given act, the rational individual is far less likely to do it. On the other hand, if the expected pleasure outweighs the expected pain, a rational person will engage in the act. Bentham listed a set of criteria that he thought would go into the decision making of a rational individual. An analogy would be an imagined two-sided balance scale in which the pros and cons of crimes are considered, and then the individual makes a rational decision about whether or not to commit the crime.

Beyond the idea of the hedonistic calculus, Bentham's contributions to the overall assumptions of classical theorizing did not revise the theoretical model in a significant way. Perhaps the most important contribution he made to the Classical School was helping to popularize the framework in Britain. In fact, Bentham became more known for his design of a prison structure, known as the

▲ **Photo 1.2**   Jeremy Bentham, often credited as the founder of University College London, insisted that his body be put on display there after his death. You can visit a replica of it today.

panopticon, which was used in several countries and in early Pennsylvania penitentiaries. This model of prisons involved using a type of "wagon wheel" design, in which a post at the center allowed 360-degree visual observation of the various "spokes": hallways that contained all of the inmate cells.

✖ The Neoclassical School of Criminology

A number of governments, including the newly formed United States, incorporated Beccaria's concepts and propositions in the development of their justice systems. The government that most strictly applied Beccaria's ideas—France, after the revolution of the late 1780s—found that it worked pretty well except for one concept. Beccaria had believed that every individual who committed a certain act against the law should be punished the same way. Although equality in punishment sounds like a good philosophy, the French realized very quickly that not everyone should be punished equally for a certain act.

The French system found that sentencing a first-time offender the same as a repeat offender did not make much sense, especially when the first-time offender was a juvenile. Furthermore, there were many circumstances in which a defendant appeared to be not malicious in doing an act, such as someone with limited mental capacity or a person who acted out of necessity. Perhaps most important, Beccaria's framework specifically dismisses the intent (i.e., mens rea) of criminal offenders, while focusing only on the harm done to society by a given act (i.e., actus reus). French society, as well as most modern societies such as the United States, deviated from Beccaria's framework in taking the intent of offenders into account, often in a very important way, such as in determining what type of charges should be filed against those accused of homicide. Therefore, a new school of thought regarding the classical/deterrence model developed, which became known as the **Neoclassical School** of criminology.

The only significant difference between the Neoclassical School and the Classical School of criminology is that the Neoclassical (*neo* means new) School takes into account contextual circumstances of the individual or situation that allow for increases or decreases in the punishment. For example, would a society want to punish a 12-year-old first-time offender the same way for shoplifting as they would punish a 35-year-old previous offender for shoplifting the same item? In addition, does a society want to punish a mentally challenged person for stealing a car one time as much as a normal person who has been convicted of stealing more than a dozen cars? The answer is, probably not—at least that is what most modern criminal justice authorities have decided, including those in the United States.

This was also the conclusion of French society, which quickly realized that, in this respect, Beccaria's system was neither fair nor effective in terms of deterrence. They came to acknowledge that circumstantial factors play an important part in how malicious or guilty a certain defendant is in a given crime, based on a number of contextual factors. The French revised their laws to take into account both mitigating and aggravating circumstances. This neoclassical concept became the standard in all Western justice systems.

The United States also followed this model and considers such contextual factors in virtually all of its charges and sentencing decisions. For example, juvenile defendants are actually processed in a completely different court. Furthermore, defendants who are first-time offenders are generally given an option for a diversion program or probation, as long as their offense is not serious.

The Neoclassical School adds an important caveat to the previously important Classical School, but it assumes virtually all other concepts and propositions of the Classical School: the social contract, due process rights, and the idea that rational beings will be deterred by the certainty, swiftness, and severity of punishment. This Neoclassical framework had, and continues to have, an extremely important impact on the world.

Loss of Dominance of Classical/Neoclassical Theory

For about 100 years after Beccaria wrote his book, the Classical/Neoclassical schools were dominant in criminological theorizing. During this time, most governments, especially those in the Western world, shifted their justice frameworks toward the neoclassical model. This has not changed, even in modern times. For example, when officials attempt to reduce certain illegal behaviors, they increase the punishments or put more effort into catching such offenders.

However, the dominance of the classical/neoclassical framework lost dominance among academics and scientists in the 19th century, especially after Darwin's publication in the 1860s of *The Origin of Species*, which introduced the concept of evolution and natural selection. This perspective shed new light on other influences on human behavior beyond free will and rational choice (e.g., genetics, psychological deficits). Despite this shift in emphasis among academic and scientific circles, it remains true that the actual workings of justice systems of most Western societies still retained the framework of the classical/neoclassical model as their model of justice, and this continues until today.

One example is the "three-strikes" laws; others include police department gang units and injunctions that condemn any observed loitering by or gathering among gang members in a specified region by listed members of established gangs. Furthermore, some jurisdictions, such as California, have created gang enhancements for sentencing; after the jury decides whether the defendant is guilty of a given crime, it then considers whether the person is a

▲ **Photo 1.3**   Charles Darwin (1809–1882), author of evolutionary theory

gang member. If a jury in California decides that the defendant is a gang member, which is usually determined by evidence provided by local police gang units, it automatically adds more time to any sentence the judge gives the individual convicted of the crime. These are just some examples of how Western justice systems still rely primarily on deterrence of criminal activity through increased enforcement and enhanced sentencing. The bottom line is that modern justice systems still base most of their policies on a classical/neoclassical theoretical framework that fell out of favor among scientists and philosophers in the late 1800s.

⊠ Policy Implications of Classical Deterrence Theory

Many policies are based on deterrence theory: the premise that increasing punishment sanctions will deter crime.[39] This is seen throughout the system of law enforcement,

---

[39]Ibid.

courts, and corrections. This is rather interesting given the fact that classical/deterrence theory has not been the dominant explanatory model among criminologists for decades. In fact, a recent poll of close to 400 criminologists in the nation ranked classical theory as 22 out of 24 theories in terms of being the most valid explanation of serious and persistent offending.[40] Still, given the dominance of classical deterrence theory in most criminal justice policies, it is very important to discuss the most common strategies.

First, the death penalty is used as a general deterrent for committing crime. As the father of deterrence theory predicted, most studies show that capital punishment has a negligible effect on criminality. A recent review of the extant literature concluded that "the death penalty does not deter crime."[41] In fact, some studies show evidence for a brutalization effect, an increase in homicides after a high-profile execution.[42] Although the evidence is somewhat mixed, it is safe to say that the death penalty is not a consistent deterrent.

Another policy flowing from classical/neoclassical models is adding more police officers to deter crime in a given area. A recent review of the existing literature concluded that simply "adding more police officers will not reduce crime."[43] Rather, it is generally up to communities to police themselves via informal factors of control (such as family, church, community ties, etc.). However, this same review did find that sometimes when police engage in problem-solving activities at a specific location, it can reduce crime, but at that point, the strategy is not based on deterrence.[44] Furthermore, a recent report concluded that proactive arrests for drunk driving have consistently been found to reduce such behavior, as does arresting offenders for domestic violence, but only if they are employed.[45]

Regarding court and correctional strategies, one example is the "scared straight" programs that became popular several decades ago.[46] These programs essentially sought to scare or deter juvenile offenders into going straight by showing them the harshness and realities of prison life. However, nearly all evaluations of these programs showed that they were ineffective, and some evaluations indicated that these programs led to

---

[40]Lee Ellis, Jonathan Cooper, and Anthony Walsh. "Criminologists' Opinions About Causes and Theories of Crime and Delinquency: A Follow-up," *The Criminologist* 33 (2008): 23–26.

[41]Samuel Walker, *Sense and Nonsense about Crime and Drugs*, 6th ed. (Belmont, CA: West/Wadsworth, 2005), 102.

[42]Bowers, "Effect of Execution."

[43]Walker, *Sense and Nonsense*, 79.

[44]Ibid., 81.

[45]Lawrence Sherman, Denise Gottfredson, Doris MacKenzie, John Eck, Peter Reuter, and Shawn Bushway, *Preventing Crime: What Works, What Doesn't, What's Promising: A Report to the United States Congress* (Washington, DC: US Department of Justice, 1997).

[46]For a review of these programs and evaluations of them, see Richard Lundman, *Prevention and Control of Juvenile Delinquency,* 2nd ed. (Oxford, UK: Oxford University Press, 1993).

higher rates of recidivism.[47] There seem to be few successful deterrent policies in the court and corrections components of the criminal justice system. A recent review found that one of the only court-mandated policies that seem promising is protection orders for battered women.[48]

The policies, programs, and strategies based on classical deterrence theory will be examined more thoroughly in the final section of this book. To sum up, however, most of these strategies don't seem to work consistently to deter. This is because such a model assumes people are rational and think carefully before choosing their behavior, whereas most research findings suggest that people often engage in behaviors that they know are irrational or that offenders tend to engage in behaviors without rational decision making,[49] which criminologists often refer to as "bounded rationality."[50] Therefore, it is not surprising that many attempts by police and other criminal justice authorities to deter potential offenders do not seem to have much effect in preventing crime. This explanation will be more fully discussed in the final section of the book.

▨ Conclusion

This section examined the earliest period of theorizing about criminological theory, which evolved from the Enlightenment. The Classical School of criminology evolved out of ideas from this Enlightenment era in the mid- to late 18th century. This school of thought emphasized free will and rational choices that individuals make, with an emphasis on making choices regarding criminal behavior based on the potential costs and benefits that could result from such behavior. This section also explored the concepts and propositions of the father of the classical school/deterrence, which built the framework on which deterrence theory is based. We also discussed the various reforms that Beccaria proposed, many of which were adopted in the formation of the U.S. Constitution and Bill of Rights. The significance of the Classical School in both theorizing about crime and in actual administration of justice in the United States cannot be overestimated. The Classical/Neoclassical School of criminology remains to this day the primary framework in which justice is administered, despite the fact that scientific researchers and academics have, for the most part, moved past this perspective to consider social and economic factors.

---

[47]Ibid.

[48]Sherman et al., *Preventing Crime.*

[49]For a review of the extant research on this topic, see Alex Piquero and Stephen Tibbetts, *Rational Choice and Criminal Behavior* (New York: Routledge, 2002).

[50]For more recent discussion on the complexity of developing policies based on deterrence/rational choice models, see Travis Pratt, "Rational Choice Theory, Crime Control Policy, and Criminological Relevance," *Criminology and Public Policy* 7 (2008): 43–52.

◪ Section Summary

- The dominant theory of criminal behavior for most of the history of human civilization used demonic, supernatural, or other metaphysical explanations of behavior.

- The Age of Enlightenment was important because it brought a new logic and rationality to understanding human behavior, especially regarding the ability of individual human beings to think for themselves. Hobbes and Rousseau were two of the more important Enlightenment philosophers, and both stressed the importance of the social contract.

- Cesare Beccaria, who is generally considered the father of criminal justice, laid out a series of recommendations for reforming the brutal justice systems that existed throughout the world in the 1700s.

- Beccaria is also widely considered the father of classical school/deterrence theory; he based virtually all of his theoretical framework on the work of Enlightenment philosophers, especially their emphasis on humans as rational beings who consider the perceived risks and benefits before committing criminal behaviors. This is the fundamental assumption of deterrence models of crime reduction.

- Beccaria discussed three key elements that punishments should have to be effective deterrents; a punishment should be certain, swift, and severe.

- Specific deterrence involves sanctioning an individual to deter that particular individual from offending in the future. General deterrence involves sanctioning an individual to deter other potential offenders by "making an example" out of the individual being punished.

- The Neoclassical School was formed because societies found it nearly impossible to punish offenders equally for a given offense. The significant difference between the Classical and Neoclassical schools is that the neoclassical model takes aggravating and mitigating circumstances into account when an individual is sentenced.

- Jeremy Bentham helped reinforce and popularize Beccaria's ideas in the English-speaking world, and he further developed the theory by proposing the "hedonistic calculus," a formula for understanding criminal behavior.

- Despite falling out of favor among most criminologists in the late 1800s, the classical/neoclassical framework remains the dominant model and philosophy of all modern Western justice systems.

KEY TERMS

Actus reus	Certainty of punishment	General deterrence
Age of Enlightenment	Classical School	Mens rea
Brutalization effect	Deterrence theory	Neoclassical School

Severity of punishment   Specific deterrence   Utilitarianism

Social contract   Swiftness of punishment

DISCUSSION QUESTIONS

1. Do you see any validity to the supernatural or religious explanations of criminal behavior? Provide examples of why you feel the way you do. Is your position supported by scientific research?

2. Which portions of Enlightenment thought do you believe are most valid in modern times? Which portions do you find least valid?

3. Of all of Beccaria's reforms, which do you think made the most significant improvement to modern criminal justice systems and why? Which do you think had the least impact and why?

4. Of the three elements of deterrence that Beccaria described, which do you think has the most important impact on deterring individuals from committing crime? Which of the three do you think has the least impact on deterring potential criminals? Back up your selections with personal experience.

5. Between general and specific deterrence, which do you think is more important for a judge to consider when sentencing a convicted individual? Why do you feel that way?

6. Provide examples of general and specific deterrence in your local community or state. Use the Internet if you can't find examples from your local community. Do you think such deterrence is effective?

7. Given the modern emphasis by the U.S. government on the definition of torture, as well as what Beccaria thought about this issue, do you think that the father of criminal justice/deterrence would agree with the interrogation policies of the Bush administration during the Iraq War, which indisputably violated the guidelines set by the Geneva Conventions? Explain your position.

8. Regarding the use of the death penalty, list and explain at least three reasons why the father of criminal justice/deterrence felt the way he did. Which of these arguments do you agree with the most? Which argument do you disagree with the most? Ultimately, are you more strongly for or against the death penalty after reading the arguments of Beccaria?

9. Regarding the Neoclassical School, which mitigating factor do you think should reduce the punishment of a criminal defendant the most? Which aggravating circumstance do you think should increase the sentence of a criminal defendant the most? Do you believe that all persons who commit the same act should be punished exactly the same, regardless of age, experience, or gender?

10. What types of policy strategies based on classical/deterrence theory do you support? Which don't you support? Why?

WEB RESOURCES

Demonic Theories of Crime

http://www.criminology.fsu.edu/crimtheory/week2.htm

http://www.salemweb.com/memorial/

Age of Enlightenment:

http://history-world.org/age_of_enlightenment.htm

http://encarta.msn.com/encyclopedia_761571679/Enlightenment_Age_of.html

Thomas Hobbes:

http://www.philosophypages.com/hy/3x.htm

http://oregonstate.edu/instruct/phl302/philosophers/hobbes.html

Rousseau:

http://www.philosophypages.com/ph/rous.htm

http://www.rousseauassociation.org/

Beccaria:

http://www.constitution.org/cb/beccaria_bio.htm

http://cepa.newschool.edu/het/profiles/beccaria.htm

http://www.criminology.fsu.edu/crimtheory/beccaria.htm

Deterrence:

http://www.associatedcontent.com/article/32600/evolution_of_deterrence_crime_theory.html

http://www.deathpenaltyinfo.org/article.php?scid=12&did=167

Bentham:

http://www.utilitarianism.com/bentham.htm

http://cepa.newschool.edu/het/profiles/bentham.htm

Neoclassical School:

http://www.answers.com/topic/neo-classical-school

READING

This entry, passages from Beccaria's *On Crimes and Punishments* (originally published in 1764), is perhaps the most important in the entire book because it provides the framework in which all societies in the Western world based their criminal justice systems. That is the model presented by the father of criminal justice, Cesare Beccaria, who published the work anonymously at first and once it became a widespread success came forward as the author. The reforms that he presents in the passages excerpted here are examples of his attempts to make justice systems more fair and rational in the time that he wrote, as a son of the Enlightenment Age, which stressed rationality, fairness, and free will among individuals.

Beccaria proposed numerous reforms for justice systems, which many of us take for granted, such as trial by peers, the right to confront and cross-examine accusers, rights to public notification of laws, and barring of torture and secret accusations. These reforms made him generally recognized as the father of criminal justice. Furthermore, Beccaria is widely considered the father of the classical school and the father of deterrence theory because he was the first to be recognized as emphasizing the importance of individuals' free will and rationale in choosing to engage in behavior after considering the perceived costs and benefits of offending. Although it is likely that he was not the first person to come to this realization, he was the first to get credit for explicitly stating this rational, decision-making process in individuals before they commit criminal acts and for identifying the three elements that make a punishment a good deterrent.

On Crimes and Punishments

Cesare Beccaria

▨ I. Origin of Punishments

Laws are the conditions by which independent and isolated men, tired of living in a constant state of war and of enjoying a freedom made useless by the uncertainty of keeping it, unite in society.[1] They sacrifice a portion of this liberty in order to enjoy the remainder in security and tranquility. The sum of all these portions of liberty sacrificed for the good of everyone constitutes the sovereignty of a nation, and the sovereign is its legitimate depository and administrator. The mere formation of this deposit, however, was not sufficient; it had to be defended against the private usurpations of each particular individual, for everyone always seeks to withdraw not only his own share of liberty from the common store, but to expropriate the portions of other men besides. Tangible motives were required sufficient to

SOURCE: Cesare Beccaria, *On Crimes and Punishments,* with notes and introduction by David Young (Indianapolis, IN: Hackett, 1985). The selection here has been abridged from the original.

dissuade the despotic spirit of each man from plunging the laws of society back into the original chaos. These tangible motives are the punishments established for lawbreakers. I say "tangible motives," since experience has shown that the common crowd does not adopt stable principles of conduct, and the universal principle of dissolution which we see in the physical and the moral world cannot be avoided except by motives that have a direct impact on the senses and appear continually to the mind to counterbalance the strong impressions of individual passions opposed to the general good. Neither eloquence nor declamations nor even the sublime truths have sufficed for long to check the emotions aroused by the vivid impressions of immediately present objects.[2]

◙ II. The Right to Punish

Every punishment which does not derive from absolute necessity, says the great Montesquieu, is tyrannical.[3] The proposition may be made general thus: every act of authority between one man and another that does not derive from absolute necessity is tyrannical. Here, then, is the foundation of the sovereign's right to punish crimes: the necessity of defending the depository of the public welfare against the usurpations of private individuals. Further, the more just punishments are, the more sacred and inviolable is personal security, and the greater is the liberty that the sovereign preserves for his subjects. Let us consult the human heart, and there we shall find the fundamental principles of the sovereign's right to punish crimes, for no lasting advantage is to be expected from political morality if it is not founded upon man's immutable sentiments. Any law that deviates from them will always encounter a resistance that will overpower it sooner or later, just as a continually applied force, however slight, eventually overcomes any violent movement communicated to a physical body.

No man freely gave up a part of his own liberty for the sake of the public good; such an illusion exists only in romances. If it were possible, each of us would wish that the agreements binding on others were not binding on himself. Every man thinks of himself as the center of all the world's affairs.

The increase in the numbers of mankind, slight in itself but too much for the means that sterile and uncultivated nature offered to satisfy increasingly interrelated needs, led the first savages to unite. These initial groups necessarily created others to resist the former, and thus the state of war was transposed from individuals to nations.[4]

It was necessity, then, that constrained men to give up part of their personal liberty; hence, it is certain that each man wanted to put only the least possible portion into the public deposit, only as much as necessary to induce others to defend it.[5] The aggregate of these smallest possible portions of individual liberty constitutes the right to punish; everything beyond that is an abuse and not justice, a face but scarcely a right. Note that the word "right" is not a contradiction of the word "force"; the former is, rather, a modification of the latter—namely, the modification most useful to the greatest number. By "justice," moreover, I do not mean anything but the bond necessary to hold private interests together. Without it, they would dissolve into the earlier state of incompatibility. All punishments that exceed what is necessary to preserve this bond are unjust by their very nature. One must beware of attaching the idea of something real to this word "justice," as though it were a physical force or a being that actually exists. It is simply a human manner of conceiving things, a manner that has an infinite influence on the happiness of everybody.[6] Most certainly I am not speaking of the other sort of justice that comes from God and that is directly related to the rewards and punishments of the life to come.

⚑ III. Consequences

The first consequence of these principles is that only the law may decree punishments for crimes, and this authority can rest only with the legislator, who represents all of society united by a social contract.[7] No magistrate (who is a part of society) can justly inflict a punishment on a member of the same society, for a penalty that exceeds the limit fixed by law is the just punishment and another besides. Thus, no magistrate may, on whatever pretext of zeal or the public good, increase the established punishment for a delinquent citizen.

The second consequence is that if every individual member is bound to society, society is likewise bound to every individual member by a contract that, by its very nature, places both parties under obligation. This obligation, which reaches from the throne to the hovel and which is equally binding on the greatest and the most wretched of men, means nothing other than that it is in everybody's interest that the contracts useful to the greatest number should be observed. Their violation, even by one person, opens the door to anarchy.* The sovereign, who represents society itself, can only establish general laws that apply to all of its members; he cannot, however, pass judgment as to whether one of them has violated the social contract, for then the nation would be divided into two parties: one, represented by the sovereign, which alleges the violation of the contract, and the other, the party of the accused, which denies it. Hence it is necessary that a third party judge the facts of the case. This is the reason that there must be a magistrate whose sentences are beyond appeal and consist of the simple assertion or denial of particular facts.[8]

The third consequence is that if extremely cruel punishments are useless, even though they were not directly opposed to the public good and to the very goal of preventing crimes, then such cruelty would nevertheless be contrary to those beneficent virtues that flow from enlightened reason, which prefers to command happy men rather than a herd of slaves who constantly exchange timid cruelties with one another; excessively severe punishments would also be contrary to justice and to the nature of the social contract itself.

⚑ IV. Interpretation of the Law[9]

There is a fourth consequence: the authority to interpret penal law can scarcely rest with criminal judges for the good reason that they are not lawmakers. When a fixed legal code that must be observed to the letter leaves the judge no other task than to examine a citizen's actions and to determine whether or not they conform to the written law, when the standard of justice and injustice that must guide the actions of the ignorant as well as the philosophic citizen is not a matter of controversy but of fact, then subjects are not exposed to the petty tyrannies of many men. Such tyrannies are all the more cruel when there is a smaller distance between the oppressor and the oppressed. They are more ruinous than the tyranny of one person, for the despotism of many can be remedied only by the despotism of a single man, and the cruelty of a despot is not proportional to his strength, but to the obstacles he encounters.[10] With fixed and immutable laws, then, citizens acquire personal security. This is just because it is the goal

---

* The word *obligation* is one of the words more frequently used in morals than in any other discipline; it is an abbreviated symbol for a chain of arguments and not for an idea. Look for an idea that corresponds to the word *obligation*, and you will not find it; reason about the matter, and you will understand and be understood.

of society, and it is useful because it enables them to calculate precisely the ill consequences of a misdeed. It is just as true that they will acquire a spirit of independence, but this will not be to shake off the laws and resist the supreme magistrates; rather, they will resist those who have dared to claim the sacred name of virtue for their weakness in yielding to their private interests or capricious opinions.

V. Obscurity of Laws

If the interpretation of laws is an evil, their obscurity, which necessarily entails interpretation, is obviously another evil, one that will be all the greater if the laws are written in a language that is foreign to the common people. This places them at the mercy of a handful of men, for they cannot judge for themselves the prospect of their own liberty or that of others. A language of this sort transforms a solemn official book into one that is virtually private and domestic. What must we think of mankind when we consider that such is the ingrained custom of a good part of cultured and enlightened Europe![11] The greater the number of people who understand the sacred law code and who have it in their hands, the less frequent crimes will be, for there is no doubt that ignorance and uncertainty concerning punishments aid the eloquence of the passions.

One consequence of these last thoughts is that, without written texts, society will never assume a fixed form of government in which power derives from the whole rather than the parts and in which the laws, which cannot be altered save by the general will, are not corrupted as they move through the crush of private interests. Experience and reason have shown us that the probability and certainty of human traditions decline the farther removed they are from their source. If there is no lasting memorial of the social contract, how will the laws resist the inevitable force of time and the passions?

From this we see how useful the printing press is. It makes the entire public, not just a few people, the depository of the sacred laws.[12] To a great extent, it has dissipated that dark spirit of cabal and intrigue that vanishes when confronted with enlightenment and learning, which its adherents affect to despise and which they really fear. This is the reason that we see the atrocity of crimes diminishing in Europe; this atrocity made our forefathers tremble and become tyrants and slaves by turns. Anyone who is acquainted with the history of the last two or three centuries, and of our own century, will be able to see how the sweetest virtues—humanity, benevolence, tolerance of human errors—sprang from the lap of luxury and easy living.[13] He will see the effects of what is erroneously called ancient simplicity and good faith: humanity cowering under implacable superstition, the avarice and ambition of a few men staining coffers of gold and royal thrones with human blood, private treasons, public massacres, every nobleman a tyrant to the common people, and ministers of the Gospels' truth soiling with blood the hands that touched the God of mercy every day. These things are not the work of this enlightened century, which some people call corrupt.[14]

VIII. Division of Crimes[15]

We have seen what the true measure of crimes is—namely, *the harm done to society*.[16] This is one of those palpable truths which one needs neither quadrants nor telescopes to discover and which are within the reach of every ordinary intellect. Through a remarkable combination of circumstances, however, such truths have been recognized with decisive certainty only by a handful of thinking men in every nation and every age. But Asiatic notions, passions bedecked with power and authority, have dissipated the simple ideas that probably formed the first philosophy of newborn societies, usually by imperceptible nudges, but sometimes by violent impressions on timid human credulity. The enlightenment of this century seems to be leading us back to these simple ideas, though with a greater firmness obtainable from a mathematically rigorous investigation, a thousand unhappy experiences, and the obstacles themselves. At this point, the proper order of presentation would lead us to distinguish all the different sorts of crimes and the way to punish them, but their changing nature in the different circumstances of various times and places would make this an immensely and tediously detailed task for us. I shall be content to call attention to the most general principles and the most pernicious and widespread errors in order to disabuse both those who, from a poorly understood love of liberty, would desire to establish anarchy and those who would like to reduce men to a cloister-like regularity.

Some crimes are immediately destructive of society or of the person who represents it; some offend against the personal security of a citizen in his life, his goods, or his honor; certain others are actions contrary to what the laws oblige everyone to do or not to do for the sake of the public good. The first, which are the greatest crimes because they do the most harm, are called lese majesty or high treason. Only tyranny and ignorance, which confound the clearest words and ideas, can assign this name (and consequently the ultimate punishment) to crimes of a different nature, thus making men, as on a thousand other occasions, the victims of a word. Every crime, however private it may be, offends society, but not every crime threatens it directly with destruction. Moral actions, like physical ones, have their limited sphere of activity and are circumscribed, like all natural movements, by time and space. Hence, only quibbling interpretation, which is usually the philosophy of slavery, can confuse what eternal truth has distinguished by immutable relations.

After high treason come those crimes which violate the security of private persons.

Since this is the chief end of every legitimate association, one cannot but assign some of the most considerable punishments established by law to the violation of the right to security which every citizen has.

⬚ XII. Purpose of Punishments

From the simple consideration of the truths expounded thus far, it is clear that the purpose of punishments is not to torment and afflict a sentient being or to undo a crime which has already been committed. Far from acting out of passion, can a political body, which is the calm agent that moderates the passions of private individuals, harbor useless cruelty, the tool of fury and fanaticism or of weak tyrants? Can the cries of a poor wretch turn back time and undo actions which have already been done? The purpose of punishment, then, is nothing other than to dissuade the criminal from doing fresh harm to his compatriots and to keep other people from doing the same. Therefore, punishments and the method of inflicting them should be chosen that, mindful of the proportion between crime and punishment, will make the most effective and lasting impression on men's minds and inflict the least torment on the body of the criminal.[17]

⬚ XV. Secret Accusations[18]

Secret accusations are an evident but time-honored abuse made necessary in many nations by the weakness of the constitution. Such a custom makes men false and dissimulating. Anyone who can suspect another person of being an informer sees in him an enemy. Men then grow accustomed to masking their personal feelings, and, through the habit of hiding them from others, they end by hiding their sentiments from themselves. How unhappy men are when they reach this point! Without clear and firm principles to guide them, they wander bewildered and aimless in the vast sea of opinions, always concerned with saving themselves from the monsters which threaten them. Their present is always embittered by the uncertainty of their future. Deprived of the lasting pleasures of tranquillity and security, only a scant few happy moments scattered here and there in their sad lives and devoured in haste and disorder console them for having been alive. And shall we make of such men the bold soldiers who defend the country and the throne? Among these men shall we find the uncorrupted magistrates who sustain and enlarge the true interests of the sovereign with free and patriotic eloquence, who bring tribute to the throne together with the love and blessings of all classes of men, giving in return to palaces and hovels alike peace, security, and the industrious hope of bettering one's lot, which is the useful leaven and the very life of states?

Who can defend himself against calumny when it is armed with tyranny's strongest shield, *secrecy*? What on earth is the form of government in which the ruler suspects every subject of being an enemy and, in order to assure the public peace, deprives each citizen of tranquillity?

What are the reasons with which people justify secret accusations and punishments? The public welfare? The security and maintenance of the established form of government? What a strange constitution it must be in which the regime that controls force and public opinion (which is even stronger than force) fears every citizen! The safety of the accuser? The laws, then, do not suffice to defend him; one must conclude that there are subjects stronger than the sovereign! The infamy of the informer? Then slander is permitted when it is secret and punished when it is public! The nature of the crime? If harmless actions, or even acts useful to the public, are deemed crimes, then accusations and judgments can never be secret enough! Can there be crimes— that is, offenses against the public—when at

the same time it is not in everyone's interest to have a public example and, hence, a public judgment? I respect every government, and I am not speaking of any one in particular. Sometimes circumstances are such that one can believe that the extirpation of an evil inherent in the system of government would mean the complete ruin of the state. But if I had to dictate laws in some deserted corner of the universe, my hand would tremble before I authorized such a custom, and I would see all posterity before my eyes. Monsieur de Montesquieu has already said that public accusations are better suited to a republic, where the public good ought to be the strongest passion of the citizens, than to a monarchy, where this passion is greatly weakened by the very nature of the government. There it is best to establish appointed commissioners who accuse lawbreakers in the name of the people.[19] But every regime, republican and monarchical alike, should inflict upon the false accuser the same punishment that the accused would have received.

⬚ XVI. Torture[20]

The torture of the accused while his trial is still in progress is a cruel practice sanctioned by the usage of most nations. Its purpose is either to make the accused confess his crime, or to resolve the contradictions into which he has fallen, or to discover his accomplices, or to purge him of infamy for some metaphysical and incomprehensible reason or other, or, finally, to find out other crimes of which he may be guilty but of which he is not accused.

A man cannot be called "guilty" before the judge has passed sentence, and society cannot withdraw its protection except when it has been determined that he has violated the contracts on the basis of which that protection was granted to him. What right, then, other than the right of force, gives a judge the power to inflict punishment on a citizen while the question of his guilt or innocence is still in doubt? This dilemma is not new: either the crime is certain, or it is not: if it is certain, then no other punishment is suitable for the criminal except the one established by law, and torture is useless because the confession of the accused is unnecessary; if the crime is uncertain, one should not torment an innocent person, for, in the eyes of the law, he is a man whose misdeeds have not been proven.[21] But I add, moreover, that one confuses all natural relationships in requiring a man to be the accuser and the accused at the same time and in making pain the crucible of truth, as though the criterion of truth lay in the muscles and fibers of a poor wretch. This is a sure way to acquit robust scoundrels and to condemn weak but innocent people. This criterion is worthy of a cannibal, and the Romans (who were themselves barbarians on more than one count) kept it only for slaves, the victims of a ferocious and over-praised "virtue."

What is the political goal of punishment? It is to intimidate others. But what justification can we give, then, for the secret and private carnage that the tyranny of custom wreaks on the guilty and the innocent? It is important that no manifest crime go unpunished, but it is useless to discover who has committed a crime that lies buried in darkness. A wrong which has already been done and for which there is no remedy cannot be punished by political society except when the failure to do so would arouse false hopes of impunity in others. If it is true that more men, whether from virtue or fear, respect the law than violate it, then the risk of torturing an innocent person should be considered all the greater when, other things being equal, the probability is greater that a man has respected the law rather than despised it.

Another ridiculous reason for torture is the purgation of infamy; that is, a man judged infamous by law must confirm his deposition with the dislocation of his bones.[22] This abuse should not be tolerated in the eighteenth century. The underlying belief is that pain, which is a sensation, purges infamy, which is simply a moral relationship. Is pain perhaps a crucible? And is infamy perhaps a mixed and impure substance? It is not difficult to go back to the origin of this ridiculous law, since the very absurdities adopted by a whole nation always have some relationship to the common and respected ideas of that nation. This custom seems to be taken from religious and spiritual ideas which have so much influence on the thoughts of men, nations, and ages. An infallible dogma assures us that the blemishes which result from human weakness and which yet have not deserved the eternal wrath of the Great Being must be purged with an incomprehensible fire. Now infamy is a civil blemish, and, since pain and fire remove spiritual and disembodied stains, will the spasms of torture not remove a civil stain, namely infamy? I believe that the confession of the criminal, which certain courts require for conviction, has an analogous origin, for, in the mysterious tribunal of penance, the confession of sins is an essential part of the sacrament. This is how men abuse the very clear light of revelation. Just as such light is the only one that still shines in times of ignorance, so docile humanity runs to it on every occasion, giving it the most absurd and far-fetched applications.[23] Infamy, however, is a sentiment that is not subject to reason or to law, but to public opinion. Torture itself causes real infamy for its victim. Hence, this method seeks to remove infamy by inflicting it.

⊠ XIX. Promptness of Punishment

The more prompt the punishment is and the sooner it follows the crime, the more just and useful it will be. I say more just, because it spares the criminal the useless and cruel torments of uncertainty, which grow with the vigor of one's imagination and the sense of one's own weakness; more just, because being deprived of one's liberty is a punishment, and this cannot precede the sentence except when necessity demands it. Imprisonment, then, simply means taking someone into custody until he is found guilty, and, as such custody is essentially punitive, it should last as short a time as possible and be as lenient as possible. The duration of imprisonment should be determined both by the time necessary for the trial and by the right of those who have been detained the longest to be tried first. The rigor of detention must not exceed what is necessary to forestall escape or the concealment of evidence. The trial itself must be completed in the shortest possible time. Can there be a more cruel contrast than the one between the indolence of a judge and the anguish of someone accused of a crime—between the comforts and pleasures of an unfeeling magistrate on the one hand, and, on the other, the tears and squalid condition of a prisoner? In general, the burden of a punishment and the consequence of a crime should have the greatest impact on others and yet be as mild as possible for the person who suffers it; for one cannot call any society "legitimate" if it does not recognize as an indisputable principle that men have wanted to subject themselves only to the least possible evils.

I have said that promptness of punishment is more useful, for the less time that passes between the misdeed and its chastisement, the stronger and more permanent is the human mind's association of the two ideas of *crime* and *punishment*, so that imperceptibly the one will come to be considered as the cause and the other as the necessary and inevitable result. It is well established that the association of ideas is the cement that shapes the whole structure of the human intellect; without it, pleasure and pain would be isolated feelings with no consequences. The farther removed men are from general ideas and universal principles—in other words, the more uneducated men are—the more they act on the basis of immediate and very familiar associations, neglecting the more remote and complicated ones. The latter are useful only to men strongly impassioned for the object after which they are striving. The light of their attention illuminates this one object only, leaving all others in darkness. Such remote and complicated associations are likewise useful to more lofty minds, for they have acquired the habit of rapidly surveying many objects at once, and they have the ability to contrast many partial sentiments with one another, so that the outcome, which is action, is less dangerous and uncertain.

The temporal proximity of crime and punishment, then, is of the utmost importance if one desires to arouse in crude and uneducated minds the idea of punishment in association with the seductive image of a certain advantageous crime. Long delay only serves to disconnect those two ideas, and whatever impression the chastisement of a crime may make, that impression will be made more as a spectacle than a punishment. Further, the impression will come only after the horror of a given crime, which ought to reinforce the feeling of punishment, has grown weak in the minds of the spectators.

Another principle serves admirably to tighten even further the connection between the misdeed and its punishment, namely, that the latter should conform as closely as possible to the nature of the crime. This analogy marvelously facilitates the contrast that should exist between the motive for a crime and the consequent impact of punishment, so that the latter draws the mind away and leads it to quite a different end than the one toward which the seductive idea of breaking the law seeks to direct it.

⊠ XXVII. Mildness of Punishments

But my train of thought has taken me away from my subject, and I hasten to return in order to clarify it. One of the greatest checks on crime is not the cruelty of punishments but their inevitability. Consequently, in order to be effective, virtues, magisterial vigilance and inexorable judicial severity must be accompanied by mild legislation. The certainty of a chastisement, even if it be moderate, will always make a greater impression than the fear of a more terrible punishment that is united with the hope of impunity; for, when they are certain, even the least of evils always terrifies men's minds, while hope, that heavenly gift that often fills us completely, always removes from us the idea of worse punishments, especially if that hope is reinforced by the examples of impunity which weakness and greed frequently accord. The very savagery of a punishment makes the criminal all the bolder in taking risks to avoid it precisely because the evil with which he is threatened is so great, so much so that he commits several crimes in order to escape the punishment for a single one of them. The countries and ages in which punishments have been most atrocious have always been the scene of the bloodiest and most inhuman actions, for the same spirit of ferocity that guided the hand of the legislator governed the hand of the parricide and the assassin. Seated on the throne, this spirit dictated iron laws for savage and slavish souls to obey; in private darkness, it moved men to destroy one tyrant in order to create another.[24]

To the degree that punishments become more cruel, men's souls become hardened, just as fluids always seek the level of surrounding objects, and the constantly active force of the passions leads to this: after a hundred years of cruel punishments, breaking on the wheel[25] occasions no more fright than imprisonment did at first. In order for a penalty to achieve its objective all that is required is that the harm of the punishment should exceed the benefit resulting from the crime. Further, the inevitability of the punishment and the loss of the anticipated advantage of the crime should enter into this calculation of the excess of harm.[26] Everything more than this is thus superfluous and therefore tyrannical. Men regulate their conduct by the repeated experience of evils which they know, not by those of which they are ignorant. Let us imagine two nations, each having a scale of punishments proportional to crimes; in one, the maximum penalty is perpetual slavery, and, in the other, breaking on the wheel. I maintain that the first nation will have as much fear of its greatest punishment as the second.[27] If for some reason the first of these nations were to adopt the more severe penalties of the second, the same reason might lead the latter to increase its punishments, passing gradually from breaking on the wheel to slower and more deliberate torments, and finally to the ultimate refinements of that science that tyrants know all too well.

Cruelty of punishments leads to two other ruinous consequences that are contrary to the very purpose of preventing crimes. The first is that it is far from easy to maintain the essential proportion between crime and punishment, for no matter how much industrious cruelty may have multiplied the forms of chastisement, they still cannot exceed the limit that the human physique and sensory capacity can endure.[28] Once this limit has been reached, it would not be possible to devise greater punishments for more harmful and atrocious crimes, and yet such punishments would be required to deter them. The second consequence is that impunity itself arises from the barbarity of punishments. There are limits to human capacities both for good and for evil, and a spectacle that is too brutal for humanity can only be a passing frenzy, never a permanent system such as the law must be. If the laws are indeed cruel, either they are changed or else fatal impunity results from the laws themselves.

Who would not tremble with horror when he reads in history books of the barbarous and useless torments that were devised and carried out in cold blood by men who were deemed wise? Who would not shudder to the depths of his being at the sight of thousands of poor wretches forced into a desperate return to the original state of nature by a misery that the law—which has always favored the few and trampled on the many—has either willed or permitted? Or at the spectacle of people accused of impossible crimes fabricated by timid ignorance? Or at the sight of persons whose only crime has been their fidelity to their own principles lacerated with deliberate formality and slow torture by men endowed with the same senses and hence with the same passions, providing a diverting show for a fanatical crowd?

▧ XXVII. The Death Penalty

This vain profusion of punishments, which has never made men better, has moved me to inquire whether capital punishment is truly useful and just in a well-organized state. By what alleged right can men slaughter their fellows? Certainly not by the authority from which sovereignty and law derive. That authority is nothing but the sum of tiny portions of the individual liberty of each person; it represents the general will, which is the aggregate of private wills. Who on earth has ever willed that other men should have the liberty to kill him? How could this minimal sacrifice of the liberty of each individual ever include the sacrifice of the greatest food of all, life itself?[29] And even if such were the case, how could this be reconciled with the principle that a man does not have the right to take his own life? And, not having this right himself, how could he transfer it to another person or to society as a whole?

The death penalty, then, is not a *right*—for I have shown that it cannot be so—but rather a war of the nation against a citizen, a campaign waged on the ground that the nation has judged the destruction of his being to be useful or necessary.[30] If I can demonstrate that capital punishment is neither useful nor necessary, however, I shall have vindicated the cause of humanity.

The death of a citizen cannot be deemed necessary except for two reasons. First, if he still has sufficient connections and such power that he can threaten the security of the nation even though he be deprived of his liberty, if his mere existence can produce a revolution dangerous to the established form of government, then his death is required. The death of such a citizen becomes necessary, then, when the nation is losing or recovering its liberty, or in times of anarchy, when disorder itself takes the place of law. Under the calm rule of law, however, and under a regime that has the full support of the nation, that is well armed against external and internal enemies with force and with public opinion (which is perhaps more effective than force itself), where only the true sovereign holds the power to command, and where riches buy pleasure and not authority, I see no necessity whatever for destroying a citizen.[31] The sole exception would be if his death were the one and only deterrent to dissuade others from committing crimes. This is the second reason for believing that capital punishment could be just and necessary.

If the experience of all ages, during which the ultimate punishments has never deterred men who were determined to harm society; if the example of the citizens of Rome; or if twenty years of the reign of the Empress Elizabeth of Muscovy, who has given the leaders of her people an illustrious example that is worth at least as much as many conquests bought with the blood of her country's sons[32]—if all this does not persuade men, who always suspect the voice of reason and heed the voice of authority, then one needs only to consult human nature in order to feel the truth of my assertion.

It is not the severity of punishment that has the greatest impact on the human mind, but rather its duration, for our sensibility is more easily and surely stimulated by tiny repeated impressions than by a strong but temporary movement. The rule of habit is universal over every sentient being, and, as man talks and walks and tends to his needs with the aid of habit, so moral ideas are fixed in his mind only by lasting and repeated blows. The most powerful restraint against crime is not the terrible but fleeting spectacle of a villain's death, but the faint and prolonged example of a man who, deprived of his liberty, has become a beast of burden, repaying the society he has offended with his labors. Each of us reflects, "I myself shall be reduced to such a condition of prolonged wretchedness if I commit similar misdeeds."[33] This thought is effective because it recurs quite frequently, and it is more powerful than the idea of death, which men always see in hazy distance.

Capital punishment makes an impression which for all its force does not offset the rapid forgetfulness that is natural to man, even in the most essential matters, and the human passions accelerate. One may posit as a general rule that violent passions grip men strongly but not for long, and thus they are apt to cause those revolutions that turn ordinary men either into Persians or else into Spartans. Under a free and tranquil regime, however, impressions should be frequent rather than strong.

The death penalty becomes an entertainment for the majority and, for a few people, the object of pity mixed with scorn. Both of these sentiments alike fill the hearts of the spectators to a greater extent than does the salutary fear that the law claims to inspire. With moderate and continuous punishments, though, such fear is the dominant sentiment because it is the only one. The limit that the legislator should assign to the rigor of punishments, then, seems to be the point at which the feeling of compassion begins to outweigh every other emotion in the hearts of those who witness a chastisement that is really carried out for their benefit rather than for the sake of the criminal.[34]

In order to be just, a penalty should have only the degree of intensity needed to deter other men from crime. Now there is no one who, on reflection, would choose the total and permanent loss of his own liberty, no matter how advantageous a crime might be. Therefore, the intensity of a sentence of servitude for life, substituted for the death penalty, has everything needed to deter the most determined spirit. Indeed, I would say more: a great many people look upon death with a tranquil and steady eye, some from fanaticism, others from vanity (a sentiment that almost always accompanies men even beyond the grave), some from a final and desperate attempt to live no longer or to leave their misery behind; but neither fanaticism nor vanity survives among fetters and chains, under the prod or the yoke, or in an iron cage, and the desperate man finds a beginning rather than an end to his troubles. Our spirit withstands violence and extreme yet fleeting pain better than it does time and unending weariness, for it can, so to speak, draw itself together for a moment to repel the former, but its elasticity is insufficient to resist the prolonged and repeated actions of the latter. With capital punishment, one crime is required for each example offered to the nation; with the penalty of a lifetime at hard labor, a single crime affords a host of lasting examples. Moreover, if it be important that men should see the power of the law frequently, judicial executions should not be separated by too great an interval; this presupposes frequent crimes. Thus, in order for this punishment to be useful, it must not make as strong an impression on men as it ought to make; in other words, it must be effective and ineffective at the same time. If someone were to say that life at hard labor is as painful as death and therefore equally cruel, I should reply that, taking all the unhappy moments of perpetual slavery together, it is perhaps even more painful, but these moments are spread out over a lifetime, and capital punishment exercises all its power in an instant. And this is the advantage of life at hard labor; it frightens the spectator more than the victim, for the former considers the entire sum of unhappy moments, and the latter is distracted from the future by the misery of the present moment. Imagination magnifies all evils, and the sufferer finds compensations and consolations unknown and unbelievable to the spectators, who substitute their own sensibility for the calloused soul of the wretch.

Here, more or less, is the line of reasoning that a thief or a murderer follows; such men have no motive but the gibbet or the wheel to keep them from breaking the law. (I am aware that developing the sentiments of one's spirit is an art that one acquires with education, but, though a thief would not express his principles well, they are no less operative for that.) "What are these laws that I must respect and that leave such a great distance between me and the rich man? He denies me the penny I ask of him, and he excuses himself by exhorting me to work, something with which he himself is unfamiliar. Who made these laws? Rich and powerful men who have never deigned to visit the squalid hovels of the poor, who have never broken a moldy crust of bread among the innocent cries of their famished children and the tears of their wives. Let us break these bonds that are so ruinous for the majority and useful to a handful of indolent tyrants; let us attack injustice at its source. I shall revert to my natural state of independence, and for a time I shall live free and happy from the fruits of my courage and industry. Perhaps I shall see the day of suffering and repentance, but that time will be brief, and, in return for a day of torment, I shall have many years of liberty and pleasure. Kings of a small band, I shall set fortune's errors right, and I shall see those tyrants grow pale and tremble in the presence of a man whom they, with insulting ostentation, respected less than their horses and dogs."[35] Then religion appears to the mind of the scoundrel, who puts everything to bad use, and, presenting him with the prospect of an easy repentance and a near certainty of eternal bliss, greatly diminishes the horror or the final tragedy.

But the man who sees before his eyes the prospect of a great many years or even a lifetime of penal servitude and suffering, exposed to the sight of his fellow citizens with whom he once lived in freedom and friendship, a slave to the laws that once protected him, will make a salutary comparison between all this, on the one hand, and uncertain success of his crimes and the brief time that he will be able to enjoy their fruits, on the other. The constant example of those whom he actually sees as victims of their own inadvertence makes a much stronger impression on him than the spectacle of punishment that hardens more than it corrects him.

Capital punishment is not useful because of the example of cruelty which it gives to men. If the passions or the necessity of war have taught people to shed human blood, the laws that moderate men's conduct ought not to augment the cruel example, which is all the more pernicious because judicial execution is carried out methodically and formally. It appear absurd to me that the laws, which are the expression of the public will and which detest and punish homicide, commit murder themselves, and, in order to dissuade citizens from assassination, command public assassination. What are the true and most effective laws? They are those pacts and conventions that everyone would observe and propose while the voice of private interest, which one always hears, is silent or in agreement with the voice of the public interest.[36] What are the sentiments of each person regarding the death penalty? We may read them in the signs of indignation and scorn with which everyone looks upon the executioner, who is, however, an innocent servant of the public will, a good citizen who contributes to the public welfare, the necessary instrument of internal security just as valorous soldiers are of external security. What, then, is the origin of this contradiction? And why is this sentiment that defies reason indelible among men? Because men, in their heart of hearts, the part of them that more than any other still retains the original form of the first nature, have always believed that one's own life should be at the mercy only of necessity, which rules the world with its iron scepter.

What must men think when they see wise magistrates and grave ministers of justices who, with tranquil indifference and slow preparation, have a criminal dragged to his death? And when they witness a judge who, with cold insensitivity and perhaps even secret satisfaction in his own authority, goes to enjoy the comforts and pleasures of life while a poor wretch writhes in his final agony, awaiting the fatal below? "Ah!" they will say, "these laws are only pretexts for violence and for the premeditated and cruel formalities of justice; they are only a conventional language for sacrificing us with greater security, like victims offered up to insatiable idol of despotism. We see assassination employed without repugnance or excitement, even though it is preached to us as horrible crime. Let us take advantage of this example. Violent death appears a terrible sight as it is described to us, but we see that it is the affair of a moment. How much less its terror will be for someone who, because he is not expecting it, is spared almost all of its pain!" These are the pernicious and fallacious arguments used more or less consciously by men disposed to crime. Among such men, as we have seen, the abuse of religion carries more weight than religion itself.

If anyone should cite against me the example of practically all ages and nations, which have assigned the death penalty to certain crimes, I shall reply that the example is annihilated in the presence of truth, against which there is no prescription, and that human history leaves us with the impression of a vast sea of errors in which a few confused and widely scattered truths are floating. Human sacrifice was common among virtually all nations, yet who will dare to excuse it? That a mere handful of societies have abstained from capital punishment for a short period only is more favorable than contrary to my case, because this is similar to the fate of great truths. They last no longer than a flash in comparison with the long dark night that surrounds humanity. The happy period has not yet arrived in which truth shall be the portion of the majority, just as error has been hitherto. Until now, only those truths that Infinite Wisdom has wished to distinguish by revealing them have been exempted from this universal law. The voice of one philosopher is too weak to overcome the hue and the cry of so many people who are guided by blind habit, but the few sages scattered across the face of the earth will echo my sentiments in their inner-most hearts. And if Truth can reach the throne of a monarch despite the infinite obstacles that separate him from her and despite his own will, let him know that she comes with the secret desires of all men; let him know that the sanguinary notoriety of conquerors will fall silent before him and that a just posterity will give his name preeminence among the peaceful trophies of the Tituses, the Antonines, and the Trajans.[37]

How happy humanity would be if laws were being given to it for the first time, now that we see beneficent monarchs seated on the thrones of Europe! They are rulers who love peaceful virtue, the sciences and the arts; they are the fathers of their people, citizens who wear the crown. The growth of their authority constitutes the happiness of their subjects because it destroys that intermediary despotism, which is all the more cruel because it is less secure, that has stifled the expression of the desire of the people.[38] Those desires are always sincere and always fortunate when they can reach the throne. If such monarchs, I say, allow ancient laws to remain, it is the result of the infinite difficulty of stripping errors of the venerable rust of many centuries. This is a reason for enlightened citizens to desire more ardently the continued increase of their authority.[39]

◼ XLVII. Conclusion

I conclude with the reflection that the magnitude of punishment ought to be relative to the condition of the nation itself. Stronger and more obvious impressions are required for the hardened spirits of a people who have scarcely emerged from a savage state. A thunderbolt is needed to fell a ferocious lion who is merely angered by a gun shot. But, to the extent that human spirits are made gentle by the social state, sensibility increases; as it increases, the severity of punishment must diminish if one wishes to maintain a constant relationship between object and feeling.[40]

From all that has been seen hitherto, one can deduce a very useful theorem, but one that scarcely conforms to custom, the usual law-giver of nations. It is this: *In order that any punishment should not be an act of violence committed by one person or many against a private citizen, it is essential that it should be public, prompt, necessary, the minimum possible under the given circumstances, proportionate to the crimes, and established by law.*

✖ Notes

1. The idea of laws as conditions of the social contract was fundamental to Rousseau. Jean-Jacques Rousseau, *Du contrat social*, in Jean-Jacques Rousseau, *Oeuvres complètes*, ed. Brenard Gagnebin and Marcel Raymond, vol. 3 (Paris, 1964), bk. 2, chap. 6, pp. 378–380.

2. This view of human nature as motivated chiefly by self-interest was common among eighteenth-century utilitarians. Helvétius had declared that all men seek to become despots and that tangible motives are necessary to check this tendency. Helvétius, *D l'esprit*, disc. 3, chap. 17, pp. 284–289. Beccaria frankly admitted that he owed a large part of his ideas to Helvétius. Beccaria to Morellet, 26 Jan. 1766, in *Opere* (Romangnoli ed.), 2:865. Kant and Hegel, of course, vehemently objected to such a theory of punishment, and it must be noted that it was not always typical of Beccaria. Immanuel Kant, *The Metaphysical Elements of Justice. Part I of the Metaphysics of Morals*, ed. and trans. John Ladd (Indianapolis, 1965), p. 100; G. W. Hegel, *Hegel's Philosophy of Right*, ed. and trans. T. M. Knox (London, 1967), p. 246.

3. Montesquieu had held that excessive and unnecessary penalties are suitable only for a despotic government. Montesquieu, *De l'esprit des lois* (Caillois ed.), bk. 6, chap. 9, 2:318–319.

4. This account of the formation of societies closely parallels the one which Montesquieu gave in his *De l'esprit des lois* (Caillois ed.), bk 1, chap. 3, 2:236–238.

5. Beccaria's account of the social contract is quite unlike the total surrender of rights of which Rousseau spoke. Rousseau, *Du contrat social* (Gagnebin and Raymond ed.), bk. 1, chap. 6, 3:360–362. It is, rather, much closer to Locke's idea that the sovereign is purely fiduciary and that the people forming a state make only a minimal surrender of their liberty. John Locke, *The Second Treatise of Government*, ed. Thomas P. Peardon (Indianapolis, 1952), chap. 3, para 21, p. 14; chap 8, paras. 95–101, pp. 54–57.

6. Beccaria's utilitarian view of justice appears very similar to the ideas expounded by Helvétius. Helvétius, *D l'esprit*, disc. 2, chap. 5, pp. 55–57; chap. 8, pp. 69–74; chap. 12, pp. 89–97.

7. Rousseau, who had an immense influence on Beccaria, insisted that only the sovereign, representing the general will, has the right to establish laws. Rousseau, *Du contrat social* (Gagnebin and Raymond ed.), bk. 1, chap. 7, 3:362–363; bk. 2, chap. 1, 3:368–369.

8. Rousseau maintained that the general will, the sovereign legislator, could lay down only general laws and could not apply them in particular cases; doing that, Rousseau declared, was the task of the magistrate. Rousseau, *Du contrat social* (Gagnebin and Raymond ed.), bk. 2, chap. 4, 3:372–375.

9. The entire chapter is a reaction against the unbridled judicial discretion characteristic of Beccaria's day. With the blend of the Roman law, local custom, royal decrees, judicial commentaries, and court precedent which constituted the legal systems in most of Europe, judges had all but total authority to decide what laws would be applied and to whom. A good account of this is to be found in Manzoni, *Opere* (Bacchelli ed.), pp. 973–989. Voltaire discovered this state of affairs in his campaign for law reform in France. Peter Gay, *Voltaire's Politics: The Poet as Realist* (New York, 1965), pp. 294–296. Conservatives, however, argued that a wide scope for interpretation reinforced a benevolent paternalistic power of the upper classes. See Facchimei, *Note ed osservazioni*, pp. 13–14, 16, 23–24.

10. Montesquieu had inveighed against the degeneration of an aristocracy into an oligarchy, deeming the latter a despotism with many despots. Montesquieu, *De l'esprit des lois* (Caillois ed.), bk. 8, chap 5, 2:353–354. Though Beccaria may have wished to elaborate on Montesquieu's theme, his target was

obviously the aristocracy, particularly the older generation of the Milanese patriciate. See Daniel M. Klang, "Reform and Enlightenment in Eighteenth-Century Lombardy," *Canadian Journal of History/Annales Canadiennes d'Histoire* 19(1984):39–70.

11. In Beccaria's day, laws were promulgated in Latin in much of Europe. Maria Theresa's criminal code of 1770 was drawn up in Latin and picturesquely named the *Nemesis Teresiana*. Some decades after Beccaria, Hegel deplored the practice of couching laws in a dead tongue. Hegel, *Philosophy of Right* (Knox ed.), p. 138.

12. This statement may be interpreted as a jab at Montesquieu, who held that a monarchy required a special intermediary body, such as the French *parlements* or Senate of Milan, to serve as a depository of law. Montesquieu, *De l'esprit des lois* (Caillois ed.), bk. 2, chap 5, 2:249.

13. In the eighteenth century, many writers defended the growth of wealth and luxury as sources of virtue. Hirschman, *Passions and Interests*, pp. 14–18. Among those with whom Beccaria was best acquainted, Helvétius, and following him, Pietro Verri, had mounted especially strong arguments in favor of luxury. Helvétius, *De l'esprit* (1759 ed.), disc. 1, chap. 3, pp. 12–24; Pietro Verri, *Meditazioni sulla economia politica*, in Pietro Verri, *Opere filsofische e di economia politica* (Milan 1835), 1:155–364; Pietro Verri, "Considerazioni sul lusso," in *Il caffe* (Romagnoli ed.), pp. 113–118.

14. Among Italian conservatives in general and older Lombard patricians in particular, it was common to deem the eighteenth century "corrupt" because of the growth of prosperity and the infusion of transalpine ideas. Venturi, *Settecento reformatore*, p. 657. Pietro Verri lampooned this outlook in his "Orazione panegirica sulla giurisprudenza milanese," written in 1763. This is printed in *Delitti* (Venturi ed.), pp. 127–146; see esp. pp. 127–129.

15. The division of crimes that Beccaria offers is similar to, and perhaps inspired by, Montesquieu's chapter on the same subject. Montesquieu, *De l'esprit des lois* (Caillois ed.), bk. 12, chap 4, 433–435.

16. Though Hegal rejected any utilitarian foundation of the right to punish, it is noteworthy that he was at one with Beccaria in declaring that the social harm of an offense is the only way of measuring its relative importance. Like Beccaria, Hegel held that the relative importance of specific crimes might vary precisely because their impact could differ in different times and places. Hegel, *Philosophy of the Right* (Knox ed.), pp. 68–72, 274.

17. Beccaria showed his most utilitarian side in discussing the purpose of punishment. In this regard, Bentham followed in his footsteps, while Kant and Hegel deplored these principles. Bentham, *Principles* (Burns and Hart ed.), chap. 13, pp. 158–164; Kant, *Justice*, p. 100; Hegel, *Philosophy of Right*, p. 246. On Beccaria's belief that punishment should achieve the maximum mental impact with the least cost in physical pain, see Michel Foucault, *Discipline and Punish: The Birth of the Prison*, trans. Alan Sheridan (New York, 1977), pp. 73–103.

18. Despite Beccaria's disclaimer, this chapter is directed especially against Venice, where the state inquisitors received secret accusations against seditious or ambitious citizens in order to foil plots against the oligarchy that controlled the Republic. This chapter was among those that prompted Facchinei, with the blessings of the Venetian authorities, to write his diatribe against Beccaria. Facchinei, *Note ed Osservazioni*, pp. 49–58. Montesquieu had written at some length on this Venetian custom, but he had praised secret accusations there, arguing that they were a necessary means of preserving the Republic's liberty. Montesquieu, *De l'esprit des lois* (Caillois ed.), bk. 2, chap 3, 2:245–246; bk. 5, chap. 8, 2:286–287; bk. 11, chap. 6, 2:397. Beccaria, a Milanese and therefore closer to Venice, had a much lower opinion of the reputed wisdom of the Republic than did Montesquieu.

19. Montesquieu wrote of the methods of accusation under various types of government in *De l'esprit des lois* (Caillois ed.), bk. 6, chap 8, 2:317.

20. Before Beccaria, Montesquieu had deplored the use of judicial torture, which was common in most places on the Continent. Montesquieu had held that this practice is suitable only in despotic states and that careful investigation, not a forced confession, is the only way to obtain evidence against a criminal. Montesquieu, *De l'esprit des lois* (Caillois ed.), bk. 2, chap 3, 2:245–246; bk. 5, chap. 8, 2:286–287; bk. 11, chap. 6, 2:397. Pietro Verri, O*sservazioni sulla tortura*, ed. Plinio Succhetto (Bologna, 1962), pp. 177–280.

21. Pietro Verri employed precisely the same argument in his "Orazione panegirica sulla giurisprudenza milanese," *Delitti* (Venturi ed.), pp. 132–133. On the efforts of Enlightenment reformers generally to introduce new criteria of judicial certainty, see Foucault, *Discipline and Punish*, pp. 38–43, 79–82.

22. As Beccaria's grandson explained, torture was originally applied to persons who were deemed disreputable and who accused other persons of crimes. Pain, it was held, gave such testimony a credibility that the

accuser's character did not. Torture, of course, was applied to suspects as well as accusers. Manzoni, *Colonno infame*, pp. 1009–1010.

23. Torture became more common from the eleventh century onward. Ecclesiastical courts held that a confession, even if extracted under torture, was essential for the salvation of the criminal's soul. But the practice was quickly taken up by secular magistrates. Pietro Verri's discussion of the matter was very similar to Beccaria's in his *Osservazioni sulla tortura*, pp. 259–263.

24. On several occasions, Montesquieu remarked that cruel punishments are suitable only in a despotic state, which is based upon fear, and that excessive penalties are most likely to be found in such a regime. Montesquieu, *Lettres persanes* (Callois ed.), vol. 1 (Paris, 1949) letter 80, pp. 252–253; Montesquieu, *De l'esprit des lois* (Caillois ed.), bk. 12, chap 4, 2:433–435.

25. Breaking on the wheel was a common form of execution in the eighteenth century. An English traveler in France described such an execution: "On the scaffold was erected a large cross exactly in the form of that commonly represented for Saint Andrew's. The executioner and his assistants then placed the prisoner on it, in such a manner that his arms and legs were extended exactly agreeable to the form of the cross, and strongly tied down; under each arm, leg, etc., was cut a notch in the wood, as a mark where the executioner might strike, and break the bone with greater facility. He held in his hand a large iron bar . . . and in the first place broke his arms, then in a moment after both his thighs; it was a melancholy, shocking sight, to see him heave his body up and down in extreme agony, and hideous to behold the terrible distortions of his face; it was a considerable time before he expired. . . ." Sacheverell Stevens, *Miscellaneous Remarks Made on . . . France, Italy, Germany, and Holland* (London, 1756) as cited in Jerry Kaplow, *The Names of Kings: The Parisian Laboring Poor in the Eighteenth Century* (New York, 1972), p. 135.

26. Bentham later elaborated on the calculation of the excess of harm over the profit of the crime, citing Beccaria in the process. Bentham, *Principles* (Burns and Hart ed.), chap. 14, pp. 165–174, esp. sec. 8, n.

27. Montesquieu had made precisely this point as early as 1721. Montesquieu, *Lettres persanes* (Callois ed.), letter 80, 1:252–253.

28. The point Beccaria is making here is based on the psychology of Helvétius, Helvétius, *D l'esprit* (1759 ed.), disc. 1, pp. 1–32; disc. 3, chaps. 1–3, pp. 187–202.

29. Beccaria's detractors were quick to seize on this argument. Facchinei, for instance, suggested that one could just as well argue that no one would ever grant the sovereign the right to punish him in any way. Facchinei, *Note ed osservazioni*, pp. 105–106. Thoroughgoing utilitarians, while perhaps sympathetic to Beccari's conclusions, deplored his contractarian arguments. See, for instance, Melchior Grimm's review of the French translation in his *Correspondance littéraire* of 1 Dec. 1765, in *Delitti* (Venturi ed.), pp. 343–344.

30. In arguing that the death penalty is an act of war, Beccaria is, in effect, saying that it is an act of annihilation, not of coercion. One may very well argue that the state has the right to coerce, but capital punishment itself (as opposed to the *threat* of capital punishment) is no coercion at all. It is certainly possible to develop arguments against the death penalty on contractarian grounds. See, for instance, Thomas W. Satre, "The Irrationality of Capital Punishment," *The Southwestern Journal of Philosophy* 6 (1975):75–87.

31. Rousseau, who may have influenced Beccaria's views on capital punishment, wrote, "There is no right to put someone to death, even in order to set an example, unless he cannot be kept alive without danger." Rousseau, *Du contrat social* (Gagnebin and Raymond ed.), bk. 2, chap. 5, 3:377.

32. Czarina Elizabeth I, who reigned from 1741 to 1762, was opposed to the death penalty and abolished it with two separate decrees in 1753 and 1754.

33. Passages such as this show Beccaria at his most utilitarian, thinking of the criminal merely as an object lesson rather than a person. Beccaria later rejected sentences of public labor for minor crimes, and he implied that public labor, since it is inherently degrading, should not be used as a punishment for any offense. Beccaria, "Brevi riflessione intorno al codice generale," in *Opere* (Romagnoli ed.), 2:709–711, 717.

34. In the eighteenth century, it was common for the disadvantaged (from whose ranks most criminals came) to feel sympathy or curiosity at the sight of a public execution. Rarely, if ever, did such a spectacle inspire "salutary fear." Foucault, *Discipline and*

Punish, pp. 54–68, 104–131; Kaplow, Names of Kings, pp. 136–137.

35. Here Beccaria states an important principle of retributivist theories of punishment, especially those of a contractarian sort: if a criminal is punished because he violated an ostensibly universally beneficial system of rules, there is no reason that he should have felt bound by those rules in the first place if they caused him to bear many burdens and receive few benefits. See esp. Herbert Morris, "Persons and Punishment," in *Punishment and Rehabilitation*, ed. Jeffrie G. Murphy (Belmont, Cal., 1973), p. 56.

36. Beccaria's description of valid laws is quite similar to Rousseau's account of the legislative general will. Rousseau, *Du contrat social* (Gagnebin and Raymond ed.), bk. 2, chap. 3, 3:371–372; chaps. 6–6, 3:378–384.

37. These Roman emperors of the first and second centuries were famous for their concern for the happiness and well-being of their people.

38. Once more, Beccaria takes aim at the aristocratic intermediary bodies, such as the French *parlements* or the Senate of Milan, which Montesquieu had praised so highly.

39. In 1792, Beccaria prepared a memorandum on the death penalty for the Austrian government. He repeated most of the arguments in this chapter, and he added a new one: unlike other punishments, the death penalty is absolutely irrevocable. Therefore, he contended, it should be inflicted only in cases of total certainty. Since humans are not infallible, however, Beccaria concluded that there could never be sufficient certainty to warrant the use of capital punishment. Cesare Beccari, "Vota . . . per la riforma del sistema criminale nell Lombardia austriaca riguardante la pena di morte" (1792) in *Opere* (Romagnoli ed.), 2:735–741. This is one of the strongest arguments against the death penalty; modern opponents of capital punishment have developed it and relied heavily upon it. See esp. Jeffrie G. Murphy, "Cruel and Unusual Punishments," in *Retribution, Justice, and Therapy*, ed. Murphy (Boston, 1979), pp. 238–244.

40. Hegel fully agreed with Beccaria on this score, arguing that punishments can and should vary according to the degree of a society's refinement or barbarism. "A penal code, then," concluded Hegel, "is primarily the child of its age and the state of civil society at the time." Hegel, *Philosophy of Right* (Knox ed.), p. 140. For a modern view substantially like Beccaria's, see Jan Gorecki, *Capital Punishment: Criminal Law and Social Evolution* (New York, 1983), esp. pp. 31–80.

REVIEW QUESTIONS

1. What two reasons did Beccaria have for arguing that punishments should be prompt or swift? Which of these relates to deterrence theory (whereas the other is a reform for fairness and due process)?

2. Of the three aspects of deterrence theory—swift, severe, and certain—which did Beccaria say was most important? Provide a quotation that summarizes his feelings regarding this belief.

3. What did Beccaria say about the death penalty? Was he for it or against it? Provide at least three reasons he took this position.

4. Discuss two or more reforms that Beccaria proposed that would make a justice system more fair.

5. Discuss how Beccaria may feel regarding how Iraqi war suspects have been tortured or held for many years without trial? Would he agree or disagree with their treatment? Why or why not?

READING

This entry from Jeremy Bentham's *Introduction to the Principles of Morals and Legislation* (1789/1962) followed Beccaria and obviously was consistent with his approach. Bentham was important in the sense that he elaborated on Beccaria's ideas and brought many of the latter's concepts and assumptions to England and other countries. Notably, it should be said that his formulations and statements were largely the same as those of Beccaria; Bentham simply provided further specification of his concepts. Still, Bentham is considered another key figure of the Classical School of criminological thought and aided in further clarification of deterrence theory in terms of his formulation of the utility and hedonistic calculation of decision making. Specifically, he provides actual criteria for determining the pains and pleasures of engaging in certain activities, which some have called the "hedonistic calculus."

Bentham further specified the idea of utility, or **utilitarianism,** which emphasizes maximization of pleasure and minimization of pain. This can be viewed on both the individual and aggregate level. On the individual level, this means that an individual will engage (or not engage) in activity that maximizes pleasure and minimizes pain, as perceived by the individual prior to the act. On the aggregate scale, utilitarianism argues that the state or government will implement laws that benefit the "greatest good for the greatest number." Although this sounds like a good policy for both levels, such a policy is often deceiving. After all, individuals often do not realize what is good in the long term and often choose what is pleasurable in the short term. Thus, they often engage in behavior that provides pleasure in the "here and now" and disregard what will happen later. In the aggregate, societies often engage in activities that appear to be pleasurable and desirable to the vast majority but are either unethical or detrimental in the long-term, such as the U.S. policy on slavery in the early 1800s or the Nazis in mid-20th-century Germany. Although most of the population agreed with the governmental policies at the time, it is quite obvious that they are often either unethical and/or detrimental to society at large.

An Introduction to the Principles of Morals and Legislation

Jeremy Bentham

⚒ Of the Principle of Utility

Nature has placed mankind under the governance of two sovereign masters, pain and pleasure. It is for them alone to point out what we ought to do, as well as to determine what we shall do. On the one hand the standard of right and wrong, on the other the chain of causes and effects, are fastened to their throne. They govern us in all we do, in all we say, in all we think: every

---

Source: Jeremy Bentham. (1764). *Introduction to the Principles of Morals and Legislation.* The selection here has been abridged from the original.

effort we can make to throw off our subjection will serve but to demonstrate and confirm it. In words a man may pretend to abjure their empire: but in reality he will remain subject to it all the while. The principle of utility recognizes this subjection and assumes it for the foundation of that system, the object of which is to rear the fabric of felicity by the hands of reason and of law. Systems which attempt to question it deal in sounds instead of sense, in caprice instead of reason, in darkness instead of light.

But enough of metaphor and declamation: it is not by such means that moral science is to be improved.

The principle of utility is the foundation of the present work: it will be proper therefore at the outset to give an explicit and determinate account of what is meant by it. By the principle of utility is meant that principle which approves or disapproves of every action whatsoever, according to the tendency which it appears to have to augment or diminish the happiness of the party whose interest is in question or, what is the same thing in other words, to promote or to oppose that happiness. I say of every action whatsoever; and therefore not only of every action of a private individual, but of every measure of government.

By utility is meant that property in any object, whereby it tends to produce benefit, advantage, pleasure, good or happiness (all this in the present case comes to the same thing) or (what comes again to the same thing) to prevent the happening of mischief, pain, evil, or unhappiness to the party whose interest is considered: if that party be the community in general, then the happiness of the community: if a particular individual, then the happiness of that individual. . . .

⬛ Value of a Lot of Pleasure or Pain, How to Be Measured

Pleasures then, and the avoidance of pains, are the ends which the legislator has in view: it behooves him therefore to understand their value. Pleasures and pains are the instruments he has to work with: it behooves him therefore to understand their force, which is again, in other words, their value. To a person considered by himself, the value of a pleasure or pain considered by itself, will be greater or less, according to the four following circumstances.

1. Its intensity.

2. Its duration.

3. Its certainty or uncertainty.

4. Its propinquity or remoteness.

These are the circumstances which are to be considered in estimating a pleasure or a pain considered each of them by itself. But when the value of any pleasure or pain is considered for the purpose of estimating the tendency of any act by which it is produced, there are two other circumstances to be taken into the account; these are,

1. Its fecundity, or the chance it has of being followed by sensations of the same kind: that is, pleasures, if it be a pleasure: pains, if it be a pain.

2. Its *purity,* or the chance it has of not being followed by sensations of the *opposite* kind: that is, pains, if it be a pleasure: pleasures, if it be a pain.

These two last, however, are in strictness scarcely to be deemed properties of the pleasure or the pain itself; they are not, therefore, in strictness to be taken into the account of the value of that pleasure or that pain. They are in strictness to be deemed properties only of the act, or other event, by which such pleasure or pain has been produced; and accordingly are only to be taken into the account of the tendency of such act or such event.

To a number of persons, with reference to each of whom the value of a pleasure or a pain is considered, it will be greater or less, according to seven circumstances: to wit, the six preceding ones; viz.

1. Its intensity.

2. Its duration.

3. Its certainty or uncertainty.

4. Its propinquity or remoteness.

5. Its fecundity.

6. Its purity.

And one other; to wit:

7. Its extent; that is, the number of persons to whom it extends; or (in other words) who are affected by it.

To take an exact account then of the general tendency of any act, by which the interests of a community are affected, proceed as follows. Begin with any one person of those whose interests seem most immediately to be affected by it: and take an account,

1. Of the value of each distinguishable pleasure which appears to be produced by it in the first instance.

2. Of the value of each pain which appears to be produced by it in the *first* instance.

3. Of the value of each pleasure which appears to be produced by it *after* the first. This constitutes the *fecundity* of the first pleasure and the *impurity* of the first *pain.*

4. Of the value of each pain which appears to be produced by it after the first. This constitutes the *fecundity* of the first pain, and the *impurity* of the first pleasure.

5. Sum up all the values of all the pleasures on the one side, and those of all the pains on the other. The balance, if it be on the side of pleasure, will give the good tendency of the act upon the whole, with respect to the interests of that *individual* person; if on the side of pain, the *bad* tendency of it upon the whole.

6. Take an account of the *number* of persons whose interests appear to be concerned; and repeat the above process with respect to each. *Sum up* the numbers expressive of the degrees of *good* tendency, which the act has, with respect to each individual, in regard to whom the tendency of it is *good* upon the whole . . . do this again with respect to each individual, in regard to whom the tendency of it is *bad* upon the whole. Take the balance; which, if on the side of *pleasure*, will give the general *good tendency* of the act, with respect to the total number or community of individuals concerned; if on the side of pain, the general *evil tendency*, with respect to the same community.

It is not to be expected that this process should be strictly pursued previously to every moral judgment, or to every legislative or judicial operation. It may, however, be always kept in view: and as near as the process actually pursued on these occasions approaches to it, so near will such process approach to the character of an exact one.

The same process is alike applicable to pleasure and pain, in whatever shape they appear: and by whatever denomination they are distinguished: to pleasure, whether it be called *good* (which is properly the cause or instrument of pleasure) or *profit* (which is distant pleasure, or the cause or instrument of distant pleasure) or *convenience,* or *advantage, benefit, emolument, happiness,* and so forth: to pain, whether it be called *evil,* (which corresponds to *good*) or *mischief,* or *inconvenience,* or *disadvantage,* or *loss,* or *unhappiness,* and so forth.

Nor is this a novel and unwarranted, any more than it is a useless theory. In all this there is nothing but what the practice of mankind, wheresoever they have a clear view of their own interest, is perfectly conformable to. An article of property, an estate in land, for instance, is valuable, on what account? On account of the pleasures of all kinds which it enables a man to produce, and what comes to the same thing the pains of all kinds which it enables him to avert. But the value of such an article of property is universally understood to rise or fall according to the length or shortness of the time which a man has in it: the certainty or uncertainty of its coming into possession: and the nearness or remoteness of the time at which, if at all, it is to come into possession. As to the *intensity* of the pleasures which a man may derive from it, this is never thought of, because it depends upon the use which each particular person may come to make of it; which cannot be estimated till the particular pleasures he may come to derive from it, or the particular pains he may come to exclude by means of it, are brought to view. For the same reason, neither does he think of the *fecundity* or *purity* of those pleasures.

REVIEW QUESTIONS

1. Of the multiple pleasures on Bentham's list, which do you think are the most important when individuals choose their behavior? Why?

2. Of the multiple pains on Bentham's list, which do you think are the most important when individuals choose their behavior? Why?

3. Provide an example of utilitarianism as being a very good policy for government. Also provide an example of utilitarianism not being a good policy for government, other than those already provided in this section.

❖

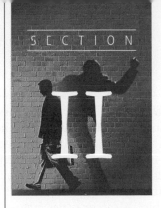

II

MODERN APPLICATIONS OF THE CLASSICAL PERSPECTIVE

Deterrence, Rational Choice, and Routine Activities/Lifestyle Theories of Crime

The introduction to this section will discuss the early aggregate studies of deterrence in the late 1960s, then the perceptual studies of the 1970s, and finally the longitudinal and scenario studies of the 1980s and 1990s to present. Other policy applications, such as increased penalties toward drunk driving, white-collar crime, and so on will also be examined. The introduction will also discuss the development of rational choice theory in economics and its later application to crime. Finally, this introduction will examine the use of routine activities/lifestyle theory as a framework for modern research and applications for reducing criminal activity.

⚓ Introduction

In the last section, we discussed the early development of the Classical/Neoclassical School of criminological thought. This theoretical perspective remained the dominant framework used by judges and practitioners in the practice of administering justice and punishment even in current times, but beginning in the late 19th century, criminological researchers dismissed the Classical/Neoclassical framework. Rather, criminological research and theorizing began emphasizing factors other than free will and deterrence. Instead, an emphasis was placed on social, biological, or other factors that go beyond free will and deterrence theory. These theories will be discussed in later sections, but first, we will examine the recent rebirth of Classical/Neoclassical theory and deterrence.

⚓ Rebirth of Deterrence Theory and Contemporary Research

As discussed above, the Classical/Neoclassical School framework fell out of favor among the scientists and philosophers in the late 19th century, largely due to the introduction of Darwin's ideas about evolution and natural selection. However, virtually all Western criminal systems retained the Classical/Neoclassical framework for their model of justice, particularly the United States. Nevertheless, the ideology of Beccaria's work was largely dismissed by academics and theorists after the presentation of Darwin's theory of evolution in the 1860s. Therefore, the Classical/Neoclassical School fell out of favor in terms of criminological theorizing for about 100 years. However, in the 1960s, the Beccarian model of offending experienced a rebirth.

In the late 1960s, several studies using aggregate measures of crime and punishment were published that used a deterrence model for explaining why individuals engage in criminal behavior. These studies revealed a new interest in the deterrent aspects of criminal behavior and further supported the importance of certainty and severity of punishment in deterring individuals from committing crime, particularly homicide. In particular, evidence was presented that showed that increased risk or certainty of punishment was associated with less crime for most serious offenses. Plus, it is a fact that most offenders who are arrested once never get arrested again, which lends some basic support for deterrence.

Many of these studies used statistical formulas to measure the degree of certainty and severity of punishment in given jurisdictions. One measure used the ratio of the crimes reported to police as compared to the number of arrests in a given jurisdiction. Another measure of certainty of punishment was the ratio of arrests to convictions, or findings of guilt, in criminal cases. Other measures were also employed. Most of the studies showed the same result: that the higher the likelihood of arrest compared with reports of crime, or the higher the conviction rate compared to the arrest rate, the lower the crime rate was in a jurisdiction. On the other hand, the scientific evidence regarding measures of severity, which such studies generally indicated by the length of sentence for comparable crimes or a similar type of measure, did not show much impact on crime.

Additional aggregate studies examined the prevalence and influence of capital punishment on the crime rate in given states.[1] The evidence showed that the states with death penalty statutes also had higher murder rates than non-death penalty states. Furthermore, the studies showed that murderers in death penalty states who were not executed actually served less time than murderers in non-death penalty states. Thus, the evidence regarding increased sanctions, including capital punishment, was mixed. Still, a review of the early deterrence studies by the National Academy of Sciences concluded that, overall, there was more evidence for a deterrent effect than against it, although the finding was reported in a tone that lacked confidence, perhaps cautious of what future studies would show.[2]

It was not long before critics noted that studies incorporating aggregate (i.e., macro-level) statistics are not adequate indicators or valid measures of the deterrence theoretical framework, largely because the model emphasizes the perceptions of individuals. Using aggregate or group statistics is flawed because different regions may have higher or lower crime rates than others, thereby creating bias in the level of ratios for certainty or severity of punishment. Furthermore, the group measures produced by these studies provide virtually no information on the degree to which individuals in those regions perceive sanctions as being certain, severe, or swift. Therefore, the emphasis on the unit of analysis in deterrence research shifted from the aggregate level to that of a more micro, individual level.

The following phase of deterrence research focused on individual perceptions of certainty and severity of sanctions, primarily drawn at one point in time, known as **cross-sectional studies**. A number of cross-sectional studies of individual perceptions of deterrence showed that perceptions of the risk or certainty of punishment were strongly associated with intentions to commit future crimes, but individual perceptions of severity of crimes were mixed. Furthermore, it readily became evident that it was not clear whether perceptions were causing changes in behavior or whether behavior was causing changes in perception. This led to the next wave of research, longitudinal studies of individual perceptions and deterrence, which measured perceptions of risk and severity, as well as behavior, over time.[3]

One of the primary concepts revealed by longitudinal research was that behavior was influencing perceptions of the risk and severity of punishment more than perceptions were influencing behavior. This was referred to as the **experiential effect**, which is appropriately named because people's previous experience highly influences their expectations regarding their chances of being caught and the resulting penalties. A common example is that of people who drive under the influence of alcohol (or other substances).

Studies show that if you ask people who have never driven drunk how likely they would be caught if they drove home drunk, most predict an unrealistically high chance of

[1] Daniel Glaser and Max S. Zeigler, "Use of the Death Penalty v. Outrage at Murder," *Crime and Delinquency* 20 (1974): 333–38); Tittle, Franklin E. Zimring and Gordon J. Hawkins, *Deterrence—The Legal Threat in Crime Control*, (Chicago: University of Chicago Press, 1973); Johannes Andenaes, *Punishment and Deterrence* (Ann Arbor, MI: University of Michigan Press, 1974); Jack P. Gibbs, *Crime, Punishment and Deterrence* (New York: Elsevier, 1975).

[2] Alfred Blumstein, Jacqueline Cohen, and Daniel Nagin, eds., *Deterrence and Incapacitation: Estimating the Effects of Criminal Sanctions on Crime Rates* (Washington, DC: National Academy of Sciences, 1978).

[3] Raymond Paternoster, Linda E. Saltzman, Gordon P. Waldo, and Theodore G. Chiricos, "Perceived Risk and Social Control: Do Sanctions Really Deter?" *Law and Society Review* 17 (1983): 457–80; Raymond Paternoster, "The Deterrent Effect of the Perceived Certainty and Severity of Punishment: A Review of the Evidence and Issues," *Justice Quarterly* 4 (1987):173–217.

getting caught. However, if you ask people who have been arrested for driving drunk, even those who have been arrested several times for this offense, they typically predict that the chance is very low. The reason for this is that these chronic drunk drivers have typically been driving under the influence for many years, mostly without being caught. It is estimated that more than 1 million miles are driven collectively by drunk drivers before one person is arrested.[4] If anything, this is likely a conservative estimate. Thus, people who drive drunk, with some doing so every day, are not likely to be deterred even when they are arrested more than once because they have done so for years. In fact, perhaps the most notable expert on the deterrence of drunk drivers, H. L. Ross and his colleagues, concluded that drunk drivers who "perceive a severe punishment if caught, but a near-zero chance of being caught, are being rational in ignoring the threat."[5] It is obvious that even the most respected scholars in the area admit that sanctions against drunk driving are nowhere near certain enough, even if they are growing in severity.

Another common example is seen with white-collar criminals. Some researchers have theorized that being caught by authorities for violating government rules, enforced by the Securities and Exchange Commission (SEC), will make these organizations less likely to commit future offenses.[6] However, business organizations have been in violation of established practices for years before getting caught, so it is likely that they will continue to ignore the rules in the future, more than organizations that have never violated the rules. As it was with drunk drivers, the certainty of punishment for white-collar violations is so low—and many would argue the severity is also quite low—that it is quite rational for businesses and business professionals to take the risk of engaging in white-collar crime.

It is interesting to note that white-collar criminals and drunk drivers are two types of offenders who are most likely to be deterred because they are mostly of the middle- to upper-level socioeconomic class. The extant research on deterrence has shown that individuals who have something to lose are the most likely to be deterred by sanctions. This makes sense: Those who are unemployed or poor or both do not have much to lose, and for them or for some minorities, incarceration may not present a significant departure from the deprived lives that they lead.

The fact that official sanctions have limitations in deterring individuals from drunk driving and white-collar crime is not a good indication of the effectiveness of deterrence-based policies. Their usefulness becomes even more questionable when other populations are considered, particularly the offenders in most predatory street crimes (e.g., robbery, burglary, etc.), in which offenders typically have nothing to lose because they come from poverty-stricken areas and are often unemployed. One recent study showed that being arrested had little effect on perceptions of the certainty of punishment; offending actually corresponded with decreases in such perceptions.[7]

---

[4]H. L. Ross, *Deterring the Drunk Driver: Legal Policy and Social Control* (Lexington, MA: Lexington Books, 1982); II. L. Ross, *Confronting Drunk Driving: Social Policy for Saving Lives* (New Haven, CT: Yale University Press, 1992); H. L. Ross, "Sobriety Checkpoints, American Style," *Journal of Criminal Justice* 22 (1994): 437–44; H. L. Ross, R. McCleary, and Gary LaFree, "Can Mandatory Jail Laws Deter Drunk Driving? The Arizona Case," *Journal of Criminal Law and Criminology* 81 (1990): 156–70.

[5]H. L. Ross et al., "Sobriety Checkpoints," 164.

[6]Sally Simpson and Christopher S. Koper, "Deterring Corporate Crime," *Criminology* 30 (1992): 347–76.

[7]Greg Pogarsky, KiDeuk Kim, and Raymond Paternoster, "Perceptual Change in the National Youth Survey: Lessons for Deterrence Theory and Offender Decision-Making," *Justice Quarterly* 22 (2005): 1–29.

Some people don't see incarceration as that much of a step down in life, given the three meals a day, shelter, and relative stability provided by such punishment. This fact epitomizes one of the most notable paradoxes we have in criminology: The individuals we most want to deter are the least likely to be deterred, primarily because they have nothing to fear. In early Enlightenment thought, Hobbes asserted that although fear was the tool used to enforce the social contract, people who weren't afraid of punishment could not be effectively deterred. That remains true in modern days.

Along these same lines, studies have consistently shown that for young male offenders—at higher risk, with low emotional/moral inhibitions, low self-control, and high impulsivity—official deterrence is highly ineffective in preventing crimes with an immediate payoff.[8] Thus, many factors go into the extent to which official sanctions can deter. As we have seen, even among those offenders who are in theory the most deterrable, official sanctions have little impact because their experience of not being caught weakens the value of deterrence.

The identification and understanding of the experiential effect had a profound effect on the evidence regarding the impact of deterrence. Researchers saw that to account for such an experiential effect, any estimation of the influence of perceived certainty or severity of punishment must control for previous behaviors and experiences engaging in such behavior. The identification of the experiential effect was the primary contribution of the longitudinal studies of deterrence, but such studies faced even further criticism.

Longitudinal studies of deterrence provided a significant improvement over the cross-sectional studies that preceded this advanced methodology. However, such longitudinal studies typically involved designs in which measures of perceptions of certainty and severity of punishment were collected at points in time that were separated by up to a year apart, including long stretches between when the crime was committed and when the offenders were asked about their perceptions of punishment. Psychological studies have clearly established that perceptions of the likelihood and severity of sanctions vary

---

[8]For a review, see Stephen Brown, Finn Esbensen, and Gilbert Geis, *Criminology,* 6th ed. (Cincinnati, OH: LexisNexis, 2007), 201–04; N. Finley and Harold Grasmick, "Gender Roles and Social Control," *Sociological Spectrum* 5 (1985): 317–30; Harold Grasmick, Robert Bursik, and Karla Kinsey, "Shame and Embarrassment as Deterrents to Noncompliance with the Law: The Case of an Antilittering Campaign," *Environment and Behavior* 23 (1991): 233–51; Harold Grasmick, Brenda Sims Blackwell, and Robert Bursik, "Changes in the Sex Patterning of Perceived Threats of Sanctions," *Law and Society Review* 27 (1993):. 679–705; P. Richards and Charles Tittle, "Gender and Perceived Chances of Arrest," *Social Forces* 59 (1981): 1182–99; G. Loewenstein, Daniel Nagin, and Raymond Paternoster, "The Effect of Sexual Arousal on Expectations of Sexual Forcefulness," *Journal of Research in Crime and Delinquency* 34 (1997): 209–28; T. Makkai and John Braithwaite, "The Dialects of Corporate Deterrence," *Journal of Research in Crime and Delinquency* 31 (1994): 347–73; Daniel Nagin and Raymond Paternoster, "Enduring Individual Differences and Rational Choice Theories of Crime," *Law and Society Review* 27 (1993): 467–96; Alex Piquero and Stephen Tibbetts, "Specifying the Direct and Indirect Effects of Low Self-Control and Situational Factors in Offenders' Decision Making: Toward a More Complete Model of Rational Offending," *Justice Quarterly* 13 (1996): 481–510; Raymond Paternoster and Sally Simpson, "Sanction Threats and Appeals to Morality: Testing a Rational Choice Model of Corporate Crime," *Law and Society Review* 30 (1996): 549–83; Daniel Nagin and Greg Pogarsky, "Integrating Celerity, Impulsivity, and Extralegal Sanction Threats into a Model of General Deterrence: Theory and Evidence," *Criminology* 39 (2001): 404–30; Alex Piquero and Greg Pogarsky, "Beyond Stafford and Warr's Reconceptualization of Deterrence: Personal and Vicarious Experiences, Impulsivity, and Offending Behavior, *Journal of Research in Crime and Delinquency* 39 (2002): 153–86; For a recent review and altered explanation of these conclusions, see Greg Pogarsky, "Identifying 'Deterrable' Offenders: Implications for Research on Deterrence," *Justice Quarterly* 19 (2002):.431–52.

significantly from day-to-day, let alone month-to-month or year-to-year.[9] Therefore, in the late 1980s and early 1990s, a new wave of deterrence research evolved, which asked study participants to estimate their immediate intent to commit a criminal act in a given situation, as well as their immediate perceptions of certainty and severity of punishment in this same situation. This wave of research was known as **scenario (vignette) research**.[10]

Scenario research (i.e., vignette design) was created to deal with the limitations of the previous methodological strategies for studying the effects of deterrence on criminal offending, specifically, the criticism that individuals' perceptions of the certainty and severity of punishment changed drastically from time to time and across different situations. The scenario method dealt with this criticism directly by providing a specific, realistic (albeit hypothetical) situation, in which a person engages in a criminal act. The participant in the study is then asked to estimate the chance that he/she would engage in such activity in the given circumstances and to respond to questions regarding perceptions of risk of getting caught (i.e., certainty of punishment) and the degree of severity of punishment they expected.

Another important and valuable aspect of scenario research was that it promoted a contemporaneous (i.e., instantaneous) response about perceptions of the risk and severity of perceived sanctions. In comparison, previous studies (e.g., aggregate, cross-sectional, longitudinal) had always relied on either group rates or individual measures of perceptions across long periods of time. However, some argue that intentions to commit crimes given a hypothetical situation are not accurate measures of what one would do in actual reality. Studies have shown an extremely high correlation between what people report doing in a given scenario and what they would do in real life.[11] A recent review of criticisms of this method of research showed that one weakness of this research was that it did not allow respondents to develop their own perceptions and costs associated with

---

[9]I. Ajzen and M. Fishbein, *Understanding Attitudes and Predicting Social Behavior* (Englewood Cliffs, NJ: Prentice Hall, 1980); M. Fishbein and I. Ajzen, *Belief, Attitude, Intention, and Behavior* (Reading, MA: Addison-Wesley, 1975); I. Ajzen and M. Fishbein, "Attitude-Behavior Relations: A Theoretical Analysis and Review of Empirical Research," *Psychological Bulletin* 84 (1977): 888–918; for a recent review, see Pogarsky et al., "Perceptual Change."

[10]Loewenstein et al., "Effect of Sexual Arousal"; Nagin and Paternoster, "Enduring Individual Differences"; Piquero and Tibbets, "Specifying the Direct"; Paternoster and Simpson, "Sanction Threats"; Ronet Bachman, Raymond Paternoster, and S. Ward, "The Rationality of Sexual Offending: Testing a Deterrence/Rational Choice Conception of Sexual Assault," *Law and Society Review* 26 (1992): 343–72; Harold Grasmick and Robert Bursik, "Conscience, Significant Others, and Rational Choice: Extending the Deterrence Model," *Law and Society Review* 24 (1990): 837–61; Harold Grasmick and D. E. Green, "Legal Punishment, Social Disapproval, and Internalization as Inhibitors of Illegal Behavior," *Journal of Criminal Law and Criminology* 71 (1980): 325–35; S. Klepper and Daniel Nagin, "The Deterrent Effects of Perceived Certainty and Severity of Punishment Revisited, *Criminology* 27 (1989):. 721–46; Stephen Tibbetts and Denise Herz, "Gender Differences in Students' Rational Decisions to Cheat," *Deviant Behavior* 18 (1996): 393–414; Stephen Tibbetts and David Myers, "Low Self-Control, Rational Choice, and Student Test Cheating," *American Journal of Criminal Justice* 23 (1999): 179–200; Stephen Tibbetts, "Shame and Rational Choice in Offending Decisions," *Criminal Justice and Behavior* 24 (1997): 234–55.

[11]D. Green, "Measures of Illegal Behavior in Individual Behavior in Individual-Level Deterrence Research," *Journal of Research in Crime and Delinquency* 26 (1989): 253–75; I. Ajzen and M. Fishbein, *Understanding Attitudes*; I. Ajzen, "From Intentions to Actions: A Theory of Planned Behavior." In *Action-Control: From Cognition to Behavior* ed. J. Kuhl and J. Beckmann (place of publication: publisher, year), 11–39; I. Ajzen and M. Fishbein, "The Prediction of Behavioral Intentions in a Choice Situation," *Journal of Experimental Psychology* 5 (1969): 400–16.

each offense.[12] Despite such criticisms, the scenario method appears to be the most accurate method that we have to date to accurately estimate the effects of individual perceptions with the likelihood of such individuals engaging in given criminal activity at a given point in time. This is something that the previous waves of deterrence research—aggregate, cross-sectional, and longitudinal studies—could not estimate.

Ultimately, the studies using the scenario method showed that participants were more affected by perceptions of certainty and less so, albeit sometimes significant, perceptions of severity. This finding supported previous methods of estimating the effects of *formal/official deterrence*, meaning the deterrent effects of three general groups: law enforcement, courts, and corrections (i.e., prisons and probation/parole). So the overall conclusion regarding the effects of official sanctions on individual decision making remained unaltered. However, one of the more interesting aspects of the scenario method research is that it helped solidify the importance of extralegal variables in deterring criminal behavior, variables that had been neglected by the previous methods.

These extralegal or informal deterrence variables, which include any factors beyond formal sanctions of police, courts, and corrections—such as employment, family, friends, or community, are typically known as informal or unofficial sanctions. These studies helped show that the deterrent effect of such informal sanctions provided most of the deterrent effect—if there was any. These findings coincided with the advent of a new model of deterrence, which became commonly known as rational choice theory.

▧ Rational Choice Theory

Rational choice theory is a perspective that criminologists adapted from economists, who used it to explain a variety of individual decisions regarding a variety of behaviors. This framework emphasizes all important factors that go into a person's decision to engage, or not engage, in a particular act. In terms of criminological research, the rational choice model emphasized both official or formal forms of deterrence, as well as the informal factors that influence individual decisions for criminal behavior. This represented a profound advance in the understanding of human behavior. After all, as studies showed, most individuals are more affected by informal factors than they are by official or formal factors.

Although there were several previous attempts to apply the rational choice model to the understanding of criminal activity, the most significant work, which brought rational choice theory into the mainstream of criminological research, was Cornish and Clarke's *The Reasoning Criminal: Rational Choice Perspectives on Offending* in 1986.[13] Furthermore, around the same time, Katz (1988) published his work, *Seductions in Crime*, which for the first time placed an emphasis on the benefits (mostly the inherent physiological pleasure) in committing crime;[14] before Katz's publication, virtually no attention had been paid to

[12]Jeffrey A. Bouffard, "Methodological and Theoretical Implications of Using Subject-Generated Consequences in Tests of Rational Choice Theory," *Justice Quarterly* 19 (2002): 747–71.

[13]D. Cornish and Ron Clarke, *The Reasoning Criminal: Rational Choice Perspectives on Offending* (New York: Springer-Verlag, 1986).

[14]Jack Katz, *Seductions of Crime* (New York: Basic Books, 1988).

the benefits of offending, let alone the "fun" that people feel when they engage in criminal behavior. A recent study showed that the publication of Cornish and Clarke's book, as well as the timing of other publications such as Katz's, led to an influx of criminological studies in the late 1980s to mid-1990s based on the rational choice model.[15]

These studies on rational choice showed that while official or formal sanctions tend to have some effect on individuals' decisions to commit crime, they almost always are relatively unimportant compared to extralegal or informal factors. The effects of people's perceptions of how much shame or loss of self-esteem they would experience, even if no one else found out that they committed the crime, was one of the most important variables in determining whether or not they would do so.[16] Additional evidence indicated that females were more influenced by such effects of shame and moral beliefs in this regard than were males.[17] Recent studies have shown that levels of personality traits, especially low self-control and empathy, are likely the reason why males and females differ so much in engaging in criminal activity.[18] Finally, the influence of peers has a profound impact on individual perceptions of the pros and cons of offending by significantly decreasing the perceived risk of punishment if people see their friends get away with crimes.[19]

Another area of rational choice research dealt with the influence that an individual's behavior would have on those around them. A recent review and test of perceived social disapproval showed that this was one of the most important variables in decisions to commit crime.[20] In addition to self-sanctions, such as feelings of shame and embarrassment, the perceived likelihood of how loved ones and friends, as well as employers, would respond is perhaps the most important factor that goes into a person's decision to engage in criminal activity. These are the people we deal with every day and the source of our livelihood, so it should not be too surprising that our perceptions of how they will react strongly affect how we behave.

Perhaps the most important finding of rational choice research was that the expected benefits, particularly the pleasure they would get from such offending, had one of the most significant effects on their decisions to offend. Many other conclusions have been

---

[15] Stephen Tibbetts and Chris Gibson, "Individual Propensities and Rational Decision-Making: Recent Findings and Promising Approaches," in *Rational Choice and Criminal Behavior*, ed. Alex Piquero and Stephen Tibbetts (New York: Routledge, 2002), 3–24.

[16] Grasmick and Bursik, "Conscience"; Pogarsky, "Identifying `Deterrable' Offenders"; Tibbetts, "Shame and Rational Choice"; Nagin and Paternoster, "Enduring Individual Differences"; Tibbetts and Herz, "Gender Differences"; Tibbetts and Myers, "Low Self-control"; Harold Grasmick, Brenda Sims Blackwell, and Robert Bursik, "Changes over Time in Gender Differences in Perceived Risk of Sanctions," *Law and Society Review* 27 (1993): 679–705; Harold Grasmick, Robert Bursik, and Bruce Arneklev, "Reduction in Drunk Driving as a Response to Increased Threats of Shame, Embarrassment, and Legal Sanctions," *Criminology* 31 (1993): 41–67; Stephen Tibbetts, "Self-Conscious Emotions and Criminal Offending," *Psychological Reports* 93 (2004): 101–31.

[17] Tibbetts and Herz, "Gender Differences"; Grasmick, Blackwell, and Bursik, "Changes in the Sex Patterning"; Finley and Grasmick, "Gender Roles"; Pogarsky et al., "Perceptual Change"; Stephen Tibbetts, "Gender Differences in Students' Rational Decisions to Cheat," *Deviant Behavior* 18 (1997): 393–414.

[18] Nagin and Paternoster, "Enduring Individual Differences"; Grasmick, Blackwell, and Bursik, "Changes Over Time"; Tibbetts, "Self-Conscious Emotions."

[19] Pogarsky et al., "Perceptual Change."

[20] Pogarsky, "Identifying 'Deterrable' Offenders."

made regarding the influence of extralegal or informal factors on criminal offending, but the ultimate conclusion that can be made is that these informal deterrent variables typically hold more influence on individual decision making regarding deviant activity than the official/formal deterrent factors that were emphasized by traditional Classical School models of behavior.

The rational choice model of criminal offending became the modern framework of deterrence. Official authorities acknowledged the influence of extralegal or informal factors, which is seen in modern efforts to incorporate the family, employment, and community in rehabilitation efforts. Such efforts are highly consistent with the current state of understanding regarding the Classical School/rational choice framework, namely that individuals are more deterred by the impact of their actions on the informal aspects of their lives, as opposed to the formal punishments they face by doing illegal acts.

▨ Routine Activities Theory

Routine activities theory is another contemporary form of the Classical School framework in the sense that it assumes a rational decision-making offender. The general model of routine activities theory was originally presented by Lawrence Cohen and Marcus Felson in 1979.[21] This theoretical framework emphasized the presence of three factors that come together in time and place to create a high likelihood for crime and victimization. These three factors are: motivated offender(s), suitable target, and lack of guardianship. Overall, the theory is appropriately named, in the sense that it assumes that most crime occurs in the daily routine of people, who happen to see—and then seize—tempting opportunities to commit crime. Studies tend to support this idea, as opposed to the idea that most offenders leave their home knowing they are going to commit a crime; the latter offenders are called "hydraulic" and are relatively rare compared to the opportunistic type.

Regarding the first factor noted as being important for increasing the likelihood of criminal activity—motivated offender—the theory of routine activities does not provide

▲ **Photo 2.1**   Marcus Felson, 1947– , Rutgers University, author of routine activity theory.

[21] Lawrence Cohen and Marcus Felson, "Social Change and Crime Rates: A Routine Activities Approach," *American Sociological Review* 44 (1979): 214–41.

much insight. Rather, the model simply assumes that some individuals tend to be motivated and leaves it at that. Fortunately, we have many other theories that can fill this notable absence. Instead, the strength of routine activities theory is in the elaboration of the other two aspects of a crime-prone environment: suitable targets and lack of guardianship.

Suitable targets can include a variety of situations. For example, a very suitable target can be a vacant house in the suburbs that the family has left for summer vacation. Data clearly show that burglaries more than double in the summer when many families are on vacation. Other forms of suitable targets range from an unlocked car to a female alone, carrying a lot of cash and credit cards and often purchased goods, at a shopping mall, which is a common place for being victimized. Other likely targets are bars or other places that serve alcohol. Offenders have traditionally targeted drunk persons because they are less likely to be able to defend themselves, in a history that extends to rolling drunks for their wallets back in the early part of the 20th century. This is only a short list of the many types of suitable targets that are available to motivated offenders in everyday life.

The third and final aspect of the routine activities model for increased likelihood of criminal activity is the lack of guardianship. Guardianship is often thought of as a police officer or security guard, which often is the case. There are many other forms of guardianship, however, such as owning a dog to protect a house, which studies demonstrate can be quite effective. Just having a car or house alarm constitutes a form of guardianship. Furthermore, the presence of an adult, neighbor, or teacher can be an effective type of guarding the area against crime. In fact, recent studies show that the presence of increased lighting in the area can prevent a significant amount of crime, with one study showing a 20% reduction in overall crime in areas randomly chosen to receive improved lighting as compared to control areas that did not.[22] Regardless of the type of guardianship, it is the absence of adequate guardianship that sets the stage for crime; on the other hand, each step taken toward protecting the place or person is likely to deter offenders from choosing the target in relation to others. Locations that have a high convergence of motivated offenders, suitable targets, and lack of guardianship are typically referred to as "hot spots."

Perhaps the most supportive evidence for routine activities theory and "hot spots" was the study of 911 calls for service during one year in Minneapolis, Minnesota.[23] This study examined all serious calls (as well as total calls) to police for a one-year period. Half of the top 10 places from which police were called were bars or locations where alcohol was served. As mentioned above, establishments that serve alcohol are often targeted by motivated offenders for their high proportion of suitable targets. Furthermore, a number of bars tend to have a low level of guardianship in relation to the number of people they serve. Readers of this book may well relate to this situation. Most college towns and cities have certain drinking establishments that are known as being a "hot spot" for crime.

Still, the Minneapolis "hot spot" study showed other types of establishments that made the top 10 rankings. These included bus depots, convenience stores, rundown motels and hotels, downtown malls and strip malls, fast-food restaurants, towing companies, and so on. The common theme linking these locations and the bars was the convergence of

---

[22]David P. Farrington and Brandon C. Welsh, "Improved Street Lighting and Crime Prevention," *Justice Quarterly* 19 (2002): 313–43.

[23]Lawrence Sherman, Patrick R. Gartin, and Michael Buerger, "Hot Spots of Predatory Crime: Routine Activities and the Criminology of Place," *Criminology* 27 (1989): 27–56.

| Figure 2.1 | Routine Activities Theory |

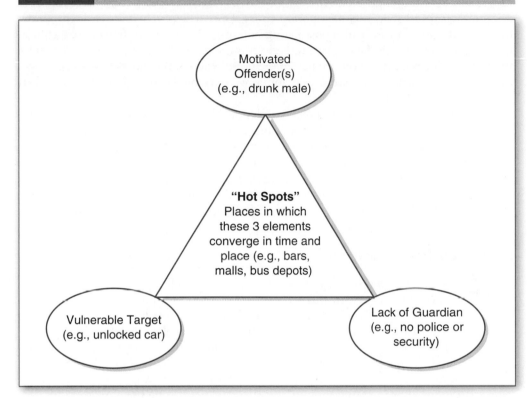

the three aspects described by routine activities theory as being predictive of criminal activity. Specifically, these are places that attract motivated offenders, largely because they have a lot of vulnerable targets and/or lack sufficient levels of security or guardianship. The routine activities framework has been applied in many contexts and places, many of them international.[24]

---

[24] Jon Gunnar Bernburg and Thorolfur Thorlindsson, "Routine Activities in Social Context: A Closer Look at the Role of Opportunity in Deviant Behavior," *Justice Quarterly* 18 (2001): 543–67; see also R. Bennett, "Routine Activity: A Cross-National Assessment of a Criminological Perspective," *Social Forces* 70 (1991): 147–63; J. Hawdon, "Deviant Lifestyles: The Social Control of Routine Activities," *Youth and Society* 28 (1996): 162–188; J. L. Massey, M. Krohn, and L. Bonati, "Property Crime and the Routine Activities of Individuals," *Journal of Research in Crime and Delinquency* 26 (1989): 378–400; Terrance Miethe, Mark Stafford, and J. Long, "Social Differences in Criminological Victimization: A Test of Routine Activities/Lifestyles Theories," *American Sociological Review* 52 (1987): 184–94; E. Mustaine and Richard Tewksbury, "Predicting Risks of Largency Theft Victimization: A Routine Activity Analysis Using Refined Lifestyle Measures," *Criminology* 36 (1998): 829–57; D. Osgood, J. Wilson, P. M. O'Malley, J. Bachman, and J. L. Johnston, "Routine Activities and Individual Deviant Behavior," *American Sociological Review* 61 (1996): 635–55; D. Roncek and P. Maier, "Bars, Blocks, and Crimes Revisited: Linking the Theory of Routine Activities to the Empiricism of Hot Spots," *Criminology* 29 (1991): 725–53; Robert Sampson and John Wooldredge, "Linking the Micro- and Macro-Level Dimensions of Lifestyle-Routine Activity and Opportunity Models of Predatory Victimization," *Journal of Quantitative Criminology* 3 (1987): 371–93.

Modern applications of routine activities theory include geographic profiling, which uses satellite positioning systems in perhaps the most attractive and marketable aspect of criminological research in contemporary times. Essentially, such research incorporates the computer software for global positioning systems (GPS) for identifying the exact location of every crime that takes place in a given jurisdiction. Such information has been used to solve or predict various crimes, to the point where serial killers have been caught because the sites where the victims were found were triangulated to show the most likely place where the killer lived.

Some theorists have proposed a theoretical model based on individuals' lifestyle, which has a large overlap with routine activities theory, as shown in previous studies reviewed.[25] It only makes sense that a person who lives a more risky lifestyle, for example, by frequenting bars or living in a high-crime area, will be at more risk by being close to various hot spots identified by routine activities theory. Although some criminologists label this phenomenon a lifestyle perspective, it is virtually synonymous with the routine activities model, because such lifestyles incorporate the same conceptual and causal factors in routine activities.

◪ Policy Implications

There are numerous policy implications that can be derived from the theories and scientific findings of this chapter. In this section, we will concentrate on some of the most important policies. First, we look at the policy of broken windows, which has many assumptions similar to those of routine activities and rational choice theories. The broken windows perspective emphasizes the need for police to "crack down" on more minor offenses to reduce more major crimes.[26] Although many cities have claimed reductions in serious crimes by using this theory (such as New York and Los Angeles), the fact is that crime was reduced by the same amount across most cities during the same time (the late 1990s to mid 2000s).

Still, other policies that can be derived from theories in this section include the "three-strikes-you're-out" policy, which assumes that offenders will make a rational choice not to commit future offenses because they could go to prison for life if they commit three; the negatives certainly outweigh the expected benefits for the third crime. For deterrence to be extremely effective, punishment must be swift, certain, and severe. Where does the policy of three-strikes fall into this equation? The bottom line is that it is much more severe than it is swift or certain. Given the theory/philosophy of Beccaria (see Section I), this policy will probably not work because it is not certain or swift. However, it is severe, in the sense that a person can be sentenced to life if they commit three felony offenses over time.

A controversial three-strikes law was passed by voter initiative in California, and other states have adopted similar types of laws.[27] It sends third-time felons to prison for the rest of their lives regardless of the nature of that third felony. California first requires

---

[25]Hawdon, "Deviant Lifestyles"; Sampson and Wooldredge, "Linking the Micro"; Brown et al., *Criminology*.

[26]James Q. Wilson and G. Kelling, "Broken Windows: The Police and Neighborhood Safety." *Atlantic Monthly*, March (1982): 29–38.

[27]D. Schichor and D. K. Sechrest, eds., *Three Strikes and You're Out: Vengeance as Social Policy* (Thousand Oaks, CA: Sage, 1996).

convictions for two "strikeable" felonies, crimes like murder, rape, aggravated assault, burglary, drug offenses, and so on. Then any third felony can trigger the life sentence. The stories about some nonviolent offenders going to prison for the rest of their lives for stealing a piece of pizza or shoplifting DVDs, while rare, are quite true.

The question we are concerned with here is, does the "three-strikes" policy work? As a specific deterrent, the answer is clearly yes; offenders who are in prison for the rest of their lives cannot commit more crimes on the streets. In that regard, three-strikes works very well. Some people feel, however, that laws like three-strikes need to have a *general* deterrent effect to be considered successful, meaning that this law should deter everyone from engaging in multiple crimes. So, is three-strikes a general deterrent? Unfortunately, there are no easy answers to this question because laws vary from state to state, the laws are used at different rates across the counties in a given state, and so forth. There is at least some consensus in the literature, however.

One study from California suggests that three-strikes reduced crime,[28] but the remaining studies show either that three-strikes has no effect on crime or that it actually *increases* crime.[29] How could three-strikes increase crime? The authors attributed an increase in homicide, following three-strikes, to the possibility that third strikers have an incentive to kill victims and any witnesses in an effort to avoid apprehension. Although this argument is tentative, it may be true.[30] This is just one of the many policy implications that can be derived from this section introduction. We expect that readers of this book will come up with many more of such policy implications, but it is vital that they examine the empirical literature in determining their usefulness in reducing criminal activity. Other policy implications regarding the theories and findings discussed in this chapter will be discussed in the final section of this book.

In a strategy that is also strongly based in the rational choice model, a number of judges have started using shaming strategies to deter offenders from recidivating.[31] They have ordered everything from publicly posting pictures of men arrested for soliciting prostitutes to forcing offenders to walk down Main Streets of towns wearing signs that announce they committed crimes. These are just two examples of an increasing trend that emphasizes the informal or community factors required to deter crime. Unfortunately, to date there have been virtually no empirical evaluations of the effectiveness of such shaming penalties, although studies of expected shame for doing an act consistently show a deterrent effect.[32]

---

[28]J. M. Shepherd, "Fear of the First Strike: The Full Deterrent Effect of California's Two- and Three-Strikes Legislation," *Journal of Legal Studies* 31 (2002): 159–201.

[29]See L. Stolzenberg and S.J. D'Alessio, "Three Strikes and You're Out: The Impact of California's New Mandatory Sentencing Law on Serious Crime Rates," *Crime and Delinquency* 43 (1997): 457–69; M. Males and D. Macallair, "Striking Out: The Failure of California's 'Three-Strikes and You're Out Law,'" *Stanford Law and Policy Review* 11 (1999): 65–72.

[30]T. B. Marvell and C. E. Moody, "The Lethal Effects of Three-Strikes Laws," *Journal of Legal Studies* 30 (2001): 89–106; see also T. Kovandzic, J. J. Sloan III, and L. M. Vieraitis, "Unintended Consequences of Politically Popular Sentencing Policy: The Homicide-Promoting Effects of 'Three Strikes' in U.S. Cities (1980-1999)," *Criminology and Public Policy* 1 (2002): 399–424. It should be noted that some of the information discussed here was taken from John Worrall's review of this section.

[31]Piquero and Tibbetts, *Rational Choice.*

[32]Tibbetts, "Gender Differences."

⊠ Conclusion

This section reviewed the more recent forms of classical/deterrence theory, such as rational choice theory, which emphasizes the effects of informal sanctions (e.g., family, friends, employment) and benefits and costs of offending, and a framework called routine activities theory, which explains why victimization tends to occur far more often in certain locations (i.e., "hot spots") due to the convergence of three key elements in time and place—motivated offender(s), vulnerable target(s), and lack of guardianship—which create an attractive opportunity for crime as individuals go about their typical, everyday activities. The common element across all of these perspectives is the underlying assumption that individuals are rational beings who have free will and thus choose their behavior based on assessments of a given situation, such as what are the possible risks versus the potential payoff. Although many studies examined in this section lend support for many of the assumptions and propositions of the classical framework, it is also clear that there is a lot more involved in explaining criminal human behavior than the individual decision making that goes on before a person engages in rule violation. After all, human beings are often not rational and often do things spontaneously without considering the potential risks beforehand, especially chronic offenders. So despite the use of the classical/neoclassical model in most systems of justice in the modern world, such theoretical models of criminal activity largely fell out of favor among experts in the mid-19th century, when an entirely new paradigm of human behavior became dominant. This new perspective became known as the Positive School, and we discuss the origin and development of this paradigm in the following Section.

⊠ Section Summary

- After 100 years of neglect by criminologists, the classical/deterrence model experienced a rebirth in the late 1960s.
- The seminal studies in the late 1960s and early 1970s were largely based on aggregate/group rates of crime, as well as group rates of certainty and severity of punishment, which showed that levels of actual punishment, especially certainty of punishment, were associated with lower levels of crime.
- A subsequent wave of deterrence research, cross-sectional surveys, which are collected at one time, supported previous findings that perceptions of certainty of punishment had a strong, inverse association with offending, whereas findings regarding severity were mixed.
- Longitudinal studies showed that much of the observed association between perceived levels of punishment and offending could be explained by the experiential effect, which is the phenomena of behavior affecting perceptions, as opposed to deterrence (i.e., perceptions affecting behavior).
- Scenario studies addressed the experiential effect by supplying a specific context through a detailed vignette and then asking what subjects would do in that specific circumstance and what their perceptions of the event were.

- Rational choice theory emphasizes not only the formal and official aspects of criminal sanctions but also the informal or unofficial aspects, such as family and community.
- Whereas traditional classical deterrence theory ignored them, rational choice theory emphasizes the benefits of offending, such as the thrill of offending, as well as the social benefits of committing crime.
- Routine activities theory provides a theoretical model that explains why certain places have far more crime than other places and why some locations have hundreds of calls to police each year, whereas other establishments have none.
- Lifestyles theoretical perspectives reveal that the way people live may predispose them to both crime and victimization.
- Routine activities theory and the lifestyles perspective are becoming key in one of the most modern approaches toward reducing or predicting crime and victimization. Specifically, Global Positioning Systems (GPS) and other forms of geographical mapping of crime events have contributed to an elevated level of research and attention given to these theoretical models, due to its importance in specifically documenting where crime occurs and in some cases predicting where future crimes will occur.
- All of the theoretical models and studies in this section were based on the classical/deterrence model, which assumes that individuals consider the potential benefits and costs of punishment and then make their decision of whether or not to engage in the criminal act.

KEY TERMS

Cross-sectional studies

Experiential effect

Rational choice theory

Routine activities theory

Scenario (vignette) research

DISCUSSION QUESTIONS

1. Do you think the deterrence model should have been reborn, or do you think it should have just been left for dead? Explain why you feel this way.

2. Regarding the aggregate level of research in deterrence studies, do you find such studies valid? Explain why or why not.

3. In the comparison of longitudinal studies versus vignette/scenario studies, which do you think offers the most valid method for examining individual perceptions regarding the costs and benefits in offending situations? Explain why you feel this way.

4. Can you relate to the experiential effect? If you can't, do you know someone who seems to resemble the behavior that results from this phenomenon? Make sure to articulate what the experiential effect is.

5. Regarding rational choice theory, would you rather be subject to formal sanctions if none of your family, friends, or employers found out that you engaged in shoplifting, or would you rather face the informal sanctions with no formal punishment (other than being arrested) for such a crime? Explain your decision.

6. As a teenager, did you or family and friends get a "rush" out of doing things that were deviant or wrong? If so, did that feeling seem to outweigh any legal or informal consequences that may have deterred you or people you know?

7. Regarding routine activities theory, which places, residences, or areas of your hometown do you feel fit this idea that certain places have more crime than others (i.e., "hot spots")? Explain how you, friends, or others (including police) in your community deal with such areas. Does it work?

8. Regarding routine activities theory, which of the three elements of the theory do you feel is the most important to address in efforts to reduce crime in the "hot spots"?

9. What type of lifestyle characteristics lead to the highest criminal/victimizing rates? List at least five factors that lead to such propensities.

10. Find at least one study that uses mapping/geographical (GPS) data and report the conclusions of that study. Do the findings and conclusions fit the routine activities theoretical framework or not? Why?

11. What types of policy strategies derived from rational choice and routine activities theories do you think would be most effective? Least effective?

WEB RESOURCES

Modern testing of deterrence:

http://www.deathpenaltyinfo.org/article.php?scid=12&did=167

http://www.jstor.org/view/00914169/ap040062/04a00040/0

Rational choice theory:

http://www.crimereduction.gov.uk/learningzone/rct.htm

http://www.answers.com/topic/rational-choice-theory-criminology

Routine activities/lifestyle theory:

http://www.indiana.edu/~theory/Kip/Routine.htm

http://home.comcast.net/~ddemelo/crime/routine.html

READING

In this study, Sorensen and his colleagues examine the extent to which capital punishment has a significant deterrent effect on murder rates in Texas, which is the state that executes far more individuals than any other in the United States. After reviewing the various designs that have been used to estimate the association between death penalty/executions on murder rates, the authors offer their own test using data from 1984 to 1997. Without giving away the findings of the study, it is notable that this test is consistent with most other studies on the deterrent effects of capital punishment and thus is a good representation of the existing empirical research to date. Readers should keep in mind that the deterrent effect of capital punishment is just one of the many issues in the debate of whether or not we should use the death penalty, just as the evidence of whether executions deter murderers is just one aspect of deterrence theory.

——— Capital Punishment and Deterrence ———

Examining the Effect of Executions on Murder in Texas

Jon Sorensen, Robert Wrinkle, Victoria Brewer, and James Marquart

The dominant approach to determining the relationship between deterrence and the death penalty involves comparing the rate of homicide or some subset of homicide and either the legal status of the death penalty or the performance of actual executions within or across particular jurisdictions. The deterrence hypothesis is supported when lower homicide rates are found within time periods or jurisdictions where the death penalty has been available or in use. If homicide rates are higher in the presence of capital punishment, the alternative, or "brutalization hypothesis," is supported (Bowers and Pierce 1980). The third possible outcome is that the death penalty is found to have no influence on homicide rates.

Empirical studies of deterrence and capital punishment are best classified by their research design. Cross-sectional designs compare homicide rates across jurisdictions. The earliest deterrence studies of this kind simply compared rates of homicide in retentionist states that have statutory provisions for the death penalty to the rates in abolitionist states without such provisions. Findings showed that retentionist states typically experienced higher rates of homicide than did abolitionist jurisdictions (Sutherland 1925). However, examination of geographical, social, and economic dissimilarities between abolitionist and retentionist states suggested that factors other than the death penalty could have influenced homicide rates.

SOURCE: Adapted from Jon Sorensen, Robert Wrinkle, Victoria Brewer, and James Marquart, "Capital Punishment and Deterrence: Examining the Effect of Executions on Murder in Texas," *Crime and Delinquency* 45, no. 4 (1999): 481–93. Copyright © 1999 Sage Publications, Inc. Reprinted by permission of Sage Publications, Inc.

Scholars then began making comparisons that are more specific between neighboring states, which were presumed to be comparable on such factors. These studies failed to support the deterrence hypothesis, finding that retentionist states most often experienced higher rates of homicide than did contiguous abolitionist states (Schuessler 1952; Sellin 1967). A new generation of cross-sectional studies has employed multiple regression analyses to predict the rate of homicide across jurisdictions while controlling for extraneous variables (Forst 1977; Passell 1975) and has consistently found executions to have no effect on murder rates (Cheatwood 1993; Peterson and Bailey 1988).

Longitudinal designs are used to study the influence of the death penalty in a single jurisdiction over time. The earliest of these studies examined homicide rates in jurisdictions before and after a legislative change in the legal status of the death penalty that either abolished or reimplemented capital punishment. The deterrence hypothesis would be supported if states experienced lower rates of homicide during retentionist periods and higher rates of homicide during abolitionist periods. These studies failed to support the deterrence hypothesis (Bedau 1967; Sellin 1967) because they found inconsistent changes in murder rates after legislative enactments.

More recently, the advent of sophisticated statistical techniques has influenced the methodology used in testing the deterrence hypothesis. Time series analysis, introduced by economist Isaac Ehrlich in 1975, has proven to be a superior means of testing the deterrent effect of the death penalty over time. Time series analyses typically concentrate on the effect of actual executions and enable the researcher to simultaneously control for the influence of alternative explanatory variables. In his initial study, Ehrlich claimed that executions carried out during 1933 through 1969 had resulted in a significant reduction in the number of homicides occurring throughout the United States. However, reanalyses of his work failed to find

support for the deterrence hypothesis; instead, researchers concluded that the reduction in homicides observed by Ehrlich was an artifact of measurement error that resulted from inappropriate design specifications and faulty statistical analysis (Baldus and Cole 1975; Bowers and Pierce 1975; Klein, Forst, and Filatov 1978). Recent time series analyses have confirmed the findings of Ehrlich's critics; they have failed to find evidence of a deterrent effect (Bailey 1983; Cochran, Chamlin, and Seth 1994; Decker and Kohfeld 1984).

Another recent advance in methodology has been to limit analyses to only those types of murder likely to be deterred by capital punishment. Because only certain instances of murder can result in the death penalty, many researchers have disaggregated the universe of homicides to limit the dependent variable to those that have death as a possible sentence. For example, Bailey and Peterson (1987) found that the likelihood of receiving a death sentence was not related to the killing of police officers in the United States during 1973 through 1984. The findings supported those of earlier studies that have failed to find a relationship between capital punishment and police killing (Cardarelli 1968; Sellin 1980). One limitation of these studies was that the researchers were unable to consider the certainty of punishment because very few executions had been carried out during that period.

In a later study that included a measure of the certainty of punishment, Peterson and Bailey (1991) analyzed the relationship between actual executions and the monthly rates of felony murder throughout the United States from 1976 through 1987. The researchers found no consistent relationship between the number of executions, the level of television publicity of these executions, and the rate of felony murder. A study following Oklahoma's return to capital punishment disaggregated homicides into felony murders and murders involving strangers (Cochran et al. 1994). Using an interrupted time-series design, Cochran and colleagues found no change in the rate of felony homicides over the 68 weeks following this highly publicized execution, but observed an increase in the rate of stranger homicides. This brutalization effect was recorded in another study that found an increase in several types of homicide in metropolitan areas after Arizona's first execution in 29 years (Thompson 1997).

After thoroughly reviewing the empirical literature, Peterson and Bailey (1998) concluded that the lack of evidence for any deterrent effect of capital punishment was incontrovertible. According to them, no credible empirical studies had ever been able to demonstrate that the severity, certainty, or celerity of capital punishment reduced the rate of homicide. However, they did envision situations that might present unique opportunities to engage the deterrence hypothesis.

One such opportunity presented itself in Texas in recent years. By far the most active death penalty state, Texas has accounted for more than a third of all executions in the United States since the reimplementation of capital punishment in the years following *Furman* v. *Georgia* (1972). In 1997 alone, Texas executed a record number of 37 capital murderers, accounting for half of the 74 U.S. executions in that year. Texas has provided an ideal natural experiment to engage the deterrence hypothesis.

One study of the effect of capital punishment on homicide rates in Texas from 1933 through 1980 found no support for the deterrence hypothesis (Decker and Kohfeld 1990). Although this study did not include the effects of any post-Furman executions in Texas, an update of their research extended the period studied through 1986. Decker and Kohfeld then found that executions were actually followed by an increase in homicide rates, supporting the brutalization hypothesis. Their studies, however, included a limited number of control variables, an aggregate measure of homicide, and the use of years as the unit of analysis. Their updated study captured few

of the executions that were to occur in the post-Furman era. The study reported here advances the work of Decker and Kohfeld by examining the deterrence hypothesis in Texas from 1984 through 1997, capturing the most active period of executions in a jurisdiction during the post-Furman period. It also simultaneously incorporates the methodological strengths of recent studies to provide one of the most compelling tests of the deterrence hypothesis completed thus far.

⊠ Data and Methods

To examine the deterrence hypothesis during the modern era, data that spanned the years from 1984 through 1997 were collected from official sources. The year 1984 was chosen as the beginning of the data collection period because of the availability of specific data on homicides and the onset of executions.[1] Because no executions took place until December 1982, the period before the onset of executions was eliminated from our analyses due to a lack of variance in the independent variable. Data collection was further limited as a result of the Houston Police Department's failure to report information on homicides in 1983 for inclusion in the *Supplemental Homicide Reports* (SHR).[2] Estimating the murder rate for 1983 would, to some unknown degree, bias measures of the deterrent effect of the lone execution in December 1982, particularly because no further executions were carried out until March 1984.[3] Because of these potential sources of bias, data collection began with the year 1984.

The number of executions served as the independent variable. The number of executions was tabulated from ledgers provided by the Texas Department of Criminal Justice—Institutional Division. The dependent variables included rates of murder and rates of felony murder. Information on the number of murders was collected from the Texas Department of Public Safety—Uniform Crime

Reporting Division. The murder rate was based on the number of murders and nonnegligent manslaughters occurring in Texas during the period studied. Excluded from this category were negligent manslaughters, accidental homicides, justifiable homicides committed by citizens and police officers, and executions performed by the state. Murders involving burglary, robbery, or sexual assault were coded as felony murders.[4]

Information on control variables that have most often been found to be related to homicide rates in previous studies was also collected. Information related to homicide in general, including the percentage of the state population living in metropolitan areas, the percentage of the population aged 18 through 34, and the unemployment rate, were culled from the *Statistical Abstracts of the United States* (see Land, McCall, and Cohen 1990; Peterson and Bailey 1991). The number of physicians per l00,000 residents was also coded from the *Statistical Abstracts of the United States* and is included as a proxy for the availability of emergency services, which could prevent an aggravated assault from turning deadly. Other variables available in the *Statistical Abstracts of the United States* that are typically included in homicide studies are the percentage of Blacks and the percentage of divorced individuals. They were excluded from this study because both were constant over the time period studied. Furthermore, these variables were not significant predictors of homicide rates in a recent deterrence study (Peterson and Bailey 1991).

Additional information was collected from alternate sources. The percentage of murders resulting in convictions was collected from the *Annual Reports of the Texas Judicial Council* as an additional measure of the certainty of punishment. The rate of incarceration per 100,000 in the state was gathered from the Bureau of Justice Statistics. The incarceration rate was included to control for possible incapacitation effects resulting from a vast increase in Texas's prison population during the time period studied. Information on the percentage of Texas residents who are on Aid to Families with Dependent Children (AFDC) was gathered from the Texas Department of Human Resources. The direction of its expected relationship to homicide is not specified herein. Although a direct relationship between welfare and homicide rates is typically expected, a recent study found that AFDC is an indicator of available resources that act to mitigate the harshness of poverty, thereby decreasing homicide rates (DeFronzo 1997).

Control variables were also calculated from the SHR data. The percentage of homicides resulting from gunshots was included as a proxy for the availability of firearms. Temporal variables were included to account for surges and lulls in the homicide rate. A high- and low-season variable specified months that were found to be significantly higher or lower in general homicide rates. High season included the months of July and August, whereas low-season included only the months of February. Because the state experienced a record number of executions in 1997, an indicator of that year was also included as a control variable. Lagged execution variables, T_1 to T_3, were also calculated.

Following Chamlin, Grasmick, Bursik, and Cochran (1992), the unit of analysis is the month. Although Chamlin and colleagues did not find significant macro-level deterrent effects when their data were aggregated at longer time intervals, they did find deterrent effects when lagging data in shorter temporal aggregations. A month was the shortest time interval available for which information about the dependent variables was recorded. Accordingly, although we aggregated the number of executions by month, control variables were typically observed on a yearly basis; thus, monthly figures were estimated using linear interpolation. These estimation procedures were appropriate because these variables were treated only as control variables, and not as alternative explanatory variables (Peterson and Bailey 1991).

⊠ Analysis and Findings

Figure 1 provides an overview of execution and murder rates during the time period encompassed by the study. This figure illustrates the episodic nature of executions. After the first execution, which was that of Charlie Brooks, was carried out in December 1982 (not included in Figure 1), the next one did not take place until James Autry was executed in March 1984. A small wave of executions, which peaked at 10 in 1986, followed. A slump in executions then occurred, with an average of four per year being carried out during 1988 through 1991. The ascendance of executions in 1992 signaled the beginning of a more substantial wave of executions, with an average of 15.5 executions per year during 1992 through 1995.

A legal challenge to Texas's procedures for speeding up the appellate processing of capital cases resulted in a moratorium on executions. With the exception of the voluntary execution of Joe Gonzalez in September 1996, executions were halted to await a decision of the Texas Court of Criminal Appeals on the legality of the new procedures. Executions resumed in February 1997, after the court's pronouncement that the expedited appellate procedures were constitutional. The next wave of executions in Texas would be of historical significance. In dispatching 37 backlogged cases, Texas reached a new record for the number of executions carried out in the state during a single year.

The rate of murder in the state from 1984 through 1991 showed no discernible trend in relation to the execution rate. Although there was a slight decrease in murder rates in 1987 through 1989 after the execution wave of the mid-1980s that could be attributed to a deterrent effect, the homicide rate only began to

| Figure 1 | Murders and Live Executions in Texas |

NOTE: This graph shows the murder rate per population of 100,000 and the number of executions per year.

increase in 1990 and 1991, which was after a 2-year lull in executions during the late 1980s. Although the increase could be attributed to the earlier lull in executions and hence support the deterrence hypothesis, the considerable lag in its increase would suggest that any deterrent effect, or lack thereof, occurred only after a considerable time and is of limited significance.

The greatest amount of support for the deterrence hypothesis is found when the decrease in murder rates is paired with the increase in executions during the 1990s. During the execution wave of the 1990s, the murder rate declined substantially in the state. In the same year that the state reached a historical high in executions, the murder rate fell below what was experienced in decades. Although this seems to provide a strong support for the deterrence hypothesis, the downward trend in homicide rates does not appear to be disturbed by the moratorium on executions in 1996, as the deterrence hypothesis would predict; instead, the downward trend continued. Although a bivariate regression model (not reported in tabular form) produced a significant equation, with executions explaining 7 percent of the variance in murder rates, the estimates were not reliable due to a high degree of serial correlation. Furthermore, murder rates have been declining throughout the United States during this same period, which suggests that factors unrelated to executions were responsible for this pattern.

To test for the influence of other causal factors, control variables were included along with executions and used to predict general murder rates and felony murder rates across the monthly series of data from 1984 through 1997. In the first model, the general murder rates were regressed on executions and the control variables. An analysis of residuals from a preliminary ordinary least squares (OLS) regression model that was used to predict murder rates indicated possible heteroscedasticity. In addition, serial autocorrelation found in the original OLS equation suggested the

need for some type of correction. The equation was recalculated using the Newey-West variance estimator, which was specifically designed to correct for these violations of OLS assumptions (Newey and West 1987; StataCorp 1997). Because the Durbin-Watson test indicated problematic levels of autocorrelation up to the third lag, the model presented in Table 1 included Newey-West variance estimates based on a model with three lags.

As shown by the coefficients presented in Table 1, the number of executions was not related to murder rates over the 14-year period that was studied. Control variables positively related to murder rates included the percentage of the population in metropolitan areas, the percentage of the population age 18 to 34, the murder conviction rate, and the high season. The low season, February, was the only variable with a significant negative relation to murder rates. Inclusion of these variables produced a high degree of fit to the data with an overall R^2 of .75.

The model presented in Table 2 limits the dependent variable to felony murders. Because the Durbin-Watson statistic did not indicate a high degree of autocorrelation and because the residuals were more normally distributed, a simple OLS regression model was employed. Because models that were run with lags, $T1$, to $T3$, showed no difference in findings, only the nonlagged model is presented below.

Just as in the model that predicted murder rates in general, the rate of felony murder was not related to the number of executions. The same variables found to be significantly related to murder rates in general were again found to be significant predictors of felony murder rates. The percentage of the population in metropolitan areas, the percentage of the population age 18 to 34, the murder conviction rate, and the high season were positively related to the felony murder rate, whereas the low season was negatively related. One additional variable having a significant positive coefficient in the felony murder model was the unemployment rate.

Table 1	Newey-West Regression Model Predicting General Murder Rates		
Variable	**b**	**SE b**	**t-value**
Number of executions	.0066	.0061	1.073
Percentage in metropolitan area	0.2166***	.0566	3.828
Percentage aged 18 to 34	.1989***	.0470	4.232
Unemployment rate	.0369	.0315	1.171
Physician rate	−.0013	.0110	−.122
Conviction rate	.0192*	.0093	2.068
Incarceration rate	.0001	.0004	.246
Percentage on AFDC	−.0419	.0596	−.702
Percentage of homicides involving guns	.0016	.0026	.617
High season	.1285***	.0323	3.983
Low season	−.1151***	.0253	−4.549
Year 1997	−.0176	.0603	−.292

NOTE: $R^2 = .750$. AFDC = Aid to Families With Dependent Children.

$*p < .05$ (one tailed). $***p < .001$ (one tailed).

Table 2	OLS Regression Model Predicting Felony Murder Rates		
Variable	**b**	**SE b**	**t-value**
Number of executions	−.0012	.0023	−.508
Percentage in metropolitan area	.0381**	.0135	2.815
Percentage aged 18 to 34	.0251*	.0131	1.916
Unemployment rate	.0153*	.0071	2.151
Physician rate	−.0001	.0024	−.060
Conviction rate	.0057*	.0028	2.028
Incarceration rate	.0000	.0001	.176
Percentage on AFDC	−.0069	.0177	−.391
Percentage of homicides involving guns	−.0004	.0006	−.635
High season	.0240***	.0069	3.503
Low season	−.0162*	.0093	−1.746
Year 1997	.0226	.0230	.982

NOTE: $R^2 = .360***$; Durbin-Watson = 1.765. AFDC = Aid to Families With Dependent Children.

$*p < .05$ (one tailed). $**p < .01$ (one tailed). $***p < .001$ (one tailed).

Conclusions

This study found that recent evidence from the most active execution state in the nation lent no support to the deterrence hypothesis. The number of executions did not appear to influence either the rate of murder in general or the rate of felony murder in particular. At the same time, no support was found for the brutalization hypothesis. Executions did not reduce murder rates; they also did not have the opposite effect of increasing murder rates. The inability to reject the null hypothesis supports findings from the vast majority of studies on deterrence and capital punishment (Peterson and Bailey 1998). From the data presented, it appears that other factors are responsible for the variations and trends in murder rates.

Once this argument is accepted, several implications can arise. Some may infer, for example, that these results suggest the repeal of the death penalty because it fails to serve the penological function that is so often offered in its defense. Others would argue that various other goals must also be taken into consideration before making this determination, such as whether the public supports its use, whether it serves the goals of retribution, whether it saves money over life imprisonment, whether it serves to provide justice to the families of victims, and whether it serves the interests of the criminal justice system in general. However, these justifications have also been challenged by research (Acker, Bohm, and Lanier 1998; Bedau 1997). Along with the steady stream of consistent findings on the failure of capital punishment in all of these areas, this study cannot help but support the abolitionist argument.

Notes

1. Detailed information on homicides, a necessity in disaggregating felony murders from more general ones, has been routinely kept by the state since 1976 in the *Supplemental Homicide Reports.*

2. In addition to being the largest jurisdiction in Texas, Houston is the most significant contributor to the number of murders, particularly felony murders, in the state.

3. Information from Houston would be crucial in estimating statewide murder rates, especially felony murder rates, for 1983, the year immediately following the first post-Furman execution,

4. Disaggregating murders into a felony murder category was imperative, because this type of murder was eligible for capital punishment. Felony-related murders have also been the category of capital murders that have most often resulted in death sentences and eventual executions in Texas (Marquart, Ekland-Olson, and Sorensen 1994). Noncapital murders, especially those occurring in the heat of passion, should not be expected to decline in response to executions because they are not punishable by death. Death-eligible homicides, particularly those involving the premeditation of felony-related murders, should reasonably be expected to decrease in response to executions if the deterrence hypothesis is correct.

References

Acker, James R., Robert M. Bohm, and Charles S. Lanier. 1998. *America's Experiment with Capital Punishment: Reflections on the Past, Present, and Future of the Ultimate Sanction.* Durham, NC: Carolina Academic Press.

Bailey, William C. 1983. "Disaggregation in Deterrence and Death Penalty Research: The Case of Murder in Chicago." *Journal of Criminal Law and Criminology* 74:827–59.

Bailey, William C. and Ruth D. Peterson. 1987. "Police Killings and Capital Punishment: The Post-Furman Period." Criminology 25:1–25.

Baldus, David C. and James W. L. Cole. 1975. "A Comparison of the Work of Thorsten Sellin and Isaac Ehrlich on the Deterrent Effect of Capital Punishment." *Yale Law Journal* 85:170–86.

Bedau, Hugo A. 1967. *The Death Penalty in America.* Rev. ed. New York: Doubleday.

———. 1997. *The Death Penalty in America: Current Controversies.* New York: Oxford University Press.

Bowers, William J. and Glenn Pierce. 1975. "The Illusion of Deterrence in Isaac Ehrlich's Research on Capital Punishment." *Yale Law Journal* 85:187–208.

———.1980. "Deterrence or Brutalization: What Is the Effect of Executions?" *Crime & Delinquency* 26:453–84.

Cardarelli, Albert P. 1968. "An Analysis of Police Killed in Criminal Action: 1961–1963." *Journal of*

Criminal Law, Criminology and Police Science 59:447–53.

Chamlin, Mitchell B., Harold G. Grasmick, Robert J. Bursik, Jr., and John K. Cochran. 1992. "Time Aggregation and Time Lag in Macro-Level Deterrence Research." *Criminology* 30:377–95.

Cheatwood, Derral. 1993. "Capital Punishment and the Deterrence of Violent Crime in Comparable Counties." *Criminal Justice Review* 18:165–79.

Cochran, John K., Mitchell B. Chamlin, and Mark Seth. 1994. "Deterrence or Brutalization? An Assessment of Oklahoma's Return to Capital Punishment." *Criminology* 32:107–34.

Decker, Scott H. and Carol W. Kohfeld. 1984. "A Deterrence Study of the Death Penalty in Illinois, 1933-1980." *Journal of Criminal Justice* 12:367–77.

Decker, Scott H. and Carol W. Kohfeld. 1990. "The Deterrent Effect of Capital Punishment in the Five Most Active Execution States: A Time Series Analysis." *Criminal Justice Review* 15:173–91.

DeFronzo, James. 1997. "Welfare and Homicide." *Journal of Research in Crime & Delinquency* 34:395–406.

Ehrlich, Isaac. 1975. "The Deterrent Effect of Capital Punishment: A Question of Life and Death." *American Economic Review* 65:397–417.

Forst, Brian. 1977. "The Deterrent Effect of Capital Punishment: A Cross-Tabular Analysis of the 1960's." *Minnesota Law Review* 61:743–67.

Furman v. Georgia, 408 U.S. 238 (1972).

Klein, Lawrence R., Brian Forst, and Victor Filatov. 1978. "The Deterrent Effect of Capital Punishment: An Assessment of Estimates." Pp. 331–60 in *Deterrence and Incapacitation: Estimating the Effects of Criminal Sanctions on Crime Rates,* edited by A. Blumstein, J. Cohen, and D. Nagin. Washington, DC: National Academy of Sciences.

Land, Kenneth C., Patricia L. McCall, and Lawrence E. Cohen. 1990. "Structural Covariates of Homicide Rates: Are There Any Invariances Across Time and

Social Space?" *American Journal of Sociology* 95:922 63.

Marquart, James W., Sheldon Ekland-Olson, and Jonathan R. Sorensen. 1994. *The Rope, the Chair, and the Needle: Capital Punishment in Texas, 1923–1990.* Austin: University of Texas Press.

Newey, Whitley and Kenneth D. West. 1987. "A Simple, Positive, Semi-Definite. Heteroskedasticity and Autocorrelation Consistent Covariance Matrix." *Econometrica* 55:703–708.

Passell, Peter. 1975. "The Deterrent Effect of the Death Penalty: A Statistical Test." *Stanford Law Review* 28:61–80.

Peterson, Ruth D. and William C. Bailey. 1988. "Murder and Capital Punishment in the Evolving Context of the Post-Furman Era." *Social Forces* 66:774–807.

———. 1991. "Felony Murder and Capital Punishment: An Examination of the Deterrence Question." *Criminology* 29:367–95.

———. 1998. "Is Capital Punishment an Effective Deterrent for Murder? An Examination of the Social Science Research." Pp. 157–82 in *America's Experiment with Capital Punishment: Reflections on the Past, Present, and Future of the Ultimate Sanction,* edited by J. R. Acker, R. M. Bohm, and C. S. Lanier. Durham, NC: Carolina Academic Press.

Schuessler, Karl F. 1952. "The Deterrent Effect of the Death Penalty." *The Annals of the American Academy of Political and Social Science* 284:54–62.

Sellin, Thorsten. 1967. *Capital Punishment.* New York: Harper & Row.

———. 1980. *The Penalty of Death.* Beverly Hills, CA: Sage.

StataCorp. 1997. Stata Statistical Software (Release 5.0) [Computer software]. College Station, TX: Author.

Sutherland, Edwin H. 1925. "Murder and the Death Penalty." *Journal of Criminal Law and Criminology* 15:522–29.

Thompson, Ernie. 1997. "Deterrence Versus Brutalization: The Case of Arizona." *Homicide Studies* 1:110–28.

REVIEW QUESTIONS

1. What do the authors conclude toward the end of their discussion of prior studies? Specifically, what did Peterson and Bailey (1998) conclude?

2. What did the authors conclude from their own findings regarding the deterrent effects of capital punishment on murder rates in Texas?

3. What did the authors conclude from their own findings regarding the "brutalization hypothesis"?

READING

This article uses the scenario design or vignettes to test the compatibility of rational choice theory with what has become the most researched and discussed theory of the last two decades: low self-control theory. Briefly mentioned in the introduction to this section, low self-control theory is a rather simple model that assumes (like other control theories, which we will cover in Section VI) that all individuals are born with a propensity for crime and that children develop self-control through socialization and discipline. However, some children's parents do not do a good job at monitoring or training their children, so these children never develop self-control and, thus, engage in crime when such opportunities present themselves. Piquero and Tibbetts review other studies that have successfully merged rational choice theory with the low self-control model and then present a test of individuals' perceptions regarding two offenses that most college students are familiar with: drunk driving and shoplifting. As you read this study, try to relate the different theoretical concepts and types of offending to people you know.

Specifying the Direct and Indirect Effects of Low Self-Control and Situational Factors in Offenders' Decision Making

Toward a More Complete Model of Rational Offending

Alex R. Piquero and Stephen G. Tibbetts

It has been argued that criminology is in a state of theoretical paralysis (Wellford 1989:119) and that its theoretical developments have stagnated (Gibbs 1987). Recently, however, theorizing in criminology has undergone two important advances. One of these was proposed by Michael Gottfredson and Travis Hirschi (1990) in *A General Theory of Crime*. Their theory concerns individual differences, or propensities, that predispose an individual toward offending; their central concept is that of low self-control. The other theoretical advancement is the rational choice perspective (Cornish and Clarke 1986, 1987). This framework emphasizes the contextual and situational factors involved in decisions to offend, as well as the "choice-structuring" properties of offenses (Cornish and Clarke 1987:935).

SOURCE: Alex R. Piquero and Stephen G. Tibbetts, "Specifying the Direct and Indirect Effects of Low Self-Control and Situational Factors in Decision Making: Toward a More Complete Model of Rational Offending, *Justice Quarterly* 13 (3): 481–510. Copyright © 1996 Routledge. Reprinted with permission.

Low self-control is established early and remains relatively stable throughout life. This is a characteristic of individuals who are more likely than others to engage in imprudent behaviors such as smoking, drinking, or gambling and commit criminal offenses such as shoplifting or assault. Gottfredson and Hirschi (1990:89) characterize low self-control as composed of elements such as immediate gratification, risk taking, orientation to the present, acts involving little skill or planning, and self-centeredness.

The rational choice framework focuses on situational inducements and impediments to offending (Cornish and Clarke 1986, 1987; Nagin and Paternoster 1993) such as the perceived costs (e.g., threat of sanctions) and benefits (e.g., pleasure) of crime. The rational choice model is consistent with a deterrence framework, especially in its focus on the perceived costs associated with committing an offense. It also includes the importance of examining an offender's perception of the benefits of offending and of informal and/or internal threats of sanction, which is absent from the traditional deterrence framework (Piliavin, Gartner, Thornton, and Matsueda 1986). Therefore the rational choice framework provides one way of looking at the influence of situational factors on offending. By the same token, this perspective is not confined to the situational determinants of (perceived) opportunity. Rational choice also examines how motivation is conditioned by situational influences and opportunities to commit crime.

Rational choice emphasizes would-be offenders' subjective perceptions of the expected rewards and costs associated with offending. From this perspective, a crime-specific focus is necessary because the costs and benefits of one crime may be quite different from those of another. This point suggests the importance of examining the choice-structuring properties of particular offenses (Cornish and Clarke 1987:935). Furthermore, the rational choice perspective suggests explanations in terms of those characteristics which promote or hinder

gratification of needs, such as low self-control, shame, moral beliefs, threat of formal sanctions, or the pleasure of offending.

Situational factors and individual propensities are related to each other in a way suggested by Harold Grasmick and his colleagues. Grasmick, Tittle, Bursik, and Arneklev (1993b) noted that situational circumstances and individual characteristics may influence the extent to which low self-control affects criminal behavior. Thus the effect of low self-control depends on the situation; that is, low self-control may condition criminal behavior. Nagin and Paternoster (1993) have examined the compatibility of these perspectives. Using scenario data from a sample of college undergraduates, they found support for the underlying propensity (low self-control) argument advocated by Gottfredson and Hirschi, as well as some support for the effect of situational factors. Attractiveness of the crime target, ease of committing the crime with minimal risk, and perceptions of the costs and benefits of committing the crime were all related significantly to offending decisions. Their analysis, however, consisted solely of examining the direct effects of exogenous variables on the dependent variable (intentions to deviate).

Our analysis builds on Nagin and Paternoster's (1993) paper. We focus on specifying low self-control in an explicit causal model while taking into account the situational factors associated with offending decisions. We believe that low self-control has a direct effect on intentions to deviate, but we also argue that low self-control has indirect effects on these intentions, which operate through a variety of situational factors. These indirect effects are an important step in understanding criminals' decision-making processes.

Whereas Gottfredson and Hirschi distinguish between crime and criminality, Birkbeck and LaFree (1993) argue that theories of crime (situational explanations) should be united with theories of criminality (stable propensities). In this paper, following suggestions emanating from the work of Birkbeck and LaFree (1993) and Nagin and Paternoster (1993), we merge theories of crime (situational factors measured by subjective perceptions) and theories of criminality (low self-control) into a more highly specified causal model of rational offending. We argue that offenders are rational decision makers who are affected by various factors. These factors include not only an individual propensity to offend (i.e., low self-control) but also situational inducements (such as the pleasure of committing the crime) and situational impediments to crime (e.g., sanction threats, shame).

Previous Research

Perceived Sanction Threats and Perceived Pleasure

Deterrence concepts have been modified and expanded (Cornish and Clarke 1986, 1987; Paternoster 1989; Piliavin et al. 1986; Stafford and Warr 1993; Williams and Hawkins 1986), and recent research conducted within the rational choice framework (Bachman, Paternoster, and Ward 1992; Klepper and Nagin 1989b; Nagin and Paternoster 1993), using factorial vignette surveys, has found support for perceptions of certainty and its negative effect on delinquent behavior. Given the consistency with which sanctions may deter certain individuals who commit certain crimes (Bachman et al. 1992; Klepper and Nagin 1989b; Nagin and Paternoster 1993; Smith and Gartin 1989), we contend that these factors are quite important in a general model of rational offending.

The rational choice framework has focused strongly on the pleasure of offending (Bachman et al. 1992; Nagin and Paternoster 1993; Piliavin et al. 1986). Most researchers have found that the perceived benefits of criminal offending are important in a would-be offender's calculations, perhaps even more important than the estimated costs (Nagin and Paternoster 1993:482). The anticipated rewards or gains

from offending may be more important than the potential costs to these individuals because the former are more immediate and more characteristic of risk taking and short-term gratification (Gottfredson and Hirschi 1990). Jack Katz (1988) argues that there are "seductions of crime," which result from the thrills and pleasures provided by committing criminal acts. Other research, however, suggests that seductions are influenced by several background factors including age, gender, and the strain associated with inadequate economic opportunities (McCarthy 1995). Almost all previous empirical tests of deterrence models neglected this beneficial dimension of offending; the few studies that have examined this construct find support for perceived pleasure (Nagin and Paternoster 1993; Piliavin et al. 1986).[1]

Shame

Thomas Scheff (1988) labeled shame as an important factor for social control. Scheff's work was followed closely by John Braithwaite's (1989) *Crime, Shame, and Reintegration,* which caused an immediate increase in the attention given to shame in criminology. Early theorizing on shame, however, tended to focus on acts of shaming by others (e.g., disintegrative/reintegrative shaming) rather than on the internal emotion of shame felt by the individual. Therefore those theorists implied that to experience shame, one must be shamed by a social audience. This assumption is not supported by the psychological literature on shame; in fact, the early researchers in this area acknowledged that most experiences of shame are not preceded by an act of shaming (H. Lewis 1971; Piers and Singer 1953). Experiences of shame are the result of a global, internal evaluation of the self in which the actor temporarily loses some of his or her self-esteem (M. Lewis 1992). Although acts of shaming by others may elicit shame in an individual, such an act need not occur to cause the person to feel that emotion (M. Lewis 1992; Piers and Singer 1953). In other words, individuals can be shamed without the presence of an audience (see Grasmick and Bursik 1990).

Despite the lack of criminological theory on the phenomenological nature of shame, researchers recently have attempted to measure the subjective experiences of shame within a rational choice framework. In these studies (Grasmick and Bursik 1990; Grasmick, Bursik, and Kinsey 1991; Grasmick, Tittle, et al. 1993b; Nagin and Paternoster 1993) respondents have been asked to describe the shame they felt, or would feel, if they had committed, or intended to commit, specific criminal offenses such as drunk driving, littering, date rape, tax evasion, or petty theft. Shame was found to have a strong inhibitory effect on the commission of all these offenses. Furthermore, for some of the offenses, shame had the strongest effect of all the variables specified in the model, including formal sanctions (Grasmick and Bursik 1990). Thus, a deterrent effect of shame seems to be strongly evident in the criminological literature.

Low Self-Control

Gottfredson and Hirschi (1990:90) contend that individuals with low self-control will tend to engage in criminal and analogous acts. Their ideas, which have met with some opposition (Akers 1991; Barlow 1991; Polk 1991), have generated a number of empirical studies (Benson and Moore 1992; Brownfield and Sorenson 1993; Gibbs and Giever 1995; Grasmick, Tittle, et al. 1993b; Keane, Maxim, and Teevan 1993; Nagin and Paternoster 1993; Polakowski 1994; Wood, Pfefferbaum, and Arneklev 1993). Although these studies generally support low self-control, some examination of this work is necessary. First, Grasmick. Tittle, et al. (1993b) developed a psychometric scale that measured low self-control, based on the criteria outlined by Gottfredson and Hirschi. The findings of their study, which examined only direct effects, indicated that low self-control was related

strongly to offending (force and fraud). Keane et al. (1993) examined the relationship between low self-control and drinking and driving. Employing a behavioral measure of self-control (use of seat belts), they found that for both males and females, low self-control was an important predictor of driving under the influence of alcohol.

Gottfredson and Hirschi (1990:90) also believe that low self-control may manifest itself in various imprudent behaviors such as smoking, drinking, and gambling. Using the same data and measures as found in Grasmick, Tittle, et al. (1993b), Arneklev, Grasnick, Tittle, and Bursik (1993) tested this proposition. The results were mixed; on one hand, the low self-control index had a direct effect on an individual's participation in various imprudent behaviors. Yet one component of that index (risk taking) was more strongly predictive than the scale as a whole. Furthermore, smoking appeared to be unaffected by low self-control.[2] Similarly, Wood et al. (1993) argued that although low self-control was a significant predictor of imprudent behaviors and some forms of delinquency, their results suggested that the low self-control measure, as well as the different dependent variables, should be disaggregated.

Gibbs and Giever (1995) examined the manifestations of low self-control on a sample of college undergraduates by creating an attitudinal measure of low self-control and examining its impact on two noncriminal behaviors, cutting class and alcohol consumption. They found that low self-control was the strongest predictor of these behaviors. Their study, however, did not include factors other than self-control, such as moral beliefs or perceived threat of sanctions.

Moral Beliefs and Prior Offending

In addition to the variables discussed above, we included two other variables in the model specification: moral beliefs and prior offending. Moral beliefs are necessary in the study of any rational choice framework because such beliefs impede criminal behavior; theorists have stressed the importance of internalized moral constraints (Bachman et al. 1992; Bishop 1984; Grasmick and Bursik 1990; Paternoster, Saltzman, Chiricos, and Waldo 1983; Tittle 1977, 1980). We also included prior offending as a control variable because it could capture the influence of other sources of stable criminality (Nagin and Paternoster 1991, 1993).

Proposed Model

The proposed model assumes that a rational human actor with low self-control encounters situational factors which push him or her toward crime (pleasure of the offense) and/or away from crime (moral beliefs, perceived risk of sanctions, and situational shame). When the push toward crime is greater than the push away from crime, an individual is more likely to choose crime. This idea is summarized by Gottfredson and Hirschi (1990:89) when they observe that a major characteristic of those with low self-control is the tendency to respond to tangible stimuli in the immediate environment and to have a concrete "here and now" orientation (also see Hirschi and Gottfredson 1993).

Although our theoretical model relies heavily on the most recent statement of control theory outlined by Gottfredson and Hirschi, it is not meant to downplay the importance of earlier control theorists, particularly Walter Reckless (1961; also see Toby 1957). In his seminal piece, Reckless noted that inner containment consists mainly of self-control, while outer containment represents the structural buffer in the person's immediate social world which is able to hold him or her within bounds (Reckless 1961:44–45). Expanding upon the idea of outer containment, one could easily infer that sanctions,

pleasure, and shame are structural buffers in an individual's immediate social world. Moreover, Block and Flynn (1956:61) state that "there are many variables in the personality of the delinquent and the delinquency-producing situation itself which the investigators may not readily discern and which themselves may constitute the critical factors involved in the delinquent act." Conceivably, then, one could argue that our theoretical model is a refinement, an extension, and an empirical test of Reckless's theory and of Block and Flynn's assertions (also see A. Cohen 1955).

Model Specification

We propose a general model of rational offending that unites low self-control with situational variables. The model assumes that low self-control is a stable personality characteristic and that prior offending and moral beliefs are characteristics antecedent to situational factors; these characteristics have direct effects on intentions to deviate and indirect effects that operate through situational characteristics associated with offending. Situational characteristics include the perceived pleasure associated with the act, the probability of sanctions as a result of committing the act, and the shame associated with committing the act. The dependent variable is an individual's intention to deviate.

Let us explain the proposed causal links in the model. The first set of variables—low self-control, moral beliefs, and prior offending—are time-stable and antecedent to the situational variables. Because most of the previous research on low self-control concentrated on direct effects, our model explicitly takes into account the indirect effects of low self-control. Much to their credit, Gottfredson and Hirschi (1990:95) anticipated these indirect effects when they argued that the impulsive or short-sighted person fails to consider the negative or painful consequences of his acts. As a result, such a person has fewer negative consequences to consider. Following this line of reasoning, we argue that low self-control will have a positive effect on perceived pleasure, and negative effects on both perceived sanctions and shame. Short-sighted individuals (i.e., those with low self-control) are more likely not to perceive shame and sanctions as important because these are long-term outcomes, whereas they perceive pleasure as a short-term result of committing an offense. In agreement with previous self-control research, we also predict that low self-control will have a positive, direct effect on intentions to deviate.

Like Gottfredson and Hirschi (1990), we believe that an individual's moral beliefs against committing an act should be associated positively both with perceived sanctions and with the shame associated with offending because belief in the morality of the law should increase the saliency of being caught as well as the shame associated with committing crime. Moral beliefs also should have a negative effect on an individual's perceived pleasure and should be related negatively to intentions to deviate. As for the third of these antecedent variables—prior offending—we hypothesize that an individual's prior offending should be related positively to perceived pleasure and intentions to deviate, and negatively to perceived sanctions and shame.

The mediating variables in the model are the situational characteristics associated with criminal offending. We propose that they have direct effects on intentions to deviate and indirect effects through other situational variables. In regard to direct effects, perceived pleasure should have a positive effect on intentions to deviate, while perceived sanctions and shame should inhibit such intentions. Similarly, perceived pleasure should exert a negative effect on perceived shame, while shame should have a positive effect on perceived sanctions and perceived sanctions should have a positive

effect on perceived pleasure, in keeping with Katz's (1988) "sneaky thrills."

☒ Methods

We collected data through a self-administered questionnaire that presented respondents with a realistic scenario describing in detail the conditions in which an actor commits a crime. The respondents were told only that the actor committed the act, not whether he or she approved of the act. Thus we focus not on the hypothetical actor's perceptions or approval of the act, but rather on the respondent's perceptions and approval. The questions were designed to measure respondents' perceptions of the costs and benefits of committing the offense described in the scenario, to estimate the probability that they would commit that offense, and to estimate the chance that their committing the offense would result in arrest and in exposure without arrest.

The scenario method differs from conventional data collection in perceptual social control/deterrence research in that it uses hypothetical, third-person scenarios of offending to elicit the dependent variable. This strategy has been used successfully in recent research on rational choice (Bachman et al. 1992; Klepper and Nagin 1989a, 1989b; Nagin and Paternoster 1993). The primary weakness of this approach is that an expressed intention to offend is not synonymous with actual offending. Fishbein and Ajzen (1975), however, argue that a person's intention to perform a particular behavior should be highly correlated with the actual performance of that behavior.[3] This proposition is supported empirically by Green (1989), whose two-wave panel design revealed a high correlation ($r = .85$) between intentions and actual performance of deviant behavior. In addition, Kim and Hunter's (1993) recent meta-analysis produced a strong relationship between attitude, intention, and behavior. In all, the scenario method is the best approach available because of its advantages, its realistic nature, and the specificity of the scenarios.[4]

The realistic and specific nature of the scenarios allows us to examine the effect of situational factors on both the intentions to offend and the anticipated risks and rewards of these behaviors. Without these contextual specifications, the respondents would impute their own details; such a situation would "undoubtedly vary across respondents and affect their responses" (Nagin and Paternoster 1993:474). Also, individuals may vary in their definition of illegal behavior. If these differences in definition vary systematically with responses measuring variables of interest, analysis of the effects of such variables on actual behavior may be misrepresented (Nagin and Paternoster 1993).

Another, perhaps more important advantage of the scenario method is its capacity to capture the "instantaneous" relationship between independent variables and the respondent's intentions to offend (Grasmick and Bursik 1990). Previous cross-sectional and panel studies on deterrence used measures of past behavior or behavior within waves to measure the dependent variable (e.g., Bishop 1984). Because perceptions of risk are unstable over time, however, this lagged type of measurement is not appropriate. These designs would tend to find lagged effects for independent variables that remained stable over time, such as moral beliefs, but no lagged effects for independent variables that are not stable, such as perceived of threats sanction (Grasmick and Bursik 1990). Therefore, because the scenario method permits the examination of "instantaneous" relationships, it is preferable to traditional designs.[5]

Sample and Scenario Design

Respondents were undergraduates at a major East Coast university, enrolled in several large introductory criminal justice courses in the fall 1993 semester. A total of 349 males and 293 females (642 in all) completed the questionnaire. Although participation was voluntary, only 4 percent of potential respondents refused to participate; given this small amount, analysis and conclusions appear not to be threatened by response bias. The respondents ranged in age from 17 to 35; the median age was 19. Because we selected introductory classes that fulfill general core requirements for the university curriculum, a substantial majority of students (69 percent) were not criminal justice majors and were currently in their freshman and sophomore years. In addition, questionnaires were administered during the second week of the semester. Therefore it is very unlikely that responses were biased by students' knowledge of deterrence or correctional concerns.[6] Listwise deletion of missing cases resulted in a sample of 604.

The Scenarios

Under an adaptation of the factorial survey methodology developed by Rossi and Anderson (1982), each student was given two scenarios—drunk driving and shoplifting—to which to respond. All of the scenarios were framed in settings familiar to these college student respondents. Selected scenario conditions were varied experimentally across persons. Respondents were asked to estimate the probability that they would commit the act specified in the scenario, to predict the chance that their commission of the offense would result in arrest, and to answer questions designed to measure their perceptions of the costs and benefits of committing the offense described in the scenario. In the present analysis, then, all respondents receive the opportunity to commit the same crimes in the same setting.[7]

⊠ Measurement of Variables

Intentions to Deviate

Separate models are estimated for each type of offense. The dependent variable is the respondent's estimate of the chance that he or she will do what the character did in the scenario. We measured intentions to offend on a scale from 0 (no chance at all) to 10 (100 percent chance). Responses were solicited for both the drunk driving (INTENTDD) and the shoplifting (INTENTSH) scenarios.

Shame

Shame is measured by two items following each scenario, which ask the respondent (1) "what is the chance" and (2) "how much of a problem" would loss of self-esteem be if he or she were to do what the actor in the scenario did, even if no one else found out. Responses to both of these items were measured on an 11-point scale (0 = no chance/no problem to 10 = 100 percent chance/very big problem). We computed shame (SHAME) by multiplying the responses of the two items; higher scores reflect a higher likelihood that the individual would feel shame if he or she were to commit the specified act.

Low Self-Control

We operationalized low self-control with a psychometric scale borrowed from Grasmick et al. (1993b), which includes 24 items intended to measure the six elements of low self-control.[8] We coded these items on a five-point Likert-type scale (1 = never to 5 = very often) and created a composite measure of self-control (SELFCONT) by summing the responses across 24 items. High scores on the scale indicate low self-control. This instrument was used in two previous studies (Grasmick et al. 1993b; Nagin and Paternoster 1993), both of which provided strong reliability and validity support for the scale. The high estimated reliability coefficient ($\alpha = .84$) gave us confidence in the internal consistency of the scale. Furthermore, the factor loadings provided by a principal-components factor analysis were comparable to those reported by Grasmick et al. (1993b).

Perceived External Sanctions

Respondents were asked to estimate the chance of arrest (*Pf:* risk of formal discovery) and the chance that others would find out if they were not arrested (*Pi:* risk of informal discovery). To measure the perceptions of the implications of discovery, we asked respondents to estimate the probability that discovery by arrest or informal exposure would result in dismissal from the university (*Pdf, Pdi*), loss of respect by close friends (*Pff, Pfi*), loss of respect by parents and relatives (*Ppf, Ppi*), and diminished job prospects (*Pjf, Pji*). Each of these perceptual measures is intended to measure the risks of informal sanctions that may threaten an individual's "stake in conformity," or bonding to the moral order. To measure the perceived risk of formal sanctions, we asked respondents to estimate the risk of jail (*Pjaf*). The drunk-driving scenario was followed by an additional item measuring the perceived chance of losing one's driver's license (*Plf*) if an arrest was made. All responses were measured on an 11-point scale (0 = no chance at all to 10 = 100 percent chance).

These measures of risk probably would have little effect on intentions unless associated with perceptions of some cost (Grasmick and Bursik 1990). Thus we asked respondents to estimate the perceived severity of each sanction. Specifically, we asked each subject to estimate "how much of a problem" each sanction would pose for them. All responses were measured on an 11-point scale (0 = no problem at all to 10 = a very big problem). To create the composite scale of perceived external sanctions, we multiplied each risk-perception response by the corresponding severity-perception response. Then we summed these separately for drunk driving and for shop-lifting (PEREXSAN); higher scores on the scale correspond to a high degree of perceived risk and cost of performing the act in question for that individual. We used the following formula:

PEREXSAN = Pi [(Pdi) (Sd) + (Pfi) (Sf) + (Ppi) (Sp) + (Pji) (Sj)] + Pf [(Pdf) (Sd) + (Pff) (Sf) + (Ppf) (Sp) + (Pjf) (Sj) + (Plf) (SI) + (Pjaf) (Sja)]

where *Sd* equals the perceived severity of sanction *d* (dismissal from university) and all other variables are as defined previously.

Moral Beliefs

To measure the perceived immorality of the behavior, we asked respondents to estimate how morally wrong they thought the incident would be if they were to commit drunk driving and shoplifting (MORALS). Response options varied on an 11-point scale (0 = not morally wrong at all to 10 = very morally wrong). Although some may contend that our respondents may not regard the behaviors under study as criminal or morally wrong, the mean moral value was 7.80 against drunk driving and 7.57 against shoplifting. These findings indicate that most of our respondents perceive even these behaviors as morally wrong.

Perceived Pleasure

To measure perceived pleasure, a single item asked respondents to estimate "how much fun or kick" it would be to commit drunk driving and shoplifting under the conditions specified in the scenarios (PLEASURE). Responses varied on an 11-point scale (0 = no fun or kick at all to 10 = a great deal of fun or kick).

Prior Offending

In addition to the variables discussed above, we included prior offending as a control in the model. We did so to capture the influence of sources of stable criminality extraneous to that of persistent individual differences due to personality traits included in the model

(such as low self-control). To measure prior offending (PRIOROFF), we included two items (one for each scenario offense) that asked the respondents how many times in the past year they had driven while drunk and how many times they had shoplifted.[9]

Hypotheses

In this paper we postulate and examine three hypotheses:

H$_1$: Low self-control has both direct and indirect effects via situational factors on intentions to deviate;

H$_2$: Situational characteristics have both direct and indirect effects on intentions to deviate and on other situational variables;

H$_3$: The model uniting the effects of low self-control and situational characteristics of crime will provide a good fit to the data.[10]

⬚ Results

We estimated models for intentions to drink and drive and to shoplift. Insignificant paths were eliminated, and we reestimated the models. Although the models for the two offenses exhibited similar results, some minor differences emerged. The analysis that follows employs the LISREL VII estimation program. LISREL's structural equation model specifies the causal relationships among the latent variables, and describes the causal effects and the amount of unexplained variance (Joreskog and Sorbom 1989).

Tables 1 and 2 report the bivariate correlations of the variables for shoplifting and for drunk driving. Because this paper focuses on constructing a model that unifies low self-control with situational factors of crime, we concentrate on the maximum-likelihood estimates provided by LISREL.[11]

Table 1	Correlation Matrix for Intentions to Shoplift ($N = 604$)						
	SELFCONT	**PEREXSAS**	**PRIOROFS**	**PLEASURS**	**MORALSSH**	**SHAME**	**INTENTSH**
SELFCONT	1.0000						
PEREXSAS	−.1416**	1.0000					
PRIOROFS	.1099*	−.0471	1.000				
PLEASURS	.2549**	−.0427	.2069**	1.0000			
MORALSSH	−.2775**	.3604**	−.1088*	−.2809	1.0000		
SHAME	−.2808**	.4827**	−1443**	−.2927	.5625**	1.0000	
INTENTSH	.3405**	−.1675**	.2927**	.4142	−4328**	−.4482	1.0000

NOTE: SELFCONT = low self-control; PEREXSAS = perceived external sanctions for shoplifting; PRIOROFS = prior offending for shoplifting; PLEASURS = perceived pleasure for shoplifting; MORALSSH = moral beliefs for shoplifting; SHAME = shame; INTENTSH = intentions to shoplift.

*$p < .01$; **$p < .001$

Table 2	Correlation Matrix for Intentions to Drive Drunk ($N = 604$)						
	SELFCONT	**PEREXSAS**	**PRIOROFD**	**PLEASURD**	**MORALSDD**	**SHAME**	**INTENTSH**
SELFCONT	1.0000						
PEREXSAD	−.1595**	1.0000					
PRIOROFD	.1698**	−.1953**	1.000				
PLEASURD	.2871**	−1292**	.0538	1.0000			
MORALSDD	−.2499**	.3663**	−.0988*	−.2169**	1.0000		
SHAME	−.2836**	.5241**	−.2129**	−.2733**	.4216**	1.0000	
INTENTDD	.3234**	−.4222**	.4222**	.3749**	−.3266**	−.4029**	1.0000

NOTE: SELFCONT = low self-control; PEREXSAD = perceived external sanctions for drunk driving; PRIOROFD – prior offending for drunk driving; PLEASURD = perceived pleasure for drunk driving; MORALSDD = moral beliefs for drunk driving; SHAME = shame; INTENTDD = intentions to drive drunk.

*$p < .01$; **$p < .001$

Analysis of Shoplifting

According to Hypothesis 1, low self-control will have a direct effect on intentions to deviate and indirect effects on intentions to deviate through situational factors. Significant maximum-likelihood estimates for shoplifting may be found in Table 3 and Figure 1. Of the four paths estimated for low self-control, three are significant. Low self-control has a direct positive effect ($b = .153$, $t = 4.438$) on intentions to shoplift and a direct positive effect ($b = .178$, $t = 4.502$) on perceived pleasure, an indication that the higher one scores on the low self-control scale, the more likely one is to intend to shoplift and to perceive pleasure from shoplifting. Low self-control has a direct negative effect ($b = −.102$, $t = −2.889$) on shame, indicating that the higher one scores on the low self-control scale, the less likely one is to experience shame due to shoplifting. The only insignificant effect is the effect of low self-control on perceived risk of sanctions.

Figure 1

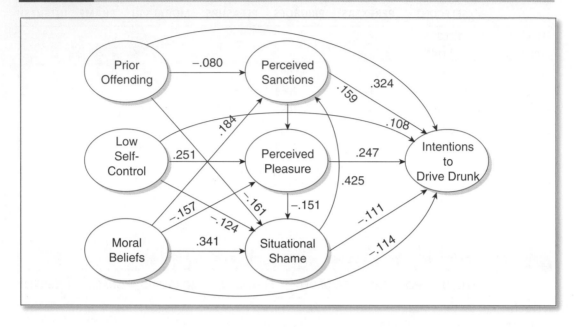

Table 3 Significant Full-Information Maximum-Likelihood Estimates for Intentions to Shoplift (N = 604)

Dependent Variables	Shame	Perceived Sanctions	Perceived Pleasure	Moral Beliefs	Prior Offending	Low Self-Control
			Independent Variables			
Intentions to Shoplift	−.214	[a]	.220	−.186	.176	.153
Shame	—[b]	—	−.173	.483	—[a]	−.102
Perceived Sanctions	.434	—	—[b]	.117	—[a]	—[a]
Perceived Pleasure	—[b]	.153	—[b]	-.267	.169	.178

NOTE: LISREL shows the effects of columns on rows.

a. Path estimated but not significant.

b. Path not established.

Therefore, low self-control not only has a direct effect on intentions to shoplift; it also indirectly affects intentions to shoplift through situational variables (pleasure and shame). These results are consistent with Gottfredson and Hirschi's (1990:95) idea that individuals with low self-control will be less likely to consider the consequences of offending.

Hypothesis 2 indicates that situational characteristics should have direct effects on intentions to shoplift and indirect effects on intentions to shoplift which operate through other situational factors. With the exception of perceived sanctions, both shame ($b = -.214$, $t = -5.372$) and preceived pleasure ($b = .220$, $t = 6.270$) have the expected effects on intentions to shoplift. The null results for perceived sanctions are not surprising: Shoplifting is a very common crime and one that can be committed with relative impunity; thus an individual's perception of being caught would likely not be salient.

As for the other effects, shame ($b = .434$, $t = 9.745$) has a positive effect on perceived sanctions, indicating that the more likely one is to perceive shame, the more likely one is to perceive the threat of sanctions as salient. Even though perceived sanctions do not affect intentions to shoplift, they affect perceived pleasure in a rather interesting manner: Perceived sanctions have a positive effect ($b = .153$, $t = 3.398$) on perceived pleasure, in keeping with Katz's (1988) notion of "sneaky thrills." It appears that among our respondents, the more one perceives the risk of sanctions as high, the more pleasure one perceives from shoplifting. Finally, perceived pleasure has a negative effect on shame ($b = -.173$, $t = -4.468$): The more one perceives pleasure from shoplifting, the less likely one is to feel shame.

Other effects include those of the other two exogenous variables, prior offending and moral beliefs. Prior offending has positive effects on intentions to shoplift ($b = .176$, $t = 5.322$) and on perceived pleasure ($b = .169$, $t = 4.421$), indicating that the more times respondents have shoplifted in the past, the more likely they are to intend to shoplift and to perceive pleasure from shoplifting. Prior behavior does not exert an effect on perceived sanctions. Moral beliefs are the only exogenous variable to be significant and consistent with all effects as predicted. Moral beliefs have the predicted negative effects on intentions to shoplift ($b = -.186$, $t = -4.669$) and on perceived pleasure ($b = -.267$, $t = -6.287$), indicating that the stronger one's moral beliefs against shoplifting, the less likely one is to intend to shoplift or to perceive pleasure from shoplifting. Likewise, moral beliefs have the predicted positive effects on shame ($b = .483$, $t = 13.599$) and on perceived sanctions ($b = .117$, $t = 2.691$), indicating that the stronger one's moral beliefs, the more likely one is to perceive shame and sanctions as important.

The indirect, direct, and total effects of the exogenous variables are displayed in Table 4. Because previous researchers did not examine the indirect effects of low self-control, we concentrate on those effects. (See Table 4 for the indirect effects of other exogenous variables.) Most of the influence of low self-control is direct, .153. Low self-control, however, also exerts important indirect effects on intentions to shoplift, totaling .065. The combined direct and indirect effects for low self-control equal .219. Of all the exogenous variables in the model, the total effects for low self-control

Table 4	Indirect, Direct, and Total Effects of Exogenous Variables on Intention to Shoplift					
	Exogenous Variables					
	Low Self-Control	Prior Behavior	Moral Beliefs	Shame	Sanctions	Pleasure
Direct Effect	.153	.176	−.186	−.215	0	.220
Indirect Effects	.065	.043	−.159	.017	.037	.034
Total Effects[a]	.219	.219	−.345	−.197	.039	.254

a. Total effects may not equal the direct and indirect effects added together because of rounding error.

rank second only to moral beliefs. These results signify the importance of specifying all types of effects of low self-control, particularly indirect effects.

To test the third hypothesis, we constructed a model that united the effects of low self-control and of situational characteristics. To determine whether the proposed model fit the data adequately, we examined the chi-square statistic of the model. Because chi-square is sensitive to sample size and to departures from normality in the data, there are alternative methods for assessing the goodness of fit of a model; one such method is the ratio of chi-square to degrees of freedom. Smith and Patterson (1985) suggest that values of 5 or less indicate an adequate fit. For this model the value is 1.01 (4.05/4), indicating an adequate fit to the data.

Analysis of Drunk Driving

The significant maximum-likelihood estimates for drunk driving are shown in Table 5

and Figure 2. For low self-control, three of the four effects are significant. Low self-control has direct positive effects on intentions to drive drunk ($b = .108$, $t = 3.167$) and on perceived pleasure ($b = .251$, $t = 6.308$), indicating that the higher one scores on the low self-control scale, the more likely one is to intend to drive drunk and to perceive pleasure from drunk driving. Low self-control exerts a negative effect on shame ($b = -.124$, $t = -3.257$), indicating that persons with low self-control are less likely to feel shame. As in the analysis of shoplifting, the effect of low self-control on perceived sanctions is insignificant.

All three situational factors have the expected effects on intentions to drive drunk. Shame ($b = -.111$, $t = -2.796$) and perceived sanctions ($b = -.159$, $t = -.4.219$) exert the expected negative effects on intentions to drink and drive, indicating that the more one perceives sanction threats and shame as important, the less likely one is to intend to drive drunk.[12] Perceived pleasure has the expected positive effect ($b = .247$, $t = 7.313$) on

Figure 2

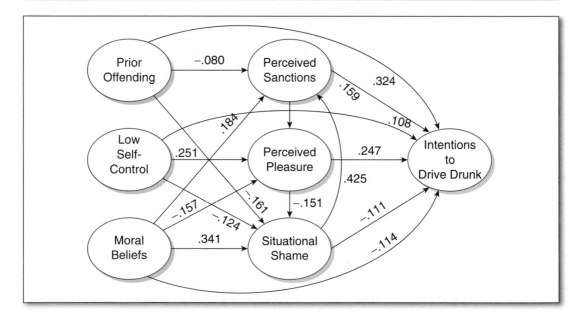

| Table 5 | Significant Full-Information Maximum-Likelihood Estimate for Intention to Drive Drunk (*N*=604) |

	Independent Variables					
Dependent Variables	**Shame**	**Perceived Sanctions**	**Perceived Pleasure**	**Moral Beliefs**	**Prior Offending**	**Low Self-Control**
Intentions to Drive Drunk	−.111	−.159	.247	−.114	.324	.108
Shame	—[b]	—[b]	−.151	.341	−.161	−.124
Perceived Sanctions	.425	—[b]	—[b]	.184	−.080	—[a]
Perceived Pleasure	—[b]	—[a]	—[b]	−.157	—[a]	.251

NOTE: LISREL shows the effects of columns on rows.

a. Path estimated but not significant.

b. Path not established.

intentions to drive drunk, indicating that the more pleasure one perceives from drunk driving, the more likely one is to intend to drive drunk. Other effects for perceived pleasure include a negative effect on shame ($b = −.151$, $t = −4.057$), indicating that the more pleasure one obtains from drinking and driving, the less likely one is to lose self-esteem. Shame has a positive effect ($b = .425$, $t = 11.123$) on perceived sanctions, indicating that the more one perceives shame as salient, the more likely one is to perceive sanction threats as also important.

Effects of the other two exogenous variables (prior offending and moral beliefs) are largely as expected. Prior offending has a negative effect on shame ($b = −.161$, $t = −4.498$) and on perceived sanctions ($b = −.080$, $t = −2.295$), indicating that the more one has driven drunk in the past, the less likely one is to feel shame and to perceive sanctions as important. In addition, prior offending has a positive effect on intentions to drive drunk ($b = .324$, $t = 9.946$), which indicates that the more one has driven drunk in the past, the more likely one is to intend to drive drunk. Prior offending has no effect on the perceived pleasure of drunk driving.

All four moral belief effects are significant. Moral beliefs have negative effects on intentions to drink and drive ($b = −.114$, $t = −3.177$)

and on perceived pleasure ($b = −.157$, $t = −3.959$), indicating that the stronger one's moral beliefs are against drunk driving, the less likely one is to intend to drive drunk and the less likely one is to derive pleasure from drinking and driving. Moral beliefs also have positive effects on shame ($b = .341$, $t = 9.269$) and on perceived sanctions ($b = .184$, $t = 4.925$), indicating that the stronger one's moral beliefs are, the more likely one is to experience shame and to perceive sanctions as important.

As in the analysis of shoplifting, low self-control not only exerts a direct effect on intentions to drink and drive but also indirectly affects intentions to drink and drive through certain situational characteristics. These findings are consistent with Gottfredson and Hirschi (1990:95) in that the effects of the perceived consequences of offending are conditioned by the individual's level of self-control. Table 6 displays the indirect, direct, and total effects of the exogenous variables for the drinking and driving model. As in the shoplifting analysis, we concentrate on the effects of low self-control. For the drunk-driving analysis, the direct effect of low self-control is .108. Possibly more intuitive and certainly more interesting is the finding that the summed indirect effects of low self-control on intentions

Table 6	Indirect, Direct, and Total Effects of Exogenous Variables on Intention to Drive Drunk					
	Low Self-Control	Prior Behavior	Moral Beliefs	Shame	Sanctions	Pleasure
Direct Effect	.108	.324	−.114	−.111	−.159	.247
Indirect Effects	.091	.041	−.133	−.068	0	.027
Total Effects[a]	.199	.366	−.247	.179	−.159	.274

a. Total effects may not equal the direct and indirect effects added together because of rounding error.

to drive drunk is .091, almost equal to the direct effect. The total effect of low self-control, with the direct and indirect effects summed, is .199, third highest among the exogenous variables.

Results concerning Hypothesis 3 in regard to drinking and driving are similar to those for shoplifting. To determine whether the model constructed for drunk driving fit the data adequately, we performed the same two tests as we conducted for shoplifting. The first test examined the ratio of chi-square to degrees of freedom. Values of less than 5 indicate an adequate fit to the data: our value was .33 (1.00/3).

✉ Conclusion

Building on the early work of Nagin and Paternoster (1993), we set out here to combine two different paths in theoretical criminology into a more complete model of offending. One path attributes crime to individual differences that are established early in life, specifically in low self-control. According to the second path, crime is the result of situational factors associated with criminal offending, such as the perceived costs and benefits of crime. As observed by Nagin and Paternoster (1993:489), these two paths have been explored separately rather than in conjunction. On the basis of our analysis, we find support for a model that integrates these two paths. The model holds after

controlling for several important factors and performs well in two different tests designed to measure the fit of the model to the data.

Aside from delineating and testing a more complete model of rational offending, this paper represents the first attempt to examine the indirect effects of low self-control. This attempt is especially important because previous research in low self-control examined only the direct effects of low self-control and rational choice characteristics (Grasmick et al. 1993b; Nagin and Paternoster 1993). Of all our findings, the indirect effects of low self-control were the most interesting. In fact, these effects were more complex than we had imagined originally. We found that low self-control had similar effects on shame and perceived pleasure across offenses, but exerted no effect on perceived sanctions in either scenario. Modeling indirect effects of low self-control is a difficult task, which we undertook with almost no previous theoretical guidance. Such effects probably depend on the offense, but currently we have too little information about the indirect effects of low self-control on offending. Additional theoretical work and further modeling of the total effects are priorities in self-control research.

The model we have presented here may be extended in the following ways. First, we would like to see future studies examine a wide array of criminal and deviant behaviors, such as drug use, sexual assault, burglary, and robbery.

Insofar as Gottfredson and Hirschi are correct, low self-control should be related to all types of criminal and deviant behaviors. Second, many variables could be interchanged with and/or added to our list of situational variables. We contend that because different offenses require different situational characteristics and circumstances, these mediating factors may change in type—but they will be situational factors nonetheless. For example, an examination of marijuana use may require inclusion of a situational variable such as the ease of obtaining marijuana, whereas an examination of breaking and entering would require situational characteristics such as the lack of capable guardians, lack of a security system, and the time of day or night. Still other examples of situational variables would include peer delinquency and peer associations. Because delinquency is overwhelmingly a group phenomenon (Reiss 1986), inclusion of such a measure has the potential to enhance the predictability of our model. This discussion should make apparent that although situational characteristics may vary in type depending on the crime, the framework of the model will remain the same: Time-stable variables such as low self-control will always precede and influence the situational variables.

▨ Notes

1. Some may argue that the pleasure associated with offending is only part of the story, and that often the more important situational factors are the amount of time and energy saved (as in drunk driving) and the value of goods stolen (as in shoplifting). Because of the lack of significant findings from Nagin and Paternoster's (1993) vignettes of these conditions, we did not vary these situational characteristics.

2. This result may be due to the average age of the sample (46.5 years). It could be that these individuals began to smoke before the effects of smoking were known to be undesirable (Arneklev et al. 1993).

3. Fishbein and Ajzen (1975) identify three criteria for maximizing the correspondence between intentions and actual behavior. The first of these criteria is the degree to which the intentions are measured with the same specificity as the behavior that is being predicted. The scenarios presented here include highly specific circumstances (see appendix). The second criterion is the stability of the expressed intention. In view of the realistic and specific conditions of the scenarios, there is no compelling reason to question the stability of these intentions. The final criterion is the degree to which the respondent can willfully carry out the intention.

4. Our scenarios were designed after those used by Nagin and Paternoster (1993) in regard to detail and contextual specificity. We achieved specificity by presenting details of the circumstance of the offense, such as naming the bar where the actor is drinking or the type of item the actor is shoplifting. The scenario approach has been used as well in research on death penalty juries (Bohm 1991).

5. We systematically varied the location of the intention questions for both the drunk driving and the shoplifting scenarios. In approximately half of these scenarios, the dependent variable item was placed directly after the scenario; other perceptual items (e.g., moral beliefs, perceived certainty) followed (this position was coded 0). In the other half, the dependent variable was located at the end of the battery of perceptual items (this position was coded 1). We adopted this procedure to examine for possible differences due to responses on the dependent variable item affecting the responses on the subsequent perceptual items. For instance, if the dependent variable item is placed directly after the scenario, the respondents may base their perceptions on their previous response to the dependent variable item. In contrast if the dependent variable item is placed after the perceptual items, respondents may respond differently on the dependent variable item if they have thought carefully about their perceptions regarding the offense. Bivariate correlations showed that the location of the dependent variable item did not have a significant effect on respondents' intentions to commit drunk driving or to shoplift ($r = .06$ and $-.05$, respectively). Therefore, we did not include this variable in the multivariate analyses.

6. The use of convenience samples in deterrence research is questionable and has drawn some criticism (Jensen, Erickson, and Gibbs 1978; Williams and Hawkins 1986). The major objection is that of representativeness. Large public universities, however, contain a moderate number of marginal offenders (Matza 1964), particularly for the kinds of offenses that are the focus of this study. In our data,

44 percent of respondents admit to having committed drunk driving in the past year (17 percent committed shoplifting in the past year). Furthermore, a Bureau of Justice Statistics Report (R. Cohen 1992) reveals that the rate of arrest for driving while under the influence of alcohol (DUI) is highest for persons between ages 21 and 24. Those in the 18-20 age range have the second-highest arrest rate for DUI. Also, a survey of 1,287 university students conducted in 1991 revealed that almost one-half were regular users of alcohol; 45 percent of these reported consuming four or more drinks at a time, and more than half reported driving within an hour after consuming their last drink (Kuhn 1992). When subjects in our sample were asked the likelihood of drinking and driving under the conditions of the scenario presented to them, only 33 percent reported "no chance." Shoplifting also has been shown to be quite common among young adults (Empey and Stafford 1991); self-reports show that shoplifting is about as common as drinking (Elliott, Ageton, Huizinga, Knowles, and Canter 1983; Hindelang, Hirschi, and Weis 1981). When subjects in our sample were asked the likelihood of committing shoplifting under the conditions of the scenario presented to them, only 37 percent reported "no chance." In addition, arrests for theft reported by the university police department totaled 1,267 for 1992; an overwhelming number of these crimes were committed by students. Given this information, one can conclude that college student populations contain frequent offenders in situations involving drunk driving and shoplifting; thus college samples are appealing for studies such as this.

7. We varied the level of risk of exposure (informal and formal) in both the shoplifting and the drunk driving scenarios. Preliminary analysis revealed no effect for these scenario-varied conditions; as a result, they were not estimated in the LISREL equations. Furthermore, we used gender as a control variable in preliminary analyses. After controlling for low self-control, the effect of gender was not significant in predicting intentions to either shoplift or drive drunk. In addition, gender had no significant effect on the other exogenous variables. Thus we did not examine gender in the LISREL models. These results confirm Gottfredson and Hirschi's (1990:144–49) predictions concerning gender, low self-control, and crime and they are consistent with previous research regarding similarity between males and females in offending behavior regarding shoplifting and drunk driving (Grasmick, Bursik, and Arneklev 1993a; Hindelang et al. 1981; Keane et al. 1993; Nagin and Paternoster 1993; Yu, Essex, and Williford 1992).

8. Persons interested in obtaining a copy of the low self-control scale can write to us or consult Grasmick et al. (1993b) or Nagin and Paternoster (1993).

9. In the models that follow, when we investigate intentions to drive drunk, we use a past behavior measure: the number of times in the past year the respondent has driven drunk. Similarly, when we examine intentions to shoplift, we use a past behavior measure of respondent's previous shoplifting. An anonymous reviewer observed correctly that a situational variable from the perspective of rational choice theory may be a dispositional variable from the perspective of self-control theory, such that one can use the drunk driving (past behavior) variables to predict shoplifting and can use the shoplifting (past behavior) variables to predict drunk driving. Insofar as dispositions rather than situations are at work, the results should be largely the same in either case. For the sake of brevity and because it is not the focus of the present analysis, we did not examine this issue here. We plan on assessing this issue, however, in future work with these data.

10. To examine the validity of this hypothesis, the LISREL computer program provides a chi-square statistic that estimates the goodness of fit of the model.

11. In using LISREL or other structural equation modeling programs, it is always good practice to provide the bivariate correlations from which the model was estimated.

12. This is the only effect for perceived sanctions and differs from the results for shoplifting. The sanction effects for drunk driving appear to be direct—not indirect, as they were for shoplifting—perhaps because of recent moral campaigns targeting drunk driving and because of the harshness of penalties that are reported by the media. This result is consistent with recent research concerning perceived sanctions on drunk driving (Grasmick, Bursik, et al. 1993a; Nagin and Paternoster 1993).

✉ References

Akers, R. 1991. "Self-Control as a General Theory of Crime." *Journal of Quantitative Criminology* 7:201–11.

Arneklev, B., H. Grasmick, C. Tittle, and R. Bursik, Jr. 1993. "Low Self-Control and Imprudent Behavior." *Journal of Quantitative Criminology* 9:225–47.

Bachman, R., R. Paternoster, and S. Ward. 1992. "The Rationality of Sexual Offending: Testing a Deterrence/ Rational Choice Conception of Sexual Assault." *Law and Society Review* 26:343–72.

Barlow, H. 1991. "Explaining Crimes and Analogous Acts, or The Unrestrained Will Grab at Pleasure Whenever They Can." *Journal of Criminal Law and Criminology* 82:229–42.

Benson, M. and E. Moore. 1992. "Are White Collar and Common Offenders the Same?" *Journal of Research in Crime and Delinquency* 29:251–72.

Birkbeck, C. and G. LaFree. 1993. "The Situational Analysis of Crime and Deviance." *Annual Review of Sociology* 19:113–37.

Bishop, D. 1984. "Legal and Extralegal Barriers to Delinquency." *Criminology* 22:403–19.

Block, H. and F. Flynn. 1956. *Delinquency: The Juvenile Offender in America Today.* New York: Random House.

Bohm, R. 1991. *Death Penalty in America: Current Research.* Cincinnati: Anderson.

Braithwaite, J. 1989. *Crime, Shame, and Reintegration.* New York: Cambridge University Press.

Brownfield, D. and A. Sorenson. 1993. "Self-Control and Juvenile Delinquency: Theoretical Issues and an Empirical Assessment of Selected Elements of a General Theory of Crime." *Deviant Behavior* 14:243–64.

Cohen, A. 1955. *Delinquent Boys.* New York: Free Press.

Cohen, R. 1992. *Drunk Driving.* Washington, DC: U.S. Department of Justice.

Cornish, D. and R. Clarke. 1986. *The Reasoning Criminal: Rational Choice Perspectives in Offending.* New York: Springer-Verlag.

———. 1987. "Understanding Crime Displacement: An Application of Rational Choice Theory." *Criminology* 25:933–47.

Elliot, D., S. Ageton, D. Huizinga, W. Knowles, and R. Canter. 1983. *The Prevalence and Incidence of Delinquent Behavior: 1976-1980.* Boulder: University of Colorado.

Empey, L. and M. Stafford. 1991. *American Delinquency: Its Meanings and Construction.* 3rd ed. Belmont, CA Wadsworth.

Fishbein, M. and I. Ajzen. 1975. *Belief, Attitudes, Intention, and Behavior.* Reading, MA: Addison-Wesley.

Gibbs, J. 1987. "The State of Criminological Theory." *Criminology* 25:821–24.

Gibbs, J. and D. Giever. 1995. "Self-Control and Its Manifestations among University Students: An Empirical Test of Gottfredson and Hirschi's General Theory." *Justice Quarterly* 12:231–55.

Gottfredson, M. and T. Hirschi. 1990. *A General Theory of Crime.* Stanford: Stanford University Press.

Grasmick, H. and R. Bursik, Jr. 1990. "Conscience, Significant Others, and Rational Choice: Extending the Deterrence Model." *Law and Society Review* 24:837–61.

Grasmick, H., R. Bursik, Jr., and B. Arneklev. 1993a. "Reduction in Drunk Driving as a Response to Increased Threats of Shame, Embarrassment, and Legal Sanctions." *Criminology* 31:41–67.

Grasmick, H., R. Bursik, Jr., and K. Kinsey. 1991. "Shame and Embarrassment as Deterrents to Noncompliance with the Law." *Environment and Behavior* 23:233–51.

Grasmick, H., C. Tittle, R. Bursik, Jr., and B. Arneklev. 1993b. "Testing the Core Implications of Gottfredson and Hirschi's General Theory of Crime." *Journal of Research in Crime and Delinquency* 30:5–29.

Green, D. 1988. "Measures of Illegal Behavior in Individual-Level Deterrence Research." *Journal of Research in Crime and Delinquency* 26:253–75.

Hindelang, M., T. Hirschi, and J. Weis. 1981. *Measuring Delinquency.* Beverly Hills: Sage.

Hirschi, T. and M. Gottfredson. 1993. "Commentary: Testing the General Theory of Crime." *Journal of Research in Crime and Delinquency* 30:47–54.

Jensen, G., M. Erickson, and J. Gibbs. 1978. "Perceived Risk of Punishment and self-Reported Delinquency." *Social Forces* 57:57–78.

Joreskog, K. and D. Sorbom. 1989. *LISREL 7 User's Guide.* Chicago: International Education Services.

Katz, J. 1988. *Seductions of Crime.* New York: Basic Books.

Keane, C, P. Maxim, and J. Teevan. 1993. "Drinking and Driving, Self-Control and Gender: Testing a General Theory of Crime." *Journal of Research in Crime and Delinquency* 30:30–46.

Kim, M. and J. Hunter. 1993. "Relationships among Attitudes, Behavioral Intentions, and Behavior: A Meta-Analysis of Past Research. Part 2." *Communications Research* 20:331–64.

Klepper, S. and D. Nagin. 1989a. "Tax Compliance and Perceptions of the Risks of Detection and Criminal Prosecution." *Law and Society Review* 23:209–40.

———. 1989b. "The Deterrent Effect of Perceived Certainty and Severity of Punishment Revisited." *Criminology* 27:721–46.

Kuhn, R. 1992. "1991 Student Drug Survey." College Park: University of Maryland, President's Committee on Alcohol and Drug Policy.

Lewis, H. 1971. *Shame and Guilt in Neurosis.* New York: International Universities Press.

Lewis, M. 1992. *Shame: The Exposed Self.* New York: Macmillan.

Matza, D. 1964. *Delinquency and Drift.* New York: Wiley.

McCarthy, B. 1995. "Not Just 'For the Thrill of It': An Instrumentalist Elaboration of Katz's Explanation of Sneaky Thrill Property Crimes." *Criminology* 33:519–38.

Nagin, D. and R. Paternoster. 1991. "On the Relationship of Past and Future Participation in Delinquency." *Criminology* 29:163–89.

———. 1993. "Enduring Individual Differences and Rational Choice Theories of Crime." *Law and Society Review* 27:467–96.

Paternoster, R. 1989. "Absolute and Restrictive Deterrence in a Panel of Youth: Explaining the Onset, Persistence/Desistence and Frequency of Delinquent Offending." *Social Problems* 36:289–309.

Paternoster, R., L. Saltzman, T. Chiricos, and G. Waldo. 1983. "Estimating Perceptual Stability and Deterrent Effects: The Role of Perceived Legal Punishment in the Inhibition of Criminal Involvement." *Journal of Criminal Law and Criminology* 74:270–97.

Piers, G. and M. Singer. 1953. *Shame and Guilt.* New York: Norton.

Piliavin, I., R. Gartner, C. Thornton, and R. Matsueda. 1986. "Crime Deterrence and Rational Choice." *American Sociological Review* 51:101–19.

Polakowski, M. 1994. "Linking Self-Control and Social Control with Deviance: Illuminating the Structure Underlying a General Theory of Crime and Its Relation to Deviant Activity." *Journal of Quantitative Criminology* 10:41–78.

Polk, K. 1991. "Review of a General Theory of Crime." *Crime and Delinquency* 37:575–81.

Reckless, W. 1961. "A New Theory of Delinquency and Crime." *Federal Probation* 25:42–46.

Reiss, A. 1986. "Co-Offender Influences on Criminal Careers." Pp. 121–60 in *Criminal Careers and "Career Criminals,"* edited by A. Blumstein, J. Cohen, J. Roth, and C. Visher. Washington, DC: National Academy Press.

Rossi, P. and A. Anderson. 1982. "The Factorial Survey Approach: An Introduction." Pp. 15–67 in *Measuring Social Judgments,* edited by P. Rossi and S. Nock. Beverly Hill, CA: Sage.

Scheff, T. 1988. "Shame and Conformity: The Deference-Emotion System." *American Sociological Review* 53:395–406.

Smith, D. and P. Gartin. 1989. "Specifying Specific Deterrence: The Influence *of* Arrest on Future Criminal Activity." *American Sociological Review* 54:94–105.

Smith, D. and B. Patterson. 1985. "Latent-Variable Models in Criminological Research: Applications and a Generalization of Joreskog's LISREL Model." *Journal of Quantitative Criminology* 1:127–58.

Stafford, M. and M. Warr. 1993. "A Reconceptualization of General and Specific Deterrence." *Journal of Research in Crime and Delinquency* 30:123–35.

Tittle, C. 1977. "Sanction Fear and the Maintenance of the Social Order." *Social Forces* 55:579–96.

———. 1980. *Sanctions and Social Deviance.* New York: Praeger.

Toby, J. 1957. "Social Disorganization and Stake in Conformity: Complimentary Factors in the Predatory Behavior of Hoodlums." *Journal of Criminal Law, Criminology and Police Science* 48:12–17.

Wellford, C. 1989. "Towards an Integrated Theory of Criminal Behavior." Pp. 119–28 in *Theoretical Integration in the Study of Deviance and Crime: Problems and Prospects,* edited by S. Messner, M. Krohn, and A. Liska. Albany: SUNY Press.

Williams, K. and R. Hawkins. 1986. "Perceptual Research on General Deterrence: A Critical Review." *Law and Society Review* 20:545–72.

Wood, P. B. Pfefferbaum, and B. Arneklev. 1993. "Risk-Taking and Self-Control: Social Psychological Correlates of Delinquency." *Journal of Crime and Justice* 16:111–30.

Yu, J., D. Essex, and W. Williford. 1992. "DWI/DWAI Offenders and Recidivism by Gender in the Eighties: A Changing Trend?" *International Journal of the Addictions* 27:637–47.

⬛ Appendix: Scenarios Used in the Study

Scenario 1: Drunk Driving/Low Risk

It's about two o'clock in the morning and Mark has spent most of Thursday night drinking with his friends at the "Vous." He decides to leave the Vous and go home to his off-campus apartment, which is about 10 miles away. Mark has had a great deal to drink. He feels drunk and wonders if he may be over the legal limit and perhaps he should not drive himself home. He knows people who have driven home drunk before, and none of them have ever gotten caught. Mark also knows that no one else will find out that he drove home drunk because only he knows how much he drank. In addition, Mark realizes that if he gets a ride home, he will have to take a bus back to the Vous in the morning to pick up his car. Mark decides to drive himself home.

Scenario 2: Drunk Driving/High Informal Risk

It's about two o'clock in the morning and Mark has spent most of Thursday night drinking with his friends at the "Vous." He decides to leave the Vous and go home to his off-campus apartment, which is about 10 miles away. Mark has had a great deal to drink. He feels drunk and wonders if he may be over the legal limit and perhaps he should not drive himself home. Mark knows that the chance of getting caught by the police is very low because he knows people who have driven home drunk, and none have ever gotten caught. But he also knows that some of his friends and members of his family may find out he drove home drunk since he saw many people he knew at the Vous. Mark realizes that if he gets a ride home, he will have to take a bus back to the Vous in the

morning to pick up his car. Mark decides to drive himself home.

Scenario 3: Drunk Drivings/High Formal Risk

It's about two o'clock in the morning and Mark has spent most of Thursday night drinking with his friends at the "Vous." He decides to leave the Vous and go home to his off-campus apartment, which is about 10 miles away. Mark has had a great deal to drink. He feels drunk and wonders if he may be over the legal limit and perhaps he should not drive himself home. Mark knows that the local police have recently implemented a "crackdown" on drunk driving, and in fact he knows two people who were arrested for drunk driving after leaving the Vous the week before. But Mark realizes that if he gets a ride home, he will have to take a bus back to the Vous in the morning to pick up his car. Mark decides to drive himself home.

Scenario 4: Shoplifting/Low Risk

It's Sunday evening, and David has gone to a small, privately owned convenience store to buy batteries for his alarm clock. He needs the batteries because he has to wake up very early the next day to take an exam for his 7:30 class. David will be studying most of the night, and he knows that if he doesn't have batteries for his alarm clock he will probably oversleep. The store is about to close when David realizes he does not have enough money to buy the batteries. The batteries are small enough to hide on himself without anyone noticing. He has enough money to buy a soda, so that no one will be suspicious of his not buying anything. David notices that he is out of sight of the only clerk, who is reading the newspaper behind the counter. He knows several people who have taken small items from the store and have not gotten caught, and in fact there seem to be no video cameras or other types of security devices in the store. Since David is alone, he knows that his friends have little chance of finding out if he takes the batteries. David decides to take the batteries.

Scenario 5: Shoplifting/High Informal Risk

It's Sunday evening. David and his best friend, Brian, have gone to a small, privately owned convenience store to buy batteries for David's alarm clock. David needs the batteries because he has to wake up very early the next day to take an exam for his 7:30 class. David will be studying most of the night, and he knows that if he doesn't have batteries for his alarm clock he will probably oversleep. The store is about to close when David realizes he does not have enough money to buy the batteries. He asks Brian if he can borrow some money, but Brian says he doesn't have any. The batteries are small enough to hide on himself without anyone noticing. David notices that he is out of sight of the only clerk, who is reading the newspaper behind the counter. He has enough money to buy a soda, so that the clerk will not be suspicious of his not buying anything. In addition, he

knows several people who have taken small items from the store and have not gotten caught, and in fact there seem to be no video cameras or other types of security devices in the store. David decides to take the batteries.

Scenario 6: Shoplifting/High Formal Risk

It's Sunday evening. David and his best friend, Brian, have gone to a small, privately owned convenience store to buy batteries for David's alarm clock. David needs the batteries because he has to wake up very early the next day to take an exam for his 7:30 class. David will be studying most of the night, and he knows that if he doesn't have batteries for his alarm clock he will probably oversleep. The store is about to close when David realizes he does not have enough money to buy the batteries. He asks Brian if he can borrow some money, but Brian says he doesn't have any. The batteries are small enough to hide on himself without anyone noticing. David knows that this store tends to prosecute shoplifters if they are caught, and in fact he read in the paper that someone was recently convicted for shoplifting a small item from this store. David takes the batteries and leaves immediately.

REVIEW QUESTIONS

1. What are some of the elements of the low self-control personality?

2. What do Piquero and Tibbetts say are some of the key concepts of the rational choice framework that go beyond traditional deterrence concepts? Which of these concepts were most supported by their own findings?

3. What finding do Piquero and Tibbetts claim is the "most interesting"?

READING

This empirical study on "hot spots" by Sherman and his colleagues is generally considered one of the contemporary classics in the literature, perhaps due to it being one of the first tests of routine activities theory to use spatial data, as measured by crime distribution by location of the crime, as opposed to individual or family victimization data.[33] Using data for a single year in Minneapolis, Minnesota, this study found that some locations—both businesses and residences—are responsible for many hundreds of calls for police each year. On the other hand, the vast majority of locations have no calls to police at all. Routine activities theory helps provide a framework for explaining this non-random distribution of criminal activity. While reading this selection, readers should consider the place that they grew up in or live in now and see if the findings of this study fit with the "hot spots" there.

Hot Spots of Predatory Crime

Routine Activities and the Criminology of Place

Lawrence W. Sherman, Patrick R. Gartin, and Michael E. Buerger

Is crime distributed randomly in space? There is much evidence that it is not. Yet there are many who suggest that it is. In a leading treatise on police innovations, for example, Skolnick and Bayley (1986: 1) observe that "we feel trapped in an environment that is like a madhouse of unpredictable violence and Quixotic threat." People victimized by crime near their homes often feel that there are no safe places and that danger lurks everywhere (Silberman, 1978: 15–16). Even many police we know, who acknowledge that some areas are more dangerous than others, often assume a random distribution of crime within areas.

For them, the practical question is not whether crime is concentrated in space, but how much.

Such analysis of variation across space is one of the basic tools of science. Many clues to the environmental causes of cancer, for example, have been revealed by the discovery of carcinogenic "hot spots": locations with extremely high rates of cancer mortality (Mason et al., 1985). Similarly, many factors associated with automobile fatalities (such as low population density and distance from emergency medicine) have been highlighted by the discovery of rural western counties with death rates 350 times higher than those in such eastern

SOURCE: Adapted from Lawrence Sherman, Patrick Gartin, and Michael Buerger, "Hot Spots of Predatory Crime: Routine Activities and the Criminology of Place," *Criminology* 27 (1989):27–56. Reprinted by permission of the American Society of Criminology.

[33]Cohen and Felson, "Social Change".

states as New Jersey (Baker et al., 1987). The methodological history of such analyses can be traced to the moral statistics tradition (Guerry, 1831; Quetelet, 1842) and the sociology of crime and deviance, which pioneered the analysis of variation in behavior across space. Durkheim's *Suicide* (1951) and Shaw et al.'s *Delinquency Areas* (1929) are two classic examples. More recently, sociologists have tested income inequality and other structural theories of crime with variation in crime rates across collectivities, at the levels of nation-states (e.g., Krahn et al., 1986; Messner, 1980), regions (e.g., Gastil, 1971; Loftin and Hill, 1974; Messner, 1983), and cities or metropolitan areas (e.g., Blau and Blau, 1982; Messner, 1982; Sampson, 1986).

⬙ Collectivities, Communities, and Places

A common problem of spatial analysis is pinpointing the locations of events. The ecological tradition in criminology has been confined to relatively large aggregations of people and space, which may mask important variation and causal properties within those aggregations. This may be especially important for within-city spatial variation.

Unlike the boundaries of nation-states and cities, the boundaries of within-city crime reporting districts do not correspond to theoretically or empirically defined collectivities, such as local communities or ethnic areas (Reiss, 1986: 26). Nor, as Reiss (1986) also points out, do official statistics on communities include many of the variables on collectivity characteristics needed to test theories of crime. The inability of community data to measure those characteristics creates major problems for community crime research (just as it does for this analysis) and leaves it vulnerable to what one sympathetic observer describes as a claim that there is little more here than "a theoretical exercise in the mapping of criminal phenomena" (Bursik, 1986: 36).

Even if collectivity characteristics can be measured at the level of community areas, those characteristics may have very different meanings and causal properties at the level of places. An independent variable like per capita alcohol consumption per hour, for example, means something very different at the street-corner level than it does at a 2-mile-square neighborhood level. It is clearly subject to a much wider range at the place level than it is at greater aggregations, with all of the effects of higher levels of consumption being concentrated on behavior in that microsocial space. Focusing on variation across smaller spaces opens up a new level of analysis that can absorb many variables that have previously been shunned as too obvious or not sufficiently sociological: the visibility of cash registers from the street, the availability of public restrooms, the readiness of landlords to evict problem tenants.

The increased range of such independent variables at a micro-place level also means that variation in crime within communities is probably greater than variations across communities (Robinson, 1950). The very meaning of the concept of a bad neighborhood is an open empirical question: whether the risk of crime is randomly or evenly distributed throughout the neighborhood, or so concentrated in some parts of the neighborhood that other parts are relatively safe.

Some recent policy research hints at the latter answer. Taylor and Gottfredson (1986: 410) conclude that there is evidence linking spatial variation in crime to the physical and social environment at the subneighborhood level of street blocks and multiple dwellings (e.g., Jacobs, 1961; Newman, 1972; Newman and Franck, 1980, 1982; but see Merry, 1981a, 1981b). Some 40 years ago, Henry McKay himself made the unpublished discovery that even within high-crime Chicago neighborhoods, entire blocks were free of offenders (Albert J. Reiss, Jr., personal communication).

Other findings suggest microlevel variation within blocks for the predatory stranger crimes of burglary, robbery, and auto theft. Salt Lake City houses with well-tended hedges were found to be less likely than other houses in the same neighborhood to be burglarized (Brown, 1983). Tallahassee apartments near the complex entrance and not facing another building were more likely to be burglarized than apartments inside the development facing other buildings (Molumby, 1976). And apartments in buildings with doormen were also less likely to be burglarized than other apartments (Repetto, 1974; Waller and Okihiro, 1978).

Microspatial variations in robbery rates also suggest nonrandom distributions. Convenience stores near vacant land or away from other places of commerce were more likely to be robbed than those in dense commercial areas (Duffala, 1976). Over a 5-year period in Gainesville, Florida, 96% of all 47 convenience stores were robbed, compared with 36% of the 67 fast-food establishments, 21% of the 71 gas stations, and 16% of the 44 liquor stores (Clifton, 1987). Conversely, over a 10-year period in Texas, gas station workers were murdered at a rate of 14.2 per 100,000 workers per year, compared with a rate of 11.9 for convenience-type store workers and 5.1 per 100,000 per year for all retail workers (Davis, 1987). Tallahassee convenience stores with the cashier visibly stationed in the middle of the store were three times more likely to have a low robbery rate as stores with the cashier set less visibly off to the side (Jeffery et al., 1987). Convenience stores with two clerks on duty may be less likely to be robbed than stores with only one (Clifton, 1987; Jeffery et al., 1987; but see Chambers, 1988).

Similar microspatial findings are reported in England. English parking lots with attendants had lower rates of auto theft than unattended parking lots (Clarke, 1983: 239). Pedestrian tunnels in downtown Birmingham, England, accounted for a negligible portion of all public space, but they produced 13% of a sample of 552 criminal attacks on persons (Poyner, 1983: 85).

Traditional collectivity theories may be appropriate for explaining community-level variation, but they seem inappropriate for small, publicly visible places with highly transient populations. Nor is it necessary to give up the explanatory task to the competing perspectives of rational choice (Cornish and Clarke, 1986) and environmental design (Jeffery, 1971; Newman, 1972). A leading recent sociological theory can address these findings, but only with a clearer definition of its unit of analysis. The routine activities approach of Cohen and Felson (1979) can be used to develop a criminology of *places*, rather than its previous restrictions to a criminology of *collectivities* or of the life-styles of victimized *individuals* (Hindelang et al., 1978; Messner and Tardiff, 1985; Miethe et al., 1987) and *households* (Massey et al., 1987).

▧ Routine Activities and Place

In their original statement of the routine activities approach, Cohen and Felson (1979: 589) attempt to account for "direct contact predatory violations," or illegal acts in which "someone definitely and intentionally takes or damages the person or property of another" (Glaser, 1971: 4). They propose that the rate at which such events occur in collectivities is affected by "the convergence in space and time of the three minimal elements of direct-contact predatory violations: (1) motivated offenders, (2) suitable targets, and (3) the absence of capable guardians against a violation" (Cohen and Felson, 1979: 589). The theory thus integrates several different vast bodies of literature: the factors affecting the supply of "motivated" offenders (e.g., Wilson and Herrnstein, 1985), the opportunity perspective on the supply of stealable property (e.g., Gould, 1969), the life-style perspective on the supply of persons vulnerable to victimization (Hindelang et al., 1978; Miethe et al., 1987), the policy research on physical "target-hardening" (e.g., Jeffery, 1971), and the literature on the deterrent threat of official and unofficial policing (e.g., Sherman, 1986) implied in the concept of guardianship.

The most important contribution of routine activities theory is the argument that crime rates are affected not only by the absolute size of the supply of offenders, targets, or guardianship, but also by the factors affecting the frequency of their convergence in space and time. The theory claims roots in social and physical ecology, and it explicitly cites Hawley's (1950: 289) space-time concepts of (1) rhythm, the regular periodicity with which events occur; (2) tempo, the number of events per unit of time, "such as the number of violations per day on a given street" (Cohen and Felson, 1979: 590); and (3) timing, the coordination of different interdependent activities, "such as the coordination of an offender's rhythms with those of a victim" (Cohen and Felson, 1979: 590)—presumably, again, at a specific place.

The major limitation of the evidence for the theory, however, is the lack of testing with ecological data on actual places where offenders, targets, and weak guardians converge. As Miethe et al. (1987: 185) point out, most tests of routine activities theory lack independent measures of the life-styles in question and substitute presumed demographic correlates for them. Although Cohen and Felson (1979: 595) do provide data on the personal risks of victimization in different places, they do not link those individual risks to variations in the amount of time individuals spend in different types of places. Rather, they explain national crime trends with national trends in presumed places of routine activities. And although Miethe et al. (1987) do measure the kinds of places in which victims and nonvictims spend their time, they take the individuals as the unit

of analysis rather than the places. Given Cohen and Felson's emphasis on the spatial and temporal ecology of crime, the most appropriate unit of analysis for the routine activities approach would seem to be places.

In this article we can go no further than prior research in testing the causal properties explaining variation in crime across places, nor do we attempt tests. Rather, our purpose is to provide a more complete description of the variation in crime across places than has been previously available in order to suggest future directions for developing a sociological criminology of place.

⊠ The Sociology of Place

The concept of place lies at the nexus of the physical and social environments, providing a unit of analysis rich in both symbolic content and social organization. We do not mean *place* in the sense of social position in a group (Goffman, 1971), nor in the broader geographical sense of a community (Cobb, 1975). Our more precise geographic concept of place can be defined as *a fixed physical environment that can be seen completely and simultaneously, at least on its surface, by one's naked eyes.*

Although the perceptual boundaries of geographic place are often ambiguous and subject to dispute, centuries of human efforts have been devoted to lessening that ambiguity through such tools as maps, survey, and street names and addresses. The variability in the social institutions of place is suggested by the absence of consecutive street addresses in Japan (Bayley, 1976: 15–16), where houses are given numbers in the order in which they are built. The geographic concept of place embraces an extraordinarily heterogeneous range of environments, from 1-room cottages to 3,000-unit hotels, from street intersections to waterfalls, from farmyards to nightclubs and banks.

The sociological concept of place can be defined as the *social organization of behavior at a geographic place.* Although a place can be coterminous with a collectivity, such as a law firm, nuclear family, or a church in a single building, the human population at places is usually too transient to constitute a collectivity. That transiency does not, however, prevent places from acquiring such variable social organizational properties as customary rules of interaction, financial wealth, forbidden and encouraged activities, prestige rank, moral value, patterns of recruitment and expulsion, legal rights and duties, and even language spoken. Compare, for example, an airport lobby and a homeless shelter, a roadside motel and a university boathouse, a hospital and a liquor store.

Places, like persons, can be seen to have routine activities subject to both formal and informal regulation. Their formal regulation has become a major arena of social and legal conflict, especially in modern America. Recent examples include historic preservation efforts to prevent demolition of buildings, community activist efforts to prevent renewal of a bar's liquor license, zoning battles over proposals to increase the human density of places through taller buildings, the "not in my back yard" (NIMBY) opposition to halfway houses for convicts and treatment centers for infants with acquired immunodeficiency syndrome (AIDS), and police attempts to stop visible street corner drug dealing. The participants in these conflicts evidence little doubt that the routine activities of places produce important aesthetic and social consequences for the quality of life in adjoining places, including crime.

Despite its conceptual richness and precedent, place has received little systematic attention. Most lacking is quantitative analysis of the causal relationships among various social and physical characteristics of place, such as the relationship of alcohol sales to vandalism controlling for the intensity of lighting, density of place population, price of alcohol, and the socioeconomic status of the patrons. The systematic study of place as a unit of analysis has been left to environmental psychologists,

architects, urban planners, and "space doctor" consultants who intervene in problem plagued public places (Hiss, 1987a, 1987b). Yet few attempts have been made to describe systematically, let alone explain, variation across different kinds of places in such behavior as crime.

There is little point in examining variation in crime by place, of course, if such variation is merely random. Criminal opportunities, it has been argued, are in some sense ubiquitous (Reiss, 1986: 6). Although previous research shows concentrations of crime in space, such concentrations could occur merely by chance— just like throwing darts at a target while blindfolded (Kinley Larntz, personal communication, 1987). The implied premise of routine activities theory is that such concentrations are not random, which some might think is obvious. But to our knowledge, that premise has never been examined across an entire city with place as the unit of analysis.

This article examines that premise. Using street addresses and intersections as an operational definition of urban places, we assess police call data as a measure of place crime in Minneapolis. We describe the distribution of crimes by place and test for the randomness of that distribution. We then consider the implications of the results for further development of a routine activities criminology of place.

✉ Police Call Data: Strengths and Weaknesses

A new source of data on crime has recently become available. With the growth of centralized police dispatching systems, three-digit emergency telephone numbers (911), and the attendant increases in calls to police, administrative data on calls to police provide a reliable indicator of time and place variations in crime (Pierce et al., 1984). Call data are relatively so precise and cast so wide a net that they some day may provide a third major indicator of crime trends—supplementing official crime reports and victimization surveys. One can even imagine a new series of federal crime statistics derived from local call data, the "Uniform Telephone Reports (UTR)." But these data have substantial limitations as well as strengths.

Limitations

Traditional measures of crime are subject primarily to underreporting, but police call data are subject to both underreporting and overreporting. Calls about crimes may either be made in error or as intentional lies, much like false fire alarms. In other cases, a single criminal event may generate more than one call. Or, if the call record is updated by additional information to the dispatcher from the police at the scene, the update may be recorded as a separate call (referred to as a mirror) rather than as replacing the earlier call.

Even a great advantage of call data— precision as to the time and place of the crime—has major limitations. There may often be a lag of many hours between the time of the crime and the computer-recorded time of the call (Kansas City Police Department, 1977; Spelman and Brown, 1981). Moreover, the Minneapolis Computer-aided Dispatch (CAD) system has data entry fields for only two locations: the location from which the call is made and the location to which a police car is to be dispatched. There is no provision for the third and most important address—the location of the event's occurrence. Moreover, a fourth location—the residence address of the caller—may sometimes be entered into one of the two available fields.

Places vary in the extent to which they suffer underreporting or overreporting. Hospitals, police stations, and public locations (e.g., gas stations and convenience stores with phone booths), to which crime victims may go in the aftermath of a crime, may suffer overreporting. This problem appears to have been compounded by rising phone bills since telephone deregulation, which police in Minneapolis and elsewhere say has caused more poor people to give up home telephone service. The locations where those crimes actually occurred suffer underreporting. This pattern is augmented by intentional lying to police dispatchers about the location of crimes by representatives of places attempting to avoid losing a license to do business, such as bars or teenage dance halls (see, e.g., Sanchez and Horwitz, 1987).

While certain commercial locations with phones may suffer overreporting of crimes, a different problem often causes underreporting from those locations. Slight variations in the descriptions of the names of commercial locations, added to the street address, are often entered into the computer. Misspellings, omission of some words (e.g., "Moby Dick's Bar" vs. "Moby's"), using the street address only, and other variations produce several distinct data files for calls at the same address—thus undercounting the total crime at that location.

Strengths

Despite these substantial limitations, police emergency telephone records do provide the widest ongoing data collection net for criminal events in the city. Unlike official crime reports, call data are virtually not screened. Police telephone operators are monitored by a continuous audio tape of all transactions, and they cannot safely fail to enter a clearly stated crime report from a clearly stated location. Comparison of the 1986 call data presented below with 1986 Uniform Crime Reports (Federal Bureau of Investigation, 1987: 85) data on Minneapolis, for example, shows that official crime reports were filed for only 66% of calls about robberies (although the Dallas police in 1986 filed reports on virtually 100% of robbery calls, which that city's officers probably reclassify from the scene more conscientiously).

For whatever error they entail, call data in some cities arguably capture many events that neither official crime reports nor victimization surveys would capture. In some cases, for example, a bystander may call about a crime in progress, yet the victim does not want to file a report. Or by the time the police arrive, the victim may be gone or refuse to provide any evidence, so no report is filed. When later asked about victimization events by a Census Bureau surveyor, the victim does not mention the crime. Thus, the criminal event is lost to all other records systems—but not from police call data.

The reliability of police call data is suggested by at least one report of a very high geographic correlation between call data and reported crimes. Data from a 1979–80 multistage stratified cluster sample of 63 street blocks in 12 Baltimore City neighborhoods found several types of reported crimes to be highly correlated with related calls for service. Calls about crimes of violence to persons, for example, were correlated with official aggravated assault reports at about $r = .80$ (Taylor et al., 1981).

As a measure of the concentration of predatory crime at specific places, the errors of overreporting appear to be counterbalanced by the errors in underreporting. Because there are no offense reports listing location of occurrence for the vast majority of all calls, we cannot even estimate how much overreporting there is by location, nor can we estimate the amount of underreporting without inspecting almost 70,000 listings of addresses that generated calls. But knowing that the two sources of error work in opposite directions gives us greater confidence in findings of high concentration than if the errors all increased overreporting.

We can also generally assume that call data about places are immune to being swamped by one-man crime waves, those occasionally visible bursts of activity by street criminals often concentrated in a small area (e.g., Iverem, 1988; Sanchez, 1988). The anecdotal evidence about such crime sprees is that they are usually spread out over several addresses, even when the offender is on foot.

As Biderman and Reiss (1967) suggest, there is no "true" count of crime events, only

different socially organized ways of counting them, each with different flaws and biases. Calls to the police provide the most extensive and faithful account of what the public tells the police about crime, with the specific errors and biases that that entails.

▧ Data Collection

Unfortunately, few if any police departments can provide researchers, or even police chiefs, with a year-long call data base ready to analyze. Computer-aided dispatch systems are designed for operational purposes, so they do not have large storage capacities. The Minneapolis system, for example, can store only about 7,000 call records on line, so the calls must be removed from the mainframe computer about once a week and stored on tape. To construct a single data file for police calls covering 1 year or longer, researchers must generally provide their own computer into which the police backup tapes are read.

Using that procedure in Minneapolis, selected data elements from each complete call record were read from all the available tapes covering the period from December 15, 1985, to December 15, 1986. Missing data were discovered for 28 days, distributed throughout the year in four blocks of about 7 days each. A total of 323,979 call records were copied into a microcomputer, after fire, ambulance, and administrative record calls from police (e.g., out to lunch) were deleted. The findings presented below are derived from those data as well as from a less precise estimate of the number of street addresses and intersections in the city.

▧ Estimating the Denominator

Although some data already exist on the distribution of crime by different types of places (Felson, 1987: 922-925; Hindelang, 1976: 300), they have been uncontrolled by any estimate of the relevant denominator: the number of such places at which crimes could possibly occur. We are unable to classify types of places, but we are at least able to estimate the total number of addresses to which crimes could possibly have been attributed.

All police calls in Minneapolis are dispatched to street addresses, intersections, or several hundred "special locations," such as parks, hospitals, and City Hall. Determining the total number of such places in a city requires great caution. The original estimate of 172,000 addresses and intersections supplied by one official source was used for a year of preliminary analysis (Sherman, 1987) until further checking revealed it to be incorrectly based on dwelling units (including individual apartments).

Further checking also revealed different estimates from different sources. The Manager of Inspections Administration for the City of Minneapolis (interview with Erin Larsen, 1987) estimates that there are 107,000 buildings in the city, plus or minus 10,000. The tax assessor's office (interview with Richard Hanson, 1987) reports that there are about 115,000 parcels of land, and about 111,000 parcels with a building on them—some of which parcels are condominium units within multiple-unit buildings. Some of the vacant lots are addresses, at which crimes could be reported, but others are slivers of land between a sidewalk and a curb; the counts do not distinguish. The Administrative Engineering Service (interview with Brian Lokkesmoe, 1987) reports that there are about 100,000 locations for garbage pickup.

The number of intersections is more consistently estimated. The Traffic Engineering Office (interview with staff, 1987) reports that a 1973 study found 6,320 intersections. A separate department, Administrative Engineering Services, reports 24,000 corners, or about 6,000 intersections. Using the estimate of 6,000 intersections, and a rough averaging of the estimated number of street addresses at

109,000, we estimate the total number of places in the city to be 115,000.

In Tables 2, 3, and 4, the data presented below have been corrected by deleting from the numerators, but not the denominators, two high-volume places at which the crimes were clearly just being reported, not occurring: City Hall and the Hennepin County Medical Center.

Hot Spots of Crime

The analysis reveals substantial concentrations of all police calls, and especially calls for predatory crime, in relatively few "hot spots." Just over half (50.4%) of all calls to the police for which cars were dispatched went to a mere 3.3% of all addresses and intersections (Table 1). A majority (60%) of all addresses generated at least one call over the course of the year, but about half of those addresses produced one call and no more. The top 5% of all locations generated an average of 24 calls each, or 1 every 2 weeks.

The number of calls per location ranged as high as 810 at a large discount store near a poor neighborhood, followed by 686 calls at a large department store, 607 calls at a corner with a 24-hours-a-day convenience store and a bar, and 479 calls at a public housing apartment building (data not displayed). To test the premise that these concentrations are not merely random clusters, we calculated a simple Poisson model of the expected frequency of locations with each level of call volume. The simple Poisson model assumes that (1) the probability of a dispatched call to police is the same for all places and (2) the probability of a call does not depend on the number of previous calls (Nelson, 1980). For a sample of 115,000 places, the frequencies of repeat calls expected by chance are significantly lower than the observed frequencies, with a maximum of 13 calls expected (and 810 observed) at any one location (Table 1).

Some of the chronic locations for calls to police are probably quite safe for the public.

Table 1	Distribution of All Dispatched Calls for Police Service in Minneapolis, by Frequency at Each Address and Intersection, December 15, 1985–December 15, 1986

No. of Calls	Observed No. of Places	Expected No. of Places	Cumulative % of Places	Cumulative % of Calls
0	45,561	6,854	100%	—
1	35.858	19,328	60.4	100.0
2	11,318	27,253	29.2	88.9
3	5,683	25,618	19.4	81.9
4	3,508	18,060	14.4	76.7
5	2,299	10,186	11.4	72.4
6	1,678	4,787	9.4	68.8
7	1,250	1,929	7.9	65.7
8	963	680	6.8	63.0
9	814	213	6.0	60.6
10	652	60	5.3	58.4
11	506	15	4.7	56.3
12	415	4	4.3	54.6
13	357	1	3.9	53.1
14	297	0	3.6	51.7
15≥	3,841	0	3.3	50.4

NOTE: Mean = 2.82; X^2 = 301,376; df = 14; p < .0001.

Twenty-four retail stores, for example, produced a total of 2,444 calls to arrest a shoplifter apprehended by store personnel, or 68% of all calls for that reason in the entire city. Other high-volume call addresses primarily generated calls about such noncriminal matters as traffic accidents, car lockouts, parking disputes, or noise complaints.

The hot spots of predatory crime are best described in Table 2, which shows the distributions of three types of predatory crime generally committed in public places: criminal sexual conduct (rape, molesting, and exposing,

designated hereafter as rape/CSC), robbery, and auto theft (also aggregates of several distinct call codes, including attempts and alarms).[1] Taken separately, each of the predatory crimes shows even greater geographic concentration than all calls for service: all 4,166 robbery calls were located at only 2.2% (against a possible 3.6%) of all places, all 3,908 auto thefts at 2.7% (against a possible 3.4%), and all 1,729 rape/CSCs at just 1.2% (against a possible 1.5%) of the places in the city. Thus, robbery had the greatest magnitude of concentration, with a 39% relative reduction in actual

Table 2	Distribution of Dispatched Police Calls for Rape, Robbery, Auto Theft, by Frequency at Each Address and Intersection in Minneapolis, December 15, 1985–December 15, 1986

No. of Crimes	Robbery		Rape/Criminal Sexual Conduct		Auto Theft		Combined	
	Observed	Expected	Observed	Expected	Observed	Expected	Observed	Expected
0	112,446	110,883	113,612	113,275	111,923	111,136	108,786	105,605
1	1,825	4,048	1,161	1,714	2,574	3,795	4,629	8,993
2	434	69	167	11	345	69	942	380
3	122	0	34	0	85	0	269	12
4	58	0	11	0	36	a	142	0
5	34	0	5	0	15	0	65	0
6	30	0	3	0	6	0	31	0
7	14	0	3	0	2	0	38	0
8	10	0	3	0	6	0	25	0
9	6	0	0	0	0	D	17	0
10	4	0	0	0	3	0	16	0
11	2	0	0	0	1	0	4	0
12	2	0	0	0	0	0	6	0
13	2	0	0	0	1	0	6	0
14	1	0	0	0	0	0	3	0
15	1	0	0	0	0	0	6	0
16	1	0	0	0	0	0	2	0
17	0	0	0	0	0	0	0	0
18	0	0	0	0	1	0	2	0
19	0	0	0	0	0	0	0	0
20 ≥	6	0	0	0	0	0	9	0
Max.	28	2	8	2	18	2	33	3
$\bar{x} =$	.036		.015		.034		.085	
$x^2 =$	3,174		2,391		1,502		8,549	
$df =$	27		7		17		32	
$p <$	.0001		.0001		.0001		.0001	

N of places = 115,000.

NOTE: Table does not include calls dispatched to Hennepin County Medical Center or City Hall.

locations from the possible number; auto theft had a 21% relative reduction, and rape/ CSC had a 20% relative reduction. There were, for example, 113 places with five or more robberies in 1 year, 35 places with five or more auto thefts, and 14 places with five or more rape/CSC calls. Conversely, 95% of the places in the city were free of any of these crimes in a 1-year period.

When the three general offense types are combined in the same places, there are 230 hot spots with 5 or more of any of the three offenses, and 54 locations with at least 10. The 6,212 actual locations with any of the three offenses constitute a relative reduction of 37% from the 9,803 possible numbers of places if there were no repeat offenses. There is also a high conditional probability of an additional offense once one has occurred, as illustrated in Figure 1. Under the simple Poisson model, each place in the city has only an 8% chance of suffering one of these crimes. But the risk of a second offense given a first shows the observed probability increasing sharply to 26%. Once a

place has had three of these offenses, the risk of recurrence within the year exceeds 50% and stays above that level at each higher level of call frequency. The exact day and time of recurrence cannot be predicted, but the fact of recurrence, given three offenses, becomes highly predictable. This predictability exceeds, for example, the 33% to 59% annual likelihood of an arrested offender being rearrested for an index offense while free (Blumstein et al., 1986: 58).

Table 2 also tests the possibility that the concentrations of predatory public crimes have occurred by chance. The simple Poisson models again show significantly lower expected frequencies of repeat offense concentrations than are actually observed. Where the Poisson distribution predicts a maximum of 2 robberies in any one place, for example, the observed maximum is actually 28.

The hot-spot pattern of concentration is not limited to crime in public places. Similar patterns were found for three types of crime commonly committed indoors (Table 3).

| Figure 1 | Conditional Probability of k + 1 Calls Given k Calls for Rape/Criminal Sexual Conduct, Robbery, and Auto Theft in Minneapolis (December 15, 1985–December 15, 1986) |

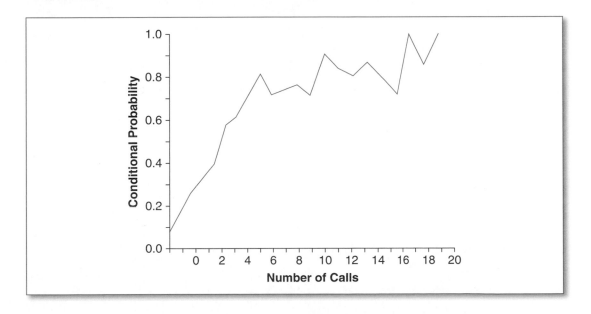

| Table 3 | Distribution of Dispatched Police Calls for Burglary, Domestic Disturbance, and Assault, by Frequency at Each Address and Intersection in Minneapolis, December 15, 1985–December 15, 1986 |

No. of Crimes	Burglary		Domestic Disturbance		Assault	
	Observed	Expected	Observed	Expected	Observed	Expected
0	97.449	94,198	105,158	92,564	106,954	98,946
1	9,032	13.745	5,519	20,090	5,245	14,870
2	1,664	1,003	1,790	2,174	1,394	1,116
3	465	44	866	161	497	58
4	170	10	468	11	293	11
5	92	0	293	0	170	0
6	55	0	208	0	86	0
7	26	0	156	0	73	0
8	13	0	97	0	43	0
9	7	0	86	0	50	0
10	6	0	56	0	29	0
11	6	0	41	0	19	0
12	1	0	48	0	17	0
13	4	0	33	0	8	0
14	1	0	18	0	8	0
15 $\geq$	7	0	161	0	112	0
Max	20	4	62	4	136	4
$N =$	109,00	115,000	115,000			
$x^2 =$	8,442	34,423	17,499			
$df =$	14	14	14			
$p < .0001$	.0001	.0001				
$\bar{x} =$	.145	.217	.150			

NOTE: Table does not include calls dispatched to Hennepin County Medical Center or City Hall.

Although their greater frequency produces somewhat less absolute concentration, domestic problems and assaults are actually more concentrated relative to the possible number of places at which those events could have occurred. All 24,928 domestic disturbances were recorded at 9% of all places (although each address can include many apartments), a 59% relative reduction from the possible locations without repeats. All 17,225 assaults were concentrated in 7% of all places, a 55% relative reduction. All 15,901 burglary calls were concentrated in 11% of the estimated 109,000 street addresses, a 27% relative reduction.

The distributions of "inside" crime are also far from random. All three observed distributions in Table 3 are significantly more concentrated than the simple Poisson model of a random pattern of crime occurrence. And as seen in Table 4, every crime we have analyzed differs from a theoretical distribution without repeat locations by a magnitude that varies from moderate to substantial.

◙ Are Dangerous Communities Generally Safe?

The data also suggest great variation in predatory crime within communities, regardless of variation across communities. This finding is based on the 5% of all places that had any predatory public crime over the course of a year, and the estimated 20% of the places in the city that are located in what local residents define as high-crime neighborhoods (interview with Alva Emerson, 1987). Assume that all of the 6,212 places with at least one rape/CSC, robbery, or auto theft were located among the 23,000 high-crime-area places (which they are not). Even then, the vast majority of places in those areas—15% of the city among the high-crime 20% of the city, or 73% of those high-crime neighborhoods—would still be free of predatory public crime over 1 year.

We do not claim that hot spots are completely unrelated to each other. To the contrary, they are clearly bunched on major thoroughfares,

Table 4	Magnitude of Concentration by Location of Six Offenses Reported to Police in Minneapolis by Telephone, December 15, 1985–December 15, 1986

| | % of City Addresses | | |
Offense	Without Repeats	Actual	Percent Reduction
Domestic Disturbance	21	8.6	59
Assault	15	7	55
Robbery	3.6	2.2	39
Burglary	14.6	10.6	27
Auto Theft	3.4	2.7	21
Rape/Criminal Sexual Conduct	1.5	1.2	20

at least in Minneapolis. They are also bunched near each other. Linnell (in press) found that the mean distance between robbery hot spots was 888 feet, and that 90% of them were located on 7 main avenues. In designing a patrol experiment affecting clusters of hot spot addresses, Sherman and Weisburd (1988) found 420 clusters of addresses of three or more hard crime calls, totaling 20 such calls, located within one-half block of each other. Moreover, 72 of those clusters had to be eliminated from the research design because they were located within two blocks of each other and would have violated the necessary independence of treatment. Yet, there were still many addresses in between those clusters about which no police calls had been received.

To be sure, most of the absence of crime by place is due to crime's overall rarity, rather than its concentration. A separate address for each of the crimes would still leave 57% of places in high-crime neighborhoods free of major predatory crimes, rather than the 73% after the actual concentration into hot spots. The relative magnitude of the concentration, then, may not be that substantial, depending on one's view of a 28% relative increase (73% over 57%) in the chance of being free of major predatory crime (assuming incorrectly, again, that all crimes are in high-crime areas). Whether from rarity or concentration, however, the general safety of places in the city's "dangerous neighborhoods" further suggests the theoretical and policy significance of the criminology of place, as distinct from the traditional criminology of neighborhoods or areas.

⬚ The Criminology of Place

Adjusting for Exposure Period and Population

Our findings must be interpreted with great caution, largely because of the enormous heterogeneity across places of both the size of the population and the periods of time at risk. The wide range of risk levels is suggested by capsule qualitative descriptions of the places with highest raw frequencies of predatory crime calls, presented in Table 5.

The nonrandom distribution of crime by place may simply be due to the nonrandom distribution of people. Geographers have long recognized the day-to-day clustering of people residing over a wide area into small "nodes" of activity (Brantingham and Brantingham, 1984: 235). If crime is concentrated in direct proportion to the concentration of people, then there may be nothing particularly criminogenic about those places. It may make sense for police to concentrate their efforts at those nodes, as an experiment in Minneapolis is now attempting to do (Sherman and Weisburd, 1988), but there is no increased per capita risk of crime for people to worry about or for theory to explain.

Two cases in point illustrate the population issue. A Dallas hotel we analyzed looked much like the worst hot spots in Minneapolis. It had a high place rate of predatory crime, with 1,245 crime reports over a 2-year period, 41 of them for violent crimes against persons. Yet the hotel covered 48 acres and had an estimated mean daily population of 3,000 guests, employees, and visitors. The per capita robbery rate was 76% lower than the per capita robbery rate for the entire city (Sherman, 1988).[2] In contrast, the bar on the Minneapolis hot-spot list (Table 5) with the highest raw frequency of predatory crime also had very high per capita rates of crime. With 25 robberies in 1 year, and an estimated mean daily population of no more than 300, Moby Dick's Bar had a robbery call rate of 83 per 1,000 persons—seven times higher than the call rate of 12 per 1,000 for the city's entire 1986 estimated population of 362,000. With 81 assaults, the per capita assault rate at the bar was 270 per 1,000 or more than 1 assault for every 4 persons in the bar over the year. Such an environment can reasonably be labeled as a dangerous place, in which individuals face substantially higher personal risks of criminal victimization than in the "average" place.

The estimates of per capita crime risk by place are further complicated by the varying time at risk. One reason the convenience stores in Gainesville had six times greater prevalence than the liquor stores of at least one armed robbery per location (Clifton, 1987) may be that the liquor stores were open for less than half the hours per week of a 24-hour, 7-day convenience store. Just as plane and automobile crash fatalities are computed per passenger mile traveled, the conceptually appropriate indicator for place crimes against persons may be crimes per person-minute spent on the premises. Property crimes would require different adjustments for time at risk. Commercial burglary, for example, should be standardized by the number of hours per week an establishment is closed, and auto theft rates might be standardized by the number of hours each car is present at a place. The violent Minneapolis bar would be even more violent if the annualized violence rates were adjusted for the limited hours per day it was occupied.

Are Places Criminogenic?

The basic theoretical problem for the criminology of place, however, is not just to account for variation in raw frequencies of place crime, or per unit/period crime target risks. The more fundamental issue is whether the routine activities of places, given their physical environment, are actually criminogenic. Do places vary in their capacity to help *cause* crime, or merely in their frequency of *hosting* crime that was going to

occur some place inevitably, regardless of the specific place? Are the routine activities of hot spots criminogenic *generators* of crime, or merely more attractive *receptors* of crime? If a crime hot spot is somehow incapacitated from producing its routine activities, would there be a corresponding net decline in total criminal events, or merely a hydraulic displacement of the same events to the next most appropriate locations?

Displacement

A routine activities criminology of place hypothesizes that crime cannot be displaced merely by displacing motivated offenders; the offenders must also be displaced *to* other places with suitable targets and weak guardianship. The findings presented above support that view. If the distribution of crime hot spots was determined solely by the concentration of offenders, then how can we explain the complete 1-year absence of predatory crimes from 73% of the places in high-crime areas in Minneapolis (compared with the expected absence from only 57%)?

Cohen and Felson (1979) support the criminogenic role of place by demonstrating temporal correlations between time spent away from home and collectivity crime rates. Other evidence suggests that variations in area-level guardianship are associated with little displacement of offending from better to more poorly guarded areas (Hakim and Rengert, 1981). The entire problem of planned reductions in criminal opportunity unintentionally producing *displacement* of crime may have been exaggerated by policymakers pessimistically resigned to the perseverance of evildoers (Clarke and Mayhew, 1988; Cornish and Clarke, 1987). But as an empirical question, the generator-versus-receptor problem is far from being resolved.

Table 5	Hot Spots in Minneapolis, December 15, 1985–December 15, 1986, with 10 or More Predatory Crimes (adjusted data before aggregation of multiple address listings)	
Rank/Description	**Robberies, Rapes, Auto Thefts**	**All Types**
1. Intersection: bars, liquor store, park	33	461
2. Bus Depot	28	343
3. Intersection: homeless shelters, bars	27	549
4. Downtown Mall	27	445
5. Intersection: adult bookstore, bars	27	431
6. Bar	25	510
7. Intersection: theater, mall, record store	25	458
8. Hotel	23	240
9. Convenience Store	22	607
10. Bar	21	219
11. Intersection: drugstore, adult theater	18	513
12. Intersection: restaurant, bar	15	445
13. Apartment Building	15	177
14. Department Store	15	449
15. Intersection: Burger King, office bldg.	15	365
16. Shopping Mall	15	305
17. Hotel	14	121
18. Bar	14	244
19. Towing Company	14	113
20. Movie Theater	14	251
21. Department Store	14	810
22. High-rise Apartment Building	14	479
23. Intersection: drugstore, adult bookstore	13	290
24. Convenience Store	13	113
25. Intersection: high concentration of bars	13	206
26. Parking Lot	13	31
27. Loring Park	13	212

Table 5	(Continued)	
Rank/Description	Robberies, Rapes, Auto Thefts	All Types
28. Restaurant	13	25
29. Apartment Building	12	142
30. Restaurant	12	198
31. Homeless Shelter	12	379
32. Detached House	11	190
33. High-rise Apartment Building	11	125
34. Apartment Building	10	233
35. Intersection: high residential area	10	156
36. High-rise Apartment Building	10	92
37. Restaurant	10	122
38. Intersection: apartments, gas stations	10	197
39. Intersection: supermarket, liquor store	10	94
40. Lake Harriet	10	171
41. Bar	10	107
42. Apartment Building	10	142
Totals	661	11,760

It seems likely, for example, that the criminogenic influence of place varies by type of offense. Crimes arising out of intimate or market relationships may be much less dependent on place than predatory stranger crimes. The concentration of domestic disturbance calls may simply indicate that certain buildings are receptors for the kinds of people most likely to experience, or at least call police about, domestic problems; such calls might occur at the same rate no matter where they lived. Some market-driven offenses, like the street sale of prostitution and illegal narcotics, may occur independently of the routine activities of places. As the recent failed crackdown on drugs in the District of Columbia suggests (Reuter et al., 1987), market crimes may create their own routine activities in otherwise relatively unorganized public places.

Yet all the literature on robbery discussed above suggests that cash business places open at night generate opportunities for robbery, the absence of which could well mean fewer robberies. The concentration of exposers in Minneapolis parks (7 of the 25 top-ranked places for sex crimes) suggest that there might be fewer exposures if there were fewer places providing both a desirable audience and abundant opportunities for concealment. Predatory stranger offenses, in particular, seem dependent on places where offenders converge with vulnerable victims and low surveillance.

Yet even predatory stranger offenses vary substantially by type of offense, as Table 4 shows, with respect to the magnitude of concentration they display relative to the number of possible locations. One can avoid robbery twice as effectively by staying away from certain places than one can avoid sex crimes or auto theft. If routine activities of places are criminogenic, they appear to be more powerfully so for some kinds of offenses than others.

Changing Places, Not People

Ironically, Cohen and Felson (1979) concluded their original analysis with an emphasis on individual life-styles as the primary aspect of routine activities affecting crime, implying the inevitability of higher crime with a more mobile life-style (cf. Hindelang et al., 1978). Focusing on the routine activities of places rather than of individual life-styles produces a different conclusion, as Felson (1987) has recently implied. On a place-specific basis, targets may be made less suitable, guardianship may be increased, and the supply of potential offenders may be reduced. Successful efforts to do so might produce net reductions in crime, holding constant the absolute size of the populations of offenders and targets. The routine activities of the person who goes to bars or convenience stores late at night does not have to change for such places to be made less criminogenic.

Many recent examples of such attempts can be found. Local ordinances passed in the late 1980s in Ohio, Florida, and New Jersey require convenience stores to have two or more clerks on duty, for the explicit purpose of reducing armed robbery through better guardianship (Clifton, 1987). A 1987 editorial in the *American Journal of Public Health* recommends that the U.S. Occupational Safety and Health Administration regulate workplace environments to reduce target suitability for robbery-homicide by requiring bulletproof barriers to protect taxi drivers

and store clerks and better placement of cash registers to increase surveillability from the street (Dietz and Baker, 1987). The mother of a boy murdered in a 1985 Orlando, Florida, convenience store robbery attempted to create a Mother against Drunk Driving (MADD)-type organization called Victims of Interstate Convenience Enterprises (VOICE) to fight for better convenience store security (Lawrence, 1986). New York police developed an Operation Padlock program to close up businesses with repeated crime problems. As part of the developing problem-oriented approach to reinventing police strategies (Eck and Spelman, 1987; Goldstein, 1979), police in Minneapolis have sought to reduce convergence of offenders and targets under weak guardianship by revoking the liquor licenses of two violent bars, based in part on the data analysis presented above (Sherman et al., 1988). Citizens in Detroit have gone as far as buying up and renovating vacant houses to prevent them from becoming crack houses (Wilkerson, 1988b), and citizens in both Detroit and Miami have burned down crack houses, with an acquittal on arson charges by one Detroit jury (Wilkerson, 1988a).

Whether such measures can produce net reductions in crime (without displacement) may be impossible to determine, given the difficulty of holding constant the collectivity supply of motivated offenders—or even of defining adequately who they are (Massey et al., 1987). But controlled experimentation may be the best means for determining the extent to which routine activities of places can be made less criminogenic. Random assignment of a large sample of clusters of hot spot addresses to different levels of guardianship by police patrol, for example, could determine (1) whether guardianship affects place crime and (2) whether crimes reduced in one place are matched by crimes increased in nearby places (Sherman and Weisburd, 1988). The convenience store industry could experiment with numbers of clerks and other guardianship measures and examine potential displacement of armed robbery to other nighttime commercial establishments in nearby jurisdictions.

At the same time, the criminology of place can be enhanced by longitudinal analysis of the characteristics associated with onset, frequency rates, seriousness, and desistance of crime in places (Blumstein et al., 1986; Wolfgang et al., 1972). For example, from 1945 (the birth year of the first Philadelphia cohort) to 1988, one liquor store in Northeast Washington under the same family management experienced 16 robberies and burglaries and 4 robbery-homicides (Mintz, 1988). How does that compare with other liquor stores? How will it compare with future rates under new management? How do liquor stores compare with other types of retail outlets, or other types of places? Such research on the "criminal careers" of places could help to specify the fertile, but still too general, routine activities concepts of target suitability, motivated offenders, and guardianship.

Like the criminology of individuals, a criminology of place could fall prey to the facile notion that getting rid of the "bad apples" will solve the problem. Neither capital punishment of places (as in arson of crack houses) nor incapacitation of the routine activities of criminal hot spots (as in revocation of liquor licenses) seems likely to eliminate crime. But since the routine activities of places may be regulated far more easily than the routine activities of persons, a criminology of place would seem to offer substantial promise for public policy as well as theory.

⬚ Notes

1. Tables 2 and 3 combine different dispatcher identifications of the same address, which the city's computer treats as different addresses if there is any variation in the description following a colon after the street address (e.g., 1414 Bouza Street: Moby's Bar). Colons and descriptions are quite rare among all addressees, but fairly common among the hot spots. The combinations were made by searching all places with 50 or more calls of all types ($n = 799$ places) for addresses with descriptions following a colon and then for each such address searching all 69,439 data records on places with any calls, for any other record beginning with the same address. (Up to six separate records were found for one address.) This procedure was unable, however, to detect misspellings of street names.

2. One possible cause for the actual "coolness" of this apparently hot spot was the much higher level of guardianship at the hotel. The ratio of patrol officers (including the hotel's security officers) to population was three times higher at the hotel than in the city as a whole, and the density of patrol presence per acre was 63 times higher at the hotel than city wide. As Felson (1987: 927) points out, growing inequality of security is characteristic of the modern metropolis. This inequality has arguably made many people into virtual prisoners of their private spaces, or "modern cliff dwellers" (Reiss. 1987: 42).

⬚ References

Baker, Susan P., R.A. Whitfield, and Brian O'Neill. 1987. Geographic variations in mortality from motor vehicle crashes. *New England Journal of Medicine* 316:1, 384–87.

Bayley, David H. 1976. *Forces of order: Police behavior in Japan and the United States.* Berkeley: University of California Press.

Biderman, Albert D., and Albert J. Reiss, Jr. 1967. On exploring the "dark figure" of crime. *Annals of the American Academy of Political and Social Sciences* 374:1–15.

Blau, Judith R., and Peter M. Blau. 1982. The cost of inequality: Metropolitan structure and violent crime. *American Sociological Review* 47:114–129.

Blumstein, Alfred, Jacqueline Cohen, Jeffrey A. Roth, and Christy Visher, eds. 1986. *Criminal careers and "career criminals."* Vol. 1. Washington, DC: National Academy Press.

Brantingham, Paul J., and Patricia L. Brantingham. 1984. *Patterns in crime.* New York: Macmillan.

Brown, Barbara, 1983. Ph.D. dissertation, Department of Psychology, University of Utah. Cited in *The New York Times,* November 17, C11.

Bursik, Robert J., Jr. 1986 Ecological stability and the dynamics of delinquency. In *Communities and crime,* edited by Albert J. Reiss, Jr. and Michael Tonry. Chicago: University of Chicago Press.

Chambers, Ray W. 1988. Gainesville convenience store security measures ordinance: A review and analysis. Unpublished manuscript.

Clarke, Ronald V. 1983. Situational crime prevention: Its theoretical basis and practical scope. In *Crime and justice: An annual review of research.* Vol. 4, edited by Michael Tonry and Norval Morris. Chicago: University of Chicago Press.

Clarke, Ronald V., and Pal Mayhew. 1988. The British gas suicide rate story and its criminological implications. In *Crime and justice: An annual review of research.* Vol. 10, edited by Michael Tonry and Norval Morris. Chicago: University of Chicago Press.

Clifton, Wayland, Jr. 1987. Convenience store robberies in Gainesville, Florida: An intervention strategy by the Gainesville police department. Paper presented at meeting of the American Society of Criminology, Montreal, November.

Cobb, Richard. 1975. *A sense of place.* London: Duckworth.

Cohen, Lawrence E. and Marcus Felson. 1979. Social change and crime rate trends: A routine activity approach. *American Sociological Review* 44:588–608.

Cornish, Derek B. and Ronald V. Clarke. 1986. *The reasoning criminal: Rational choice perspectives on offending.* New York: Springer-Verlag.

———. 1987. Understanding crime displacement: An application of rational choice theory. *Criminology* 25:933–48.

Davis, Harold. 1987. Workplace homicides of Texas males. *American Journal of Public Health* 77:1, 290–93.

Dietz, Park Elliott, and Susan P. Baker. 1987. Murder at work. *American Journal of Public Health* 77:273–74.

Duffala, Dennis C. 1976. Convenience stores, armed robbery, and physical environmental features. *American Behavioral Scientist* 20:227–46.

Durkheim, Emile. [1897] 1951. *Suicide: A study in sociology.* 1897. New York: Free Press.

Eck, John, and William Spelman. 1987 *Problem-solving.* Washington, DC: Police Executive Research Forum.

Federal Bureau of Investigation. 1987 *Crime in the United States, 1986.* Washington, DC: Government Printing Office.

Felson, Marcus. 1987. Routine activities and crime prevention in the developing metropolis. *Criminology* 25:911–32.

Gastil, Raymond D. 1971. Homicide and a regional culture of violence. *American Sociological Review* 36:412–17.

Glaser, Daniel. 1971. *Social deviance.* Chicago: Markham.

Goffman, Erving. 1971. The insanity of place. In *Relations in public: Microstudies of the public order.* London: Penguin.

Goldstein, Herman. 1979. Improving policing: A problem-oriented approach. *Crime and Delinquency* 25:236–38.

Gould, Leroy. 1969. The changing structure of property crime in an affluent society. *Social Forces* 48:50–59.

Guerry, A. M. 1831. *Essai sur la statistique morale de la France.* Paris: Chez Corchard.

Hakim, Simon, and George F. Rengert, eds. 1981. *Crime spillover.* Beverly Hills, CA: Sage.

Hawley, Amos. 1950. *Human ecology.* New York: Ronald Press.

Hindelang, Michael. 1976. *Criminal victimization in eight American cities.* Cambridge, MA: Ballinger.

Hindelang, Michael, Michael Gottfredson, and James Garofalo. 1978. *Victims of personal crime.* Cambridge, MA: Ballinger.

Hiss, Tony. 1987a. Experiencing places-I. *The New Yorker,* June 22, 45–68.

———. 1987b. Experiencing places-II. *The New Yorker,* June 29: 73–86.

Iverem, Esther. 1988 A teen-age addict is held in the killings of 5 in East Harlem. *The New York Times,* January 10, 1.

Jacobs, Jane. 1961 *The death and life of great American cities.* New York: Vintage.

Jeffery, C. Ray. 1971. *Crime prevention through environmental design.* Beverly Hills, CA: Sage.

Jeffery, C. Ray, Ronald Hunter, and Jeffery Griswold. 1987 Crime analysis, computers, and convenience store robberies. Appendix D to Wayland Clifton, Jr., Convenience store robberies in Gainesville, Florida: An intervention strategy by the Gainesville police department. Paper presented at meeting of the American Society of Criminology, Montreal, November.

Kansas City (Mo.) Police Department. 1977. *Response time analysis,* 4 vols. Kansas City: Author.

Krahn, Harvey, Timothy F. Hartnagel, and John W. Gartrell. 1986. Income inequality and homicide rates: Cross-national data and criminological theories. *Crimonology* 24: 269–95.

Lawrence, Donna. 1986. Mad mother fights for tighter C-store security. *C-Store Digest,* October 20, 1

Linnell, Deborah. In Press. The geographic distribution of hot spots of robbery, rape, and auto theft in Minneapolis. M.A. thesis, University of Maryland, Institute of Criminal Justice and Criminology.

Loftin, Colin, and Robert Hill. 1974. Regional subculture and homicide: An examination of the Gastil-Hackney hypothesis. *American Sociological Review* 39:714–24.

Mason, T. J., F. W. McKay, R. Hoover, W. J. Blot, and J. F. Fraumeni, Jr. 1985. *Atlas of cancer mortality for U.S. counties: 1950-69,* DHEW Publication (NIH) 75-780. Bethesda, MD: National Cancer Institute.

Massey, James L., Marvin D. Krohn, and Lisa Bonati. 1987 The routine activities of individuals and property crime. Paper presented at meeting of the American Society of Criminology, Montreal.

Merry, Sally E. 1981a. Defensible space undefended: Social factors in crime prevention through environmental design. *Urban Affairs Quarterly* 16: 397–422.

———. 1981b. *Urban danger: Life in a neighborhood of strangers.* Philadelphia: Temple University Press.

Messner, Steven. 1980. Income inequality and murder rates: Some cross-national findings. *Comparative Social Research* 3: 185–98.

———. 1982. Poverty, inequality and the urban homicide rate. *Criminology* 20:103–14.

———. 1983. Regional differences in the economic correlates of the urban homicide rate: Some evidence on the importance of cultural context. *Criminology* 21:477–88.

Messner, Steven, and Kenneth Tardiff. 1985. The social ecology of urban homicide: An application of the "routine activities" approach. *Criminology* 23:241–67.

Miethe, Terance D.. Mark C. Stafford, and J. Scott Long. 1987. Social differentiation in criminal victimization: A test of routine activities lifestyle theories. *American Sociological Review* 52: 1984–94.

Mintz, John. 1988. NE store owners retiring with stock of memories. *The Washington Post,* October 17, CI.

Molumby, Thomas. 1976. Patterns of crime in a university housing project. *American Behavioral Scientist* 20:247–59.

Nelson, James F. 1980. Multiple victimization in American cities: A statistical analysis of rare events. *American Journal of Sociology* 85: 870–91.

Newman, Oscar. 1972. *Defensible space: Crime prevention through urban design.* New York: Macmillan.

Newman, Oscar, and K. A. Franck. 1980. *Factors affecting crime and instability in urban housing developments.* Washington, DC Government Printing Office.

———. 1982. The effects of building size on personal crime and fear of crime. *Population and Environment* 5: 203–20.

Pierce, Glen L., Susan A. Spaar, and LeBaron R. Briggs IV. 1984. The character of police work: Implications for the delivery of services. Center for Applied Social Research, Northeastern University, Boston.

Poyner, Barry. 1983. *Design against crime: Beyond defensible space.* London: Butterworth.

Quetelet, L. Adolphe J. 1842. *A treatise on man and the development of his faculties.* Edinburgh: Chambers.

Reiss, Albert J., Jr. 1986 Why are communities important in understanding crime? In *Communities and crime,* ed. Albert J. Reiss, Jr., and Michael Tonry. Chicago: University of Chicago Press.

———. 1987. The legitimacy of intrusion into private space. In *Sage criminal justice systems annual series.,* Vol. 23, *Private policing.* Newbury Park, CA: Sage.

Reppetto, Thomas A. 1974. *Residential crime.* Cambridge, MA: Ballinger.

Reuter, Peter, John Haaga, Patrick Murphy, and Amy Praskac. 1987. Drug use and policy in the Washington metropolitan area: An assessment. Draft. Washington, DC: Rand.

Robinson, William S. 1950. Ecological correlations and the behavior of individuals. *American Sociological Review* 15:351–57.

Sampson, Robert. 1986 Crime in cities: Formal and informal social control. In *Communities and crime,* ed. Albert J. Reiss, Jr., and Michael Tonry. Chicago: University of Chicago Press.

Sanchez, Rene. 1988. Suspect in rape arrested: 5 assaults occurred near Dupont Circle. *The Washington Post,* January 8.

Sanchez, Rene, and Sari Horwitz. 1987. Transcript disputes go-go assault account: 911 caller, police differ on stabbing site. *The Washington Post,* November 18, CI.

Shaw, Clifford R., Henry D. McKay, Frederick Zorbaugh, and Leonard S. Cottrell. 1929. *Delinquency areas.* Chicago: University of Chicago Press.

Sherman, Lawrence W. 1986 Policing communities: What works? In *Communities and crime,* ed. Albert J. Reiss, Jr., and Michael Tonry, Chicago: University of Chicago Press.

———. 1987. Repeat calls to police in Minneapolis. *Crime Control Reports,* No. 5. Washington, DC: Crime Control Institute.

———. 1988. Violent stranger crime at a large hotel: A case study in risk assessment methods. Unpublished manuscript, Crime Control Institute, Washington, DC.

Sherman, Lawrence W., and David Weisburd. 1988. Policing the hot spots of crime: A redesign of the Kansas City preventive patrol experiment. Unpublished manuscript, Crime Control Institute, Washington, DC.

Sherman, Lawrence, Michael E. Buerger, and Patrick R. Gartin, 1988. Beyond dial-a-cop: Repeat call addresses policing. Unpublished manuscript, Crime Control Institute, Washington, DC.

Silberman, Charles. 1978. *Criminal violence, criminal justice.* New York: Simon and Schuster.

Skolnick, Jerome, and David Bayley. 1986. *The new blue line.* New York: Free Press.

Spelman, William, and Dale K. Brown. 1981. *Calling the police: Citizen reporting of serious crime.* Washington, DC: Police Executive Research Forum.

Taylor, Ralph B., Steven D. Gottfredson, and S. Brower. 1981. *Informal social control in the residential urban environment.* Baltimore, Md.: Johns Hopkins University Press, Center for Metropolitan Planning and Research.

Taylor, Ralph B., and Steven Gottfredson. 1986. Environmental design, crime, and prevention: An examination of community dynamics. In *Communities and crime* ed. Albert J. Reiss, Jr., and Michael Tonry. Chicago: University of Chicago Press.

Waller, Irvin, and Norman Okihiro. 1978. *Burglary: The victim and the public.* Toronto: University of Toronto Press.

Wilkerson, Isabel. 1988a. Crack house fire: Justice or vigilantism? *The New York Times,* October 22.

———. 1988b. Detroit citizens join with church to rid community of drugs. *The New York Times,* June 29.

Wilson, James Q., and Richard Herrnstein. 1985. *Crime and human nature.* New York: Simon and Schuster.

Wolfgang, Marvin E., Robert Figlio, and Thorsten Sellin. 1972. *Delinquency in a birth cohort.* Chicago: University of Chicago Press.

REVIEW QUESTIONS

1. What measure did Sherman et al. use to measure how crime was distributed? What are the strengths and weaknesses of this measure?

2. Regarding the more serious crimes that Sherman et al. measure, what types of establishments appear the most in the Top 10? Why do such locations appear to epitomize the various elements of routine activities theory?

3. Why would fast-food restaurants or towing companies appear as "hot spots" for crime? What types of characteristics do these locations have that would predispose them to crime/victimization?

READING

Like the authors of the previous reading, LaGrange uses a routine activities theoretical framework for her study of crime in Edmonton, Alberta, and goes a bit further by examining the influence of environmental predictors of why crime tends to be higher in certain areas. Furthermore, she uses more advanced mapping software for her analysis, using a relatively new program called MapInfo, which is based on coordinates and geocoding of crime activity. In addition, she reviews the influence of the theory of "broken windows" that was originally presented by Wilson and Kelling.[34] That theory argues that in certain areas, a broken window or other physical eyesore will be fixed very quickly, whereas in other areas, one broken window quickly leads to other broken windows and additional dilapidation because the residents and business owners do not care much about the area or do not have the financial means to pay for such improvements. Subsequently, this leads to higher crime rates in the area because motivated offenders feel they have less chance of getting caught in run-down areas where residents do not have the inclination or resources to secure the area. This often becomes a serious feedback loop, and physical and crime conditions continue to decline in a rapid cycle. While reading the following selection, readers should consider the importance that environmental planning and the layout of cities have on the opportunities for offenders to commit crime.

The Impact of Neighborhoods, Schools, and Malls on the Spatial Distribution of Property Damage

Teresa C. LaGrange

Analysis of the geographic distribution of crime has had a long history in criminology, dating from the work of sociologists at the Chicago School. Most of this research has focused on index crimes such as homicide, robbery, rape, or burglary, and it has linked rates of offending to the social and residential characteristics of the neighborhoods where these crimes occur. Minor crimes, however, such as property damage and vandalism, occur far more frequently in any city than the more widely studied index offenses. Recent research

SOURCE: Teresa C. LaGrange, "The Impact of Neighborhoods, Schools, and Malls on the Spatial Distribution of Property Damage," *Journal of Research in Crime and Delinquency* 36, no. 4 (1999): 393–422. Copyright © 1999 Sage Publications, Inc. Reprinted by permission of Sage Publications, Inc.

[34]Wilson and Kelling, "Broken Windows".

has suggested that these crimes may be significant factors in the occurrence of urban crime in a more general sense, through processes described as the "broken windows" effect or "spiral of decay" (Felson 1998: 131; Skogan 1990:40; see also Kelling and Coles 1996). According to these perspectives, the accumulation of minor property damage sends a subtle signal to potential offenders that guardianship is low and that crimes may be carried out undetected. Thus, crimes of all types are more likely to be committed in run-down areas of a city. An understanding of the geographic distribution of minor offenses like property damage and vandalism, therefore, has the potential to shed light on the spatial patterns of a much broader range of criminal activity.

Although the causal variables that are usually examined in relation to spatial variations in crime rates have been those associated with structural theories of offending (Bursik 1988; Park and Burgess 1933; Shaw and McKay 1942; Snodgrass 1976), the broken windows phenomenon emphasizes that the immediate situational context may also be an important factor in the occurrence of crime. A would-be offender needs a suitable victim or target, and must encounter it in circumstances that permit a crime to be carried out without interruption. Recognition of these fundamental observations has led to the recent development of perspectives such as routine activities theory, which looks beyond the attributes of residents of high-crime areas to consider those social and environmental factors that make such areas conducive to crime (Cohen and Felson 1979; Felson 1986, 1987, 1994, 1998; Felson and Cohen 1980). As formulated and developed by Felson and Cohen, routine activities theory portrays crime as the convergence of the following three elements: motivated offenders, potential victims or targets, and unguarded access (Felson 1998:53). Thus, features of the urban environment that contribute to the convergence

of these factors by increasing the concentration of offenders and victims, or reducing guardianship, will result in increased crime rates. Such factors may include characteristics of neighborhoods, such as the age composition of residents or the proportion of rental housing. They may include physical structures in the environment such as bars (Block and Block 1995; Roncek and Maier 1991), schools (Roncek and Faggiani 1985; Roncek and LoBosco 1983), or shopping malls (Engstad 1980) that serve as crime attractors or generators (Brantingham and Brantingham 1991, 1994, 1995; Jarvis 1972).

The primary objective of the current study is to apply these concepts, derived from routine activities theory and previous research on the spatial distribution of crime, to the occurrence of minor property damage throughout one city during a single year. The occurrence of damage incidents as recorded by three different city departments or agencies is considered in relationship to neighborhood characteristics, residential composition, and environmental structures (schools and malls) that contribute to the convergence of likely offenders and reduced guardianship. The research has a number of implications. The link between cumulative property damage and more serious crimes means that the analysis of these crimes can make an important contribution to the understanding of urban crime patterns. Furthermore, because property damage and vandalism tend to be directed primarily toward public or impersonal spaces and facilities, an understanding of their etiology may provide a useful model for research into other forms of anonymous environmental damage (Skogan 1990:37). Finally, this research extends the principles of routine activities theory to previously understudied types of criminal activity, and thus sheds further light on the way in which guardianship and opportunity intersect to permit crimes to occur.

⬚ Theory and Previous Research

It is well known that crimes tend to cluster in certain areas of a city, so that some areas have higher crime rates than others (Brantingham and Brantingham 1984, 1993; Dunn 1984; Figlio, Hakim, and Rengert 1986; Georges-Abeyie and Harries 1980): This is one of the least disputed facts about crime, and it has been repeatedly supported by numerous studies spanning several decades, different cities, and diverse offenses. Previous research has examined, for example, the locations of homicides in Cleveland (Bensing and Schroeder 1960), Chicago (Block 1976), and Houston (Bullock 1955); robberies in Seattle (Schmid 1960); and burglaries in Washington, D.C. (Scarr 1973). Furthermore, it was the early observation of spatial variations that initially contributed to the development of the classical perspectives on crime. Early theorists equated areas of the city with neighborhoods having enduring and distinctive features; this conceptualization led to the development of theories that attempted to explain the criminogenic nature of these areas (Kornhauser 1978). Factors associated with higher crime rates included the proportion of unemployed persons, the amount of rental housing, the overall residential density, and the length of time residents remained in the area. Various mechanisms have been identified that link these area characteristics to the criminal inclinations of residents—blocked opportunities and the absence of legitimate pursuits, the attraction of illegitimate opportunities, or a general lack of neighborhood social control (Allan and Steffensmeier 1989; Bursik 1988; Park and Burgess 1933; Sampson and Grove 1989; Shaw and McKay 1942; Snodgrass, 1976).

Contemporary research on when and where crimes occur, however, has been strongly influenced by the recognition that human activities, including crimes, take place within a specific social and physical environment (Hawley 1950, 1971). Thus, contemporary studies have taken into consideration the situational context that surrounds the occurrence of criminal events (Sacco and Kennedy 1994), rather than focusing exclusively on offender motivation. Even highly motivated offenders require certain conditions to complete their crimes—conditions that include such factors as the vulnerability of victims or targets and the presence or absence of witnesses (Sherman, Gartin, and Buerger 1989). Analyses of situational factors that include the routine activities of both likely offenders and potential victims have demonstrated that these requisite elements of crime converge nonrandomly; that is, some areas provide all the components for crime to occur more frequently and more regularly than others. The circumstances, under which crimes occur, from this perspective, are functions of social and structural phenomena that allow people to translate their criminal inclinations into action (Felson 1986, 1994, 1998).

Therefore, the social characteristics of residents of specific areas may not be a direct cause of the crimes rates in those areas (Allan and Steffensmeier 1989). Some, if not all, of the crimes may be committed by outsiders who gravitate to these areas rather than by residents themselves (Costanzo, Halperin, and Gale 1986:74). Crime rates might be high, not because of the criminal inclinations of residents, but because of the criminal opportunities that such areas provide (Dunn 1984; Felson and Cohen 1980). The nature and extent of guardianship in an area is directly related to whether people are at home or away from home and whether there are many people about during different times of day. It is affected by whether residents know their neighbors, and are capable of both recognizing and responding to events that appear out of the ordinary. Social characteristics such as the number of homemakers, single adults, or retired persons living in the area contribute to different patterns of local activity, dictating whether the majority of residents are at home during the day, during the evenings, or only rarely; and whether residents come and go regularly or at all hours.

Residents in their teens or early adulthood are likely to be absent from their homes more frequently, as are those who are single (Felson and Cohen 1980; Hindelang, Gottfredson, and Garofalo 1978). In areas where a large proportion of residents fall into these categories, guardianship may be substantially reduced. Similarly, areas with a high proportion of rental units have a less permanent resident population than well-established owner-occupied housing tracts (Bursik 1988). In areas with a high population turnover, it is less likely that residents will know their neighbors and know who does or does not have a legitimate reason for being in the area. Residents may therefore be unable to exercise informal surveillance by direct observation and by questioning of strangers or suspicious activities (Sampson and Grove 1989). Consistent with a routine activities perspective that emphasizes opportunity; higher crime rates in such areas may be directly related to the inability of residents to exercise suitable guardianship.

The population density of an area, by contrast, may have the opposite effect. Although high density might be hypothesized to increase the convergence of potential offenders and suitable targets, and has been linked by some research to higher crime rates (Hartnagel and Lee 1990), the sheer number of people in an area may make it more difficult for potential offenders to commit a crime without being observed. In general, density may lead to lower rates for crimes that involve secrecy (burglary, for example) and higher rates for crimes that arise from proximity (larceny and muggings) (Decker, Shichor, and O'Brien 1982:52–53; Felson 1998:29; see also Roncek and Faggiani 1985; Roncek and LoBosco 1983; Shlomo 1968).

Physical Structures in the Environment

In addition to the social environment provided by residents and neighborhoods, the urban landscape is shaped by physical structures that influence human activities. Two types of public structures, shopping malls and high schools, have been identified by previous studies as being significantly related to higher crime rates in adjacent areas (Engstad 1980; Roncek and Faggiani 1985; Roncek and LoBosco 1983). Both serve to attract a large number of nonresidents into an area who come and go with little formal supervision—a situation that not only brings potential offenders into the area, but also reduces guardianship.

Shopping malls. Shopping malls exist for the express purpose of attracting potential customers to the shops and services that they shelter. The large number of people who come and go in the streets around a mall works to reduce effective guardianship, because distinguishing between legitimate patrons and persons who are simply loitering may be difficult. It might be predicted that this combination of factors would result in higher crime rates in the areas immediately adjacent to malls. Consistent with this prediction, Engstad (1980) found significantly higher frequencies of auto crimes, thefts, and miscellaneous offenses in three urban census tracts, each of which contained a major shopping center,[1] when compared to surrounding areas with similar social and demographic characteristics. Areas with shopping centers had from 2.1 to 6.5 times as many offenses as the averages recorded for adjacent areas, and 1.5 to 3.7 times as many offenses as the maximum recorded for adjacent areas (Engstad 1980:210–11).

High schools. High schools also contribute to increased traffic and activity in the immediate, surrounding vicinity. Furthermore, this population consists of persons in their teens, who are implicated in higher rates of offending than other demographic groups. Thus, research into the impact of high schools has consistently demonstrated that their presence is associated with higher crime rates in surrounding neighborhoods. In the first of two studies conducted in medium-sized American cities, Roncek and LoBosco (1983) examined crimes occurring in a relatively new, affluent city that ranked lower in crime rates than other cities of comparable size. The authors report higher rates of several types of index crimes[2] in one-block areas immediately adjacent to public high schools,[3] controlling for social, housing, and demographic composition of the areas. Although the size of a school's enrollment was evaluated as a possible predictor, it was found to be nonsignificant.

Routine Activities and Minor Crimes

Although the impact of routine activities of residents and environmental structures such as schools and malls has been investigated for serious crimes such as robbery, burglary, or rape, these factors have not been considered in relation to minor offenses like property damage. One of the primary justifications for the emphasis on serious crimes, without question, is the potential such offenses have for grave social harm. Such crimes typically and understandably arouse the greatest fear and concern among the public, and they receive the greatest attention and resources from criminal justice agencies. Minor crimes do not appear to pose the same sort of immediate social threat, and hence are usually considered to be of secondary concern. Yet, the rate of occurrence for minor crimes far surpasses, in any city, the rates for more serious offenses. Such crimes as vandalism are a constantly occurring and ever present problem in most contemporary North American cities, and their cumulative impact can prove very costly (Bell, Bell, and Godefroy 1988; Skogan 1990). More significant, the

accumulation of such incidents has been implicated in the spiral of decay (Felson 1998:131) that leads to the devaluation of urban neighborhoods. The impact is more than economic; the occurrence of minor crimes in an area, especially those that involve visible property damage, may elicit other criminal activity (Chalfant 1992; Challinger 1987; Kelling and Coles 1996).

The assertion that behavior can be formed by the circumstances surrounding it has been criticized as mechanistic at best and environmental determinism at worst (Clarke 1978). It is based, however, on the fundamental tenets of learning theory (Marongiu and Newman 1997; Wortley 1997:66). Specific situational factors can act as eliciting stimuli or behavioral cues to engage in or restrain certain behaviors. For example, in an influential and widely cited study, Zimbardo (1970) demonstrated that "releaser cues," in the form of existing damage, led to the rapid destruction of seemingly abandoned vehicles on the street. The view that criminal events are contingent on the situation in which they occur is consistent with rational theories that attribute actions, including criminal ones, to a balancing of costs and benefits (Clarke and Felson 1993; Cornish and Clarke 1986). Benefits may be perceived as more likely if there is seemingly a low risk of apprehension. Thus, the perception of opportunity may in fact contribute to the occurrence of crime. Risks, on the other hand, particularly the perceived risks of being caught, may work to suppress such behavior. Areas where minor offenses such as property damage or vandalism occur and accumulate may convey a subtle signal to potential offenders that guardianship and social control are low. This perception may be taken, consciously or unconsciously, as an indication that the risks of detection and apprehension for criminal activity are negligible. That perception, in turn, may lead to an escalation in the rate of other, potentially more serious crimes (Felson 1998:131; Wortley 1997:67).

Social characteristics of areas that contribute to these elements may therefore be associated with higher rates of property damage. Areas that contain many unemployed persons, more young males, more rental than owned housing, and a larger number of transient residents will have greater movement and activity in and around the neighborhood, both during the course of the daily routine and over longer periods of time. These factors, in turn, make it more difficult to distinguish between strangers and residents and to determine the nature of their activities, thereby weakening guardianship. On the other hand, areas of high density and areas where more residents are home-makers or retired, and therefore home for more hours during the day, may not afford anonymous access to unguarded targets. Because potential offenders may feel constrained by the possibility of observation, less damage may occur in such areas.

Beyond the social characteristics of neighborhoods, physical features of the environment, such as malls and schools, that contribute to the convergence of offenders in an area might be expected to result in increased rates of minor crimes, just as with more serious crimes. Malls serve to draw large numbers of people into an area, some of whom may be potential offenders. Malls also impede effective guardianship because of the difficulty in distinguishing between legitimate and illegitimate visitors to the area. The presence of secondary schools within an area can also be expected to result in more crimes of this type because, like malls, such facilities increase local human traffic and thus interfere with guardianship. Furthermore, both malls and schools serve to draw together young people in their teens—the specific age groups most likely to be involved in minor property crime (Beaulieu 1982; Erickson and Jensen 1977; Gladstone 1978; Gold 1970).

The Current Research

In keeping with the foregoing discussion, this study examines the geographic distribution of minor property crime recorded as mischief and vandalism in a medium-sized Canadian city during a 1-year period.[4] Previous studies of the spatial distribution of crime have, for the most part, relied on official crime data as the most widely available and consistent information on crimes throughout all parts of a given geographic area. These statistics usually reflect police activity in response to crime. It is axiomatic that such records undercount crime for a variety of reasons, including the fact that many crimes go unreported and therefore fail to show up in official records, However, under-reporting is especially likely to be a problem for official counts of very minor crimes such as property damage and vandalism, because victims may view such incidents as too minor to report. To address the potential methodological issues that arise from reliance on official reports for minor crimes, this study uses three types of data obtained from three departments in the city that served as the study site: the city's Department of Parks and Recreation, its transit department, and its police service.[5] Using census enumeration areas as a unit of aggregation,[6] the geographic patterns of these three types of damage are evaluated using ordinary least square (OLS) regression in relationship to two categories of predictors; the social characteristics of residents of each area, taken from municipal census data, and the presence or absence of two types of environmental structures, shopping malls and secondary schools.

Method

Data for this research was collected in Edmonton, Alberta, a western Canadian city with a population of approximately 600,000. The three agencies cooperating in this research—the police service, the city transit department, and the Department of Parks and Recreation—provided records of the damage done to the facilities under their supervision during the calendar year of 1992. Each of these data sources was unique due to the nature of the targets, the types of incidents that could occur, and the way in which incidents were recorded; thus, merging them into a single composite index was deemed to be unsuitable. Each measure was therefore retained as a separate indicator of damage.

Mischief. Records were obtained from the Edmonton Police Service on all mischief incidents reported during 1992. Under Canadian law, the offence category of mischief refers to willful, malicious damage or public behavior. Most such incidents are property-related vandalism. Because many of them are relatively minor, however, police data are potentially biased, both by underreporting and by differential response. Research indicates that in Canada, as elsewhere, the extent to which citizens report crime varies. For very minor offenses, and particularly where there is little likelihood of identifying and arresting an offender, reporting is low (Griffiths and Verdun-Jones 1989; Silverman, Teevan, and Sacco 1996). Canadian victimization surveys reveal that most respondents do not report minor property crimes, even when the property is their own. Typically, less than half of such incidents are reported. Most respondents blame their failure to report on the trivial nature of incidents (60 percent) or on the lack of benefit expected from formal police action (47 percent) (Gartner and Doob 1996). Hence, the majority of incidents involving minor property damage does not show up in official police records. In addition, police responses to citizen complaints may vary according to the seriousness of the crime, considerations of the immediate workload, and time pressures. An officer's perception of the neighborhood where the report comes from may also influence the response (Hagan, Gillis, and Chan 1978). These factors can be expected to substantially restrict the number of incidents of minor property damage that show up in official police records, a reduction

that may vary considerably for different neighborhoods.

For this study, additional data were obtained from two other contributing agencies. These data consisted of maintenance records to damaged structures, avoiding the difficulties associated with behavior and selective recording of damage. Where damage was observed, maintenance personnel for each department filed a report indicating their assessment of the cause, whether it was an accident, normal wear and tear, theft, or deliberate damage. In contrast to police data, which provided information on single incidents, both of these agencies kept periodic maintenance records that include an unknown amount of cumulative damage.

Transit vandalism. Records obtained from the transit department reported damage to bus stops and shelters throughout the city.[7] These records were compiled on the basis of reports from staff maintenance personnel who made regular visits to each shelter. Given the nature of these structures, assessments on the type of damage were deemed to be fairly accurate—their construction is designed to resist most inadvertent damage and discourage theft. The majority of incidents involved damage to the glass or plexiglass panels from which shelters are constructed; graffiti was excluded from the reports. However, maintenance personnel only visited the shelters on their route on a monthly basis. Any damage that was recorded at that time therefore incorporated an unknown number of actual incidents (D. Kowalchuk, personal communication, June 30, 1993).

Park vandalism. Reports on damage to facilities maintained by the Department of Parks and Recreation reflected periodic visits by maintenance personnel. Unlike the transit department records, however, which listed only a relatively restricted type of damage to specific types of structures, records from the Department of Parks and Recreation included a wide variety of different incidents, reflecting the greater variation in the types of structures and grounds for which the department had responsibility. In addition to all parks within the municipal limits, the Department of Parks and Recreation maintained recreational facilities, cemeteries, the grounds of public buildings, and tracts of public landscaping. Hence, there were potentially many different types of damage done to lawns, flowerbeds, buildings, pools, and other structures. This greater variety, however, renders the issue of determining the exact nature of damage and judging whether it was deliberate more problematic. Although maintenance personnel attempted to discriminate deliberate damage from accidental, the extent to which these judgments were correct is unknown. In addition, there is no way to ascertain whether similar criteria were used in evaluating incidents in different locations. Decisions were made by individual workers who filed reports on damages and classified them according to the type of incident (W. Gorman, personal communication, June 2, 1993).

Census Enumeration Areas

Data on population and housing characteristics in the city were obtained from the 1992 City of Edmonton Municipal Census, which reported aggregate information for each census enumeration area. Enumeration areas are subdivisions of the permanent statistical units established by Statistics Canada for the national census (Lalu 1989:1). The smaller enumeration areas do not provide as much demographic detail as the larger census tracts. Their smaller size, however, allows for greater variation in population characteristics.

The routine activities perspective that forms the theoretical basis for this research does not equate the concept of neighborhood with that of community or that of the distinctive sociocultural attributes inherent in this term as developed in the earlier ecological

tradition. Nevertheless, some minimal assumption about neighborhood is implicit in the expectation that people can potentially come to know their neighbors, that they may become familiar with their neighbors' children and habitual routines, and that these developments are important in informal guardianship. Although the placement of boundaries along naturally occurring lines of demarcation like main roadways, parks, ravines, and commercial strips may appear to be arbitrary, such divisions also tend to create effective barriers that limit guardianship. Residents preoccupied with daily activities are less likely to take note of events that occur across the park, or across the railroad tracks, than those next door or on their own block. It is in this limited sense that the use of census enumeration boundaries provides a basis for comparing areas.

Area characteristics. Area characteristics that are used as predictors of vandalism and mischief for this research included a measure of residents who had lived 1 year or less at their current address (new residents), calculated as a percentage of the total population in each area. Additional population measures included the percentage of the total population identified as homemakers, as retired, and as unemployed at the time of the enumeration. The two following specific segments of the population were included: males between the ages of 10 and 19 and males between the ages of 20 and 24. Both were calculated as a percentage of the total number of males in the area.[8] Housing characteristics were calculated as a percentage of the total number of housing units in the area. These included the percentage of renters, rooming houses, and vacant housing units. A final measure, child density, was calculated as the average number of children per household with ages from 5 to 19.[9]

Environmental structures. Malls included in this study were the 17 largest malls in the city. Commercial strips and smaller neighborhood malls were excluded.[10] Secondary schools included the four following types of schools: the junior and senior high schools of the public school district, and the junior and senior high schools of the Catholic school district. Several schools in the Catholic school district served students at more than one level, including three junior-senior high schools; these schools were coded as high schools in the analysis. Each of these facilities was coded 0 (absence) or 1 (presence). Previous research on the impact of schools (Roncek and Faggiani 1985; Roncek and LoBosco 1983) had identified a single-block radius as the extent of significant differences. Therefore, schools were introduced as variables only within the enumeration area in which they were located. The impact of malls, however, was expected to be wider, due in part to their greater use of land within a given area and in part to the greater amount of human traffic that they generate. Therefore, malls were introduced both for their presence within an area and for their presence in an adjacent area.[11]

Mapping of Crime Incidents

To obtain area measures of crime rates that might be evaluated in relation to area-level predictors, crime data were spatially located within the city using a computerized mapping program (MapInfo) that assigns x and y coordinates to a reference map.[12] For this study, two base maps were used: a computerized street map containing streets and block numbers for all city addresses, and a map of the boundaries of the census enumeration areas. Mischief incidents as reported by the police department were recorded as specific points, in most cases, a street address, and they were geocoded as such. Of the original 13,537 incidents reported to the police, 97 percent (13,131) were successfully geocoded.[13] Locations for the remaining 259 cases (1.9 percent) could not be identified. Transit vandalism incidents were recorded as occurring at intersections, with route direction information provided; route direction was used to determine assignment to one of the corners of the intersection. Of the 1,337 incidents, 1,325 were successfully geocoded. Park vandalism incidents were recorded according to the park or facility where they occurred rather than a street address. Each record also indicated a structure or type of structure, such as a pool house or field house; this portion of the record was used to refine the point of location further. Out of the 402 incidents recorded during 1992, all but 9 were successfully geocoded.[14]

All incidents from each source were separately aggregated according to the census enumeration area boundaries, resulting in a total figure for each type of damage within each area. These figures represented the actual count of incidents recorded in each enumeration area. In resident populations, areas ranged from 0 (for six areas of the city) to 3,201 (for the most densely populated). They ranged in size from .027 square kilometers to over 55 square kilometers. These two factors are inversely related; that is, the largest areas were low in population, whereas the smaller areas were high.

Area amounts of crime are generally calculated as a population-based rate, obtained by dividing the total number of crimes by the total population and then multiplying by a constant (1,000 or 100,000). Crime rates provide a useful way of comparing units such as cities, states, and even countries, because it is plausible to assume that a larger population would contain a larger number of offenders, even if the proportion of such offenders within the population remained constant. For this analysis, however, population-based rates present a number of shortcomings. Although it is logical to calculate a population-based rate of crime for larger geographical units such as a city in comparison to other cities, because most of the crime in a city can logically be attributed to residents, it is less reasonable to make the same assumption about smaller geographic units such as enumeration areas.

Although it may be true that offenders are likely to select targets close to their own residences, the question remains of how close (Costanzo et al. 1986). Five blocks might be considered close; yet, this is a sufficient distance to place a criminal event two enumeration areas away from the offender's home territory. Following a similar argument, Harries asserts that "most if not all the incidents may be attributable to outsiders . . . theoretically, zero events might be 'blamed' on residents, again making nonsense of the rate concept" (Harries 1993:4). This observation is underscored by the fact that crimes can and do occur in areas where there is a very small (or sometimes zero) population. The use of a population-based rate would inflate the amount of crime for these areas.[15]

Difficulties also arise in relation to the physical size of an area. The incidents examined in this research involve crimes, not against individual victims, but against the physical environment. Furthermore, although these crimes are often directed at structures in populated areas such as residences and stores, damage can also occur in parks, ravines, and in industrial or commercial areas. It could therefore be argued that a geographic rate (crimes per square kilometers) would provide a better measure, because larger areas would provide more opportunities for crime. However, the opportunities inherent in available targets, by themselves, are not sufficient to predict crime; it is the convergence of potential offenders with these targets in the absence of suitable guardianship that provides the conditions under which crime is likely to occur. These factors, in turn, are linked to the movements of human activity. Although the empty land around a city's perimeter may theoretically be damaged, areas that are populated, either residentially or because they are activity centers, are likely to experience a greater convergence of all of the components contributing to crime.

These arguments suggest that both the physical size of an area and the number of persons who live there are important considerations; one alone cannot provide an adequate basis for constructing a rate to control for their effects. For this reason, the actual counts of the three types of incidents that occur within each area are retained as measures. To control for the effects of area size and population, these measures are introduced as independent variables in the multiple regression analysis (Agresti and Finlay 1997; Bollen and Ward 1980).[16]

◤ Results

Edmonton, Alberta, had almost three-quarter million residents within its metropolitan limits during 1992, the year for which census data used in this study was gathered. The total area occupied by the city was 690.74 square kilometers (see Table 1).[17] Schools included 13 senior high schools and 26 junior high schools in the public school district, and 9 senior high schools and 11 junior high schools in the Catholic school district. There were, in addition, 17 shopping malls. The municipal enumeration had identified 662 areas. Listwise elimination of missing cases and the removal of one extreme outlier from the analysis[18] resulted in 654 areas that were used in the subsequent analysis. The mapping and aggregating of incidents of mischief and vandalism resulted in positively skewed distributions. Extreme scores at the upper end of the distribution for mischief and transit vandalism were recoded at the 90th percentile (Nagin and Smith 1990).[19] Incidents of recoded mischief ranged from 0 to 37, with a mean of 17.72; recoded transit vandalism ranged from 0 to 5, with a mean of 1.63; and park vandalism ranged from 1 to 29, with a mean of .59.

Ordinary least squares (OLS) regression was used to examine relationships between predictors and dependent variables. The results are summarized in Table 2. Of the residential characteristics, the percentage of unemployed ($b = .57, p < .01$) and the percentage of residents who were males aged 20 to

Table 1	Descriptive Statistics: Means, Standard Deviations, Minimums, and Maximums for Variables by Census Enumeration Areas ($N = 654$)				
Variable	**Minimum**	**Maximum**	**Total**	**M**	**SD**
Area per square kilometer	.03	55.03	690.74	1.04	4.32
Total population	0	3,201	617,299	933	358
New residents (%)	.00	56.41		20.39	11.43
Resident characteristics (%)					
Homemakers	.00	19.62		6.06	2.90
Retired	.00	98.30		11.76	10.12
Unemployed	.00	66.67		5.55	5.11
Males age 10 to 19	.00	92.05		11.84	5.91
Males age 20 to 24	.00	50.82		9.39	4.98
Residences					
Total	0	1,246	252,325	381	122
Renters (%)	.00	100.00		44.33	28.85
Rooming houses (%)	.00	100.00		.40	3.94
Vacant (%)	.00	28.68		4.32	4.01
Child density	.00	9.50		.49	.45
Catholic high school			9		
Catholic junior high			11		
Public high school			13		
Public junior high			26		
Mall in area			17		
Mall adjacent			102		
Mischief incidents[a][b]	.00	37.00	11,712	17.72	10.05
Transit vandalism[b][c]	.00	5.00	1,079	1.63	1.91
Park vandalism[d]	.00	29.00	394	.59	2.42

SOURCE: City of Edmonton Municipal Census (1992) unless otherwise indicated.

a. Source: City of Edmonton Police Service.
b. Scores above 90th percentile recoded.
c. Source: City of Edmonton Transportation Department.
d. Source: City of Edmonton Parks and Recreation Department.

Table 2	Regression Coefficients for Mischief Incidents, Transit Damage, and Park Damage on Area Characteristics and Presence of Malls and Schools ($N = 654$)					
	Mischief		**Transit**		**Park**	
Variable	B	β	B	β	b	β
Area per square kilometer	−.11	−.05	−.01	−.01	.00	.00
Total population	.01**	.21	.00**	.12	.00	−.01
New residents (%)	−.13**	−.15	−.02	−.10	−.01	−.05
Resident characteristics (%)						
Homemakers	.24	.07	.04	.08	−.02	−.03
Retired	.00	.05	.01	.09	.01	.05
Unemployed	.57**	.29	.04*	.11	.08**	.18
Males age 10 to 19	.00	.14	.01	.08	−.01	−.01
Males age 20 to 24	.27**	.13	.04	.01	−.01	−.09
Residences (%)						
Renters	01*	.15	.01*	.13	.01	.04
Rooming houses	.22*	.09	.02	.01	−.01	−.01
Vacant	.41**	.16	−.05	−.01	−.01	−.05
Child density	−1.66	−.08	−.30	−.07.	.08	.01
Catholic high school	4.13	.05	−.98	−.06	2.52**	.12
Catholic junior high	2.63	.03	.34	.02	.29	.02
Public high school	8.32**	.11	1.15*	.08	5.35**	.30
Public junior high	2.03	.04	.90*	.09	−.02	.00
Mall in area	11.19**	.18	1.61**	.13	2.91**	.19
Mall adjacent	2.58**	.09	.21	.04	−.37	−.06
R^2	.26		.10		.21	

$^*p < .05.$ $^{**}p < .01.$ Two-tailed tests.

24 ($b = .27$, $p < .01$) were found to be statistically significant predictors of increased mischief. Only the percentage unemployed, however, was identified as a statistically significant predictor for all three crime measures. For transit vandalism, it predicted a small

increase ($b = .04$, $p < .05$), and for park vandalism, a somewhat greater increase ($b = .08$, $p < .01$). Other measures of residential characteristics, however, had little consistent impact. Those few that were statistically significant for one type of measure proved to be nonsignificant for the other two types. The percentage of new residents was significantly related to mischief ($b = -.13$, $p < .01$), but it predicts lower levels instead of the expected higher levels. For transit and park vandalism, the percentage of new residents was nonsignificant. The percentage of area residences that were rooming houses ($b = .22$, $p < .05$) and the percentage that were vacant ($b = .41$, $p < .01$) were also significant predictors of increased mischief, but not of transit or park damage. The percentage of renters was related to a small but statistically significant increase in both mischief and transit vandalism ($b = .01$, $p < .05$ for both), but it was nonsignificant for park vandalism.

In contrast, the presence or absence of public high schools and shopping malls within an area were consistent and robust predictors for increased damage of all types. Because all these facilities had been coded as dummy variables, the reported coefficients represent the difference between two conditional means—one for those areas with such a structure, and one for those without. The presence of a high school predicted a substantial and statistically significant increase in mischief incidents ($b = 8.32$), transit incidents ($b = 1.15$), and park vandalism ($b = 5.35$). Effects for a mall in a given area were of greater magnitude, with $b = 11.19$ for mischief, $b = 1.61$ for transit, and $b = 2.91$ for park vandalism. Contrary to expectations, having a mall in an adjacent area did not predict a significant increase in transit or park vandalism, although the reported effect ($b = 2.58$) was both robust and statistically significant for mischief. Consistent with previous literature, the significant relationship identified for high schools applied only to those of the public school system. Catholic schools were associated with a significant

increase in park vandalism ($b = 2.52$, $p < .01$), but they had little consistent relationship to other types of damage. In contrast to the prominent effect identified for high schools, the presence of a junior high school had a negligible impact. Although these schools were statistically significant for predicting transit vandalism, they did not predict a corresponding increase in mischief or park vandalism.

These differences in incidences reported for high schools and malls cannot be attributed to differences in area characteristics, as measured by the variables included in this study. As reported in Table 3, which summarizes comparisons between areas with and without schools and malls, and is based on independent samples t tests, areas containing high schools did not differ significantly from those without in terms of population characteristics. The same conclusion is apparent for areas containing malls, when they are compared to other areas.[20] These types of structural facilities are dispersed throughout the city in neighborhoods of all kinds. The fact that property crime is consistently higher in surrounding areas appears to be related more to their presence than to any distinctive differences in the residential environment.

⊠ Discussion

Damage or vandalism was defined, measured, and counted in widely divergent ways by the agencies that provided data for this research. These three measures of damage, however, yielded very similar results when their spatial patterns throughout the city were examined. In spite of the expectation that residential and neighborhood characteristics would influence the amount of property crime that occurred in an area, these variables had little consistent impact for all types of damage. The unemployment rate, however, was a significant predictor of increased levels of mischief and transit and park vandalism. Other characteristics that

| Table 3 | Comparison of Means for Areas with and without Schools and Malls: Independent Sample *t* Test (*N* = 654) | | | | | | | | |

		Resident Characteristics (in percentage)					Damage Incidents		
Structure	n	New Residents	Unemployed	Males Age 10 to 19	Males Age 20 to 24	Renters	Mischief	Transit	Park
Catholic school									
High School									
Present	9	18.07	4.68	11.52	6.94	39.40	23.22	1.00	3.44
Absent	645	20.45	5.57	11.86	9.41*	44.39	17.67	1.64	.56
Junior high									
Present	11	17.20	6.91	11.73	7.93	37.01	21.73	2.72	1.09
Absent	643	20.48	5.54	11.86	9.40*	44.45	17.67	1.61*	.59
Public School									
High school									
Present	12	20.32	4.78	9.97	6.67	54.78	29.16	2.80	6.75
Absent	642	20.42	5.57	11.98	9.39	44.13	17.52**	1.16*	.48
Junior high									
Present	26	14.39	4.08	11.97	7.83	37.49	19.88	2.80	.84
Absent	628	20.67**	5.62**	11.85	9.44**	44.61	17.65	1.59**	.59
Mall									
In Area									
Present	17	24.44	6.55	10.18	9.84	59.86	31.11	3.59	4.41
Absent	637	20.31	5.53	11.90	9.37	43.91	17.38**	1.59**	.50**
Adjacent									
Present	102	21.37	6.32	11.92	9.69	50.01	21.76	2.00	.59
Absent	552	20.24	5.42	11.84	9.32	43.28*	17.00**	1.57*	.60

* $p < .05$. ** $p < .01$. Two-tailed tests.

were expected to increase the likelihood of property damage were either nonsignificant (the percentage of teen males living in an area) or were predictors of one type of damage but not another (the percentage of local housing that was renter occupied). Those areas with high schools and malls, however, were found to have consistently higher rates of all three types of damage, controlling for differences in social, residential, and demographic characteristics. These results are consistent with a routine activities interpretation of the circumstances in which damage occurs. Routine activities theory argues that crimes will be committed when potential offenders are confronted with the opportunities afforded by available targets in situations of reduced guardianship. Based on that model, the observation that mischief and vandalism are concentrated in certain locations can be interpreted as reflecting differences in the way those areas and their surroundings permit the convergence of these factors. High schools are built throughout the populated areas of a city, in prosperous neighborhoods as well as poorer ones. Similarly, shopping malls are scattered throughout the urban landscape. Both schools and malls are public structures that generate a steady flow of traffic into and out of an area, and draw in a high volume of nonresidents. The increased traffic that results not only contributes to the convergence of potential offenders, it also impedes guardianship by making it more difficult to distinguish loiterers with an illegitimate purpose from those with a legitimate one. Thus, it might be argued that these structures provide situational incentives to commit crime.

The way in which unemployment affects the amount of property damage occurring in some areas, however, is open to several alternative interpretations. Areas with high unemployment are typically the poorer areas of a city (Kornhauser 1978).[21] Are residents of these areas more inclined to damage and destroy their surroundings than residents in better neighborhoods? It might be, as some writers have argued, that the urban poor lack social ties and experience marginalization; therefore, they destroy the symbols of a society that has largely excluded them. The physical damage inflicted on their surroundings, according to this interpretation, represents a form of "nonverbal communication . . . the mutilation of objects

and environments for which the perpetrator does not feel any code fellowship" (Roos 1992:75, 81; see also Cohen 1973).

This explanation is consistent with classic structural theories that have their roots in the concept of social disorganization. By implication, higher crime rates in these areas are attributed to the activities of local residents. Although the argument may be a valid one, its accuracy cannot be determined based on aggregate data (Robinson 1950). Furthermore, the link between unemployment and property crime could occur due to factors unrelated to the motivation or inclinations of area residents. Areas with high unemployment tend to be, as previously noted, poorer areas of the city, and these areas often contain a housing mix characterized by multifamily dwellings, low-rise apartment buildings, and row housing. Areas where these multiple-family housing facilities dominate may be deficient in structurally imposed forms of guardianship (Crowe 1991; Crowe and Zahm 1994; Hough and Mayhew 1980; Wilson 1978). These facilities often include common areas in parking lots, pathways, and open spaces between buildings. Public areas of this type render informal surveillance and supervision more difficult, and the lack of ownership implicit in such land use may discourage residents from intervening in questionable activities (Felson 1998; Geason and Wilson 1990; see also Newman 1972). Furthermore, the routine daily activities of residents are frequently less predictable in areas of high unemployment. Lacking the imposed constraints of a daily work schedule, unemployed persons are more likely to enter and depart their premises according to idiosyncratic circumstances rather than according to a repetitive and habitual routine. Residents of these areas are thus less likely to be able to assess the legitimacy or illegitimacy of neighborhood traffic, and they may therefore fail to challenge activities or persons about whom they are uncertain. Hence, some or all of the effects identified for unemployment as a predictor of increased property crime may be related to the reduced guardianship in these areas rather than differences in motivation.

Routine activities theory, and the notion that crimes occur in certain places because of the criminal opportunities provided by the physical environment of those places, provides an explanation for the patterns of minor property crime analyzed in this research. However, there is still the question of whether the opportunities presented by circumstances in certain areas prompt the motivation to act, or whether the crimes that occur there would have occurred anyway (Sherman et al. 1989). This issue is relevant to this study of property crime, but it is also germane to the question of whether the occurrence of such crimes creates an atmosphere of low guardianship, which then leads to more serious crimes. If crimes occur in certain locations because motivated offenders actively seek out opportunities to wreak their damage, then identifying criminogenic areas or hot spots and the circumstances that generate them is simply an effort in description. Furthermore, increasing guardianship in such locations will accomplish little other than displacement.

Felson and Cohen argue that motivations may remain static, whereas opportunities to commit crime fluctuate according to variations in patterns of routine human activities (1980:397). Similarly, Gottfredson and Hirschi (1990) have suggested that offenders have an inherent, unvarying criminality; the occurrence of specific crimes, however, is shaped largely by immediate opportunities. These explanations of offending portray human behavior as the product of the weighing of risks versus benefits, and they are analogous to more traditional versions of control theory, which assert that deviant behavior is an omnipresent vulnerability (Hirschi 1969; Kornhauser 1978:24). Such models of offending suggest that opportunities implicit in reduced guardianship might do more than simply provide the context for events that would inevitably have happened somewhere. Instead,

such opportunities might act to stimulate the behavior, and thus act as proximate causes.

Conclusion

This research extends the geographic analysis of crime patterns, which has previously focused on serious index crimes, to study the occurrence of minor property crimes in a Canadian city during a 1-year period. As with the index crimes examined in previous research, the results of this study reveal a marked concentration of criminal incidents in certain areas—specifically, those containing high schools or malls, and those with higher unemployment. Reasons for the concentration of mischief and vandalism in these areas have been framed in terms of routine activities theory's model of crime as arising from the convergence of offenders and targets in the absence of effective guardianship. The results support the conclusion that situational opportunities presented by urban ecological features can account for variations in crime patterns. Some urban areas, due to their residential composition, may be particularly attractive to vandals and to those inclined to minor property crime because they afford inherently low guardianship. Further weakening of guardianship is brought about by the presence of facilities like schools and malls that bring a large number of nonresidents into the vicinity on a daily basis. Furthermore, these facilities attract the segments of population most likely to engage in minor damage and vandalism—people in their teens. Although replication is necessary to confirm these findings, the results of this analysis have implications for crime control. If minor property damage and vandalism are seen as the product of routine activities that arise primarily from the convergence of offenders and reduced guardianship, then prevention may most effectively be focused on disrupting the way in which these factors intersect—through an increase in both active and passive guardianship. Furthermore, the relevance of these findings goes beyond the damage itself. Crimes such as vandalism and mischief are often trivial as single events, but they are collectively significant in their impact on the perception of guardianship, and hence on potential offenders' weighing of the risks and benefits associated with more serious and socially harmful criminal behavior. The accumulation of property crime in an area may provide a signal to the criminally inclined that there is little likelihood of apprehension; therefore, such areas become likely sites for more serious criminal activities. In light of recent research into what has been described as the broken windows effect, an understanding of the circumstances surrounding minor property damage sheds further light on urban crime patterns more generally.

▧ Notes

1. The cited research uses the term *shopping center* to describe the retail facilities studied, in contrast to the current usage of the term *mall*. Although there is a semantic distinction between the two terms based on whether facilities have a common nonretail area (the mall itself), the city in which the current research was conducted does not make such a distinction. The two terms are therefore used interchangeably in this study.

2. The offence categories examined were murder, rape, assault, robbery, burglary, grand theft, and auto theft (Roncek and LoBosco 1983).

3. A similar effect was not observed for private high schools. The authors speculated that "the grounds of public high schools are public property and legitimately available for use by anyone while the grounds of private schools are not" (Roncek and LoBosco 1983).

4. *Vandalism* is defined variously as "intentional acts aimed at damaging or destroying" (Moser 1992); "intentional hostile behavior aimed at damaging environmental objects"; "willful or malicious destruction, injury, disfigurement, or defacement" (Federal Bureau of Investigation 1994; see also Cohen 1973; Levy-Leboyer 1984). Although the term vandalism is commonly used in Canada, it is not contained in the Canadian Criminal Code. Instead, instances of vandalism are recorded and prosecuted under the statutes for mischief.

5. Incidents involving property damage and vandalism are recorded under the Canadian Criminal Code as mischief more than or less than $1,000 (Rodrigues 1990:210–211, section 430). Although offences recorded under this section usually refer to property damage, some incidents may refer to other behavior. Section 430.1 of the Canadian Criminal Code states that "Everyone commits mischief who willfully (a) destroys or damages property; (b) renders property dangerous, useless, inoperative or ineffective; (c) obstructs, interrupts or interferes with the lawful use, enjoyment or operation of property" (Rodrigues 1990:210–211). Section 430.1.1 adds that anyone who "destroys or alters data," "renders data meaningless, useless, or ineffective," or otherwise interferes with data is also guilty of mischief (Rodrigues 1990:210–211). Actions falling within the latter subsection would not be considered property damage. It is, however, impossible to determine from the data source whether police records on mischief include offenses of this nature, and, if so, how many.

6. Enumeration areas are subdivisions of the permanent census tracts established by Statistics Canada, and they represent the smallest unit of census aggregation. There are typically several enumeration areas within a tract. Their boundaries are intended to define an area as homogenous as possible in terms of socioeconomic characteristics and to follow, where feasible, well-established natural boundaries (Lalu 1989:1).

7. Although the transit department maintained additional records on damage to buses and Light Rail Transit trains, they were excluded from this study, because they could not be spatially located.

8. Age groupings were chosen to allow a measure of residents in their teens or early adulthood. Actual cutoffs between age groups, however, reflect the limitations of the municipal census data, which recorded age by gender in categories rather than in any substantive or legal distinctions (in Canada, the cutoff between juvenile and adult status is 18 years).

9. Child density was included instead of broader measures of residential or neighborhood density; this was based on previous research that has identified this measure as predictive of increased levels of minor property damage (see Wilson 1978).

10. Commercial strips and smaller neighborhood malls were excluded because an exhaustive, valid measure of their presence or absence in a given area could not be developed for the city in which data were

gathered. Many areas contained small, neighborhood strip malls that were identifiable as separate corporate entities. Others had commercial strips of very similar composition (typically centering on fast food outlets or convenience stores) that reflected the clustering of separate facilities within a municipal commercial zone. These latter areas were not identified as malls, although they effectively functioned as such.

11. An adjacent area was operationally defined as one sharing one or more common boundaries with an area containing a mall.

12. More detailed information on the technique of geocoding used for this study is available from the author.

13. This number includes three types of records: (1) those which had been recorded as a specific street address; (2) those recorded as occurring at a particular named building or facility, which in turn had a street address; and (3) those recorded as occurring at an intersection. For the latter group of incidents, a specific street address could not be identified, nor could it be determined on which of four potential corners an incident had occurred. All incidents of this nature were therefore geocoded to the northwest corner of the intersection.

14. Missing data for the three types of property damage incidents included cases in which an address or specific location could not be identified, cases in which the indicated address or location did not exist (recording agency errors), and cases in which the indicated address or location corresponded to two or more potential locations.

15. The use of a population figure as the denominator in calculating a rate for the dependent variable may lead to spurious positive results if the same population figure is used to compute independent variables. Such a situation might arise in this analysis because predictors include the percentage of area residents falling into certain demographic categories (Bollen and Ward 1980: 61).

16. A further source of potential bias arises from the nature of the data and the possibility of spatial autocorrelation. Multivariate analyses such as the ordinary least squares (OLS) regression models employed in this study are based on the expectation that error terms are independent and do not vary systematically—a requirement that is rarely met with spatial data (Upton and Fingleton 1985:371). Instead, such data are likely to exhibit organized patterns or systematic spatial variation in values across a map (Cliff and Ord

1981:6; Upton and Fingleton 1985:151). High unemployment rates in one area, for example, do not abruptly drop at the border of that area; they tend to continue into neighboring areas. Such facilities as shopping malls, by contrast, will almost universally be absent from any area next to one in which they are present. However, the impact of such potential sources of bias is reduced as n increases (Upton and Fingleton 1985:365), and this study employs a fairly large n of over 650. Although spatial statistics have been developed to correct for such problems (see Anselin 1990a, 1990b; Anselin et al. 1996; Blommestein and Koper 1997), they involve more complex models than OLS regression and thus are appropriate if it appears that autocorrelation contributes to a significant distortion of coefficients. For this analysis, a post-hoc analysis of the differences between theoretically derived (expected) values and the corresponding observed values suggested little consistent spatial patterning that would seriously bias results, and hence regression models were retained (Cliff and Ord 1981:76).

17. The dimensions are those calculated by summing the area per square kilometer for all enumeration areas defined by the City of Edmonton Municipal Census (1992) using mapping software. The size varies somewhat from that reported by other sources.

18. A single enumeration area was excluded from the analysis as an outlier. There were no statistically significant differences in population characteristics when this area was compared to others in the city. All measures were within one standard deviation of the mean for the city as a whole. It differed, however, in that it contains an architectural phenomenon touted as "the world's largest shopping mall." Although the relative size of malls was not included as a variable in this study, the enormity of this structure places it in a category by itself, so that it could not be treated as equivalent to other malls. It should be noted, however, that mischief and vandalism rates recorded for this area were extremely high, consistent with the assumption that such a facility would predict increased amounts of crime.

19. Recoding of park vandalism would have restricted this measure to three categories, which was deemed to provide inadequate variation. This measure was thus retained as recorded in the data.

20. The only exception is size; areas containing malls and schools are significantly smaller than those without. This finding is consistent with the observation, made previously, that there was an inverse relationship

between area size and population. Larger areas were the more sparsely populated outlying districts, reflecting their smaller number of residents; they were less likely to contain schools and malls. Densely populated areas closer to the midzones of the city, by contrast, were smaller, but they were more likely to contain such facilities.

21. The data used to analyze residential characteristics in this study did not contain measures of average income or median housing values, which would have permitted a more direct measure of socioeconomic status.

≋ References

Agresti, A., and B. Finlay. 1997. *Statistical Methods for the Social Sciences,* 3d ed. Saddle River, NJ: Prentice-Hall.

Allan, E. A., and D. J. Steffensmeier. 1989. "Youth, Underemployment, and Property Crime: Differential Effects of Job Availability and Job Quality on Juvenile and Young Adult Arrest Rates." *American Sociological Review* 54:107–23.

Anselin, Luc. 1990a. "Some Robust Approaches to Testing and Estimation in Spatial Econometrics." *Regional Science and Urban Economics* 29:141–63.

———. 1990b. "Spatial Dependence and Spatial Structural Instability in Applied Regression Analysis." *Journal of Regional Science* 30:185–207.

Anselin, Luc, Anil K. Bera, Raymond Florax, and Mann J. Yoon. 1996. "Simple Diagnostic Tests for Spatial Dependence." *Regional Science and Urban Economics* 26:77–104.

Beaulieu, L. 1982. *Vandalism; Responses and Responsibilities.* Report of the Task Force on Vandalism. Ontario, Canada: Queen's Printer.

Bell, M. M., M. M Bell, and K. Godefroy. 1988. "The Impact of Graffiti on Neighborhoods and One Community's Response." Presented at the International Symposium on Vandalism: Research, Prevention, and Social Policy, April 20–23, Seattle, WA.

Bensing, R. C, and O. Schroeder, Jr. 1960. *Homicide in an Urban Community.* Springfield, IL: Charles Thomas.

Block, R. 1976. "Homicide in Chicago: A Nine-Year Study (1965–1973)." *Journal of Criminal Law and Criminology* 66:510.

Block, R. L., and C. R. Block. 1995. "Space, Place, and Crime: Hot Spot Areas and Hot Places of Liquor-related Crime." Pp. 145–83 in *Crime Prevention Studies,* Vol. 4, *Crime and Place,* edited by J. E. Eck and D. Weisburd. Monsey, NY: Criminal Justice Press.

Blommestein, Hans J., and Nick A. M. Koper. 1997. "The Influence of Sample Size on the Degree of Redundancy in Spatial Lag Operators." *Journal of Econometrics* 82:317–33.

Bollen, K. A., and S. Ward. 1980, "Ratio Variables in Aggregate Data Analysis." Pp. 60–79 in *Aggregate Data: Analysis and Interpretation,* edited by E. F. Borgatta and D. J. Jackson. Beverly Hills, CA: Sage.

Brantingham, P. J., and P. L. Brantingham, 1991. *Environmental Criminology.* Prospect Heights, IL: Waveland.

———. 1994. "Mobility, Notoriety, and Crime: A Study in Crime Patterns of Urban Nodal Points." *Journal of Environmental Systems* 11:89–99.

Brantingham, P. L., and P. J. Brantingham. 1984. *Patterns in Crime.* New York: Macmillan.

———. 1993. "Environment, Routine and Situation: Toward a Pattern Theory of Crime." Pp. 259–94 in *Routine Activity and Rational Choice: Advances in Criminological Theory,* edited by R. V. Clarke and M. Felson. New Brunswick, NJ: Transaction Books.

———. 1995. "Criminality of Place: Crime Generators and Crime Attractors." *European Journal of Criminal Policy and Research* 3:5–26.

Bullock, H. A. 1955. "Urban Homicide in Theory and Fact." *Journal of Criminal Law, Criminology, and Police Science* 45:565–75.

Bursik, R. J., Jr. 1988. "Social Disorganization and Theories of Crime and Delinquency: Problems and Prospects." *Criminology* 26:519–51.

Chalfant, H. 1992. "No One Is in Control." Pp. 4–11 in *Vandalism: Research, Prevention and Social Policy,* edited by H. H. Chistensen, D. R. Johnson, and M. H. Brookes. Portland, OR: Department of Agriculture Forest Service.

Challinger, D. 1987. *Preventing Property Crime: Proceedings of a Seminar.* Canberra: Australian Institute of Criminology.

City of Edmonton Municipal Census. 1992. Census Summary Data File. Edmonton, Canada: City of Edmonton Computing Resources.

Clarke, R.V.G. 1978. *Tackling Vandalism.* Home Office Research Study No. 47. London, UK: HMSO.

Clarke, R.V.G., and M. Felson. 1993. *Routine Activity and Rational Choice.* New Brunswick, NJ: Transaction Press.

Cliff, Andrew D., and J. K. Ord. 1981. *Spatial Process: Models and Applications.* London, UK: Pion.

Cohen, L. E., and M. Felson. 1979. "Social Change and Crime Rate Trends: A Routine Activity Approach." *American Sociological Review* 44:588–608.

Cohen, S. 1973. "Property Destruction: Motives and Meanings." Pp. 23–54 in *Vandalism,* edited by Colin Ward. New York: Van Nostrand Reinhold.

Cornish, D. B., and R. V. Clarke. 1986. *The Reasoning Criminal: Rational Choice Perspectives on Offending.* New York: Springer-Verlag.

Costanzo, C. M., W. C. Halperin, and N. Gale. 1986. "Criminal Mobility and the Directional Component in Journeys to Crime." Pp. 73-95 in *Metropolitan Crime Patterns,* edited by R. M. Figlio, S. Hakim, and G. F. Rengert. Monsey, NY: Willow Tree Press.

Crowe, T. D. 1991. *Crime Prevention Through Environmental Design: Applications of Architectural Design and Space Management Concepts.* Boston, MA: Butterworth-Heinemann.

Crowe, T. D., and D. Zahm. 1994. "Crime Prevention Through Environmental Design." *Land Management* 7:220-27.

Decker, David L., David Shichor, and Robert M. O'Brien. 1982. *Urban Structure and Victimization.* Lexington, MA: Lexington Books/D. C. Heath and Company.

Dunn, C. S. 1984. "Crime Area Research." Pp. 5–25 in *Patterns in Crime,* edited by P. L. Brantingham and P. J. Brantingham. New York: Macmillan.

Engstad, P. A. 1980. "Environmental Opportunities and the Ecology of Crime." Pp. 206–22 in *Crime in Canadian Society,* 2nd ed., edited by R. A. Silverman and J. J. Teevan, Jr. Toronto, Canada: Butterworths.

Erickson, M. L., and G. F. Jensen. 1977. "Delinquency Is Still Group Behavior!: Toward Revitalizing the Group Premise in the Sociology of Deviance." *Journal of Criminal Law and Criminology.* 68:262–73.

Federal Bureau of Investigation. 1994. *Uniform Crime Reports.* Washington, DC: Government Printing Office.

Felson, M 1986 "Routine Activities, Social Controls, Rational Decisions and Criminal Outcomes." Pp. 119–28 in *The Reasoning Criminal,* edited by D. Cornish and R. V. Clarke. New York: Springer-Verlag.

———. 1987. "Routine Activities and Crime Prevention in the Developing Metropolis." *Criminology* 25:11–31.

———. 1994. *Crime and Everyday Life.* Thousand Oaks, CA: Pine Forge Press.

———. 1998. *Crime & Everyday Life.* 2nd ed. Thousand Oaks, CA: Pine Forge Press.

Felson, M., and L. E. Cohen. 1980. "Human Ecology and Crime: A Routine Activity Approach." *Human Ecology* 8:389–406.

Figlio, R. M., S. Hakim, and G. F. Rengert. 1986. *Metropolitan Crime Patterns.* Monsey, NY: Willow Tree Press.

Gartner, R., and A. N. Doob. 1996. "Trends in Criminal Victimization: 1988-1993." Pp. 90–104 in *Crime in Canadian Society* (5th ed.), edited by R. A. Silverman, J. J. Teevan Jr., and V. F. Sacco. Toronto, Canada: Butterworths.

Geason, S., and P. R. Wilson. 1990. *Preventing Graffiti and Vandalism.* Canberra: Australian Institute of Criminology.

Georges-Abeyie, D. E., and K. D. Harries. 1980. *Crime: A Spatial Perspective.* New York: Columbia University Press.

Gladstone, F. J. 1978. "Vandalism Amongst Adolescent Schoolboys." Pp. 19–39 in *Tackling Vandalism,* Home Office Research Study No. 47, edited by R. V. G. Clarke. London, UK: HMSO.

Gold, M. 1970. *Delinquent Behavior in an American City.* Belmont, CA: Brooks-Cole.

Gottfredson, M. R., and T. Hirschi. 1990. *A General Theory of Crime.* Stanford, CA: Stanford University Press.

Griffiths, C. T., and S. N. Verdun-Jones. 1989. *Canadian Criminal Justice.* Toronto, Canada: Butterworths.

Hagan, John, A. R. Gillis, and J. Chan. 1978. "Explaining Official Delinquency: A Spatial Study of Class Conflict and Control." *Sociological Quarterly* 19:386–98.

Harries, K. 1993. "The Ecology of Homicide and Assault: Baltimore City and County, 1989-1991." Presented at the American Society of Criminology Annual Meetings, November 4, Phoenix, AZ.

Hartnagel, T. F, and G, Won Lee. 1990. "Urban Crime in Canada." *Canadian Journal of Criminology* (October):591–606.

Hawley, A. H. 1950. *Human Ecology: A Theory of Community Structure.* New York: Ronald Press.

———. 1971. *Urban Society: An Ecological Approach.* New York: Ronald Press.

Hindelang, M. J., M. R. Gottfredson, and J. Garofalo. 1978. *Victims of Personal Crime: An Empirical Foundation for a Theory of Personal Victimization.* Cambridge, MA: Ballinger.

Hirschi, T. 1969. *Causes of Delinquency*. Berkeley: University of California Press.

Hough, M., and P. Mayhew. 1980. *Crime and Public Housing: Proceedings of a Workshop*. London, UK: Home Office.

Jarvis, G. K. 1972. "The Ecological Analysis of Juvenile Delinquency in a Canadian City." Pp. 195–211 in *Deviant Behavior and Society Reaction*, edited by C. L. Boydell, C. F. Grindstaff, and P. C. Whitehead. Toronto, Canada: Holt, Rinehart and Winston.

Kelling, G. L., and C. Coles. 1996. *Fixing Broken Windows: Restoring Order and Reducing Crime in Our Communities*. New York: Free Press.

Kornhauser, R. R. 1978. *Social Sources of Delinquency*. Chicago, IL: University of Chicago Press.

Lalu, N. M. 1989. *Changing Profiles of Edmonton Census Tracts*. Edmonton, Canada: University of Alberta Population Research Laboratory.

Levy-Leboyer, C. 1984. *Vandalism: Behavior and Motivations*. New York: North-Holland.

Marongiu, P., and G. Newman. 1997. "Situational Crime Prevention and the Utilitarian Tradition." Pp. 115–35 in *Rational Choice and Situational Crime Prevention: Theoretical Foundations*, edited by G. Newman, R. V. Clarke, and S. G. Shoham. Dartmouth, VT: Ashgate Publications.

Moser, G. 1992. "What Is Vandalism? Towards a Psycho-social Definition and its Implications." Pp. 49–70 in *Vandalism: Research, Prevention, and Social Policy*, edited by H. H. Chistensen, D. R. Johnson, and M. H. Brookes. Portland, OR: Department of Agriculture Forest Service,

Nagin, D. S., and D. A. Smith. 1990. "Participation in and Frequency of Delinquent Behavior: A Test for Structural Differences." *Quantitative Criminology* 6:335–65.

Newman, O. 1972. *Defensible Space: Crime Prevention Through Urban Design*. New York: Macmillan.

Park, R. E., and E. W. Burgess. 1933. *Introduction to the Science of Sociology*. 2nd ed. Chicago, IL: University of Chicago Press.

Robinson, W. S. 1950. "Ecological Correlation and the Behavior of Individuals." American *Sociological Review* 15:351–57.

Rodrigues, Gary P., ed. 1990. *Canadian Criminal Code: Pocket Criminal Code*. Toronto, Canada: Carswell.

Roncek, D. W., and D. Faggiani. 1985. "High Schools and Crime: a Replication." *The Sociological Quarterly* 26:491–505.

Roncek, D. W., and A. LoBosco. 1983. "The Effect of High Schools on Crime in Their Neighborhoods." *Social Science Quarterly* 64:598–613.

Roncek, D. W., and P. Maier. 1991. "Bars, Blocks, and Crimes Revisited: Linking the Theory of Routine Activities to the Empiricism of Hot Spots." *Criminology* 29:725–53.

Roos, H. 1992. "Vandalism as a Symbolic Act in 'Free-zones.'" Pp. 71–87 in *Vandalism: Research, Prevention, and Social Policy*, edited by H. H. Christensen, D. R. Johnson, and M. H. Brookes. Portland, OR: Department of Agriculture Forest Service.

Sacco, V. F., and L. W. Kennedy. 1994. *The Criminal Event*. Scarborough, Canada: Nelson.

Sampson, R. J., and W. B. Grove. 1989. "Community Structure and Crime: Testing Social Disorganization Theory." *American Journal of Sociology* 94:774–802.

Scarr, H. A. 1973. *Patterns of Burlgary*. Washington, DC: Law Enforcement Assistance Administration, National Institute of Law Enforcement and Criminal Justice.

Schmid, C. F. 1960. "Urban Crime Areas, Part I & Part II." *American Sociological Review* 25:527–43, 655–78.

Shaw, C. R., and H. D. McKay, 1942. *Juvenile Delinquency and Urban Areas*. Chicago, IL: University of Chicago Press.

Sherman, L. W., P. R. Gartin, and M. E. Buerger. 1989. "Hot Spots of Predatory Crime: Routine Activities and the Criminology of Place." *Criminology* 17:17–49.

Shlomo, Angel. 1968. "Discouraging Crime Through City Planning." Working Paper No. 75, Space Sciences Laboratory and the Institute of Urban and Regional Development, University of California, Berkeley, CA.

Silverman, R. A., J. J. Teevan, Jr., and V. F. Sacco. 1996. *Crime in Canadian Society*. 5th ed. Toronto, Canada: Harcourt Brace.

Skogan, W. G. 1990. *Disorder and Decline: Crime and the Spiral of Decay in American Neighborhoods*. New York: Free Press.

Snodgrass, J. 1976. "C. R. Shaw and H. D. Mckay: Chicago Criminologists." *The British Journal of Criminology* 16:1–19.

Upton, Graham J. G., and Bernard Fingleton. 1985. *Spatial Data Analysis by Example*. New York: Wiley.

Wilson, S. 1978. "Vandalism and 'Defensible Space' on London Housing Estates." Pp. 41–65 in *Tackling Vandalism*. Home Office Research Study No. 47, edited by R. V. G. Clarke. London, UK: HMSO.

Wortley, R. 1997. "Reconsidering the Role of Opportunity in Situational Crime Prevention." Pp. 65–81 in *Rational Choice and Situational Crime Prevention: Theoretical Foundations,* edited by G. Newman, R. V. Clarke, and S. G. Shoham. Dartmouth, VT: Ashgate Publications.

Zimbardo, Philip. 1970. "The Human Choice: Individuation, Reason, and Order Versus Deindividuation, Impulse, and Chaos." Pp. 237–307 in *Nebraska Symposium on Motivation,* edited by W. Arnold and D. Levine. Lincoln: University of Nebraska Press.

REVIEW QUESTIONS

1. Of the three environmental factors that LaGrange examines, which appears to be the one that most affects high crime levels?

2. To what extent do the findings by LaGrange apply to the broken windows theory?

3. Have you observed such influences of environmental predictors in the place where you live or grew up? Provide details regarding how the proximity of such establishments or demographic characteristics have influenced the crime levels near your residences.

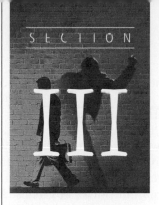

III

EARLY POSITIVE SCHOOL PERSPECTIVES OF CRIMINALITY

The introduction for this section will discuss the dramatic differences in assumptions between the Classical and Positive Schools of criminological thought. It will also discuss the pre-Darwinian perspectives of human behavior (e.g., phrenology), as well as the influence that Darwin had on perspectives of all social sciences, particularly criminology. Finally, the introduction will discuss the theories and methods used by early positivists, particularly Lombroso, IQ theorists, and body-type researchers, with an emphasis on the criticisms of these perspectives, methodology, and resulting policies.

☒ Introduction

After many decades of dominance by the Classical School (see Sections I and II), academics and scientists were becoming aware that the deterrence framework did not explain the distribution of crime. Their restlessness led to new explanatory models of crime and behavior. Most of these perspectives focused on the fact that certain individuals or groups tend to offend more than others and that such "inferior" individuals should be controlled or even eliminated. This ideological framework fit a more general stance towards **eugenics**, which is the study of and policies related to the improvement of the human race via control over reproduction, which we will see was explicitly mandated for certain groups. Thus, the conclusion was that there must be notable variations across individuals and groups that can help determine who are most at risk of offending.

So in the early to mid-1800s, several perspectives were offered regarding how to determine which individuals or groups were most likely to commit crime. Many of these theoretical frameworks were based on establishing a framework for distinguishing the more superior individuals or groups from the more inferior individuals or groups. Such intentions were likely related to the increased use of slavery in the world during the 1800s, as well as the fight of imperialism over rebellions at that time. For example, slavery was at its peak in the United States during this period, and many European countries controlled many dozens of colonies, which they were trying to retain for profit and domain.

Perhaps the first example of this belief was represented by craniometry. **Craniometry** was the belief that the size of the brain or skull represented the superiority or inferiority of certain individuals or ethnic/racial groups.[1]

The size of the brain and the skull were considered because at that time it was believed that a person's skull perfectly conformed to the brain structure; thus, the size of the skull was believed to reflect the size of the brain. Modern science has challenged this assumption, but there actually is a significant correlation between the size of the skull and the size of the brain. Still, even according to the assumptions of the craniometrists, it is unlikely that much can be gathered from the overall size of the brain, and certainly the skull, from simple measurement of mass.

The scientists who studied this model would measure the various sizes or circumferences of the skull if they were dealing with living subjects. If they were dealing with recently dead subjects, then they would actually measure the weight or volume of the brains of the "participants." When dealing with subjects who had died long before, the craniometrists would pour seeds into the skull area and then measure the volume of the skull by pouring the seeds into a graduated cylinder. Later, when these scientists realized that seeds were not a valid measure of volume, they moved toward using buckshot or ball bearings.

Most studies by the craniometrists tended to show that the subjects of White, Western European descent were superior to other ethnic groups in terms of larger circumference or volume in brain or skull size. Furthermore, the front portion of the brain (i.e., genu) was thought to be larger in the superior individuals or groups, and the hind portion of the brain or skull (i.e., splenium) was predicted to be larger in the lesser individuals or groups. Notably, these researchers typically knew which brains or skulls belonged to which ethnic or racial group before measurements were taken, making for an unethical and improper methodology. Such biased measurements continued throughout the 19th century and into the early 1900s.[2] These examinations were largely done with the intention of

---

[1]For a review, see Nicole Rafter, "The Murderous Dutch Fiddler," *Theoretical Criminology* 9. no. 1 (2005): 65–96.

[2]For example, see the comparison made between Robert Bean, "Some Racial Peculiarities of the Negro Brain," *American Journal of Anatomy* 5 (1906): 353–432, which showed a distinct difference in the brains across race when brains were identified by race before comparison, and F. P. Mall, "On Several Anatomical Characters of the Human Brain, Said to Vary According to Race and Sex, with Especial Reference to the Weight of the Frontal Lobe," *American Journal of Anatomy* 9 (1909): 1–32, which showed virtually no differences among the same brains when comparisons were made without knowing the races of the brains prior to comparison; see discussion in Stephen Jay Gould, *The Mismeasure of Man*, 2nd ed. (New York: Norton, 1996).

furthering the assumptions of eugenics, which aimed to prove under the banner of science that certain individuals and ethnic or racial groups are inferior to others. The fact that this was their intent is underscored by subsequent tests using the same subjects but without knowing which skulls/ brains were from certain ethnic or racial groups; these later studies showed only a small correlation between size of the skull or brain and certain behaviors or personality.[3]

Furthermore, once some of the early practitioners of craniometry died, their brains had a volume that was less than average or average. The brain of of K. F. Gauss, for example, was relatively small but more convoluted, in terms of more gyrus fissures in the brain. Craniometrists then switched their postulates to say that more convoluted or complex structure of brains, with more fissures and gyrus, indicated superior brains.[4] However, this argument was even more tentative and vague than the former hypotheses of craniometrists and thus did not last long. The same was true of craniometry, thanks to its noticeable lack of validity. However, it is important to note that modern studies show that people who have significantly larger brains tend to score higher on intelligence tests.[5]

Despite the failure of craniometry to explain the difference between criminals and noncriminals, scientists were not ready to give up the assumption

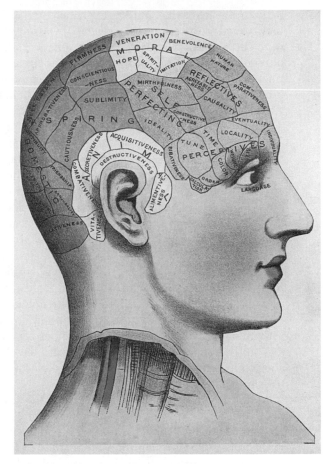

▲ **Photo 3.1**   This diagram shows the sections of the brain, as detailed by 19th-century phrenologists, who believed that each section was responsible for a particular human personality trait. If a section were enlarged or shrunken, the personality would be likewise abnormal. Doctors, particularly those doing entry examinations at American prisons, would examine the new inmate's head for bumps or cavities to develop a criminal profile. For example, if the section of brain responsible for "acquisitiveness" was enlarged, the offender probably was a thief. Lombroso and his school combined phrenology with other models that included external physical appearance traits that could single out criminals from the general population.

---

[3]See Mall, "On Several Anatomical Characters"; much of the discussion in this section is taken from Gould, *The Mismeasure.*

[4]E. A. Spitska, "A Study of the Brains of Six Eminent Scientists and Scholars Belonging to the Anthropological Society, Together with a Description of the Skull of Professor E. D. Cope," *Transactions of the American Philosophical Society* 21 (1907): 175–308.

[5]Stanley Coren, *The Left-Hander Syndrome* (New York: Vintage, 1993); James Kalat, *Biological Psychology*, 8th ed. (New York: Wadsworth, 2004).

that criminal behavior could be explained by visual differences in the skull (or brain), and they certainly weren't ready to give up the assumption that certain ethnic or racial groups were superior or inferior relative to others. Therefore, the experts of the time created phrenology. **Phrenology** is the science of determining human dispositions based on distinctions (e.g., bumps) in the skull, which are believed to conform to the shape of the brain.[6] Readers should keep in mind that much of the theorizing by the phrenologists still aimed to support the assumptions of eugenics and show that certain individuals and groups of people were inferior or superior to others.

It is important to keep in mind that, like the craniometrists, phrenologists assumed that the shape of the skull conformed to the shape of the brain. Thus, a bump or other abnormality on the skull directly related to an abnormality in the brain at that spot. Such an assumption has been refuted by modern scientific evidence, so it is not surprising that phrenology fell out of favor in criminological thought rather quickly. Like its predecessor, however, phrenology got some things right. Certain parts of the brain are indeed responsible for specific tasks.

For example, in the original phrenological map, destructiveness was indicated by abnormalities above the left ear. Modern scientific studies show that the most vital part of the brain in terms of criminality associated with trauma is the left temporal lobe, the area above the left ear.[7] Furthermore, phrenologists were right in other ways. Most readers know that specific portions of the brain govern the operation of different physical activities; one area governs the action of our hands, whereas other areas govern our arms, legs, and so on. So the phrenologists had a few things right but were completely wrong about the extent to which bumps on the skull could indicate who would be most disposed to criminal behavior.

Once phrenology fell out of favor among scientists, the researchers and society did not want to depart from the assumption that certain individuals or ethnic groups were inferior to others. Therefore, another discipline known as physiognomy became popular in the mid-1800s. **Physiognomy** is the study of facial and other bodily aspects to indicate developmental problems, such as criminality. Not surprisingly, the early physiognomy studies focused on contrasting various racial or ethnic groups to prove that some were superior or inferior to others.[8]

Using modern understandings of science, it should be obvious from the illustrations that physiognomy did not last long as a respected scientific perspective of criminality. At any time other than the late 1800s, their ideas would not have been accepted for long, if at all. However, the timing of such a theory was on its side. Specifically, Darwin published his work, *The Origin of Species*, in the late 1800s and made a huge impact on societal views regarding the rank order of groups in societies.

Darwin's model outlined a vague framework suggesting that humans had evolved from more primitive beings and that the human species (like all others) evolved from a number of adaptations preferred by natural selection. In other words, some species are

---

[6]Orson S. Fowler, *Fowler's Practical Phrenology: Giving a Concise Elementary View of Phrenology* (New York: O. S. Fowler, 1842).

[7]For a review, see Adrian Raine, *The Psychopathology of Crime* (San Diego, CA: Academic Press, 1993).

[8]J. C. Nott and G. R. Gliddon, *Types of Mankind* (Philadelphia, PA: Lippincott, Grambo, and Company, 1854); J. C. Nott and G. R. Gliddon, *Indigenous Races on Earth* (Philadelphia, PA: Lippincott, 1868).

selected by their ability to adapt to the environment, whereas others do not adapt as well and die off, or at least become inferior in terms of dominance. This assumption of Darwin's work, which was quickly and widely accepted by both society and scientists throughout the world, falsely led to an inclination to believe that certain ethnic or racial groups are inferior or superior to groups.

Darwin was not a criminologist, so he is not considered the father or theorist of any major schools of thought. However, he did set the stage for what followed in criminological thought. Specifically, Darwin's theory laid the groundwork for what would become the first major scientific theory of crime, namely Cesare Lombroso's theory of born criminals, which also tied together the assumptions and propositions of craniometry, phrenology, and physiognomy.

▧ Lombroso's Theory of Atavism and Born Criminals

Basing his work on Darwin's theory of natural selection, Cesare Lombroso (1835–1909) created what is widely considered the first attempt toward scientific theory in criminological thought. Most previous theorists were not scientists; Beccaria, for example, was trained in law and never tested his propositions. Unlike the craniometrists and phrenologists, Beccaria's goal was not to explain levels of criminality. However, Lombroso was trained in medical science, and he aimed to document his observations and use scientific methodology. Furthermore, timing was on his side in the sense that Darwin's theory was published 15 years prior to the publication of Lombroso's major work and in that time had become immensely popular with both scientists and the public.

Lombroso's Theory of Crime. The first edition of Lombroso's work, *The Criminal Man*, was published in 1876 and created an immediate response in most Western societies, influencing both their ideas and their policies related to crime

▲ **Photo 3.2**   Cesare Lombroso (1836–1909)

and justice.[9] In this work, Lombroso outlined a theory of crime that largely brought together the pre-Darwinian theories of craniometry, phrenology, and physiognomy. Furthermore, Lombroso thought that certain groups and individuals were *atavistic* and

---

[9]Cesare Lombroso, *The Criminal Man (L'uomo Delinquente)*, 1st ed. (Milan: Hoepli, 1876); 2nd ed. (Turin: Bocca, 1878).

likely to be born to commit crime. **Atavistic or atavism** means that a person or feature of an individual is a throwback to an earlier stage of evolutionary development. In other words, he thought serious criminals were lower forms of humanity in terms of evolutionary progression. For example, Lombroso would likely suggest that chronic offenders are more like earlier stages of humankind, like the "missing link," than they are like modern humans.

Lombroso noted other types of offenders, such as mentally ill and "criminaloids," who committed minor offenses due to external or environmental circumstances, but he argued that the "*born criminals*" should be the target in addressing crime, insisting that they were the most serious and violent criminals in any society. These are what most criminologists now refer to as chronic offenders. Furthermore, Lombroso claimed that born criminals cannot be stopped from their natural tendencies to be antisocial.

On the other hand, Lombroso claimed that although it was their nature to commit crime, societies could prevent or reduce the crimes that they were inevitably going to commit. According to Lombroso, societies could identify born criminals, even early in life, through their stigmata. **Stigmata** are physical manifestations of the atavism of an individual, features that indicate a prior evolutionary stage of development.

Lombroso's List of Stigmata. According to Lombroso, more than five stigmata indicate that an individual is atavistic and inevitably will be a born criminal. Understandably, readers may be wondering what these stigmata are, given their importance. This is a great question, but the answer varies. In the beginning, this list was largely based on Lombroso's work as a physician; it included such things as large eyes, large ears, and so on. Lombroso changed this list as he went along, however, which might be considered poor science, even in the last edition of his book published well into the 1900s.

For the most part, stigmata consisted of facial and bodily features that deviated from the norm. In other words, abnormally small or large noses, abnormally small or large ears, abnormally small or large eyes, abnormally small or large jaws—almost anything that went outside the bell curve on normal human physical development. Lombroso also threw in some extraphysiological features, such as tattoos and a family history of epilepsy and other disorders.[10] Although tattoos may be somewhat correlated to crime and delinquency, is it likely that they caused such antisocial behavior? Given Lombroso's model that people are born criminal, it is quite unlikely that these factors are causally linked to criminality. How many babies are born with tattoos? Ignoring the illogical nature of many of the stigmata, Lombroso professed that people who had more than five of these physical features were born criminals and that something should be done to prevent the inevitable in their future offending career.

As a physician working for the Italian army, Lombroso examined the bodies of war criminals who were captured and brought in for analysis. According to Lombroso,[11] he first came to the realization of the nature of the criminal when a particular war criminal was brought in for him to examine:

---

[10]Gould, *The Mismeasure.*

[11]Lombroso, *Criminal Man,* as cited and discussed by Ian Taylor, P. Walton, and J. Young, *The New Criminology: For a Social Theory of Deviance* (London: Routledge, 1973), 41.

This was not merely an idea, but a flash of inspiration. At the sight of that skull, I seemed to see all of a sudden, lighted up as a vast plain under a flaming sky, the problem of the nature of the criminal—an atavistic being who reproduces in his person the ferocious instincts of primitive humanity and the inferior animals.

This was Lombroso's first exposure to such a criminal and his acknowledgment of the theory that he created. He further expanded on this theory by specifying some of the physical features he observed from this individual:[12]

Thus were explained anatomically the enormous jaws, high cheek bones . . . solitary lines in the palms, extreme size of the orbits, handle-shaped ears found in criminals, savages and apes, insensibility to pain, extremely acute sight, tattooing, excessive idleness, love of orgies, and the irresponsible craving of evil for its own sake, the desire not only to extinguish life in the victim, but to mutilate the corpse, tear its flesh and drink its blood.

Although most people may now laugh at his words, at the time he wrote this description, it would have rung true to most readers, which is likely why his book was the dominant text for many decades in the criminological field. In this description, Lombroso incorporates many of the core principles of his theory: the idea of criminals being atavistic or biological throwbacks in evolution, as well as the stigmata that became so important in his theoretical model.

A good example of the popular acceptance of Lombroso's "scientific" stigmata was that Bram Stoker used it in the 1896 novel, *Dracula*, a character based on Lombrosian traits of the villain, such as the high bridge of the thin nose, arched nostrils, massive eyebrows, pointed ears, and so on. This novel was published in the late 1800s, when Lombroso's theory was highly dominant in society and science. Lombroso's ideas became quite popular among academics, scientists, philosophers, fiction writers, and those responsible for criminal justice policy.

Beyond identifying born criminals by their stigmata, Lombroso said he could associate the stigmata with certain types of criminals, for example, anarchists, burglars, murderers, shoplifters, and so on. Of course, his work is quite invalid by modern research standards.

Lombroso as the Father of Criminology and the Father of the Positive School. Lombroso's theory came a decade and a half after Darwin's work had been published and had spread rapidly throughout the Western world. Also, Lombroso's model supported what were then the Western world's views on slavery, deportation, and so on. Due to this timing and the fact that Lombroso became known as the first individual who actually tested his hypotheses through observation, Lombroso is widely considered the father of criminology. This title does not indicate respect for his theory, which has been largely rejected, or for his methods, which are considered highly invalid by modern standards. It is deserved, however, in the sense that he was the first person to gain recognition in testing his theoretical propositions. Furthermore, his theory coincided with a political movement that became popular at that time: the Fascist and Nazi movements of the early 1900s.

---

[12]Taylor et al., *New Criminology*, 41–42.

Beyond being considered the father of criminology, Lombroso is also considered the father of the **Positive School** of criminology because he was the first to gain prominence in identifying factors beyond free will or free choice, which the Classical School said were the sole cause of crime. Although previous theorists had presented perspectives that went beyond free will, such as craniometrists and phrenologists, Lombroso was the first to gain widespread attention. Lombroso's perspective gained almost immediate support in all developed countries of that time, which is the most likely reason for why Lombroso is considered the father of the Positive School of criminology.

It is important to understand the assumptions of positivism, which most experts consider somewhat synonymous with the term determinism. **Determinism** is the assumption that most human behavior is determined by factors beyond free will and free choice. In other words, determinism (i.e., the Positive School) assumes that human beings do not decide how they will act by logically thinking through the costs and benefits of a given situation. Rather, the Positive School attributes all kinds of behavior, especially crime, on biological, psychological, and sociological variables.

Many readers probably feel they chose their career paths and made most other key decisions in their lives. However, scientific evidence shows otherwise. For example, studies clearly show that far more than 90% of the world's population have adopted the religious affiliation (e.g., Baptist, Buddhist, Catholic, Judaism, etc.) of their parents or caretakers. Therefore, what most people consider an extremely important decision, the choice of what they believe regarding a higher being or force, is almost completely determined by the environment in which they were brought up. Almost no one sits down and studies various religions before deciding which one suits him or her the best. Rather, in almost all cases, religion is determined by their culture, and this finding goes against the Classical School's assumption that free will rules. The same type of argument can be made about the clothes we wear, the food we prefer, and the activities that give us pleasure.

Another way of distinguishing positivism and determinism from the Classical School lies in the way scientists view human behavior, which can be seen best as an analogy with chemistry. Specifically, a chemist assumes that if a certain element is subjected to certain temperatures, pressures, or mixtures with other elements, it will result in a predicted response to such influences. In a highly comparable way, a positivist assumes that when human beings are subjected to poverty, delinquent peers, low intelligence, or other factors, they will react and behave in a predictable way. Therefore, there is virtually no difference in how a chemist feels about particles and elements and how a positivistic scientist feels about how humans react when exposed to biological and social factors.

In Lombroso's case, the deterministic factor was the biological makeup of the individuals. However, we shall see in the next several sections that positivistic theories focus on a large range of variables, from biology to psychology to social aspects. For example, many readers may believe that bad parenting, poverty, and associating with delinquent peers are some of the most important factors in predicting criminality. If you believe that such variables have a significant influence on decisions to commit crime, then you are likely a positive theorist; you believe that crime is caused by a factor above and beyond free choice or free will.

Lombroso's Policy Implications. Beyond the theoretical aspects of Lombroso's theory of criminality, it is important to realize that his perspective had profound consequences for policy. Lombroso was called to testify in numerous criminal processes and trials to determine the guilt or innocence of suspects. Under the banner of science (comparable to what we consider DNA or fingerprint analysis in modern times), Lombroso was asked to specify whether or not a suspect had committed a crime.[13] Lombroso based such judgments on the visual stigmata that he could see among suspects.[14] Lombroso documented many of his experiences as an expert witness at criminal trials. Here is one example: "[one suspect] was, in fact, the most perfect type of the born criminal: enormous jaws, frontal sinuses, and zygomata, thin upper lip, huge incisors, unusually large head . . . [h]e was convicted."[15]

When Lombroso was not available for such "scientific" determinations of the guilty persons, his colleagues or students (often referred to as lieutenants) were often sent. Some of these students, such as Enrico Ferri and Raphael Garrofalo, became quite active in the Fascist regime of Italy in the early 1900s. This model of government, like the Nazi Party of Germany, sought to remove the "inferior" groups from society.

Another policy implication in some parts of the world was identifying young children on the basis of observed stigmata, which often become noticeable in the first 5 to 10 years of life. This led to tracking or isolating certain children, largely based on physiological features. Although many readers may consider such policies ridiculous, modern medicine has supported the identification, documentation, and importance of what are termed **minor physical anomalies (MPAs),** which it holds may indicate high risk of developmental problems. Some of these MPAs include:[16]

- Head circumference out of the normal range
- Fine "electric" hair
- More than one hair whorl
- Epicanthus, which is observed as a fold of skin from the lower eyelids to the nose, which appears as droopy eyelids
- Hypertelorism (orbital), which represents an increased interorbital distance
- Malformed ears
- Low-set ears
- Excessively large gap between the first and second toes
- Webbing between toes or fingers
- No ear lobes
- Curved fifth finger
- Third toe is longer than the second toe
- Asymmetrical ears
- Furrowed tongue
- Simian crease

---

[13]See Gould, *The Mismeasure.*

[14]For a review of such identifications, see Gould, *The Mismeasure.*

[15]Cesare Lombroso, *Crime: Its Causes and Remedies* (Boston, MA: Little & Brown, 1911), 436.

[16]Taken from Waldrop and Halverson, (1971), as cited and reviewed by Diana Fishbein, *Biobehavioral Perspectives on Criminology* (Belmont, CA: Wadsworth, 2001).

Given that such visible physical aspects are still correlated with developmental problems, including criminality, it is obvious that Lombroso's model of stigmata in predicting antisocial problems has implications to the present day. Such implications are more accepted by modern medical science than they are in the criminological literature. Furthermore, some modern scientific studies have shown that being unattractive predicts criminal offending, which somewhat supports Lombroso's theory of crime.[17]

About three decades after Lombroso's original work was published, and after a long period of dominance, criminologists began to question his theory of atavism and stigmata. Furthermore, it became clear that more was involved in criminality than just the way people looked, such as psychological aspects of individuals. However, scientists and societies were not ready to depart from theories like Lombroso's, which assumed that certain people or groups of people were inferior to others, so they simply chose another factor to emphasize: intelligence or IQ.

⌧ The IQ Testing Era

Despite the evidence against Lombroso that was presented, his theorizing remained dominant until the early 1900s, when criminologists realized that his stigmata, and the idea of a born criminal was not valid. However, even at that time, theorists and researchers were not ready to give up on the eugenics assumption that certain ethnic or racial groups were superior or inferior to others. Thus, a new theory emerged based on a more quantified measure that was originated with benevolent intentions in France by Alfred Binet. This new measure was IQ, short for intelligence quotient. At that time, intelligence quotient was calculated as chronological age divided by mental age, which was then multiplied by 100, with average scores being 100. This scale was changed enormously over time, but the bottom line was using the test to determine whether someone was above or below the base score of average (100).

As mentioned above, Binet had good intentions: He created IQ scores to identify youth who were not performing up to par on educational skills. Binet was explicit in stating that IQ could be changed, which is why he proposed a score to identify slow learners so that they could be trained to increase their IQ.[18] However, when Binet's work was brought over to the United States, his basic assumptions and propositions were twisted. One of the most prominent individuals who used Binet's IQ test in the United States for purposes of deporting, incapacitating, sterilizing, and otherwise ridding society of low IQ individuals was H. H. Goddard.

Goddard is generally considered the leading authority on the use and interpretation of IQ testing in the United States.[19] He adapted Binet's model to examine the immigrants who were coming into the United States from foreign lands. It is important to note that Goddard proposed quite different assumptions regarding intelligence or IQ than did Binet. Goddard asserted that IQ was static or innate, meaning that such levels could not be changed, even with training. His assumption was that intelligence was passed from generation to generation, inherited from parents.

---

[17]Robert Agnew, "Appearance and Delinquency," Criminology 22 (1984): 421–40.

[18]Gould, *The Mismeasure.*

[19]Again, most of this discussion is taken from Gould, *The Mismeasure*, because his review is perhaps the best known in the current literature.

Goddard labeled low IQ **feeblemindedness,** which actually became a technical, scientific term in the early 1900s meaning those who had significantly below average levels of intelligence. Of course, being a scientist, Goddard specified certain levels of feeblemindedness, which were ranked based on the degree to which the score of the individual was below average. Ranking from the best to the lowest intelligence, the first group was the *morons,* the second-lowest group was the *imbeciles,* and the lowest-intelligence group was the *idiots.*

According to Goddard, from a eugenics point of view, the biggest threat to the progress of humanity was not the idiots but the morons. In Goddard's words: "The idiot is not our greatest problem. . . . He does not continue the race with a line of children like himself. . . . It is the moron type that makes for us our great problem."[20] Thus, the moron is the group of the three categories of the feebleminded that is smart enough to slip through the cracks and reproduce.

Goddard received many grants to fund his research to identify the feebleminded. Goddard took his research team to the major immigration center at Ellis Island in the early 1900s to identify the feebleminded as they attempted to enter the United States. Many members of his team were women, who he felt were better at distinguishing the feebleminded by sight:

> The people who are best at this work, and who I believe should do this work, are women. Women seem to have closer observation than men. It was quite impossible for others to see how . . . women could pick out the feeble-minded without the aid of the Binet test at all.[21]

▲ **Photo 3.3** Ellis Island

---

[20]H. H. Goddard, *The Kallikak Family, a Study of the Heredity of Feeble-Mindedness* (New York: MacMillan, 1912).

[21]H. H. Goddard, "The Binet Tests in Relation to Immigration," *Journal of Psycho-Asthenics* 18 (1913): 105–07, quote taken from p. 106; as cited in Gould, *The Mismeasure.*

Goddard was proud of the increase in the deportation of potential immigrants to the United States, enthusiastically reporting that deportations for the reason of mental deficiency increased by 350% in 1913 and 570% in 1914 over the average of the preceding five years.[22] However, over time, Goddard realized that his policy recommendations of deportation, incarceration, and sterilization were not based on accurate science.

After consistently validating his IQ test on immigrants and mental patients, Goddard finally tested his intelligence scale on a relatively representative cross-section of American citizens, namely draftees for military service during World War I. The results showed that many of these recruits would score as feebleminded (i.e., lower than a mental age of 12) on the IQ test. Therefore, Goddard lowered the criterion of what determined a person of feeblemindedness from a mental age of 12 to a mental age of 8. Although this appears to be a clear admission that his scientific method was inaccurate, Goddard continued to promote his model of the feebleminded for many years, and societies used his ideas.

However, toward the end of his career, Goddard admitted that intelligence could be improved, despite his earlier assumptions that it was innate and static.[23] In fact, Goddard actually claimed that he had "gone over to the enemy."[24] However, despite Goddard's admission that his assumptions and testing were not determinant of individuals' intelligence, the snowball had been rolling for too long and had gathered too much strength to fight even the most notable theorist's admonishment of the perspective.

Sterilization of individuals, mostly females, continued in the United States based on scores of intelligence tests. Often the justification was not the person's intelligence scores but those of his or her mother or father. Goddard had proclaimed that the "germ-plasm" determining feeblemindedness was passed on from one generation to the next, so it inevitably resulted in offspring being feebleminded as well. Thus, the U.S. government typically sterilized individuals, typically women, based on IQ scores of their parents.

The case of *Buck v. Bell,* brought to the U.S. Supreme Court in 1927, dealt with the issue of sterilizing individuals who had scored, or whose parents had scored, as mentally deficient on intelligence scales. The majority opinion, written by one of the court's most respected jurists, Oliver Wendell Holmes, Jr., stated that:

> We have seen more than once that the public welfare may call upon the best citizens for their lives. It would be strange if it could not call upon those who already sap the strength of the state for these lesser sacrifices. . . . Three generations of imbeciles are enough.[25]

Thus, the highest court in the United States upheld the use of sterilization for the purposes of limiting reproduction among individuals who were deemed as being "feebleminded" according to an IQ score. Such sterilizations continued until the 1970s, when the practice was finally halted. Governors of many states, such as North Carolina,

---

[22]Gould, *The Mismeasure*, p. 198.

[23]H. H. Goddard, "Feeblemindedness: A Question of Definition," *Journal of Psycho-Asthenics* 33 (1928): 219–27, as discussed in Gould, *The Mismeasure.*

[24]H. H. Goddard, "Feeblemindedness," 224.

[25]As quoted in Gould, *The Mismeasure*, 365.

Virginia, and California, have given public apologies for what was done. For example, in 2002, the governor of California, Gray Davis, apologized for the state law passed almost a century earlier that had resulted in the sterilization of about 19,000 women in California.

Although this aspect of U.S. history is often hidden from the public, it did occur, and it is important to acknowledge this blot on our nation's history, especially at a time when we were fighting abuses of civil rights by the Nazis and other regimes. Ultimately, the sterilizations, deportations, and incarcerations based on IQ testing are an embarrassing episode in the history of the United States.

For decades, the issue of IQ was not researched or discussed much in the literature. However, in the 1970s, a very important study was published in which Travis Hirschi and Michael Hindelang examined the effect of intelligence on youths.[26] Hirschi and Hindelang found that among youth of the same race and social class, intelligence had a significant effect on delinquency and criminality among individuals. This study, as well as others, showed that the IQ of delinquents or criminals is about 10 points lower than the scores of noncriminals.[27]

This study led to a rebirth in research regarding intelligence testing in criminological research. A number of recent studies have shown that certain types of intelligence are more important than others. For example, several studies have shown that having a low verbal intelligence has the most significant impact on predicting delinquent and criminal behavior.[28]

This tendency makes sense because verbal skills are important for virtually all aspects of life, from everyday interactions with significant others to filling out forms at work to dealing with people via employment. In contrast, most people do not have to perform advanced math or quantitative skills at their job, or in day-to-day experiences, let alone spatial and other forms of intelligence that are more abstract. Thus, low verbal IQ is the type of intelligence that represents the most direct prediction for criminality, and this is most likely due to the general need for such skills in routine daily activities. After all, people who lack such communication skills will likely find it hard to obtain or retain employment, as well as deal with family and social problems, due their lack of being able to communicate in daily life.

This rebirth in studies regarding the link between intelligence and criminality seemed to reach a peak with the publication of Richard Herrnstein and Charles Murray's *The Bell Curve* in 1994.[29] Although this publication changed the terminology of *moron, imbecile,* and *idiot* to relatively benign terms (e.g., "cognitively disadvantaged"), their argument

---

[26]Travis Hirschi and Michael Hindelang, "Intelligence and Delinquency: A Revisionist Review," *American Sociological Review* 42 (1977): 571–87.

[27]For a review, see Raymond Paternoster and Ronet Bachman, *Explaining Criminals and Crime* (Los Angeles, CA: Roxbury, 2001).

[28]For a review, see Chris L. Gibson, Alex R. Piquero, and Stephen G. Tibbetts, "The Contribution of Family Adversity and Verbal IQ to Criminal Behavior," *International Journal of Offender Therapy and Comparative Criminology* 45 (2001): 574–92; see also, Hirschi and Hindelang, "Intelligence and Delinquency"; Terrie Moffitt, "The Neuropsychology of Delinquency: A Critical Review of Theory and Research," in *Crime and Justice: An Annual Review of Research*, Vol. 12, ed. M. Tonry and N. Morris (Chicago: University of Chicago Press, 1990), pp. 99–169,; Terrie Moffitt and B. Henry, "Neuropsychological Studies of Juvenile Delinquency and Juvenile Violence," in *The Neuropsychology of Aggression*, ed. J. S. Miller (Boston, MA: Kluwer, 1991), 67–91; also see conclusion by Paternoster and Bachman, *Explaining Criminals*, 51.

[29]Richard J. Herrnstein and Charles Murray, *The Bell Curve: Intelligence and Class Structure in the United States* (New York: Free Press, 1994).

was consistent with that of the feeblemindedness researchers of the early 20th century. Herrnstein and Murray argued that people with low IQ scores are somewhat destined to be unsuccessful in school, become unemployed, produce illegitimate children, and commit crime. They also suggest that IQ or intelligence is primarily innate, or genetically determined, with little chance of improving it. These authors also noted that African Americans tend to score the lowest, whereas Asians and Jewish people tend to score highest, and they offer results from social indicators that support their argument that these intelligence levels result in relative success in life in terms of group-level statistics.

This book produced a public outcry, resulting in symposiums at major universities and other venues in which the authors' postulates were largely condemned. As noted by other reviews of the impact of this work, some professors at public institutions were sued in court because they used this book in their classes.[30] The book received blistering reviews from fellow scientists.[31] However, none of these scientific critics has fully addressed the undisputed fact that African Americans consistently score low on intelligence tests and that Asians and Jews score higher on these examinations. Furthermore, none have adequately addressed the issue that even within these populations, low IQ scores (especially on verbal tests) predict criminality. For example, in samples of African Americans, the group that scores lowest on verbal intelligence consistently commits more crime and is more likely to become delinquent or criminal. So despite the harsh criticism of *The Bell Curve*, it is apparent that there is some validity to the authors' arguments.

With the popularity of intelligence testing and IQ scores in the early 20th century, it is not surprising that this was also the period when other psychological models of deviance and criminality became popular. However, one of the most popular involved body-type theories.

◪ Body Type Theory: Sheldon's Model of Somatotyping

Although there were numerous theories based on body types in the late 1800s and early 1900s, such as Lombroso's and those of others who called themselves criminal anthropologists, none of these perspectives had a more enduring impact than that of William Sheldon. In the mid-1940s, a new theoretical perspective was proposed that merged the concepts of biology and psychology. Sheldon claimed that in the embryonic and fetal stages of development, individuals tend to have an emphasis on the development of certain tissue layers.[32] According to Sheldon, these varying degrees of emphasis were largely due to heredity, which led to the development of certain body types and temperaments or personalities. This became the best-known body-type theory, also known as **somatotyping**.

According to Sheldon, all embryos must develop three distinct tissue layers, and this much is still acknowledged by perinatal medical researchers. The first layer of tissue is the

---

[30]See Stephen Brown, Finn Esbensen, and Gilbert Geis, *Criminology* 6th ed. (Cincinnati, OH: LexisNexis, 2007) , 260.

[31]J. Blaine Hudson, "Scientific Racism: The Politics of Tests, Race, and Genes," *Black Scholar* 25 (1995): 1–10; David Perkins, *Outsmarting IQ: The Emerging Science of Learnable Intelligence* (New York: The Free Press, 1995); Francis Cullen, Paul Gendreau, G. Roger Jarjoura, and John P. Wright, "Crime and the Bell Curve: Lessons From Intelligent Criminology," *Crime and Delinquency* 43 (1997): 387–411; Robert Hauser, "Review of the Bell Curve," *Contemporary Sociology* 24 (1995): 149–53; Howard Taylor, "Book Review, *The Bell Curve*," *Contemporary Sociology* 24 (1995): 153–58; Gould, *The Mismeasure*.

[32]William Sheldon, E. M. Hartl, and E. McDermott, *Varieties of Delinquent Youth* (New York: Harper, 1949).

endoderm, which is the inner layer of tissues and includes the internal organs, such as the stomach, large intestine, small intestine, and so on. The middle layer of tissue, called the **mesoderm,** includes the muscles, bones, ligaments, and tendons. The **ectoderm** is the outer layer of tissue, which includes the skin, capillaries, and much of the nervous system's sensors.

Sheldon used these medical facts regarding various tissue layers to propose that certain individuals tend to emphasize certain tissue layers relative to others, typically due to inherited dispositions. In turn, Sheldon believed that such emphases led to certain body types in an individual, such that people who had a focus on their endoderm in development would inevitably become **endomorphic,** or obese (see Figure 3.1a). Individuals who had an emphasis on the middle layer of tissue typically became **mesomorphic,** or of an athletic or muscular build (see Figure 3.1b). Finally,

Figure 3.1a

Endomorph. *Physical traits*: soft body, underdeveloped muscles, round shape, overdeveloped digestive system. *Associated personality traits*: love of food, tolerance, evenness of emotions, love of comfort, sociability, good humor, relaxed mood, need for affection.

Figure 3.1b

Mesomorph. *Physical traits*: hard, muscular body; overly mature appearance; rectangular shape; thick skin; upright posture. *Associated personality traits*: love of adventure, desire for power and dominance, courage, indifference to what other think or want, assertive mien, boldness, zest for physical activity, competitive nature, love of risk and chance.

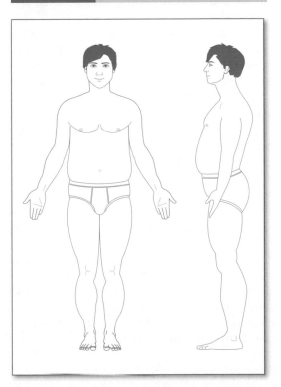

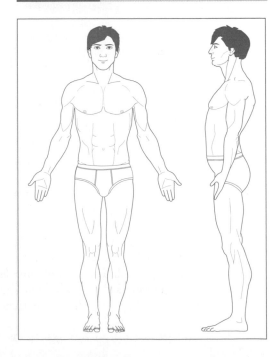

Figure 3.1c

Ectomorph. *Physical traits*: thin, flat chest, delicate build, young appearance, lightly muscled, stoop-shouldered, large brain. *Associated personality traits*: self-conscious, preference for privacy, introverted, inhibited, socially anxious, artistic, mentally intense, emotionally restrained.

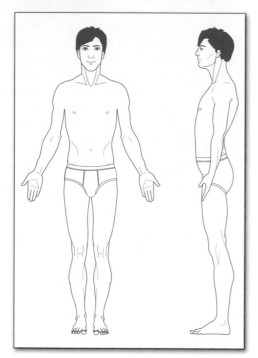

someone who had an emphasis on the other layer of tissue in embryonic development would end up being an **ectomorphic** build, or thin (see Figure 3.1c).

Sheldon and his research team graded each subject on three dimensions corresponding to these body types. Each body type was measured on a scale of 1 to 7, with 7 being the highest score. Obviously, no one could score a zero for any body type because all tissue layers are needed for survival; we all need our internal organs, bone/muscular structure, and outer systems (e.g., skin, capillaries, etc.). Each somatotype always had the following order: endomorphy, mesomorphy, ectomorphy.

So the scores on a typical somatotype might be 3-6-2, which would indicate that this person scored a 3 (a little lower than average) on endomorphy, a 6 (high) on mesomorphy, and a 2 (relatively low) on ectomorphy. According to Sheldon's theory, this hypothetical subject would be a likely candidate for criminality, because of the relatively high score on mesomorphy. In fact, the results from his data, as well as all studies that have examined the association of body types and delinquency/criminality, would support this prediction.

Perhaps most important, Sheldon proposed that these body types matched personality traits or temperaments. Individuals who were endomorphic (obese), Shelden claimed, tended to be more jolly or lazy. The technical term for this temperament is referred to as **viscerotonic**. In contrast, people who were mesomorphic (muscular) typically had a temperament of being risk-taking and aggressive, called **somotonic**. Last, individuals who were ectomorphic (thin) tended to have a personality of being introverted and shy, which is referred to as **cerebrotonic**. Obviously, members of the middle group, the mesomorphs, had the highest propensity toward criminality because they were disposed toward a risk-taking and aggressive personality.

Interestingly, many politicians were subjects in Sheldon's research. Most entering freshmen at Ivy League schools, especially Harvard, were asked to pose for photos for Sheldon's studies. The Smithsonian Institution still retains a collection of nude photos, which includes George W. Bush, Hillary Rodham Clinton, and many other notable figures.[33]

Sheldon used poor methodology to test his theory. He based his measures of subjects' body types on what he subjectively judged from viewing three perspectives of each

---

[33]See discussion in Brown et al., op cit., p. 246.

subject and often from only three pictures taken of each subject in the three poses. He also had his trained staff view many of the photos and make their determination of how these individuals scored on each category of body type. The evidence of the reliability among these scorings was shown to be weak, meaning that the trained staff did not tend to agree with Sheldon or among themselves on the somatotype for each participant.

This is not surprising given the high level of variation in body types and the fact that Sheldon and his colleagues did not employ the technology that is used today, such as caliper tests and submersion in water tanks, which provide the information for which he was searching. Readers may have altered their weight, going from an ecto-morphic or mesomorphic build to a more endomorphic form, or vice versa. Presented with the argument that individuals often alter their body type via diet or exercise, Sheldon responded that he could tell what the "natural" body type of each individual was from the three pictures that were taken. Obviously, this position is not a strong one, which is backed up by the poor inter-rater reliability shown by his staff. Therefore, Sheldon's methodology is questionable, which casts doubt on the entire theoretical framework.

Despite the problematic issues regarding his methodology, Sheldon clearly showed that mesomorphs, or individuals who had muscular builds and tended to be more risk-taking, were more delinquent and criminal than individuals who had other body types/temperaments.[34] Furthermore, other researchers, even those who despised Sheldon's theory, found the same associations between mesomorphy and criminality, as well as related temperaments (i.e., somotonic) and criminality.[35] Subsequent studies showed that mesomorphic boys were far more likely to have personality traits that predicted criminality, such as aggression, short temper, self-centeredness, and impulsivity.

Recent theorists have also noted the link between an athletic, muscular build and the highly extroverted, aggressive personality that is often associated with this body type.[36] In fact, some recent theorists have gone so far as to claim that chronic offenders, both male and female, can be identified early in life by the their relatively "v-shaped" pelvic structure, as opposed to a "u-shaped" pelvic structure.[37] The v-shaped is said to indicate relatively high level of androgens (male hormones, like testosterone) in their system, which predisposes individuals toward crime. On the other hand, a more u-shaped pelvis indicates relatively low levels of such androgens and therefore lower propensity toward aggression and criminality. Using this logic, it may be true that more hair on an individual's arms (whether they be male or female) is predictive of a high likelihood for committing crime. However, no research exists regarding this factor.

Regarding the use of body types and characteristics in explaining crime, many of the hard-line sociologists who have attempted to examine or replicate Sheldon's results have never been able to refute the association between mesomorphs and delinquency/criminality, nor the association between mesomorphy and somotonic temperament of being risk

---

[34]Sheldon et al., *Varieties.*

[35]Sheldon Glueck and Eleanor Glueck, *Physique and Delinquency* (New York: Harper and Row, 1956); see also, Emil Hartl, *Physique and Delinquent Behavior* (New York: Academic Press, 1982); Juan Cortes, *Delinquency and Crime* (New York: Seminar Press, 1972); for the most recent applications, see Hans J. Eysenck and G. H. Gudjonsson, *The Causes and Cures of Criminality* (New York: Plenum, 1989).

[36]James Q. Wilson and Richard Herrnstein, *Crime and Human Nature* (New York: Simon and Schuster, 1985).

[37]Eysenck and Gudjonsson, *Causes and Cures.*

taking or aggressive.[38] Thus, the association between being muscular or athletically built and engaging in criminal activities is now undisputed and assumed to be true. Still, sociologists have taken issue with what is causing this association.

Whereas Sheldon claimed it was due to inherited traits for a certain body type, sociologists argue that this association is due to societal expectations: Muscular male youth would be encouraged to engage in risk-taking and aggressive behavior. For example, a young male with an athletic build will be encouraged to join sports teams and engage in high-risk behavior by peers. Who would gangs most desire as members? The more muscular, athletic individuals would be better at fighting and performing acts that require a certain degree of physical strength and stamina.

Ultimately, it is now established that mesomorphs are more likely to commit crime.[39] Furthermore, the personality traits that are linked to having an athletic or muscular build are dispositions toward risk-taking and aggressiveness, and few scientists dispute this correlation. No matter which theoretical model is adopted, whether it be the biopsychologists or the sociologists, the fact of the matter is that mesomorphs are indeed more likely to be risk-taking and aggressive and, thus, to commit more crime than individuals of other body types.

However, whether it is biological or sociological is a debate that shows the importance of theory in criminological research. After all, the link between mesomorphy and criminality is now undisputed; the explanation of why this link exists has become a theoretical debate. Is it the biological influence or the sociological influence? Readers may make their own determination, if not now then later.

Our position is that both biology and social environment are likely to interact with one another in explaining this link. Thus, it is most likely that both nature and nurture are in play in this association between mesomorphy and crime, and both Sheldon and his critics may be correct. A middle ground can often be found in theorizing on criminality. It is important to keep in mind that theories in criminology, as a science, are always considered subject to falsification and criticism and can always be improved. Therefore, our stance on the validity and influence of this theory, as well as others, should not be surprising.

⬛ Policy Implications

Many policy implications can be derived from the theories presented in this chapter. First, one could propose more thorough medical screening at birth and in early childhood, especially regarding minor physical anomalies (MPAs). The studies reviewed in this chapter obviously implicate numerous MPAs in developmental problems (mostly in the womb). These MPAs are a red flag signaling problems, especially in cognitive abilities, which are likely to have a significant impact on criminal behavior.[40]

---

[38]Cortes, *Delinquency and Crime*; Glueck and Glueck, *Physique and Delinquency*; for a review, see Eysenck and Gudjonsson, *Causes and Cures*.

[39]For a review, see Lee Ellis, *Criminology: A Global Perspective* (supplemental tables and references) (Minot, ND: Pyramid Press, 2000).

[40]D. Fishbein, *Biobehavioral Perspectives*.

Other policy implications derived from the theories and findings of this chapter involve having same-sex classes for children in school because they focus on deficiencies that have been shown for both young boys and girls. Numerous school districts now have policies that specify same-sex math courses for female children. This same strategy might be considered for male children in English or literature courses because males have a biological disposition for a lower aptitude level than females regarding this area of study. Furthermore, far more screening should be done regarding the IQ and aptitude levels of young children to identify which children require extra attention because studies show that such early intervention can make a big difference in improving their IQ/aptitude.

A recent report that reviewed the extant literature regarding what types of programs work best for reducing crime noted the importance of diagnosing early head trauma and further concluded that one of the most consistently supported programs for such at-risk children are those that involve weekly infant home visitation.[41] Another obvious policy implication derived from biosocial theory is mandatory health insurance for pregnant mothers and children, which is quite likely the most efficient way to reduce crime in the long term.[42] Finally, all youth should be screened for abnormal levels hormones, neurotransmitters, and toxins (especially lead).[43] These and other policy strategies will be discussed in the last section of the book.

▨ Conclusion

In this introduction, we discussed the development of the early Positive School of criminology. The Positive School can be seen as the opposite of the Classical School perspective, which we covered in Sections I and II, because positivism assumes that individuals have virtually no free will; rather, criminal behavior is the result of determinism, which means that factors other than rational decision making, such as poverty, intelligence, bad parenting, and unemployment, influence us and determine our behavior.

The earliest positivist theories, such as craniometry and phrenology, were developed in the early 1800s but did not become popular outside of scientific circles, likely because they were presented prior to Charles Darwin's theory of evolution. In the 1860s, Darwin's theory became widely accepted, which set the stage for the father of criminology, Cesare Lombroso, to propose his theory of born criminals. Lombroso's theory was based on Darwin's theory of evolution and argued that the worst criminals are born that way, being biological throwbacks to an earlier stage of evolution. Unfortunately, Lombroso's theory leads to numerous policies that fit the philosophy and politics of Fascism, which found useful a theory that said certain people were inferior to others. However, Lombroso and many of his contemporaries became aware that the field should shift to a more multifactor approach, such as the environment and social factors interacting with physiological influences.

---

[41]Lawrence Sherman, Denise Gottfredson, Doris MacKenzie, John Eck, Peter Reuter, and Shawn Bushway, *Preventing Crime: What Works, What Doesn't, What's Promising: A Report to the United States Congress* (Washington, DC: U.S. Department of Justice, 1997).

[42]John Wright, Stephen Tibbetts, and Leah Daigle, *Criminals in the Making: Criminality Across the Life Course* (Thousand Oaks, CA: Sage, 2008).

[43]Ibid.

We also discussed theories regarding low IQ scores, traditionally known as feeble-mindedness. Although most recent studies show that there is a correlation between crime and low IQ, this association is not quite as strong as thought in the early 1900s. Modern studies show consistent evidence that low verbal IQ is related with criminality,[44] especially when coupled with sociological factors, such as weak family structure. This is the state of the criminological field today, and it will be discussed in the next section.

Finally, we explored the theories and evidence regarding body types in predisposing an individual toward criminality. Studies have shown that the more athletic/mesomorphic an individual is, the higher the probability that this individual will be involved in criminality. This relationship is likely based on hormonal levels, and this type of association will be explored in the next section.

Ultimately, we have examined a variety of physiological and psychological factors that predict criminal offending, according to empirical research. Still, the existence of such factors is largely conditional based on environmental and social factors.

⊠ Section Summary

+ The Positive School of criminology assumes the opposite of the Classical School. Whereas the Classical School assumes that individuals commit crime because they freely choose to act after rationally considering the expected costs and benefits of the behavior, the Positive School assumes that individuals have virtually no free will or choice in the matter; rather, their behavior is determined by factors outside of free will, such as poverty, low intelligence, bad child rearing, and unemployment.

+ The earliest positive theories, such as craniometry and phrenology, emphasized measuring the size and shape of the skull/brain. These perspectives did not become very popular because they preceded Charles Darwin's theory of evolution.

+ Cesare Lombroso, the father of criminology, presented a theoretical model that assumed the worst criminals are born that way. Highly influenced by Darwin, Lombroso claimed that born criminals are evolutionary throwbacks that are not as highly developed as most people.

+ Lombroso claimed these born criminals could be identified by physical features called stigmata. This lead to numerous policy implications that fit with the societal beliefs at that time, such as Fascism.

+ In the early 1900s, the IQ test was invented in France and was quickly used by American researchers in their quest to identify the "feebleminded." This lead to massive numbers of deportation, sterilizations, and institutionalizations across the United States and elsewhere.

+ Modern studies support a link between low verbal IQ and criminality, even within a given race, social class, or gender.

---

[44]Gibson et al., "The Contribution of Family Adversity."

♦ Merging elements of the early physiological and psychological perspectives were the body-type theories. The best-known of these was somatotyping, which was proposed by William Sheldon. Sheldon found that an athletic/muscular build (i.e., mesomorphy) was linked to an aggressive, risk-taking personality, which is in turn was associated with higher levels of crime.

♦ Despite the methodological problems with Sheldon's body-type theory, many propositions and associations of the perspective hold true in modern studies.

♦ The early Positive School theories set the stage for most of the other theories we will be covering in this book because they placed an emphasis on using the scientific method for studying and explaining criminal activity.

KEY TERMS

Atavistic/Atavism	Endomorphic	Physiognomy
Cerebrotonic	Eugenics	Positive School
Craniometry	Feeblemindedness	Somotonic
Determinism	Mesoderm	Somatotyping
Ectoderm	Mesomorphic	Stigmata
Ectomorphic	Minor physical anomalies (MPAs)	Viscerotonic
Endoderm	Phrenology	

DISCUSSION QUESTIONS

1. What characteristics distinguish the Positive School from the Classical School regarding criminal thought? Which of these schools do you lean toward in your own perspective of crime and why?

2. Name and describe the various early schools of positivistic theories that existed in the early to mid-1800s (pre-Darwin), as well as the influence that they had on later schools of thought regarding criminality. Do you see any validity in these approaches (because modern medical science does) and articulate why?

3. What was the significant reason(s) that these early schools of Positivistic theories did not gain much momentum in societal popularity? Does this lack of popularity relate to the neglect of biological perspectives of crime in modern times?

4. What portion of Lombroso's theory of criminality do you find least valid? Which do you find most valid?

5. Most readers have taken the equivalent of an IQ test (e.g., SAT or ACT tests). Do you believe that this score is a fair representation of your knowledge as compared to others and why? Do your feelings reflect the criticisms of experts regarding the use of IQ in identifying potential offenders, such as feeblemindedness theory?

6. In light of the scientific findings that show that verbal IQ is shown to be a consistent predictor of criminality among virtually all populations/samples, can you provide evidence from your personal experience for why this occurs?

7. Regarding Sheldon's body-type theory, what portion of this theory do you find most valid? What do you find least valid?

8. If you had to give yourself a somatotype (e.g., 3-6-2), what would it be? Explain why your score would be the one you provide, and note whether this would make you likely to be a criminal in Sheldon's model.

9. Provide a somatotype of five of your family members or best friends. Does the somatotype have any correlation with criminality according to Sheldon's predictions? Either way, describe your findings.

10. Ultimately, do you believe some of the positive theoretical perspectives presented in this section are valid, or do you think they should be entirely dismissed in terms of understanding/predicting crime? Either way, state your case.

11. What types of policies would you implement, if you were in charge, given the theories and findings in this chapter?

WEB RESOURCES

Phrenology/Craniometry

http://pages.britishlibrary.net/phrenology/

http://www.phrenology.org/index.html

http://skepdic.com/cranial.html

http://library.thinkquest.org/C0121653/craniometry.htm

Lombroso

http://www.museocriminologico.it/lombroso_1_uk.htm

IQ Testing/Feeblemindedness

http://www.vineland.org/history/trainingschool/history/eugenics.htm

http://www.vineland.org/history/trainingschool/history/eugenics.htm

Body Type Theories/Somatotyping

http://www.innerexplorations.com/psytext/shel.htm

http://www.teachpe.com/multi/somatotyping_eating_disorders_performance.htm

READING

In this selection, Nicole Rafter discusses the key contributions that were made by phrenology and other early positive perspectives. Although they are often laughed at and dismissed by contemporary criminologists, few understand that without such developments and research offered by phrenology and other early biological theories of crime, neither modern medical science nor criminology would be as advanced as it is now. While reading this selection, try to avoid judging the early perspectives by today's standards. After all, a century from now, many of the theories we currently believe in will likely be laughed at and dismissed.

The Murderous Dutch Fiddler

Criminology, History, and the Problem of Phrenology

Nicole H. Rafter

Phrenology—the early 19th-century system of reading character from the contours of the skull—produced one of the most radical reorientations in ideas about crime and punishment ever proposed in the Western world. In the area of jurisprudence, its practitioners worked to re-establish criminal law on a new philosophical basis; to overhaul ideas about criminal responsibility; and, in a retributivist age, to develop a rehabilitative rationale for sentencing. In the area of penology, phrenologists opposed capital punishment and proposed innovations in prisoner management that influenced criminal justice for the next 150 years. But it was in the area of criminology that phrenologists proved themselves most innovative, as they developed the first comprehensive explanation of criminal behavior.

On the basis of their understanding of the brain as an aggregation of independent organs or "faculties," phrenologists could explain every form of criminal behavior, from petty theft to wife-beating to homicide. They had guidelines for distinguishing between sane and insane criminals; they introduced the idea that people vary in their propensity to crime; and they could account for differences in crime rates by age, nationality, race and sex. Phrenologists could even explain the behavior of criminals whom we today would call serial killers and psychopaths, as in this case from one of phrenology's basic texts:

Source: Adapted from Nicole Rafter, "The Murderous Dutch Fiddler: Criminology, History, and the Problem of Phrenology," *Theoretical Criminology* 9(1): 65–96. Copyright © 2005 Sage Publications Ltd. Reprinted by permission of Sage Publications Ltd. and the author.

At the beginning of the last century several murders were committed in Holland, on the frontiers of the province of Cleves. For a long time the murderer remained unknown; but at last an old fiddler, who was accustomed to play on the violin at country weddings, was suspected in consequence of some expressions of his children. Led before the justice, he confessed thirty-four murders, and he asserted that he had committed them without any cause of enmity, and without any intention of robbing, but only because he was extremely delighted with bloodshed. (Spurzheim, 1815: 308)

At a time when most people would have explained the Dutch fiddler's behavior in terms of sin, phrenologists attributed it to innate biological defect. Their criminological ambition and scope—their desire to develop a science of criminal behavior—excited progressive thinkers on both sides of the Atlantic.

Phrenologists' writings on criminal jurisprudence, penology and criminology were part of a much broader, all-encompassing biosocial system that aimed at scientifically explaining not only criminal behavior but *all* human behavior (and a great deal of animal behavior as well). Their system rested on five fundamental assumptions:

1. The brain is the organ of the mind.

2. The brain is an aggregation of about 30[1] separate organs or faculties, such as Combativeness, Covetiveness, and Destructiveness, that function independently.

3. The more active an organ, the larger its size.

4. The relative size of the organs can be estimated by inspecting the contours of the skull.

5. The relative size of the organs can be increased or decreased through exercise and self-discipline.[2]

These fundamental ideas, all but the last of them formulated about 1800 by the Viennese physician Franz Joseph Gall, became the basis of an international movement to develop a science of phrenology and spread its gospel. The movement fell into two stages: a scientific phase, from about 1800 to 1830, when the phrenological system was developed, mainly by physicians and psychiatrists; and an overlapping popularizing stage, from about 1820 to 1850, during which phrenology became a fad, complete with marketers, clubs and hucksters. But the timing and duration of these phases differed by place. Although phrenology itself underwent little development after the 1840s, its ideas segued into the theory of degeneration that underpinned concepts of deviance into the 1920s. Moreover, some phrenological societies remained active into the 20th century.

Like other very early students of social behavior, phrenologists adopted the previously developed methods of the natural sciences, assuming that the social world could be studied using the same procedures. They collected data, formulated hypotheses and made positivist assumptions about the possibility of direct, objective apprehension of social phenomena. During its scientific phase, phrenology intersected with a range of other scientific endeavors, including anatomy, anthropology, physiology, psychology and psychiatry, and it used a range of scientific procedures, including empirical observation, induction and deduction. (Some phrenologists also claimed to use the experimental method; but their failure to experiment rigorously proved to be their scientific Achilles' heel.) Phrenology constituted an ambitious and complex effort to break with older metaphysical and theological explanations of behavior and replace them with an empirical science.

Today, phrenology is remembered primarily for the popular culture of its second stage:

the manufacture of inkwells and caneheads shaped like phrenological skulls, with the organs marked out for study; the calling in of phrenological experts to examine the heads of job applicants; the quackery of itinerant practitioners of "bumpology." It has been dismissed as a medical cult, discredited science, dead science, pathological science and pseudo-science. These refusals to take phrenology seriously as an early scientific discourse place criminologists in an awkward position. Phrenology constituted an important episode in the history of criminology and criminal justice; but to recall that history is to risk seeming ridiculous. The problem becomes: How can criminologists relate to this apparently embarrassing forerunner?

Criminologists have essentially three choices when confronted with phrenology:

1. *Ignore it.* Traditionally, historians of criminology and criminal justice have chosen this route. It is difficult to find an extended discussion of phrenology in any standard history of criminology or criminal justice other than Fink (1938/1962) and Savitz et al. (1977).[3] Histories of phrenology itself sometimes include a chapter on penology, but even they slight what phrenologists said about the causes of crime. It may well be that historians of phrenology have simply been unaware of the doctrine's significance in the evolution of criminology. More difficult to explain is the marginalization of phrenology by historians of criminology and criminal justice. While historians of insanity have thoroughly explored the phrenological model of mental disturbance and its impact on the development of neuroscience and treatment of the mentally ill, for crime-and-justice historians the rule has been to ignore it.

2. *Make it "relevant."* This approach would involve mining phrenological doctrine for material resembling today's research on the role of brain dysfunction in criminal behavior and then treating phrenology as a precursor science. Given phrenology's disrepute, it is

unlikely that any present-day PET-scanner of criminals' brains would claim phrenological ancestry.[4] However, an example of this approach can be found in an article on the ways in which phrenology anticipated later ideas in American psychology (Bakan, 1966).[5] The trouble with this kind of approach is that it reduces the past to anticipations of the present, denying it value in its own right. Equally misguided would be any effort, such as that cited by Shapin (1982: 157), to fold phrenology into a sociology of error or mistake. Phrenology was indeed erroneous, but it was neither an error nor a mistake; it was an early science of the mind, and to reduce it to something else is no more respectful of the past than the first alternative of ignoring it entirely.

3. *Come to terms with it.* This approach would acknowledge phrenology as an episode in the history of criminology and criminal justice, evaluate its influence and significance and attempt to establish some sort of relationship with it.

Aside from the two exceptions cited above, the third approach seems not to have been tried. This article aims at implementing it.

Our attitudes toward phrenology will depend on our conception of the criminological enterprise and ultimately on how we define science. If we conceive of criminology as an independent and free-floating subject, a set of truths about crime that it is the job of criminologists using scientific methods to discover, then we must agree that phrenologists failed, and we can safely ignore them. The history of criminology will become a chronicle of the stockpiling of currently acceptable scientific techniques and knowledge (see Kuhn, 1970: 2). If, however, we conceive of criminology and other sciences as discourses formulated in time and space, shaped by their social contexts and by scientists' own backgrounds, then we can open the historical door to phrenology. We can view it as a discourse on the human brain that greatly advanced understandings of mind-behavior relationships (Carlson, 1958; Cantor, 1975; Cooter, 1976/1981, 1984; Shapin, 1982; Young, 1990), that advocated scientific methods but failed in some respects to meet the scientific criteria of its own day, and that formed the first coherent explanation of criminality.

Phrenology: Context and Substance

Phrenology emerged out of the Enlightenment drive to replace metaphysical and theological explanations with scientific accounts of natural and social phenomena. "One fact is to me more positive and decisive than a thousand metaphysical opinions," declared Johann Gaspar Spurzheim, one of phrenology's founders, in a phrase much admired by his followers (*Phrenological Journal and Miscellany*, 1834–6: title page). Whereas churchmen interpreted the world in terms of divine creation and insisted on religious authority, in the late 18th- and early 19th-centuries laymen were growing interested in less authoritarian, more rational approaches to understanding. The new emphasis on observation and human reasoning as sources of knowledge was reinforced by democratic revolutions in North America and France—vivid demonstrations of the possibility of breaking free of older systems. With democracy came the ideal of universal education and the bold notion that any educated person might at least dabble in the study of natural phenomena. Phrenology grew out of the Enlightenment's enthusiasm for scientific explanation and its democratic impulses. Insanity and criminality, previously interpreted as signs of sin, now seemed as though they might be comprehensible in scientific terms. At the same time, the fall of authoritarian regimes, their gradual replacement by bourgeois industrial societies and the growing distaste for older, retributivist punishments of the body created a demand for new methods of social ordering and discipline (Foucault, 1977;

McLaren, 1981). This, very roughly, was the situation about 1800, when phrenology made its first appearance.

To understand why it was phrenology and not some other science that became the basis for the first fully developed theory of crime, we need to look at the way the personal interests and research skills of phrenology's founders intersected with the scientific context in which they worked and the social and cultural contexts in which their doctrine took root. The scientific context was one of widespread interest in applying scientific methods to the study of social phenomena, and researchers had no reason to doubt that natural science methods would also work for the investigation of social and psychological events. Earlier, the utilitarians Jeremy Bentham and Cesare Beccaria had offered a rational choice framework for explaining criminal behavior, but utilitarianism was hardly a full-blown psychological or social theory, and in any case crime rates cast doubt on its sidecar theory of deterrence. The work of A. M. Guerry (1833), Adolphe Quetelet (1835), and Siméon-Denis Poisson (1837) on crime statistics—another sign of the hunger for social science—lay in the future (albeit the very near future), and none of these men was particularly interested in the causes of crime. However, Lavater had recently elaborated his system for reading the face for signs of trouble; and moral insanity theorists, drawing on faculty psychology, were already starting to develop a biological-fault theory of criminality. Thus the very first efforts to study deviance scientifically had already set this type of work on an explanatory trajectory heading toward the brain.

Explanations of Crime

Spurzheim's chapter on "The Organ of the Propensity to Destroy, or of Destructiveness" illustrates both his methods and phrenology's applicability to the study of crime. He begins by observing that animals vary in their propensity to kill, even within species and breeds.

Gall had a little dog which had this propensity in so high a degree, that he would sometimes watch several hours for a mouse, and as soon as it was killed he left it; notwithstanding repeated punishment he had also an irresistible propensity to kill birds. (Spurzheim, 1815: 305)

In man, too, Spurzheim continues, the destructive propensity manifests itself with different degrees of intensity: some people are merely indifferent to animals' pain; others enjoy seeing animals killed; and still others experience "the most irresistible desire to kill" (1815: 306). Spurzheim gives many examples, including that of the Dutch fiddler, and explores their implications. The examples seem to demonstrate that "the propensity to kill is a matter independent of education and training" (1815: 310), a function of mental organization alone. Spurzheim also reports on the related research of Philippe Pinel, the French psychiatrist who at about the same time was observing in madmen a similarly "fierce impulsion to destroy," and he gives many of Pinel's examples (1815: 312–15).

To Spurzheim, the conclusion seemed inescapable: there must be an organ of the brain that determines the propensity to kill and it must function independently of other propensities, which continue to work normally even in extreme cases like that of the Dutch fiddler. Gall (1825, vol. 4: 64) had earlier identified an organ of Murder, having found a well-developed protuberance at the same spot in the skulls of two murderers. However, Spurzheim objects to naming an organ "according to its abuse" and therefore changes the name of Murder to Destructiveness, attributing to it the propensity, not only to kill, but also

to pinch, scratch, bite, cut, break, pierce, devastate. . . . We are convinced, by a great number of observations, that the seat of this organ is on the side of

the head immediately above the ears. . . . It is commonly larger in men than in women; yet there are exceptions from this rule. (1815: 317–18)

In sum, on the basis of numerous examples, Spurzheim has identified the primary cause of homicide: overdevelopment of the organ of Destructiveness, which is the seat of both negative and useful forms of destruction.

Moreover, because post-Gall phrenologists conceived of the brain as plastic, malleable and capable of change, they were able to combine their determinism with an optimistic, rehabilitative approach to crime and other social problems without a sense of contradiction. Conceiving of character traits as heritable but not fixed, they could simultaneously argue that criminals are not responsible for their crimes *and* that, with treatment, they can be cured of criminality.

In practice, most phrenologists dodged the full implications of their doctrine for free will by developing a typology of mankind according to degree of criminal responsibility.

While phrenologists' classifications of criminals by degrees of responsibility and free will differed in their particulars, together they formed the nucleus of the idea that later flowered in Lombroso's hierarchical typology of Criminal Man (Lombroso-Ferrero, 1911/1972).

Although the major phrenological texts on issues of crime and justice were produced by professional men like Simpson and George Combe, anyone could add to the store of phrenological knowledge about crime. From Sydney to Stockholm, York to Heidelberg, and Rochester, New York, to Lexington, Kentucky, amateur phrenologists studied the heads of living and dead criminals, mailed their findings to phrenological journals, and reported them at meetings of phrenological societies. The 1834–6 volume of the *Phrenological Journal and Miscellany,* for example, included a reader's article on a tame ram with unusually well-developed Destructiveness who violently butted adults and terrorized children. The same issue carried speculations on the relation of Benevolence and Destructiveness in a certain pirate (he had become a buccaneer to revenge himself on Spaniards for some cruelty) and character analyses of the head casts of recently executed murderers. Relatedly, when in 1848 an explosion sent a metal rod flying through the head of a Vermont railroad worker named Phineas Gage, the local physician wrote up the case in such a way as to support phrenology, and on the basis of hearsay the *American Phrenological Journal* reported that:

> after the man recovered, and while he was recovering, he was gross, profane, coarse, and vulgar. . . . Before the injury he was quiet and respectful . . . [T]he iron rod passed through the regions of BENEVOLENCE and VENERATION, which left these organs without influence in his character, hence his profanity, and want of respect and kindness; giving the animal propensities absolute control in his character. (quoted in Barker, 1995: 678)

Thus phrenology enabled ordinary people to contribute to scientific knowledge, including knowledge about the causes of crime.

◪ The Achievements of Phrenology

Acceptance at first came slowly to phrenology. The Austrian emperor, alarmed by Gall's radical materialism and its implicit denial of free will, expelled Gall from the country; Paris, to which Gall and Spurzheim moved to carry on their research, proved only slightly less hostile (McLaren, 1981). Breaking with Gall and relocating to England, Spurzheim again encountered skepticism and ridicule (e.g., Combe, 1831), but once he learned how to make Gall's doctrine accessible through his books and

attractive through his teaching about the potential for human change, phrenology enjoyed greater success. Lavater's physiognomy had primed the public for a theory of head-character relationships, with the title of Spurzheim's *Physiognomical System of Drs. Gall and Spurzheim* suggesting the ease of the transition between the two doctrines. Gall and Spurzheim were both outstanding brain anatomists, a skill that gave them threshold credentials and authority in their new science based on claims about the anatomy of the brain. Moreover, in a scientisitic age, phrenology made a show of applying scientific methods and achieving positivist results, and it had remarkable internal consistency. Such factors helped phrenology overcome the initial resistance.

Once the doctrine began to take root, its social context helped it to thrive. In a period when social reforms seemed both imperative and achievable, phrenology provided a sturdy platform on which to erect major programs of change. In a century when—to an extent difficult to comprehend today—ordinary people lived in fear of becoming insane, phrenology showed how insanity might be staved off through the cultivation of certain faculties. Equally important to psychologists and philosophers struggling to make sense of body-mind relationships, phrenology offered a way out of the mazes of Cartesian dualism by holding, simply, that the mind is not separate from the body but rather a function of the brain. And in an era of intense debate over the application of the insanity defense, the phrenological image of independent faculties in the brain offered a relatively clear way to conceptualize the new category of partial or moral insanity—a breakdown of a single organ while the rest continued to function normally.

Gall and Spurzheim had sketchily indicated their doctrine's implications for understanding and reforming criminal behavior (Gall, 1825, vol. 1: 336–69; Spurzheim, 1828: 273–327). Later phrenologists who built on this foundation worked mainly in the interstices between phrenology's two major stages, after the basics had been established but before disrepute set in among intellectuals. Catching phrenology at its peak of plausibility, they were able to achieve major reorientations in ways of thinking about social problems.

What, then, did phrenology accomplish in the area of crime and justice?

In criminology:

- Phrenology helped establish the idea that criminal behavior can and should be studied scientifically. It introduced scientific methods into the study of criminal behavior and inaugurated what became the positivist tradition in criminology.
- Phrenology produced the first systematic and comprehensive theory of criminal behavior, although it did not conceptualize its project in these terms.
- Breaking with the utilitarian model of Beccaria and Bentham, who were not much concerned with differences among criminals, phrenologists introduced the idea that people vary in their degree of criminal responsibility and in their propensity to commit crime.
- Phrenologists consolidated and advanced the medical model of criminal behavior,[6] according to which criminals (or at least some criminals) are not bad but sick. This concept of crime as a disease profoundly influenced later analyses of criminal behavior.
- By explaining criminality in terms of defective brain organization, phrenology established a biological foundation on which later criminologists built, including late 19th-century degenerationists and criminal anthropologists. It also laid the foundation for eugenic criminology (Rafter, 1997). The idea that the cause of crime may lie in brain defects (or genes that lead to brain

defects) seems today to be making a comeback (e.g., Raine et al., 1995).

In criminal jurisprudence:

- Phrenology rationalized jurisprudence. At the dawn of the 19th century, on the threshold of the urban industrial world, it helped reorient criminal jurisprudence away from the principles of retribution and deterrence on which it had long rested, and toward more systematic, proactive measures for reformation and social defense.
- The doctrine raised questions about criminal responsibility that in time led to new approaches to criminal insanity and new ways of conceptualizing dangerousness.
- Phrenologists proposed indefinite and indeterminate sentencing. In addition, they hinted at (without clearly articulating) the idea of sentencing according to biological fitness.

In penology:

- The first to propose a systematic program for reforming criminals, phrenologists advocated rehabilitative measures that shaped the course of "corrections" until the 1970s.
- While rudimentary prisoner classification had been practiced in early lockups such as Philadelphia's Walnut Street Jail, phrenologists introduced the idea of studying convicts at the point of admission to prison and then dividing them into treatment groups according to intelligence and character. They also introduced the idea of classifying prisons and designating one for incorrigibles.
- More generally, phrenologists helped the next generation of prison administrators conceive of penology as a

science that might professionalize prison management and medicalize work with convicts (Wines, 1871; Boies, 1901).

In sum, phrenology put into circulation powerful new concepts about crime and justice that eventually became part of the broader culture. The results lived on long after the husk of organology had fallen by the wayside.

By the 1830s, phrenology had begun to lose its plausibility among intellectuals and professionals. Some close students of the doctrine, like the English surgeon John Abernethy, had asked tough questions from the beginning. How, Abernethy had demanded (1821: 66–7), were the organs coordinated? "By committees of the several organs, and a board of control"? Abernethy also worried about negative labeling: "[S]uppose a man to have large knobs on his head which are said to indicate him to be a knave and a thief, can he expect assistance and confidence from any one?" (1821: 8).) The social philosopher Auguste Comte, the psychiatrist Isaac Ray and others who had begun as converts to phrenology gradually lost faith (Carlson, 1958: 536; McLaren, 1981: 19–20). Still others, of course, had never seen anything in the doctrine but blasphemy and sympathy with criminals. While phrenology remained popular through the mid-century and phrenologists continued to gather empirical proofs of their doctrine, to the scientifically inclined it was increasingly clear that almost any evidence could be regarded as confirmation of such a multi-faceted theory (Cantor, 1975: 211–18). Nor did phrenologists conduct experiments to see if their doctrine could be refuted. The aspect of phrenology that may have harmed it most, scientifically, was its redundancy: even advocates eventually realized that one could reach the same conclusions about the nature of human behavior without recourse to organology.[7] In any case, by 1850, social problems were starting to appear less tractable than they had

seemed in the sunnier light of the century's early decades. Alarmed, social theorists and policy makers were no longer able to start from phrenology's premise that most people are biologically normal and naturally good. Those who had endorsed phrenology now abandoned it in favor of the newer doctrine of degeneration, with its more hereditarian cast and implications for more coercive measures of social control (Pick, 1989).

The great exception to this rule was Cesare Lombroso, the Italian psychiatrist who founded criminal anthropology with his book on *Criminal Man* (orig. 1876/2005). While contemporaries stampeded to adopt the degenerationist explanation of criminality, Lombroso clung to his Darwinian, anthropological explanation of the criminal as an atavism or reversion to an earlier evolutionary stage. A major emerging question in Lombroso studies asks why Lombroso swam against the degenerationist current until very late in his career. The definitive answer may become clear when we gain access to all five editions of *Criminal Man* (Lombroso, 1876/2005; see also Wolfgang, 1972; Pick, 1989), but a preliminary response is that Lombroso was so hugely influenced by phrenology that he found it difficult to let go (so to speak) of the cranium. One suggestive bit of evidence can be found in his office, which still exists in its original state in the Lombroso museum in Turin, Italy: on the desk, next to the skull of the brigand Villella, sits a gigantic phrenological head.

⊠ Criminology and the Bad Science Issue

Phrenology is not the only disreputable ancestor in criminology's genealogy. Criminologists also have to come to terms with such forerunners as criminal anthropology, the feeblemindedness theory of crime and Earnest A. Hooton's 1930 attempts to revivify eugenic criminology. To ignore a now discredited science like phrenology on the grounds that it was wrong is to miss an important opportunity to see how science is shaped by its social context and by the circumstances of those who generated it. In fact, some science historians argue that it is

> almost more useful sometimes to learn something of the misfires and the mistaken hypotheses of early scientists, . . . and even to pursue these courses of scientific development which led into a blind alley, but which still had their effect on the progress of science in general. What is wrong in the history of science as in all other forms of history is to keep the present day always before one's mind as the basis of reference. (Kragh, 1987: 81, quoting H. Butterfield, 1950)

For the study of the nature and history of science, then, discredited sciences can be especially useful.

Moreover, to proceed as though there were a bright line between good and bad science is to ignore the fact that social and historical factors shape the acceptance of *all* science—good, bad, anti-, pseudo-, pathological, partly right, Greek, Renaissance and presumably authoritative 21st-century science. No scientific activity occurs in a vacuum, insulated from its social context, and thus it is futile to look for a pure, totally objective science. Even if science could be vacuum-packed, one could not easily distinguish between good science and pseudoscience. Finding ways to differentiate between sound and flawed science has been a major preoccupation of recent philosophers of science. Some have challenged the Enlightenment view of science as a rational, systematic, progressive activity (Kuhn, 1970;

Feyerabend, 1975; Lakatos, 1978; Merton and Barber, 2003). They do not speak with one voice, of course, but individually or collectively they have argued that the scientific method itself is something of a myth, since many scientific discoveries occur serendipitously, even anarchically, bypassing the step-by-step process enshrined in the just-so story of scientific methodology. The findings of even the physical and life sciences may be historically relative, in the view of some theorists, while others maintain that although science can produce change, it does not produce progress. One need not swallow these critiques whole to recognize what they imply: to dismiss phrenology on the grounds that it was bad science is to take a naïve, outmoded view of science. Perhaps only the passage of time can teach us which large-scale scientific research programs (and phrenology certainly fits this category) lead to truth or falsity (Lakatos, 1973; see also Lakatos and Feyerabend, 1999).

Sociologists and criminologists may continue to disregard phrenology on the grounds that it offered a *biological* theory of crime. Whereas in actuality it offered a biosocial theory, one that pictured a constant interaction between the faculties and environment, it did have a strong biological component; and so for the sake of argument let us suppose for a moment that phrenology was an exclusively biological theory. Reflexive mistrust of biological theories per se, while it is historically and ethically understandable, is becoming increasingly suspect. As the phrenology example itself shows, biological theories are not necessary bigoted or conservative. Phrenology did biologize difference, but in its own context it was a progressive, even radical theory. One might well keep the liberalism and indeed progressivism of phrenology in mind today as biological theories make their comeback, even while we also guard against their tendency to reach eugenic conclusions. Phrenology can help us remember that biological theories are no more inherently reactionary than sociological theories are inherently bias-free.

The history of criminology is generally an underdeveloped field, one to which Americans, in particular, have paid little attention. Thanks to David Garland (1985, 2002), Paul Rock (1994, 1998) and Neil Davie (2004), British criminologists have a relatively clear overview of their own disciplinary evolution, a solid scaffolding on which to construct more detailed studies. U.S. criminologists, in contrast, have a shakier sense of their field's origins and development, and in fact, the U.S.-based criminologist who has produced some of the best historical work—Piers Beirne (1993, 1994)— was born and schooled in England. Americans' greater disinterest can be explained, at least in part, by conclusions reached by the historian Dorothy Ross in her *Origins of American Social Science* (1991). "American social science," Ross observes, "bears the distinctive mark of its national origins":

> Its liberal values, practical bent, shallow historical vision, and technocratic confidence are recognizable features. . . . To foreign and domestic critics, these characteristics make American social science ahistorical and scientistic, lacking in appreciation of historical difference and complexity. . . . What is so marked about American social science is the degree to which it is modeled on the natural rather than the historical sciences. (Ross, 1991: xiii)

Ross ties the ahistorical nature of American social science to the experience of settling a new continent and untouched spaces; Americans "could relegate history to the past while they acted out their destiny in the realm of nature . . . they could develop in space rather than time" (1991: 25). Ross urges American social scientists to give more recognition to history in order to relativize their work and become more keenly aware that social science developed through human choices. While Ross's analyses pertain specifically to U.S. social science, some of her conclusions are relevant to British as well as American criminologists.

Social science, as Ross recognizes, is constituted by activities as well as findings and results. It is not a constant but rather an ongoing process. It is contingent on verification, of course, but it is also contingent on what is defined as scientifically interesting at any point in time and what methods of proof and disproof are available. Even our idea of what science is depends on the past. Our current understandings of the social roles of criminology, criminal jurisprudence and "corrections" were shaped partly by phrenology. From today's perspective, phrenologists were wrong scientifically—the bumps of the skull do not reflect one's character—but they left a powerful legacy, and if we try to ignore their work, we avoid part of ourselves as well.

⊠ Notes

1. The system's founder, Franz Joseph Gall, identified 27 organs; his closest follower, Johann Gaspar Spurzheim, identified 33.

2. It was not Gall himself but Spurzheim who introduced the hopeful idea that the faculties might be modified by exercise and other forms of treatment.

3. Both Fink (1938/1962) and Savitz et al. (1977) contain significant factual errors. The latter is particularly unsophisticated and superficial.

4. For an overview and links to detailed reports on imaging research on criminals' brains, see Abbott (2001); for an especially derogatory comparison to phrenology, see Uttal (2001). But it is also noteworthy that evolutionary psychologists are starting to

talk in terms of "mental modules" and "mental organs" (Gander, 2003), language that closely echoes phrenology.

5. A related example is to be found in Schlag (1997), comparing law with phrenology in order to disparage the scientific claims of both. In this instance, phrenology is invoked, not as a precursor but an analog. But again, the author is not interested in phrenology in its own right but rather in using it to make a point about another field.

6. The medical model itself had been introduced slightly earlier by moral insanity theorists.

7. For more details on the scientific failings of phrenology, see Young (1990).

◼ References

Abbott, Alison (2001) "Into the Mind of a Killer," *Nature* 410(15 March): 296–8. Available at http://www.nature.com/nature/journal/v.410/

Abernethy, John (1821) *Reflections on Gall and Spurzheim's System of Physiognomy and Phrenology.* London: Longman, Hurst, Rees, Orme & Brown.

Bakan, David (1966) "The Influence of Phrenology on American Psychology," *Journal of the History of the Behavioral Sciences* 2(3): 200–20.

Barker, F.G., II (1995) "Phineas among the Phrenologists: The American Crowbar Case and 19th-Century Theories of Cerebral Localization," *Journal of Neurosurgery* 82(4): 672–82.

Beirne, Piers (1993) *Inventing Criminology: Essays on the Rise of Homo Criminalis.* Albany, NY: State University of New York Press.

Beirne, Piers (ed.) (1994) *The Origins and Growth of Criminology: Essays on Intellectual History, 1760–1945.* Aldershot: Dartmouth.

Boies, Henry M. (1901) *The Science of Penology.* New York: Putnam's.

Cantor, G.N. (1975) "The Edinburgh Phrenology Debate: 1803–1828," *Annals of Science* 32(3): 195–218.

Carlson, Eric T. (1958) "The Influence of Phrenology in Early American Psychiatric Thought," *American Journal of Psychiatry* 115: 535–8.

Combe, Andrew (1831) *Observations on Mental Derangement.* Edinburgh: John Anderson.

Cooter, Roger (1976/1981) "Phrenology and British Alienists, ca. 1825–1845," in Andrew Scull (ed.) *Madhouses, Mad-Doctors, and Madmen,* pp. 58–104. London: The Athlone Press.

Cooter, Roger (1984) *The Cultural Meaning of Popular Science: Phrenology and the Organization of Consent in Nineteenth-Century Britain.* Cambridge: Cambridge University Press.

Davie, Neil (2004) *Les Visages de la Criminalité: À la Recherche d'une Théorie Scientifique du Criminel Type en Angleterre (1860–1914).* Paris: Éditions Kimé To be published in English (2005) as *Tracing the Criminal: The Rise of Scientific Crimonology in Britain, 1860–1918.* Oxford: Blackwell Press.

Feyerabend, Paul K. (1975) *Against Method.* London: Verso.

Fink, Arthur E. (1938/1962) *Causes of Crime: Biological Theories in the United States, 1800–1915.* New York: A.S. Barnes, Perpetua Edition.

Foucault, Michel (1977) *Discipline and Punish.* Trans. A. Sheridan. New York: Pantheon.

Gall, F.G. (1825) *Sur les Fonctions du Cerveau et sur Celles de Chacune de Ses Parties* (6 vols). Paris: J.-B. Ballière.

Gander, Eric M. (2003) *On Our Minds: How Evolutionary Psychology Is Reshaping the Nature-versus-Nurture Debate.* Baltimore, MD: The Johns Hopkins University Press.

Garland, D. (1985) *Punishment and Welfare: A History of Penal Strategies.* Aldershot, England: Gower Publishing Co.

Garland, David (2002) "Of Crimes and Criminals: The Development of Criminology in Britain," in Mike Maguire, Rod Morgan and Robert Reiner (eds) *The Oxford Handbook of Criminology,* 3rd edn, pp. 7–50. New York: Oxford University Press.

Kragh, Helge (1987) *An Introduction to the Historiography of Science.* Cambridge: Cambridge University Press.

Kuhn, Thomas (1970) *The Structure of Scientific Revolutions* (2nd edn). Chicago, IL: University of Chicago Press.

Lakatos, Imre (1973) *Science and Pseudoscience.* Transcript of a talk broadcast in 1973. Available at http://www.lse.ac.uk/collections/lakatos/science And PseudoscienceTranscript.htm

Lakatos, Imre (1978) *The Methodology of Scientific Research Programmes* (vol. 1). Ed. John Worrall and Gregory Currie. Cambridge: Cambridge University Press.

Lakatos, Imre and Paul Feyerabend (1999) *For and Against Method.* Ed. and with an introduction by

Matteo Motterlini. Chicago, IL: University of Chicago Press.

Lavater, J.C. (n.d.) *Essays on Physiognomy*. Abridged from Mr Holcroft's translation. London: Printed for G.G.J. & J. Robinson.

Lombroso, Cesare (2005) *Criminal Man*. Trans. and with a new introduction by Mary Gibson and Nicole Hahn Rafter. Durham, NC: Duke University Press.

Lombroso-Ferrero, Gina (1911/1972) *Criminal Man According to the Classification of Cesare Lombroso*. Repr. Montclair, NJ: Patterson, Smith.

McLaren, Angus (1981) "A Prehistory of the Social Sciences: Phrenology in France," *Comparative Studies in Society and History* 23 (Jan): 3–22.

Merton, Robert K. and Elinor Barber (2003) *The Travels and Adventures of Serendipity: A Study in Sociological Semantics and the Sociology of Science*. Princeton, NJ: Princeton University Press.

Pick, Daniel (1989) *Faces of Degeneration: A European Disorder, c. 1848–1918*. Cambridge: Cambridge University Press.

Rafter, Nicole Hahn (1997) *Creating Born Criminals*. Urbana, IL: University of Illinois Press.

Raine, Adrian, Todd Lencz and Sarnoff A. Mednick (eds) (1995) *Schizotypal Personality*. Cambridge: Cambridge University Press.

Rock, Paul (1994) *History of Criminology*. Aldershot, England: Dartmouth Publishing Co.

Rock, Paul (ed.) (1998) *History of British Criminology*. Oxford: Clarendon Press.

Ross, Dorothy (1991) *The Origins of American Social Science*. Cambridge: Cambridge University Press.

Savitz, L., S.H. Turner and T. Dickman (1977) "The Origins of Scientific Criminology: Franz Gall as the First Criminologist," in R.F. Meier (ed.) *Theory in Criminology*, pp. 41–56. Beverly Hills, CA: Sage Publications.

Schlag, Peter (1997) "Law and Phrenology," *Harvard Law Review* 110(4): 877–921.

Shapin, Steven (1982) "History of Science and Its Sociological Reconstruction," *History of Science* 20: 157–211.

Spurzheim, J.G. (1815) *The Physiognomical System of Drs. Gall and Sptrrzheim* (2nd ed.). London: Printed for Baldwin, Cradock & Joy.

Spurzheim, J.G. (1828) *A View of the Elementary Principles of Education: Founded on the Study of the Nature of Man* (2nd edn). London: Treuttel, Würtz & Richter.

Uttal, W.R. (2001) *The New Phrenology*. Massachusetts: MIT Press.

Wines, E.C. (ed.) (1871) *Transactions of the National Congress on Penitentiary and Reformatory Discipline Held at Cincinnati, Ohio, Oct. 12–18, 1870*. Albany, NY: Weed, Parsons & Company.

Wolfgang, Marvin (1972) "Cesare Lombroso, 1835–1909," in Hermann Mannheim (ed.) *Pioneers in Criminology*, 2nd edn, pp. 232–91. Montclair, NJ: Patterson Smith.

Young, R.M. (1990) *Mind, Brain and Adaptation in the Nineteenth Century*. Oxford: Clarendon.

REVIEW QUESTIONS

1. Which of the five fundamental assumptions of phrenology do you find the most valid? The least valid?

2. List at least five personality traits that were governed by various portions of the brain. Where were some of those traits believed to be located?

3. What types of punishments and penology philosophies did the phrenologists recommend? What types did they recommend against using? Why?

READING

Lombroso is generally considered the father of criminology as well as the father of the positive school, primarily because he was the first to get credit for testing his theoretical propositions through observation. He emphasized using the scientific method in examining criminal behavior. This selection includes some of the key portions of his evolution-influenced theory of criminals, particularly the ones he referred to as "born criminals." He also reviews many examples of stigmata, or physical manifestations he believed were signs of evolutionary inferiority. As you read this selection, keep in mind that although most of these ideas have been discredited, at the time, they became very popular among both social scientists and society at large. Furthermore, Lombroso's theory was the dominant theory of criminality for several decades, which was longer than virtually any other theoretical model in over a century.

The Criminal Man (L'uomo delinquente)

Cesare Lombroso (as translated by Mary Gibson and Nicole H. Rafter)

▨ 2. Anthropometry and Physiognomy of 832 Criminals

To many, my attempt to conclude anything at all about the cranial dimensions of the criminal man from a few measurements of cadavers will seem futile and rash. Fortunately, however, I have been able to compare these measurements with those taken from 832 live specimens of criminals, thanks to the help of colleagues who are prison directors and prison physicians.

In terms of height, criminals reproduce their regional types. In Italy, they are very tall in the Veneto (1.69 meters), fairly tall in Umbria and Lombardy (1.66 m), less tall in Emilia, Calabria, and Piedmont (1.63 m), slightly shorter in Naples, Sicily, and the Marches, and shortest of all in Sardinia (1.59 m).[1] Compared with healthy men in the army, criminals appear to be taller than the average Italian, especially in the Veneto, Umbria, Lombardy, Sicily, and Calabria. In the Marches, Naples, and Piedmont, criminals are the same height as healthy men.

These findings, however, are skewed by the preponderance of robbers and murderers in my sample, and thus they conflict with the conclusions of Thomson and Wilson.[2] Robbers and murderers are taller than rapists, forgers, and especially

---

thieves.[a] As for weight, we can compare the findings on 1,331 soldiers, studied by me and Dr. Franchini, with the average for criminals from each region. In the Veneto, healthy men weighed an average of 68 kilograms, while criminals weighed 62.5kg.[3] But in most other regions, most notably Naples, Sicily, and Piedmont, criminals' average weight exceeded that of healthy men.

There are many erroneous ideas in circulation about the physiognomy, or facial expressions, of criminals. Novelists turn them into frightening-looking men with beards that go right up to their eyes, penetrating ferocious gazes, and hawklike noses. More serious observers, such as Casper, err on the other extreme, finding no differences between criminals and normal men.[4] Both are wrong. It is certainly true that there are criminals with notably large cranial capacity and beautifully formed skulls, just as there are those with perfectly regular physiognomy, particularly among adroit swindlers and gang leaders. Lavater and Polli wrote about a murderer whose face resembled one of the angels painted by Guido (*Saggio di Fisiognomia,* 1837).[5] But criminals whose handsome features make a strong impression can be misleading precisely because they contradict our expectations. They are usually individuals of uncommon intelligence, a trait associated with gracefulness of form.

When, on the other hand, one ignores those rare individuals who form the oligarchy of the criminal world to study the entire spectrum of these wretches, as I have done in various prisons, one has to conclude that while offenders may not look fierce, there is nearly always something strange about their appearance. It can even be said that each type of crime is committed by men with particular physiognomic characteristics, such as lack of a beard or an abundance of hair; this may explain why the overall appearance is neither delicate nor pleasant.

In general, thieves are notable for their expressive faces and manual dexterity, small wandering eyes that are often oblique in form, thick and close eyebrows, distorted or squashed noses, thin beards and hair, and sloping foreheads. Like rapists, they often have jug ears. Rapists, however, nearly always have sparkling eyes, delicate features, and swollen lips and eyelids. Most of them are frail; some are hunchbacked. Pederasts are often distinguished by a feminine elegance of the hair and feminine clothing, which they insist on wearing even under their prison uniforms.

Habitual murderers have a cold, glassy stare and eyes that are sometimes bloodshot and filmy; the nose is often hawklike and always large; the jaw is strong, the cheekbones broad; and their hair is dark, abundant, and crisply textured. Their beards are scanty, their canine teeth very developed, and their lips thin. Often their faces contract, exposing the teeth. Among nearly all arsonists, I have observed a softness of skin, an almost childlike appearance, and an abundance of thick straight hair that is almost feminine. One extremely curious example from Pesaro, known as "the woman," was truly feminine in appearance and behavior.

Nearly all criminals have jug ears, thick hair, thin beards, pronounced sinuses, protruding chins, and broad cheekbones. Dumollard, a rapist and murderer, had a deformed upper lip and very thick black hair.[6] The rapist Mingrat had a low forehead, jug ears, and an enormous square jaw. Archaeologists have established that the cruelest of the Caesars—Commodius, Nero, and Tiberius—had jug ears and swollen temples.

But anthropology needs numbers, not isolated, generic descriptions, especially for use in forensic medicine. Thus I will provide statistics on 390 criminals from the regions of Emilia, the Marehes, and southern Italy. Table 1 compares the hair color of these 390 criminals with that of 868 Italian soldiers from the same regions and

---

a. Thomson found an average weight of 151 pounds among 423 Scottish criminals, 106 pounds among 147 Irish criminals, and 149 pounds among 55 English criminals. The average height was 5 feet 6.9 inches for the Scottish criminals, 5 feet 6.2 inches for the English, and 5 feet 6.6 inches for the Irish.

Table 1	Hair Color of Soldiers and Criminals							
	Brown (percent)		Black (percent)		Blond (percent)		Red (percent)	
Region	Soldiers	Criminals	Soldiers	Criminals	Soldiers	Criminals	Soldiers	Criminals
Sicily	51	41	25	54	17	0	0	3
Calabria	39	50	20	33	15	0	0	0
Naples	50	50	28	40	22	5	0.3	0
Central Italy	56	66	20	33	21	5	1	5
Piedmont	47	35	13	35	34	29	0	0
Lombardy	38	33	16	33	32	33	0	0
Insane from Pavia	83	–	12	–	4	–	0	–

NOTE: The last category, the insane from Pavia, is not consistent with Lombroso's regional categories but seems to be an attempt to compare the insane with soldiers (normal men) and criminals.

90 insane from Pavia. These figures show that hair color of criminals replicates typical regional characteristics, but only up to a certain point.

Jug ears are found on 28 percent of criminals, but the proportion varies by region: 47 percent of Sicilian criminals have jug ears, as do 33 percent from Piedmont, 11 percent from Naples, 33 percent from the Romagna, 9 percent from Sardinia, and 36 percent from Lombardy. Nine percent of all criminals have very long ears, although that proportion rises to 10 percent in Lombardy and the Romagna and 18 percent in Sicily and Piedmont.

It is difficult to determine the muscular force of criminals even with the best dynamometers because the subjects are completely out of condition after long periods of detention and inertia.[7] The problem is often compounded by the malignant spirit that characterizes prisoners' whole existence. They pretend to be weaker than they really are and do not put much effort into pushing the dynamometer. In this regard, it is noteworthy that, as I was able to verify in the penitentiary of Ancona, prisoners are more energetic when they work continuously than in institutions that permit them to be idle. Rapists, brigands, and arsonists are the strongest and thieves and forgers the weakest, based on measurements of traction. Murderers and pickpockets differ in strength only by a slight fraction.

Criminal Women

At this point little can be said about female criminals because I have been allowed to examine only twenty-one of them and did so with much less ease than in the case of men. But this does not pose a complete obstacle because, first, I do not have enough information on normal women to make a comparison with criminal women; and, second, Parent-Duchatelet offers us numerous and reliable statistics on prostitutes, a class of women almost identical to criminal women in moral terms.[8] In addition, the esteemed Dr. Soresina has provided me with measurements on fifty-four prostitutes who were patients in the Milan lock hospital for venereal disease.[9]

The average cranial capacity of the twenty-one female criminals in my sample was 1,442 cc, slightly less than that of twenty insane women without dementia (1,468 cc), and above that of nineteen idiots with dementia (1,393 cc).

Table 2 presents these differences more clearly. Female criminals, and especially prostitutes, have oversized heads, but these are found only in a fraction of insane women. The rate of microcephaly, or a cranial circumference of forty-eight centimeters, among prostitutes is four times greater among the insane, who in turn have a rate double that of criminal women.

The only conclusion about the physiognomy of criminal women that I can draw from my sample is that female criminals tend to be masculine. (Among prostitutes even the voice often seems virile.) The only exception was a poisoner. Two out of twenty-one criminal women closely resembled the insane with their protruding and asymmetrical ears. Where criminal women differ most markedly from the insane is in the rich luxuriance of their hair. Not a single woman in my sample was bald, and only one showed precociously graying hair. Thomson, too, has noted rich manes of hair among female criminals.

Summing up in a few words that which scientific exigencies oblige me to express with arid numbers,[10] I conclude:

— The criminal is taller than the normal individual and even moreso than the insane, with a broader chest, darker hair and (with the exception of Venetians) greater weight.

— In head volume, the criminal presents a series of submicrocephalic craniums, double the number for normal men, but fewer than in the case of the insane.

— Criminals, especially forgers, also exceed the insane in large-volume heads. But the average head size of criminals never reaches the size of healthy men.

— The cephalic index, or shape of the criminal skull, varies with ethnicity

Table 2	Cranial Circumference of Insane Women, Prostitutes, and Criminal Women					
	Insane Women (86)		Prostitutes (54)		Criminal Women (21)	
Cranial Circumference (cm.)	total	percent	total	percent	total	percent
48	2	0.01	8	4.3		–
49	16	1.04	9	4.8	1	0.2
50	23	1.09	8	4.8	2	0.2
51	20	1.07	15	8.1	4	0.8
52	20	1.07		–	7	1.4
53	3	0.02	3	1.6	6	1.2
54	1	0.05	1	0.5	1	0.2
55			2	1.0		–
56			2	1.0		–

NOTE: The number of insane women in the second column adds to 85 rather than 86; the total number of prostitutes adds to 48 rather than 54. The percentage columns have no relationship to the other data in the table, illustrating Lombroso's statistical ineptitude or general carelessness with detail.

but tends to be brachycephalic, or short-headed, particularly among robbers.[11] Criminal skulls present frequent asymmetry, although less often than among the insane.

— Compared to the insane, criminals have more traumatic lesions of the head and oblique eyes. But they less frequently display degeneration of the temple arteries, abnormalities of the ear, thin beards, tics, virility of appearance (if they are female), dilated pupils, and, still less often, graying hair and baldness.

— Criminals and the insane show equal rates of prognathism, unequally sized pupils, distorted noses, and receding foreheads.

— Measured on the dynamometer, criminals reveal greater weakness than normal men, though they are not as weak as the insane.

— More often than in the healthy population, criminals have brown or dark eyes and thick black hair. Such hair is most frequently found among robbers.

— Hunchbacks are extremely rare among murderers but are more common among rapists, forgers and arsonists.

— Arsonists, and even more so thieves, tend to have gray irises; members of both groups are always shorter, lighter, weaker, and smaller in cranial capacity than pickpockets, who are in turn shorter, lighter, and weaker than murderers.

Among criminal women, one thing that can be said with certainty is that, like their male counterparts, they are taller than the insane. Yet they are shorter, and, perhaps with the exception of prostitutes, lighter than healthy women. All three female groups are identical in their average cranial circumference. Prostitutes show both a greater than average number of large heads and more microcephaly. In prostitutes, extremely small heads are four times more common than among the mad, and the rates among the mad are twice as high as those of the criminal. Prostitutes have dark, thick hair, and in Lombardy, but not in France, they frequently have dark irises. Female criminals are weaker than the insane and more often masculine looking.

Prognathism, thick and crisp hair, thin beards, dark skin, pointed skulls, oblique eyes, small craniums, overdeveloped jaws, receding foreheads, large ears, similarity between the sexes, muscular weakness—all these characteristics confirm the findings from autopsies to demonstrate that European criminals bear a strong racial resemblance to Australian aborigines and Mongols.

3. Tattoos

One of the most singular characteristics of primitive men and those who still live in a state of nature is the frequency with which they undergo tattooing. This operation, which has both its surgical and aesthetic aspects, derives its name from an Oceanic language. In Italy, the practice is known as *marea* [mark], *nzito*, *segno* [sign], and *devozione* [devotion]. It occurs only among the lower classes—peasants, sailors, workers, shepherds, soldiers, and even more frequently among criminals. Because of its common occurrence among criminals, tattooing has assumed a new and special anatomico-legal significance that calls for close and careful study. But first we must examine its frequency among normal individuals for the sake of comparison.

To do this, I will use data on 7,114 individuals: 4,380 were soldiers and 2,734 criminals, prostitutes, and criminal soldiers. For this information, I have to thank that valiant student of forensic medicine, Tarehini Bonfanti, as well as the illustrious Dr. Baroffio, Cavaliere Alborghetti of Bergamo, Professor Gamba of Turin, Dr. Soresina of Milan, and Professor De Amicis of Naples. Table 3 summarizes the results of the survey.

In Italy, as among savages, tattooing is infrequent among women. Among noncriminal men, the use of tattoos is decreasing, with rates ten times lower in 1873 than in 1863. On the other hand, the custom persists and indeed reaches enormous proportions among the criminal population, both military and nonmilitary, where among 1,432 individuals examined, 115 or 7.9 percent sported tattoos. The most common place for tattoos is the smooth part of the forearm, followed by the shoulder, the chest (sailors), and the fingers (miners), on which they take the form of a ring. Only men who had been to Oceania or were in prison had tattoos on their backs or pubic regions.

The symbols and meanings of tattoos are generally divisible into the categories of love, religion, war, and profession. Tattoos are external signs of beliefs and passions predominant among working-class men. After a careful study of the designs chosen by 102 criminals, I found that several not only appear frequently but also carry a particular significance. In 2 out of the 102 cases, tattoos marvelously reveal a nature that is violent, vindictive, or divided by conflicting intentions. For example, an old Piedmontese sailor, who had been a swindler and committed murder as part of a vendetta, had inscribed on his chest between two fists the sad phrase "I swear to revenge myself." A Venetian thief who was a recidivist bore the following lugubrious words on his chest: "I will come to a miserable end." It is said that criminals know their own fate and engrave it on their skin.

Other tattoos are obscene, either from their design or the region of the body where they appear. Only a few soldiers, mostly deserters released from prison, had obscene tattoos in the genital region. More significant results are obtained from my direct study of 102 tattooed

Table 3	Tattoos in Soldiers, Criminals, and Prostitutes		
Year	**Group (total number)**	**Medical Examiner**	**Percentage With Tattoos**
1863	Soldiers (1,147)	Dr. Lombroso	11.6
1873	Soldiers (2,739)	Dr. Baroffio	1.4
	Criminal soldiers (150)	Dr. Baroffio	8.6
1872	Criminals (500)	Dr. Lombroso	6.0
1873	Criminals (134)	Dr. Alborghetti	15.0
	Criminals (650)	Dr. Tarehini	7.0
	Criminal women (300)	Dr. Gamba	1.6
1866–1873	Prostitutes (1,000)	Dr. Soresina	—
1871	Prostitutes (small number)	—	a few
1874	Prostitutes (small number)	De Amicis	a few

NOTE: In this eclectic table, Lombroso has combined studies covering different parts of Italy and different years.

criminals, of whom four had obscene tattoos.[b] One had a figure of a nude woman traced over the length of his penis. A second had the face of a woman drawn on the glans, in such a way that the woman's mouth was the penis's orifice; higher up on the penis was the Savoyard flag. A third had the initials of his lover on his penis; a fourth had a bouquet of flowers. All of this reveals not only a shamelessness but also insensitivity, given that the sexual organs are extremely sensitive to pain. Even savages avoid tattooing these areas.

The study of tattoos sometimes helps us track individuals to criminal organizations. Many members of the Camorra, a Neapolitan criminal organization, have tarantula drawn on their arms; and three young arsonists from Milan sported tattoos with the same initials. Even tattoos that do not seem to have anything criminal about them and resemble those of farmers, shepherds, and sailors from the same region can be useful to the legal system and forensic medicine: they may reveal an individual's identity, his origins, and the important events of his life. Criminals, too, are aware of the advantages to the legal system offered by these involuntary revelations; thus the cleverest of them avoid tattoos or try to remove those they already have.

It would be interesting for anthropologists to ponder why such an apparently disadvantageous custom is maintained, given the discomfort and damage it causes. Here are some hypotheses:

— Religion, which tends to preserve ancient habits and customs, certainly perpetuates the practice of tattooing. Those who are devoted to particular saints believe that having their name inscribed on their flesh signifies their devotion and affection.

— A second reason is imitation.[12] Proof of this curious influence is the fact that members of an entire regiment will often sport the same design, such as a heart.

— Laziness also plays a part. For this reason tattoos are often found on deserters, prisoners, and sailors. In one prison, I found that twenty-five out of forty-one inmates had obtained their tattoos while in custody. Idleness is more painful than pain itself.

— Equally if not more influential is vanity. Even those who are not psychiatrists are aware of the way this powerful sentiment, found at all social levels and possibly even among animals, can prompt the most bizarre and damaging behavior. This is why savages who walk around naked have designs on their chests, while those who wear clothes choose to have tattoos in places that are visible or easily revealed, such as the forearm, and more often on the right forearm than on the left.

— Also influential is the sense of camaraderie, and perhaps even, as the initials of those Milanese arsonists suggest, a sense of "sect." I would not be surprised if some *camorristi* adopted, in addition to tattoos of frogs or tarantulas, other primitive ornamentations to distinguish their sect, such as wearing rings, pins, little chains, or sporting a certain style of whiskers. African tribes distinguish themselves by scarring their faces.

— Up to a certain point, noble human passions are also involved. The rites of the paternal village, the image of the patron saint, scenes of infancy, and

---

b. Tardieu has written about a coach driver and an iron smith who had boots tattooed on their penises.

depictions of distant friends—naturally one does not want to forget these things. Tattoos bring memories to life in the poor soldier's mind; the "prick of remembrance" overcomes distance, deprivations, and dangers.

— The passion of love, or rather, eroticism, is also important, as indicated by the obscene figures (4 out of 102 cases) or the initials of lovers (10 out of 102 cases) found on our criminals, and similar images on lesbians and prostitutes. Tribal women tattoo themselves to show they are unmarried. In men, too, tattoos often indicate virility. As Darwin puts it somewhat exaggeratedly, tattoos are both a sign and a means of sexual selection.[13]

Among Europeans, the most important reason for tattooing is atavism and that other form of atavism called traditionalism both of which characterize primitive men and men living in a state of nature. There is not a single primitive tribe that does not use tattooing. It is only natural that a custom widespread among savages and prehistoric peoples would reappear among certain lower-class groups. One such group is sailors, who display the same temperament as savages with their violent passions, blunted sensitivity, puerile vanity, and extreme laziness. Sailors even adopt the habits, superstitions, and songs of primitive peoples. In their nudity, prostitutes, too, recall savage customs.

The foregoing should suffice to demonstrate to judges and practitioners of forensic medicine that tattoos can signify a previous incarceration. Criminals' predilection for this custom is sufficient to distinguish them from the insane. Both groups experience forced internment, strong emotion, and long periods of boredom. Although madmen resort to such strange pastimes as rolling stones, snipping their clothes and even their flesh, scribbling on

walls, and filling entire reams of paper, they very rarely make designs on their skin. This is yet another proof of the influence of atavism on tattooing because madness is almost never a congenital illness and rarely atavistic.

Forensic medicine should recognize that in the case of the criminal man, who is in constant struggle against society, tattoos—like scars—are professional characteristics.

⬚ 4. Emotions of Criminals

The unusual predilection of criminals for something as painful as tattooing suggests that they have less sensitivity to pain than ordinary men, a phenomenon that can also be observed in the insane, especially those with dementia. Except in cases of idiocy, what appears as insensitivity to pain is actually the dominance of certain passions. Thus lesbian prostitutes, to reach their lovers in hospital, use red-hot irons to give themselves blisters that resemble pustular eruptions. I once saw two murderers who had for a long time hated one another throw themselves at each other during exercise hour, remaining embroiled for some minutes, one biting the other's lip, the other tearing out the hair of his adversary. When they were finally separated, they were more concerned about their unfinished brawl than their wounds, which became seriously infected.

More generally, criminals exhibit a certain moral insensitivity. It is not so much that they lack all feelings, as bad novelists have led us to believe. But certainly the emotions that are most intense in ordinary men's hearts seem in the criminal man to be almost silent, especially after puberty. The first feeling to disappear is sympathy for the misfortunes of others, an emotion that is, according to some psychologists, profoundly rooted in human nature. The murderer Lacenaire confessed that he had never felt the slightest sense of regret seeing any cadaver except that of his cat: "The sight of agony has no effect on me whatsoever. I would kill a man as easily I would drink a glass of water."[14] Complete indifference to their victims and to the bloody traces of their crimes is a constant characteristic of all true criminals, one that distinguishes them from normal men.

An executioner, Pantoni, told me that nearly all robbers and murderers go to their death laughing. A thief from Voghera, shortly before his execution, ordered a boiled chicken and ate it with gusto. Another inmate wanted to choose his favorite among the three executioners and called him his "professor." While being taken to the gallows, the assassin Valle from Alessandria, who had killed two or three of his companions out of pure caprice, loudly sang the well-known song "It's Not True That Death Is the Worst of All Evils."

The criminal's insensitivity is further proved by the frequency with which an assassin's accomplices will return to murdering people just after he is put to death. Also instructive are the joking words that in criminal jargon are used to name the executioner and his instruments, as well as the tales that are told in prison, where hanging is the favorite theme.[c] This provides one of the strongest arguments for the abolition of the death penalty, which clearly dissuades only a very small number of these wretches from committing crimes. Instead, it may encourage crime thanks to that law of imitation that so dominates among the vulgar classes and to the horrendous prestige that accrues to the person of the condemned. His criminal companions are made envious by the lugubrious and solemn ritual of execution before a crowd of spectators.[15]

Insensitivity to their own and other's pain explains why some criminals can commit acts that seem to be extraordinarily courageous. Thus Holland, Doincau, Mottino, Fieschi, and Saint-Clair had previously won medals for valor on the battlefield. Coppa threw himself into the midst of a battalion firing his gun and came out unharmed. These apparent acts of courage are really only the effect of criminals' insensitivity and infantile impetuousness, which prevent them from recognizing even certain danger. It makes them blind when they have a goal or a passion to satisfy.

Insensitivity combined with precipitous passions explains the lack of logic in crimes and the disjuncture between the gravity of a deed and the motive. For example, a prisoner killed a fellow inmate because he snored too loudly and would not or could not stop (Lauvergne, p. 108). Another, from Alessandria, killed his fellow inmate because he would not polish his shoes. Such moral insensitivity among criminals explains a paradox: the frequent cruelty of criminals who at other times seem capable of kindness.

Criminals' feelings are not always completely gone; some may survive while others disappear. Troppman, a killer of women and children, cried on hearing the name of his mother. D'Avanzo, who roasted and ate a man's calf muscles, later, wrote poetry. Immediately after committing murder, Feron ran to his girlfriend's children and gave them sweets. Holland confessed to a murder committed because he wanted to obtain money for his family, saying, "I did it for my poor child."

Parent-Duchatelet shows that while many prostitutes lose all ties to their own family, others use their ill-gotten gains to provide for their children, their parents, and even friends. They are excessively passionate about their lovers, so much so that even violence does not detach them. One unfortunate prostitute, after breaking her leg trying to escape her pimp's beatings, returned to him. Assaulted once again, she suffered a broken arm but lost nothing of her intense affection.

In most criminals, the nobler sentiments tend to be abnormal, excessive, and unstable.

---

c. Fregier. *Des classes dangéreuses*, 1841, p. 111. In German slang, to be hanged is *Heimgangen*, going back home. In Italian there are various terms for being hanged, including "to grimace" or "to squint."

Mabille, to entertain a friend he had made one night in a restaurant, committed a murder. A certain Maggin said to me, "The cause of my crimes is that I fall into friendships too easily; I cannot see a companion offended without putting a hand on my dagger to revenge him."

A few tenacious passions dominate criminals in place of their absent or unstable social and family feelings. First among these is pride, or rather, an excessive sense of self-worth, which seems to grow in inverse proportion to merit. It is almost as if the psyche is dominated by the same law that governs reflex reactions, which grow stronger as the nervous system weakens. But in the case of criminals, this disequilibrium reaches gigantic proportions. The vanity of criminals exceeds that of artists, the literati, and flirtatious women. A death sentence did not worry Lacenaire nearly as much as criticism of his dreadful poetry and fear of public ridicule. As he put it, "I do not mind being hated, but do mind being mocked" for verses like the following:

The storm leaves a track

While the humble flower passes without a trace.

The most common motive for modern crimes is vanity, the need to shine in society, which is sadly known as "cutting a fine figure." At the Pallanza prison, a criminal told me, "I killed my sister-in-law because our family was too big, and it was thus difficult to make a show in the world." Denaud killed his wife, and his lover killed her husband, so that they could marry and save their reputation. When an infamous thief adopted a certain type of waistcoat and tie, his fellows imitated him, dressing themselves in the same style. Thus Inspector Vidocq found, among twenty-two thieves caught in one single day, twenty who wore the same color waistcoat.[16] They were vain about their strength, their looks, their courage, their ill-gotten and short-lived gains, and, most distressing of all, their ability to commit crimes.

Criminals resemble prostitutes, who always believe they belong to the highest grade of their profession. To prostitutes, the phrase "you are a one-lira woman" is a deep offense. Male inmates who have stolen thousands of lire laugh at petty thieves; and murderers consider themselves superior to thieves and swindlers.

The excessive vanity of criminals explains why they discuss their crimes before and after committing them, showing incredible lack of foresight and providing the justice system with the best possible weapon for finding and sentencing them. Shortly before killing his father, the patricide Marcellino said, "When my father returns from the fields, he will remain here forever." Berard, before going to commit the last of his crimes—the murder of three rich women—was heard to say: "I want to be connected to something big; oh, how I will be talked about!" But the clearest and most curious example of criminals' incredible vanity is a photograph discovered by the police in Ravenna showing three villains who, after having killed a companion, had themselves portrayed in the positions they had as aimed while striking their blows. They felt the need to immortalize this strange moment at the risk of being reported and apprehended, which in fact happened.

A natural consequence of criminals' limitless vanity and inordinate sense of self is an inclination toward revenge for even the pettiest motives, as in the example of the inmate who killed someone for refusing to polish his shoes. Prostitutes exhibit the same tendency. "They frequently become enraged for the smallest reason, for a comment about being ugly, for example. In this regard they are more childish than their own children. Prostitutes would consider themselves dishonored if they did not react" (Parent-Duchatelet, p. 152). Such violent passions also lie behind the ferocity of ancient and savage peoples, although today they are so rare as to appear monstrous.

Once criminals have experienced the terrible pleasure of blood, violence becomes an uncontrollable addiction. Strangely, criminals are not ashamed of their bloodlust, but treat it with a sort of pride. Thus while he lay dying, Spadolino lamented having killed ninety-nine men instead of reaching a hundred.

Everyone agrees that the few violent women far exceed men in their ferocity and cruelty. The brigand women of southern Italy and the female revolutionaries of Paris invented unspeakable tortures. It was women who sold the flesh of policemen; who forced a man to eat his own roasted penis; and who threaded human bodies on a pike. Thus Shakespeare depicts Lady Macbeth as more cruel and cold than her male accomplice.

After the delights of taking revenge and satisfying his vanity, the criminal finds no greater pleasures than those offered by drinking and gambling. However, the passion for alcohol is very complex, being both a cause and an effect of crime. Indeed, alcohol is a triple cause of crime. First, the children of alcoholics often become criminals, and second, inebriation gives cowards the courage to undertake their dreadful deeds, as well as providing them with a future justification for their crimes. Furthermore, precocious drinking seduces the young into crime.[17] Third, the tavern is a common gathering place for accomplices, where they hatch their plots and upend the proceeds. For many it is their only home. The innkeeper is their banker, with whom criminals deposit their ill-gotten gains.

Few criminals fail to feel a lively passion for gambling, which explains a continual contradiction in the life of the malefactor: he greatly desires the belongings of others but at the same time squanders the money he has stolen, possibly because it was so easily acquired. Love of gambling also explains why most criminals end up poor despite possessing at times enormous sums.

Rarely does a male criminal feel true love for women. His is a carnal love, which almost always takes place in brothels (especially in London, where two-thirds of these are dens of criminals) and which develops at a very young age.[d] Prostitutes experience lesbian love, which distinguishes them from normal women, and they are passionate about flowers, dancing, and dining.

But the pleasures of gambling, eating, sex, and even revenge are nothing but intermediate steps to criminals' predominate passion, that of the orgy. Even though criminals shy away from regular society, they crave a kind of social life all their own, veritable orgies in which they enjoy the jubilant, tumultuous, riotous, and sensuous companionship of other offenders and even police spies.

I will not discuss criminals' many other passions, which vary according to their habits and intelligence. These range from the most terrible, such as pederasty, to the most noble, such as music; collecting books, paintings, and medals; and the enjoyment of flowers—a particular enthusiasm of the prostitute. One can find unusual emotions among criminals as among healthy people; what distinguishes criminal passions is instability, impetuousness, and often violence. In the quest to satisfy themselves, they give no thought to the consequences.

In many of these characteristics, criminals resemble the insane, who also exhibit not only violent and unstable passions but also insensitivity to pain, an exaggerated egotism, and (less frequently) a craving for alcohol. But the insane rarely show a predilection for gambling and orgies, and more than criminals, they will suddenly start to hate those who have hitherto been dearest to them. While the criminal cannot

---

d. Of 3,287 homicides and assaults in Italy, 299 were caused by sexual jealousy and 47 for prostitution and loose behavior. Of 41,454 crimes, 1,499 involved illicit loves. In England, of 10,000 persons sentenced, 3,608 were prostitutes; in Italy, of 383 women sentenced, 12 were prostitutes. Of 208 crimes committed for reasons of love according to Descruret, 91 were for adultery, 96 for concubinage, and 13 for jealousy. Of 10,000 crimes of violence in France and England, 1,477 were for reasons of love (Guerry). Of 10,899 suicides in France, 981 were for love.

live without his companions and seeks them out even at his peril, the mad prefer to be alone, fleeing from every association with others. Plotting is as rare in the mental hospital as it is common in the prison.

In their emotional intensity, criminals closely resemble not the insane but savages. All travelers know that among the Negroes and savages of America, sensitivity to pain is so limited that the former laugh as they mutilate their hands to escape work, while the latter sing their tribe's praises while being burned alive. During puberty initiations, the young savages of America submit without complaint to cruel tortures that would kill a European. For example, they hang themselves upside down on butchers' hooks over dense columns of smoke. Such insensitivity encourages painful tattooing—something few Europeans can bear—and customs like cutting the lips and the fingers or pulling out teeth during funeral ceremonies.

Even moral sensitivity is weak or nonexistent among savages. The Caesars of the yellow races, called Tamerlanes, made their monuments out of pyramids of dried human heads. The emperor Nero was infamous for his barbarity, but even it paled in comparison to the cruel rites of the Chinese. Savages and criminals are further alike in the impetuosity and instability of their passions. Savages, as Lubbock tells us, have quick and violent emotions; while their strength and passions are those of adults, in character they are children. Similarly, Schaffhausen reports that "in many respects savages are like children: they feel strongly and think little; they love games, dancing and ornaments; and they are sometimes curious and timid. They are unaware of danger, but deep down they are cowards, vengeful and cruel in their vendettas."[e] A Cacique, returning from a failed hunting expedition, was greeted by his young son, who ran between his legs; to vent his rage, the father picked him up by the leg and dashed him against some rocks.

✎ Notes

1. These heights in meters are equivalent to 5 feet 4.4 inches; 5 feet 3.6 inches; 5 feet 2.8 inches; 5 feet 2.5 inches; and 5 feet 1.8 inches, respectively.

2. James Bruce Thomson (1810–73), chief physician at Scotland's Perth Prison, published influential articles based on degenerationist theory and arguing that many criminals are born criminals; these included "The Psychology of Criminals" in the 1870 issue of the *Journal of Mental Science*. Lombroso was fascinated by Thomson's work and cited it as evidence for his own theory. The Wilson to whom Lombroso refers was probably the British craniometrist George Wilson. The work of both men is discussed in Davie 2004, 2005.

3. These weights in kilograms are equivalent to 150 pounds and 138 pounds, respectively.

4. Johann Ludwig Casper (1787–1853) was a prominent professor of forensic pathology in Berlin.

5. Johann Kasper Lavater (1741–1801) attempted to turn physiognomy, or the reading of character from facial expressions, into a science. In the early nineteenth century, physiognomy flowed into phrenology, and later in the century, phrenology flowed into criminal anthropology. The Guido to whom Lombroso here refers was probably the Italian baroque painter Guido Reni (1575–1642).

6. The French criminal Martin Dumollard, famous for drinking his victims' blood, was executed in 1862.

7. The dynamometer, or strength-testing machine, used by Lombroso was oval in shape with a dial to record results. Subjects had to compress the oval to test "compressive strength" and pull at the oval to test "tractive strength."

8. Alexandre J. B. Parent-Duchatelet (1805–59) was the first European to conduct a large-scale study of prostitution, *De la prostitution dans la ville de Paris* (*Prostitution in the City of Paris*, 1836). Lombroso relied on Parent-Duchatelet's work for most of his early data on female crime. In this first edition of

---

e. *Uber den Zustand der Wilden*, 1868.

Criminal Man, Lombroso begins to establish his theory that prostitution—that is sexual deviancy—constitutes the typical and most widespread form of female crime.

9. Lock hospitals (or *sifilicomi* in Italian) were nineteenth-century institutions for the internment of prostitutes with venereal disease. Although hospitals in name, they resembled prisons to which prostitutes were admitted forcibly by police. Lock hospitals provided criminologists like Lombroso with a captive female population for physical and psychological examinations.

10. In this summary of his conclusions on the anthropometry and physiognomy of criminals, Lombroso is referring (with only a few exceptions) to men. Some of his conclusions are based on material that we cut from this chapter.

11. The cephalic index was an important tool for nineteenth-century criminal anthropologists in their attempts to categorize races and identify born criminals. They obtained the cephalic index by multiplying the width of the skull by one hundred and dividing the result by the length of the skull. The resulting numbers enabled them to classify skulls into various categories including brachycephalic (or short-headed) and dolichocephalic (or long-headed). Born criminals fell into both categories; Lombroso considered their cephalic indices to be abnormal when the numbers deviated from the norm for their geographical region.

12. Imitation became an important criminological concept in the late nineteenth century, especially in explanations of crowd behavior. The French jurist Gabriel Tarde (1843–1904) is best known for applying the concept of imitation to the etiology of crime in his 1890 book *La philosophie pénale* (*Penal Philosophy*). In this first edition of *Criminal Man*, Lombroso anticipates Tarde as well as later criminological debates over imitation. See Barrows 1981.

13. Charles Darwin (1809–82), the British naturalist, elaborated the theory of evolution in his famous work, *The Origin of Species* (1859). For Lombroso, Darwin's theory confirmed his intuition that criminals were atavistic or throwbacks on the evolutionary scale. In the third edition of *Criminal Man*, Lombroso aspires to do for criminals what Darwin had done for plant and animal species.

14. The French thief and poet Pierre-François Lancenair (1800–36) both shocked and fascinated public opinion with his murder of the widow Chardon and her son in 1835. The following year, he was executed by guillotine for his crime.

15. Lombroso was ambivalent about the death penalty, but in general he went from opposition to grudging acceptance of it as a means of social defense against recidivist born criminals and particularly violent members of organized crime like *mafiosi* and brigands. Yet he muffled his support of capital punishment in deference to the opposition of the majority of his positivist colleagues and the public in general. In 1889, Italy abolished the death penalty in its new criminal code.

16. The notorious French criminal Eugène François Vidocq (1775–1857) turned informer and finally became head of the French urban police force in 1811. He is the model for both Jean Valjean and Inspector Javert in Victor Hugo's novel *Les misérables* (1862).

17. Lombroso often employs the word *precocity* to mean the premature development of physical or psychological characteristics or early indulgence in adult behaviors. In all cases, precocity signals abnormality for Lombroso.

REVIEW QUESTIONS

1. What are some of the characteristics of the face of habitual murderers, according to Lombroso? What about thieves? Rapists?

2. What does Lombroso claim are key characteristics of female criminals? Do these differ from male criminals?

3. Review some of the tattoos that Lombroso observed on criminals. How do these tattoos relate to the emotions of criminals?

❖

READING

In this selection, Gibson, Piquero, and Tibbetts review the existing research on the link between low verbal IQ and criminality and provide an empirical study to further advance the state of knowledge on this link. They present a test of the interaction between verbal IQ with an environmental factor: family adversity. Thus, this theoretical model assumes a "nature via nurture" perspective (as opposed to "nature versus nurture"), which we will review in more detail in Section IV. While reading this selection, you should keep in mind the contrast between how IQ is viewed and being studied now, as opposed to how it was studied in the early 1900s.

The Contribution of Family Adversity and Verbal IQ to Criminal Behavior

Chris L. Gibson, Alex R. Piquero, and Stephen G. Tibbetts

Since the early 1990s, there has been a progressive effort to identify risk and protective factors that contribute to increasing and decreasing the odds of differential offending behavior (e.g., early onset of offending, violent offending, etc.). Farrington (2000) referred to this model as the *risk factor prevention paradigm*. Although recently introduced to the field of criminology by scholars such as Hawkins and Catalano (1992) and Farrington (2000), the risk factor prevention paradigm has been widely applied in the fields of public health and medicine to successfully address and prevent life-threatening illnesses such as cancer and heart conditions. As it pertains to criminology, this paradigm is simple in that its goal is to identify the important risk and protective factors of

offending behaviors and then, based on the identification of such factors, implement prevention techniques designed to minimize the risk factors and maximize the protective factors. As noted by Farrington (2000), this relatively new and simple approach has been advocated and adopted in the United States (Loeber & Farrington, 1998) and expanded to several industrialized countries such as the United Kingdom (Nutall, Goldblatt, & Lewis, 1998) and the Netherlands (Junger-Tas, 1997).

The risk factor prevention paradigm consists of both risk factors and protective factors. A risk factor by definition is a variable that predicts later involvement in offending (Farrington, 2000; Kazdin, Kraemer, Kessler, Kupfer, & Offord, 1997). Researchers have adopted several ways of assessing the effects of

SOURCE: Chris L. Gibson, Alex R. Piquero, and Stephen G. Tibbetts, "The Contribution of Family Adversity and Verbal IQ to Criminal Behavior," *International Journal of Offender Therapy and Comparative Criminology* 45, no. 5 (2001): 574–92. © 2001 Sage Publications, Inc. Used by permission of Sage Publications, Inc.

risk factors on later offending; thus, oftentimes a risk factor is referred to as an extreme category of an independent variable (Farrington, 2000). Such risk factors tend to co-occur, making it difficult to disentangle their effects on future offending behavior. Given the complex nature of how risk factors may contribute to the explanation of later offending, Farrington (2000) suggested that multiplicative interactions of such factors should be explored in an attempt to understand how they are linked to offending behaviors (e.g., prevalence, early onset, etc.). Protective factors are another component of the risk factor prevention paradigm. Although the definition and existence of protective factors are controversial, Rutter (1985) stated that one possible definition of a protective factor is a variable that mediates the likelihood that a risk factor will increase later offending behavior.

The majority of studies that have been conducted to assess risk factors for offending have used prospective longitudinal data (see Denno, 1990; Farrington & Loeber, 1999; Moffitt, 1993). Most of these studies have traditionally focused on individual, family, peer, school, and socioeconomic factors measured in childhood and/or adolescence that predict the later development of offending and violent offending (for a review, see Farrington, 1998). Two well-known longitudinal studies conducted in London and Pittsburgh have identified several comparative risk factors that predict subsequent delinquency and youth violence. These studies concluded that individual factors such as hyperactivity, poor concentration, and low achievement measured at ages 8 through 10 were significantly related to court referrals between ages 10 and 16 (Farrington & Loeber, 1999). Another important individual risk factor is low intelligence (Farrington, 1998; Moffitt, 1993), which was found to be a significant risk factor for court convictions and self-report offending in London. In addition, both studies also concluded that family adversity measures such as an antisocial father, large family size, low family income, a broken family, poor parental supervision, and parental disharmony measured at ages 8 to 10 were significant predictors of court referrals between ages 10 and 16 (Farrington & Loeber, 1999).

Although several longitudinal investigations (Denno, 1990; Farrington & Loeber, 1999; Moffitt, 1994) have identified important risk factors that increase the odds of subsequent offending behavior, there have been limited empirical efforts to assess how such risk factors interact with one another to increase/decrease the odds of offending. This is especially true with regard to specific types of offending such as adolescent-limited versus life-course persistent offending (Moffitt, 1993). Furthermore, Farrington (2000) added that another key yet underresearched issue is whether the strength of the relationship between such risk factors and outcomes is similar or variable across distinct groups of offenders. Given that many longitudinal studies assessing risk and protective factors have used samples of Caucasian youth, it makes it difficult to generalize such findings to other groups of individuals (e.g., African American, inner-city youth).

The present study builds on the risk factor prevention paradigm by investigating the interactive effect of verbal IQ and family adversity as they relate to the prevalence of offending as well as an early onset of offending in a sample of 987 African American, inner-city youth born and raised in Philadelphia. This investigation is important due to the limited empirical knowledge on how such risk factors interact with one another to predict different types of offending, especially among a sample of urban, inner-city, African American youth. Herein, we apply Moffitt's (1993) international hypothesis of antisocial behavior as a theoretical guide for investigating the joint contribution of both risk factors in explaining criminal behavior. First, we hypothesize that the Verbal IQ × Family Adversity interaction will not be predictive of whether an individual is an offender by age 18 because Moffitt (1993) suggested that the interaction will only be a risk factor for a certain type of offending (i.e., life-course persistent offending). Second, we examine Moffitt's (1993) hypothesis that the Verbal IQ × Family Adversity interaction will be important in weeding out specific types of offenders within the age-crime curve (i.e., the Verbal IQ × Family Adversity interaction should be a risk factor for early onset of offending, which has been shown to be an important indicator of life-course persistent styles of offending).

The review of the literature is as follows. First, we discuss the importance of establishing known risk factors for early onset and how early onset is related to life-course persistent offending patterns. Second, we present Moffitt's (1993) theoretical framework, emphasizing her international hypothesis. Third, we discuss the extant empirical evidence of the relationship between verbal IQ and criminal behavior and the link between family adversity and criminal behavior.

⊠ Correlates of Early Onset and Persistent Offending

Research has shown that adult persistent offending is rooted in early childhood behavioral problems. In concordance with the strong and positive association observed among past and future offending (Gottfredson & Hirschi, 1990; Nagin & Paternoster, 1991; Robins, 1966, 1978; Wilson & Herrnstein, 1985), the age at which a first offense occurs (i.e., onset) has similarly been found to be strongly correlated with future offending, especially serious, habitual, and violent offending (Blumstein, Cohen, Roth, & Visher, 1986; Dunford & Elliott, 1984; Farrington, 1986, 1998; Farrington et al., 1990; LeBlanc & Frechette, 1989; Loeber & LeBlanc, 1990; Patterson, Crosby, & Vuchinich, 1992; Reiss & Roth, 1993; Sampson & Laub, 1993; Tolan,

1987; Wolfgang, 1983). In addition, relationships have been observed between early onset, conduct disorder, and oppositional defiant disorder (Wasserman & Miller, 1998).

Although early onset is important to the understanding of homotypic and heterotypic continuity, the risk and protective factors that increase and/or decrease the odds of an early onset is an entirely different question. Acquiring knowledge of the risk and protective factors that increase and/or decrease the probability of an early onset would allow policy makers and researchers to more accurately identify individuals at risk for early onset and persistent offending styles, thus allowing for early implementation of prevention strategies (Farrington, 1998, 2000; Moffitt, 1993). Although few question this agenda and its profound policy implications, reviews of the literature pertaining to early onset suggest that there is limited evidence of risk factors associated with early onset (Farrington, 1998; Farrington et al., 1990).

Research has assessed the influence of several independent risk factors on the development of early offending behavior. Factors such as neuropsychological and psychosocial deficiencies (Gorenstein, 1990; Moffitt, 1990), poor psychomotor skills (Farrington & Hawkins, 1991), as well as family adversity and economic/social deprivation have been identified as predictors of an early onset (Farrington & Hawkins, 1991; Moffitt, 1990). A small number of studies have assessed biosocial interactions between some of the previously mentioned risk factors. Such studies have assessed psychological and environmental factors coupled with various biological and/or physiological traits (e.g., low birth weight, heart rate, and pre/perinatal complications) in predicting early onset of offending behavior (Moffitt, 1990; Raine, Brennan, & Mednick, 1994; Tibbetts & Piquero, 1999) as well as violent offending patterns (Brennan, Mednick, & Raine, 1997; Kandel & Mednick, 1991; Piquero & Tibbetts, 1999; Reiss & Roth, 1993). Although studies that have investigated interactions between individual differences (neuropsychological and personality traits) and environmental factors on offending outcomes are limited, Moffitt (1993) stated, "It is now widely acknowledged that personality and behavior are shaped in large measure by the interaction between the person and the environment" (p. 682).

Moffitt's Interaction Hypothesis

Moffitt (1993), among others (Wolfgang, Thornberry, & Figlio, 1987), suggested that a relatively small group of offenders (6% to 10%) exhibit criminal behavior early in life and are likely to be chronic in their offending patterns. Specifically, Moffitt claimed that the small groups of individuals who incur an early onset are likely to possess two specific risk factors early in life, neuropsychological problems in early childhood and disadvantaged environments and/or family adversity. Moffitt stated that these risk factors interact to predict early onset of offending behavior, leading to a life-course persistent style of offending.

Moffitt, Lynam, and Silva (1994) claimed that problem behavior begins early in childhood because neuropsychological deficiencies disrupt normal development, and it is these deficits that increase vulnerability to the criminogenic aspects of disadvantaged rearing environments. In agreement with studies on low verbal and communication skills (Tarter, Hegedus, Alterman, & Katz-Garris, 1983; Tarter, Hegedus, Winsten, & Alterman, 1984), Moffitt (1993) noted that children with varying degrees of neuropsychological deficiencies evoke a challenge to the most well-prepared parents. Specifically, these deficiencies may elicit increasingly more physical punishment from caregivers, especially if the family is living in a disadvantaged or distressed environment (Moffitt, 1997).

Preliminary findings from Moffitt's (1990) ongoing New Zealand study showed that young boys who scored low on neuropsychological tests and lived in adverse home environments had a mean aggression score more than four times greater than that of boys with either neuropsychological deficiencies or adverse home environments alone (also see Moffitt et al., 1994). More recently, Moffitt (1997) used data from the Pittsburgh Youth Study to assess the relationship between neuropsychological test scores and delinquency among two groups of African American youth living in "good" and "disadvantaged" urban neighborhoods. The interactive effect of neuropsychological test scores and environmental factors on boys' delinquent behaviors were assessed by categorizing the boys on their mean score, either below or above, on a measure of cognitive impulsivity. The boys were also categorized into good and disadvantaged neighborhoods based on census tract data. Results show that there was a marginally significant interaction effect on delinquency. Moffitt (1997) stated that the results indicate that neuropsychological deficit and delinquency coexist among individuals in disadvantaged inner-city environments. Furthermore, regardless of the type of neighborhood, boys with neuropsychological problems were more likely to be more delinquent.

Moffitt theorized that a variety of factors disrupt the central nervous system of the fetus/infant such as prenatal and perinatal complications. Importantly, these deficits manifest themselves in various ways such as temperament difficulties, lower executive functioning, and poor verbal test scores. Such cognitive deficiencies and temperamental deficits among children are found more pervasively in unsupportive or adverse environments (Moffitt, 1993). In agreement with empirical research (Alexander & Comely, 1987; Caldwell, 1981; Greenberg, 1983), Moffitt (1997) suggested that neuropsychologically vulnerable children are found at higher rates in inner cities because prenatal care is scarce, premature births are more common, infant malnutrition is problematic, and the possibility for exposure to toxic and infectious agents is greater.

Moffitt (1993) suggested that the neuro-environmental interactions should not be capable of distinguishing offenders from nonoffenders, but should distinguish between types of offenders (i.e., early onset and late onset). In particular, the interaction between neuropsychological deficiencies and disadvantaged environments should predict early onset.

Verbal IQ and Criminal Behavior

Moffitt's (1993) theoretical framework suggests that verbal functioning is one of the two sorts of neuropsychological deficits that are most empirically associated with an early onset, which has been argued to be one manifestation of life-course persistent offending (see Gibson, Piquero, & Tibbetts, 2000). Moffitt (1993) suggested that "the verbal deficits of antisocial children are pervasive, affecting receptive listening and reading, problem solving, expressive speech and writing, and memory" (p. 680). Research has shown that cognitive deficits and criminal behavior share variation that is independent of the effects of social class, race, test motivation, and academic achievement (Lynam, Moffitt, & Stouthamer-Loeber, 1993; Moffitt, 1990). Moreover, existing empirical evidence supports the conclusion that the association between verbal deficiencies and offending behavior is one of the largest and most robust effects in the investigation of criminal behavior (see Hirschi & Hindelang, 1977; Moffitt, 1990; Moffitt & Henry, 1991; Moffitt & Silva, 1988). The consistency of these findings gives support to the notion that delinquents have a language manipulation deficit and that individuals with such neuropsychological deficiencies tend to be involved in offending at an early age. However, there is limited empirical evidence showing that childhood verbal IQ test scores interact with one's social environment to minimize and/or maximize their likelihood of early offending behavior (Moffitt, 1993, 1997).

Family Adversity and Criminal Behavior

Moffitt (1993) emphasized that children with cognitive and temporal deficiencies are often born into nonsupportive families that are oftentimes saturated with family adversity. For Moffitt (1993), vulnerable children "are disproportionately found in environments that will not be ameliorative because many sources of neural development co-occur with family disadvantage" (p. 681).

Although Moffitt (1993) used the term *criminogenic environment* throughout her theoretical argument, she implied that this is synonymous with family adversity. In defining family adversity, several empirical investigations have employed socioeconomic status (SES), single parenting, age of mother at birth, and multiple family transitions as risk factors that independently and collectively measure family adversity (Loeber & Farrington, 1998; Moffitt, 1990; Nagin, Pogarsky, & Farrington, 1997; Tibbetts & Piquero, 1999).

The aforementioned family adversity risk factors have all been shown to be independently related to several subsequent types of criminal offending. For example, empirical evidence has shown that maternal age at birth is a risk factor for offending behavior for mother's offspring (Nagin et al., 1997). To account for this linkage, scholars argue that young mothers are less likely to have well-developed parenting/role model skills, which oftentimes leads to neglect and poor supervision of offspring, particularly when other family adversity risk factors such as low SES and parental separation are involved (Nagin et al., 1997). Other empirical investigations have documented the adversities of being a young mother. Young mothers are more likely to engage in problem behavior (Elster, Ketterlinus, & Lamb, 1990), head a single-parent household, live in poverty (Grogger & Bronars, 1993), and fail to finish high school (Ahn, 1994), all of which have been used as indicators of family adversity and subsequently have been found to be related to offspring criminal behavior.

Many studies show that children from single-parent families are exposed to an increased risk for behavioral problems. Such studies have shown that broken homes and early separation of parents predict criminal offending (Farrington, 1992, 1993; McCord, 1982). In the New-Castle Thousand Family study, Kolvin and colleagues (Kolvin, Miller, Scott, Gatzanis, & Fleeting, 1990) found a significant association between parental divorce or separation before age 5 and later convictions up to age 33. The importance of the effects of single-parent homes is also shown in the English national longitudinal survey of more than 5,000 children born in 1946 (Wadsworth, 1979). Specifically, boys from broken homes due to divorce or separation were significantly more likely of being convicted or officially cautioned up to age 21. Furthermore, children from single-parent families have been found to have a variety of adverse problems such as conduct disorder and substance abuse (Blum, Boyle, & Offord, 1988; Boyle & Offord, 1986).

Current Effort

In this article, we build on prior research in two ways. First, we examine two different outcome variables that are hypothesized to be related to various risk factors as implicated by Moffitt's theory and other empirical research (Farrington & Loeber, 1999). Second, we more directly measure the role of neuropsychological risk by explicating measures of verbal IQ and expand prior measurement of family adversity by integrating several other risk factors that have been related to offending. Toward this end, we examine two hypotheses from Moffitt's developmental taxonomy that center around the role of neuropsychological risk and family adversity. The following two outcome variables are employed: (a) whether the subject is an offender (no/yes) and (b) given that the subject

is an offender, whether he or she exhibited an early onset of offending.

Moffitt hypothesized that the Neuropsychological Risk × Family Adversity interaction would not be predictive of who does/does not offend. For example, several research efforts have shown that there are important differences (in degree and kind) within the offending population such that offenders are not a homogenous population (D'Unger, Land, McCall, & Nagin, 1998; Nagin, Farrington, & Moffitt, 1995; Nagin & Land, 1993; Piquero et al., 2001). Furthermore, the Neuropsychological Risk × Family Adversity interaction is not meant to distinguish between offenders and nonoffenders; instead, it best predicts a "special sort of delinquency" that is related to styles of life-course persistent offending (Moffitt, 1997). For Moffitt's theory, the Neuropsychological Risk × Family Adversity interaction should be predictive of an early but not late onset of offending because adolescence-limited offenders do not suffer from any individual-level deficits in neuropsychological risk or self-control (whereas the life-course persistent offenders do) (Moffitt, 1993; Moffitt et al., 1994; Moffitt, Caspi, Dickson, Silva, & Stanton, 1996).

Two hypotheses are investigated herein. First, the Neuropsychological Risk × Family Adversity interaction will not be a significant predictor of whether or not individuals offend. Second, the Neuropsychological Risk × Family Adversity interaction will be able to distinguish among the offending population. In particular, the interaction will be a significant predictor of early but not late onset of offending. Because family adversity has independently been found to be related to whether individuals offend or refrain from offending, we expect it to be related to whether subjects offend or do not, but we do not expect it to be a significant discriminator between those who exhibit an early as opposed to late onset of offending. In sum, the key focus of the current analysis lies in understanding the ways in which neuropsychological risk interacts with social and environmental conditions to increase the probability that certain forms of offending will occur (Moffitt, 1997).

Data

Data used to examine these hypotheses are drawn from the Philadelphia portion of the National Collaborative Perinatal Project (NCPP) (Denno, 1990). Designed as a health and development study, the NCPP followed prospectively the course of more than 56,000 pregnancies enrolled between 1959 and 1966 at several university-affiliated medical schools in the United States (Niswander & Gordon, 1972). Pregnancies for the Philadelphia site came from Pennsylvania Hospital.

A wide variety of variables were collected, including events of gestation, labor, and delivery as well as children's mental, motor, sensory, and physical development to 7 years of age. Major findings from the NCPP have been detailed elsewhere (Broman, Nichols, & Kennedy, 1975; Nichols & Chen, 1981), and specific criminological investigations have also been undertaken, primarily with the Philadelphia (Denno, 1990; Piquero, 2000a, 2000b, 2000c; Piquero & Tibbetts, 1999; Tibbetts & Piquero, 1999) and Providence (Lipsett, Buka, & Lipsitt, 1990; Piquero & Buka, 2001) cohorts.

Archived data from the Philadelphia cohort of the NCPP consisted of 987 subjects. All members of this subsample were African American, and the majority was of middle to lower class in socioeconomic status. Detailed criminal history information for this cohort was collected from the Philadelphia Police Department apart from the larger NCPP project by researchers at the University of Pennsylvania in the early 1980s when the cohort was 18 years of age. Several criteria were established for inclusion in the Philadelphia subsample. All subjects were born and raised until young adulthood in Philadelphia, received very similar medical treatment early in life, attended Philadelphia public schools, and shared a predominantly lower to lower-middle socioeconomic status (Denno, 1990).

The Philadelphia NCPP data provide a unique opportunity to study the risk factors associated with the development of criminal offending. In fact, Moffitt (1997) regarded these data as among the best for the study of neuropsychological and cognitive risk factors and their relation to criminal offending.

⊠ Variables

Dependent Variables

As stated earlier, we examine the following two distinct outcome variables: (a) whether the participant was/was not an offender by age 18 and (b) if the participant was an offender, the age at first police contact (i.e., early onset). The former variable was measured by the presence of a police contact with the Philadelphia Police Department by age 18. Of the 987 subjects, 220 incurred at least one police contact by age 18. This variable is coded as 0 (nonoffender) and 1 (offender). The latter variable was measured as the age at the first police contact among those individuals incurring at least one police contact by age 18. Following previous research (Patterson et al., 1992; Simons, Wu, Conger, & Lorenz, 1994; Tibbetts & Piquero, 1999), an early onset of offending is measured as onset prior to age 14 (coded 1, $n = 151$), whereas a late onset of offending is measured as onset at or after age 14 (coded 0, $n = 69$). In the Philadelphia data, age 14 marks the peak onset age and occupies the highest hazard, thus providing empirical support for selecting this cutoff.

Independent Variables

Verbal IQ. Verbal IQ is measured by the verbal IQ score on the Wechsler Intelligence Scale for Children (WISC). According to Friedes (1972), the WISC is "the best available test purported to measure intelligence in children," and researchers have established a strong link between verbal IQ and delinquency (Hirschi & Hindelang, 1977; Moffitt, 1997; Wilson & Herrnstein, 1985). Moreover, Caspi and his colleagues (Caspi, Harkness, Moffitt, & Silva, 1996) argued that the WISC is the "most psychometrically trustworthy measure of intellectual performance." The verbal IQ score is a summary measure of verbal ability based on a composite of the following four subtests: (a) information, (b) comprehension, (c) digit span, and (d) vocabulary.

Family adversity. Using the data available in the Philadelphia portion of the NCPP, we measured family adversity in a similar manner to the Dunedin Multidisciplinary Health and Development Study (Pryor & Woodward, 1996; also see Rutter, 1978). All of the variables that comprise the family adversity scale have independently been related to criminal offending (see Loeber & Farrington, 1998; Nagin et al., 1997; Tibbetts & Piquero, 1999). The family adversity scale is composed of the summation of the following four dichotomous items: (a) low socioeconomic status (coded 1 for lowest 25th percentile, coded 0 otherwise), (b) single mother at child birth and at age 7 (coded 1 if single at both ages, 0 otherwise), (c) age of mother at birth of child (younger than 21 coded 1, older

than 21 coded 0), and (d) the number of family transitions through age 7 (coded 0 for zero transitions, 1 for one or more transitions). Higher scores indicate more family adversity.

Birth weight. Birth weight was measured immediately on delivery by hospital staff members and ranged from 3 to 12 pounds. Following the literature on low birth weight (Paneth, 1995; World Health Organization, 1950), this variable was recoded to less than 6 pounds (1) and equal to/greater than 6 pounds (0). Research has shown birth weight to be an important correlate of a variety of negative sequelae, including criminal behavior (Tibbetts & Piquero, 1999).

Gender. This variable was coded as male (1) and female (2).

Interaction. Following Jaccard, Turrisi, and Wan (1990), we created an interaction between family adversity and verbal IQ that was based on the multiplicative of the mean-centered scores of these two items. Mean centering is undertaken to avoid problems associated with multicollinearity among the two variables comprising the interaction as well as the interaction itself (descriptive statistics for all variables may be found in Table 1).

Table 1	Descriptive Statistics			
Variable	Mean	Standard Deviation	Minimum	Maximum
Gender	1.50	0.50	1	2
Low birth weight	0.34	0.47	0	1
Verbal IQ	91.92	11.38	57	133
Family adversity	1.17	1.01	0	4
Offender	0.22	0.42	0	1
Early onset	0.36	0.48	0	1

NOTE: Based on $n = 220$.

Hypotheses

Two key hypotheses are investigated in this article. The first is that the Verbal IQ × Family Adversity interaction will not predict whether an individual is/is not an offender by age 18. We employ logistic regression for this analysis where the outcome variable is nonoffender (0) and offender (1). The second is that the Verbal IQ × Family Adversity interaction will be predictive of the age at first police contact. Specifically, the interaction should be predictive of an early but not late onset of offending. We employ logistic regression for this analysis where the outcome variable is late (0) and early onset (1).

Results

Prior to examining the logistic regression results, we first present some preliminary bivariate analyses to establish the baseline association among the key variables in our analysis. Three ANOVAs were conducted where the family adversity scale was examined with the following three outcome variables: (a) verbal IQ, (b) whether the subject was an offender, and (c) among offenders, whether the subject exhibited an early onset.

Results (not shown) suggest that those individuals scoring highest on the family adversity scale (i.e., more risk factors) scored the lowest on verbal IQ, whereas those individuals scoring lowest on the family adversity scale (i.e., zero risk factors) scored the highest on verbal IQ ($F = 4.85$, $p < .05$). In terms of predicting offending, results once again showed the expected effect; that is, those individuals scoring highest on the family adversity scale were the most likely to become offenders, whereas those individuals scoring lowest on the family adversity scale were the least likely to become offenders ($F = 3.2S$, $p < .05$). The family adversity scale was not related to early onset ($F = .93$, $p > .05$).

Predicting Offending

In Table 2, the estimates for predicting offending are presented. As can be observed, two variables are statistically significant, gender and family adversity. Specifically, compared to males, females ($B = -1.10$) are significantly less likely to become offenders. In terms of family

Table 2	Logistic Regression Predicting Offending			
Variable	**B**	**SE(B)**	**Wald**	**Exp(B)**
Gender	−1.100	.171	41.253*	0.332
Low birth weight	0.046	.174	0.071	1.047
Verbal IQ	−0.003	.007	0.243	0.996
Family adversity	0.245	.080	9.400*	1.278
Interaction	0.003	.007	0.235	1.003
Constant	0.281	.245	1.309	

Chi-square/df 55.19/5
*$p < .05$

adversity, these results suggest a positive relationship; that is, the higher the score on family adversity (B = .24), the higher the probability of the subject becoming an offender. Birth weight was not a significant predictor of offending. Most importantly, and in accord with Moffitt's hypothesis, neither the additive effect of verbal IQ nor the interaction between verbal IQ and family adversity were significant predictors of whether a subject was an offender. These results are consistent with Moffitt's hypothesis because the outcome variable speaks only to the prevalence of offending and not to the different types of offenders categorized by their age of onset.

Predicting Early Onset

Since Moffitt anticipated that the Verbal IQ × Family Adversity interaction will be important in differentiating distinct offenders within the age-crime curve, we examine whether such an interaction is present for the prediction of the age at first police contact. Recall that Moffitt's theory hypothesizes that life-course persisters are more likely than adolescent-limited offenders to exhibit an early age of onset (Moffitt et al., 1994, 1996).

As can be observed from Table 3, two of the five coefficients in this model are predictive of early onset. Specifically, individuals incurring a low birth weight (B = .72) are significantly more likely to incur an early age of onset. As expected by Moffitt, the interaction between verbal IQ and family adversity is a significant predictor of early onset (B = −.039). The sign of this interaction requires some discussion.

The interaction sign is negative, implying that higher verbal IQ scores serve to inhibit the deleterious consequences of family adversity; thus, individuals who are at most risk for an early onset of offending are those individuals who have the lowest verbal IQ scores and who have the highest family adversity scores.

Thus, when verbal IQ is at its lowest level (57), the effect of family adversity on early onset is −2.213. When verbal IQ is at its highest level (133), the effect of family adversity is at its lowest level (−5.177). This result suggests that subjects who are proficient in verbal abilities are able to ward off the detrimental consequences of family adversity. These results are consistent with those obtained by other research in that exceptionally strong verbal skills can be an asset for resisting the effects of

Table 3	Logistic Regression Predicting Early Onset			
Variable	B	SE(B)	Wald	Expt(B)
Gender	−.109	.335	0.106	0.896
Low birth weight	.724	.316	5.249*	2.062
Verbal IQ	−003	.015	0.054	0.996
Family adversity	.010	.158	0.004	1.010
Interaction	−.039	.015	6.284*	0.961
Constant	−.843	.471	3.207	

Chi-square/*df* 12.72/5
*p < .05.

adverse familial environments (Kandel et al., 1988; White, Moffitt, & Silva, 1989). Furthermore, this finding is in concordance with Rutter's (1985) definition of a protective factor in that individuals who have high verbal IQ scores are able to minimize the detrimental effects of family adversity on early onset.

◤ Discussion

This study provides one of the few attempts to assess the prospective link between the interactive effect of two known risk factors, verbal IQ deficiencies and family adversity, on two different forms of criminal offending among a longitudinal sample of urban, inner-city, African American youth. Given the understudied nature of such interactions (Brennan et al., 1997; Raine, 1993; Tibbetts & Piquero, 1999), this assessment should be seen as an important contributor to the risk factor prevention paradigm. At the same time, our findings should be regarded as an empirical contribution to Moffitt's theoretical model as it pertains to the risk factors for differential types of criminal behavior. Not only do our results lend support to Moffitt's interactional hypothesis concerning the correlates of early onset of offending, but our results show that the negative effect of biosocial interactions on early offending are underway or rooted in early childhood.

Our results reveal that the two risk factors under investigation had differential effects on two types of offending behavior (offending prevalence and early onset of offending). First, regarding the model predicting the prevalence of offending, our results are consistent with Moffitt's theory. The interaction of Verbal IQ × Family Adversity did not distinguish between whether or not a youth had committed an offense before age 18, thus lending further support to Moffitt's claim that the interaction cannot distinguish between offenders and nonoffenders. Second, as predicted by Moffitt, our findings show that low verbal IQ scores at

age 7 interact with family adversity to predict early onset of offending. Specifically, the odds of incurring an early onset were increased when there was a co-occurrence of both low verbal IQ scores and high family adversity. Further analysis of the interaction term revealed that high verbal IQ scores act as a protective factor because high verbal IQ scores were shown to minimize the deleterious effect that family adversity had on early onset. Neither verbal IQ nor family adversity exerted an independent effect on early onset of offending. In sum, our findings suggest that there may be different risk factors across different types of offending measures.

Although we feel that our findings have made an important contribution, the data used for this investigation suffer from some limitations. First, this study used only WISC verbal IQ scores as an indicator of neuropsychological risk. Although Moffitt (1993) clearly stated that verbal deficiencies are the best risk indicators of neuropsychological problems, future studies should attempt to examine other proxies (e.g., minor physical anomalies, maternal drug use during pregnancy, birth/delivery complications, etc.) or more direct measures (e.g., positron emission tomography scans and magnetic resonance imaging) of neuropsychological risk (Raine, 1993) in concert with criminogenic or disadvantaged environments in predicting early onset. Second, our analysis focused only on one manifestation of life-course persistent offending. Other dimensions such as chronicity, seriousness, and violence were not examined. Future efforts may wish to examine the extent to which the interactions studied herein apply to related life-course persistent offending dimensions. Third, Moffitt's (1993) strategy for research, along with others (Farrington, 1998), suggests that "reports of antisocial behavior should be gathered from multiple sources to tap pervasiveness across circumstances" (p. 694). Due to the original data collection protocol, we were only able to employ official data records. Although many researchers have used official measures for identifying early onset (Moffitt et al., 1994; Simons et al., 1994), such indices would probably best be used in conjunction with multiple measures such as self-reports, teacher reports, and parental reports of early adverse behavioral problems. Different operationalizations of early onset may possibly increase the validity and confidence of the results in this study. Fourth, due to the fact that the sample consisted of only inner-city, African American youth, and the generalizability of our findings may be limited. Therefore, the significant effect of the interaction of risk factors in our study may be diminished (or enhanced) in other populations (e.g., rural Caucasian youth). Finally, future efforts may wish to explore how Moffitt's interaction hypothesis relates to sociopathy. This would be a fruitful research agenda because estimates suggest that sociopaths comprise a small portion of the male population (between 3% and 4%) yet account for 33% to 80% of the chronic offender population (Cohen & Vila, 1996). Information is needed on the extent to which this group of offenders resembles the most extreme end of the life-course persistent continuum.

Given that the present study's results suggest that predictors of life-course persistent types of offending patterns may be underway in early childhood, it would be fruitful to initiate prevention and intervention strategies early in the life course. Programs should place an emphasis on the development of social skills and cognitive tasks when training children to generate multiple alternative solutions to problems, whereas preschool enrichment programs may wish to target the social/cognitive and behavioral correlates of early antisocial behaviors. Continued research into the etiology of life-course persistent styles of offending is likely to provide much more needed information on prevention tactics across various settings and sample compositions.

⊠ References

Ahn, N. (1994). Teenage childbearing and high school completion: Accounting for individual heterogeneity. *Family Planning Perspectives, 26,* 17–21.

Alexander, G. R., & Comely, D. A. (1987). Racial disparities in pregnancy outcome role of prenatal care utilization and maternal risk status. *American Journal of Preventive Medicine, 3,* 254–261.

Blum, H. M., Boyle, M. H., & Offord, D. R. (1988). Single-parent families: Child psychiatric disorder and school performance. *Journal of the American Academy of Child and Adolescent Psychiatry, 27,* 214–219.

Blumstein, A., Cohen, J., Roth, J., & Visher, C. (1986). *Criminal careers and "career criminals."* Washington, DC: National Academy Press.

Boyle, M. H., & Offord, D. R. (1986). Smoking, drinking and use of illicit drugs among adolescents in Ontario: Prevalence, patterns of use and socio-demographic correlates. *Canadian Medical Association Journal, 135,* 1113–1121.

Brennan, P., Mednick, S., & Raine, A. (1997). Biosocial interactions and violence: A focus on perinatal factors. In A. Raine, P. Brennan, D. Farrington, & S. Mednick (Eds.), *Biosocial basis of violence* (pp. 163–174). New York: Plenum.

Broman, S. H., Nichols, P. L., & Kennedy, W. A. (1975). *Preschool IQ: Prenatal and early developmental correlates.* New York: John Wiley.

Caldwell, J. (1981). Maternal education as a factor in child mortality. *World Health Organization, 75*–79.

Caspi, A., Harkness, A. R., Moffitt, T. E., & Silva, P. A. (1996). Intellectual performance: Continuity and change. In P. A. Silva & W. R. Stanton (Eds.), *From child to adult: The Dunedin multidisciplinary health and development study* (pp. 59–74). Oxford, UK: Oxford University Press.

Cohen, L., & Vila, B. (1996). Self-control and social control: An exposition of the Gottfredson-Hirschi/Sampson-Laub debate. *Studies on Crime and Crime Prevention, 5,* 125–150.

Denno, D. (1990). *Biology and violence: From birth to adulthood.* Cambridge, UK: Cambridge University Press.

Dunford, F., & Elliott, D. (1984). Identifying career offenders with self-reported data. *Journal of Research in Crime and Delinquency, 21,* 57–86.

D'Unger, A. V., Land, K. C, McCall, P. L., & Nagin, D. S. (1998). How many latent classes of delinquent/criminal careers? Results from mixed poisson regression analyses. *American Journal of Sociology, 103,* 1593–1630.

Elster, A. B., Ketterlinus, R., & Lamb, M. E. (1990). Association between parenthood and problem behavior in a national sample of adolescents. *Pediatrics, 85,* 1044–1055.

Farrington, D. (1986). Age and crime. In M. Tonry & N. Morris (Eds.), *Crime and justice: An annual review of research* (Vol. 7, pp. 189–250). Chicago: University of Chicago Press.

Farrington, D. (1992). Explaining the beginning, progress, and ending of antisocial behavior from birth to adulthood. In J. McCord (Ed.), *Facts, frameworks, and forecasts: Advances in criminological theory* (Vol. 3, pp. 253–286). New Brunswick, NJ: Transaction Books.

Farrington, D. (1993). Childhood origins of teenage antisocial behaviour and adult social dysfunction. *Journal of the Royal Society of Medicine, 86,* 13–70.

Farrington, D. (1998). Predictors, causes, and correlates of male youth violence. In M. Tonry & N. Morris (Eds.), *Crime and justice: A review of research* (Vol. 7, pp. 421–475). Chicago: University of Chicago Press.

Farrington, D. (2000). Explaining and preventing crime: The globalization of knowledge—The American Society of Criminology 1999 presidential address. *Criminology, 38,* 1–24.

Farrington, D., & Hawkins, J. (1991). Predicting participation, early onset, and later persistence in officially recorded offending. *Criminal Behaviour Mental Health, 1,* 1–33.

Farrington, D., & Loeber, R. (1999). Transatlantic replicability of risk factors in the development of delinquency. In P. Cohen, C. Slomkowski, & L. N. Robins (Eds.), *Historical and geographical influences on psychopalhology* (pp. 299–329). Hillsdale, NJ: Lawrence Erlbaum.

Farrington, D., Loeber, R., Elliott, D., Hawkins, J., Kandel, D., Klein, M., McCord, L, Rowe, D., & Tremblay, R.(1990). Advancing knowledge about the onset of delinquency and crime. In B. Lahey & A. Kazdin (Eds.), *Advances in clinical and child psychology* (Vol. 13, pp. 283–342). New York: Plenum.

Friedes, D. (1972). Review of the Wechsler Intelligence Scale for Children. In O. K. Buros (Ed.), *The seventh mental measurements yearbook* (pp. 430–432). Highland Park, NJ: Gryphon.

Gibson, C. L., Piquero, A. R., & Tibbetts, S. G. (2000). Assessing the relationship between maternal cigarette smoking during pregnancy and age at first police contact. *Justice Quarterly, 17,* 519–542.

Gorenstein, E. (1990). Neuropsychology of juvenile delinquency. *Forensic Reports, 3,* 15–48.

Gottfredson, M., &Hirschi, T. (1990). *The general theory of crime.* Stanford, CA: Stanford University Press.

Greenberg, R. S. (1983). The impact of prenatal care in different social groups. *American Journal of Obstetrics and Gynecology, 145,* 797–801.

Grogger, J., & Bronars, S. (1993). The socioeconomic consequences of teenage childbearing: Findings from a natural experiment. *Family Planning Perspectives, 25,* 156–161

Hawkins, D. J., & Catalano, R. F. (1992). *Communities that care.* San Francisco: Jossey-Bass.

Hirschi, T., & Hindelang, M. J. (1977). Intelligence and delinquency: A revisionist review. *American Sociological Review, 42,* 571–587.

Jaccard, J., Turrisi, R., & Wan, C. K. (1990). *Interaction effects in multiple regression.* Newbury Park, CA: Sage.

Junger-Tas, J. (1997). *Jeugd en gezin* [Youth and family]. The Hague, the Netherlands: Ministry of Justice.

Kandel, E., & Mednick, S. (1991). Perinatal complications predict violent offending. *Criminology, 29,* 519–530.

Kandel, E., Mednick, S. A., Kirkegaard-Sorenson, L., Hutchings, B., Knop, J., Rosenberg, J. R., & Schulsinger, F. (1988). IQ as a protective factor for subjects at high risk for antisocial behavior. *Journal of Consulting and Clinical Psychology, 56,* 224–226.

Kazdin, A. E., Kraemer, H. C, Kessler, R. S., Kupfer, D. J., & Offord, D. R. (1997). Contributions of risk-factor research to developmental psychopathology. *Clinical Psychology Review, 17,* 375–406.

Kolvin, I., Miller, F.J.W., Scott, D. M., Gatzanis, S.R.M., & Fleeting, M. (1990). *Continuities of deprivation?* Aldershot, UK: Avebury.

LeBlanc, M., & Frechette, M. (1989). *Male criminal activity from childhood through youth.* New York/ Berlin: Springer-Verlag.

Lipsett, P. D., Buka, S. L., & Lipsitt, L. P. (1990). Early intelligence scores and subsequent delinquency: A prospective study. *American Journal of Family Therapy, 18,* 197–208.

Loeber, R., & Farrington, D. (1998). *Serious and violent juvenile offenders: Risk factors and successful interventions.* Thousand Oaks, CA: Sage.

Loeber, R., & LeBlanc, M. (1990). Toward a developmental criminology. In M. Tonry & N. Morris (Eds.), *Crime and justice; An annual review of research* (Vol. 12, pp. 375–473). Chicago: University of Chicago Press.

Lynam, D., Moffitt, T. E., & Stouthamer-Loeber, M. (1993). Explaining the relationship between IQ and delinquency: Class, race, test motivation, school failure and self-control. *Journal of Abnormal Psychology, 102,* 187–196.

McCord, J. (1982). A longitudinal view of the relationship between paternal absence and crime. In J. Gunn & D. P. Farrington (Eds.), *Abnormal offenders, delinquency, and the criminal justice system* (pp. 113–128). Chichester, UK: Wiley.

Moffitt, T. E. (1990). The neuropsychology of delinquency: A critical review of theory and research. In M. Tonry & N. Morris (Eds.), *Crime and justice: An annual review of research* (Vol. 12, pp. 99–169). Chicago: University of Chicago Press.

Moffitt, T. E. (1993). Adolescence-limited and life-course persistent antisocial behavior: A developmental taxonomy. *Psychological Review, 100,* 674–701.

Moffitt, T. E. (1994). Natural histories of delinquency. In E. Weitekamp & H. Kerner (Eds.), *Cross national longitudinal research on human development and criminal behavior* (pp. 3–61). Boston: Kluwer.

Moffitt, T. E. (1997). Neuropsychology, antisocial behavior, and neighborhood context. In J. McCord (Ed.), *Violence and childhood in the inner city* (pp. 116-170). Cambridge, UK: Cambridge University Press.

Moffitt, T. E., Caspi, A., Dickson, N., Silva, P., & Stanton, W. (1996). Childhood-onset versus adolescent-onset antisocial conduct problems in males: Natural history from ages 3 to 18 years. *Development and Psychopathology, 8,* 399–424.

Moffitt, T. E., & Henry, B. (1991). Neuropsychological studies of juvenile delinquency and juvenile violence. In J. S. Milner (Ed.), *The neuropsychology of aggression* (pp. 67–91). Boston: Kluwer.

Moffitt, T. E., Lynam, D., & Silva, P. A. (1994). Neuropsychological tests predicting persistent male delinquency. *Criminology, 32,* 277–300.

Moffitt, T. E., & Silva, P. A. (1988). Neuropsychological deficit and self-reported delinquency in an unselected birth cohort. *Journal of the American Academy of Child and Adolescent Psychiatry, 27,* 233–240.

Nagin, D. S., Farrington, D. P., & Moffitt, T. E. (1995). Life-course trajectories of different types of offenders. *Criminology, 33,* 111–139.

Nagin, D. S., & Land, K. C. (1993). Age, criminal careers, and population heterogeneity: Specification and estimation of a nonparametric, mixed poisson model. *Criminology, 31,* 327–362.

Nagin, D. S., & Paternoster, R. (1991). On the relationship of past and future offending. *Criminology, 29,* 163–189.

Nagin, D. S., Pogarsky, G., & Farrington, D. (1997). Adolescent mothers and the criminal behaviors of their children. *Law and Society Review, 31,* 137–162.

Nichols, P. L., & Chen, T. C. (1981). *Minimal brain dysfunction: A prospective study.* Hillsdale, NJ: Lawrence Erlbaum.

Niswander, K., & Gordon, M. (1972). *The women and their pregnancies.* Washington, DC: Department of Health, Education, and Welfare.

Nutall, C. P., Goldblatt, P., & Lewis, C. (1998). *Reducing offending: An assessment of research evidence on ways of dealing with offending behaviour.* London: HMSO.

Paneth, N. S. (1995). The problem of low birth weight. *The Future of Children: Low Birth Weight* (Center for the Future of Children), 5, 613–618.

Patterson, G., Crosby, L., & Vuchinich, S. (1992). Predicting risk for early police arrest. *Journal of Quantitative Criminology, 8,* 335–355.

Piquero, A. (2000a). Are chronic offenders the most serious offenders? Exploring the relationship with special attention on offense skewness and gender differences. *Journal of Criminal Justice, 28,* 103–116.

Piquero, A. (2000b). Frequency, specialization and violence in offending careers. *Journal of Research in Crime and Delinquency, 37,* 392–418.

Piquero, A. (2000c). Testing Moffitt's neuropsychological variation hypothesis for the prediction of life-course persistent offending. *Psychology, Crime, and Law, 7,* 193–216.

Piquero, A., Blumstein, A., Brame, R., Haapanen, R., Mulvey, E. P., & Nagin, D. S. (2001). Assessing the impact of exposure time and incapacitation on longitudinal trajectories of criminal offending. *Journal of Adolescent Research, 16,* 54–74.

Piquero, A., & Buka, S. (2001). Investigating race and gender differences in specialization in violence. In B. Cohen, R. A. Silverman, & T. P. Thornberry (Eds.), *Criminology at the millennium.* Boston: Kluwer.

Piquero, A., & Tibbetts, S. (1999). The impact of pre/perinatal disturbances and disadvantaged familial environments in predicting criminal offending. *Studies on Crime and Crime Prevention, 8,* 52–70.

Pryor, J., & Woodward, L. (1996). Families and parenting. In P. A. Silva & W. R. Stanton (Eds.), *From child to adult: The Dunedin multidisciplinary health and development study* (pp. 247–258). Oxford, UK: Oxford University Press.

Raine, A. (1993). *The psychopathology of crime: Criminal behavior as a clinical disorder.* San Diego: Academic Press.

Raine, A., Brennan, P., & Mednick, S. (1994). Birth complications combined with early maternal rejection at age 1 year predispose to violent crime at age 18 years. *Archives of General Psychiatry, 51,* 984–988.

Reiss, A., & Roth, J. (1993). *Understanding and preventing violence.* Washington, DC: National Academy Press.

Robins, L. (1966). *Deviant children grown up.* Baltimore: Williams and Wilkins.

Robins, L. (1978). Sturdy childhood predictors of adult antisocial behavior: Replications from longitudinal studies. *Psychological Medicine, 8,* 611–622.

Rutter, M. (1978). Family, area, and school influences in the genesis of conduct disorders. In L. A. Hersov, M. Berger, & D. Shaffer (Eds.), *Aggression and antisocial behavior in childhood and adolescence* (pp. 95–113). Elmsford, NY: Pergamon.

Rutter, M. (1985). Resilience in the face of adversity: Protective factors and resistance to psychiatric disorders. *British Journal of Psychiatry, 147,* 598–611.

Sampson, R. J., & Laub, J. H. (1993). *Crime in the making: Pathways and turning points through life.* Cambridge, MA: Harvard University Press.

Simons, R., Wu, C, Conger, R., & Lorenz, F. (1994). Two routes to delinquency: Differences between early and late starters in the impact of parenting and deviant peers. *Criminology, 32,* 247–276.

Tarter, R. E., Hegedus, A., Alterman, A. L., & Katz-Garris, L. (1983). Cognitive capabilities of juvenile violent, nonviolent and sexual offenders. *Journal of Nervous and Mental Disease, 171,* 564–567.

Tarter, R. E., Hegedus, A. M., Winsten, N. E., & Alterman, A.I. (1984). Neuropsychological, personality and familial characteristics of physically abused delinquents. *Journal of the American Academy of Child Psychiatry, 23,* 668–674.

Tibbetts, S., & Piquero, A. (1999). The influence of gender, low birth weight, and disadvantaged environment in predicting early onset of offending: A test of Moffitt's interaction hypothesis. *Criminology, 37,* 843–877.

Tolan, P. (1987). Implications of age of onset for delinquency risk. *Journal of Abnormal Child Psychology, 15,* 47–65.

Wadsworth, M. (1979). *Roots of delinquency.* London: Martin Robertson.

Wasserman, G., & Miller, L. (1998). The prevention of serious and violent juvenile offending. In R. Loeber & D. P. Farrington (Eds.), *Serious and violent juvenile offenders* (pp. 197–247). Thousand Oaks, CA: Sage.

White, J., Moffitt, T. E., & Silva, P. A. (1989). A prospective replication of the protective effects of IQ in subjects at high risk for juvenile delinquency. *Journal of Clinical and Consulting Psychology, 57,* 719–724.

Wilson, J. Q., & Herrnstein, R. (1985). *Crime and human nature.* New York: Simon & Schuster.

Wolfgang, M. (1983). Delinquency in two birth cohorts. In K. Van Dusen & S. Mednick (Eds.), *Prospective studies of crime and delinquency* (pp. 7–17). Boston: Kluwer.

Wolfgang, M., Thornberry, T., & Figlio, R. (1987). *From boy to man, from delinquency to crime.* Chicago: University of Chicago Press.

World Health Organization. (1950). *Public health aspects of low birth weight* (WHO Technical Report Series, No. 27). Geneva, Switzerland: Author.

REVIEW QUESTIONS

1. How exactly did Gibson et al. measure family adversity? What type of IQ test did they use? Do these measures reflect the theoretical concepts they are meant to measure?

2. What type of sample did Gibson et al. utilize for this study? What are the strengths and weaknesses in using that type of sample, in terms of supporting the theoretical propositions/hypotheses of this study?

3. What did Gibson et al. conclude based on their findings? Did any of the findings or conclusions surprise you? Do any go against the theoretical assumptions you have regarding criminal behavior?

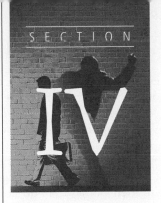

IV

MODERN BIOSOCIAL PERSPECTIVES OF CRIMINAL BEHAVIOR

The introduction to this section will discuss the more modern biological studies of the 20th century. We will begin with the early 1900s, particularly studies that sought to emphasize the influence on criminality of biological factors, such as families, twins, and children who were adopted in infancy. Virtually all of these studies have shown a significant biological effect in the development of criminal propensities. Then we will examine the influence of a variety of physiological factors, including chromosomal mutations, hormones, neurotransmitters, brain trauma, and other dispositional aspects in individuals' nervous systems A special emphasis will be placed on showing the consistent evidence that has been found for an interaction between the physiological and environmental factors (i.e., biosocial factors).

⬛ Introduction

This section will examine a variety of perspectives that deal with interactions between physiological and environmental factors, which is currently the dominant model explaining criminal behavior. First, we will discuss the early studies that attempted to place an emphasis on the biological aspects of offending: family, twin, and adoption studies. All of these studies show that biological influences are more important than social and environmental factors, and most also conclude that when both negative biological and disadvantaged environmental variables are combined, these individuals are by far the most likely to offend in the future, which fully supports the interaction between nature and nurture factors.

Later in this section, we will examine other physiological factors, such as hormones and neurotransmitters. We will see that chronic, violent offenders tend to have significantly different levels of hormones and other chemicals in their bodies than do normal individuals. Furthermore, we will examine brain trauma and activity among violent offenders, and we will see that habitual violent criminals tend to have lower levels of brain wave patterns and anxiety levels than normal persons. Ultimately, we will see in this section that numerous physiological distinctions can be made between chronic violent offenders and others, but that these differences are most evident when physical factors are combined with being raised in poor, disadvantaged environments.

◼ Nature Versus Nurture: Studies Examining the Influence of Genetics and Environment

At the same time that Freud was developing his perspective of psychological deviance, other researchers were busy testing the influence of heredity versus environment to see which of these two components had the strongest effect on predicting criminality. This type of testing produced four waves of research: family studies, twin studies, adoption studies, and in recent years, studies of identical twins separated at birth. Each of these waves of research, some more than others, contributed to our understanding of how much criminality is inherited from our parents (or other ancestors) versus how much is due to cultural norms, such as family or community. Ultimately, all of the studies have shown that the interaction between these two aspects—genetics and environment—is what causes crime among individuals and groups in society.

Family Studies

The most notable **family studies** were done in the early 1900s by Dugdale, in his study of the Jukes family, and the previously discussed researcher Goddard, who studied the Kallikak family.[1] These studies were supposed to test the proposition that criminality is more likely to be found in certain families, which would indicate that crime is inherited. Due to the similarity of the results, we will focus here on Goddard's work on the Kallikak family.

This study showed that a much higher proportion of children from that family became criminal. Furthermore, Goddard thought that many of the individuals (often children) from the Kallikak family actually looked like criminals, which fit Lombroso's theory of stigmata. In fact, Goddard had photographs made of many members of this family to back up these claims. However, follow-up investigations of Goddard's research show that many of these photographs were actually altered to make the subjects appear

[1]Stephen Jay Gould, *The Mismeasure of Man* (New York: Norton, 1981).

more sinister or evil (fitting Lombroso's stigmata) by altering their facial features, most notably their eyes.[2]

Despite the despicable methodological problems with Goddard's data and subsequent findings, two important conclusions can be made from the family studies that were done in the early 1900s. The first is that criminality is indeed more common in some families; in fact, no study has ever shown otherwise. However, this tendency cannot be shown to be a product of heredity or genetics. After all, individuals from the same family are also products of a similar environment—often a bad one—so this conclusion of the family studies does little in advancing knowledge regarding the relative influence of nature versus nurture in terms of predicting criminality.

The second conclusion of family studies was more insightful and interesting. Specifically, the family studies showed that criminality by the mother (or head female caretaker) had a much stronger influence on the future criminality of the children than did the father's criminality. This is likely due to two factors. The first is that the father is often absent most of the time while the children are being raised. But perhaps more important, it takes much more for a woman to transgress social norms and become a convicted offender, which indicates that the mother is highly antisocial; this gives some (albeit limited) credence to the argument that criminality is somewhat inherited. Despite this conclusion, it should be apparent from the weaknesses in the methodology of family studies that this finding did not hold much weight in the nature versus nurture debate. Thus, a new wave of research soon emerged that did a better job of measuring the influence of genetics versus environment, which was twin studies.

Twin Studies

After family studies, the next wave of tests done to determine the relative influence on criminality between nature and nurture involved **twin studies,** the examination of identical twin pairs versus fraternal twin pairs. Identical twins are also known as **monozygotic twins** because they come from a single (hence *mono*) egg *(zygote);* they are typically referred to in scientific literature as MZ (monozygotic) twins. Such twins share 100% of their genotype, meaning they are identical in terms of genetic makeup. On the other hand, fraternal twins are typically referred to as **dizygotic twins** because they come from two (hence *di*) separate eggs *(zygotes);* they are known in the scientific literature as DZ (dizygotic) twins. Such DZ twins share 50% of the genes that can vary (all humans share approximately 99% of genetic makeup, leaving about 1% that can vary), which is the same amount that any siblings from the same two parents share. DZ twins can be of different gender and may look and behave quite differently, as many readers have probably observed.

The goal of the twin studies was to examine the **concordance rates** between MZ twin pairs versus that of DZ twin pairs regarding delinquency. Concordance is a count based on whether two people (or a twin pair) share a certain trait (or lack of the trait); for our purposes, the trait is criminal offending. Regarding a count of concordance, if one twin is an offender, then we look to see if the other is also an offender. If that person is, then it would be concordant given the fact that the first twin was a criminal offender. Also, if

---

[2]For an excellent discussion of the alteration of Goddard's photographs, see Gould, *The Mismeasure.*

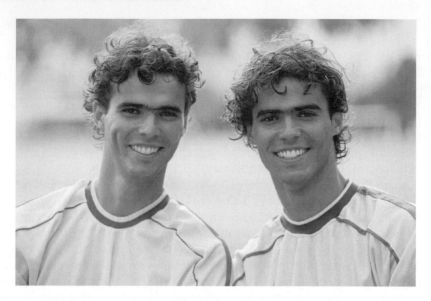

▲ **Photo 4.1**   Identical Twins

neither of the two twins were offenders, then that also would be concordant because they both lack the trait. However, if one twin is a criminal offender and the other twin of the pair is not an offender, then this would be discordant, in the sense that one has a trait that the other does not.

Thus, the twin studies focused on comparing the concordance rates of MZ twin pairs versus those of DZ twin pairs, with the assumption that any significant difference in concordance can be attributed to the similarity of the genetic makeup of the MZ twins (which is 100%) versus the DZ twins, which is significantly less (i.e., 50%). If genetics plays a major role in determining the criminality of individuals, then it would be expected that MZ twins would have a significantly higher concordance rate for being criminal offenders than would DZ twins. It is assumed that both MZ twin pairs and DZ twin pairs in these studies were raised in more or less the same environment, if they were brought up in the same families at the same time.

A number of studies were performed in the early and mid-1900s that examined the concordance rates between MZ and DZ twin pairs. These studies clearly showed that identical twins had far higher concordance rates than did fraternal (DZ) twins, with most studies showing twice as much concordance or more for MZ twins, even for serious criminality.[3]

However, the studies regarding the comparisons between the twins were strongly criticized for reasons that most readers see on an everyday basis. Specifically, identical twins, who look almost exactly alike, are typically dressed the same by their parents and treated the same by the public. In addition, they are generally expected to behave the same way.

---

[3]See review in Adrian Raine, *The Psychopathology of Crime* (San Diego, CA: Academic Press, 1993); see also Juan B. Cortes, *Delinquency and Crime* (New York: Seminar Press, 1972); K. O. Christiansen, "Seriousness of Criminality and Concordance among Danish Twins," in *Crime, Criminology, and Public Policy*, ed, R. Hood (New York: The Free Press, 1974).

However, this is not true for fraternal twins, who often look very different and quite often are of different genders.

This produced the foundation for criticisms of the twin studies, which was a very valid argument that the higher rate of concordance among MZ twins could have been due to the extremely similar way they were treated or expected to behave by society. Another criticism of the early twin studies had to do with the accuracy in determining whether twins were fraternal or identical, which was often done by sight in the early tests.[4] Although these criticisms were seemingly valid, the most recent meta-analysis, which examined virtually all of the twin studies done up to the 1990s, concluded that the twin studies showed evidence of a significant hereditary basis for criminality.[5] Still, the criticisms of such studies were quite valid; therefore, researchers in the early- to mid-1900s involved in the nature versus nurture debate attempted to address these valid criticisms by moving on to another methodological approach for examining this debate: adoption studies.

Adoption Studies

Due to the valid criticisms leveled at twin studies in determining the relative influence of nature (biological) or nurture (environmental), researchers in this area moved on to **adoption studies**, which examined the predictive influence of the biological parents versus that of the adoptive parents, who raised the children from infancy to adulthood. It is important to realize that in such studies, the adoptees were typically given up for adoption prior to six months of age, meaning that the biological parents had relatively no interaction with their natural children; rather, they were almost completely raised from infancy by the adoptive parents.

Perhaps the most notable of the adoption studies was done by Sarnoff Mednick and his colleagues, in which they examined male children born in Copenhagen between 1927 and 1941 who had been adopted early in life.[6] This study and virtually all others that have examined adoptees in this light found that by far, the highest predictability for future criminality was found for adopted youth who had *both* biological parents and adoptive parents who were convicted criminals. However, the Mednick study also showed that the criminality of biological parent(s) had a far greater predictive effect on future criminality of offspring than did the criminality of the adoptive parents. Still, the adopted children who were least likely to become criminal had no parent with a criminal background. In light of this last conclusion, readers realize that these findings support the major contentions of the authors of this textbook, in the sense that they fully back up the "nature *via* nurture" argument as opposed to the "nature *versus* nurture" argument, in the sense that they support the idea that it is both biological *and* environmental factors that contribute to the future criminality of youth.

Unfortunately, the researchers who performed these studies focused on the other two groups of youth, those who had either only criminal biological parents or only criminal adoptive parents. Thus, the adoption studies have found that the adoptees who had only

---

[4]Adrian Raine, *Psychopathology of Crime*.

[5]Glenn Walters, "A Meta-Analysis of the Gene-Crime Relationship," *Criminology* 30 (1992): 595–613.

[6]Barry Hutchings and Sarnoff A. Mednick, "Criminality in Adoptees and Their Adoptive and Biological Parents: A Pilot Study," in *Biosocial Bases of Criminal Behavior*, ed. S. Mednick and K. O. Christiansen (New York: Gardner Press, 1977), 127–41.

biological parents who were criminal had a much higher likelihood of becoming criminal as compared to the youths who had only adoptive parents who were criminal. Obviously, this finding supports the genetic influence in predisposing people toward criminality. However, this methodology was subject to criticism.

Perhaps the most notable criticism of adoption studies was that adoption agencies typically incorporated a policy of **selective placement,** in which adoptees are placed with adoptive families similar in terms of demographics and background to their biological parents. Such selective placement could bias the results of the adoption studies. However, recent analyses have examined the impact of such bias, concluding that even when accounting for the influence of selective placement, the ultimate findings of the adoption studies are still somewhat valid.[7] Children's biological parents likely have more influence on their future criminality than the adoptive parents who raise them from infancy to adulthood. Still, the criticism of selective placement was strong enough to encourage a fourth wave of research in the "nature versus nurture" debate, which became studies on identical twins separated at birth.

Twins Separated at Birth

Until recently, studies of identical twins separated at birth were virtually impossible because it was so difficult to get a high number of identical twins who were indeed separated early. But since the early 1990s, **twins separated at birth studies** have been possible. Readers should keep in mind that in many of the identical twin pairs studied for these investigations, the individuals did not even know that they had a twin. Furthermore, the environments in which they were raised were often extremely different; one twin might be raised by a very poor family in an urban environment while the other twin was raised by a middle- to upper-class family in a rural environment.

These studies, the most notable being done at the University of Minnesota, found that the twin pairs often showed extremely similar tendencies for criminality, sometimes more than those seen in concordance rates for identical twins raised together.[8] This finding obviously supports the profound influence of genetics and heredity, which is not surprising to most well-read scientists, who now acknowledge the extreme importance of inheritance of physiological and psychological aspects regarding human behavior. Perhaps more surprising was why separated identical twins who never knew that they had a twin, and were often raised in extremely different circumstances, had just as similar or even more similar concordance rates than identical twins who were raised together.

The leading theory for this phenomenon is that identical twins who are raised together actually go out of their way to deviate from their natural tendencies to form an identity separate from their identical twin, with whom they have spent their entire life. As for criticisms of this methodology, no significant criticism has been presented. Thus, it is somewhat undisputed at this point in the scientific literature that the identical twins separated at birth studies have shown that genetics has a significant impact on human behavior, especially regarding criminal activity.

---

[7]See James Q. Wilson and Richard J. Herrnstein, *Crime and Human Nature* (New York: Simon and Schuster, 1985); Walters, "Gene-Crime Relationship."

[8]T. J. Bouchard, D. T. Lykken, M. McGue, N. L. Segal, and A. Tellegen, "Sources of Human Psycological Differences: The Minnesota Study of Twins Reared Apart," *Science* 250 (1990): 223–28.

Ultimately, taking all of the "nature versus nurture" methodological approaches and subsequent findings together, the best conclusion that can be made is that genetics and heredity both have a significant impact on criminality. Environment simply cannot account for all of the consistent results seen in the comparisons between identical twins and fraternal twins, those of identical twins separated at birth, and those of adoptees with criminal biological parents versus those who did not have such parents. Despite the taboo nature and controversial response from the findings of such studies, it is quite clear that when nature and nurture are compared, biological rather than environmental factors tend to have the most influence on the criminality of individuals. Still, the authors of this book hope that readers will emphasize the importance of the interaction between nature and nurture (better stated as nature *via* nurture). Ultimately, we hope that we have shown quite convincingly through scientific study that it is the interplay between biology and the environment that is most important in determining human behavior.

Perhaps in response to this "nature versus nurture" emphasis debate, a new theoretical perspective was offered in the mid-1900s that merged biological and psychological factors in explaining criminality. Although leaning more toward the "nature" side of the debate, critics would use this same perspective to promote the "nurture" side, so this framework was useful in promoting the interaction between biology and sociological factors.

Cytogenetic Studies: The XYY Factor

Beyond the body-type theories, another theory was proposed in the early 1900s regarding biological conditions that predispose individuals toward crime: cytogenetic studies. **Cytogenetic studies** of crime focus on the genetic makeup of individuals, with a specific focus on abnormalities in their chromosomal makeup, and specifically, chromosomal abnormalities that occur randomly in the population. Many of the chromosomal mutations that have been studied (such as XYY) typically result not from heredity but from random mutations in chromosomal formation.

The normal chromosomal makeup for women is XX, which represents an X from the mother and an X from the father. The normal chromosomal makeup for men is XY, which represents an X from the mother and a Y from the father. However, as in many species of animals, there are often genetic mutations, which we see in human beings. Consistent with evolutionary theory, virtually all possible variations of chromosomes that are possible have been found in the human population, such as XXY, XYY, and many others. We will focus our discussion on the chromosomal mutations that have been most strongly linked to criminality.

One of the first chromosomal mutations recognized as a predictor of criminal activity was the mutation of XYY. In 1965, the first major study showed that this mutation was far more common in a Scottish male population of mental patients than in the general population.[9] Specifically, in the general population, XYY occurs in about 1 of every 1,000 males.

The first major study that examined the influence of XYY sampled about 200 men in the mental hospital; one occurrence would have been predicted, assuming what was known about the general population. The study, however, found 13 individuals who were XYY, which suggested that individuals who have mental disorders are more likely to have

---

[9]P. A. Jacobs, M. Brunton, M. M. Melville, R. P. Brittian, and W. F. McClemmot, "Aggressive Behavior, Mental Sub-Normality, and the XYY Male," *Nature* 208 (1965): 1351–52.

XYY than those who do not. Males who have XYY have at least 13 times (or 1300%) the chance of having behavioral disorders compared with those who do not have this chromosomal abnormality. Subsequent studies have not been able to dismiss the effect of XYY on criminality, but they have concluded that this mutation is more often linked with property crime than with violent crime.[10]

Would knowing this relationship help in policies toward crime? Probably not, considering the fact that there were still 90% of the male mental patients who were not XYY. Still, this study showed the importance of looking at chromosomal mutations as a predictor of criminal behavior.

Such mutations include numerous chromosomal abnormalities, such as XYY, which is a male who is given an extra Y chromosome, making him more "male-like." These individuals are often very tall but slow in terms of social and intelligence skills. Another type of mutation is XXY, which is otherwise known as Klinefelter's syndrome; it results in more feminine males (homosexuality has been linked to this mutation). Many other types of mutations have been observed, but it is the XYY mutation that has been the primary focus of studies, largely due to the higher levels of testosterone produced by this chromosomal mutation (see Figure 4.1).

| Figure 4.1 | Hypothetical Scattergram Relating Masculinity/Androgen Level (Designated by Karyotype) to Deviance |

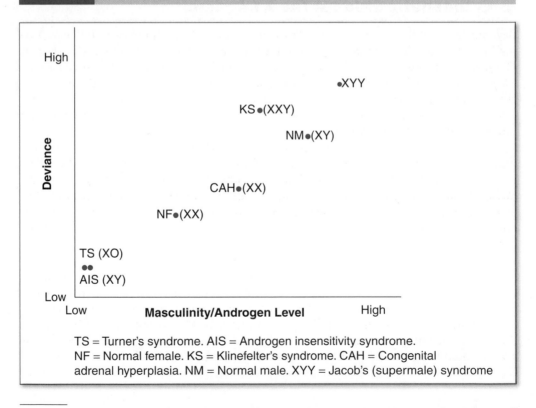

---

[10]See review in Raymond Paternoster and Ronet Bachman, *Explaining Criminals and Crime*, (Los Angeles, CA: Roxbury, 2001), 53.

One study examined the relative criminality and deviance of a group of individuals in each of these groups of chromosomal mutations (see Figure 4.1).[11] This study found the more male hormones produced by the chromosomal mutation, the more likely people with the mutation were to commit criminal and deviant acts. On the other hand, the more feminine hormones produced by the chromosomal mutation, the less likely the individuals were to commit criminal activity. Ultimately, all of these variations in chromosomes show that there is a continuum of degrees of "femaleness" and "maleness," and that the more male-like the individual is in terms of chromosomes, the more he is likely to commit criminal behavior.

Ultimately, the cytogenetic studies showed that somewhat random abnormalities in an individual's genetic makeup can have profound influence on his or her level of criminality. Whether or not this can or should be used in policy related to crime is another matter, but the point is that genetics does indeed contribute to dispositions toward whether someone is likely to commit criminal acts. The extent to which male hormones or androgens are increased by the mutation is an important predictor of criminalistic traits.

▨ Hormones and Neurotransmitters: Chemicals That Determine Criminal Behavior

Various chemicals in the brain and the rest of the body determine how we think, perceive, and react to various stimuli. Hormones, such as testosterone and estrogen, carry chemical signals to the body as they are released from certain glands and structures. Some studies have shown that a relatively excessive amount of testosterone in the body is consistently linked to criminal or aggressive behavior, with most studies showing a moderate relationship.[12] This relationship is seen even in early years of life.[13] On the other side of the coin, studies have also shown that hormonal changes in females can cause criminal behavior. Specifically, studies have shown that a high proportion of the women in prison for violent crimes committed their crimes during their premenstrual cycle, at which time women experience a high level of hormones that make them more "male-like" during that time, due to relatively low levels of estrogen as compared to progesterone.[14]

Anyone who doubts the impact of hormones on behavior should examine the scientific literature regarding performance on intelligence tests at different times of day. Virtually everyone performs better on spatial and mathematical tests early in the day, when people have relatively higher levels of testosterone and other male hormones in their bodies; on the other hand, virtually everyone performs better on verbal tasks in the afternoon or evening, when people have relatively higher levels of estrogen or other female hormones

---

[11]Anthony Walsh, "Genetic and Cytogenetic Intersex Anomalies: Can They Help Us to Understand Gender Differences in Deviant Behavior?" *International Journal of Offender Therapy and Comparative Criminology* 39 (1995): 151–66.

[12]Alan Booth and D. Wayne Osgood, "The Influence of Testosterone on Deviance in Adulthood: Assessing and Explaining the Relationship," *Criminology* 31 (1993): 93–117; H. Soler, P. Vinayak, and D. Quadagno, "Biosocial Aspects of Domestic Violence," *Psychoneuroendocrinology* 25 (2000): 721–73; for a review, see Lee Ellis and Anthony Walsh, *Criminology: A Global Perspective* (Boston: Allyn and Bacon, 2000).

[13]J. R. Sanchez-Martin, E. Fano, L. Ahedo, J. Cardas, P. F. Brain, and A. Azpiroz, "Relating Testosterone Levels and Free Play Social Behavior in Male and Female Preschool Children," *Psychoneuroendocrinology* 25 (2000): 773–83.

[14]Diane Halpern, *Sex Differences in Cognitive Abilities* (Mahwah, NJ: Lawrence Erlbaum, 2000).

in their system.[15] Furthermore, studies have shown that individuals who are given shots of androgens (male hormones) before math tests tend to do significantly better on spatial and mathematics tests than they would do otherwise. Scientific studies show the same is true for people who are given shots of female hormones prior to verbal/reading tests.

It is important to realize that this process of differential levels of hormones begins at a very early age, specifically at about the fifth week after conception. At that time, the Y chromosome of the male tells the developing fetus that it is a male and stimulates production of higher levels of testosterone. So even during the first few months of gestation, the genes on the Y chromosome significantly alter the course of genital, and thus hormonal development.[16]

This level of testosterone alters the genitals of the fetus/embryo through gestation, as well as later changes in the genital area, and produces profound increases in testosterone in the teenage and early adult years. This produces not only physical differences but also huge personality and behavioral alterations.[17] High levels of testosterone and other androgens tend to "masculinize" the brain toward risk-taking behavior, whereas the lower levels typically found in females tend to result in the default feminine model.[18] High levels of testosterone have numerous consequences, such as lowered sensitivity to pain, enhanced seeking of sensory stimulation, and a right-hemisphere shift of dominance in the brain, which has been linked to higher levels of spatial aptitude but lower levels of verbal reasoning and empathy. These consequences have profound implications for criminal activity and are more likely to occur in males than females.[19]

Ultimately, hormones have a profound effect on how individuals think and perceive their environment. All criminal behavior comes down to cognitive decisions in our three-pound brain. So it should not be surprising that hormones play a highly active role in this decision-making process. Nevertheless, hormones are probably secondary compared to levels of **neurotransmitters,** which are chemicals in the brain and body that help transmit electric signals from one neuron to another.

Neurotransmitters can be distinguished from hormones in the sense that hormones carry a signal in themselves that is not electric, whereas the signals that neurotransmitters carry are indeed electric. Neurotransmitters are chemicals that are released when a neuron, the basic unit of our nervous system, wants to send an electric message to a neighboring neuron(s). Sending such a message requires the creation of neural pathways,

---

[15]Ibid.

[16]Lee Ellis, "A Theory Explaining Biological Correlates of Criminality," *European Journal of Criminality* 2 (2005): 287–315.

[17]See Ellis, "A Theory."

[18]C. Burr, "Homosexuality and Biology," *Atlantic Monthly* 271(1993): 47–65; Lee Ellis and M. A. Ames, "Neurohormonal Functioning and Sexual Orientation: A Theory of Homosexuality-Heterosexuality," *Psychological Bulletin* 101 (1987): 233–258; Ellis, "A Theory."

[19]M. Reite, C. M. Cullum, J. Stocker, P. Teale, and E. Kozora, "Neuropsychological Test Performance and MEG-based Brain Lateralization: Sex Differences," *Brain Research Bulletin,* 32 (1993): 325–28; J. Moll, R. Oliveira-Souza, P. J. Eslinger, B. E. Bramanti, J. Mourao-Miranda, P. A. Andreiuolo, and L. Pessoa, "The Neural Correlates of Moral Sensitivity: A Functional Magnetic Resonance Imaging Investigation of Basic and Moral Emotions," *Journal of Neuroscience* 22 (2002): 2730–36; K. Badger, R. Simpson Craft, and L. Jensen, "Age and Gender Differences in Value Orientation Among American Adolescents," *Adolescence* 33 (1998): 591–96.

which means that neurotransmitters must be activated in processing the signal. At any given moment, healthy levels of various neurotransmitters are needed to pass messages from one neuron to the next across gaps between them, gaps called synapses.

Although there are many types of neurotransmitters, the most studied neurotransmitters in relation to criminal activity are dopamine and serotonin. **Dopamine** is most commonly linked to feeling good. For example, dopamine is the chemical that tells us when we are experiencing good sensations, such as good food, sex, or experiences. Most illicit drugs elicit a pleasurable sensation by enhancing the level of dopamine in our systems. Cocaine and methamphetamine, for example, tell the body to produce more dopamine and inhibit the enzymes that typically "mop up" the dopamine in our system after it is used.

Although a number of studies show that low levels of dopamine are linked to high rates of criminality, other studies show no association with—or even a positive link to—criminal behavior.[20] However, the relationship between dopamine and criminal behavior is probably curvilinear, such that both extremely high and extremely low levels of dopamine are associated with deviance. Unfortunately, no conclusion can be made at this point, due to the lack of scientific evidence regarding this chemical.

On the other hand, a clear conclusion can be made about the other major neurotransmitter that has been implicated in criminal offending: **serotonin.** Studies have consistently shown that low levels of serotonin are linked with criminal offending.[21] Serotonin is important in virtually all information processing, whether it be learning or emotional processing; thus, it is vital in most aspects of interactions with the environment. Those who have low levels of serotonin are likely to have problems in everyday communication and life in general. Therefore, it is not surprising that low levels of serotonin are strongly linked to criminal activity.

⊠ Brain Injuries

Another area of physiological problems associated with criminal activity is that of trauma to the brain. As mentioned before, the brain is only three pounds, but it is responsible for every criminal act that an individual commits, so any problems related to this structure have profound implications regarding behavior, especially deviance and criminal activity.

Studies have consistently shown that damage to any part of the brain increases the risk of crime by individuals in the future. However, trauma to certain portions of the brain tends to have more serious consequences than injury to other areas. Specifically, damage to the frontal or temporal lobes (particularly those on the left side) appears to

---

[20]For reviews see Raine, *Psychopathology of Crime*, and Diana H. Fishbein, *Biobehavioral Perspectives in Criminology* (Belmont, CA: Wadsworth/Thomson Learning, 2001).

[21]For a review, see Ellis, "A Theory"; see also; R. Blumensohn, G. Ratzoni, A. Weizman, M. Israeli, N. Greuner, A. Apter, S. Tyano, and A. Biegon, "Reduction in Serotonin 5HT Receptor Binding on Platelets of Delinquent Adolescents," *Psychopharmacology* 118 (1995): 354–56; E. F. Coccaro, R. J. Kavoussi, T. B. Cooper, and R. L. Hauger, "Central Serotonin Activity and Aggression," *American Journal of Pyschiatry* 154 (1997): 1430–35; M. Dolan, W. J. F. Deakin, N. Roberts, and I. Anderson, "Serotonergic and Cognitive Impairment in Impulsive Aggressive Personality Disordered Offenders: Are There Implications for Treatment?" *Psychological Medicine* 32 (2002): 105–17; H. Davidson, K. M. Putnam, and C. L. Larson, "Dysfunction in the Neural Circuitry of Emotion Regulation: A Possible Prelude to Violence," *Science* 289 (2000): 591–94.

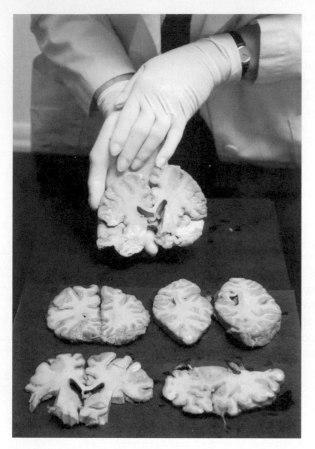

▲ **Photo 4.2** Harkening back to the 19th century, when postmortem examinations of the brains of criminals were a frequent phenomenon, the brain of serial killer John Wayne Gacy was dissected after his execution. The attempt to locate an organic explanation of his monstrous behavior was unsuccessful.

have the most consistent associations with criminal offending.[22] These findings make sense, primarily because the **frontal lobes** (which includes the prefrontal cortex) are the area of the brain where the realm of higher-level problem solving and "executive" functioning takes place.[23] Thus, the frontal lobes, and especially the left side of the frontal lobes, process what we are thinking and inhibit us from doing what we are emotionally charged to do. Thus, any moral reasoning relies on this executive area of the brain because it is the region that considers long-term consequences.[24] If people suffer damage to the frontal lobe, they will be far more inclined to act on their emotional urges without any logical inhibitions imputed from the frontal lobe region.

In a similar vein, the temporal lobe region is highly related to the memory and emotional structures in our brain. To clarify, the **temporal lobes** cover and communicate almost directly with certain structures of our brain's limbic system in the brain. Certain limbic structures largely govern our memories (hippocampus) and emotions (amygdale). Any damage to the temporal lobe, which is generally located above the ear, is likely to damage these structures or the effective communication of these structures to other portions of the brain. Therefore, it is understandable why trauma to the temporal region of the brain is linked to future criminality.

[22]For a review, see Raine, *Psychopathology of Crime*, 129–54; see also, J. M. Tonkonogy, "Violence and Temporal Lobe Lesion: Head CT and MRI Data," *Journal of Neuropsychiatry* 3 (1991): 189–96; P. Wright, J. Nobrega, R. Langevin, and G. Wortzman, "Brain Density and Symmetry in Pedophilic and Sexually Aggressive Offenders," *Annals of Sex Research* 3 (1990): 319–28.

[23]Ellis, "A Theory," 294; see also D. Fishbein, "Neuropsychological and Emotional Regulatory Processes in Antisocial Behavior," in *Biosocial Criminology: Challenging Environmentalism's Supremacy*, ed. A. Walsh and L. Ellis (New York: Nova Science, 2003), 185–208.

[24]B. J. Anderson, M. D. Holmes, and E. Ostresch, "Male and Female Delinquents' Attachments and Effects of Attachments on Severity of Self-Reported Delinquency," *Criminal Justice and Behavior* 26 (1999): 435–52.

⊠ Central and Autonomic Nervous System Activity

The brain is a key player in two different types of neurological systems that have been linked to criminal activity. The first is the **central nervous system (CNS)**, which largely involves our brain and spinal column, and governs our voluntary motor activities. For example, the fact that you are actually reading this sentence means that you are in control of this brain-processing activity. Empirical studies of the influence of CNS functioning on criminality have traditionally focused on brain wave patterns, with most using electroencephalograms (EEGs). Although EEGs do not do a good job in describing which areas of the brain are active or inactive, they do reveal how much the brain as an entire organ is performing at certain times.

Studies have compared brain wave patterns of known chronic offenders (i.e., psychopaths, repeat violent offenders) to those of "normal" people (i.e., those who have never been charged with a crime).[25] These studies consistently show that the brain wave patterns of chronic offenders are abnormal as compared to the normal population, with most studies showing slower brain wave patterns in psychopaths.[26] Four types of brain wave patterns are found, from slowest to fastest: delta, theta, alpha, and beta.[27] Delta waves are often seen when people sleep, whereas theta is typically observed in times of lower levels of being awake, such as drowsiness. Relatively, alpha waves (which tend to be divided into "slow alpha" and "fast alpha" wave patterns, as are beta waves) tend to be related to more relaxed wakefulness, and beta waves are observed with high levels of wakefulness, such as in times of extreme alertness and particularly in times of excited activity.

The studies that have compared brain wave patterns among chronic offenders and normals have shown significant differences. Psychopaths tend to have more activity in the theta (or sometimes "slow alpha") patterns, whereas normals tend to show more activity in the "fast alpha" or beta types of waves. These consistent findings reveal that the cortical arousal of chronic offenders tends to be significantly slower than that of people who do not typically commit crimes. Thus, it is likely that chronic offenders typically do not have the mental functioning that would be disposed toward accurate assessments regarding the consequences of committing criminal behavior. Consistent with these relatively low levels of cortical activity in the central nervous system is that of the findings of studies examining the autonomic nervous systems of individuals.

The second area of the nervous system that has been most linked to criminal behavior is the **autonomic nervous system (ANS)**, which is primarily responsible for involuntary

---

[25]A. Raine and P. H. Venables, "Enhanced P3 Evoked Potentials and Longer P3 Recovery Times in Psychopaths," *Psychophysiology* 25 (1988): 30–38; L. O. Bauer, "Frontal P300 Decrements, Childhood Conduct Disorder, Family History, and the Prediction of Relapse among Abstinent Cocaine Abusers," *Drug and Alcohol Dependence* 44 (1997): 1–10; for a review, see Ellis, "A Theory."

[26]For a review, see Raine, *Psychopathology of Crime*, 174–78; also see R. D. Hare, *Psychopathy: Theory and Practice* (New York: Wiley, 1970); J. Volavka, "Electroencephalogram among Criminals," in *The Causes of Crime: New Biological Approaches*, ed. S. A. Mednick, T. E. Moffitt, and S. Stack (Cambridge, UK: Cambridge University Press, 1987), 137–45; P. H. Venables, "Psychophysiology and Crime: Theory and Data," in *Biological Contributions to Crime Causation*, ed. T. E. Moffitt and S. A. Mednick (Dordrecht, Holland: Martinus Nijhoff, 1988); P. H. Venables and A. Raine, "Biological Theory," in *Applying Psychology to Imprisonment: Theory and Practice*, ed. B. McGurk, D. Thornton, and M. Williams (London: Her Majesty's Stationery Office, 1987), 3–28.

[27]For further discussion and explanation, see Raine, *Psychopathology of Crime*, 174–77.

motor activities, such as heart rate, dilation of pupils, and electric conductivity in the skin. This is the type of physiological activity that is measured by polygraph measures, or lie detector tests. Such measures capitalize on the inability of individuals to control physiological responses to anxiety, which occurs in most normal persons when they lie, especially regarding illegal behavior. However, such measures are not infallible because the individuals who are most at risk of being serious, violent offenders are the most likely to pass such tests even though they are lying.

Consistent with the findings regarding CNS arousal levels, studies have consistently shown that individuals who have a significantly low level of ANS functioning are far more likely to commit criminal acts.[28] For example, studies consistently show that chronic violent offenders tend to have much slower resting heartbeats than normal people, with a number of studies estimating this difference as much as 10 heartbeats per minute slower for the offenders.[29] This is a highly significant gap that cannot be explained away by alternative theories—for example, the explanation that offenders are just less excited in laboratory tests.

Furthermore, people who have such low levels of ANS arousal tend to experience what is known in the psychological literature as "stimulus hunger." Stimulus hunger means that individuals with such a low level of ANS arousal may constantly seek out experiences and stimuli that are risky and thus often illegal. Readers may recall children they have known who can never seem to get enough attention, with some even seeming to enjoy being spanked or other forms of harsh punishment. In addition, people with a low level of ANS arousal may feel no anxiety about punishment, even corporal punishment, and thus they do not adequately learn right from wrong through normal forms of discipline. This is perhaps one of the reasons why children who are diagnosed with the disorder of Attention Deficit-Hyperactivity Disorder (ADHD) have a higher likelihood of becoming criminal than their peers.

Because people who are accurately diagnosed with ADHD have a neurological abnormality—a significantly low functioning ANS level of arousal—doctors prescribe stimulants (e.g., Ritalin) for such youth. It may seem counterintuitive to prescribe a hyperactive person a stimulant, however, the medication boosts the individual's ANS functioning to a normal level of arousal. This makes such individuals experience a healthy level of anxiety, which they would not normally experience from wrongdoing. Assuming that the medication is properly prescribed and at the correct dosage, children who are treated tend to become more attuned to the discipline that they face if they engage in rule violation.

---

[28]For a review, see Raine, *Psychopathology of Crime*, 159–73; also see the following reviews and studies: P. H. Venables, "Childhood Markers for Adult Disorders," *Journal of Child Psychology and Psychiatric and Allied Disorders* 30 (1989): 347–364; A. Raine and P. H. Venables, "Skin Conductance Responsivity in Psychopaths to Orienting, Defensive, and Consonant-Vowel Stimuli," *Journal of Psychophysiology* 2 (1988): 221–25; T. Bice, "Cognitive and Psychophysiological Differences in Proactive and Reactive Aggressive Boys," Doctoral dissertation, Department of Psychology, University of Southern California (Ann Arbor, MI: UMI, 1993).

[29]See E. Mezzacappa, R. E. Tremblay, D. Kindlon, J. P. Saul, L. Arseneault, J. Seguin, R. O. Pihl, and F. Earls, "Anxiety, Antisocial Behavior, and Heart Rate Regulation in Adolescent Males," *Journal of Psychiatry* 38, (1997): 457–69; G. A. Rogeness, C. Cepeda, C. A. Macedo, C. Fischer, and W. R. Harris, "Differences in Heart Rate and Blood Pressure in Children with Conduct Disorder, Major Depression, and Separation Anxiety," *Psychiatry Research* 33 (1990): 199–206; D. J. Kindlon, R. E. Tremblay, E. Mezzacappa, F. Earls, D. Laurent, and B. Schaal, "Longitudinal Patterns of Heart Rate and Fighting Behavior in 9- through 12-year-old Boys," *Journal of the American Academy of Child and Adolescent Psychiatry* 34 (1995): 371–77.

Children who do not fear punishment at all—in fact, some of them do not feel anxiety, even when being physically punished (e.g., spanking)—are likely to have a lower than average ANS level of functioning. Such individuals are likely to become chronic offenders if this disorder is not addressed because they will not respond to discipline or consider the long-term consequences of their risky behavior. If people didn't fear punishment or negative consequences from their behavior, they might be more likely to engage in selfish, greedy behavior. Thus, it is important to address this issue when it becomes evident. On the other hand, children will be children, and ADHD and other forms of disorders have been overly diagnosed in recent years. A well-trained physician should decide whether an individual has such a low level of ANS functioning that medication and/or therapy is required to curb deviant behavior.

Individuals who have significantly lower ANS arousal are likely to pass lie detector tests because they feel virtually no or little anxiety when they lie; many of them lie all the time. Thus, it is ironic but the very people that lie-detecting measures are meant to capture are the most likely to pass such tests, which is probably why they are typically not allowed to be used in court. Only through medication and/or cognitive-behavioral therapy can such individuals learn to consider the long-term consequences of the decisions that they make regarding their behavior.

Individuals with low levels of ANS functioning are not always destined to become chronic offenders. Some evidence has shown that people with low ANS arousal often become successful corporate executives, decorated military soldiers, world-champion athletes, and high-level politicians. Most of these occupations require people who constantly seek out exciting, risky behaviors, and others require constant and convincing forms of lying to others. So there are many legal outlets and productive ways to use the natural tendencies of individuals with low levels of ANS functioning. Such individuals could perhaps be steered toward such occupations and opportunities when they present themselves. It is clearly a better option than committing antisocial acts against others in society.

Ultimately, low levels of cortical arousal in both the CNS and ANS are clearly linked to a predisposition toward criminal activity. However, modern medical research and societal opportunities exist to help such individuals divert their tendencies toward more prosocial outlets.

⬛ Biosocial Approaches Toward Explaining Criminal Behavior

Perhaps the most important, and most recent perspective of how criminality is formed is that of biosocial approaches toward explaining crime. Specifically, if there is any conclusion that can be made regarding the previous theories and research in this section, it is that both genetics *and* environment influence behavior, particularly the interaction between the two. Even the most fundamental aspects of life can be explained by these two groups of factors.

For example, if we look at the height of individuals, we can predict with a great amount of accuracy how tall a person will be by looking at the individual's parents and other ancestors because much of height is determined by a person's genotype. However, even for something as physiological as height, the environment plays a large role. As many readers will observe, individuals who are raised in poor, underdeveloped areas (e.g., Mexico, Asia) are smaller in height than children raised in the United States. However, individuals who descend from parents and relatives in these underdeveloped areas but are raised in the United States tend to be just as tall (if not taller) than children born here. This is largely due to diet, which obviously is an environmental factor.

In other words, our genotype provides a certain range or "window" that determines the height of an individual, which is based on ancestral factors. But the extent to which an individual grows to the maximum or minimum, or somewhere in between, is largely dependent on what occurs in the environment as they develop. This is why biologists make a distinction between genotype, which is directly due to genetics, and **phenotype**, which is a manifestation of genetics interacting with the environment. The same type of biosocial effect is seen on criminal behavior.

Furthermore, over the last decades, a number of empirical investigations have examined the extent to which physiological variables interact with environmental variables, and the findings of these studies have shown consistent predictions regarding criminality. Such studies have been more accurate than those that rely on either physiological and genetic variables or environmental factors separately. For example, findings from a cohort study in Philadelphia showed that individuals who had low birth weight were more likely to commit crime, but that was true primarily if they were raised in a lower income family or a family with weak social structure.[30] Those who were raised in a relatively high-income household or a strong family structure were unlikely to become criminals. It was the coupling of both a physiological deficiency (i.e., low birth weight) and an environmental deficit (i.e., weak family structure or income) that had a profound effect on criminal behavior.

In addition, recent studies have shown that when incarcerated juveniles were assigned to diets with limited levels of simple carbohydrates (e.g., sugars), their reported levels of violations during incarceration declined by almost half (45%).[31] Furthermore, other studies have reported that various food additives and dyes, such as those commonly found in processed foods, can also have a significant influence on criminal behavior. Thus, the old saying "You are what you eat" appears to have some scientific weight behind it, at least regarding criminal behavior. Additional studies have found that high levels of certain toxins, particularly lead and manganese, can have a profound effect on behavior, including criminality. Recent studies have found a consistent, strong effect on high levels of lead in predicting criminal behavior. Unfortunately, medical studies have also found many subtle forms of ingesting high levels of lead, such as the fake jewelry that many children wear as toys. Also unfortunate is that the individuals who are most vulnerable to high levels of lead, as with virtually every toxin, are children, and yet they are the most likely to be exposed to high levels of lead. Even more unfortunate is that the populations (e.g., poor, urban, etc.) most susceptible to biosocial interactions are most likely to experience high levels of lead, largely due to the old paint, which often contained lead, and other household products that contained dangerous toxins.[32]

---

[30]Stephen Tibbetts and Alex Piquero, "The Influence of Gender, Low Birth Weight, and Disadvantaged Environment in Predicting Early Onset of Offending: A Test of Moffitt's Interactional Hypothesis," *Criminology* 37 (1999): 843–78.

[31]S. J. Shoenthaler, *Improve Your Child's IQ and Behavior* (London, UK: BBC Books, 1991).

[32]John Wright, Stephen Tibbett, and Leah Daigle, *Criminals in the Making* (Thousand Oaks, CA: Sage, 2008).

Consistently, other studies have shown that pre- and perinatal problems alone do not predict violence very well. However, when such perinatal problems are considered along with environmental deficits, such as weak family structure, this biosocial relationship often predicts violent but not property crime.[33] Other studies have shown the effects on criminality of a biosocial interaction between the impact of physiological factors within the first minute of life, called Apgar scores, and environmental factors, exposure to nicotine.[34] Additional studies have also found that the interaction of maternal cigarette smoking and fathers' absence from the household was associated with criminal behavior, especially early in life, which is one of the biggest predictors of chronic offending in the future.[35] One of the most revealing studies showed that although only 4 percent of a sample of 4,269 individuals had both birth complications and maternal rejection; this relatively small group of people accounted for more than 18 percent of the total violent crimes committed by the whole sample.[36] So studies have clearly shown that the interaction between biological factors and environmental deficiencies has the most consistent effect on predicting criminality.

▧ Policy Implications

The theories in this section have plenty of policy implications; a few of the primary interventions are discussed here. First, there should be universal, funded preschool for all children. This early life stage is important not only for developing academic skills, but also for fostering healthy social and disciplinary skills, which children who do not attend preschool often fail to develop.[37] In addition, there should be funded mental health and drug counseling for all young children and adolescents who exhibit symptoms of mental disorders or drug problems.[38] There should also be universal funding for health care for all expectant mothers, especially those who have high risk factors (e.g., poor, inner city, etc.). Also, and perhaps most important, there should be far more thorough examinations of children's physiological makeup, in terms of hormones, neurotransmitters, brain formation/functioning, and genetic design, so that earlier interventions can take place. It has been shown empirically that the earlier interventions take place, the better the outcome.[39] These and other policy implications based on modern biosocial theories will be discussed in the final section of the book.

---

[33]Alex Piquero and Stephen Tibbetts, "The Impact of Pre/Perinatal Disturbances and Disadvantaged Familial Environment in Predicting Criminal Offending," *Studies on Crime and Crime Prevention* 8 (1999): 52–71.

[34]Chris Gibson and Stephen Tibbetts, "Interaction Between Maternal Cigarette Smoking and Apgar Scores in Predicting Offending," *Psychological Reports* 83 (1998): 579–86.

[35]Chris Gibson and Stephen Tibbetts, "A Biosocial Interaction in Predicting Early Onset of Offending," *Psychological Reports* 86 (2000): 509–18.

[36]Adrian Raine, P. A. Brennan, and S. A. Mednick, "Birth Complications Combined with Early Maternal Rejection at Age 1 Year Predispose to Violent Crime at Age 18 Years," *Archives of General Psychiatry* 51 (1994): 984–88.

[37]John P. Wright, Stephen Tibbetts, and Leah Daigle, *Criminals in the Making* (Thousand Oaks, CA: Sage).

[38]Ibid.

[39]Wu, S. S., C. Ma, and R. L. Carter, "Risk Factors for Infant Maltreatment: A Population-Based Study," *Child Abuse & Neglect* 28 (2004): 1253–264. Tremblay, R., "Tracking the Origins of Criminal Behavior: Back to the Future," *The Criminologist* 31 (2006): 1, 3–7.

◪ Conclusion

This section has examined a large range of explanations of criminal behavior that place most of the blame on biological and/or psychological factors, which are typically intertwined. These types of explanations were primarily popular in the early years of the development of criminology as a science, but they have also been shown in recent years to be quite valid as significant factors in individual decisions to commit crime. This section examined the influence of genetics and environment in family studies, twin studies, adoption studies, and studies of identical twins separated at birth. These studies ultimately showed the consistent influence of inheritance and genetics in predisposing individuals toward criminal activity. This was supported by the influence of hormones (e.g., testosterone) in human behavior, as well as the influence of variations in chromosomal mutations (e.g., XYY). Recent research has supported both of these theories in showing that people with high levels of male androgens are far more likely to commit crime than those who do not have high levels of these hormones. The link between brain trauma was also discussed, with an emphasis on the consistent association between damage to the left and/or frontal parts of the brain. We also examined theories regarding variations in levels of functioning of the CNS and ANS; all empirical studies have shown that low levels of functioning of these systems have links to criminality. Finally, we explored the extent to which the interaction between physiological factors and environmental variables contribute to the most consistent prediction of criminal offending. Ultimately, it is interesting that the very theories that were key in the early years of development of criminology as a science are now showing strong support in studies as being a primary influence on criminal behavior.

◪ Section Summary

- Early studies that examined the influence of biology focused on case studies of certain families. These studies showed that criminality was indeed clustered among certain families, but such studies did not separate biology from environment.
- The next stage of studies examined the concordance rates of identical twins versus non-identical twins. These studies led to a conclusion that genetic makeup was very important, but critics called their conclusions into question.
- The following stage of research examined adoptees, namely the predictive nature of which parents had more influence in their future criminal behavior. These studies revealed that biological parents (whom they never knew) had far more influence than the adoptive parents who raised them. However, there were criticisms of these studies, so the findings were questioned.
- The final stage of the biology versus environment influence was that of identical twins separated at birth as compared to identical twins raised together. These studies showed that the twins who were separated at birth are just as similar, if not more so, than the twins who are reared together. To date, there are no criticisms of this method of study. Thus, it appears that all four waves of study are consistent in showing that biological influences are vitally important in explaining the criminality of individuals.
- This section also examined chromosomal mutations, such as the XYY mutation, which has consistently shown associations with criminality. Much, if not most, of this link is believed to be due to the increased male androgens (e.g., testosterone) produced by individuals who have this XYY chromosomal mutation.

- Studies have consistently shown that individuals with higher levels of testosterone and other androgens are more disposed toward criminality; for example, normal males are far more likely than normal females to engage in violent crimes.

- This section reviewed findings from studies that show that abnormal levels of certain neurotransmitters, such as serotonin, are far more likely to engage in crime than those who have normal levels of these chemicals in the brain/body.

- Studies show criminality is more likely among individuals who have experienced brain trauma or have lower levels of brain functioning, especially in certain regions of the brain, such as the frontal and temporal lobes, which are the regions that largely govern the higher level, problem-solving functions.

- This section also reviewed the disposition of individuals regarding two aspects of the nervous system, specifically central and autonomic nervous systems. Those who have significantly slower brain waves and lower anxiety levels are far more likely to commit crimes.

KEY TERMS

Adoption studies	Dizygotic twins	Phenotype
Autonomic nervous system/ ANS	Dopamine	Selective placement
	Family studies	Serotonin
Central nervous system/ CNS	Frontal lobes	Temporal lobes
Concordance rates	Monozygotic twins	Twins separated at birth studies
Cytogenetic studies	Neurotransmitters	Twin studies

DISCUSSION QUESTIONS

1. Is there any validity to family studies in determining the role of genetics in criminal behavior? Explain why or why not.

2. Explain the rationale of studies that compare the concordance rates of identical twins and fraternal twins who are raised together. What do most of these studies show regarding the influence of genetics on criminal behavior? What are the criticisms of these studies?

3. Explain the rationale of studies that examine the biological and adoptive parents of adopted children. What do most of these studies show regarding the influence of genetics on criminal behavior? What are the criticisms of these studies?

4. What are the general findings of identical twins who are separated at birth? What implications do these findings have for the importance of genetics or heritability regarding criminal behavior? Can you find a criticism for such findings?

5. Explain what cytogenetic disorders are and describe the related disorder that is most linked to criminal behavior. What characteristics of this type of disorder seems to be driving the higher propensity toward crime?

6. What types of hormones have been shown by scientific studies to be linked to criminal activity? Give specific examples that show this link to be true.

7. Explain what neurotransmitters are, and describe which neurotransmitters are key in predicting criminal offending. Provide support from previous scientific studies.

8. Which areas of the brain, given trauma, have shown the greatest vulnerability regarding criminal offending. Does the lack of healthy functioning in these areas/lobes make sense? Why?

9. How do the brain wave patterns differ among chronic, violent criminals versus normal people? Does this make sense in biosocial models of criminality?

10. How does the autonomic nervous system differ among chronic, violent criminals versus normal people? Does this make sense in biosocial models of criminality?

11. What types of policy implications would you support based on the information provided by empirical studies reviewed in this chapter?

WEB RESOURCES

Family/Twin/Adoption Studies

http://www.criminology.fsu.edu/crimtheory/dugdale.htm

http://www.bookrags.com/research/twin-studies-wog/

http://www.nuffieldbioethics.org/go/browseablepublications/geneticsandhb/report_412.html

Cytogenetic Studies

http://www.asa3.org/ASA/PSCF/1969/JASA6-69Anderson.html

Hormones and Neurotransmitters

http://www.gender.org.uk/about/06encrn/63faggrs.htm

http://faculty.ncwc.edu/TOConnor/301/301lect05.htm

http://serendip.brynmawr.edu/biology/b103/f02/web1/kamlin.html

Autonomic/Central Nervous Systems

http://campus.umr.edu/police/cvsa/hamilton.htm

http://www.mentalhelp.net/poc/view_doc.php?type=doc&id=263

http://www.annalsnyas.org/cgi/content/abstract/794/1/46

Brain Trauma and Crime

http://www.auseinet.com/journal/vol3iss1/chanhudsonparmenter.pdf

http://www.crimetimes.org/

READING

In this selection, Glenn Walters presents a meta-analysis of several dozen studies that examined the influence of biological and environmental factors regarding criminality. It should be mentioned that a meta-analysis is a study that combines findings of previous studies and estimates a diagnostic, statistical measure for making a conclusion based on these previous findings. Walters examined virtually all of the studies from previous family, twin, and adoption studies, and he provides a conclusion about the influence of biology based on these previous studies. However, it should be noted that Walters did not include the most valid and important form of studies of this genre, namely identical twins separated at birth, which is not surprising given the fact that most of the studies investigating this type of methodology were not published in 1992, when he wrote this piece. If such studies had been available, the findings would likely have been slightly different in terms of degree. Readers should consider the strengths and weaknesses of the various methods of each of these approaches as they read this selection.

A Meta-Analysis of the Gene-Crime Relationship

Glenn D. Walters

Research addressing the putative relationship between heredity and human behavior has generated controversy, debate, and strong opinion from both sides. Nowhere is this more evident than in studies exploring the possibility of a gene-crime nexus. While the majority of studies on this topic intimate that certain forms of criminal activity may be influenced by genetic factors, Walters and White (1989) conclude that many of those studies are so methodologically flawed that they defy singular interpretation. Responding to this summation, Brennan and Mednick (1990) question the objectivity and specificity of many of Walters and White's criticisms, which Walters (1990) counters by reasserting that the problem lies with genetic research studies on crime, not the Walters and White review. Since this controversy shows no signs of abating, it would appear that the time has come for a more objective analysis of the relationship presumed to exist between genetic inheritance and criminal conduct.

One way by which such an appraisal might be accomplished is through application of the meta-analytic research technique. Meta-analysis allows for the collection of aggregate data from a series of research reports, which are

SOURCE: Glenn Walters, "A Meta-analysis of the Gene-Crime Relationship," *Criminology* 30, no. 4 (1992): 595–613. Reprinted by permission of the American Society of Criminology.

then converted to a common scale and reanalyzed using various statistical techniques. In contrast to narrative reviews of the literature, which consider the qualitative side of a particular research question, meta-analyses conceptualize outcomes derived from individual studies as data points amenable to quantitative analysis. A common statistic, the effect size, is calculated for each study and then averaged, summarized, and analyzed using standard statistical procedures. (The phi coefficient [Φ], calculated from 2×2 tables of criminal outcomes [present, absent], serves as the effect size measure in this investigation.) The individual strengths and weaknesses of the meta-analytic technique have been discussed at length by other investigators (Glass et al., 1981; Hedges and Olkin, 1985; R. Rosenthal, 1984; Wilson and Rachman, 1983), although there is sufficient evidence to suggest that the meta-analytic technique provides yet another avenue through which the gene-crime question might be addressed.

Three primary methodologies have been employed in research on the heritability of criminal conduct: family studies, twin studies, and adoption studies. Family studies are grounded in the knowledge that family members are more genetically similar than non-family members. Initially, an index group comprising proband (persons who display the trait or behavior under study) and control (persons who do not display the trait or behavior under study) subjects is identified. Next, the prevalence of a target trait or behavior (e.g., criminality, delinquency) is measured in the relatives, often first-degree relatives (parents, siblings, children), of the proband and control subjects. A possible genetic effect is indicated when the trait or behavior is found more frequently in the relatives of proband subjects than in the relatives of control subjects. However, because family members share many of the same environmental experiences, family studies cannot distinguish between the individual contributions of heredity and common environment.

Twin studies address the question of heritability by comparing monozygotic (single-egg) and dizygotic (dual-egg) twins, armed with the knowledge that monozygotic (MZ), or identical, twins possess greater genetic similarity than do dizygotic (DZ), or fraternal, twins. This is because MZ twins share the same genetic inheritance, whereas DZ twins share only half their genes. A genetic effect is suggested when the concordance rate (presence of the trait or behavior in both twins) is higher in MZ twins than in DZ twins. Like family studies, however, the twin method suffers from a potential confounding of genetic and environmental influences because research suggests that MZ twins tend to be treated more similarly by others, spend significantly more time together (Kidd and Matthysee, 1978), and share a greater sense of mutual identity (Dalgard and Kringlen, 1976) than do DZ twins.

Because the family and twin methods were found to be less than ideal for isolating genetic and environmental contributions to criminal outcome, adoption studies were introduced. Adoption studies cross-lag and compare the behavior of biological and adoptive parents with the behavior of index subjects adopted away from their biological homes at a relatively early age. The advantage of the adoption method is that the shared environmental effect of being raised in a particular home and the effect of biological parentage can be studied separately due to the fact that the subject was not raised by his or her biological parents. Although the adoption method of investigating the gene-crime relationship is superior to the family and twin methods, potential problems of interpretation arise when the adoption does not take place shortly after birth or the adoption agency follows a practice of matching the biological and adoptive homes on such potentially important characteristics as family income or socioeconomic status (see Walters and White, 1989). It is possible to combine these individual methods, as exemplified by studies comparing MZ twins reared apart and reared together, although mixed-method studies are rarely found in the gene-crime literature.

The decision to employ a case-to-case statistical model, such as is provided by the phi coefficient, rather than a case-to-base rate model, such as has been suggested by Gottesman and Carey (1983) in their use of the tetrachoric coefficient procedure, is central to the logic of this meta-analysis. A procedure that allows for direct comparisons of subjects from the same sample—whether this involves contrasting subjects with and without a family history of criminality or examining the relative concordance of MZ and DZ twins—seems to capture more clearly the spirit of gene-crime research than a procedure that pits subject groups against some estimated population base rate. After all, the purpose of the twin method is to compare MZ and DZ twins with each other rather than with an estimate of the population base rate, which research suggests, may vary by as much as 20 percentage points from one study to the next.

Researchers in the area of behavior genetics often make use of the heritability coefficient (h), in which the concordance rate is doubled to compensate for the fact that genetic studies do not typically compare genetically identical pairs of subjects with genetically unrelated pairs of subjects. Twin studies, as noted, compare MZ twins, who share 100% of their genetic inheritance, with DZ twins, who share 50% of their genetic inheritance. Adoption studies, on the other hand, contrast child-biological parent pairs, who share 50% of their genes in common, with child-adoptive parent pairs, who are genetically unrelated. Consequently, researchers in the field of behavior genetics estimate the heritability coefficient by doubling the concordance rate or simple correlations obtained in family, twin, or adoption studies. The heritability coefficient was not used to estimate effect sizes in this investigation because (1) data sufficient to calculate this coefficient were not always available, (2) the case-by-case analysis employed

seemed more compatible with other major meta-analyses of criminology-related issues (cf. Andrews et al., 1990; Tittle et al., 1977), and (3) the heritability coefficient, regardless of how scientific it may sound, is no more immune from interpretation problems introduced by a research design that confounds genetic and environmental influences than any other statistic (see Trasler, 1987).

This investigation was undertaken to document the existence, strength, and magnitude of a heredity-crime relationship. The principal null hypothesis was that various indices of genetic inheritance and criminal outcome would not correlate. In addition to assessing the overall relationship between heredity and crime, the effect sizes for family, twin, and adoptive studies were calculated individually. Finally, the influence of several potentially important moderating variables (subject gender, nationality of the research sample, publication date of the referenced article, quality of the research design) on the gene-crime relationship was gauged. The respective null hypotheses predicted that each of the four moderating variables (gender, nationality, publication date, design quality) would fail to modify the gene-crime relationship significantly. The rationale for selecting these variables was based on their facility of operationality, prior inclusion in other meta-analyses, and potential relevance to the criminal justice field.

☒ Method

Sample of Studies

Thirty-eight family, twin, and adoption studies yielding data pertinent to the proposed affiliation of heredity and crime indicators were subjected to meta-analysis. The studies were identified through an exhaustive review of the genetic literature on crime, and all major studies addressing the gene-crime relationship were included. Several of the studies, however, used overlapping (or even identical) samples. Data from overlapping studies were only included in the meta-analysis if they shed new light on the gene-crime question by way of variations in procedure, methodology, or the measurement of criminal outcome.

Procedure

The 38 studies included in this meta-analysis produced 54 2 × 2 contingency tables relevant to the hypothesized link between heredity and crime. In studies following a family methodology, the gene-crime correlation was estimated by comparing the family criminal backgrounds of proband and control index subjects. This is a relatively weak test of the gene-crime hypothesis in that family studies are incapable of distinguishing between genetic and environmental influences. With regard to twin studies, the gene-crime association was calculated on the basis of a comparison of concordance rates for MZ and DZ twins. The genetic status-criminal outcome connection was quantified in adoption studies by contrasting the rate of biological parent criminality observed in the backgrounds of criminal, adopted-away offspring with the rate of biological parent criminality observed in the backgrounds of noncriminal controls.

The phi coefficient was used to measure the strength and direction of the putative association between heredity and crime. In line with recommendations made by R. Rosenthal (1984), these phi coefficients were transformed into Fisher z_r's for the purpose of combining and comparing the results of individual studies. These z_r values were also used to construct a mean Z estimate by which the statistical significance of the average effect sizes for family, twin, adoption, and combined sample studies was evaluated. The mean z_r values were subsequently converted back to phi coefficients following completion of the data analysis phase of this study. In addition to a standard, unweighted mean estimate of phi, a weighted mean estimate, based on the value of z_r and

degrees of freedom ($N = 3$) for the particular study in question, was also calculated.

Four potential moderating variables were also examined: gender, nationality, year of publication, and quality of the research design. Gender of the research sample was coded male, female, or (in studies that failed to provide separate rates for males and females) both. The nationality of the research sample was coded according to whether the study was conducted in the United States or in a foreign (usually European) nation. The year 1975 served as the cutoff point for the publication date measure; studies published prior to 1975 were put into one category and studies published in 1975 or thereafter were put into a second category. The year 1975 was selected as the cutoff because it was the year in which D. Rosenthal published one of the first critical reviews of genetic studies on crime with clear implications and recommendations for future research.

The quality of a study's research design was rated adequate (+) when it satisfied three (four in the case of twin studies) basic criteria: use of a clearly defined outcome measure of criminality, use of a control group, at least 10 subjects per research cell, and for twin studies, a reliable measure of zygosity. Studies failing to satisfy any one of the three (or for twin studies, four) criteria were judged to have had problems with their research design and were given a rating of (−) on this measure. A second rater provided independent ratings of the design quality measure for all 54 gene-crime comparisons, the results of which indicated a respectable degree of interrater reliability for the two sets of ratings (K = .86). The four gene-crime comparisons on which there was disagreement were discussed and a consensus rating was derived.

☒ Results

The studies included in this meta-analysis of the gene-crime literature are summarized in Tables 1 (family studies), 2 (twin studies), and

Table 1 Family Studies Addressing the Gene-Crime Relationship

Study	Location	Index Subjects			Family Subjects				Outcome (%)		Φ	Moderating Var.			
		N	Sex	Criteria	N	Sex	Relation	Criteria	Proband	Control		Sex	Nat	Pub	Des
Glueck & Glueck (1930)	Concord, Mass.	402	M	Delinquency	2,404	M&F	1st-degree relatives	Arrest of family members	53.2	26.7	.27	M	USA	Pre	−
Glueck & Glueck (1934)	Framingham, Mass.	426	F	Delinquency	3.165	M&F	1st-degree relatives	Arrest of family members	45.5	28.6	.25	F	USA	Pre	−
Robins & Lewis (1966)	St. Louis, Mo.	67	M	Delinquency	134	M&F	Parents	Delinquency/ criminality	28.0	13.0	.20	M	USA	Pre	+
Guze et al. (1967)	St. Louis, Mo.	93	M	Felony conviction	519	M&F	1st-degree relatives	Sociopathy	9.0	2.7	.13	M	USA	Pre	−
Cloninger & Guze (1973)	St. Louis, Mo.	66	F	Sociopathy	288 / 106	M&F / M&F	1st-degree relatives	Arrest / Sociopathy	26.0 / 19.8	3.8 / 2.8	.31 / .27	F / F	USA / USA	Pre / Pre	− / −
Cloninger et al. (1975)	St. Louis, Mo.	58 / 28	M / F	Sociopathy	272 / 115	M&F / M&F	1st-degree relatives	Sociopathy	10.3 / 27.8	2.2 / 4.3	.17 / .32	M / F	USA / USA	Post / Post	− / −
Robins et al. (1975)	St. Louis, Mo.	86 / 59	M / F	Delinquency	172 / 118	M&F / M&F	Parents	Delinquency/ criminality	43.0 / 23.0	0.0 / 0.0	.41 / .32	M / F	USA / USA	Post / Post	+ / +
Stewart & Leone (1978)	Iowa City, Iowa	36	M	Unsocialized aggr. reaction	72	M&F	Parents	Antisocial personality	26.5	2.6	.35	M	USA	Post	−
Farrington (1979)	London, England	409	M	Self-reported delinquency	818	M&F	Parents	Criminal conviction	43.8	26.5	.21	M	For	Port	+
Lewis et al. (1983)	St. Louis, Mo.	131 / 281	M / F	Antisocial personality		M&F / M&F	1st-degree relatives	Antisocial acts	18.2 / 41.3	5.2 / 14.1	.20 / .26	M / F	USA / USA	Post / Post	− / −
Farrington et al. (1988)	London, England	63	M	Felony conviction	126	M&F	Parents	Criminal conviction	78.3	41.2	.35	M	For	Post	+

NOTE: Outcome: Proband = percentage of criminal outcome index subjects with one or more criminal outcome relatives. Control = percentage of noncriminal index subjects with one or more criminal outcome relatives. Because all of the index subjects in the Glueck and Glueck (1930, 1934), Guze et al. (1967), Cloninger and Guze (1973), and Cloninger et al. (1975) studies had criminal outcome, the control concordance for these studies was determined from a base-rate estimate of criminal outcome calculated on the basis of data collected by Cloninger et al. (1975) and organized according to the sex (male, female) and race (white, nonwhite) of family subjects in the proband condition.

Moderating variables: Sea = gender of index subjects—M = males, F = females, and B = mixed samples of males and females; Nat = nationality of the subject sample—USA = a US-based sample and For = a European- or Asian-based sample; Pub = year of publication—Pre = a publication date earlier than 1975 and Post = a publication date of 1975 or later; Des = whether the design of the study satisfies three criteria, i.e., a clearly defined outcome, use of a control group, and an adequate sample size (+ vs −).

271

Table 2 Twin Studies Addressing the Gene-Crime Relationship

Study	Location	Criteria	Monozygotic Twins			Dizygotic Twins			Φ	Moderating Var.			
			N	Sex	Concord (%)	N	Sex	Concord. (%)		Sex	Nat	Pub	Des
Lange (1930)	Germany	Criminal arrest	13	M	76.9	17	M	11.8	.66	M	For	Pre	−
LeGras (1932)	Holland	Imprisonment	4	M&F	100.0	5	M&F	0.0	.99	B	For	Pre	−
Rosanoff et al. (1934)	United States	Criminal conviction	38	M	76.3	23	M	21.7	.53	M	USA	Pre	−
			7	F	85.7	4	F	25.0	.59	F	USA	Pre	−
Kranz (1936)	Germany	Criminal arrest	31	M&F	64.5	43	M&F	53.5	.11	B	For	Pre	+
Stumpfl (1936)	Germany	Registered criminality	15	M	60.0	17	M	41.2	.19	M	For	Pre	−
			3	F	66.7	2	F	0.0	.67	F	For	Pre	−
Slater (1938)	United Kingdom	Criminal arrest	2	M&F	50.0	10	M&F	30.0	.14	B	For	Pre	−
Borgstrom (1939)	Finland	Registered criminality	4	M&F	75.0	5	M&F	40.0	.36	B	For	Pre	−
Yoshimasu (1961)	Japan	Criminal arrest	28	M	60.7	18	M	11.1	.49	M	For	Pre	−
Hayashi (1967)	Japan	Delinquency	15	M	80.0	4	M	75.0	.06	M	For	Pre	−
Christiansen (1970)	Denmark	Registered criminality	67	M	35.8	114	M	12.3	.28	M	For	Pre	+

			Monozygotic Twins			Dizygotic Twins				Moderating Var.			
Study	Location	Criteria	N	Sex	Concord (%)	N	Sex	Concord. (%)	Φ	Sex	Nat	Pub	Des
Dalgard & Kringlen (1976)	Norway	Legal violations	49	M	22.4	89	M	18.0	.05	M	For	Post	+
		Felonies	31	M	25.8	54	M	14.8	.14	M	For	Post	+
Shields (1977)	United Kingdom	Delinquency	5	M	80.0	9	M	77.8	.03	M	For	Post	–
Rowe (1983)*	United States	Self-reported antisocial acts	61	M	73.8	38	M	68.4	.06	M	USA	Post	+
			107	F	75.7	59	F	67.8	.09	F	USA	Post	–
Gurling et al. (1984)	United Kingdom	Criminal conviction	14	M&F	7.1	14	M&F	14.3	–.11	B	For	Post	–

NOTE: For monozygotic and dyzygotic twins: Concord = percentage of criminal outcome index twins with criminal outcome co-twin. Because all of the index subjects in the Lange (1930), LeGras (1932), Kranz (1936), Stumpfl (1936), Slater (1938), Borgstrom (1939), Rosanoff et al. (1934), Yoshimasu (1961), and Shields (1977) studies had criminal outcomes, the control concordance for these studies was determined by the base rate of criminality calculated according to sex (male, female) using rates provided by Christiansen (1970, 1974).

Moderator Variables: Sex = gender of index subjects M = males, F = females, and B = mixed samples of males and females; Nat = nationality of the subject sample—USA = a US-based sample and For = a European- or Asian-based sample; Pub = year of publication—Pre = a publication date earlier than 1975 and Post = a publication date of 1975 or later; Des = whether the design of the study satisfies four criteria, i.e., a clearly defined outcome, a reliable procedure for determining zygosity, use of a control group, and an adequate sample size (+ vs –).

*Cutting scores were derived separately for male and females based on the median number of antisocial acts reported by each group (males = 18, females = 6).

Table 3 Adoption Studies Addressing the Gene-Crime Relationship

		Criterion Behavior		Proband Subjects			Control Subjects				Moderating Var.			
Study	Location	Adoptee	Bio. Parent	N	Sex	Outcome (%)	N	Sex	Outcome (%)	Φ	Sex	Nat	Pub	Des
Bohman (1972)	Sweden	School adjustment problems	Registered criminality (father)	118	M	26.3	117	M	35.9	−.10	M	For	Pre	−
				59	F	47.4	140	F	31.4	.15	F	For	Pre	−
Crowe (1972)	United States	Arrest	Incarceration (mother)	52	M&F	15.4	52	M&F	3.8	.20	B	USA	Pre	−
Schulsinger (1972)	Denmark	Psychopathy	Psychopathy	54	M&F	9.3	56	M&F	1.8	.15	B	For	Pre	−
Crowe (1974)	United States	Antisocial personality	Incarceration (mother)	46	M&F	28.2	46	M&F	4.3	.32	B	USA	Pre	−
Hutchings & Mednick (1975)	Denmark	Criminal conviction	Criminal conviction	164	M	48.8	807	M	33.8	.17	M	For	Post	+
Bohman (1978)	Sweden	Registered criminality	Registered criminality	169	M	12.5	812	M	13.0	−.01	M	For	Post	+
				32	F	3.1	1,167	F	2.8	.01	F	For	Post	+
Cadoret (1978)	United States	Antisocial personality	Antisocial personality	9	M	11.1	38	M	7.9	.04	M	USA	Post	−
				3	F	33.3	34	F	2.9	.32	F	USA	Post	−
Cloninger et al. (1982)	Sweden	Registered criminality	Registered criminality	76	M	15.8	786	M	3.4	.17	M	For	Post	+

| Study | Location | Criterion Behavior | | Proband Subjects | | | Control Subjects | | | Moderating Var. | | | | |
		Adoptee	Bio. Parent	N	Sex	Outcome (%)	N	Sex	Outcome (%)	Φ	Sex	Nat	Pub	Des
Mednick et al. (1984)	Denmark	Criminal convictions: 3 or more convictions	Criminal convictions: 3 or more convictions	1,226	M	20.2	2,492	M	13.0	.09	M	For	Post	+
				147	M	25.2	3,571	M	10.7	.09	M	For	Post	+
Cadoret et al. (1985)	United States	Antisocial personality	Antisocial personality	21	M	57.1	106	M	26.4	.25	M	USA	Post	+
				9	F	33.3	78	F	20.5	.09	F	USA	Post	−
Cloninger & Gottesman (1987)	Sweden	Criminality without alcohol	Registered criminality	39	M	20.5	823	M	8.3	.09	M	For	Post	−
				16	F	50.0	897	F	14.5	.13	F	For	Post	−
Moffitt (1987)	Denmark	Nonviolent conviction; Violent conviction	Criminal conviction; Criminal conviction;	1,252	M	16.0	2,555	M	10.6	.08	M	For	Post	+
				164	M	3.5	3,643	M	3.4	.00	M	For	Post	+
Baker et al. (1989)	Denmark	Property offense	Property offense	408	M	35.5	3,222	M	23.8	.08	M	For	Post	+
				88	F	42.0	3,834	F	25.8	.05	F	For	Post	+

NOTE: Proband subjects = adoptees who satisfied the criminal outcome criterion; control subjects = adoptees who failed to satisfy the criminal outcome criterion. Outcome = percentage of index subjects with at least one criminal outcome biological parent. Moderating variables are coded as in Table 1.

3 (adoption studies). The three tables provide information on each study's location, basic subject characteristics (sample size, gender), and the results obtained for proband and control subjects. In each case the criterion behavior is the definition of criminality employed in the study, be it a global definition like delinquency or a more specific delineation such as documented convictions or results from a formal self-report measure of past criminal activity. The rates of familial criminality attained by proband and control index subjects are also provided in each table—familial being defined as biological parents or first-degree relatives in family studies, co-twins in twin studies, and biological parents in adoption studies. The status of the four moderating variables and the phi coefficient effect sizes for each of the 54 comparisons generated by the meta-analysis are also provided in these tables.

Characteristics of the overall relationship between heredity and crime, as well as the characteristics of the effect sizes for family, twin, and adoption studies are provided in Table 4. As is readily discernible from a cursory inspection of this table, the unweighted mean

phi coefficient is modest ($\Phi = .25$), and the weighted mean phi coefficient falls in the low-modest range ($\Phi = .09$).* Nevertheless, the mean overall, family, twin, and adoption study effect sizes attained statistical significance using a procedure (Z transformation) that takes into account the size of the subject sample (see Table 4). A nonparametric binomial test of the proportion of positive to negative effect sizes revealed a pattern of results in favor of positively valenced effect sizes ($p < .001$). It should be noted, however, that the mean effect sizes obtained with adoption studies (weighted mean, $\Phi = .07$; unweighted mean, $\Phi = .11$), perhaps the strongest of the three methods commonly used to investigate the heritability of crime, are low.

Because there were statistically significant between-group differences in the standard error variance estimates for three of the four moderating variables included in the meta-analysis, separate error variance estimates were used in calculating the t-score values for moderating variable effects. Subsequent analyses revealed statistically significant differences for year of publication, $t (52) = 2.26$, $p < .05$, and

Table 4   Effect Sizes for Studies Addressing the Gene-Crime Relationship

	Overall Effect	Family Studies	Twin Studies	Adoption Studies
Number of Φ Estimates	54	15	18	21
Maximum Φ	.99	.41	.99	.32
Median Φ	.17	.27	.16	.09
Minimum Φ	−.11	.13	−.11	−.10
Weighted Mean (Φ)	.09	.26	.21	.07
Unweighted Mean (Φ)	.25	.27	.30	.11
T-Test of Mean Z	9.42*	7.87*	4.18*	5.54*

* $p < .001$ (one-tailed test).

* That the unweighted mean effect size is nearly three times the size of the weighted mean effect size can be explained by the fact that the adoption studies, by virtue of their large sample sizes, contributed disproportionally to the mean weighted effect size.

quality of the research design, $t(52) = 2.11$, $p < .05$, but not for gender, $t(52) = 1.19$, $p > .10$, or nationality, $t(52) = 0.22$, $p > .10$. Overall moderating variable effects, as well as moderating variable effects for each of the three types of studies, are displayed in Table 5.

A multiple regression analysis was then performed with the four moderating variables

Table 5	Effects of Moderating Variables on the Gene-Crime Relationship							
	Subject Gender		Nationality		Year of Publication		Design Quality	
	Male	Female	USA	Non-USA	Pre-75	75-on	+	−
Overall Effect								
Number of Φ Estimates	31	15	23	31	23	31	22	32
Maximum Φ	.66	.67	.59	.99	.99	.41	41	.99
Median Φ	.17	.26	.26	.13	.27	.09	.10	.22
Minimum Φ	−.10	.01	.04	−.11	−.10	−.11	−.01	−.11
Mean of Φ Distribution	.19	.27	.26	.24	.38	.15	.15	.33
SD of Φ Distribution	.19	.21	.15	.45	.49	.13	.12	.48
Family Studies								
Number of Φ Estimates	9	6	13	2	6	9	5	10
Maximum Φ	.41	.32	.41	.35	.31	.41	.41	.35
Median Φ	.21	.29	.27	.28	.26	.32	.32	.26
Minimum Φ	.13	.25	.13	.21	.13	.17	.20	.13
Mean of Φ Distribution	.26	.29	.27	.28	.24	.29	.30	.25
SD of Φ Distribution	.10	.03	.08	.11	.07	.09	.08	.07
Twin Studies								
Number of Φ Estimates	10	3	4	14	12	6	6	12
Maximum Φ	.66	.67	.59	.99	.99	.14	.28	.99
Median Φ	.16	.59	.31	.16	.42	.06	.10	.42
Minimum Φ	.03	.09	.06	−.11	.06	−.11	.05	−.11
Mean of Φ Distribution	.27	.48	.34	.43	.54	.04	.12	.49
SD of Φ Distribution	.26	.36	.32	.60	.60	.08	.09	.63
Adoption Studies								
Number of Φ Estimates	12	6	6	15	5	16	11	10
Maximum Φ	.25	.32	.32	.17	.32	.32	.25	.32
Median Φ	.08	.11	.22	.09	.15	.09	.08	.14
Minimum Φ	−.10	.01	.04	−.10	−.10	−.01	−.01	−.10
Mean of Φ Distribution	.08	.12	.20	.08	.14	.10	.09	.14
SD of Φ Distribution	.09	.11	.12	.08	.16	.09	.08	.13

NOTE: The mean of Φ distribution is the unweighted mean.

serving as predictor measures and the phi coefficient serving as the criterion measure. This analysis produced a multiple correlation of .46 ($R^2 = .21$) and individual beta weights of −.37 for year of publication, −.16 for nationality, −.07 for design quality, and .02 for gender. Since the year of publication and design quality measures were significantly correlated ($\Phi = .48$, $p < .001$), a separate series of regression analyses were calculated, the results of which indicated that the multiple correlation fell from .46 to .33 when the year of publication was dropped from the regression equation but remained reasonably stable at .45 when design quality was removed from the equation.

◪ Discussion

The results of this meta-analysis of the proposed gene-crime nexus reveal a consistent and statistically significant association between various indices of heredity and crime. With a mean unweighted effect size of .25, median effect size of .17, and mean weighted effect size of .09, there would appear to be guarded optimism for a genetic interpretation of certain facets of criminal behavior. However, several factors limit the significance of the overall relationship observed in this investigation. Due, in part, to the large samples employed by several of the investigators in this area of research, the gene-crime connection was shown to be highly statistically significant. However, the actual magnitude of the relationship was modest and the practical significance of the findings uncertain. Further, higher quality studies and those published in 1975 and later provided less support for the gene-crime hypothesis than lower quality studies and those published prior to 1975. Finally, the mean effect size produced by studies using an adoptive methodology, possibly the strongest of the three primary strategies used to investigate gene-behavior relationships, was less favorable to the gene-crime hypothesis than those produced by family and twin studies.

In considering the three primary methodologies used to examine the gene-crime relationship, a few comments are in order. First, since the family method is incapable of distinguishing between the individual contributions of heredity and common environment, the effect sizes observed in the family studies should be viewed as an "upper limit" estimate of genetic influence. Consequently, the actual heritability of criminal conduct is somewhere between zero and this upper limit, and, in all likelihood, resides at a level substantially lower than that displayed in Table 1 and Table 4. Second, twin studies were introduced in an effort to resolve the heredity-environment confound problem that hinders interpretation of family study data on the strength of the assumption that environmental similarity is largely equivalent across sets of identical and fraternal twins. However, several investigators have uncovered the presence of significant MZ-DZ differences in twin influence (Dalgard and Kringlen, 1976; Rowe, 1985). which when controlled, place the heritability of criminal conduct at a level commensurate with that obtained using the more conservative adoption method and considerably lower than that traditionally associated with the outcome of twin research (Carey, 1992). It is worth noting that adoption studies not only provide the most accurate test of the gene-crime hypothesis, but also yield the lowest heritability estimates.

Although the heritability coefficient was not used as the effect size measure in this investigation, it is still possible to estimate this coefficient by taking the weighted and unweighted mean phi correlations, partialling out the variance attributable to the four moderating variables (i.e., 21%), and then multiplying that figure by two (because index group differences in genetic similarity for the studies included in this meta-analysis vary by 50% rather than 100%). This procedure yields heritability estimates of 41 to 43% for family studies, 33 to 47% for twin studies, and 11 to 17% for adoption studies. These heritability estimates can then be used to partition variation in criminal outcomes into the three component sources (genetic, common environmental, specific environmental) of the behavior genetic model. The genetic component would seem to be best captured by the heritability estimate from the adoption studies (namely, 11 to 17%). Since family studies contain both genetic and common environmental sources of variance, the common environmental component (experiences shared by family members) might be estimated by subtracting the heritability estimate obtained in adoption studies from the heritability estimate obtained in family studies. This results in a common environmental component estimate of 24 to 32%. The remaining 51 to 65% of variance in criminal outcome would appear to comprise specific environmental influences (experiences unique to the individual) and measurement error.

Besides differences in outcome occurring as a result of variations in the research method employed (family studies vs. twin studies vs. adoption studies), this investigation suggests that at least two features of the study itself (year of publication and quality of the research design) modify the observed relationship between heredity and crime. Studies published prior to 1975, many of which were of poorer methodological quality than studies published in 1975 and later, tend to yield results more favorable to the genetic hypothesis than more recently conducted and better designed investigations. However, the multiple regression results obtained when year of publication and design quality were systematically removed from the moderating variable equation suggest that year of publication, not design quality, was the most salient moderating condition in this meta-analysis. Although merely speculative at this point, this finding may indicate an important shift in the etiological foundations of crime over time, with heredity playing a more significant role in the criminal behavior of subjects raised during the early part of this century than is the case with more contemporary

samples of subjects. In a related vein, a shift in the economic-crime relationship was observed between the nineteenth and twentieth centuries whereby crime became less closely tied to adverse economic conditions and began to take on the appearance of being more clearly motivated by opportunity and self-centered goals (Wilson and Herrnstein, 1985). Further study is obviously required for a fuller understanding of how year of publication impacts on the gene-crime relationship. However, this investigation suggests that there is something about the recentness of publication, independent of the quality of the research design, that is important in defining support for the gene-crime hypothesis.

While the meta-analytic technique provides many advantages over the traditional literature review, it is not without certain problems and limitations. A major concern voiced by critics of this technique is that studies of differing quality are mixed and given equal weight. One might answer this admonition by pointing out that the quality of a study's research design was one of the moderating variables considered in this investigation and that higher quality designs tended to produce results that were less supportive of the crime-gene hypothesis than lower quality designs. Interpretation of the overall effect size, however, is hindered by the fact that the procedure for twin studies on the one hand, and for family and adoption studies on the other, differed slightly. Whereas the phi coefficients for family and adoption studies were based on the rate of familial/parental criminality among subjects possessing and failing to possess criminal records, the phi coefficients obtained from twin studies were based on a comparison of MZ and DZ twins. For this reason, moderating variable effects for family, twin, and adoption studies were presented separately in Table 5.

Perhaps one of the reasons why genetic research on crime has not contributed more to an understanding of criminal behavior is that researchers have become preoccupied with the either-or thinking that sometimes dominates discussions of the nature-nurture question. Rather than presupposing a genetic or environmental explanation for a particular human behavior like crime, the science of criminology might be better served by investigations that search for a workable integration of genetic and environmental concerns. In so doing, researchers will have to determine exactly what is being inherited by persons who are at increased biological risk for later criminality. A careful reading of several of the articles published on the topic of heredity and crime reveals that some authors appear to propose a single and/or direct link between heredity and crime that may even be specific for certain categories of criminal offense (see Cloninger et al., 1982). A more realistic and scientifically defensible interpretation of gene-crime data, however, might be found in Rowe and Osgood's (1984) approach wherein genetic factors are viewed as contributing to certain individual differences that in turn interact with specific sociological and environmental conditions to bring about criminal and delinquent outcomes. Dividing the variance obtained from their analyses of twins into its genetic, common environmental, and specific environmental components, Rowe and Osgood were able to show how individual variations in genetic background may contribute to delinquency outcomes, but also how genetic findings can be effectively integrated with conventional sociological theory on peer associations. Rowe and Osgood's study also points out the importance of considering the gene-environment interaction in addition to the individual contributions of nature and nurture in understanding criminal behavior.

Since it seems unlikely that a direct genetic link for crime exists anywhere but in the minds of a handful of investigators, future research in this area should probably be directed at exploring the personality/behavioral characteristics that likely bridge the modest gene-crime association observed in this study. Variables potentially

capable of explaining the observed relationship between heredity and criminal behavior include intelligence (Hirschi and Hindelang, 1977), temperament (Olweus, 1980), and physiological reactivity (Venables, 1987), all of which are significantly affected by heredity and have been shown to correlate meaningfully with criminal outcome. Investigators in this area of research endeavor could enhance the relevance of their studies further by developing behaviorally oriented definitions of criminality, rather than relying exclusively on legal criteria, and including designs that highlight the gene-environment interaction (see Cadoret et al., 1983) rather than focusing on heredity and environment as if they were independent and mutually exclusive entities.

The task, then, is to formulate a coherent theory of gene-crime interrelationships that effectively integrates person-oriented considerations like heredity with the more popular environmental interpretations of crime and delinquency. Given the paucity of meaningful outcomes generated by research on the gene-crime hypothesis, one might wonder why heredity has been selected from the large audience of potential person-based correlates of crime to receive so much attention from certain investigators and scholars (see Rushton, 1987; Wilson and Herrnstein, 1985). Perhaps this reveals a reluctance to reject the biological positivistic roots of our criminologic forebears, or maybe it simply denotes the natural human tendency to look for easy answers to intricate and complex questions. Many criminologists, on the other hand, seem inclined to reject genetic explanations of criminal involvement before they have even had a chance to examine the pertinent data. As students of crime we must come to realize that to develop a comprehensive understanding of crime and criminal behavior we will have to consider many variables, including genetic factors, common environmental influences, and a host of specific environmental correlates, in our explanatory equation. But most important, we must be willing to examine how these individual components of the behavior genetic model interact to bring about criminal and delinquent outcomes.

⊠ References

Andrews, D. A., Ivan Zinger, Robert D. Hoge, James Bonta, Paul Gendreau, and Francis T. Cullen. 1990. "Does Correctional Treatment Work? A Clinically Relevant and Psychologically Informed Meta-Analysis." *Criminology* 28:369–404.

Baker, Laura A., Wendy Mack, Terrie E. Moffitt, and Sarnoff A. Mednick. 1989. "Sex Differences in Property Crime in a Danish Adoption Cohort." *Behavior Genetics* 19:355–70.

Bohman, Michael. 1972. "A Study of Adopted Children, Their Background, Environment, and Adjustment." *Acta Paediatrica Scandinavica* 61:90–97.

_____. 1978. "Some Genetic Aspects of Alcoholism and Criminality: A Population of Adoptees." *Archives of General Psychiatry* 35:269–76.

Borgstrom, C. A. 1939. "Eine serie von kriminellen zwilligen. Archiv fur Rassen-und." *Gesellschafts-biologie* 33:334–43.

Brennan, Patricia A., and Sarnoff A. Mednick. 1990. "A Reply to Walters and White: 'Heredity and crime.'" *Criminology* 28:657–61.

Cadoret, Remi J. 1978. "Psychopathology in Adopted Away Offspring of Biologic Parents with Antisocial Behavior." *Archives of General Psychiatry* 35:176–84.

Cadoret, Remi J., Colleen Cain, and Raymond Crowe. 1983. "Evidence for a Gene-Environment Interaction in the Development of Adolescent Antisocial Behavior." *Behavior Genetics* 13:301–10.

Cadoret, Remi J., Thomas O'Gorman, Ed Troughton, and Ellen Heywood. 1985. "Alcoholism and Antisocial Personality: Interrelationships, Genetic and Environmental Factors." *Archives of General Psychiatry* 42:161–67.

Carey, Gregory. 1992. "Twin Imitation for Antisocial Behavior: Implications for Genetic and Family Environment Research." *Journal of Abnormal Psychology* 101:18–25.

Christiansen, Karl O. 1970. "Crime in a Danish Twin Population." *Acta Geneticae Medicae Gemellologiae: Twin Research* 19:323–26.

_____. 1974. "Seriousness of Criminality and Concordance among Danish Twins." In *Crime, Criminology, and Public Policy*, ed. R. Hood. London: Heinemann.

Cloninger, C. Robert, and Irving I. Gottesman. 1987. "Genetic and Environmental Factors in Antisocial Behavior Disorders." In *The Causes of Crime: New Biological Approaches*, ed. Sarnoff A. Mednick, Terrie E. Moffitt, and Susan A. Stack. New York: Cambridge University Press.

Cloninger, C. Robert, and Samuel B. Guze. 1973. "Psychiatric Illness in the Families of Female Criminals: A Study of 288 First-Degree Relatives." *British Journal of Psychiatry* 122:697–703.

Cloninger, C. Robert, Theodore Reich, and Samuel B. Guze. 1975. "The Multifactorial Model of Disease Transmission: II. Sex Differences in the Familial Transmission of Sociopathy (Antisocial Personality)." *British Journal of Psychiatry* 127:11–22.

Cloninger, C. Robert, Soren Sigvardsson, Michael Bohman, and Anne-Liis von Knorring. 1982. "Predisposition to Petty Criminality in Swedish Adoptees: II. Cross-Fostering Analysis of Gene-Environment Interaction." *Archives of General Psychiatry* 39:1242–47.

Crowe, Raymond R. 1972. "The Adopted Offspring of Women Criminal Offenders: A Study of Their Arrest Records." *Archives of General Psychiatry* 27:600–03.

_____. 1974. "An Adoption Study of Antisocial Personality." *Archives of General Psychiatry* 31:785–91.

Dalgard, Odd S., and Einar Kringlen. 1976. "A Norwegian Twin Study of Criminality." *British Journal of Criminology* 16:213–32.

Farrington, David P. 1979. "Environmental Stress, Delinquent Behavior, and Conviction." In *Stress and Anxiety*, Vol. 6, ed. I. G. Sarason and Charles D. Spielberger. Washington, DC: Hemisphere.

Farrington, David P., Bernard Gallagher, Lynda Morley, Raymond J. St. Ledger, and Donald J. West. 1988. "Are There Any Successful Men from Criminogenic Backgrounds?" *Psychiatry* 51:116–30.

Glass, Gene V., Barry McGaw, and Mary L. Smith. 1981. *Meta-Analysis in Social Research*. Beverly Hills, CA: Sage.

Glueck, Sheldon, and Eleanor Glueck. 1930. *Five Hundred Criminal Careers*. New York: Knopf.

_____. 1934. *Five Hundred Delinquent Women*. New York: Knopf.

Gottesman, Irving I., and Gregory Carey. 1983. "Extracting Meaning and Direction from Twin Data." *Psychiatric Developments* 1:35–50.

Gurling, H. M. D., B. E. Oppenheim, and R. M. Murray. 1984. "Depression, Criminality, and Psychopathology Associated with Alcoholism: Evidence from a Twin Study." *Acta Geneticae Medicae Gemellologiae: Twin Research* 33:333–39.

Guze, Samuel B., Edwin D. Wolfgram, Joe K. McKinney, and Dennis P. Cantwell. 1967. "Psychiatric Illness in the Families of Convicted Criminals: A Study of 519 First-Degree Relatives." *Diseases of the Nervous System* 28:651–59.

Hayashi, S. 1967. "A Study of Juvenile Delinquency in Twins." *Bulletin of the Osaka Medical School* 12:373–78.

Hedges, Larry V., and Ingram Olkin. 1985. *Statistical Methods for Meta-Analysis.* Orlando, FL: Academic Press.

Hirschi, Travis, and Michael J. Hindelang. 1977. "Intelligence and Delinquency: A Revisionist Review." *American Sociological Review* 42:571–87.

Hutchings, Barry, and Sarnoff A. Mednick. 1975. "Registered Criminality in the Adoptive and Biological Parents of Registered Male Criminal Adoptees." In *Genetic Research in Psychiatry*, ed. R. R. Fieve, D. Rosenthal, and H. Brill. Baltimore: John Hopkins University Press.

Kidd, Kenneth K., and Steven Matthysee. 1978. "Research Designs for the Study of Gene-Environment Interactions in Psychiatric Disorders: Report of a Foundations Fund for Research in Psychiatry Panel." *Archives of General Psychiatry* 35:925–32.

Kranz, Heinrich. 1936. *Lebenschieksale Krimineller Zwillinge.* Berlin: Springer-Verlag.

Lange, Johannes. 1930. *Crime and Destiny.* Trans. C. Haldane. New York: Charles Boni.

LeGras, Auguste M. 1932. *Psychose en Criminaliteit bij Tweelingen.* Utrecht: University of Utrecht.

Lewis, Collins E., John Rice, and John E. Helzer. 1983. "Diagnostic Interactions: Alcoholism and Antisocial Personality." *Journal of Nervous and Mental Disease* 171:105–13.

Mednick, Sarnoff A., William F. Gabrielli, and Barry Hutchings. 1984. "Genetic Influence in Criminal Convictions: Evidence from an Adoption Cohort." *Science* 224:891–94.

Moffitt, Terrie E. 1987. "Parental Mental Disorder and Offspring Criminal Behavior: An Adoption Study." *Psychiatry* 50:346–60.

Olweus, Daniel. 1980. "Familial and Temperamental Determinants of Aggressive Behavior in Adolescent Boys: A Causal Analysis." *Developmental Psychology* 16:644–60.

Robins, Lee N., and Ruth G. Lewis. 1966. "The Role of the Antisocial Family in School Completion and Delinquency: A Three-Generation Study." *Sociological Quarterly* 7:500–14.

Robins, Lee N., Patricia A. West, and Barbara L. Herjanic. 1975. "Arrests and Delinquency in Two Generations: A Study of Black Urban Families and Their Children." *Journal of Child Psychology and Psychiatry* 16:125–40.

Rosanoff, Aaron J., Leva M. Handy, and Isabel A. Rosanoff. 1934. "Criminality and Delinquency in Twins." *Journal of Criminal Law and Criminology* 24:923–24.

Rosenthal, David. 1975. "Heredity in Criminality." *Criminal Justice and Behavior* 2:3–21.

Rosenthal, Robert. 1984. *Meta-Analytic Procedures for Social Research.* Beverly Hills, CA: Sage.

Rowe, David C. 1983. "Biometrical Genetic Models of Self-Reported Delinquent Behavior: A Twin Study." *Behavior Genetics* 13:473–89.

———. 1985. "Sibling Interaction and Self-Reported Delinquent Behavior: A Study of 265 Twin Pairs." *Criminology* 23:223–40.

Rowe, David C., and D. Wayne Osgood. 1984. "Heredity and Sociological Theories of Delinquency: A Reconsideration." *American Sociological Review* 49:526–40.

Rushton, J. Philippe. 1987. Population differences in rule-following behaviour: Race evolution and crime. Paper presented at the 39th Annual Meeting of the American Society of Criminology, Montreal.

Schulsinger, Fini. 1972. "Psychopathy: Heredity and Environment." *International Journal of Mental Health* 1:190–206.

Shields, James. 1977. "Polygenetic Influences." In *Child Psychiatry: Modern Approaches*, ed. Michael Rutter and L. Hersov. Oxford, UK: Blackwell.

Slater, Eliot. 1938. "Zur erbpathologie des manisch-depressiven irreseins: Die eltern und kinder von manisch-depressiven." *Zeitschrift Fuer Die Gesamte Neurologie und Psychiatrie* 163:1–147.

Stewart, Mark A., and Loida Leone. 1978. "A Family Study of Unsocialized Aggressive Boys." *Biological Psychiatry* 13:107–17.

Stumpfl, Friedrich. 1936. *Die Ursprunge des Verbrechens am Lebenslauf von Zwillingen.* Leipzig: Georg Thieme Verlag.

Tittle, Charles R., Wayne J. Villimez, and Douglas A. Smith. 1977. "The Myth of Social Class and Criminality: An Empirical Assessment of the Empirical Evidence." *American Sociological Review* 43:643–656.

Trasler, Gordon. 1987. "Some Cautions for the Biological Approach to Crime Causation." In *The Causes of Crime: New Biological Approaches*, ed. Sarnoff A. Mednick, Terrie E. Moffitt, and Susan A. Stack. New York: Cambridge University Press.

Venables, Peter H. 1987. "Autonomic Nervous System Factors in Criminal Behavior." In *The Causes of Crime: New Biological Approaches*, ed. Sarnoff A. Mednick, Terrie E. Moffitt, and Susan A. Stack. New York: Cambridge University Press.

Walters, Glenn D. 1990. "Heredity, Crime, and the Killing-the-Bearer-of-Bad-News Syndrome: A Reply to Brennan and Mednick." *Criminology* 28:663–67.

Walters, Glenn D., and Thomas W. White. 1989. "Heredity and Crime: Bad Genes or Bad Research?" *Criminology* 27:455–85.

Wilson, G. Terence, and S. J. Rachman. 1983. "Meta-Analysis and Evaluation of Psychotherapy Outcome: Limitations and Liabilities." *Journal of Consulting and Clinical Psychology* 51:54–64.

Wilson, James Q., and Richard J. Herrnstein. 1985. *Crime and Human Nature.* New York: Simon & Schuster.

Yoshimasu, Shufu. 1961. "The Criminological Significance of the Family In Light of the Study of Criminal Twins." *Acta Criminologiae et Medicinae Legalis Japanica* 27:117–41.

REVIEW QUESTIONS

1. Which of the three types of methodological approaches seemed to show the most support for a gene-crime link? Which of the three showed the least support for this link?

2. What types of characteristics of these studies seemed to modify the level of support for the gene-crime link?

3. Although he does not discuss theoretical models at length, what types of concepts does Walters mention as potential explanations for the heredity and crime association?

4. Do you agree or disagree with what Walters concludes in the last paragraph of the article? Why?

READING

In Lee Ellis's piece, we see an attempt at integrating many of the factors that we have explored in this section, as well as Section III, into an explanatory model of offending. In this selection, we will see elements of evolution, early development problems, neurology, androgens, nervous systems, IQ/intelligence, emotions, and other factors as leading to various deviant or offending behaviors. This is the most current synthesis and use of the existing biosocial literature in developing a theoretical model of criminality. Although this model has not been tested or incorporated into modern criminological theoretical development, it is a theory that shows promise due to its incorporation of many of the biosocial aspects that have been supported by empirical studies. While reading this piece, readers

SOURCE: Lee Ellis, "A Theory Explaining Biological Correlates of Criminality." *European Journal of Criminology* 2, no. 3 (2005): 287–315. Copyright © 2005 Sage Publications Ltd. Reprinted by permission of Sage Publications Ltd. and the author.

should consider how many of these factors fit together in a larger framework that helps explain human behavior, specifically that of criminal activity.

A Theory Explaining Biological Correlates of Criminality

Lee Ellis

Despite growing evidence that biology plays an important role in human behavior, most theories of criminal behavior continue to focus on learning and social environmental variables. This article proposes a biosocial theory of criminality that leads one to expect variables such as age, gender and social status will be associated with offending in very specific ways. According to the theory, androgens (male sex hormones) have the ability to affect the brain in ways that increase the probability of what is termed competitive/victimizing behavior (CVB). This behavior is hypothesized to exist along a continuum, with "crude" (criminal) forms at one end and "sophisticated" (commercial) forms at the other. Theoretically, individuals whose brains receive a great deal of androgen exposure will be prone toward CVB. However, if they have normal or high capabilities to learn and plan, they will transition rapidly from criminal to non-criminal forms of the behavior following the onset of puberty. Individuals with high androgen exposure and poor learning and planning capabilities, on the other hand, often continue to exhibit criminality for decades following the onset of puberty.

⊠ The Evolutionary Neuroandrogenic Theory of Criminal Behavior

The theory to be presented is called the evolutionary neuroandrogenic theory (ENA). The main types of offenses it attempts to explain are those that harm others, either by injuring them physically or by depriving them of their property. Two main propositions lie at the heart of ENA theory. The first addresses evolutionary issues by asserting that the commission of victimful crimes evolved as an aspect of human reproduction, especially among males. The second is concerned with identifying the neurochemistry responsible for increasing the probability of criminality among males relative to females. The theory maintains that sex hormones alter male brain functioning in ways that promote CVB, which is hypothesized to include the commission of violent and property crimes.

The concept of CVB is illustrated in Table 1. At one end of the continuum are acts that intentionally and directly either injure others or dispossess them of their property. In all societies with written laws, these obviously harmful acts are criminalized. At the other end of the CVB continuum are acts that make no profits on the sale of goods or services, although those who administer and maintain the organizations under which they operate usually receive much higher wages than do those who provide most of the day-to-day labor. In a purely socialist economy, the latter type of minimally competitive activities is all that is allowed; all other forms are criminalized. A capitalist economy, on the other hand, will permit profit-making commerce and often even tolerate commerce that involves significant

Table 1	Continuum of Victimizing Behavior (Reflecting Competitive/Victimizing Tendencies)			
The Continuum	very crude _____	intermediate _____		very sophisticated
Probability of Being Criminalized	virtually certain _____	intermediate _____		exceedingly unlikely
Examples	Violent and property offenses ("street crime")	Embezzlement, fraud ("white collar crime")	Deceptive business practices, price gouging	Profit-making commerce · Nonprofit-making commerce

degrees of deception. With the concept of CVB in mind, the two propositions upon which the theory rests can now be described.

▧ The Evolutionary Proposition

Throughout the world, males engage in victimful crimes (especially those involving violence) to a greater extent than do females. To explain why, ENA theory maintains that female mating preferences play a pivotal role. The nature of this mating preference is that females consider social status criteria much more than males do in making mate choices, a pattern that has been documented throughout the world (Ellis, 2001). From an evolutionary standpoint, this female preference has served to increase the chances of females mating with males who are reliable provisioners of resources, allowing females to focus more of their time and energy on bearing offspring. Another consequence has been that female choice has made it possible for males who are status strivers to pass on their genes at higher rates than males who are not. Such female preferences are found in other mammals, as evidenced by their mating more with dominant males than with subordinate males.

According to ENA theory, female preferences for status-striving males have caused most males to devote considerable time and energy to competing for resources, an endeavor that often victimizes others. In other words, natural selection pressure on females to prefer status-striving mates has resulted in males with an inclination toward CVB. ENA theory maintains that the brains of males have been selected for exhibiting competitive/victimizing behavior to a greater extent than the brains of females, and that one of the manifestations of this evolved sex difference is that males are more prone than females toward victimful criminality.

Theoretically, the same natural selection pressure that has resulted in the evolution of CVB has also favored males who flaunt and even exaggerate their resource-procuring capabilities. More unpleasant consequences of the female bias for resource provisioning mates are male tendencies to seek opportunities to circumvent female caution in mating by using deceptive and even forceful copulation tactics. This implies that rape will always be more prevalent among males than among females. ENA theory also leads one to expect complex social systems to develop in order to prevent crime victimization. In evolutionary terms, these systems are known as *counterstrategies*. An example of a counter-strategy to crude forms of CVB is the evolution of the criminal justice system.

As with any theory founded on neo-Darwinian thinking, ENA theory assumes that genes are responsible for substantial proportions of the variation in the traits being investigated. In the present context, the average male is assumed to have a greater genetic propensity toward CVB than is true for the average female. However, this assumption must be compromised with the fact that males and females share nearly all of their genes. Consequently, the only possible way for the theory to be correct is for some of the genes that promote criminality (along with other forms of CVB) to be located on the one chromosome that males and females do not share—the Y-chromosome.

⊠ The Neuroandrogenic Proposition

The second proposition of ENA theory asserts that three different aspects of brain functioning affect an individual's chances of criminal offending by promoting CVB. Two additional neurological factors help to inhibit offending by speeding up the acquisition of sophisticated forms of CVB. Testosterone's ability to affect brain functioning in ways that promote CVB is not simple, but most of the complexities will not be considered here. The main point to keep in mind is that testosterone production occurs in two distinct phases: the organizational (or perinatal) phase and the activational (or postpubertal) phase. Most of the permanent effects of testosterone occur perinatally. If levels of testosterone are high, the brain will be masculinized; if they are low, the brain will remain in its default feminine mode.

ENA theory asserts that androgens increase the probability of CVB by decreasing an individual's sensitivity to adverse environmental consequences resulting from exhibiting CVB. This lowered sensitivity is accomplished by inclining the brain to be *suboptimally aroused*. Suboptimal arousal manifests itself in terms of individuals seeking elevated levels of sensory stimulation and having diminished sensitivity to pain.

The second way androgens promote CVB according to ENA theory is by inclining the limbic system to seizure more readily, especially under stressful conditions. At the extreme, these seizures include such clinical conditions as epilepsy and Tourette's syndrome. Less extreme manifestations of limbic seizuring are known as

episodic dyscontrol and *limbic psychotic trigger*. These latter patterns include sudden bursts of rage and other negative emotions, which often trigger forceful actions against a perceived provocateur.

Third, ENA theory asserts that androgen exposure causes neocortical functioning to be less concentrated in the left (language-dominated) hemisphere and to shift more toward a right hemispheric focus. As a result of this so-called *rightward shift in neocortical functioning*, males rely less on language-based reasoning, emphasizing instead reasoning which involves spatial and temporal calculations of risk and reward probabilities. Coinciding with this evidence are intriguing new research findings based on functional magnetic resonance imaging (fMRI) which suggest that empathy-based moral reasoning occurs primarily in the left hemisphere. Predictably, empathy-based moral reasoning seems to be less pronounced in males than in females. Such evidence suggests that empathy-based moral reasoning is more likely to prevent victimful criminality than so-called justice-based moral reasoning.

Theoretically, the three androgen-enhanced brain processes just described have evolved in males more than in females because these processes contribute to CVB. Furthermore, competitive/victimizing behavior has evolved in males more than in females because it facilitates male reproductive success more than it facilitates female reproductive success.

⊠ Inhibiting Criminal Forms of Competitive/ Victimizing Behavior

Regarding the inhibiting aspects of brain functioning, two factors are theoretically involved. One has to do with learning ability and the other entails foresight and planning ability. According to ENA theory, the ability to learn will correlate with the rapidity of male transitioning from crude to sophisticated forms of

CVB. This means that intelligence and other measures of learning ability should be inversely associated with persistent involvement in criminal behavior. Likewise, neurological underpinnings of intelligence such as brain size and neural efficiency should also correlate negatively with persistent offending. These predictions apply only to persistent victimful offending, with a much weaker link to occasional delinquency and possibly none with victimless criminality.

The frontal lobes, especially their prefrontal regions, play a vital role in coordinating complex sequences of actions intended to accomplish long-term goals. These prefrontal regions tend to keenly monitor the brain's limbic region, where most emotions reside. Then the prefrontal regions devise plans for either maximizing pleasant emotions or minimizing unpleasant ones. In other words, for the brain to integrate experiences into well-coordinated and feedback-contingent strategies for reaching long-term goals, the frontal lobes perform what has come to be called *executive cognitive functioning*. Moral reasoning often draws heavily on executive cognitive functioning since it often requires anticipating the long-term consequences of one's actions.

Factors that can impact executive cognitive functioning include genetics, prenatal complications, and various types of physical and chemical trauma throughout life. According to ENA theory, inefficient executive cognitive functioning contributes to criminal behavior. Similar conclusions have been put forth in recent years by several other researchers.

To summarize, ENA theory asserts that three aspects of brain functioning promote competitive/victimizing behavior, the crudest forms of which are victimful crimes. At least partially counterbalancing these androgen-promoted tendencies are high intelligence and efficient executive cognitive functioning. These latter two factors affect the speed with which individuals quickly learn to express their competitive/victimizing tendencies in sophisticated rather than crude ways. Sophisticated expressions are less likely to elicit retaliation by victims, their relatives, and the criminal justice system than are crude ones. Males with low intelligence and/or with the least efficient executive cognitive functioning will therefore exhibit the highest rates of victimful criminal behavior.

✉ Correlates of Criminal Behavior

Twelve biological correlates of crime with special relevance to ENA theory (testosterone, mesomorphy, maternal smoking during pregnancy, hypoglycemia, epilepsy, heart rate, skin conductivity, cortisol, serotonin, monoamine oxidase, slow brainwave patterns, and P300 amplitude) are discussed below.

Testosterone. ENA theory predicts that correlations will be found between testosterone and CVB. However, the nature of these correlations will not involve a simple one-to-one correspondence between an individual's crime probability and the amount of testosterone in his/her brain at any given point in time. Earlier, a distinction was made between the organizational and activational effects of testosterone on brain functioning, and that the most permanent and irreversible effects of testosterone occur perinatally. For this reason alone, testosterone levels circulating in the blood stream or in saliva following puberty may have little direct correlation with neurological levels, especially within each sex. Therefore, one should not expect to find a strong correlation between blood or saliva levels of testosterone among, say, 20-year-old males and the number of offenses they have committed even though testosterone levels in the brain at various stages in development are quite influential on offending probabilities.

Numerous studies have investigated the possible relationship between blood levels or saliva levels of testosterone and involvement in criminal behavior, and most have found modest positive correlations (Maras et al. 2003). Additional evidence of a connection between testosterone and aggressive forms of criminality involves a recent study of domestic violence, where offending males had higher levels of saliva testosterone than did males with no history of such violence (Soler et al. 2000).

Overall, it is safe to generalize that circulating testosterone levels exhibit a modest positive association with male offending probabilities, particularly in the case of adult violent offenses. According to ENA theory, males are more violent than females, not because of cultural expectations or sex role training, but mainly because of their brains being exposed to much higher levels of testosterone than the brains of females.

Mesomorphy. Body types exist in three extreme forms. These are sometimes represented with a bulging triangle. Most people are located in the center of the triangle, exhibiting what is termed a basically balanced body type. At one corner of the triangle are persons who are extremely muscular, especially in the upper body, called mesomorphs. Ectomorphs occupy a second corner. Individuals with this body type are unusually slender and non-muscular. In the third corner, one finds endomorphs, individuals who are overweight and have little muscularity.

Studies have consistently revealed that offending probabilities are higher among individuals who exhibit a mesomorphic body type than either of the two other extreme body types (e.g., Blackson & Tarter 1994). ENA theory explains this relationship by noting that testosterone affects more than the brain; it also enhances muscle tissue, especially in the upper part of the body.

Maternal Smoking During Pregnancy. There is considerable evidence that maternal smoking may lead to an elevated probability of offspring becoming delinquent (e.g., Räsänes et al. 1999). ENA theory assumes that fetal exposure to

carbon monoxide and other neurotoxins found in cigarette smoke disrupt brain development in ways that adversely affects IQ or executive cognitive functioning, thereby making it more difficult for offspring to maintain their behavior within prescribed legal boundaries. However, it is possible that genes contributing to nicotine addiction may also contribute to criminal behavior. In fact, a recent study reported that the link between childhood conduct disorders (a frequent precursor to later criminality) and maternal smoking was mainly the result of mutual genetic influences (Maughan et al. 2004).

Hypoglycemia. Glucose, a type of natural sugar, is the main fuel used by the brain. The production of glucose is largely regulated by the pancreas in response to chemical messages from a portion of the brain called the hypothalamus. When the hypothalamus senses that glucose levels are becoming too high or too low, it sends chemical instructions to the pancreas to either curtail or increase production of glucose by regulating the amount of insulin released into the blood system. In most people, this feedback regulatory process helps to maintain brain glucose at remarkably stable levels. For a variety of reasons, some people have difficulty stabilizing brain glucose levels. These people are said to be hypoglycemic. Dramatic fluctuations in brain glucose can cause temporary disturbances in thoughts and moods, with the most common symptoms being confusion, difficulty concentrating, and irritability.

Studies have indicated that hypoglycemia is associated with an elevated probability of crime, especially of a violent nature (e.g., Virkkunen 1986). To explain such a connection, ENA theory draws attention to the importance of maintaining communication between the various parts of the brain in order to control emotionality. In particular, if the frontal lobes receive distorted signals from the limbic system, bizarre types of behavioral responses sometimes result, including responses that are violent and antisocial.

Epilepsy. Epilepsy is a neurological disorder typified by seizures. These seizures are tantamount to "electrical storms' in the brain. While people vary in genetic susceptibilities, seizures are usually induced by environmental factors such as physical injuries to the brain, viral infections, birth trauma, and exposure to various chemicals.

The main behavioral symptoms of epilepsy are known as *convulsions* (or *fits*), although not all epileptics have full-blown convulsive episodes. Mild epileptic episodes may manifest themselves as little more than a momentary pause in an on-going activity accompanied by a glazed stare. Seizures that have little to no noticeable debilitating effects on coordinated movement are called *subconvulsive* (or *subclinical*) *seizures*. Studies of human populations have shown that epilepsy affects only about one in every 150 to 200 persons. In prison populations, however, the prevalence of epilepsy is around one in 50, at least three times higher than in the general population (e.g., Mendez et al. 1993).

ENA theory can explain the links between epilepsy and offending by noting that very basic and primitive emotional responses sometimes emanate from the limbic region of the brain. While seizures in motor control centers are most likely to receive a diagnosis of epilepsy, seizures in the limbic region could provoke very basic survival instincts.

Resting Heart and Pulse Rates. Heart and pulse rates rise in response to strenuous exercise along with stressful and frightening experiences. Studies have shown that on average, the resting heart rate and pulse rate of convicted offenders are lower than those of persons in general (e.g., Mezzacappa et al. 1997:463). ENA theory would account for these relationships by stipulating that both low heart and low pulse rates are physiological indicators of suboptimal arousal. Such arousal levels should incline individuals to seek more intense stimulation and to tolerate unpleasant

environmental feedback to a greater extent than individuals with normal or superoptimal arousal under most circumstances.

Skin Conductivity (Galvanic Skin Response). Sweat contains high concentrations of sodium, which is a good electrical conductor. A device called a Galvanic Skin Response (GSR) meter was developed nearly a century ago to monitor palm sweat. The GSR works by measuring electrical impulses passing through our bodies from one electrode to another. Thus, by putting one's fingers on two unconnected electrodes of a GSR device, one completes an electrical circuit through which imperceptible amounts of electricity flows. Temperature obviously affects how much people sweat, but so too do emotions. The more intense one's emotions become (especially those of fear and anger), the more one will sweat, and thus the stronger will be the readings on the GSR meter.

Numerous studies have examined the possibility that persons with the greatest propensities toward criminal behavior have distinctive skin conductivity patterns. These studies suggest that offenders exhibit lower skin conductivity under standard testing conditions than do people in general (e.g., Buikhuisen et al. 1989; Raine et al. 1996). As in the case of heart and pulse rates, ENA theory can account for such findings by hypothesizing that low GSR readings especially under stressful testing conditions are another indication of suboptimal arousal.

Cortisol. So-called stress hormones are secreted mainly by the adrenal glands during times of anxiety, stress, and fear. The stress hormone that has been investigated most in connection with criminality is cortisol. Most of these studies have suggested that offenders have below normal levels (e.g., Lindman et al. 1997). As with heart rates and skin conductivity, one could anticipate a low cortisol high criminality relationship by assuming that low cortisol production even in the face of stress is another indicator of suboptimal arousal. This would suggest that offenders are less intimidated by threatening aspects of their environments than are persons in general.

Serotonin. Serotonin is an important neurotransmitter. When serotonin is relatively active in the synaptic regions connecting adjacent nerve cells, people typically report feeling a sense of contentment and calm. Several drugs that have been designed to treat depression and anxiety disorders operate by either prolonging the presence of serotonin in the synaptic gaps between neurons or by facilitating the ability of receptor sites on the dendrites to bond to the serotonin that is available. Low serotonin activity has been linked to crime by numerous studies, especially impulsive crimes (e.g., Virkkunen et al. 1996; Matykiewicz et al. 1997). Explaining the link between serotonin and criminality from the perspective of ENA theory draws attention to serotonin pathways connecting the brain's prefrontal areas with the emotion-control centers in the limbic system. Serotonin may facilitate the sort of executive cognitive functioning required to restrain impulsive behavior, especially regarding rage and persistent frustration.

Monoamine Oxidase. Monoamine oxidase (MAO) is an enzyme found throughout the body. Within the brain, MAO helps to break down and clear away neurotransmitter molecules (including serotonin), portions of which often linger in the synaptic gap after activating adjacent nerve cells. Studies indicate that MAO activity is unusually low among offenders (e.g., Alm et al. 1996; Klinteberg 1996). ENA draws attention to the fact that low MAO activity seems to be related to high levels of testosterone. Furthermore, low MAO brain activity may interfere with the brain's ability to manufacture or utilize serotonin.

Brain Waves and Low P300 Amplitude. Brain waves are measured using electrodes placed on the scalp. These electrodes can detect electrical activity occurring close to the surface of the brain fairly clearly. Despite their complexity, brain waves can be roughly classified in terms of ranging from being rapid and regular (alpha brain waves) to being slow and irregular (delta brain waves). Most studies based on electroencephalographic (EEG) readings have found that offenders have slower brain waves than do persons in general (e.g., Petersen et al. 1982).

Unlike traditional brain wave measurement, modern computerized brain wave detection is able to average responses to dozens of identical stimuli presented to subjects at random intervals. This reveals a distinctive brain wave pattern or "signature" for each individual. Nearly everyone exhibits a noticeable spike in electrical voltage, interrupted by a "dip" approximately one-third of a second following presentation of test stimuli. This is called the P300 amplitude of an event-related evoked potential. From a cognitive standpoint, the P300 amplitude is thought to reflect neurological events central to attention and memory.

While research has been equivocal thus far in the case of criminality, several studies have found a greater dip in P300 responses by individuals diagnosed with antisocial personality disorder than is true for general populations (see Costa et al. 2000). ENA theory can account for slower EEG patterns among offenders and a P300 decrement among persons with antisocial behavior by again focusing on suboptimal arousal. From a neurological standpoint, both slow brain waves and a tendency toward a greater than normal P300 decrement can be considered symptomatic of suboptimal arousal. If ENA theory is correct, both of these conditions will be found associated with elevated brain exposure to testosterone.

◪ Summary and Conclusions

Unlike social environmental theories, the evolutionary neuroandrogenic (ENA) theory can account for statistical associations between biological variables and criminal behavior. Furthermore, ENA theory predicts the universal concentration of offending among males between the ages of 13 and 30, patterns that strictly environmental theories have always had difficulty explaining. As its name implies, ENA theory rests on two over-arching assumptions. The first assumption is an extension of Darwin's theory of evolution by natural selection. It maintains that males on average exhibit CVB more than females because females who prefer to mate with such males increase their chances of having mates who are competent provisioners of resources. These female biases have evolved because females who have had the assistance of competent provisioners have left more offspring in subsequent generations than other females. No comparable reproductive advantage comes to males who select mates based on resource procurement capabilities.

Some forms of CVB are crude in the sense of requiring little learning, nearly all of which are either assaultive or confiscatory in nature. Other forms are sophisticated in the sense that they require complex learning and involve much more subtle types of "victimization." A major expression of sophisticated competitive/victimizing behavior involves profitable business ventures and/or the management of large organizations. In most societies, these expressions are tolerated and even encouraged. However, the vast majority of people in all societies condemn the crudest expressions of CVB, and, in all literate societies, the criminal justice system has evolved to punish such behavior.

The theory's second assumption is that genes on the Y-chromosome have evolved which cause male brains to exhibit higher rates of competitive/victimizing behavior than female brains. These genes operate in part by causing would-be ovaries to develop instead into testes early in fetal development. Once differentiated, the testes produce testosterone and other sex hormones, which have three hypothesized effects upon brain functioning, all of which

promote CVB. The three effects are termed *suboptimal arousal*, *seizuring proneness*, and *a rightward shift in neocortical functioning*. Furthermore, two neurological processes are hypothesized to help individuals shift from crude to sophisticated forms of competitive/victimizing behavior. These are learning ability (or intelligence) and executive cognitive functioning (or planning ability). The better one's learning ability or executive functioning, the quicker he/she will transition from crude to sophisticated forms of the behavior.

☒ References

Alm, P. O., af Klinteberg, B., Humble, K., Leppert, J., Sorensen, S., Thorell, L. H., et al. (1996). Psychopathy, platelet MAO activity and criminality among former juvenile delinquents. *Acta Psychiatrica Scandinavica, 94*, 105–111.

Blackson, T. C., & Tarter, R. E. (1994). Individual, family, and peer affiliation factors predisposing to early-age onset of alcohol and drug use. *Alcoholism: Clinical and Experimental Research, 18*, 813–821.

Buikhuisen, W., Eurelings-Bontekoe, E. H. M., & Host, K. B. (1989). Crime and recovery time: Mednick revisited. *International Journal of Law and Psychiatry, 12*, 29–40.

Costa, L., Bauer, L., Kuperman, S., Porjesz, B., O'Connor, S., & Hesselbrock, V. M. (2000). Frontal P300 decrements, alcohol dependence, and antisocial personality disorder. *Biological Psychiatry, 47*, 1064–1071.

Ellis, L. (2001). The biosocial female choice theory of social stratification. *Social Biology, 48*, 297–319.

Klinteberg, A. (1996). Biology, norms, and personality: A developmental perspective: Psychobiology of sensation seeking. *Neuropsychobiology, 34*(3), 146–154.

Lindman, R. E., Aromaki, A. S., & Eriksson, C. J. P. (1997). Sober-state cortisol as a predictor of drunken violence. *Alcohol and Alcoholism, 32*, 621–626.

Maras, A., Laucht, M., Gerdes, D., Wilhelm, C., Lewicka, S., Haack, D., et al. (2003). Association of testosterone and dihydrotestosterone with externalizing behavior in adolescent boys and girls. *Psychoneuroendocrinology, 28*, 932–940.

Matykiewicz, L., La Grange, L., Vance, P., Wang, M., & Reyes, E. (1997). Adjudicated adolescent males: Measures of urinary 5-hydroxyindoleacetic acid and reactive hypoglycemia. *Personality and Individual Differences, 22*, 327–332.

Maughan, B., Taylor, A., Caspi, A., & Moffitt, T. E. (2004). Prenatal smoking and early childhood conduct problems: Testing genetic and environmental explanations of the association. *Archives of General Psychiatry, 61*, 836.

Mendez, M. F., Doss, R. C., & Taylor, J. (1993). Interictal violence in epilepsy: Relationship to behavior and seizure variables. *The Journal of Nervous and Mental Disease, 181*, 566–569.

Mezzacappa, E., Tremblay, R. E., Kindlon, D., Saul, J. P., Arseneault, L., Seguin, J., et al. (1997). Anxiety, antisocial behavior, and heart rate regulation in adolescent males. *Journal of Psychiatry, 38*, 457–469.

Petersen, K. G. I., Matousek, M., Mednick, S. A., Volovka, J., & Pollock, V. (1982). EEG antecedents of thievery. *Acta Psychiatrica Scandinavica, 65*, 331–338.

Raine, A., Venables, P. H., & Williams, M. (1996). Better autonomic conditioning and faster electrodermal half-recovery time at age 15 years as possible protective factors against crime at age 29 years. *Developmental Psychology, 32*, 624–630.

Rasanes, P., Hakko, H., Isohanni, M., Hodgins, S., Jarvelin, M.-R., & Tiihonen, J. (1999). Maternal smoking during pregnancy and risk of criminal behavior among adult male offspring in the Northern Finland 1966 birth cohort. *American Journal of Psychiatry, 156*, 857–862.

Soler, H., Vinayak, P., & Quadagno, D. (2000). Biosocial aspects of domestic violence. *Psychoneuroendocrinology, 25*, 721–739.

Virkkunen, M. (1986). Reactive hypoglycemic tendency among habitually violent offenders. *Nutrition Reviews, 44*(Supplement), 94–103.

Virkkunen, M., Eggert, M., Rawlings, R., & Linnoila, M. (1996). A prospective follow-up study of alcoholic violent offenders and fire setters. *Archives of General Psychiatry, 53*, 523–529.

REVIEW QUESTIONS

1. What does Ellis call the theoretical model he presents in this piece? Explain why this name is appropriate given the primary propositions of the theory?

2. Of the twelve biological correlates of crime that Ellis reviews, which three do you feel are the most important or valid? Which three do you feel are least important/valid?

3. What three hypothesized effects on brain functioning are predicted by Ellis's theory? Do you agree with all three? Which do you think is most important or valid in predicting most criminal behavior?

❖

READING

Jana Bufkin and Vickie Luttrell are two of the leading researchers in the area of brain function in criminal behavior. In this selection, they reveal their findings regarding the significant differences that they have observed in their neuroimaging studies of violent criminals. It will be clear that there are notable differences in the three-pound organ of the brain that

SOURCE: Jana L. Bufkin and Vickie R. Luttrell, "Neuroimaging Studies of Aggressive and Violent Behavior," *Trauma, Violence, and Abuse* 6, no. 2 (2005): 176–91. Copyright © 2005 Sage Publications, Inc. Used by permission of Sage Publications, Inc.

distinguish most violent offenders from normal people (or even nonviolent offenders). While reading this selection, readers should keep in mind the fact that many things can go wrong in the functioning of the brain and that criminologists must be aware of these problems if they wish to understand why some individuals engage in criminal behavior.

Neuroimaging Studies of Aggressive and Violent Behavior

Current Findings and Implications for Criminology and Criminal Justice

Jana L. Bufkin and Vickie R. Luttrell

Aggressive and/or violent behaviors persist as significant social problems. In response, a substantial amount of research has been conducted to determine the roots of such behavior. Case studies of patients with neurological disorders or those who have suffered traumatic brain injury provide provocative insights into which brain regions, when damaged, might predispose to irresponsible, violent behavior. Psychophysiological and neuropsychological assessments have also demonstrated that violent offenders have lower brain functioning than controls, including lower verbal ability and diminished executive functioning. However, until recently it has been impossible to determine which brain areas in particular may be dysfunctional in violent individuals. With the availability of new functional and structural neuroimaging techniques, such as single-photon emission computed tomography (SPECT), positron emission tomography (PET), magnetic resonance imaging (MRI) and functional MRI (fMRI), it is now possible to examine regional brain dysfunction with a higher sensitivity and accuracy than was possible with previous techniques. This newfound ability to view the brain "in action"

has broadened our understanding of the neural circuitry that underlies emotional regulation and affiliated behaviors. In particular, evidence suggests that individuals who are vulnerable to faulty regulation of negative emotion may be at increased risk for aggressive and/or violent behavior.

In this review, we evaluate the proposed link between faulty emotion regulation and aggressive or violent behavior. We define *aggression* as any threatening or physically assaultive behavior directed at persons or the environment. *Violence* refers to behaviors that inflict physical harm in violation of social norms. Specifically, we (a) discuss briefly the neurobiology of emotion regulation and how disruptions in the neural circuitry underlying emotion regulation might predispose to impulsive aggression and violence; (b) summarize the results of modern neuroimaging studies that have directly assessed brain functioning and/or structure in aggressive, violent, and/or antisocial samples and evaluate the consistency of these findings in the context of negative emotion regulation; and (c) discuss theoretical and practical implications for criminology and criminal justice.

Emotion Regulation and Theoretical Links to Impulsive Aggression and Violence

Emotion is regulated by a complex neural circuit that involves several cortical areas, including the prefrontal cortex, the anterior cingulate cortex (ACC), the posterior right hemisphere, and the insular cortex, as well as several subcortical structures, such as the amygdala, hippocampus, and thalamus. These cortical and subcortical areas are intricately and extensively interconnected. In this article, we focus on three key elements of this neural circuitry: the prefrontal cortex, the ACC, and the amygdala.

The prefrontal cortex is a histologically heterogeneous region of the brain and has several (somewhat) functionally distinct sectors, including the ventromedial cortex and the orbitofrontal cortex (OFC). Damage to the ventromedial cortex and its behavioral affiliations have been assessed in case studies of individuals who experienced traumatic brain injury, either during childhood or adulthood, and in large, systematic studies on cohorts of war veterans with head injury.

Studies have found that patients with early-onset ventromedial lesions experience an insensitivity to future consequences, an inability to modify so-called risky behaviors even when more advantageous options are presented, and defective autonomic responses to punishment contingencies. Studies have also demonstrated that patients with adult-onset ventromedial damage show defects in real-life decision making, are oblivious to the future consequences of their actions, seem to be guided by immediate prospects only, and fail to respond autonomically to anticipated negative future outcomes.

The OFC, also a part of the prefrontal circuit, receives highly processed sensory information concerning a person's environmental experiences. The OFC is hypothesized to play a role in mediating behavior based on social context and appears to play a role in the perception of social signals, in particular, facial expressions of anger. Blair et al. (1999), using PET scans, assessed 13 male volunteers as they viewed static images of human faces expressing varying degrees of anger. They found that increasing the intensity of angry facial expressions was associated with enhanced activity in participants' OFC and the ACC. Dougherty et al. (1999) used functional neuroimaging and symptom provocation techniques to study the neurobiology of induced anger states and found that imaginal anger was associated with enhanced activation of the left OFC, right ACC (affective division), and bilateral anterior temporal regions. Also using imaginal scenarios, Pietrini et al. (2000) found that functional deactivation of OFC areas was strongest when participants were instructed to express unrestrained aggression toward assailants rather than when they tried to inhibit this imaginal aggression. Taken together, these lines of evidence support the suggestion that heightened activity in the left OFC may prevent a behavioral response to induced anger.

Based on these findings, and consistent with fearlessness theories of human aggression, a logical prediction is that OFC and ACC activity in response to provocation may be attenuated in certain individuals, predisposing them to aggression and violence. Consistent with this prediction, patients with OFC damage tend to exhibit poor impulse control, aggressive outbursts, verbal lewdness, and a lack of interpersonal sensitivity, which may increase the probability of sporadic so-called crimes of passion and encounters with the legal system. In contrast, evidence suggests that the ACC plays a role in processing the affective aspects of painful stimuli, such as the perceived unpleasantness that accompanies actual or potential tissue damage.

In addition to the prefrontal cortex and the ACC, another hypothesized neural component of emotion regulation is the amygdala, a subcortical structure, which is located on the medial margin of the temporal lobes. Similar to the OFC, the amygdala appears to play a role in extracting emotional content from environmental stimuli and may also play a role in individuals' ability to regulate negative emotion. However, neuroimaging studies have found that the amygdala is activated in response to cues that connote threat, such as facial expressions of fear (instead of anger), and that increasing the intensity of fearful facial expressions is associated with an increased activation of the left amygdala.

Davidson, Putnam, et al. (2000) suggested that individuals can typically regulate their negative affect and can also profit from restraint-producing cues in their environment, such as others' facial expressions of fear or anger. Information about behaviors that connote threat (e.g., hostile stares, threatening words, or lunging postures) is conveyed to the amygdala, which then projects to other limbic structures, and it is there that information about social context derived from OFC projections is integrated with one's current perceptions. The OFC, through its connections with other prefrontal sectors and with the amygdala, plays an important role in inhibiting impulsive outbursts because prefrontal activations that occur during anger arousal constrain the impulsive expression of emotional behavior.

Davidson, Putnam, et al. (2000) also proposed that dysfunctions in one or more of these regions and/or in the interconnections among them may be associated with faulty regulation of negative emotion and an increased propensity for impulsive aggression and violence. First, people with prefrontal and/or amygdalar dysfunction might misinterpret environmental cues, such as the facial expressions of others, and react impulsively, as a preemptive strike, to a misperceived threat. The perception of whether a stimulus is threatening is decisive in the cognitive processing leading to the aggressive behavior. Evidence suggests that individuals vary considerably in their ability to suppress

negative emotion. Therefore, individuals with decreased prefrontal activity may have greater difficulty suppressing negative emotions than those individuals who have greater prefrontal activation. Finally, although prefrontal activity helps one to suppress negative emotion, this negative emotion is generated by subcortical structures, including the amygdala. Therefore, an individual may be more prone to violence in general, and impulsive violence, in particular, if prefrontal functioning is diminished in relation to subcortical activity.

Research on individuals who have suffered traumatic head injury is of key importance in understanding the neural substrates of aggressive and/or violent behavior; however, Brower and Price (2001) noted many limitations of head injury studies, such as inadequate controls for known risk factors, including prior history of aggressive or violent behavior, socioeconomic status, stability of employment, and substance abuse. Research of behavior following head injury is also one step removed from the question of whether aggressive and/or violent individuals (who may have no history of head trauma) have neurological dysfunction localized to specific areas in the brain.

Studies of aggressive, violent, and/or antisocial offenders using functional (SPECT and PET) and structural (MRI) neuroimaging are beginning to reveal abnormalities in these groups (Raine, Lencz, et al., 2000). Specifically, 17 neuroimaging studies have been conducted on samples derived from forensic settings, prisons, psychiatric hospitals, and on violent offenders who are noninstitutionalized. Our review of these works reveals four consistent patterns:

(a) prefrontal dysfunction is associated with aggressive and/or violent behavioral histories;

(b) temporal lobe dysfunction, particularly left-sided medial-temporal (subcortical) activity, is associated with aggression and/or violence;

(c) the relative balance of activity between the prefrontal cortex and the subcortical structures is associated with impulsive aggression and/or violence; and

(d) the neural circuitry underlying the regulation of emotion and its affiliated behaviors is complex. Each of these patterns is described in theoretical context below.

Prefrontal Dysfunction Is Associated With Aggressive and/or Violent Behavioral Histories

Of the 17 studies reviewed, 14 specifically examined possible links between frontal lobe pathology and aggressive and/or violent behavior. In the 10 SPECT and PET studies, 100% reported deficits in either prefrontal (8 of 10 studies) or frontal (2 of 10 studies) functioning in aggressive, violent, and/or antisocial groups compared to nonaggressive patients or healthy controls. Analyses of specific regions in the medial prefrontal cortex revealed that individuals who were aggressive and/or violent had significantly lower prefrontal activity in the OFC (4 of 10 studies), anterior medial cortex (5 of 10 studies), medial frontal cortex (2 of 10 studies), and/ or superior frontal cortex (1 of 10 studies). In the four MRI studies, 50% (2 of 4 studies) reported decreased grey matter volume in prefrontal or frontal regions, and 25% (1 of 4 studies) reported nonspecific white matter abnormalities, not localized to the frontal cortex.

The consistency with which prefrontal disruption occurs across studies, each of which investigated participants with different types of violent behaviors, suggests that prefrontal dysfunction may underlie a predisposition to violence. Evidence is strongest for an association between prefrontal dysfunction and an impulsive subtype of aggressive behavior. Empirical findings concerning the regulation of negative emotion suggest that prefrontal sectors, such as the OFC, appear to play a role in the interpretation of environmental stimuli and the potential for danger. Consequently, disruptions in prefrontal functioning may lead individuals who are impulsive and aggressive to misinterpret situations as threatening and potentially dangerous, which in turn increases the probability of violent behavior against a perceived threat.

Nevertheless, four caveats are noteworthy. First, although prefrontal disruption was consistently related to aggressive and/or violent behavior, this association may reflect a predisposition only, requiring other environmental, psychological, and social factors to enhance or diminish this biological risk. Second, prefrontal dysfunction has also been documented in a wide variety of psychiatric and neurological disorders not associated with violence, and it may be argued that frontal hypometabolism is a general, nonspecific finding associated with a broad range of conditions. However, Drevets and Raichle (1995) reported that although frontal deficits have been observed in conditions, such as major depression, schizophrenia, and obsessive-compulsive disorder, the neurological profile for individuals who are aggressive and/or violent is different from these other groups. For example, while murderers exhibit widespread bilateral prefrontal dysfunction, individuals with depression tend to have disruptions localized in the left hemisphere only and to the left dorsolateral prefrontal cortex, in particular. (See Raine, Buchsbaum, et al., 1997, for a discussion of alterations in brain functioning across a variety of psychiatric conditions.)

Of the 10 SPECT and PET studies reviewed, 70% reported temporal lobe dysfunction in aggressive and/or violent groups, with reductions in left temporal lobe activity in 6 of 7 studies. Examination of the medial-temporal lobe, which includes subcortical structures, such as the amygdala, hippocampus, and basal ganglia, revealed that subcortical disruptions also characterized individuals who were aggressive and/or violent (4 of 7 studies). In the six MRI studies that examined the possibility of temporal lobe abnormalities, 100% (6 of 6 studies) reported temporal irregularities, including asymmetrical gyral patterns in the temporal-parietal region, decreases in anterior-inferior temporal lobe volume (including the amygdala-hippocampal region or adjacent areas), increases in left temporal lobe volume, or pathologies specific to the amygdala.

It is important to note that excessive right subcortical activity or abnormal temporal lobe structure was most common in patients with a history of intense violent behavior, such as that seen in those with intermittent explosive disorder rather than in patients who had aggressive personality types or who had high scores on an aggression scale. In humans, right-hemisphere activation has been suggested to play a role in the generation of negative affect. Therefore, increased subcortical activity in the right hemisphere could lead an individual to experience negative affect that promotes aggressive feelings and acts as a general predisposition to aggression and violence. These findings are generally consistent with current conceptions of emotion regulation and its purported relationship to impulsive violence, in particular.

The Relative Balance of Activity Between the Prefrontal Cortex and the Subcortical Structures Is Associated With Impulsive Aggression and/or Violence

Previous research has suggested that individuals may be predisposed to impulsive violence if prefrontal functioning is diminished relative to subcortical activity. Raine, Meloy, et al. (1998) found that reduced prefrontal functioning relative to subcortical functioning was characteristic of those who commit impulsive acts of aggression and/or violence. By contrast, aggression and/or violence of a predatory nature was not related to reduced prefrontal and/ or subcortical ratios. They also

suggest that although most biological studies of aggression and/or violence have not distinguished between impulsive and premeditated aggression, this distinction is likely relevant to understanding the neuroanatomical and functional underpinnings of these behaviors.

An additional line of evidence that lends support to the impulsive and/or predatory distinction comes from investigations regarding the mechanism underlying the suppression of negative emotion. The neurochemical link mediating prefrontal and/or subcortical interactions is purportedly an inhibitory serotonergic connection from the prefrontal cortex to the amygdala. The prefrontal cortex is a region with a high density of serotonin receptors, which sends efferents to the brainstem where most of the brain's serotonin-producing neurons originate. The prefrontal cortex, amygdala, and hippocampus also receive serotonergic innervation. Therefore, it is logical that dysfunction in the prefrontal and/or subcortical regions disrupts serotonergic activity in the brain. Consistent with this hypothesis, the serotonergic system has been shown to be dysfunctional in victims of violent suicide attempts, impulsive violent offenders, impulsive arsonists, violent offender and arson recidivists, children and adolescents with disruptive behavior disorders, and "acting out" hostility in normal volunteers. In all those studies, low serotonin levels were strongly related to the maladaptive behaviors noted.

◪ Implications for Criminology and Criminal Justice

Historically, paradigms guiding criminological programs of study have tended to bypass complex webs of interconnections that produce and reproduce criminality, favoring instead an emphasis on one dimension or level of analysis. The trend has been to maintain a specialized focus, often within the confines of a sociological or a legalistic model. Attempts to expand the image of crime through theory integration, which have surfaced quite frequently since the mid-1970s, shift attention to different realities of crime. The general emphasis, however, has been on the integration of ideas within and/or across the two dominant paradigms rather than on a broader, interdisciplinary strategy.

Resistance to interdisciplinarity or disciplinary cross-fertilization has not been inconsequential. Failure to incorporate interdisciplinary insights has stifled exploration of the intersections among structure, culture, and the body, leaving a knowledge void where provocative social facts "merely hang in space as interesting curiosities" (Pallone & Hennessy, 2000, chap. 22, p. 11), and critical questions go unanswered. More specifically, lack of an imagination of how nurture and nature interact to affect behavior, or what may be understood as biography in historical context, has resulted in an incapacity to either deal with variability or deal with it well. Some male individuals socialized in a patriarchal society rape and some adolescents from poor, urban, single-families chronically offend; however, most do not. In other words, there is individual variation within social contexts, and those differences may be better understood if criminologists begin to consider all pertinent angles or dimensions.

Although the studies in our review may appear to be firmly planted in the tradition of specialization and unidimensional thinking, they should be interpreted within the framework of Barak's (1998) interdisciplinary criminology, where knowledges relevant to a behavioral outcome are treated as complements in an image expansion project. Understanding that each perspective offers a reality of behavior from a different, though interrelated angle, the objective is to develop a logical network of theories that will capture the most dimensions and provide the most accurate information about phenomena of interest. In this vein, our appraisal of knowledge from the field of neuroscience

intends to elucidate the image of aggression and/or violence without supplanting other perspectives and paradigms. Our desire is not to reduce aggression and/or violence to brain functioning but to inform of advances in neurological analyses of emotion regulation and their importance to studies of that behavior.

It should be noted that the more comprehensive, interdisciplinary paradigm has been embraced by some criminologists linked to the biological sciences. In the 17 studies reviewed, researchers attempted to examine (or at least statistically controlled for) a variety of biological, psychological, and social correlates of aggressive and/or violent behavior and, in some cases, analyzed biopsychosocial interactions affecting behavior. Across studies, biological variables included history of head injury, substance use/abuse/dependence, diseases of the nervous system, left-handedness, body weight, height, head circumference, and sex. Psychological variables included the presence of psychological disorders (such as schizophrenia), indices of intellectual functioning (such as IQ scores), and performance-related motivational differences. Social variables included indices of psychosocial deprivation (such as physical and/or sexual abuse, extreme poverty, neglect, foster home placement, being raised in an institution, parental criminality, parental physical fights, severe family conflict, early parental divorce), family size, and ethnicity.

The benefits of integrating ideas or investigating an image from several angles in a research design is demonstrated in the neuroimaging studies provided. Raine, Lencz, et al. (2000), for example, found that prefrontal and autonomic deficits contributed substantially to the prediction of group membership (antisocial personality disorder vs. control group) over and above 10 demographic and psychosocial measures. The 10 demographic and psychosocial variables accounted for 41.3% of the variance. After the addition of three biological variables into the regression equation (prefrontal gray matter, heart rate, and skin conductance), amount of variance explained increased significantly to 76.7%, and the prediction of group membership increased from 73% to 88.5% classified correctly. These findings suggest that a more contextualized theoretical grasp of aggression and/or violence is possible when this behavior is conceptualized as multi-dimensional. When nature-nurture dichotomies are countered by interdisciplinary image expansion, clues about individual variability emerge, and criminologists come closer to understanding the complexity of aggression and/or violence.

The compelling evidence about this behavior revealed in the reviewed neuroimaging studies is valuable, then, not because it allows for completely reliable predictions of behavioral outcomes, but because it makes the image of aggression and/or violence a little less murky. Moreover, when merged with existing knowledges, particularly ideas about social structures and social psychology (sociological model) and rational choice (legalistic model), such findings may spawn new visions of justice centered on prevention and treatment. Within an interdisciplinary framework that values neuroscience, virtually every essential sociological factor elaborated by criminologists, structural and processual, acquires a greater potential to explain aggression and/or violence and influence policy making. According to the works in our review, as well as other research in this area, all forms of child abuse and neglect, direct exposure to violence (including media violence), an unstable family life, poor parenting, lack of prenatal and perinatal services, individual drug use, maternal drug use during pregnancy, poor educational and employment structures, poverty, and even exposure to racism play a vital role in the production of aggression and/or violence. Thus, the inclusion of insights from neuroscience further legitimizes prevention strategies touted by advocates of the sociological paradigm, from social disorganization theory to self-control theory.

When aggression and/or violence is not prevented, the criminal justice system is granted responsibility for social control. . . . Drawing from empirical findings across disciplines and levels of analysis, a vision of therapeutic justice encourages the development of holistic treatment regimens that hold offenders to "scientifically rational and legally appropriate degree[s] of accountability" (Nygaard, 2000, chap. 23, p. 12). The potential for this approach to replace the utilitarian model lies in its continued ability to unveil the often-perplexing ways in which choice is structured. This facilitates an awareness that the legally appropriate and the scientifically rational are in unity. Human creativity is not ignored in this paradigm; however, the clearer image it provides points to an amalgam of limitations. When an individual is brought into the criminal justice system, an inter-disciplinarian seeks to examine those restrictions on behavior and to tailor treatments accordingly.

With varying levels of success, criminologists have sought to qualify choice and diminish the impact of legalistic factors on conceptions of justice since the advent of positivism in the 19th century. Assessments of measures associated with social psychology, psychology, and psychiatry, along with input implicating structural concerns, such as unemployment, have been utilized, and a plethora of interventions have evolved. Thus, cracks in the utilitarian mold of justice have accrued, laying the foundation for interdisciplinarity in thought and in treatment. Applied to aggression and/or violence, this translates into the implementation of treatment plans with multidimensional components, to include neurological techniques that address how brain dysfunction affects choice. Although not the sole neurological strategy, the intervention most consistently promoted is drug therapy. Several types of drugs, such as anti-convulsants, psychostimulants, and serotonergic agents, have been successful in reducing aggressive behavior. Inter-disciplinary thinkers should not be hesitant to consider using these pharmacological remedies when biopsychosocial indicators overwhelmingly suggest that an individual is at risk to violently recidivate, for it is a step in the direction of therapeutic justice.

Other than paradigmatic preferences disallowing an interdisciplinary consideration of aggressive and/or violent crimes and lack of funding, the largest obstacle in attaining therapeutic justice is the inability to predict future behavior. When informed by neuroscience, classification and prediction instruments are fine-tuned. To illustrate, Robinson and Kelley (2000) discovered that, among probationers, indicators of brain dysfunction correlated with repeat violent offending, as opposed to repeat nonviolent offending and first-time offending. Birth complications, family abuse, head injury, parental drug use, abnormal interpersonal characteristics, and offender substance abuse were found to be risk factors for recidivism within this group. Given that it is estimated that less comprehensive prediction models reap false positives in approximately two thirds of all cases, added precision is welcome.

Still, prediction is not foolproof. Shortcomings in this area lead some to conclude that drug therapy and other invasive strategies are unwarranted. Before throwing in the towel, it should be acknowledged that pharmacological remedies already abound in the criminal justice system, along with many other intrusions. Knowledge from neuroscience merely allows for the targeted distribution of services to appropriate populations, a fruitful strategy given the scarcity of resources at the system's disposal. Prevention strategies directed at alleviating environmental conditions that increase the probability for aggression and/or violence are optimal; however, criminologists should not dismiss neuroscientific individual-level interventions in cases where patterns of aggressive and/or violent criminality are detected. Converging lines of evidence suggest that those patterns are produced by a unique combination of external and internal risk factors, each of which is integral to the construction of treatment regimens intended to effect therapeutic justice.

Blind spots in the image of aggression and/or violence should not deter interventions where they hold promise for enhancing quality of life. It is unfair and unjust to those processed in the criminal justice system and to society at large for criminologists to ignore this evidence and the control strategies proposed.

⬛ Conclusion

Functional and morphometric neuroimaging has enhanced our understanding of the distributed neural networks that subserve complex emotional behaviors. Research emanating from affective, behavioral, and clinical neuroscience paradigms is converging on the conclusion that there is a significant neurological basis of aggressive and/or violent behavior over and above contributions from the psychosocial environment. In particular, and consistent with modern theories of emotion regulation, reduced prefrontal and/or subcortical ratios may predispose to impulsive aggression and/or violence. Further progress in the study of these behaviors will require a forensically informed, interdisciplinary approach that integrates neuropsychological and psychophysiological methods for the study of the brain, emotional processing, and behavior.

As this line of interdisciplinary research unfolds, it is vital that criminology and criminal justice begin to incorporate what is known about human behavior into their explanatory models, as well as classrooms. Evidence suggests that brain structure and brain functioning do affect behavior, particularly aggressive and/or violent behavior. It is also the case that neuroscience offers means for curbing aggression and/or violence. Traditional criminology and criminal justice paradigms tend to sidestep these issues because of aversions to less dominant knowledges, especially biological programs of study. Biological insights are often dubbed Lombrosian, suggesting that some behavioral scientists retain notions of a born criminal easily identifiable using some magic test. Continued aversion to anything biological on these grounds is anachronistic and will hamper the development of theory and policy.

The problem is that neurobiological discovery has carried on with little to no input from criminology and criminal justice, and there is every reason to believe that the research will progress. There is also reason to believe that the functioning of the criminal justice system will be affected by the findings produced. The general public is already being widely exposed to such advances through numerous television news clips and articles appearing in newspapers and weekly periodicals. If criminology and criminal justice want to be relevant in more than a historical sense when it comes to theorizing about aggressive and/or violent behavior and formulating policies accordingly, it is imperative that the field embrace the interdisciplinary model.

▨ Implications for Practice, Policy, and Research

Practice

Bridging the gap between nature and nurture, a biopsychosocial model for understanding aggression enhances the explanatory capacity of sociologically based criminological theories by accounting for individual variability within social contexts.

Insights derived from a biopsychosocial model offer the most promise in the realm of crime prevention, which entails devising holistic treatment strategies for those exposed to numerous risk factors.

Policy

The accuracy of risk classification devices, used extensively throughout the criminal justice process, may be enhanced by incorporating what is known about negative emotion regulation.

Research

Research reveals that other cortical and subcortical structures likely play a role in emotion regulation through their inextricable link to the prefrontal and medial-temporal regions. The complexity of this neural circuitry must be explored with greater precision.

▨ References

Barak, G. (1998). *Integrating criminologies.* Needham Heights, MA: Allyn & Bacon.

Blair, R. J. R., Morris, J. S., Frith, C. D., Perrett, D. I., & Dolan, R. J. (1999). Dissociable neural responses to facial expressions of sadness and anger. *Brain, 122,* 883–893.

Brower, M. C., & Price, B. H. (2001). Neuropsychiatry of frontal lobe dysfunction in violent and criminal behavior: A critical review. *Journal of Neurology, Neurosurgery, and Psychiatry, 71,* 720–726.

Davidson, R. J., Putnam, K. M., & Larson, C. L. (2000). Dysfunction in the neural circuitry of emotion regulation: A possible prelude to violence. *Science, 289*(5479), 591–594.

Dougherty, D. D., Shin, L. M., Alpert, N. M., Pitman, R. K., Orr, S. P., Lasko, M., et al. (1999). Anger in healthy men: A PET study using script-driven imagery. *Biological Psychiatry, 46,* 466–472.

Drevets, W. C., & Raichle, M. E. (1995). Positron emission tomographic imaging studies of human emotional disorders. In M. S. Gazzaniga (Ed.), *The cognitive neurosciences* (pp. 1153–1164). Cambridge, MA: MIT Press.

Nygaard, R. L. (2000). The dawn of therapeutic justice. In D. H. Fishbein (Ed.), *The science, treatment, and prevention of antisocial behaviors: Application to the criminal justice system* (chap. 23, pp. 1–18). Kingston, NJ: Civic Research Institute.

Pallone, N. J., & Hennessy, J. J. (2000). Indifferent communication between social science and neuroscience: The case of "biological brain-proneness' for criminal aggression. In D. H. Fishbein (Ed.), *The science, treatment, and prevention of antisocial behaviors: Application to the criminal justice system* (chap. 22, pp. 1–13). Kingston, NJ: Civic Research Institute.

Pietrini, P., Guazzelli, M., Basso, G., Jaffe, K., & Grafman, J. (2000). Neural correlates of imaginal aggressive behavior assessed by positron emission tomography in healthy subjects. *American Journal of Psychiatry, 157*(11), 1772–1781.

Raine, A., Buchsbaum, M., & LaCasse, L. (1997). Brain abnormalities in murderers indicated by positron emission tomography. *Biological Psychiatry, 42,* 495–508.

Raine, A., Lencz, T., Bihrle, S., LaCasse, L., & Colletti, P. (2000). Reduced prefrontal gray matter volume and reduced autonomic activity in antisocial personality disorder. *Archives of General Psychiatry, 57,* 119–127.

Raine, A., Meloy, J. R., Bihrle, S., Stoddard, J., LaCasse, L., & Buchsbaum, M. S. (1998). Reduced prefrontal and increased subcortical brain functioning assessed using positron emission tomography in predatory and affective murderers. *Behavioral Sciences and the Law, 16,* 319–332.

Robinson, M., & Kelley, T. (2000). The identification of neurological correlates of brain dysfunction in offenders by probation officers. In D. H. Fishbein (Ed.), *The science, treatment, and prevention of antisocial behaviors: Application to the criminal justice system* (chap. 12, pp. 12-1–12-20). Kingston, NJ: Civic Research Institute.

REVIEW QUESTIONS

1. According to Bufkin and Luttrell, what are four consistent patterns revealed from their review of 17 neuroimaging studies? Which of these do you feel is most valid? Least valid?

2. What types of implications for criminological theory do Bufkin and Luttrell discuss?

3. What types of implications for criminal justice do they discuss?

❖

READING

In this selection, McGloin and her colleagues present a study that examines the nature of the effects of maternal cigarette smoking during pregnancy. They point out the problems of previous studies, namely concentrating on measures that are not directly based on neuropsychological problems, but rather on IQ tests. Furthermore, they base their study on a highly respected and influential model of criminal development presented by Terrie Moffitt,[40] which claimed that the most serious, chronic offenders are those that experience both a disadvantaged environment (e.g., bad neighborhood, bad parenting, etc.) and early neuropsychological problems (e.g., due to maternal cigarette smoking or other toxins, pregnancy/delivery complications, etc.). Although other studies have supported Moffitt's developmental model (which will be discussed at length in Section X), this may be the only study that has directly tested and supported her model. While reading this selection, readers should consider that what happens in the womb may have an effect on what happens much later in life.

Source: Jean McGloin, Travis Pratt, and Alex Piquero, "A Life-Course Analysis of the Criminogenic Effects of Maternal Cigarette Smoking During Pregnancy," *Journal of Research in Crime and Delinquency* 43, no. 4 (2006): 412–26. Copyright © 2006 Sage Publications, Inc. Used by permission of Sage Publications, Inc.

[40]Terrie Moffitt, "Adolescence Limited and Life Course Persistent Antisocial Behavioral: A Developmental Taxonomy," *Psychological Review* 100 (1993): 674–701.

A Life-Course Analysis of the Criminogenic Effects of Maternal Cigarette Smoking During Pregnancy

A Research Note on the Mediating Impact of Neuropsychological Deficit

Jean Marie McGloin, Travis C. Pratt, and Alex R. Piquero

Research from a variety of disciplines indicates that maternal cigarette smoking (MCS) during pregnancy is associated with an array of problematic behavioral outcomes in offspring (Cornelius and Day 2000; Wakschlag et al. 2002). Of particular interest to criminologists, this may include such traditional criminological outcomes as violent, persistent, and early-onset offending (Brennan, Grekin, and Mednick 1999; Gibson, Piquero, and Tibbetts 2000; Rasanen et al. 1999). Furthermore, Cotton's (1994) assertion that 20 to 25 percent of pregnant women who smoke continue to do so throughout their pregnancies suggests that this risk factor is worthy of criminologists' attention.

Nevertheless, the question of how MCS risk manifests into criminal behavior still remains. On one hand, some studies have used Moffitt's (1993) developmental taxonomy as a framework for empirical investigation (see Gibson et al. 2000; Gibson and Tibbetts 2000; Piquero et al. 2002). Indeed, MCS fits nicely within the battery of the various congenital risks that serve as the hypothesized roots of life-course-persistent (LCP) offending. Even so, previous investigations that have drawn on Moffitt's framework have not truly tested the developmental pathway specified by her theory. Instead, research has largely focused on the direct relationship between MCS and various measures of LCP offending, which treat MCS as a proxy for neuropsychological deficit (Gibson et al. 2000; Gibson and Tibbetts 2000; Piquero et al. 2002).

The problem with this approach is that Moffitt (1993) did not suggest that MCS is a measure of neuropsychological deficit but rather that MCS is a precursor to such deficits. She specified a mediating relationship in which congenital risks, such as MCS, increase the likelihood of neuropsychological deficits occurring in children, which in turn increase the probability that such youths will eventually engage in LCP offending. Thus, empirical research has yet to be conducted that addresses whether the relationship between MCS and LCP offending is mediated by more direct measures of neuropsychological deficit.

In this research, we addressed this void by determining whether neuropsychological deficit does in fact mediate the connection between MCS and LCP offending. We assessed this relationship while using a number of controls for other early biological risk factors (e.g., low birth weight [LBW]) as well as indicators of social disadvantage at multiple points in time. Our broader purpose, therefore, was to determine whether criminologists should continue to think about MCS in the context of theories that specify neuropsychological deficits—as opposed to, say, parenting effects (i.e., is it simply that mothers who smoke while pregnant also turn out to be inept at shaping prosocial behavior in their kids?)—as a key predictor of criminal behavior.

Despite this goal, we recognize that MCS is, by any reasonable estimation, a distal criminogenic risk factor. To be sure, other variables, such as self-control (Pratt and Cullen 2000), deviant peer influences, and antisocial attitudes (Andrews and Bonta 2001), have all been found to consistently predict antisocial behavior more consistently and robustly than MCS (Wakschlag et al. 2002). Nevertheless, the MCS-crime link provides criminologists with a unique opportunity to examine the relative validity of some of the claims made by the dominant theoretical traditions in the field. Indeed, a complete explanation of the causal processes underlying the link between MCS and crime or deviance may end up telling us a lot about the compatibility, or lack thereof, of theories that specify biological versus social-psychological causes of crime. Again, we took an initial first step in this process by testing the degree to which the link between MCS and LCP offending is mediated by measures of neuropsychological deficit.

⊠ Theoretical Context

Moffitt (1993) argued that two offending trajectories with distinct etiologies are obscured within the aggregate age-crime curve. Individuals on the LCP pathway begin offending at an early age and continue throughout the life course, engaging in an array of deviance, including criminal behavior. Individuals on the adolescence-limited trajectory instead start offending at a relatively older age, have a transitory offending time frame typically defined by minor rebellious offending, and desist on the transition to adulthood (Moffitt 1993; Piquero 2000; Piquero et al. 1999). LCP offending occurs in a small proportion of the population and develops out of an evocative interaction between neuropsychological deficits and a disadvantaged environment. Adolescence-limited offending, in contrast, is found in the majority of the population and develops out of social mimicry during the maturational gap of adolescence.

Accordingly, when criminologists concentrate on MCS as a risk factor, they typically do so with a focus on the LCP etiological pathway.

According to Moffitt (1993), developmental disturbances of the fetal brain, which can be caused by various pre- and perinatal risk factors, such as exposure to toxins, poor maternal nutrition, and MCS, produce deficits in the central nervous system. Lynam, Moffitt, and Stouthamer-Loeber (1993) stated that "deficits in the neuropsychological abilities referred to as 'executive functions' interfere with a person's ability to monitor and control his or her own behavior" (p. 188). In short, the hypothesis is that congenital, and therefore biological, risks produce neuropsychological deficits, which can manifest behaviorally in a bad temperament and, later, criminal behavior.

Indeed, neuropsychological deficit, which has also been measured through proxies such as low cognitive ability (Denno 1990; Moffitt 1990), has emerged as an important discriminating factor between offenders and nonoffenders (see Hirschi and Hindelang 1977; Wolfgang, Figlio, and Sellin 1972) as well as a predictor of more severe markers of offending within criminal populations (McGloin and Pratt 2003; Piquero and White 2003). Moreover, despite criticism about the validity of articulating only two trajectories (Nagin, Farrington, and Moffitt 1995; White, Bates, and Buyske 2001), the main hypothesis that offenders, especially serious offenders, suffer from neuropsychological and/or biological deficits is well supported empirically (Moffitt. Lynam, and Silva 1994; Piquero 2001; Piquero and White 2003).

Even so, in studying the link between MCS and crime or deviance, existing criminological research has used MCS as a proxy of neuropsychological deficit rather than as a precursor to it. For example, Gibson et al. (2000) found a significant relationship between MCS and age at first police contact net of statistical controls,[1] a finding that was echoed by Gibson and Tibbetts[2] (2000) and by Piquero et al. (2002). Although these studies offered some support for the association between MCS and LCP offending, none of them specified the purported mediating mechanism between MCS and neuropsychological deficit.

A substantial gap in the research therefore continues to exist. To be sure, the precise mechanism whereby MCS operates as a criminogenic risk factor is still unknown. Moffitt (1993) argued that neuropsychological deficit is a result of some injury to, or disturbance of, the proper fetal developmental process. She did not suggest that such risks, in this case MCS during pregnancy, should themselves be treated as a proxy for minimal brain dysfunction but rather that they are assumed to represent, among others, the primary causes of neuropsychological deficits. Accordingly, this research explicitly examined the mechanism whereby MCS produces a criminogenic risk (i.e., through its effect on neuropsychological deficits) for LCP offending.

▧ Methods

Data

Data for this project came from both the original Philadelphia portion of the National Collaborative Perinatal Project (NCPP) and a recent criminal history search conducted on the original cohort of 987 youths born to African American mothers who participated in the NCPP (see Piquero et al. 2002). Moffitt (1997) considered the NCPP data to be particularly well suited to analyses addressing issues associated with neuropsychological deficits, thereby making the use of the NCPP particularly relevant for the questions under investigation in this study. Given the focus on LCP offending, analyses were conducted on the offender subsample ($n = 220$) of the original 987 subjects.[3]

In the early 1980s, information was collected related to school functioning and criminal histories, including all police contacts through age 17 (see Denno 1990).[4] Adult criminal history data, in the form of convictions, are available through age 36 for those sample members born into the 1962 cohort and through age 39 for those born into the 1959 cohort.[5] Some might suggest that conviction data from the adult follow-up are not as reliable as self-report, police contact, or arrest data. All sources of criminal justice data, however, are subject to limitations (see Hindelang et al. 1981; Lauritsen 1998; Wolfgang et al. 1972), and extant theory does not anticipate that certain relationships would be found only when analyzing certain types of outcome data.[6] Moreover, many empirical investigations have used various sources of criminal history data, including self-report and conviction data (Farrington et al. 1996; Ge, Donnellan, and Wenk 2001; Moffitt et al. 1994). To this end, Farrington (1989) showed that self-reports and official conviction data produce "comparable and complementary results on such important topics as prevalence, continuity, versatility, and specialization in different types of offenses" (p. 418).[7]

Dependent Variable

LCP offending. Individuals who exhibited an early onset (prior to age 14) and accumulated at least two adult convictions during the follow-up period (after age 18) were coded 1, and all other sample members were coded 0 (the same approach with these data was taken by Gibson et al. 2000, Gibson and Tibbetts 2000, McGloin and Pratt 2003, Piquero 2001, and Tibbetts and Piquero 1999; see also Piquero and White 2003).

Independent Variables

Neuropsychological deficit. To measure neuropsychological deficit, we used the total battery score of the California Achievement Test (CAT), a measure that has also been used in past criminological research (see Ge et al. 2001; McGloin and Pratt 2003; Piquero and White 2003). The CAT yields total scores in the academic domain of reading, arithmetic, and language (Tiegs and Clark 1970: 14). The total battery score for Grades 7 and 8 (ages 12 to 14), which we used in this study, reflects a student's standing in terms of total achievement level. The CAT is, in general, highly praised in terms of its validity, comprehensive test and interpretive materials, reliability, and standardization procedure (Denno 1990:173).[8]

We used the CAT because we recognized the potential conceptual, empirical, and ideological concerns with using IQ to assess neuropsychological functioning within a sample of economically disadvantaged African American youth. Moreover, Moffitt (1990) argued that the array of cognitive tests is so highly interrelated that interchanging them is acceptable. One could also argue that research on LCP offending should consider measures other than IQ to establish convergent validity (McGloin and Pratt 2003). This is especially true with this data set, which has been subject to much empirical investigation under Moffitt's (1993) framework and has largely relied on IQ as the measure of neuropsychological deficit. Nevertheless, for those who prefer IQ as a measure, we also estimated the models with the verbal subscale of the Wechsler Intelligence Scale for Children (WISC), which was assessed at ages 7 to 8.[9]

Sex. Sex was coded 1 for male and 2 for female. Of these 220 offenders, approximately 70 percent were male.

Low birth weight. LBW was associated with MCS and has been shown to have a relationship with criminal outcomes (see Gibson et al. 2000; McGloin and Pratt 2003; Tibbetts and Piquero 1999). Following designation made by the World Health Organization and used in past research, LBW is a dichotomous variable indicating its presence (1) or absence (0). The cutoff for LBW is five pounds, eight ounces.[10] Of the 220 offenders, 19 percent were LBW.

Risk at birth. Our risk-at-birth composite was a summed index of three dichotomous items

measured at birth from the mother: birth complications, marital status, and age at childbirth. Following Farrington and Loeber (2000), these variables were dichotomized to reflect the risk-factor paradigm. Consistent with Nagin, Farrington, and Pogarsky (1997), mother's age at childbirth was coded 0 (18 and older) or 1 (under 18), and mother's marital status at childbirth was coded 0 (married) or 1 (single). Finally, birth complications were coded 0 (for no birth complications during pregnancy) or 1 (for one or more birth complications during pregnancy). All of these items have been considered indicators of maternal disadvantage and have been related to important offspring problems, including crime (see Farrington and Loeber 2000; Raine 1993).

Risk at age 7. Socioeconomic status was originally measured in a standardized method by all sites of the NCPP with a single-item score, ranging from 0 to 100, that was a composite measure of three indicators collected at age 7 for each child: education of the head of the household, income of the head of the household, and occupation of the head of the household (Myrianthopoulos and French 1968). Those individuals scoring in the lowest 25th percentile (very low socioeconomic status) were coded 1, and all others were coded 0.

Maternal cigarette smoking. MCS was assessed during pregnancy. Mothers were asked to self-report the average number of cigarettes they smoked each day. Although this variable was originally coded continuously, we followed the coding procedure outlined by Brennan et al. (1999) and replicated by others: 0, 1 to 2, 3 to 10, 11 to 20, and 20 or more. This measure, then, assessed a potential dose-response relationship.[11]

Analytic Strategy

The analysis focused on determining whether neuropsychological deficit mediates the relationship between MCS and LCP offending. Given the dichotomous dependent variable of LCP offending, the main analyses estimated the multivariate models via logistic regression techniques. To establish a mediating relationship, three criteria must be satisfied. First, MCS must predict the mediating mechanism of interest, neuropsychological deficit. Second, it must be established that MCS in fact predicts LCP offending. Finally, the inclusion of the potential mediator, neuropsychological deficit, should eliminate much (if not all) of the significance of MCS in the multivariate model.

With this in mind, the first step in the analysis determined whether MCS predicted neuropsychological deficit, net of statistical controls. This analysis relied on ordinary least squares regression techniques, given the continuous measure of neuropsychological deficit. Next, three models were estimated for the prediction of LCP offending. Model 1 included the MCS measure along with the control variables to gauge the relationship between MCS and LCP offending. Model 2 included the measure of neuropsychological deficit to determine if it stripped MCS of its significance and acted as a mediator. Finally, model 3 included the composite of risk at age 7, which was added separately from the other controls given its temporal distinction.

▧ Results

Table 1 presents the results from the ordinary least squares regression model predicting neuropsychological deficit. The findings suggest that the offspring of mothers who smoked (many) cigarettes during pregnancy were more likely to experience lower scores on the CAT.[12]

For the next step in the analysis, Table 2 presents the three logistic regression models predicting LCP offending. Model 1 revealed that MCS did predict LCP offending, net of statistical controls.[13] In particular, it was the most severe category of smoking that held

Table 1	Slopes and Standard Errors for the Ordinary Least Squares Regression Model Predicting Neuropsychological Deficit ($n = 220$)

Variable	B	SE
Sex	1.022	2.972
Low birth weight	−2.599	3.642
Risk at birth	2.736	1.660
Maternal cigarette smoking	−2.661**	1.161
Constant	20.399	
F	2.406	
R^2	.044	

**$p < .0.5$.

significance: The offspring of mothers who smoked 20 or more cigarettes per day, compared with mothers who did not smoke, were significantly more likely to manifest LCP offending. This suggests that the damage of MCS exists on a spectrum. Others have also confirmed the importance of allowing for a potential dose-response relationship, finding a growing risk for a certain type of offending as the number of cigarettes a mother smoked during pregnancy increased (Brennan et al. 1999; Gibson et al. 2000; Piquero et al. 2002).

Model 2 included the CAT variable to determine whether introducing an indicator of neuropsychological deficit mediated the effect of the MCS measure. Although the CAT emerged as a significant predictor of LCP offending, it did not affect the significance level of MCS. Thus, neuropsychological deficit did not appear to mediate the relationship between MCS and LCP offending. The same conclusion was reached even with the inclusion of risk at age 7 in model 3, which showed that adding this additional risk factor to the model did not eliminate the significance of the MCS variable.

Table 2	Logit Coefficients and Standard Errors for the Logistic Regression Models Predicting Life-Course-Persistent Offending ($n = 220$)					
	Model 1		**Model 2**		**Model 3**	
Variable	B	SE	B	SE	B	SE
Sex	−1.213**	.550	−1.240**	.563	−1.241**	.564
Low birth weight	.326	.485	.344	.498	.345	.498
Risk at birth	.155	.236	.279	.247	.276	.253
Maternal cigarette smoking (cigarettes per day)						
1 to 2	.590	.872	.553	.904	.556	.907
3 to 10	.250	.449	.082	.461	+.083	.462
11 to 20	.921	.679	.720	.694	.721	.694
More than 20	1.635*	.884	1.681*	.913	1.686*	.924
Neuropsychological deficit			−.036**	.015	−0.36**	.015
Risk at age 7					.018	.447
Constant	−.660		−.073		−.077	
−2 log likelihood	175.982		167.633**		167.631**	

$*p < .10. **p < .05.$

◼ Discussion

The primary focus of this research was to determine the mechanism whereby MCS manifests itself as a criminogenic risk factor. Despite the fact that previous research has used Moffitt's (1993) theoretical framework, no study has specified this mediating mechanism articulated by Moffitt. The results presented here show that although MCS is a significant precursor to neuropsychological deficit and that neuropsychological deficit significantly predicts LCP offending, this is not the mediating mechanism at work in the relationship between MCS and LCP offending. Indeed, a significant relationship remains between MCS and LCP offending that operates independent of neuropsychological dysfunction.

This finding, although somewhat inconsistent with Moffitt's (1993) theoretical predictions, is not necessarily inconsistent with earlier empirical work. Previous research in

this area that has used IQ as a control variable has shown that MCS retained its ability to predict criminal outcomes (Gibson et al. 2000; Piquero et al. 2002). Thus, these models provided a hint that Moffitt's theoretical framework was not explaining the nature of the relationship between MCS and LCP offending. The findings offered here, though in a decidedly more explicit fashion, suggest that the risk of MCS does not operate (solely) through the indirect pathway of neuropsychological deficit. At a minimum, these results call into question using an MCS measure as a proxy for neuropsychological deficit.

The question now, therefore, concerns what is in fact the mediating mechanism between MCS and LCP offending. Accordingly, there are two primary potential mediators that should be addressed in future research. First, parenting may play an important role in explaining the empirical relationship between MCS and offending. A mother who smokes cigarettes during pregnancy, particularly a "high-rate" smoker, may illustrate a "foreshadowing" propensity to put her immediate, hedonistic tendencies and desires before long-term considerations for her child's health. Perhaps it is not the toxins inherent to cigarettes that are so damaging to a child (although such toxins certainly provide no known benefit). Rather, it may be that smoking is predictive of poor parenting practices. To be sure, a variety of criminological perspectives highlight the importance of parenting. For example, social learning theorists argue that parents who serve as models of deviance and/or create reinforcement contingencies supportive of such behavior can essentially mold a delinquent child (Akers 1998). Others note that the probability of delinquency increases when parents fail to consistently provide their children with affirming social support (Cullen 1994; Wright and Cullen 2001). Control perspectives, from Hirschi's (1969) original social-bond perspective to Gottfredson and Hirschi's (1990) self-control theory, also rely heavily on the notion of parental efficacy. Although some researchers have highlighted this potential pathway (Gibson et al. 2000), it still remains a hypothesized rather than an empirically confirmed developmental process.

Second, MCS is a known risk for temperamental and conduct problems in childhood. For example, Sadowski and Parish (2005) noted that MCS consistently predicts attention-deficit hyperactivity disorder (ADHD), even when controlling for other important factors, such as maternal socioeconomic status. When combined with the finding that ADHD predicts delinquency (Pratt et al. 2002), mainly through its influence on self-control (Unnever, Cullen, and Pratt 2003), it becomes clear that this is another potential mediating pathway. It is worth noting that these two mediating mechanisms may also act in concert. Given that low self-control is endogenous to ADHD, and that Gottfredson and Hirschi (1990) allowed for variation in temperament, which

makes some youth more vulnerable to ineffective parenting practices, a child born to a mother who engages in MCS may have conduct problems and be subject to poor parenting practices. Thus, future research should also investigate a potential interaction between these two pathways.

It is also worth noting a potential genetic explanation, in which the connection between MCS and ADHD in offspring may reflect inherited biological predispositions. For example, recent research suggests that ADHD is highly heritable, with some estimates claiming that up to 80 percent of the variation in the disorder is genetic or biological (see the discussion by Pratt et al. 2002). Thus, the link between MCS and ADHD may potentially reflect mothers who have ADHD and engage in excessive smoking as self-medication. Given the recent focus on genetics with regard to self-control (Wright and Beaver 2005), it would also be wise to consider the role of heredity as another pathway through which the criminogenic effects of MCS may operate.

In the end, though the scope of the present study was modest, our results highlight some important issues. In particular, it should be recognized that the purported link between MCS and LCP offending has been embedded within the larger assumption that damage to a developing fetus is what starts this developmental pathway. The findings presented here, however, question this assumption and instead indicate that our attention should shift elsewhere, perhaps to parenting and/or self-control. Indeed, it seems that the criminogenic risk of MCS, which was supposedly "owned" by one theory, may belong under the heading of another.

▧ Notes

1. This association disappeared in the female subsample with controls, but Gibson et al. argued that this may have been due to the small sample size, because the odds ratio for the women was actually larger than that for the men.

2. Gibson and Tibbetts (2000) did not simply specify a direct relationship. They also included an interaction term with MCS and the absence of a father or husband in the household to specify the neuropsychological deficit and disadvantage environment interaction of which Moffitt wrote.

3. This subsample was defined as those individuals who had at least one official police contact by the age of 18 years.

4. *Police contact* refers to whether a juvenile had contact with police that resulted in either an official arrest or a remedial disposition. This measure of juvenile offending has also been used in the Philadelphia (see Wolfgang et al. 1972; Tracy, Wolfgang, and Figlio 1990) and Racine (see Shannon 1991) cohort studies. More generally, police contacts are positively and moderately correlated with both self-reported offending estimates and other official records of offending such as arrests and convictions (Hindelang, Hirschi, and Weis 1981). Furthermore, Smith and Gartin (1989) noted that "among the domain of official measures, police contacts provide a closer approximation of the true level of offending than arrests or convictions" (p. 102).

5. The criminal history data included offenses committed and officially processed in Philadelphia only. As was the case in another Philadelphia birth cohort study, Tracy and Kempf-Leonard (1996) knew that the sample members were residents of Philadelphia through age 17 but had to assume that the same was true for the adult period as well. In addition, data for women may have been compromised by the fact that women may have married and changed their surnames. Although the original names and social security numbers of the sample subjects were known in the present study, several women changed both. Thus, the analysis may have underestimated the criminal offending of female sample members.

6. Recall that Moffitt et al. (1994) found neuropsychological scores to be related to police, self-report, and court records in very similar ways.

7. As a practical matter, the city of Philadelphia, following the Pennsylvania Crime Code, expunges all arrests from the criminal history database for all of its citizens if the arrests do not lead to convictions within three years; therefore, the use of arrest data would also be limited in the sense that a true arrest for an offense may have in fact occurred but was deleted from the criminal history database because of the three-year rule.

8. Our use of the CAT as a proxy for neuropsychological deficit, although not necessarily ideal (relative to a more direct measure), is consistent with Moffitt's (1990) view that measures such as the CAT serve as adequate proxies for neuropsychological deficit. Our interest in this research, therefore, was to be as consistent as possible to Moffitt's work, because we were assessing the relative merits of her explanation of the relationship between MCS and crime or delinquency.

9. It is true that measures of executive functioning would also capture neuropsychological deficit. Even so, Moffitt (1990) argued strongly that verbal IQ is a valid measure of neuropsychological deficit.

10. Because LBW is endogenous to MCS, some may view its inclusion in the models as unnecessary. The two variables did not, however, evidence multicollinearity, and failing to include LBW could have overestimated the impact of MCS. Thus, we erred on the conservative side and retained this measure. Additional analyses illustrated that removing LBW did not substantively alter the findings.

11. There were no indications of multicollinearity among the independent variables.

12. When we reestimated this model with the entire sample $(N = 987)$, MCS still predicted neuropsychological deficit. It also was a significant predictor when using the WISC measure rather than the CAT.

13. Given the limited regressor space provided by the relatively small sample size $(n = 220)$ with fully specified models, we should note that the significance levels for the MCS variable (the category of 20 or more cigarettes per day) were consistently at $p = .064$, below the .10 cutoff but above the "industry standard" of .05. Because p-level cutoff points are both theoretically arbitrary and highly dependent on sample size (Tabachnick and Fidell 2001), we maintain that the significance levels reported here indicate a substantively meaningful relationship in the multivariate models.

◢ References

Akers, Ronald L. 1998. *Social Learning and Social Structure: A General Theory of Crime and Deviance.* Boston: Northeastern University Press.

Andrews, D. A., and James Bonta. 2001. *The Psychology of Criminal Conduct.* 3rd ed. Cincinnati, OH: Anderson.

Brennan, Patricia A., Emily R, Grekin, and Sarnoff Mednick. 1999. "Maternal Cigarette Smoking During Pregnancy and Adult Male Criminal Outcomes." *Archives of General Psychiatry* 56:215–19.

Cornelius, Marie D. and Nancy Day. 2000. "The Effects of Tobacco Use During and After Pregnancy on Exposed Children." *Alcohol Research and Health* 24:242–49.

Cotton, P. 1994. "Smoking Cigarettes May Do Developing Fetus More Harm Than Ingesting Cocaine, Some Experts Say." *JAMA* 271:576–77.

Cullen, Francis T. 1994, "Social Support as an Organizing Concept for Criminology: Presidential Address to the Academy of Criminal Justice Sciences." *Justice Quarterly* 11:527–59.

Denno, Deborah J. 1990. *Biology and Violence.* Cambridge, UK: Cambridge University Press.

Farrington, David P. 1989. "Self-Reported and Official Offending from Adolescence to Adulthood." Pp. 399–423 in *Cross-National Research in Self-Reported Crime and Delinquency,* edited by M. W. Klein. Boston: Kluwer Academic.

Farrington, David P., and Rolf Loeber. 2000. "Some Benefits of Dichotomization in Psychiatric and Criminological Research." *Criminal Behaviour and Mental Health* 10:100–22.

Farrington, David P., Rolf Loeber, Magda Stouthamer-Loeber, W. B. Van Kammen, and L. Schmidt. 1996. "Self-Report Delinquency and a Combined Seriousness Scale Based on Boys, Mothers, and Teachers: Concurrent And Predictive Validity." *Criminology* 34:493–517.

Ge, Xiaojia, M. Brent Donnellan, and Ernst Wenk. 2001. "The Development of Persistent Criminal Offending in Males." *Criminal Justice and Behavior* 28:731–55.

Gibson, Chris L., Alex R. Piquero, and Stephen G. Tibbetts. 2000. "Assessing the Relationship between Maternal Cigarette Smoking During Pregnancy and Age at First Police Contact." *Justice Quarterly* 17:519–41.

Gibson, Chris L., and Stephen G. Tibbetts. 2000. "A Biosocial Interaction in Predicting Early Onset of Offending." *Psychological Reports* 86:509–18.

Gottfredson, Michael R., and Travis Hirschi. 1990. *A General Theory of Crime.* Stanford, CA: Stanford University Press.

Hindelang, Michael J., Travis Hirschi, and Joseph Weis. 1981. *Measuring Delinquency.* Beverly Hills, CA: Sage.

Hirschi, Travis. 1969. *Causes of Delinquency.* Berkeley: University of California Press.

Hirschi, Travis, and Michael J. Hindelang. 1977. "Intelligence and Delinquency: A Revisionist Review." *American Sociological Review* 42:571–87.

Lauritsen, Janet L. 1998. "The Age-Crime Debate: Assessing the Limits of Longitudinal Self-Report Data." *Social Forces* 77:127–55.

Lynam, Donald R., Terrie E. Moffitt, and Magda Stouthamer-Loeber. 1993. "Explaining the Relationship Between IQ and Delinquency: Class, Race, Test Motivation, School Failure or Self-Control?" *Journal of Abnormal Psychology* 102:187–96.

McGloin, Jean M., and Travis C. Pratt. 2003. "Cognitive Ability and Delinquent Behavior among Inner-City Youth: A Life-Course Analysis of Main, Mediating, and Interaction Effects." *International Journal of Offender Therapy and Comparative Criminology* 47:253–71.

Moffitt, Terrie E. 1990. "The Neuropsychology of Juvenile Delinquency: A Critical Review." Pp. 99–170 in *Crime and Justice: A Review of the Research, Vol. 12,* edited by M. Tonry and N. Morris. Chicago: University of Chicago Press.

———. 1993. "Adolescence-Limited and Life-Course-Persistent Antisocial Behavior: A Developmental Taxonomy." *Psychological Review* 100:674–701.

———. 1997. "Neuropsychology, Antisocial Behavior, and Neighborhood Context." Pp. 116–70 in *Violence,* edited by J. McCord. New York: Cambridge University Press.

Moffitt, Terrie E., Donald Lynam, and Phil A. Silva. 1994. "Neuropsychological Tests Predicting Persistent Male Delinquency." *Criminology* 32:277–300.

Myrianthopoulos, Ntinos C., and K. S. French, 1968. "An Application of the U.S. Bureau of the Census Socioeconomic Index to a Large, Diversified Patient Population." *Social Science Medicine* 2:283–99.

Nagin, Daniel S., David P. Farrington, and Terrie E. Moffitt. 1995. "Life-Course Trajectories of Different Types of Offenders." *Criminology* 33:111–39.

Nagin, Daniel, Daniel P. Farrington, and Greg Pogarsky. 1997. "Adolescent Mothers and the Criminal Behavior of Their Children." *Law and Society Review* 31:137–62.

Piquero, Alex R. 2000. "Frequency, Specialization, and Violence in Offending Careers." *Journal of Research in Crime and Delinquency* 37:392–418.

———. 2001, "Testing Moffitt's Neuropsychological Variation Hypothesis for the Prediction of Life-Course Persistent Offending." *Psychology, Crime and Law* 7:193–215.

Piquero, Alex R., Chris L. Gibson, Stephen G. Tibbetts, Michael G. Turner, and Solomon H. Katz. 2002. "Maternal Cigarette Smoking During Pregnancy and Life Course Persistent Offending." *International Journal of Offender Therapy and Comparative Criminology* 46:231–48.

Piquero, Alex R., Raymond Paternoster, Paul Mazerolle, Robert Brame, and Charles W, Dean. 1999. "Onset Age and Offense Specialization." *Journal of Research in Crime and Delinquency* 36:275–99.

Piquero, Alex R., and Norman A White. 2003. "On the Relationship between Cognitive Abilities and Life Course-Persistent Offending among a Sample of African Americans: A Longitudinal Test of Moffitt's Hypothesis." *Journal of Criminal Justice* 31:399–409.

Pratt, Travis C., and Francis T. Cullen. 2000. "The Empirical Status of Gottfredson and Hirschi's General Theory of Crime: A Meta-Analysis." *Criminology* 38:931–64.

Pratt, Travis C., Francis T. Cullen, Kristie R. Blevins, Leah Daigle, and James D. Unnever. 2002. "The Relationship of Attention Deficit Hyperactivity Disorder to Crime and Delinquency: A Meta-Analysis." *International Journal of Police Science and Management* 4:344–60.

Raine, Adrian. 1993. *The Psychopathology of Crime.* San Diego, CA: Academic Press.

Rasanen, Pirkko, Helina Hakko. Matti Isohanni, Sheilagh Hodgins, Marjo-Ritta Jarvelin, and Jari Tihonen. 1999. "Maternal Smoking during Pregnancy and Risk of Criminal Behavior among Adult Male Offspring in a Northern Finland 1966 Cohort." *American Journal of Psychiatry* 156:857–62.

Sadowski, Kelly, and Thomas G. Parish. 2005. "Maternal Smoking Contributes to the Development of Childhood ADHD." *Internet Journal of Allied Health Sciences and Practice* 3(1). Available at http://ijahsp.nova.edu/articles/vol3num1/sadowski .htm

Shannon, Lyle W. 1991. *Changing Patterns of Delinquency and Crime: A Longitudinal Study in Racine.* Boulder, CO: Westview.

Smith, Douglas A., and Patrick R. Gartin. 1989. "Specifying Specific Deterrence: The Influence of Arrest on Future Criminal Activity." *American Sociological Review* 54:94–105.

Tabachnick, Barbara G., and Linda S. Fidell. 2001. *Using Multivariate Statistics.* 4th ed. Boston: Allyn &. Bacon.

Tibbetts, Stephen G., and Alex R. Piquero, 1999. "The Influence of Gender, Low Birth Weight, and Disadvantaged Environment in Predicting Early Onset of Offending: A Test of Moffitt's Inter-actional Hypothesis." *Criminology* 37:843–77.

Tiegs, Ernest W., and Willis W. Clark. 1970. *Examiner's Manual and Test Coordinator's Handbook: California Achievement Texts.* New York: McGraw-Hill.

Tracy, Paul E., and Kimberley Kempf-Leonard. 1996. *Continuity and Discontinuity in Criminal Careers.* New York: Plenum.

Tracy, Paul E., Marvin E. Wolfgang, and Robert M. Figlio. 1990. *Delinquency Careen in Two Birth Cohorts.* New York: Plenum.

Unnever, James D., Francis T. Cullen, and Travis C. Pratt. 2003. "Parental Management, ADHD, and Delinquent Involvement: Reassessing Gottfredson and Hirschi's General Theory." *Justice Quarterly* 20:471–500.

Wakschlag, Lauren S., Kate E. Pickett, Edwin Cook, Neal L. Benowitz, and Bennett Leventhal. 2002. "Maternal Smoking during Pregnancy and Severe Antisocial Behavior on Offspring: A Review." *American Journal of Public Health* 92:966–74.

White, Helene R., Marsha E, Bates, and Steven Buyske. 2001. "Adolescence-Limited versus Persistent Delinquency: Extending Moffitt's Hypothesis into Adulthood." *Journal of Abnormal Psychology* 110:600–09.

Wolfgang, Marvin E., Robert Figlio, and Thorstin Sellin. 1972. *Delinquency in a Birth Cohort.* Chicago: University of Chicago Press.

Wright. John P., and Kevin M. Beaver. 2005. "Do Parents Really Matter in Creating Self-Control in Their Children? A Genetically Informed Test of Gottfredson and Hirschi's Theory of Low Self-Control." *Criminology* 43:1169–1202.

Wright. John Paul, and Francis T. Cullen. 2001. "Parental Efficacy and Delinquent Behavior: Do Control and Support Matter?" *Criminology* 39:677–705.

REVIEW QUESTIONS

1. In terms of Moffitt's theoretical framework, name and briefly describe the two distinct offending trajectories, as well as what causes each of them.

2. How did McGloin, Piquero, and Pratt measure neuropsychological deficit? Do you agree that this measure is a valid measure of this concept? Why?

3. One of the major findings of this study is inconsistent with Moffitt's theoretical predictions. What finding is this? What potential explanations do the authors present for this finding?

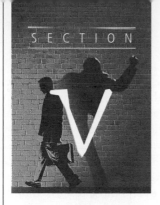

EARLY SOCIAL STRUCTURE AND STRAIN THEORIES OF CRIME

The introduction for this section will review the development of anomie/strain theory, starting with early social structure theorists, such as Durkheim, then Merton, and on to the most modern versions of strain theory (e.g., general strain theory). We will also examine the empirical research findings on this perspective, which remains one of the dominant theoretical explanations of criminal behavior today. We will finish discussing the policy implications of this research.

⬛ Introduction

This introduction reviews explanations of criminal conduct that emphasize the differences among the varying groups in societies, particularly in the United States. Such differences are easy to see in everyday life, and many theoretical models place the blame for crime on observed inequalities and/or cultural differences between groups. In contrast to the theories presented in previous sections, social structure theories disregard any biological or psychological variations across individuals. Instead, social structure theories assume that crime is caused by the way societies are structurally organized.

These social structure theories vary in many ways, most notably in what they propose as the primary constructs and processes in causing criminal activity. For example, some structural models place an emphasis on variations in economic or academic success, whereas others focus on differences in cultural norms and values. Still others concentrate on the actual breakdown of the social structure in certain neighborhoods and the resulting social disorganization that occurs from this process, a topic we will reserve for Section VI. Regardless of their differences, all of the theories examined in this section emphasize a common theme: Certain groups of individuals are more likely to break the law because of disadvantages or cultural differences resulting from the way society is structured.

The most important distinction between these theories and those discussed in previous sections is that they emphasize group differences instead of individual differences. Structural models tend to focus on the macro level of analysis, as opposed to the micro level. Therefore, it is not surprising that social structure theories are commonly used to explain the propensity of certain racial/ethnic groups for committing crime, as well as the overrepresentation of the lower class in criminal activity.

As you will see, these theoretical frameworks were presented as early as the 1800s and reached prominence in the early to mid-1900s, when the political, cultural, and economic climate of society was most conducive to such explanations. Although social structural models of crime have diminished in popularity in recent decades,[1] there is much validity to their propositions in numerous applications to contemporary society.

⬚ Early Theories of Social Structure: Early to Mid-1800s

Most criminological and sociological historians trace the origin of social structure theories to the research done in the early to mid-1800s by a number of European researchers, with the most important including Auguste Comte, Andre-Michel Guerry, and Adolphe Quetelet.[2] It is important to understand why structural theories developed in 19th-century Europe. The Industrial Revolution, defined by most historians as beginning in the mid-1700s and ending in the mid-1800s, was in full swing at the turn of the century and continued throughout most of the 1800s, so societies were quickly transitioning from primarily agriculturally based economies to industrial-based economies. This transition inevitably brought people from rural farmlands to dense urban cities, with a resulting enormous increase in social problems. These social problems ranged from failure to properly dispose of waste and garbage, to constantly losing children and not being able to find them, to much higher rates of crime (which urban areas continue to show today, as compared to suburban and rural areas).

The problems associated with such fast urbanization, as well as the shift in economics, led to a drastic change in basic social structures in Europe, as well as the United States. At the same time, other types of revolutions were also having an effect. Both the American (1776) and French Revolutions (1789) occurred in the last quarter of the 18th century. These two revolutions, inspired by the Enlightenment movement (see Section I), shared an ideology that rejected tyranny and insisted that people should have a voice in how they were

---

[1]Anthony Walsh and Lee Ellis, "Political Ideology and American Criminologists' Explanations for Criminal Behavior," *The Criminologist* 24 (1999): 1, 4; Lee Ellis and Anthony Walsh, "Criminologists' Opinions about Causes and Theories of Crime and Delinquency," *The Criminologist* 24 (1999): 1–4.

[2]Much of the discussion of the development of structural theories of the 19th century is drawn from James W. Vander Zanden, *Sociology: The Core*, 2nd ed. (New York: McGraw-Hill, 1990), 8–14.

governed. Along with the Industrial Revolution, these political revolutions affected intellectual theorizing on social structures, as well as their impact on crime, throughout the 1800s.

Auguste Comte

One of the first important theorists in the area of social structure theory was Auguste Comte (1798–1857), who is widely credited with coining the term *sociology*.[3] Comte distinguished the concepts of social statics and social dynamics. **Social statics** are aspects of society that relate to stability and social order; they allow societies to continue and endure. **Social dynamics** are aspects of social life that alter how societies are structured and pattern the development of societal institutions. Although such conceptualization seems elementary by today's standards, it had a significant influence on sociological thinking at the time. Furthermore, the distinction between static and dynamic societal factors was incorporated in several criminological theories in decades to come.

Between 1851 and 1854, Comte published a four-volume work entitled *A System of Positive Polity* that encouraged the use of scientific methods to observe and measure societal factors.[4] Although we tend to take this for granted in modern times, the idea of applying such methods to help explain social processes was rather profound at the time; probably for this reason, he is generally considered the founder or father of sociology. Comte's work set the stage for the positivistic perspective, which emphasized social determinism and rejected the notion of free will and individual choice that was common up until that time.

Andre-Michel Guerry and Adolphe Quetelet

After the first modern national crime statistics were published in France in the early 1800s, a French lawyer named Andre-Michel Guerry (1802–1866) published a report that examined these statistics and concluded that property crimes were higher in wealthy areas, but violent crime was much higher in poor areas.[5] Some experts have claimed that this report likely represents the first study of scientific criminology;[6] it was later expanded and published as a book. Ultimately, Guerry concluded that the explanation was opportunity: The wealthy had more to steal, and that is the primary cause of property crime. Interestingly, this conclusion is supported by recent U.S. Department of Justice statistics, which show that compared to lower class households, property crime is just as common—and maybe more—in middle- to upper-class households, but violent crime is not. [7]As Guerry stated centuries ago, there is more to steal in the wealthier areas, and poor individuals take the opportunity to steal such goods and currency from those households/establishments.

---

[3]Vander Zanden, *The Core*, 8–9.

[4]Auguste Comte, *A System of Positive Polity*, trans. J. H. Bridges (New York: Franklin, 1875).

[5]For more thorough discussions of Guerry and Quetelet, see the sources from which I have drawn the information presented here: Piers Beirne, *Inventing Criminology* (Albany: SUNY Press, 1993); George B. Vold, Thomas J. Bernard, and Jeffrey B. Snipes, *Theoretical Criminology*, 5th ed. (New York: Oxford University Press, 2002).

[6]Terrence Morris, *The Criminal Area* (New York: Routledge, 1957), 42–53, as cited in Vold et al., *Theoretical Criminology*, 22.

[7]U.S. Department of Justice, Bureau of Justice Statistics, *Sourcebook of Criminal Justice Statistics, 2000*, NCJ-190251 (Washington, DC: USGPO, 2001), Table 3.26, p. 202 ; U.S. Department of Justice, Bureau of Justice Statistics, *Sourcebook of Criminal Justice Statistics, 2000*, NCJ-190251 (Washington, DC: USGPO, 2001), Table 3.13, p. 194.

Adolphe Quetelet (1796–1874) was a Belgian researcher who, like Guerry, examined French statistics in the mid-1800s. Besides showing relative stability in the trends of crime rates in France, such as in age distribution and female-to-male ratios of offending, Quetelet also showed that certain types of individuals were more likely to commit crime.[8] Specifically, young, male, poor, uneducated, and unemployed individuals were more likely to commit crime than their counterparts,[9] a finding also supported by modern research. Like Guerry, Quetelet concluded that opportunities had a lot to do with where crime was concentrated, in addition to the demographic characteristics.

However, Quetelet added a special component: Greater inequality or gaps between wealth and poverty in the same place tends to excite temptations and passions, he said. This is a concept referred to as **relative deprivation**, a quite distinct condition from simple poverty.

For example, a number of deprived areas in the United States do not have high rates of crime, likely because virtually everyone is poor, so people are generally content with their lives relative to their neighbors. However, in areas of the country where very poor people live in close proximity to very wealthy people, this causes animosity and feelings of being deprived. Studies have supported this hypothesis.[10] It may well explain why Washington, D.C., perhaps the most powerful city in the world and one with many neighborhoods that are severely rundown and poor, has a higher crime rate than any other jurisdiction in the country.[11] Modern studies have also showed a clear linear association between higher crime rates and relative deprivation. For example, David Sang-Yoon Lee found that crime rates were far higher in cities that had wider gaps in income: the larger the gap between the 10th and 90th percentiles, the greater the crime levels.[12]

In addition to the concept of relative deprivation, Quetelet also showed that areas with the most rapidly changing economic conditions also showed high crime rates. He is perhaps best known for his comment, "The crimes . . . committed seem to be a necessary result of our social organization. . . . Society prepares the crime, and the guilty are only the instruments by which it is executed."[13] This statement makes it clear that crime is a result of societal structure, and not the result of individual propensities or personal decision making. Thus, it is not surprising that Quetelet's position was controversial at the time in which he wrote, and he was rigorously attacked by critics for removing all decision-making capabilities from his

---

[8]Beirne, *Inventing Criminology*, 78–81.

[9]Vold et al., *Theoretical Criminology*, 23–26.

[10]Velmer Burton and Frank Cullen, "The Empirical Status of Strain Theory," *Journal of Crime and Justice* 15 (1992): 1–30; Nikos Passas, "Continuities in the Anomie Tradition," in *Advances in Criminological Theory*, Vol. 6: *The Legacy of Anomie Theory*, ed. Freda Adler and William S. Laufer (New Brunswick: Transaction, 1995); Nikos Passas, "Anomie, Reference Groups, and Relative Deprivation," in *The Future of Anomie Theory*, ed. Nikos Passas and Robert Agnew (Boston, MA: Northeastern University Press, 1997).

[11]U.S. Department of Justice, Bureau of Justice Statistics, *Sourcebook of Criminal Justice Statistics, 2000*, NCJ-190251 (Washington, DC: USGPO, 2001), Table 3.124, p. 290.

[12]David Sang-Yoon Lee, "An Empirical Investigation of the Economic Incentives for Criminal Behavior," B.A. thesis in economics (Boston, MA: Harvard University, 1993), as cited in Richard B. Freeman, "The Labor Market," in *Crime*, ed. James Q. Wilson and Joan Petersilia (San Francisco, CA: ICS Press, 1995), 171–92.

[13]Beirne, *Inventing Criminology*, 88, as cited in Vold et al., *Theoretical Criminology*, 25.

model of behavior. In response, Quetelet argued that his model could help lower crime rates by leading to social reforms that address the inequalities due to the social structure.[14]

One of the essential points of the Guerry's and Quetelet's work is the positivistic nature of their conclusions: that the distribution of crime is not random but rather the result of certain types of people committing certain types of crimes in particular places, largely due to the way society is structured and the way it distributes resources. This perspective of criminality strongly supports the tendency of crime to be clustered in certain neighborhoods, as well as among certain people. Such findings support a structural, positivistic perspective of criminality, in which criminality is seen as being deterministic and, thus, caused by factors outside of an individual's control. In some ways, early structural theories were a response to the failure of the classical approach to crime control. As the 19th century drew to a close, classical and deterrence-based perspectives of crime fell out of favor while social structure theories and other positivist theories of crime, such as the structural models by Guerry and Quetelet, attracted far more attention.

Durkheim and the Concept of Anomie

Although influenced by earlier theorists (e.g., Comte, Guerry and Quetelet), Emile Durkheim (1858–1916) was perhaps the most influential theorist in the state of modern structural perspectives on criminality.[15] Like most other social theorists of the 19th century, he was strongly affected by the American and French revolutions and the Industrial Revolution. In his doctoral dissertation (1893) at the University of Paris, the first sociological dissertation at that institution, Durkheim developed a general model of societal development largely based on the economic/labor distribution, in which societies are seen as evolving from a simplistic mechanical society toward an multilayered organic society (see Figure 5.1).

Figure 5.1	Durkheim's Continuum of Development From Mechanical to Organic Societies

Mechanical Societies	**Organic Societies**
⟶ Industrialization	⟶
Primitive	Modern
Rural	Urban
Agricultural-based economy	Industrial-based economy
Simple division of labor (few divisions)	Complex division of labor (many specialized divisions)
Law used to enforce conformity	Law used to regulate interactions among divisions
Typically stronger collective conscience	Typically weaker collective conscience

---

[14]Vold et al., *Theoretical Criminology*, 25–26.

[15]Much of this discussion of Durkheim is taken from Vold et al., *Theoretical Criminology*, Chapter 6, as well as Vander Zanden, *The Core*, 11–13.

As outlined in this dissertation, entitled *The Division of Labor in Society*, in primitive **mechanical societies,** all members essentially perform the same functions, such as hunting (typically males) and gathering (typically females). Although there are a few anomalies (e.g., medicine man), virtually everyone essentially experiences the same daily routine. Such similarities in work, as well as constant interaction with like members of the society, leads to a strong uniformity in values, which Durkheim called the **collective conscience**. The collective conscience is the degree to which individuals of a society think alike, or as Durkheim put it, the totality of social likenesses. The similar norms and values among people in these primitive mechanical societies creates "mechanical solidarity," a very simple-layered social structure with a very strong collective conscience. In mechanical societies, law functions to enforce the conformity of the group.

However, as societies progress toward more modern, **organic societies** in the industrial age, the distribution of labor becomes more highly specified. An "organic solidarity" arises in which people tend to depend on other groups because of the highly specified division of labor, and laws have the primary function of regulating the interactions and maintaining solidarity among the groups.

For example, modern researchers at universities in the United States tend to be trained in extremely narrow topics—one might be an expert on something as specific as the antennae of certain species of ants. On the other hand, some individuals are still gathering trash from the cans on the same streets every single day. The antennae experts probably have little interaction with and not much in common with the garbage collectors, other than to pay them. According to Durkheim, to move from the universally shared roles of mechanical societies to the extremely specific roles of organic societal organization results in huge cultural differences and giant contrasts in normative values and attitudes across groups. Thus, the collective conscience in such societies is weak, largely because there is little agreement on moral beliefs or opinions. The preexisting solidarity among members breaks down, and the bonds are weakened, which creates a climate for antisocial behavior.

Durkheim was clear in stating that crime is not only normal, but necessary in all societies. As a result, his theory is often considered a good representation of **structural functionalism**. He claimed that all social behaviors, especially crime, provided essential functions in a society. Durkheim thought crime served several functions. First, it defines the moral boundaries of societies. Few people know or realize what is against the societal laws until they see someone punished for a violation. This reinforces their understanding of both what the rules are and what it means to break the rules. Furthermore, the identification of rule-breakers creates a bond among the other members of the society, perhaps through a common sense of self-righteousness or superiority.

In later works, Durkheim said the resultant bonding is what makes crime so necessary in a society. Given a community that has no law violators, society will change the legal definitions of what constitutes a crime to define some of its members as criminals, he thought. Examples of this are prevalent, but perhaps the most convincing is that of the Salem witch trials, in which hundreds of individuals were accused and tried for an almost laughable offense, and more than a dozen were executed. Durkheim would say this was inevitable because crime was so low in the Massachusetts Bay Colony, as historical records confirm, that society had to come up with a fabricated criterion for defining certain members of the population as offenders.

Other examples are common in everyday life. The fastest way to have a group of strangers bond is to give them a common enemy, which often means forming into cliques and ganging up on others in the group. In a group of three or more college roommates, for example, two or more of the individuals will quickly join together and complain about the others. This is an inevitable phenomenon of human interaction and group dynamics, which has always existed throughout the world across time and place. As Durkheim said, even in "a society of saints. . . . Crimes . . . will there be unknown; but faults which appear venial to the layman will create there the same scandal that the ordinary offense does in ordinary consciousnesses . . . this society . . . will define these acts as criminal and will treat them as such."[16]

Law enforcement should always be cautious in "cracking down" on gangs, especially during relatively inactive periods, because it may make the gang stronger. Like all societal groups, when a common enemy appears, gang members—even those who do not typically get along—will come together and "circle the wagons" to protect themselves via strength in numbers. This very powerful bonding effect is one that many sociologists, and especially gang researchers, have consistently observed.[17]

Traditional, mostly mechanical societies could count on relative consensus about moral values and norms, and this sometimes led to too much control and a stagnation of creative thought. Durkheim thought that progress in society typically depends on deviating from established moral boundaries, especially if the society is in the mechanical stage. The many examples include virtually all religious icons. Jesus, Buddha, and Mohammed were persecuted as criminals for deviating from societal norms in the time they preached. Political heroes have also been prosecuted and incarcerated as criminals, such as Gandhi in India, Mandela in South Africa, and Dr. King in the United States. In one of the most compelling cases, scientist and astronomer Galileo proposed a theory that Earth was not the center of the universe. Even though he was right, he was persecuted for his belief in a society that strictly adhered to its beliefs many centuries ago. Durkheim was clearly accurate in saying that the normative structure in some societies is so strong it hinders progress and that crime is the price societies pay for progress.

In contrast to mechanical societies, modern societies do not have such extreme restraint against deviations from established norms. Rather, almost the opposite is true; there are too many differences across groups because the division of labor is highly specialized. Thus, the varying roles in society, such as farmers versus scientific researchers, lead to extreme differences in the cultural values and norms of the various groups. There is a breakdown in the collective conscience because there is really no longer a "collective" nature in society. Therefore, law focuses not on defining the norms of society but on governing the interactions that take place among the different classes. According to Durkheim, law provides a service in regulating such interactions as societies progress toward more organic (more industrial) forms.

Durkheim emphasized that human beings, unlike other animal species who live according to their spontaneous needs, have no internal mechanism to signal when their

---

[16]Emile Durkheim, *The Rules of the Sociological Method*, trans. Sarah A. Solovay and John H. Mueller, ed. G. E. G. Catlin (New York: Free Press, 1965), as cited in Vold et al., *Theoretical Criminology*.

[17]See discussion in Malcolm Klein, "Street Gang Cycles," in *Crime*, ed. James Q. Wilson and Joan Petersilia (San Francisco, CA: ICS Press, 1995), 217–36.

needs and desires are satiated. Therefore, the selfish desires of humankind are limitless; the more an individual has, the more he or she wants. People are greedy by nature, and without something to tell them what they need and deserve, they will never feel content.[18] According to Durkheim, society provides the mechanism for limiting this insatiable appetite, having the sole power to create laws that set limits.

Durkheim also noted that in times of rapid change, society fails in this role of regulating desires and expectations. This rapid change can be due to numerous factors, such as war or social movements (like the changes seen in the United States in the 1960s). The transitions Durkheim likely had in mind when he wrote were the American and French revolutions and the Industrial Revolution. When society's ability to serve as a regulatory mechanism breaks down, the selfish, greedy tendencies of individuals are uncontrolled, causing a state Durkheim called **anomie**, or "normlessness." Societies in such anomic states experience increases in many social problems, particularly criminal activity.

Durkheim was clear that whether the rapid change was for good or bad, it would have negative effects on society. For example, whether the U.S. economy was improving, as it did during the late 1960s, or quickly tanking, as it did during the Depression of the 1930s, criminal activity would increase due to the lack of stability in regulating human expectations and desires. Interestingly, these two periods of time experienced the greatest crime waves of the 20th century, particularly for murder.[19]

Another fact that supports Durkheim's predictions is that middle- and upper-class individuals have higher suicide rates than those from lower classes. This is consistent with the idea that it is better to have stability, even if it means always being poor, than it is to have instability at higher levels of income. In his best known work, *Suicide*, Durkheim took an act that would seem to be the ultimate form of free choice or free will and showed that the decision to take one's own life was largely determined by external social factors. He argued that suicide was a "social fact," a product of meanings and structural aspects that result from interactions among people.

Durkheim showed that the rate of suicide was significantly lower among individuals who were married, younger, and practiced religions that were more interactive and communal (e.g., Jewish). All of these characteristics boil down to one aspect: The more social interaction and bonding with the community, the less the suicide. So Durkheim concluded that variations in suicide rates are due to differences in social solidarity or bonding to society. Examples of this are still seen today, as in recent reports of high rates of suicide among people who live in remote areas, like Alaska (which has the highest rate of juvenile suicide), Wyoming, Montana, and the northern portions of Nevada. Another way of looking at the implications of Durkheim's conclusions is that social relationships are what makes people happy and fulfilled. If they are isolated or have weak bonds to society, they will likely be depressed and not content with their lives.

Another reason that Durkheim's examination of suicide was important was that he showed that suicide rates increased in times of rapid economic growth or rapid decline.

---

[18]For more details on these issues, see Emile Durkheim's works: *The Division of Labor in Society* (New York: Free Press, 1893/1965); *Suicide* (New York: Free Press, 1897/1951).

[19]U.S. Department of Justice, Bureau of Justice Statistics, *Violent Crime in the United States*, (Washington, DC: Bureau of Justice Statistics, 1996), as illustrated in Joseph P. Senna and Larry G. Siegel, *Introduction to Criminal Justice*, 8th ed. (Belmont, CA: West/Wadsworth, 1999).

Although researchers later argued that crime rates did not always follow this pattern,[20] Durkheim used quantified measures to test his propositions as the positivistic approach recommended. At the least, Durkheim created a prototype of how theory and empirical research could be combined in testing differences across social groups. This theoretical framework was drawn on heavily for one of the most influential and accepted criminological theories of the 20th century: strain theory.

⌧ Strain Theories

All forms of **strain theory** share an emphasis on frustration as a factor in crime causation, hence the name "strain" theories. Although the theories differ regarding what exactly is causing the frustration—and the way individuals cope (or don't cope) with stress and anger— they all hold that strain is the primary causal factor in the development of criminality. Strain theories all trace their origin to Robert K. Merton's seminal theoretical framework.

Merton's Strain Theory

Working in the 1930s, Merton drew heavily on Durkheim's idea of anomie in developing his own theory of structural strain.[21] Although Merton altered the definition of anomie, Durkheim's theoretical framework was a vital influence in the evolution of strain theory. Combining Durkheimian concepts and propositions with an emphasis on American culture, Merton's structural model became one of the most popular perspectives in criminological thought in the early 1900s and remains one of the most cited theories of crime in criminological literature.

Cultural Context and Assumptions of Strain Theory

Some have claimed that Merton's seminal piece in 1938 was the most influential theoretical formulation in criminological literature and one of the most frequently cited papers in sociology.[22] Although its popularity is partially due to its strong foundation in previous structural theories, Merton's strain theory also benefited from the timing of its

---

[20]William J. Chambliss, "Functional and Conflict Theories of Crime," in *Whose Law? What Order?* ed. William J. Chambliss and Milton Mankoff (New York: John Wiley, 1976), 11–16.

[21]For reviews of Merton's theory, see both the original and more recent works by Merton himself: Robert K. Merton, "Social Structure and Anomie," *American Sociological Review* 3 (1938): 672–82; Robert K. Merton, *Social Theory and Social Structure* (New York: Free Press, 1968); Robert K. Merton, "Opportunity Structure: The Emergence, Diffusion, and Differentiation as Sociological Concept, 1930s-1950s," in *Advances in Criminological Theory: The Legacy of Anomie Theory,* ed. Freda Adler and William Laufer (New Brunswick, NJ: Transaction Press, 1995), Vol. 6. For reviews by others, see the following: Ronald L. Akers and Christine S. Sellers, *Criminological Theories: Introduction, Evaluation, and Application,* 4th ed. (Los Angeles, CA: Roxbury, 2004), 164–68; Stephen E. Brown, Finn-Aage Esbensen, and Gilbert Geis, *Criminology: Explaining Crime and Its Context,* 5th ed. (Cincinnati, OH: Anderson, 2004), 297–307; Marshall B. Clinard, "The Theoretical Implications of Anomie and Deviant Behavior," in *Anomie and Deviant Behavior,* ed. M. B. Clinard (New York: Free Press, 1964), 1–56; Thomas J. Bernard, "Testing Structural Strain Theories," *Journal of Research in Crime and Delinquency* 24 (1987): 262–80.

[22]Clinard, "Theoretical Implications"; also see discussion in Brown et al., *Criminology,* 297.

publication. Virtually every theory discussed in this book became popular because it was well suited to the political and social climate, fitting current perspectives of how the world works. Perhaps no other theory better represents this phenomenon than strain theory.

Most historians would agree that, in the United States, the most significant social issue of the 1930s was the economy. The influence of the Great Depression, largely a result of a stock market crash in 1929, affected virtually every aspect of life in the United States. Unemployment and extreme poverty soared, along with suicide rates and crime rates, particularly murder rates.[23] American society was fertile ground for a theory of crime that placed virtually all of the blame on the U.S. economic structure.

On the other side of coin, Merton was highly influenced by what he saw happening to the country during the Depression: how much the economic institution impacted almost all other social factors, particularly crime. He watched how the breakdown of the economic structure drove people to kill themselves or others, not to mention the rise in property crimes, such as theft. Many once-successful individuals were now poor, and some felt driven to crime for survival. Notably, Durkheim's hypotheses regarding crime and suicide were supported during this time of rapid change, and Merton apparently realized that the framework simply had to be brought up-to-date and supplemented.

One of the key assumptions that distinguishes strain theory from Durkheim's perspective is that Merton altered his version of what anomie means. Merton focused on the nearly universal socialization of the "American Dream" in U.S. society, the idea that as long as people work very hard and pay their dues, they will achieve their goals in the end. According to Merton, the socialized image of the goal is material wealth, whereas the socialized concept of the means of achieving the goal is hard work (e.g., education, labor). The conventional model of the American Dream was consistent with the Protestant work ethic, which called for working hard for a long time knowing that you will be paid off in the distant future.

Furthermore, Merton thought that nearly everyone was socialized to believe in the American Dream, no matter what the economic class of their childhood. There is some empirical support for this belief, and it makes sense. Virtually all parents, even if they are poor, want to give their children hope for the future if they are willing to work hard in school and at a job. Parents and society usually use celebrities as examples of this process: individuals who started off poor and rose to become wealthy. Modern examples include former Secretary of State Colin Powell, owner of basketball's Dallas Mavericks Mark Cuban, Oscar winner Hillary Swank, and Hollywood director/screenwriter Quentin Tarantino, not to mention Arnold Schwarzenegger's amazing rise from teenage immigrant to Mr. Olympia and California governor.

These stories epitomize the American Dream, but parents and society do not also teach the reality of the situation. As Merton points out, a small percentage of people rise from the lower class to become materially successful, but the vast majority of poor children don't have much chance of ever obtaining such wealth. This near-universal socialization of the American Dream—which turns out not to be true for most people—causes most of the strain and frustration in American society, Merton said. Furthermore, he thought that most of the strain and frustration was due not to the failure to achieve conventional goals (i.e., wealth) but rather to the differential emphasis placed on material goals and the de-emphasis on the importance of conventional means.

---

[23]Joseph J. Senna and Larry J. Siegel, *Introduction to Criminal Justice,* 8th ed. (Belmont, CA: Wadsworth, 1999), 44.

Merton's Concept of Anomie and Strain. Merton claimed that in an ideal society there would be an equal emphasis on the conventional goals and means. However, in many societies, one is emphasized more than the other. Merton thought the United States epitomized the type of society that emphasized the goals far more than the means. According to Merton, the disequilibrium in emphasis between the goals or means of societies is what he called anomie. So like Durkheim, Merton's anomie was a negative state for society; however, the two men had very different ideas of how this state of society was caused. Whereas Durkheim believed that anomie was primarily caused by a society transitioning too fast to maintain its regulatory control over its members, for Merton, anomie represented too much focus on the goals of wealth in the United States, at the expense of the conventional means.

Some hypothetical situations will illustrate. Which of the following two men would be more respected by youth (or even adults) in our society: (1) John, who has his PhD in physics and lives in a one-bedroom apartment because his job as a postdoctoral student pays $25,000 a year; or (2) Joe, who is a relatively successful drug dealer and owns a four-bedroom home, drives a Hummer, dropped out of school in the 10th grade, and makes about $90,000 a year? In years of asking such a question to our classes, the answer is usually Joe, the drug dealer. After all, he appears to have obtained the American Dream, and little emphasis is placed on how he achieved it.

Still another way of supporting Merton's idea that America is too focused on the goal of material success is to ask you, the reader, to think about why you are taking the time to read this section and/or to attend college. Specifically, the question for you is: If you knew that studying this book—or earning a college degree—would not lead to a better employment position, would you read it anyway just to increase your own knowledge? In over a decade of putting this question to about 10,000 students in various university courses, one of the authors of this book has found that only about 5% (usually fewer) of respondents said "yes." Interestingly, when asked why they put all of this work into attending classes, many of them said it was for the partying or social life. Ultimately, it seems that most college students would not be engaging in the hard work it takes to educate themselves if it weren't for some payoff at the end of the task. In some ways, this supports Merton's claim that there is an emphasis on the goals, with little or no intrinsic value being placed on the work itself (i.e., means). This phenomenon is not meant to be a disheartening or a negative statement; it is meant only to exhibit the reality of American culture. It is quite common in our society to place an emphasis on the goal of financial success, as opposed to hard work or education.

Merton thought that individuals, particularly those in the lower class, eventually realize that the ideal of the American Dream is a lie, or at least a false illusion for the vast majority. This revelation will likely take place when people are in their late-teenage to mid-twenties years, and according to Merton, this is when the frustration or strain is evident. That is consistent with the age-crime peak of offending at the approximate age of 17. Learning that hard work won't necessarily provide rewards, some individuals begin to innovate ways that they can achieve material success without the conventional means of getting it. Obviously, this is often through criminal activity. However, not all individuals deal with strain in this way; most people who are poor do not resort to crime. To Merton's credit, he explained that individuals deal with the limited economic structure of society in different ways. He referred to these variations as adaptations to strain.

Adaptations to Strain. There are five **adaptations to strain** according to Merton. The first of these is **conformity**, in which people buy into the conventional goals of society and the conventional means of working hard in school or labor.[24] Like conformists, most readers of this book would probably like to achieve material success and are willing to use conventional means of achieving success through educational efforts and doing a good job at work. Another adaptation to strain is **ritualism.** Ritualists do not pursue the goal of material success, probably because they know they don't have a realistic chance of obtaining it. However, they do buy into the conventional means, in the sense that they like to do their job or are happy with just making ends meet through their current position. For example, studies have shown that some of the most contented and happy people in society are those who don't hope to become rich and are quite content with their blue-collar job and often have a strong sense of pride in their work, even if it is sometimes quite menial. To these people, work is a type of ritual, performed without a goal in mind; rather, it is a form of intrinsic goal, in and of itself. Ultimately, conformists and ritualists tend to be low risks for offending, which is in contrast to the other adaptations to strain.

The other three adaptations to strain are far more likely to engage in criminal offending: **innovation, retreatism,** and **rebellion.** Perhaps most likely to become predatory street criminals are the innovators, who Merton claimed greatly desire the conventional goals of material success but are not willing to engage in conventional means. Obviously, drug dealers and professional auto thieves, as well as many other variations of chronic property criminals (e.g., bank robbers, etc.) would fit this adaptation. They are innovating ways to achieve the goals without the hard work that is usually required. However, innovators are not always criminals. In fact, many of them are the most respected individuals in our society. For example, some entrepreneurs have used the capitalistic system of our society to produce useful products and services (e.g., the men who designed Google for the Internet) and have made a fortune at a very young age without advanced college education or years of work at a company. Other examples are successful athletes who sign multimillion-dollar contracts at age 18. So it should be clear that not all innovators are criminals.

The fourth adaptation to strain is retreatism. Such individuals do not seek to achieve the goals of society, and they also do not buy into the idea of conventional hard work. There are many varieties of this adaptation: people who become homeless by choice or who isolate themselves in desolate places without human contact. A good example of a retreatist is Ted Kaczinsky, the Unabomber, who left a good position as a professor at University of California at Berkeley to live in an isolated cabin in Montana, which had no running water or electricity; he did not interact with humans for many months at a time. Other types of retreatists, perhaps the most likely to be criminal, are those that are heavy drug users who actively disengage from social life and try to escape via psychologically altering drugs. All of these forms of retreatists seek to drop out of society altogether, thus not buying into the means or goals of society.

Finally, the last adaptation to strain according to Merton is rebellion, which is the most complex of the five adaptations. Interestingly, rebels buy into the idea of societal goals and means, but they do not buy into those currently in place. Most true rebels are criminals by definition, largely because they are trying to overthrow the current societal structure. For example, the founders of the United States were all rebels because they actively fought the governing state—English rule—and clearly committed treason in the process. Had they

---

[24]Merton, *Social Theory.*

lost or been caught during the American Revolution, they would have been executed as the criminals they were by law. However, because they won the war, they became heroes and presidents. Another example is Karl Marx. He bought into goals and means of society, just not those of capitalistic societies. Rather, he proposed socialism/communism as a means to the goal of utopia. So there are many contexts in which rebels can be criminals, but sometimes they end up being heroes.

Merton also noted that one individual can represent more than one adaptation. Perhaps the best example is that of the Unabomber, who obviously started out as a conformist, in that he was a respected professor at University of California, Berkeley, who was well on his way to tenure and promotion. He then seemed to shift to a retreatist, in that he isolated himself from society. Later, he became a rebel who bombed innocent people in his quest to implement his own goals and means, which he described in his Manifesto and which he coerced several national newspapers to publish. This subsequently resulted in his apprehension when his brother read it and informed authorities he thought his brother was the author.

Finally, some have applied an athletic analogy to these adaptations.[25] Assuming a basketball game is taking place, conformists will play to win, but they will always play by the rules and won't cheat. Ritualists will play the game just because they like to play; they don't care about winning. Innovators will play to win, and they will break any rules they can to triumph in the game. Retreatists don't like to play and obviously don't care about winning. Finally, rebels will not like the rules on the court, so they will try to steal the ball and go play by their own rules on another court. Although this is a somewhat simplistic analogy, it is likely to help readers remember the adaptations and perhaps enable them to apply these ways of dealing with strain to everyday situations, such as resorting to criminal activity.

Evidence and Criticisms of Merton's Strain Theory. Although Merton's framework, which emphasized the importance of the economic structure, appeared to have a high degree of face validity during the Great Depression, many scientific studies showed mixed support for strain theory. Although research that examined the effects of poverty on violence and official rates of various crimes has found relatively consistent support (albeit weaker effects than strain theory implies), a series of studies of self-reported delinquent behavior found little or no relationship between social class and criminality.[26] Furthermore, the idea that unemployment drives people to commit crime has received little support.[27]

---

[25]Vold et al., *Theoretical Criminology*, 140.

[26]For examples and reviews of this research, see F. Ivan Nye, *Family Relationships and Delinquent Behavior* (New York: Wiley, 1958); Ronald L. Akers, "Socio-economic Status and Delinquent Behavior: A Retest," *Journal of Research in Crime and Delinquency* 1 (1964): 38–46; Charles R. Tittle and Wayne J. Villemez, "Social Class and Criminality," *Social Forces* 56 (1977): 474–503; Charles R. Tittle, Wayne J. Villemez, and Douglas A. Smith, "The Myth of Social Class and Criminality: An Empirical Assessment of the Empirical Evidence," *American Sociological Review* 43 (1978): 643–56; Michael J. Hindelang, Travis Hirschi, and Joseph C. Weis, "Correlates of Delinquency: The Illusion of Discrepancy Between Self-Report and Official Measures," *American Sociological Review* 44 (1979): 995–1014; Michael J. Hindelang, *Measuring Delinquency* (Beverly Hills, CA: Sage, 1980); Terence P. Thornberry and Margaret Farnworth, "Social Correlates of Criminal Involvement," *American Sociological Review* 47 (1982): 505–517; Gregory R. Dunaway, Francis T. Cullen, Velmer S. Burton, and T. David Evans, "The Myth of Social Class and Crime Revisited: An Examination of Class and Adult Criminality," *Criminology* 38 (2000): 589–632; for one of the most thorough reviews, see Charles R. Tittle and Robert F. Meier, "Specifying the SES/Delinquency Relationship," *Criminology* 28 (1990): 271–99.

[27]Gary Kleck and Ted Chiricos, "Unemployment and Property Crime: A Target-Specific Assessment of Opportunity and Motivation as Mediating Factors," *Criminology* 40 (2000): 649–80.

On the other hand, some experts have argued that Merton's strain theory is primarily a structural model of crime that is more a theory of societal groups, not individual motivations.[28] Therefore, some modern studies have used aggregated group rates (i.e., macro-level measures) to test the effects of deprivation, as opposed to using individual (micro-level) rates of inequality and crime. Most of these studies provide some support for the hypothesis that social groups and regions with higher rates of deprivation and inequality have higher rates of criminal activity.[29] In sum, there appears to be some support for Merton's strain theory when the level of analysis is the macro-level, and official measures are being used to indicate criminality.

However, many critics have claimed that these studies do not directly measure perceptions or feelings of strain, so they are only indirect examinations of Merton's theory. In light of these criticisms, some researchers have focused on the disparity in what individuals aspire to in various aspects of life (e.g., school, occupation, social life) versus what they realistically expect to achieve.[30] The rationale of these studies is that if an individual has high aspirations (i.e., goals) but low expectations of actually achieving the goals due to structural barriers, then that individual is more likely to experience feelings of frustration and strain. Furthermore, it was predicted that the larger the gap between aspirations and expectations, the stronger the sense of strain. Of the studies that examined discrepancies between aspirations and expectations, most did not find evidence linking a large gap between these two levels with criminal activity. In fact, several studies found that for most antisocial respondents, there was virtually no gap between aspirations and expectations. Rather, most of the subjects (typically young males) who reported highest levels of criminal activity tended to report low levels of both aspirations and expectations.

Surprisingly, when aspirations were high, it seemed to inhibit offending, even when expectations to achieve those goals were low. One interpretation of these findings is that individuals who have high goals will not jeopardize their chances even if those are slim. On the other hand, individuals who don't have high goals are likely to be indifferent to their future and, in a sense, have nothing to lose. So without a stake in conventional society, this predisposes them to crime. While this conclusion supports social control theories, it does not provide support for strain theory.

Some critics have argued that most studies on the discrepancies between aspirations and expectations have not been done correctly. For example, Farnworth and Leiber

---

[28]Bernard, "Testing Theories"; Steven F. Messner, "Merton's 'Social Structure and Anomie': The Road Not Taken," *Deviant Behavior* 9 (1988): 33–53; also see discussion in Burton and Cullen, "Empirical Status."

[29]For a review of these studies, see Kenneth C. Land, Patricia L. McCall, and Lawrence E. Cohen, "Structural Covariates of Homicide Rates: Are There Any Invariances Across Time and Social Space?" *American Journal of Sociology* 95 (1990): 922–63.

[30]For examples and reviews of these types of studies, see Travis Hirschi, *Causes of Delinquency* (Berkeley, CA: University of California Press, 1969); Allen E. Liska, "Aspirations, Expectations, and Delinquency: Stress and Additive Models," *Sociological Quarterly* 12 (1971): 99–107; Margaret Farnworth and Michael J. Leiber, "Strain Theory Revisited: Economic Goals, Educational Means, and Delinquency," *American Sociological Review* 54 (1989): 263–74; Burton and Cullen, "Empirical Status"; Velmer S. Burton, Francis T. Cullen, T. David Evans, and R. Gregory Dunaway, "Reconsidering Strain Theory: Operationalization, Rival Theories, and Adult Criminality," *Journal of Quantitative Criminology* 10 (1994): 213–39; Robert F. Agnew, Francis T. Cullen, Velmer S. Burton, T. David Evans, and R. Gregory Dunaway, "A New Test of Classic Strain Theory," *Justice Quarterly* 13 (1996): 681–704; also see discussion of this issue in Akers and Sellers, *Criminological Theories*, 173–75.

claimed that it was a mistake to examine differences between educational goals and expectations, or differences between occupational goals and expectations, which is what most of these studies did.[31] Rather, they proposed testing the gap between economic aspirations (i.e., goals) and educational expectations (i.e., means of achieving the goals). This makes sense, and Farnworth and Leiber found support for a gap between these two factors and criminality. However, they also found that people who reported low economic aspirations were more likely to be delinquent, which supports the previous studies that they criticized. Another criticism of this type of strain theory studies is that simply reporting a gap between expectations and aspirations may not mean that the individuals actually feel strain; rather, researchers have simply, and perhaps wrongfully, assumed that a gap between the two measures indicates feelings of frustration.[32]

Other criticisms of Merton's strain theory include some historical evidence and its failure to explain the age-crime curve. Regarding the historical evidence, it is hard to understand why some of the largest increases in crime took place during a period of relative economic prosperity, namely the late 1960s. Crime increased more than ever before (in times of recorded measures) between 1965 and 1973, which were generally good economic years in the United States. Therefore, if strain theory is presented as the primary explanation for criminal activity, it would probably have a hard time explaining this historical era. On the other hand, given the growth in the economy in the 1960s and early 1970s, this may have caused more disparity between rich and poor, thereby producing more relative deprivation.

The other major criticism of strain theory is that it does not explain one of the most established facts in the field: the age-crime curve. In virtually every society in the world, across time and place, predatory street crimes (e.g., robbery, rape, murder, burglary, larceny, etc.), tend to peak sharply in the teenage years to early twenties and then drop off very quickly, certainly before age 30. However, most studies show that feelings of stress and frustration tend not to follow this pattern. For example, suicide rates tend to be just as high or higher as one gets older, with middle-aged and elderly people having much higher rates of suicide than those in their teens or early twenties.

On the other hand, it can be argued that crime rates go down even though strain can continue or even increase with age because individuals develop coping mechanisms for dealing with the frustrations they feel. But even if this is true regarding criminal behavior, apparently this doesn't seem to prevent suicidal tendencies. General strain theory emphasized this concept. However, before we cover general strain theory, we will discuss two other variations of Merton's theory that were both developed between 1955 and 1960 to explain gang formation and behavior using a structural strain framework.

Variations of Merton's Strain Theory: Cohen and Cloward and Ohlin

Cohen's Theory of Lower-Class Status Frustration and Gang Formation. In 1955, Albert Cohen presented a theory of gang formation that used Merton's strain theory as a basis for why individuals resort to such group behavior.[33] In Cohen's model, young lower class

---

[31]Farnworth and Leiber, "Strain Theory."

[32]Agnew et al., "A New Test."

[33]Albert Cohen, *Delinquent Boys: The Culture of the Gang* (New York: Free Press, 1955).

males are at a disadvantage in school because they lack the normal interaction, socialization, and discipline instituted by educated parents of the middle-class. This is in line with Merton's original framework of a predisposed disadvantage for underclass youth. According to Cohen, such youths are likely to experience failure in school because they are unprepared to conform with middle-class values and fail to meet the "middle-class measuring rod," which emphasizes motivation, accountability, responsibility, deferred gratification, long-term planning, respect for authority and property, controlling emotions, and so on.

Like Merton, Cohen emphasized the youths' internalization of the American Dream and fair chances for success, so failure to be successful according to this middle-class standard is very frustrating for them. The strain that they feel as a result of failure in school performance and lack of respect among their peers is often referred to as "status frustration." It leads them to develop a system of values opposed to middle-class standards and values. Some have claimed that this represents a Freudian defense mechanism known as **reaction formation**, which involves adopting attitudes or committing behaviors that are opposite of what is expected as a form of defiance and so that they will feel less guilt for not living up to the standards they cannot achieve. Specifically, instead of abiding by middle-class norms of obedience to authority, school achievement, and respect for authority, these youth change their normative beliefs to value the opposite characteristics: malicious, negativistic, and nonutilitarian delinquent activity.

Delinquent youths will begin to value destruction of property and skipping school, not because these behaviors lead to a payoff or success in the conventional world, but simply because they defy the conventional order. In other words, they turn the middle-class values upside down and consider activity that violates the conventional norms and laws as good, thereby psychologically and physically rejecting the cultural system that has been imposed on them without preparation and fair distribution of resources. Furthermore, Cohen claimed that while these behaviors do not appear to have much utility or value, they are quite valuable and important from the perspective of the strained youth. Specifically, they do these acts to gain respect from their peers.

Cohen stated that he believed that this tendency to reject middle-class values was the primary cause of gangs, a classic example of "birds of a feather flock together." Not all lower-class males resort to crime and join a gang in response to this structural disadvantage. Other variations, beyond that of the **delinquent boy** which is described above, include the **college boy** and the **corner boy.** The "college boy" responds to his disadvantaged situation by dedicating himself to overcoming the odds and competing in the middle-class schools despite the unlikely chances for success. The "corner boy" responds to the situation by accepting his place as a lower-class individual who will somewhat passively make the best of life at the bottom of the social order.

As compared to Merton's original adaptations, Cohen's delinquent boy is probably best seen as similar to rebellion because he rejects the means and goals (middle-class values and success in school) of conventional society, substituting new means and goals (negativistic behaviors and peer respect in the gang). Some would argue that delinquent boys should be seen as innovators because their goals are ultimately the same: peer respect; however, the actual peers involved completely change, so we argue that through the reaction formation process, the delinquent boy actually creates his own goals and means that go against the conventional, middle-class goals and means. Regarding the college boy, the adaptation that seems to fit the best is that of conformity, because the college boy continues to believe in the conventional goals (i.e., financial success/achievement) and means (i.e., hard work via education/labor) of middle-class society. Finally, the corner boy probably best fits the adaptation of ritualism, because he knows that he likely will never achieve the goals of society but resigns himself to not obtaining financial success; at the same time, he does not resort to predatory street crime, but rather holds a stable blue-collar job or makes ends meet in other typically legal ways. Some corner boys end up simply collecting welfare and give up working altogether; they may actually become more like the adaptation of retreatism because they have virtually given up on the conventional means (hard work) of society, as well as the goals.

At the time that Cohen developed his theory, official statistics showed that virtually all gang violence, and most violence for that matter, was concentrated among lower-class male youth. However, with the development of self-report studies in the 1960s, Cohen's theory was shown to be somewhat overstated: Middle-class youth were well represented among those who commit delinquent acts.[34] Other studies have also been critical of Cohen's theory, particularly the portions that deal with his proposition that crime rates would increase after youths drop out of school and join a gang. Although the findings are mixed, many studies have found that delinquency is often higher before the youth drops out of school and may actually decline once they drop out and become employed.[35] Some critics have pointed out that such findings discredit Cohen's theory, but this is not necessarily true. After all, delinquency may be peaking right before the youth drops out because they feel most frustrated and strained, whereas delinquency may be decreasing after they drop out because some are raising their self-esteem by earning a wage and taking pride in having a job.

Still, studies have clearly shown that lower-class youth are far more likely to have problems in school and that school failure is consistently linked to criminality.[36] Furthermore, there is little dispute that much of delinquency represents malicious, negativistic, and nonutilitarian activity. For example, what do individuals have to gain from destroying mail boxes or spraying graffiti on walls? These acts will never gain much in money or any payoff other than peer respect. So, ultimately, it appears that there is some face validity to what Cohen proposed, in the sense that some youth engage in behavior that has no other value than earning peer respect, even though that behavior is negativistic and nonutilitarian according to the values of conventional society. Regardless of some criticisms of Cohen's model, he provided an important structural strain theory of the development of gangs and lower-class delinquency.

---

[34]Tittle et al., "Myth of Class"; Hindelang et al., "Correlates of Delinquency."

[35]See Bernard, "Testing Theories"; Merton, "Opportunity Structure"; Delbert Elliott and Harwin Voss, *Delinquency and Dropout* (Lexington, MA: D. C. Health, 1974); Terence P. Thornberry, Melanie Moore, and R. L. Christenson, "The Effect of Dropping Out of High School on Subsequent Criminal Behavior," *Criminology* 23 (1985): 3–18; G. Roger Jarjoura, "Does Dropping Out of School Enhance Delinquent Involvement? Results from a Large-Scale National Probability Sample," *Criminology* 31 (1993): 149–72; G. Roger Jarjoura, "The Conditional Effect of Social Class on the Dropout Delinquency Relationship," *Journal of Research in Crime and Delinquency* 33 (1996): 232–55; see discussion in Donald J. Shoemaker, *Theories of Delinquency*, 5th ed. (New York: Oxford University Press, 2005).

[36]Alexander Liazos, "School, Alienation, and Delinquency," *Crime and Delinquency* 24 (1978): 355–70; Joseph W. Rogers and G. Larry Mays, *Juvenile Delinquency and Juvenile Justice* (New York: Wiley, 1987); Clarence E. Tygart, "Strain Theory and Public School Vandalism: Academic Tracking, School Social Status, and Students' Academic Achievement," *Youth and Society* 20 (1988): 106–18; Hirschi, *Causes of Delinquency*.

Cloward and Ohlin's Theory of Differential Opportunity. Five years after Cohen published his theory, Cloward and Ohlin presented yet another structural strain theory of gang formation and behavior.[37] Like Merton and Cohen, Cloward and Ohlin assumed in their model that all youth, including those in the lower class, are socialized to believe in the American Dream, and when individuals realize that they are blocked from conventional opportunities, they become frustrated and strained. What distinguishes Cloward and Ohlin's theory from that of the previous strain theories is that they identified three different types of gangs based on the characteristics of the neighborhood's social structure. They thought the nature of gangs varied according to the availability of illegal opportunities in the social structure. So whereas previous strain theories focused only on lack of legal opportunities, Cloward and Ohlin's model emphasized *both legal and illegal* opportunities; the availability or lack of these opportunities largely determined what type of gang would form in that neighborhood, hence the name differential opportunity theory. Furthermore, the authors acknowledged Edwin Sutherland's influence on their theory (see Section VII), and it is evident in their focus on the associations made in the neighborhood. According to differential opportunity theory, the three types of gangs that form are criminal gangs, conflict gangs, and retreatist gangs.

Criminal gangs form in lower-class neighborhoods that have an organized structure of adult criminal behavior. Such neighborhoods are so organized and stable that the criminal networks are often known and accepted by the conventional individuals in the area. In these neighborhoods, the adult gangsters mentor the youth and take them under their wing. This can pay off for the adult criminals, too, because youth can often be used to do the "dirty work" for the criminal enterprises in the neighborhood without risk of serious punishment if they are caught. The successful adult offenders supply the youth with the motives and techniques for committing crime. So while members of criminal gangs were blocked from legal opportunities, they are offered ample opportunities in the illegal realm.

Criminal gangs tend to reflect the strong organization and stability of such neighborhoods. Therefore, criminal gangs primarily commit property or economic crimes, with the goal of making a profit through illegal behavior. These crimes can range from "running numbers" as local bookies to "fencing" stolen goods to running businesses that are a front for vice crimes (e.g., prostitution, drug trading). All of these businesses involve making a profit illegally, and there is often a system or structure in which the criminal activity takes place. Furthermore, these criminal gangs are most like the Merton adaptation of innovation because the members still want to achieve the goals of conventional society (financial success). Because of their strong organizational structure, these gangs favor members who have self-control and are good at planning over individuals who are highly impulsive or uncontrolled.

Examples of criminal gangs are seen in movies depicting highly organized neighborhoods often consisting of primarily one ethnicity, such as *The Godfather, The Godfather II, A Bronx Tale, State of Grace, Sleepers, New Jack City, Clockers, Goodfellas, Better Luck Tomorrow,* and many others that were partially based on real events. All of these depictions

---

[37]Richard A. Cloward and Lloyd E. Ohlin, *Delinquency and Opportunity: A Theory of Delinquent Gangs* (New York: The Free Press, 1960). For discussion and theoretical critique of the model, see Bernard, "Testing Theories," and Merton, "Opportunity Structure."

involve a highly structured hierarchy of a criminal enterprise, which is largely a manifestation of the organization of the neighborhood. The Hollywood motion pictures also involve stories about older criminals taking younger males from the neighborhood under their wing and training them in the ways of the criminal network. Furthermore, virtually all ethnic groups offer examples of this type of gang/neighborhood; the list of movies above includes Italian American, Irish American, African American, and Asian American examples. Thus, criminal gangs can be found across the racial and ethnic spectrum, largely because all groups have certain neighborhoods that exhibit strong organization and stability.

Conflict gangs were another type of gang that Cloward and Ohlin identified. Conflict gangs tend to develop in neighborhoods that have weak stability and little or no organization. In fact, the neighborhood often seems to be in a state of flux with people constantly moving in and out. Because the youth in the neighborhood do not have a solid crime network or adult criminal mentors, they tend to form as a relatively disorganized gang, and they typically lack the skills and knowledge to make a profit through criminal activity. Therefore, the primary illegal activity of conflict gangs is violence, which is used to gain prominence and respect among themselves and the neighborhood. Due to the disorganized nature of the neighborhood, as well as the gang itself, conflict gangs never quite achieve the respect and stability that criminal gangs typically achieve. The members of conflict gangs tend to be more impulsive and lack self-control as compared to members of criminal gangs, largely because there are no adult criminal mentors to control them.

According to Cloward and Ohlin, conflict gangs are blocked from both legitimate and illegitimate opportunities. If applying Merton's adaptations, conflict gangs would probably fit the category of rebellion, largely because none of the other categories fits well, but it can be argued that conflict gangs have rejected the goals and means of conventional society and implemented their own values, which emphasize violence. Examples of motion pictures that depict this type of breakdown in community structure and a mostly violent gang culture are *Menace to Society, Boyz in the Hood, A Clockwork Orange, Colors, The Outsiders,* and others, which all emphasized the chaos and violence that results when neighborhood and family organization is weak.

Finally, if an individual is a "double failure" in both the legitimate and illegitimate worlds, meaning that they can't achieve success in school or status in their local gangs, they join together to form retreatist gangs. **Retreatist gangs** are made up of these individuals who have failed to succeed in the conventional world and also could not achieve status in the criminal or conflict gangs of their neighborhoods. Because members of retreatist gangs are no good at making a profit from crime or using violence to achieve status, the primary form of offending in retreatist gangs is usually drug usage. Like Merton's retreatist adaptation to strain, the members of the retreatist gangs often want simply to escape from reality. Therefore, the primary activity of the gang when they get together is usually just to get high, which is well represented by Hollywood in such movies as *Trainspotting, Drug-Store Cowboy,* and *Panic in Needle Park.* In all of these movies, the only true goal of the gangs was getting stoned to escape from the worlds where they had failed.

There are a number of empirical studies and critiques of Cloward and Ohlin's theory, with the criticisms being similar to those of Merton's strain theory. Specifically, the critics argue that there is little evidence that gaps between what lower-class youth aspire to and what they expect to achieve produce frustration and strain, nor do such gaps appear predictive of gang membership or criminality.[38] Another criticism of Cloward and Ohlin's theory is the inability to find empirical evidence that supports their model of the formation of three types of gangs and their specialization in offending. While some research supports the existence of gangs that appear to specialize in certain forms of offending, many studies find that the observed specialization of gangs is not exactly the way that Cloward and Ohlin proposed.[39] Additional studies have shown that many gangs tend not to specialize, but rather engage in a wider variety of offending behaviors.

Despite the criticisms of Cloward and Ohlin's model of gang formation, their theoretical framework inspired policy largely due to the influence their work had on Attorney General Robert Kennedy, who had read their book. In fact, Kennedy asked Lloyd Ohlin to assist in developing federal policies regarding delinquency,[40] which resulted in the Juvenile Delinquency Prevention and Control Act of 1961. Cloward and Ohlin's theory was a major influence on the Mobilization for Youth project in New York City, which along with the federal legislation, stressed creating education and work opportunities for youth. Although evaluations of this program showed little effect on reducing delinquency, it was impressive that such theorizing about lower-class male youths could have such a large impact on policy interventions.

Ultimately, the variations of strain theory presented by Cohen, as well as Cloward and Ohlin, provided additional revisions that seemed at the time to advance the validity of strain theory. However, most of these revisions were based on official statistics that showed lower-class male youth committed most crime, which was later shown by self-reports to be exaggerated.[41] Once scholars realized that most of the earlier models were not empirically valid for most criminal activity, strain theory became unpopular for several decades. But during the 1980s, another version of strain was devised by Robert Agnew, who rejuvenated the interest in strain theory by devising a way to make the theory more general and applicable to a larger variety of crimes and forms of deviance.

General Strain Theory

In the 1980s, Robert Agnew proposed **general strain theory,** which covers a much larger range of behavior by not concentrating on simply the lower class and which provides a more applicable model for the frustrations that all individuals feel in everyday

---

[38]James F. Short, "Gang Delinquency and Anomie," in *Anomie and Deviant Behavior*, ed. Marshall B. Clinard (New York: Free Press, 1964), 98–127; James F. Short, Ramon Rivera, and Ray A. Tennyson, "Perceived Opportunities, Gang Membership, and Delinquency," *American Sociological Review* 30 (1965): 56–67; James F. Short and Fred L. Strodtbeck, *Group Processes and Gang Delinquency* (Chicago: University of Chicago Press, 1965); Liska, "Aspirations"; Hirschi, *Causes of Delinquency;* see discussion in Shoemaker, *Theories of Delinquency*, 121–30.

[39]Irving Spergel, *Racketville, Slumtown, and Haulburg* (Chicago: University of Chicago Press, 1964); Short and Strodtbeck, *Group Processes*; Paul E. Tracy, Marvin E. Wolfgang, and Robert M. Figlio, *Delinquency Careers in Two Birth Cohorts* (New York: Plenum, 1990).

[40]Lamar T. Empey, *American Delinquency,* 4th ed. (Homewood, IL: Dorsey, 1991).

[41]Tittle et al., "Myth of Class"; Hindelang et al., "Correlates of Delinquency."

life.[42] Unlike other strain theories, general strain theory does not rely on assumptions about the frustration arising when people realize that the American Dream is a false promise to those of the lower classes. Rather, this theoretical framework assumes that people of all social classes and economic position deal with frustrations in routine daily life.

Previous strain theories, such as the models proposed by Merton, Cohen, and Cloward and Ohlin, focused on individuals' *failure to achieve positively valued goals* that they had been socialized to work to obtain. General strain theory also focuses on this source of strain; however, it identifies two additional categories of strain: *presentation of noxious stimuli* and *removal of positively valued stimuli*. In addition to the failure to achieve one's goals, Agnew claimed that the presentation of noxious stimuli (i.e., bad things) in one's life could cause major stress and frustration. Examples of noxious stimuli would include things like an abusive parent, a critical teacher, or an overly demanding boss. These are just some of the many negative factors that can exist in one's life—the number of examples is endless.

Figure 5.2	Model of General Strain Theory

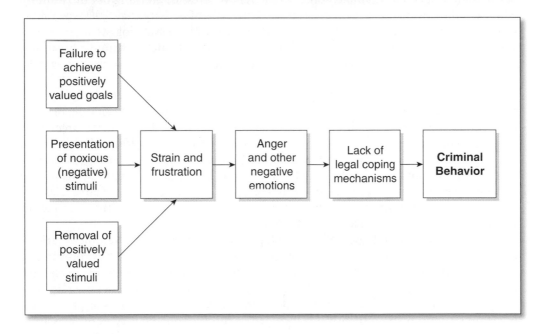

---

[42]For Agnew's works regarding this theory, see Robert Agnew, "A Revised Strain Theory of Delinquency," *Social Forces* 64 (1985): 151–67; Robert Agnew, "Foundation for a General Strain Theory of Crime and Delinquency," *Criminology* 30 (1992): 47–88; Robert Agnew and Helene Raskin White, "An Empirical Test of General Strain Theory," *Criminology* 30 (1992): 475–500; Robert Agnew, "Controlling Delinquency: Recommendations from General Strain Theory," in *Crime and Public Policy: Putting Theory to Work*, ed. Hugh Barlow (Boulder, CO: Westview Press, 1995), 43–70; Robert Agnew, Francis Cullen, Velmer Burton, T. David Evans, and R. Gregory Dunaway, "A New Test of General Strain Theory," *Justice Quarterly* 13 (1996): 681–704; Robert Agnew, "Building on the Foundation of General Strain Theory: Specifying the Types of Strain Most Likely to Lead to Crime and Delinquency," *Journal of Research in Crime and Delinquency* 38 (2001): 319–61.

The other strain category Agnew identified was the removal of positive stimuli, which is likely the largest cause of frustration. Examples of removal of positively valued stimuli include the loss of a good job, loss of the use of a car for a period of time, or the loss of a loved one(s). As with the other two sources of strain, a number of examples may have varying degrees of influence, depending on the individual. One person may not feel much frustration in losing a job or divorcing a spouse, whereas another person may experience severe anxiety or depression from such events.

Ultimately, general strain theory proposes that these three categories of strain (failure to achieve goals, noxious stimuli, and removal of positive stimuli) will lead to stress and that this results in a propensity to feel *anger*. Anger can be seen as a primary mediating factor in the causal model of the gender strain framework. It is predicted that to the extent that the three sources of strain cause feelings of anger in an individual, he or she will be predisposed to commit crime and deviance. However, Agnew was clear in stating that if an individual can somehow cope with this anger in a positive way, then such feelings do not necessarily have to result in criminal activity. These coping mechanisms vary widely across individuals, with different strategies working for some people better than others. For example, some people relieve stress by working out or running, whereas others watch television or a movie. One type of activity that has shown relatively consistent success in relieving stress is laughter, which psychologists are now prescribing as a release of stress. Another is yoga, which largely includes simple breathing techniques, such as taking several deep breaths, which are physiological shown to enhance release of stress.

Although he did not originally provide details on how coping mechanisms work or explore the extant psychological research on these strategies, Agnew specifically pointed to such mechanisms in dealing with anger in prosocial ways. The primary prediction regarding coping mechanisms is that individuals who find ways to deal with their stress and anger in a positive way will no longer be predisposed to commit crime, whereas individuals who do not find a healthy, positive outlet for their anger and frustrations will be far more likely to commit crime. Obviously, the goal is to reduce the use of antisocial and negative coping to strain, such as drug usage or aggression, which are either criminal in themselves or increase the likelihood of offending.

Recent research and theoretical development has more fully examined various coping mechanisms and their effectiveness in reducing anger and, thus, preventing criminal activity. Obviously, in focusing on individuals' perceptions of stress and anger, as well as their personal abilities to cope with such feelings, general strain theory places more emphasis on the micro level of analysis. Still, due to its origins deriving from structural strain theory, it is included in this section and is typically classified as belonging to the category of strain theories that includes the earlier, more macro-level oriented theories. In addition, recent studies and revisions of the theory have attempted to examine the validity of general strain theory propositions at the macro, structural-level.[43]

---

[43]Timothy Brezina, Alex Piquero, and Paul Mazerolle, "Student Anger and Aggressive Behavior in School: An Initial Test of Angew's Macro-Level Strain Theory," *Journal of Research in Crime and Delinquency* 38 (2001): 362–86.

Since general strain theory was first proposed in the mid-1980s, there has been a vast amount of research examining various aspects of the general strain theory.[44] For the most part, studies have generally supported the model. Most studies find a link between the three categories of strain and higher rates of criminality, as well as a link between the sources of strain and feelings of anger or other negative emotions (e.g., anxiety, depression).[45] However, the theory, and especially the way it has been tested, have also been criticized.

It is important for strain research to measure subjects' perceptions and feelings of frustration, not simply the occurrence of certain events themselves. Unfortunately, some studies have only looked at the latter, and the validity of such findings is questionable.[46] Other studies, however, have directly measured subjective perceptions of frustration, as well as personal feelings of anger.[47]

Such studies have found mixed support for the hypothesis that certain events lead to anger,[48] but less support for the prediction that anger leads to criminality, and this link is particularly weak for nonviolent offending.[49] On the other hand, the most recent studies have found support for the links between strain and anger, as well as anger and criminal

---

[44]Agnew and White, "Empirical Test"; Agnew et al., "A New Test"; Raymond Paternoster and Paul Mazerolle, "General Strain Theory and Delinquency: A Replication and Extension," *Journal of Research in Crime and Delinquency* 31 (1994): 235–63; Timothy Brezina, "Adapting to Strain: An Examination of Delinquent Coping Responses," *Criminology* 34 (1996): 39–60; John P. Hoffman and Alan S. Miller, "A Latent Variable Analysis of General Strain Theory," *Journal of Quantitative Criminology* 14 (1998): 83–110; John P. Hoffman and Felicia Gray Cerbone, "Stressful Life Events and Delinquency Escalation in Early Adolescence," *Criminology* 37 (1999): 343–74; Paul Mazerolle and Alex Piquero, "Linking Exposure to Strain with Anger: An Investigation of Deviant Adaptations," *Journal of Criminal Justice* 26 (1998): 195–211; Paul Mazerolle, "Gender, General Strain, and Delinquency: An Empirical Examination," *Justice Quarterly* 15 (1998): 65–91; John P. Hoffman and S. Susan Su, "Stressful Life Events and Adolescent Substance Use and Depression: Conditional and Gender Differentiated Effects," *Substance Use and Misuse* 33 (1998): 2219–62; Lisa M. Broidy, "A Test of General Strain Theory," *Criminology* 39 (2001): 9–36; Nicole Leeper Piquero and Miriam Sealock, "Generalizing General Strain Theory: An Examination of an Offending Population," *Justice Quarterly* 17 (2000): 449–84; Paul Mazerolle and Jeff Maahs, "General Strain and Delinquency: An Alternative Examination of Conditioning Influences," *Justice Quarterly* 17 (2000): 753–78; Paul Mazerolle, Velmer Burton, Francis Cullen, T. David Evans, and Gary Payne, "Strain, Anger, and Delinquent Adaptations: Specifying General Strain Theory," *Journal of Criminal Justice* (2000): 89–101; Stephen W. Baron and Timothy F. Hartnagel, "Street Youth and Labor Market Strain," *Journal of Criminal Justice* 30 (2002): 519–33; Carter Hay, "Family Strain, Gender, and Delinquency," *Sociological Perspectives* 46 (2003): 107–35; Sung Joon Jang and Byron R. Johnson, "Strain, Negative Emotions, and Deviant Coping Among African Americans: A Test of General Strain Theory," *Journal of Quantitative Criminology* 19 (2003): 79–105; Paul Mazerolle, Alex Piquero, and George E. Capowich, "Examining the Links Between Strain, Situational and Dispositional Anger, and Crime: Further Specifying and Testing General Strain Theory," *Youth and Society* 35 (2003): 131–57; Stephen W. Baron, "General Strain, Street Youth and Crime: A Test of Agnew's Revised Theory," *Criminology* 42 (2004): 457–84.

[45]A recent review of this research can be found in Baron, "General Strain," 457–67.

[46]For examples, see Paternoster and Mazerolle, "General Strain Theory"; Hoffman and Cerbone, "Stressful Life Events"; Hoffman and Su, "Stressful Life Events."

[47]Broidy, "A Test"; Baron, "General Strain."

[48]See Brezina, "Adapting to Strain"; Broidy, "A Test"; Mazerolle and Piquero, "Linking Exposure."

[49]Mazerolle and Piquero, "Linking Exposure"; Piquero and Sealock, "Generalizing Theory"; Mazerolle et al., "Strain, Anger."

behavior, particularly when coping variables are considered.[50] Still, many of the studies that examine the effects of anger use time-stable "trait" measures, as opposed to incident-specific "state" measures that would be more consistent with the situation-specific emphasis of general strain theory.[51] This is similar to the methodological criticism that has been leveled against studies of self-conscious emotions, particularly shame and guilt; when it comes to measuring emotions such as anger and shame, criminologists should choose their measures carefully and make sure the instruments are consistent with the theory they are testing. Thus, future research on general strain theory should employ more effective, subjective measures of straining events and situational states of anger.

Regardless of the criticisms of general strain theory, it is hard to deny its face validity. After all, virtually everyone can relate to reacting differently to similar situations based on what type of day they are having. For example, we all have days in which everything seems to be going great—it's a Friday, you receive accolades at work, and you are looking forward to a nice weekend with your friends or family. If someone says something derogatory to you or cuts you off in traffic, being in such a good mood, you are probably inclined to let it go. On the other hand, we also all have days in which everything seems to be going horribly—it's Monday, you get blamed for mishaps at work, and you have a fight with your spouse or significant other. At this time, if someone yells at you or cuts you off in traffic, you may be more inclined to respond aggressively in some way. Or perhaps more commonly, you will overreact and snap at a loved one or friend when they really didn't do much to deserve it; this is often a form of displacement in which a cumulative buildup of stressors results in taking it out on another individual(s). In many ways, this supports general strain theory, and therefore, it is very prevalent and easy to see in everyday life.

Summary of Strain Theories

The common assumption that is found across all variations of strain theory is that crime is far more common among individuals who are under a great degree of stress and frustration, especially those who can't cope with stress in a positive way. The origin of most variations of strain theory can be traced to Durkheim's and Merton's concept of anomie, which essentially means a state of chaos or normlessness in society due to a breakdown in the ability of societal institutions to regulate human desires, thereby resulting in feelings of strain.

Although different types of strain theories were proposed and gained popularity at various periods of time throughout the 20th century, they all became accepted during eras that were politically and culturally conducive to such perspectives, especially regarding the differences across the strain models. For example, Merton's formulation of strain in the 1930s emphasized the importance of the economic institution, which was developed

---

[50]Baron, "General Strain"; Mazerolle, Piquero, and Capowich, "Examining the Links."

[51]For examples, see Mazerolle et al., "Strain, Anger"; Baron, "General Strain"; see discussion in Mazerolle, Piquero, and Capowich, "Examining the Links"; Akers and Sellers, *Criminological Theories*, 180–82.

and became very popular during the Great Depression. Then in the late-1950s, two strain theories that focused on gang formation were developed, by Cohen and by Cloward and Ohlin; they became popular among politicians and society due to the focus on official statistics suggesting most crime at that time was being committed by lower-class, inner-city male youth, many of whom were gang members. Finally, Agnew developed his general strain model in the mid- to late 1980s, during a period in which a number of general theories of crime were being developed (e.g., Gottfredson and Hirschi's low self-control theory and Sampson and Laub's developmental theory), so such models were popular at that time, particular those that emphasized personality traits (such as anger) and experiences of individuals. So all of the variations of strain, like all of the theories discussed in this book, were manifestations of the period in which they were developed and became widely accepted by academics, politicians and society.

⊠ Policy Implications

Although this chapter dealt with a wide range of theories regarding social structure, the most applicable policy implications are those suggested by the most recent theoretical models of this genre. Thus, we will focus on the key policy factors in the most modern versions of this perspective. The factors that are most vital for policy implications regarding social structure theories are those regarding educational and vocational opportunities and programs that develop healthy coping mechanisms to deal with stress.

Empirical studies have shown that intervention programs are needed for high-risk youths that focus on educational and/or vocational training and opportunities because developing motivation for such endeavors can have a significant impact on reducing their offending rates.[52] Providing individuals with a job, or the preparation for one, is key to building a more stable life, even if it is not a high-paying position. As a result, the individual is less likely to feel stressed or "strained". In modern times, people are lucky to have a stable job, and this must be communicated to our youth, and hopefully, they will find some intrinsic value in the work they do.

Another key area of recommendations from this perspective involves developing healthy coping mechanisms. Everyone deals with stress virtually every day. The key is not avoiding stress or strain because that is inevitable. Rather, the key is to develop healthy, legal ways to cope with such strain. Many programs have been created to train individuals on how to handle such stress without resorting to antisocial behavior. There has been some success in such "anger management" programs, particularly the ones that take a cognitive-behavioral approach and often involve role-playing to teach individuals to think before they act.[53]

---

[52]John P. Wright, Stephen G. Tibbetts, and Leah Daigle, *Criminals in the Making: Criminality Across the Life Course* (Thousand Oaks, CA: Sage, 2008), 204–7.

[53]Patricia Van Voorhis, Michael Braswell, and David Lester, *Correctional Counseling and Rehabilitation*, 3rd ed. (Cincinnati, OH: Anderson, 1997).

⊠ Conclusion

This introduction examined the theories that emphasize the inequitable social structure as the primary cause of crime. We examined early perspectives, which established that societies vary in the extent to which they are stratified and looked at the consequences of inequalities and complexities of such structures. Our examination of strain theories explored theoretical models proposing that individuals and groups who are not offered equal opportunities for achieving success can develop stress and frustration and, in turn, dispositions for committing criminal behavior. Finally, we examined the policy recommendations suggested by this theoretical model, which included the need to provide individuals with educational and job opportunities and help them develop healthy coping mechanisms.

⊠ Section Summary

- ◆ First, we discussed the primary distinction between social structure theories and other types of explanations (e.g., biological or social process theories).
- ◆ We examined the importance of the early sociological positivists, particularly Guerry, Quetelet, and Durkheim, and their contributions to the study of deviance and crime.
- ◆ We explored reasons why strain theory was developed and became popular in its time and discussed the primary assumptions and propositions of Merton's strain theory.
- ◆ We identified, defined, and examined examples of all five adaptations to strain.
- ◆ We discussed the variations of strain theory that were presented by Cohen, Cloward and Ohlin, and Agnew, as well as the empirical support that has been found regarding each.

KEY TERMS

Adaptations to strain	Delinquent boy	Retreatism
Anomie	General strain theory	Retreatist gangs
Collective conscience	Innovation	Ritualism
College boy	Mechanical societies	Social dynamics
Conflict gangs	Organic societies	Social statics
Conformity	Reaction formation	Strain theory
Corner boy	Rebellion	
Criminal gangs	Relative deprivation	

DISCUSSION QUESTIONS

1. How does sociological positivism differ from biological or psychological positivism?

2. Which of the early sociological positivism theorists do you think contributed the most to the evolution of social structure theories of crime? Why? Do you think their ideas still hold up today?

3. Can you think of modern examples of Durkheim's image of mechanical societies? Do you think such societies have more or less crime than modern organic societies?

4. What type of adaptation to strain do you think you fit the best? The least? What adaptation do you think best fits your professor? Your postal delivery worker? Your garbage collector?

5. Do you know school friends who fit Cohen's model of status frustration? What did they do in response to the feelings of strain?

6. How would you describe the neighborhood where you or others you know grew up in terms of Cloward and Ohlin's model of organization/disorganization? Can you relate to the types of gangs they discussed?

WEB RESOURCES

Emile Durkheim

http://www.emile-durkheim.com

Strain Theory

http://www.umsl.edu/~keelr/200/strain.html

http://www.iejs.com/Criminology/anomie_and_strain_theory.htm

http://www.apsu.edu/oconnort/crim/crimtheory11.htm

READING

In this selection, Robert Merton, who is one of the most cited and respected sociologists of the 20th century, explains his theory for why individuals commit crime. Specifically, he asserted that the primary cause of crime was the economy. It plays a large part in determining criminal offending, as well as the many adaptations to strain that Merton proposes. Although strain theory, particularly Merton's framework, emphasizes the macro- or group-level of analysis, the theory also describes five stereotypical ways that individuals deal with such frustrations or strain.

SOURCE: Robert Merton, "Social Structure and Anomie," *American Sociological Review* 3, no. 3 (2002): 672–82. Selection taken from S. Cote, ed., *Criminological Theories: Bridging the Past to the Future* (Thousand Oaks, CA: Sage Publications, 2002), 95–103. (Original publication 1938)

Social Structure and Anomie

Robert K. Merton

There persists a notable tendency in socio-logical theory to attribute the malfunctioning of social structure primarily to those of man's imperious biological drives which are not adequately restrained by social control. In this view, the social order is solely a device for "impulse management" and the "social processing" of tensions. These impulses which break through social control, be it noted, are held to be biologically derived. Nonconformity is assumed to be rooted in original nature. Conformity is by implication the result of a utilitarian calculus or unreasoned conditioning. This point of view, whatever its other deficiencies, clearly begs one question. It provides no basis for determining the non-biological conditions which induce deviations from prescribed patterns of conduct. In this paper, it will be suggested that certain phases of social structure generate the circumstances in which infringement of social codes constitutes a "normal" response.

The conceptual scheme to be outlined is designed to provide a coherent, systematic approach to the study of socio-cultural sources of deviate behavior. Our primary aim lies in discovering how some social structures *exert a definite pressure* upon certain persons in the society to engage in nonconformist rather than conformist conduct. The many ramifications of the scheme cannot all be discussed; the problems mentioned outnumber those explicitly treated.

Among the elements of social and cultural structure, two are important for our purposes. These are analytically separable although they merge imperceptibly in concrete situations. The first consists of culturally defined goals, purposes, and interests. It comprises a frame of aspirational reference. These goals are more or less integrated and involve varying degrees of prestige and sentiment. They constitute a basic, but not the exclusive, component of what Linton aptly has called "designs for group living." Some of these cultural aspirations are related to the original drives of man, but they are not determined by them. The second phase of the social structure defines, regulates, and controls the acceptable modes of achieving these goals. Every social group invariably couples its scale of desired ends with moral or institutional regulation of permissible and required procedures for attaining these ends. These regulatory norms and moral imperatives do not necessarily coincide with technical or efficiency norms. Many procedures which from the standpoint of particular *individuals* would be most efficient in securing desired values, e.g., illicit oil-stock schemes, theft, fraud, are ruled out of the institutional area of permitted conduct. The choice of expedients is limited by the institutional norms.

To say that these two elements, culture goals and institutional norms, operate jointly is not to say that the ranges of alternative behaviors and aims bear some constant relation to one another. The emphasis upon certain goals may vary independently of the degree of emphasis upon institutional means. There may develop a disproportionate, at times, a virtually exclusive, stress upon the value of specific goals, involving relatively slight concern with the institutionally appropriate modes of attaining these goals. The limiting case in this direction is reached when the range of alternative procedures is limited only by technical rather than institutional considerations. Any and all devices which promise attainment of the all important goal would be permitted in this hypothetical polar case. This constitutes one type of cultural malintegration. A second polar type is found in groups where activities originally conceived as instrumental are transmuted into ends in themselves. The original purposes are forgotten and ritualistic adherence to institutionally prescribed conduct becomes virtually obsessive. Stability is largely ensured while change is flouted. The range of alternative behaviors is severely limited. There develops a tradition-bound, sacred society characterized by neophobia. The occupational psychosis of the bureaucrat may be cited as a case in point. Finally, there are the intermediate types of groups where a balance between culture goals and institutional means is maintained. These are the significantly integrated and relatively stable, though changing groups.

An effective equilibrium between the two phases of the social structure is maintained as long as satisfactions accrue to individuals who conform to both constraints, viz., satisfactions from the achievement of the goals and satisfactions emerging directly from the institutionally canalized modes of striving to attain these ends. Success, in such equilibrated cases, is twofold. Success is reckoned in terms of the product and in terms of the process, in terms of the outcome and in terms of activities. Continuing satisfactions must derive from sheer participation in a competitive order as well as from eclipsing one's competitors if the order itself is to be sustained. The occasional sacrifices involved in institutionalized conduct must be compensated by socialized rewards. The distribution of statuses and roles through competition must be so organized that positive incentives for conformity to roles and adherence to status obligations are provided for every position within the distributive order. Aberrant conduct, therefore, may be viewed as a symptom of dissociation between culturally defined aspirations and socially structured means.

Of the types of groups which result from the independent variation of the two phases of the social structure, we shall be primarily concerned with the first, namely, that involving a disproportionate accent on goals. This statement must be recast in a proper perspective. In no group is there an absence of regulatory codes governing conduct, yet groups do vary in the degree to which these folkways, mores, and institutional controls are effectively integrated with the more diffuse goals which are part of the culture matrix. Emotional convictions may cluster about the complex of socially acclaimed ends, meanwhile shifting their support from the culturally defined implementation of these ends. As we shall see, certain aspects of the social structure may generate countermores and antisocial behavior precisely because of differential emphases on goals and regulations. In the extreme case, the latter may be so vitiated by the goal-emphasis that the range of behavior is limited only by considerations of technical expediency. The sole significant question then becomes, which available means is most efficient in netting the socially approved value? The technically most feasible procedure, whether legitimate or not, is preferred to the institutionally prescribed conduct. As this process continues, the integration of the society becomes tenuous and anomie ensues.

Thus, in competitive athletics, when the aim of victory is shorn of its institutional trappings and success in contests becomes construed as "winning the game" rather than "winning through circumscribed modes of activity," a premium is implicitly set upon the use of illegitimate but technically efficient means. The star of the opposing football team is surreptitiously slugged; the wrestler furtively incapacitates his opponent through ingenious but illicit techniques; university alumni covertly subsidize "students" whose talents are largely confined to the athletic field. The emphasis on the goal has so attenuated the satisfactions deriving from sheer participation in the competitive activity that these satisfactions are virtually confined to a successful outcome. Through the same process, tension generated by the desire to win in a poker game is relieved by successfully dealing oneself four aces, or, when the cult of success has become completely dominant, by sagaciously shuffling the cards in a game of solitaire. The faint twinge of uneasiness in the last instance and the surreptitious nature of public delicts indicate clearly that the institutional rules of the game *are known* to those who evade them, but that the emotional supports of these rules are largely vitiated by cultural exaggeration of the success-goal. They are microcosmic images of the social macrocosm.

Of course, this process is not restricted to the realm of sport. The process whereby exaltation of the end generates a *literal demoralization*, i.e., a deinstitutionalization, of the means is one which characterizes many groups in which the two phases of the social structure are not highly integrated. The extreme emphasis upon the accumulation of wealth as a symbol of success in our own society militates against the completely effective control of institutionally regulated modes of acquiring a fortune. Fraud, corruption, vice, crime, in short, the entire catalogue of proscribed behavior, becomes increasingly common when the emphasis on the *culturally induced* success-goal becomes divorced from a coordinated institutional emphasis. This observation is of crucial theoretical importance in examining the doctrine that antisocial behavior most frequently derives from biological drives breaking through the restraints imposed by society. The difference is one between a strictly utilitarian interpretation which conceives man's ends as random and an analysis which finds these ends deriving from the basic values of the culture.

Our analysis can scarcely stop at this juncture. We must turn to other aspects of the social structure if we are to deal with the social genesis of the varying rates and types of deviate behavior characteristic of different societies. Thus far, we have sketched three ideal types of social orders constituted by distinctive patterns of relations between culture ends and means. Turning from these types of *culture patterning*, we find five logically possible, alternative modes of adjustment or adaptation by *individuals* within the culture-bearing society or group. These are schematically presented in Figure 1, where (+) signifies "acceptance," (−) signifies "elimination" and (±) signifies "rejection and substitution of new goals and standards."

Our discussion of the relation between these alternative responses and other phases of the social structure must be prefaced by the observation that persons may shift from one alternative to another as they engage in different social activities. These categories refer to role adjustments in specific situations, not to personality *in toto*. To treat the development of this process in various spheres of conduct would introduce a complexity unmanageable within the confines of this paper. For this reason, we shall be concerned primarily with economic activity in the broad sense, "the production, exchange, distribution and consumption of goods and services" in our competitive society, wherein wealth has taken on a highly symbolic cast. Our task is to search out

Figure 1		
	Culture Goals	**Institutionalized Means**
I. Conformity	+	+
II. Innovation	+	−
III. Ritualism	−	+
IV. Retreatism	−	−
V. Rebellion	±	±

some of the factors which exert pressure upon individuals to engage in certain of these logically possible alternative responses. This choice, as we shall see, is far from random.

In every society, Adaptation I (conformity to both culture goals and means) is the most common and widely diffused. Were this not so, the stability and continuity of the society could not be maintained. The mesh of expectancies which constitutes every social order is sustained by the modal behavior of its members falling within the first category. Conventional role behavior oriented toward the basic values of the group is the rule rather than the exception. It is this fact alone which permits us to speak of a human aggregate as comprising a group or society.

Conversely, Adaptation IV (rejection of goals and means) is the least common. Persons who "adjust" (or maladjust) in this fashion are, strictly speaking, *in* the society but not *of* it. Sociologically, these constitute the true "aliens." Not sharing the common frame of orientation, they can be included within the societal population merely in a fictional sense. In this category are some of the activities of psychotics, psychoneurotics, chronic autists, pariahs, outcasts, vagrants, vagabonds, tramps, chronic drunkards and drug addicts. These have relinquished, in certain spheres of activity, the culturally defined goals, involving complete aim-inhibition in the polar case, and their adjustments are not in accord with institutional norms. This is not to say that in some cases the source of their behavioral adjustments is not in part the very social structure which they have in effect repudiated nor that their very existence within a social area does not constitute a problem for the socialized population.

This mode of "adjustment" occurs, as far as structural sources are concerned, when both the culture goals and institutionalized procedures have been assimilated thoroughly by the individual and imbued with affect and high positive value, but where those institutionalized procedures which promise a measure of successful attainment of the goals are not available to the individual. In such instances, there results a twofold mental conflict insofar as the moral obligation for adopting institutional means conflicts with the pressure to resort to illegitimate means (which may attain the goal) and inasmuch as the individual is shut off from means which are both legitimate *and* effective. The competitive order is maintained, but the frustrated and handicapped individual who cannot cope with this order drops out.

Defeatism, quietism, and resignation are manifested in escape mechanisms which ultimately lead the individual to "escape" from the requirements of the society. It is an expedient which arises from continued failure to attain the goal by legitimate measures and from an inability to adopt the illegitimate route because of internalized prohibitions and institutionalized compulsives, *during which process the supreme value of the success-goal has as yet not been renounced.* The conflict is resolved by eliminating *both* precipitating elements, the goals and means. The escape is complete, the conflict is eliminated and the individual is asocialized.

Be it noted that where frustration derives from the inaccessibility of effective institutional means for attaining economic or any other type of highly valued "success," that Adaptations II, III, and V (innovation, ritualism and rebellion) are also possible. The result will be determined by the particular personality, and thus, the *particular* cultural background, involved. Inadequate socialization will result in the innovation response whereby the conflict and frustration are eliminated by relinquishing the institutional means and retaining the success-aspiration; an extreme assimilation of institutional demands will lead to ritualism wherein the goal is dropped as beyond one's reach but conformity to the mores persists; and rebellion occurs when emancipation from the reigning standards, due to frustration or to marginalist perspectives, leads to the attempt to introduce a "new social order."

Our major concern is with the illegitimacy adjustment. This involves the use of conventionally proscribed but frequently effective means of attaining at least the simulacrum of culturally defined success—wealth, power, and the like. As we have seen, this adjustment occurs when the individual has assimilated the cultural emphasis on success without equally internalizing the morally prescribed norms governing means for its attainment. The question arises, Which phases of our social structure predispose toward this mode of adjustment? We may examine a concrete instance, effectively analyzed by Lohman, which provides a clue to the answer. Lohman has shown that specialized areas of vice in the near north side of Chicago constitute a "normal" response to a situation where the cultural emphasis upon pecuniary success has been absorbed, but where there is little access to conventional and legitimate means for attaining such success. The conventional occupational opportunities of persons in this area are almost completely limited to manual labor. Given our cultural stigmatization of manual labor, and its correlate, the prestige of white collar work, it is clear that the result is a strain toward innovational practices. The limitation of opportunity to unskilled labor and the resultant low income cannot compete *in terms of conventional standards of achievement* with the high income from organized vice.

For our purposes, this situation involves two important features. First, such antisocial behavior is in a sense "called forth" by certain conventional values of the culture *and* by the class structure involving differential access to the approved opportunities for legitimate, prestige-bearing pursuit of the culture goals. The lack of high integration between the means-and-end elements of the cultural pattern and the particular class structure combine to favor a heightened frequency of antisocial

conduct in such groups. The second consideration is of equal significance. Recourse to the first of the alternative responses, legitimate effort, is limited by the fact that actual advance toward desired success-symbols through conventional channels is, despite our persisting open-class ideology, relatively rare and difficult for those handicapped by little formal education and few economic resources. The dominant pressure of group standards of success is, therefore, on the gradual attenuation of legitimate, but by and large ineffective, strivings and the increasing use of illegitimate, but more or less effective, expedients of vice and crime. The cultural demands made on persons in this situation are incompatible. On the one hand, they are asked to orient their conduct toward the prospect of accumulating wealth and on the other, they are largely denied effective opportunities to do so institutionally. The consequences of such structural inconsistency are psycho-pathological personality, and/or antisocial conduct, and/or revolutionary activities. The equilibrium between culturally designated means and ends becomes highly unstable with the progressive emphasis on attaining the prestige-laden ends by any means whatsoever. Within this context, Capone represents the triumph of amoral intelligence over morally prescribed "failure," when the channels of vertical mobility are closed or narrowed *in a society which places a high premium on economic affluence and social ascent for all its members.*

This last qualification is of primary importance. It suggests that other phases of the social structure besides the extreme emphasis on pecuniary success must be considered if we are to understand the social sources of antisocial behavior. A high frequency of deviate behavior is not generated simply by "lack of opportunity" or by this exaggerated pecuniary emphasis. A comparatively rigidified class structure, a feudalistic or caste order, may limit such opportunities far beyond the point which obtains in our society today. It is only when a system of cultural values extols, virtually above all else, certain common symbols of success for the population at large while its social structure rigorously restricts or completely eliminates access to approved modes of acquiring these symbols for a considerable part of the same population, that antisocial behavior ensues on a considerable scale. In other words, our egalitarian ideology denies by implication the existence of noncompeting groups and individuals in the pursuit of pecuniary success. The same body of success-symbols is held to be desirable for all. These goals are held to transcend class lines, not to be bounded by them, yet the actual social organization is such that there exist class differentials in the accessibility of these common success-symbols. Frustration and thwarted aspiration lead to the search for avenues of escape from a culturally induced intolerable situation; or unrelieved ambition may eventuate in illicit attempts to acquire the dominant values. The American stress on pecuniary success and ambitiousness for all thus invites exaggerated anxieties, hostilities, neuroses and antisocial behavior.

This theoretical analysis may go far toward explaining the varying correlations between crime and poverty. Poverty is not an isolated variable. It is one in a complex of interdependent social and cultural variables. When viewed in such a context, it represents quite different states of affairs. Poverty as such, and consequent limitation of opportunity, are not sufficient to induce a conspicuously high rate of criminal behavior. Even the often mentioned "poverty in the midst of plenty" will not necessarily lead to this result. Only insofar as poverty and associated disadvantages in competition for the culture values approved for *all* members of the society is linked with the assimilation of a cultural emphasis on monetary accumulation as a symbol of success is antisocial conduct a "normal" outcome. Thus, poverty is less highly correlated with crime in southeastern Europe than in the United States.

The possibilities of vertical mobility in these European areas would seem to be fewer than in this country, so that neither poverty *per se* nor its association with limited opportunity is sufficient to account for the varying correlations. It is only when the full configuration is considered, poverty, limited opportunity and a commonly shared system of success symbols, that we can explain the higher association between poverty and crime in our society than in others where rigidified class structure is coupled with *differential class symbols of achievement.*

In societies such as our own, then, the pressure of prestige-bearing success tends to eliminate the effective social constraint over means employed to this end. "The-end-justifies-the-means" doctrine becomes a guiding tenet for action when the cultural structure unduly exalts the end and the social organization unduly limits possible recourse to approved means. Otherwise put, this notion and associated behavior reflect a lack of cultural coordination. In international relations, the effects of this lack of integration are notoriously apparent. An emphasis upon national power is not readily coordinated with an inept organization of legitimate, i.e., internationally defined and accepted, means for attaining this goal. The result is a tendency toward the abrogation of international law, treaties become scraps of paper, "undeclared warfare" serves as a technical evasion, the bombing of civilian populations is rationalized, just as the same societal situation induces the same sway of illegitimacy among individuals.

The social order we have described necessarily produces this "strain toward dissolution." The pressure of such an order is upon outdoing one's competitors. The choice of means within the ambit of institutional control will persist as long as the sentiments supporting a competitive system, i.e., deriving from the possibility of outranking competitors and hence enjoying the favorable response of others, are distributed throughout the entire system of activities and are not confined merely to the final result. A stable social structure demands a balanced distribution of affect among its various segments. When there occurs a shift of emphasis from the satisfactions deriving from competition itself to almost exclusive concern with successful competition, the resultant stress leads to the breakdown of the regulatory structure. With the resulting attenuation of the institutional imperatives, there occurs an approximation of the situation erroneously held by utilitarians to be typical of society generally wherein calculations of advantage and fear of punishment are the sole regulating agencies. In such situations, as Hobbes observed, force and fraud come to constitute the sole virtues in view of their relative efficiency in attaining goals, which were for him, of course, not culturally derived.

It should be apparent that the foregoing discussion is not pitched on a moralistic plane. Whatever the sentiments of the writer or reader concerning the ethical desirability of co-ordinating the means-and-goals phases of the social structure, one must agree that lack of such coordination leads to anomie. Insofar as one of the most general functions of social organization is to provide a basis for calculability and regularity of behavior, it is increasingly limited in effectiveness as these elements of the structure become dissociated. At the extreme, predictability virtually disappears and what may be properly termed cultural chaos or anomie intervenes.

This statement, being brief, is also incomplete. It has not included an exhaustive treatment of the various structural elements which predispose toward one rather than another of the alternative responses open to individuals; it has neglected, but not denied the relevance of, the factors determining the specific incidence of these responses; it has not enumerated the various concrete responses which are constituted by combinations of specific values of the analytical variables; it has omitted, or included only by implication, any consideration of the social functions performed by illicit responses; it has not tested the full explanatory power of the analytical scheme by examining a large number of group variations in the frequency of deviate and conformist behavior; it has not adequately dealt with rebellious conduct which seeks to refashion the social framework radically; it has not examined the relevance of cultural conflict for an analysis of culture-goal and institutional-means malintegration. It is suggested that these and related problems may be profitably analyzed by this scheme.

REVIEW QUESTIONS

1. Why do you think Merton's theory became so popular when it did? Why?

2. Which adaptation to strain do you most identify with?

3. What types of criminal behavior does Merton's theory have a more difficult problem in explaining? Which types of offending does it do the best in explaining?

READING

In this selection, Craig Hemmens uses the lyrics of Bruce Springsteen as an analogy for Merton's causes and adaptations to strain and frustration regarding the economical structure to explain and understand the stereotypical experiences of certain individuals. The hypothetical (and/or observed) nature of the strain or frustration that people experience when they are subject to such problems provides a direct analogy to the lyrics of this poet/ songwriter, and this provides a good illustration of "real life" intersecting with traditional criminological theory. This entry is an example for how criminological theory can be applied to practical examples of actual human behavior and feelings, outside of academic theorizing.

There's a Darkness on the Edge of Town

Merton's Five Modes of Adaptation in the Lyrics of Bruce Springsteen

Craig Hemmens

Poor man wanna be rich, rich man wanna be king

And a king ain't satisfied till he rules everything

—Bruce Springsteen (*Badlands,* 1978)

⊠ Introduction

Criminologists have long struggled to explain crime and deviance. A number of sociological theories of crime causation have been posited, including differential association, social control, and labeling. One which has received enormous attention is Robert Merton's "strain" theory. Merton expanded upon Durkheim's concept of "anomie" (Durkheim 1951) and applied it to the modern, capitalist society, as exemplified by America in the mid-twentieth century. Merton's article "Social Structure and Anomie" (1938) is considered one of the most influential works in criminology (Williams & McShane 1994). According to Merton, anomie, or "strain," results when legitimate opportunities to attain success goals are blocked by structural obstacles.

Source: Craig Hemmens, "There's a Darkness on the Edge of Town: Merton's Five Modes of Adaptation in the Lyrics of Bruce Springsteen," *International Journal of Comparative and Applied Criminal Justice* 23, no. 1 (1999): 127–36. Reprinted with permission.

It is often said that art imitates life. Nathaniel Hawthorne's (1962) novels vividly portray the repressed sexuality of the Puritan period, William Faulkner's (1929) work depicts the racial conflict in the South during the early twentieth century, and Arthur Miller's (1949) writing presents the cultural malaise of America in the 1950s. Music is also an art form which often represents life. Rock music is a relatively new rnusical form, adapted from a number of sources including the blues and gospel music (Barlow 1989; Palmer 1981). While its popularity has eclipsed that of any other popular music form, often it has been criticized by musical scholars for its meaningless lyrics (Bayles 1994; Pollock 1983, 1993). The music of Bruce Springsteen is an exception to the notion that rock music neglects the realities of everyday life. Much of his work deals with the lack of equity and fairness in American society and the consequent strain experienced by many individuals.

Academic writing is not infrequently criticized as being intended solely for an academic audience. While strain theory is intuitively appealing, the bulk of discussion of the theory has been limited to the pages of academic journals. Bruce Springsteen is a musician, not a criminologist. The significance of his work lies in his ability to articulate, on a popular level, many of the concepts expressed by academicians such as Robert Merton. This article provides a content analysis of the lyrics of one Springsteen album, *Darkness on the Edge of Town* (1978) and discusses how his world view reflects, in popular form, the tenets of Merton's "strain" theory.

☒ Strain Theory

In his articulation of strain theory, Merton focused on the disjuncture between goals and means. Building upon Durkheim's (1951) study of suicide and anomie, Merton saw American society as promulgating the common goal of material success, while failing to provide equal opportunities for all in society to achieve this goal. The lower socio-economic classes had fewer opportunities to get an education and a good job, which Merton saw as the keys to achieving material success. Consequently, the lower classes were more likely to be frustrated in their efforts to achieve the sort of wealth and success that American culture taught them to want. The result, according to Merton, was anomie, or "strain" (Hirschi 1969). This strain could cause the individual to react to it in one of five ways, what Merton referred to as "modes of adaptation" (Merton 1938).

Merton suggested that there were five possible responses, or "modes of adaptation" to the frustration created by blocked opportunities: conformity, innovation, retreatism, rebellion, and ritualism. Conformists maintain support for both the goals and means of society. This is the largest of the five groups and includes most of mainstream society. Innovators accept the goals but not the means of achieving those goals; instead they find alternative means of achieving them, some of which may be illegal. Retreatists abandon both the goals and means of society and create their own, which may include unlawful activity. Drug addicts are an example. Rebels substitute new goals and means for the original ones endorsed by the wider society. An example is a revolutionary such as Lenin. Ritualists reject society's goals but continue to support the means. An example is treating a job as a form of security rather than a means of achieving success.

Strain theory suggests that the structure of society affects the action of individuals. It is what sociologists refer to as a macro or structural level theory, as it seeks to explain behavior in the aggregate (Farnworth & Leiber 1989). Its proponents do not attempt to make direct predictions about individual behavior (Vold & Bernard 1986). It is also limited in scope. Merton's anomie theory is intended to explain differences in crime by class (Burton 1991). While the theory suggests that massive changes in American social structure might be necessary to remove the barriers to success which lead to strain and deviance, strain theory is essentially conservative. The theory does not reject capitalism per se; it merely challenges the structure of American capitalism. It does not require an overthrow of the government but instead a reworking of the institution (Burton & Cullen 1992).

By the 1970s a sizable portion of the criminological community had turned away from strain theory. The theory was criticized for inadequate operationalization of its concepts and a lack of empirical support (Hirschi 1969; Kornhauser 1978). It has been forcefully argued, however, that such criticisms were based on an imprecise formulation of the theory, one which truncated strain theory and made it susceptible to misinterpretation and inaccurate empirical testing (Burton & Cullen 1992). Despite its apparent limitations, strain theory remains a popular explanation for deviant behavior. Indeed, strain theory has recently received renewed attention from criminologists (Agnew 1985, 1989, 1992).

While strain theory has impacted policy making and continues to provide fertile ground for academic research, its appeal stretches beyond the walls of academia and the halls of government. Part of its success can be attributed to its intuitive appeal. Strain theory, for many, simply makes sense as an explanation for deviant behavior in a capitalistic, materialistic society in which one's reach is always likely to be exceeded by one's goals. Images of strain theory abound in popular culture, from art to literature to film. Perhaps the most forceful exponent of strain theory in popular media is Bruce Springsteen.

◪ The Lyrics of Bruce Springsteen

Bruce Springsteen released his first record, *Greetings From Asbury Park,* in 1973. He was signed to Columbia Records by John Hammond,

who also discovered Bob Dylan. Hammond saw Springsteen as a singer-songwriter in the same vein as Dylan, but Springsteen preferred to work within the context of a full rock and roll band (Marsh 1979). His first album and his second, *The Wild, the Innocent,* and *The E Street Shuffle,* also released in 1973, garnered critical acclaim but few sales. Both albums contained songs which celebrated the freedom and exuberance of youth, but also hinted at the difficulty of growing up poor in the modern world. The writing was unpolished, but the lyrics contained a number of powerful images.

In 1975 Springsteen gained a national audience with the release of *Born to Run.* The title song became a staple of rock radio. The album displayed a newfound maturity and songwriting polish. The tone of the album was also darker than the previous albums. While the song "Born to Run" celebrated the possibilities of the open road, it also contained an acknowledgment that the characters in the song had no choice but to flee their humble upbringing:

> Baby this town rips the bones from your back
>
> It's a death trap, It's a suicide rap
>
> We gotta get out while we're young
>
> 'Cause tramps like us, baby we were born to run
>
> *—Born to Run*

The theme of the album was escape, trying to get out and get what you want and deserve. Escape was possible, but only via a limited number of avenues, some legitimate, others not:

> Kids flash guitars just like switchblades
>
> Hustling for the record machine
>
> The hungry and the hunted
>
> Explode into rock'n roll bands
>
> *—Jungleland*

And all we gotta do is hold up our end

Here stuff this in your pocket

It'll look like you're carrying a friend

And remember, just don't smile

Change your shirt, 'cause tonight we got style

—*Meeting Across the River*

Springsteen's songwriting had decidedly taken a turn. The emphasis was on individuals who felt trapped, either in dead end jobs, failed relationships, or dying industrial towns. The album sold well, and after Springsteen appeared simultaneously on the covers of *Time* and *Newsweek,* he seemed destined to be rock's next superstar (Marsh 1979).

Springsteen's next album, *Darkness on the Edge of Town* (1978), was not released for three years, an unusually long time between albums. The delay was attributable in part to a lawsuit between Springsteen and his producer, which was eventually settled (Marsh 1987). The lawsuit was only part of the reason for the delay, however, as Springsteen spent an enormous amount of time in the studio recording the new album. The cover of the album is a clear indication that the album is different from what had come before. Where Springsteen had posed laughing with his saxophone player on the cover of *Born to Run,* he stands alone on the cover of *Darkness on the Edge of Town,* staring blankly into the camera. Instead of posing in front of a plain white backdrop, as on the previous album, Springsteen is photographed standing in a room, leaning up against a wall covered in fading wallpaper. 1978 was the height of the disco music craze, with its emphasis on outlandish clothes, a style which many rock acts embraced. Springsteen, by contrast, is dressed in a plain white T-shirt and a black leather jacket. He looks more like a day laborer than a rock star.

Darkness on the Edge of Town continues the exploration of the themes first examined on *Born to Run,* but the tone is different. Where *Born to Run* was dominated by saxophone and piano, *Darkness on the Edge of Town* is driven by guitars. The change in the instrumentation gives the new music a sense of urgency and power. Where *Born to Run* had a tone of frustration and sadness, *Darkness on he Edge of Town* has an angry tone. Lyrically, Springsteen continues to explore the ways that individuals dealt with the inequities of modem society. According to the title song:

Some folks are born into a good life,

Other folks get it anyway, anyhow.

—*Darkness on the Edge of Town*

The entire album is populated with the same characters found in *Born to Run.* But where the individuals in *Born to Run* believed that escape was possible, the protagonists in *Darkness on the Edge of Town* have no illusions:

Through the mansions of fear, through the mansions of pain,

I see my daddy walking through them factory gates in the rain,

Factory takes his hearing, factory gives him life,

The working, the working, just the working life

—*Factory*

Many of the characters on the album are trapped in dead-end jobs, unable to leave but unwilling to quit dreaming of a better life. The hero of "Racing in the Street" has a wife and a job, but neither holds his interest. The only meaning in his life is provided by his car, "a sixty-nine Chevy with a 396." He spends his night racing cars as a means of escaping the drudgery of everyday life:

Some guys they just give up living

And start dying little by little, piece by piece,

Some guys come home from work and wash up,

And go racin' in the street

—*Racing in the Street*

The division between the "haves" and the "have-nots" is clear to the protagonists in these songs, and this class division is viewed as illegitimate:

Poor man wanna be rich,

Rich man wanna be king,

And a king ain't satisfied till he rules everything

—*Badlands*

While some of the protagonists appear to accept responsibility for their failed lives, others look elsewhere, placing the blame on their upbringing:

You're born into this life paying

for the sins of somebody else's past,

Daddy worked his whole life, for nothin' but the pain,

Now he walks these empty rooms, looking for something to blame

—*Adam Raised a Cain*

Some of the protagonists find what they are looking for, an outlet for their rage and frustration:

End of the day, factory whistle cries,

Men walk through these gates with death in their eyes,

And you just better believe, boy,

Somebody's gonna get hurt tonight

—*Factory*

Darkness on the Edge of Town is an angry album. The lyrics are full of rage, and the instrumentation conveys this anger. The protagonists of the album are not criminals, however. Most of them hold jobs and have families. But they are angry, and the anger is caused by their frustration with being unable to attain what they believe they have been taught to strive for:

Everybody's got a hunger, a hunger they can't resist,

There's so much that you want, you deserve much more than this,

But if dreams came true, oh wouldn't that be nice,

But this ain't no dream we're livin' through tonight

—*Prove It All Night*

Some of Springsteen's characters turn to deviance as a response to stressful life events which anger them. The anger of the speaker in "The Promised Land" is palpable and clearly directed outward:

Sometimes I feel so weak I just want to explode

Explode and tear this whole town apart

Take a knife and cut this pain from my heart

Find somebody itching for something to start

—*The Promised Land*

Most of Springsteen's characters do not become deviant as a consequence of their feelings of strain, however. This is in accord with Merton's writings, which stress that deviance is only one possible response to strain. The angry characters are most likely to become deviant. But others who experience strain, particularly the strain associated with the disjuncture between goals and means, either conform, ritualize, retreat, or rebel.

Discussion

The notion that feelings of societal inequity can create deviance-producing strain is clearly reflected in Springsteen's music, as is the recognition that strain does not necessarily lead to deviance. Many of Springsteen's characters are drawn from the lower classes, from segments of society that seem to have been abandoned by larger society, or who are trapped living lives of "quiet desperation," to paraphrase Thoreau (1854). These characters are faced with stressful life events ranging from the loss of a job to the loss of a loved one. Some respond with anger, others resign themselves to their lot in life, and others attempt to find a new way. It is those who respond to strain with anger who are most likely to deviate. Springsteen has acknowledged the impact his upbringing in a lower class family in a dying post-industrial town had on him and his songwriting:

> My memory is of my father trying in find work, what that does to you, and how that affects your image of your manhood, as a provider. The loss of that role is devastating. I write coming from that Spot—the spot of disaffection, of loners, outsiders. But not outlaws. It's about people trying to find their way in, but somebody won't let them in. Or they can't find their way in. And what are the actions that leads to? (Corn 1996)

Springsteen's lyrics are filled with images of social injustice and its impact on average working class Americans. And he has made it clear in interviews that social inequality is a major concern of his. Discussing the problems of contemporary society, he said:

> It would take a tremendous concentration of national will, on the order of a domestic Marshall Plan, to do the things that need to be done to achieve a real kind of social justice and equality. (Corn 1996)

Strain theory, for all the blame it places on American society as a cause of deviant behavior, is in many respects a conservative theory. Strain theory does not assert that it is capitalism which is the root cause of crime, as Marxist criminology does (Taylor et al. 1973). Strain theory does not imply that American society is irreparable. According to strain theory, it is the emphasis on material success to the exclusion of all other goals that inevitably leads to a disjunction between goals and means. Recent efforts have been made to demonstrate how strain can be reduced by creating other equally valued societal goals.

This is all in accord with the values Springsteen promotes in his work. Springsteen's lyrics stress the value of community, shared responsibility, and interconnectedness. He has gone to some length to promote these values, even using the concert stage as a vehicle. During the tour in support of his biggest selling album, *Born in the USA* (1984), he spoke during each show about the importance of community and then made a large public donation to a local food bank and encouraged concert attendees to become involved in similar local charities.

Conclusion

Popular music has been the subject of academic ridicule and public disdain since long before Elvis ever stepped into his blue suede shoes. Blues was dismissed as the illiterate ramblings of poor Southern blacks (Gillette 1983; Guralnick 1986). Country and bluegrass music were disparagingly termed "hillbilly music" (Goldrosen & Beecher 1986). In recent years, however, as the first generation raised almost exclusively on popular music has come of age and moved into academia, the reevaluation process has begun. It is now chic to praise the verbal wordplay of country songwriters (Tichi 1994) and the intricacies of blues phraseology (Guralnick 1986, 1994). Rock music has been accorded some credit, but even when it is applauded as technically skillful, its substance is still often dismissed as teenage pabulum.

This paper argues that rock music is sometimes much more than that. Like any art form, rock music is both a creation and a reflection of the culture from which it comes. Its potential impact is undeniably substantial—it reaches a far larger audience than do the works of authors or artists, and it certainly reaches a far greater audience than do the works of academicians. While Robert Merton's strain theory may provide an excellent explanation of how Americans respond to the strain of everyday life, many more Americans listen to the songs of Bruce Springsteen. Thus it is of some interest and import that his songwriting embodies the tenets of strain theory. This is not to suggest that what Bruce Springsteen has to say by itself has resulted in any change in society. It merely indicates that there is some sociological value to his work, and that strain theory, by whatever name it goes or manner in which it is presented, has enormous intuitive appeal, appeal which cuts across layers of discourse.

Finally, it is perhaps strain theory itself which explains why Springsteen's sociological observations would find outlet in popular song rather than academia, for as the Rolling Stones song "Street Fighting Man" (covered by Springsteen during the *Born in the USA* tour) says: "What can a poor boy do—'cept sing in a rock and roll band?"

⊠ References

Agnew, R. 1985. "A Revised Strain Theory of Delinquency." *Social Forces* 64: 151–67.

——. 1989. "A Longitudinal Test of the Revised Strain Theory." *Journal of Quantitative Criminology* 5: 373–387.

——. 1992. "Foundation for A General Strain Theory of Crime and Delinquency." *Criminology* 30: 47–86.

Barlow, W. 1989. *Looking up at Down: The Emergence of Blues Culture*. Philadelphia: Temple University Press.

Bayles, M. 1994. *Hole in Our Soul: The Loss of Beauty and Meaning in American Popular Music*. New York: The Free Press.

Burton, V. S., Jr. 1991. "Explaining Adult Criminality: Testing Strain, Differential Association, and Control Theories." Unpublished dissertation, University of Cincinnati. Ann Arbor, Michigan: University Microfilms.

Burton, V. S., Jr., and F. T. Cullen. 1992. "The Empirical Status of Strain Theory." *Journal of Crime and Justice* 15: 1–30.

Corn, D. 1996. "Springsteen Tells the Story of the Secret America." *Mother Jones*.

Durkheim, E. 1951. *Suicide: A Sociological Study*. New York: The Free Press.

Farnworth, M., and M. J. Leiber. 1989. "Strain Theory Revisited: Economic Goals, Educational Means, and Delinquency." *American Sociological Review* 54: 259–79.

Faulkner, W. 1929. *The Sound and the Fury*. New York: Random House.

Gillette, C. 1983. *The Sound of the City: The Rise of Rock and Roll*. New York: Pantheon.

Goldrosen, J., and J. Beecher. 1986. *Remembering Buddy: The Definitive Biography of Buddy Holly*. New York: Penguin Books.

Guralnick, P. 1986. *Sweet Soul Music: Rhythm and Blues and the Southern Drama of Freedom*. New York: Harper and Row.

——. 1994. *Last Train to Memphis: The Rise of Elvis Presley*. Boston: Little, Brown.

Hawthorne, N. 1962. *The Scarlet Letter*. New York: Knopf.

Hirschi, T. 1969. *Causes of Delinquency*. Berkeley: University of California Press.

Kornhauser, R. R. 1978. *Social Sources of Delinquency*. Chicago: University of Chicago Press.

Marsh, D. 1979. *Born to Run: The Bruce Springsteen Story*. New York: Doubleday.

—— 1987. *Glory Days*. New York: Pantheon.

Merton, R. K. 1938. "Social Structure and Anomie." *American Sociological Review* 3: 672–82.

Miller, Arthur. 1949. *Death of a Salesman.* New York: Knopf.

Palmer, R. 1981. *Deep Blues.* New York; Viking.

Pollock, B. 1983. *When the Music Mattered: Rock in the 1960's.* New York: Holt, Rinehart, and Winston.

——. 1993. *Hipper Than Our Kids: A Rock and Roll Journal of the Baby Boom Generation.* New York; Macmillan.

Taylor, I, P. Walton, and J. Young. 1973. *The New Criminology.* New York: Harper and Row.

Thoreau, H. D. 1854. *Walden, or Life in the Woods.* Boston: Harvard University Press.

Tichi, C. 1994. *High Lonesome: The American Culture of Country Music,* Chapel Hill: University of North Carolina Press.

Void, G. B., and T. J. Bernard. 1986. *Theoretical Criminology.* New York: Oxford University Press.

Williams, F. P., and M. D. McShane. 1994. *Criminological Theory.* Englewood Cliffs, NJ: PrenticeHall.

Albums Cited

Springsteen, Bruce. (1973). *Greetings from Asbury Park.* New York: Columbia Records.

Springsteen, Bruce. (1973). *The wild, the innocent, and the E Street shuffle.* New York: Columbia Records.

Springsteen, Bruce. (1975). *Born to run.* New York: Columbia Records.

Springsteen, Bruce. (1978). *Darkness on the edge of town.* New York: Columbia Records.

REVIEW QUESTIONS

1. Given Hemmens's examination of Springsteen's lyrics, do you feel that you have a better understanding of Merton's ideas regarding strain? Does this differ from the understanding or feelings you had from Merton's original paper? If so, why?

2. Which portion of Springsteen's lyrics do you find most compelling or applicable in terms of Merton's ideas about strain or adaptations to strain? Which adaptation(s) to strain do you find are best represented in Springsteen's lyrics?

3. Can you think of lyrics from other musicians (or movies, books) that would better, or similarly, explain Merton's concept of strain or adaptations to strain?

READING

In this selection, Messner and Rosenfeld present a more macro-level approach to Merton's strain theory, with a special emphasis on the social structure of the United States. The authors emphasize that while virtually all individuals are socialized to believe in the American Dream, most individuals will not achieve financial success, which creates strain or feelings of frustration, consistent with Merton's original formulation. However, these authors advance the original strain theory by Merton by incorporating the work of Currie, particularly regarding the concepts of devaluation of noneconomic goals and functions, as

Source: Adapted from Steven F. Messner and Richard Rosenfeld, *Crime and the American Dream,* 3rd ed. (Belmont, CA: Wadsworth, 2001). Copyright © 2001 Wadsworth, a part of Cengage Learning, Inc. Reproduced by permission. www.cengage.com/permissions.

well as the intrusion of economic forces into virtually all social institutions, which they refer to as "penetration." This selection is an often cited and well-respected piece of scholarship, and it has inspired many other scholars to engage in studies based on the framework the authors present.

Crime and the American Dream

Steven F. Messner and Richard Rosenfeld

The Virtues and Vices of the American Dream

The thesis of this book is that the American Dream itself and the normal social conditions engendered by it are deeply implicated in the problem of crime. In our use of the term "the American Dream," we refer to a broad cultural ethos that entails a commitment to the goal of material success, to be pursued by everyone in society, under conditions of open, individual competition.

The American Dream has both an evaluative and a cognitive dimension associated with it. People are socialized to accept the desirability of pursuing the goal of material success, and they are encouraged to believe that the chances of realizing the Dream are sufficiently high to justify a continued commitment to this cultural goal. These beliefs and commitments in many respects define what it means to be an enculturated member of our society. The ethos refers quite literally to the *American* dream. . . .

The Value Foundations of the American Dream

What sets the United States apart from other modern industrial nations, according to Merton, is the cultural ethos of the American Dream. Merton himself does not provide a formal definition of the American Dream, but it is possible to formulate a reasonably concise characterization of this cultural orientation based on Merton's discussion of American culture in general, his scattered references to the American Dream, and the commentary of others on Merton's work.[1] Our definition . . . is as follows: The American Dream refers to a commitment to the goal of material success, to be pursued by everyone in society, under conditions of open, individual competition.

The American Dream is a powerful force in our society because it embodies the basic value commitments of the culture: its achievement orientation, individualism, universalism, and peculiar form of materialism that has been described as the "fetishism of money" (Taylor, Walton, and Young, 1973, p. 94). Each of these value orientations contributes to the anomic character of the American Dream: its strong emphasis on the importance of realizing cultural goals in comparison with its relatively weak emphasis on the importance of using the legitimate means to do so.

Before examining the value complex underlying the American Dream, we caution against an overly simplistic interpretation of American culture. The United States is a complex and, in many respects, culturally pluralistic society. It neither contains a single, monolithic value system nor exhibits complete consensus surrounding specific value orientations. Historically, certain groups have been completely excluded from the American Dream.

An obvious example is that of enslaved African-Americans in the antebellum South. In addition, cultural prescriptions and mandates are filtered through prevailing gender roles. Indeed, we argue later in this chapter that the interpretation of the American Dream differs to some extent for men and women. We nevertheless concur with Jennifer Hochschild's claim that the American Dream has been, and continues to be, a "defining characteristic of American culture," a cultural ethos "against which all competitors must contend" (Hochschild, 1995, p. xi). An adequate understanding of the crime problem in the United States, therefore, is impossible without reference to the cluster of values underlying the American Dream: achievement, individualism, universalism, and materialism.

Achievement

A defining feature of American culture is its strong achievement orientation. People are encouraged to make something of themselves, to set goals, and to strive to reach them. At the same time, personal worth tends to be evaluated on the basis of the outcome of these efforts. Success, in other words, is to a large extent the ultimate measure of a person's value. . . . Given such a value orientation, the failure to achieve is readily equated with a failure to make any meaningful contribution to society at all. The cultural pressures to achieve at any cost are thus very intense. In this way, a strong achievement orientation, at the level of basic cultural values, is highly conducive to the mentality that "it's not how you play the game; it's whether you win or lose" (Orru, 1990, p. 234).

Individualism

A second basic value orientation at the core of American culture is individualism. Americans are deeply committed to individual rights and individual autonomy. . . . This obsession with the individual, when combined with the strong achievement orientation in American culture, exacerbates the tendency toward anomie. In the pursuit of success, people are encouraged to "make it" on their own. Fellow members of society thus become competitors and rivals in the struggle to achieve social rewards and, ultimately, to validate personal worth. . . .

Universalism

A third basic value orientation in American culture is universalism. Socialization into the cultural goals of American society has a decidedly democratic quality. With few exceptions, everyone is encouraged to aspire to social ascent, and everyone is susceptible to evaluation on the basis of individual achievements. An important corollary of this universal entitlement to dream about success is that the hazards of failure are also universal. . . .

The "Fetishism" of Money

Finally, in American culture, success is signified in a distinctive way: by the accumulation of monetary rewards. Money is awarded special priority in American culture. As Merton observes, "In some large measure, money has been consecrated as a value in itself, over and above its expenditure for articles of consumption or its use for the enhancement of power." The point to emphasize here is not that Americans are uniquely materialistic, for a strong interest in material well-being can be found in most societies. Rather, the distinctive feature of American culture is the preeminent role of money as the "metric" of success. . . . There is an important implication of the signification of achievement with reference to monetary rewards. Monetary success is inherently open-ended. It is always possible in principle to have more money. Hence, the American Dream offers "no final stopping point." It requires "never-ending achievement" (Merton, 1968, p. 190; Passas, 1990, p. 159).

In sum, the dominant value patterns of American culture—specifically, its achievement orientation, its competitive individualism, its universalism in goal orientations and evaluative standards, when harnessed to the preeminent goal of monetary success—crystallize into the distinctive cultural ethos of the American Dream. . . .

Cultural forces thus play a prominent role in our explanation of the high levels of crime in American society. However, a complete sociological explanation of crime must extend beyond features of culture and incorporate social structural factors as well. Culture does not exist in isolation from social structure but rather is expressed in, reproduced by, and occasionally impeded by, social structure. Any comprehensive explanation that emphasizes "culture" as a cause of crime must therefore also consider the relevant range of structural conditions through which the cultural sources of crime are enacted. In our view, the most important of these structural conditions are the institutional arrangements of society.

◩ The Nature and Functioning of Social Institutions

Social institutions are the building blocks of whole societies. As such, they constitute the basic subject matter of macrolevel analysis. . . . The functions of institutions in social systems have been compared with the functions of instincts in biological organisms: both channel behavior to meet basic system needs. . . . The basic social needs around which institutions develop include the need to (1) adapt to the environment, (2) mobilize and deploy resources for the achievement of collective goals, and (3) socialize members to accept the society's fundamental normative patterns (Downes & Rock, 1982; Parsons, 1951).

Adaptation to the environment is the primary responsibility of economic institutions. The *economy* consists of activities organized around the production and distribution of goods and services. It functions to satisfy the basic material requirements for human existence, such as the need for food, clothing, and shelter.

The political system, or *polity,* mobilizes and distributes power to attain collective goals. One collective purpose of special importance is the maintenance of public safety. Political institutions are responsible for "protecting members of society from invasions from without, controlling crime and disorder within, and providing channels for resolving conflicts of interest" (Bassis, Gelles, & Levine, 1991, p. 142). As part of the polity, agencies of the civil and criminal justice systems have major responsibility for crime control and the lawful resolution of conflicts.

The institution of the *family* has primary responsibility for the regulation of sexual activity and for the replacement of members of society. These tasks involve establishing and enforcing the limits of legitimate sexual relations among adults, the physical care and nurturing of children, and the socialization of children into the values, goals, and beliefs of the dominant culture. Families also bear much of the responsibility for the care of dependent persons in society more generally (for example, caring for the infirm and the elderly). In addition, a particularly important function of the family in modern societies is to provide emotional support for its members. To a significant degree, the family serves as a refuge from the tensions and stresses generated in other institutional domains. . . .

The institution of *education* shares many of the socialization functions of the family. Like the family, schools are given responsibility for transmitting basic cultural standards to new generations. In modern industrial societies, schools are also oriented toward the specific task of preparing youth for the demands of adult roles and, in particular, occupational roles. In addition, education is intended to enhance personal adjustment, facilitate the development of individual human potential, and advance the general knowledge base of the culture.

These four social institutions—the economy, the polity, the family, and education—are the central focus of our analysis of crime. They do not, of course, exhaust the institutional structure of modern societies, nor are they the only institutions with relevance to crime. Religion and mass communications, for example, have been the subjects of important criminological research (Stark, Kent, & Doyle, 1982; Surette, 1992). However, the economy, the polity, the family, and education are, in our view, central to what may be called an "institutional understanding" of crime. . . . Any given society therefore will be characterized by a distinctive arrangement of social institutions that reflects a balancing of the sometimes competing claims and requisites of the different institutions, yielding a distinctive institutional balance of power. . . .

⊠ The American Dream and the Institutional Balance of Power

The core elements of the American Dream—a strong achievement orientation, a commitment to competitive individualism, universalism, and, most important, the glorification of material success—have their institutional underpinnings in the economy. The most important feature of the economy of the United States is its capitalist nature. The defining characteristics of any capitalist economy are private ownership and control of property and free-market mechanisms for the production and distribution of goods and services.

These structural arrangements are conducive to and presuppose certain cultural orientations. For the economy to operate efficiently, the private owners of property must be profit oriented and eager to invest, and workers must be willing to exchange their labor for wages. The motivational mechanism underlying these conditions is the promise of financial returns. The internal logic of a capitalist economy thus presumes that an attraction to monetary rewards as a result of achievement in the marketplace is widely diffused throughout the population (Passas, 1990, p. 159; Polanyi, [1944] 1957).

A capitalist economy is also highly competitive for all those involved, property owners and workers alike. Firms that are unable to adapt to shifting consumer demands or to fluctuations in the business cycle are likely to fail. Workers who cannot keep up with changing skill requirements or who are unproductive in comparison with others are likely to be fired. This intense competition discourages economic actors from becoming wedded to conventional ways of doing things and instead encourages them to substitute new techniques for traditional ones if these techniques offer advantages in meeting economic goals. Therefore, a capitalist economy naturally cultivates a competitive, innovative spirit.

What is distinctive about the United States, however, is the *exaggerated* emphasis on monetary success and the *unrestrained* receptivity to innovation. The goal of monetary success overwhelms other goals and becomes the principal measuring rod for achievements. The resulting proclivity and pressures to innovate resist any regulation that is not justified by purely technical considerations. The obvious question arises: Why have cultural orientations that express the inherent logic of capitalism evolved to a particularly extreme degree in American society? The answer, we submit, lies in the inability of other social institutions to tame economic imperatives. In short, the institutional balance of power is tilted toward the economy.

The historical evidence suggests that this distinctive institutional structure has always existed in the United States. In his analysis of American slavery, the historian Stanley Elkins observed that capitalism emerged "as the principle dynamic force in American society," free to develop according to its own institutional logic without interference from "prior traditional institutions, with competing claims of their own." Whereas capitalism developed in European societies (and later in Japan) within powerful preexisting institutional frameworks, the institutional structure of American society emerged simultaneously with, and was profoundly shaped by, the requirements of capitalist development. American capitalism thus took on a "purity of form" unknown in other capitalist societies (Elkins, 1968, p. 43). Moreover, other institutions were cast in distinctly subsidiary positions in relation to the economy. . . .

Robert Heilbroner writes that "*American* capitalism, not American *capitalism*" is responsible for the features of our society that distinguish it, for better or worse, from other capitalist societies (Heilbroner, 1991, pp. 539–540). . . . [S]erious crime rates in the United States are unusually high when compared with those in other modern capitalist societies. These differences, therefore, cannot be accounted for by capitalism alone. Variation in levels of crime and other aspects of these nations is rooted . . . in their contrasting institutional settings. . . . We accept the basic argument that capitalism developed in the United States without the institutional restraints found in other societies. As a consequence, the economy assumed an unusual dominance in the institutional structure of society from the very beginning of the nation's history, and this distinctive institutional arrangement has continued to the present.

Our notion of economic dominance in the institutional balance of power is similar to Elliott Currie's concept of a "market society" as distinct from a "market economy." According to Currie, in a market society "the pursuit of private gain becomes the organizing principle of all areas of social life—not simply a mechanism that we may use to accomplish certain circumscribed ends" (Currie, 1991, p. 255). Economic dominance characteristic of the American market society is manifested, we argue, in three interrelated ways:

1. devaluation of noneconomic institutional functions and roles,

2. accommodation to economic requirements by other institutions, and

3. penetration of economic norms into other institutional domains.

Devaluation

Noneconomic goals, positions, and roles are devalued in American society relative to the ends and means of economic activity. An example is the relative devaluation of the distinctive functions of education and of the social roles that fulfill these functions. Education is regarded largely as a means to occupational attainment, which in turn is valued primarily insofar as it promises economic rewards. Neither the acquisition of knowledge nor learning for its own sake is highly valued. . . .

Similar processes are observed in the context of the family, although the tendency toward devaluation is perhaps not as pronounced as in other institutional arenas. There is a paradox here, because "family values" are typically extolled in public rhetoric. Nevertheless, family life has a tenuous position in American culture. It is the home*owner* rather than the home*maker* who is widely admired and envied—and whose image is reflected in the American Dream. . . .

The distinctive function of the polity, providing for the collective good, also tends to be devalued in comparison with economic functions. The general public has little regard for politics as an intrinsically valuable activity and confers little social honor on the role of the politician. Indeed, the label "politician" is commonly used in a disparaging way. . . .

Interestingly, one distinctive function of the polity does not appear to be generally devalued, namely, crime control. There is widespread agreement among the American public that government should undertake vigorous efforts to deal with the crime problem. If anything, Americans want government to do more to control crime. Fifty-four percent of Americans think the government spends too little on law enforcement; only 8 percent think the government spends too much.[2] Yet this apparent exception is compatible with the claim of economic dominance. Americans' obsession with crime is rooted in fears that crime threatens, according to political analyst Thomas Edsall [1992], "their security, their values, their rights, and their livelihoods and the competitive prospects of their children." . . . Because crime control bears directly on the pursuit of the American Dream, this particular function of the polity receives high priority.

Accommodation

A second way in which the dominance of the economy is manifested is in the *accommodations* that emerge in those situations where institutional claims are in competition. Economic conditions and requirements typically exert a much stronger influence on the operation of other institutions than vice versa. For example, family routines are dominated by the schedules, rewards, and penalties of the labor market. Whereas parents worry about "finding time" for their families, few workers must "find time" for their jobs. On the contrary, many feel fortunate that the economy has found time for them. Consider the resistance to parental leave in the United States. Most industrialized nations mandate paid maternity or parental leave by law to enable parents to care for infants at home without threat of job loss. . . .

Educational institutions are also more likely to accommodate to the demands of the economy than is the economy to respond to the requirements of education. The timing of schooling reflects occupational demands rather than intrinsic features of the learning process

or personal interest in the pursuit of knowledge. People go to school largely to prepare for "good" jobs. And once they are in the labor market, there is little opportunity to pursue further education for its own sake. When workers do return to school, it is almost always to upgrade skills or credentials to keep pace with job demands, to seek higher-paying jobs, or to "retool" during spells of unemployment. . . .

The polity likewise is dependent on the economy for financial support. To run effective campaigns, politicians and political parties rely on private donations. Even if money does not guarantee the outcome of an election, any candidate who hopes to win must attract significant financial support from private sources. . . .

Penetration

A final way in which the dominance of the economy in the institutional balance of power is manifested is in the *penetration* of economic norms into other institutional areas. Learning takes place within the context of individualized competition for external rewards, and teaching becomes oriented toward testing. Schools rely on grading as a system of extrinsic rewards, like wages, to ensure compliance with goals. . . .

Education itself is increasingly viewed as a commodity, no different from other consumer goods. Economic terminology permeates the very language of education, as in the emphasis on the "customer-driven classroom . . . accountability" conceptualized in terms of the "value added" to students in the educational production process, and the emphasis on students themselves as "products."[3] . . . Within the polity, a "bottom-line" mentality develops. Effective politicians are those who deliver the goods. Moreover, the notion that the government would work better if it were run more like a business continues to be an article of faith among large segments of the American public. Many Americans in fact seem to prefer business leaders over public officials to perform key political functions. . . .

The family has probably been most resistant to the intrusion of economic norms. Yet even here, pressures toward penetration are apparent. Contributions to family life tend to be measured against the all-important breadwinner role, which has been extended to include women who work in the paid labor force. No corresponding movement of men into the role of homemaker has occurred. Here again, shifts in popular terminology are also instructive. Husbands and wives are "partners" who "manage" the household "division of labor" in accordance with the "marriage contract." We are aware of few comparable shifts in kin-based terminology, or primary group norms, from the family to the workplace.[4]

In sum, the social organization of the United States is characterized by a striking dominance of the economy in the institutional balance of power. As a result of this economic dominance, the inherent tendencies of a capitalist economy to orient the members of society toward an unrestrained pursuit of economic achievements are developed to an extreme degree. These tendencies are expressed at the cultural level in the preeminence of the competitive, individualistic pursuit of monetary success as the overriding goal—the American Dream—and in the relative deemphasis placed on the importance of using normative means to reach this goal—anomie. The anomic nature of the American Dream and the institutional structure of American society are thus mutually supportive and reinforcing. In the next section, we turn to the implications of this type of social organization for crime.

▧ Anomie and the Weakening of Institutional Control

Both of the core features of the social organization of the United States—culture and institutional structure—are implicated in the genesis of high levels of crime. At the cultural level, the dominant ethos of the American Dream stimulates criminal motivations and at the same time promotes a weak normative environment (anomie). At the institutional level, the dominance of the economy in the institutional balance of power fosters weak social control. And, as just explained, both culture and institutional structure are themselves interdependent. These interconnections between culture, social structure, and crime are presented schematically in Figure 1.

The cultural stimulation of criminal motivations derives from the distinctive content of the American Dream. Given the strong, relentless pressure for everyone to succeed, understood in terms of an inherently elusive monetary goal, people formulate wants and desires that are difficult, if not impossible, to satisfy within the confines of legally permissible behavior. This feature of the American Dream helps explain criminal behavior with an instrumental character, behavior that offers monetary rewards. This type of behavior includes white-collar offenses, street crimes such as robbery and drug dealing, and other crimes that occur as a consequence of these activities.

At the same time, the American Dream does not contain within it strong injunctions against substituting more effective, illegitimate means for less effective, legitimate means in the pursuit of monetary success. To the contrary, the distinctive cultural message accompanying the monetary success goal in the American Dream is the devaluation of all but the most technically efficient means. This anomic orientation leads not simply to high levels of crime in general but to especially violent forms of economic crime, for which the United States is known throughout the industrial world, such as mugging, carjacking, and home invasion.

Of course, the American Dream does not completely subsume culture. Other elements of culture affirm the legitimacy of legal prohibitions and the desirability of lawful behavior.

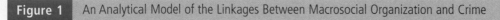

Figure 1   An Analytical Model of the Linkages Between Macrosocial Organization and Crime

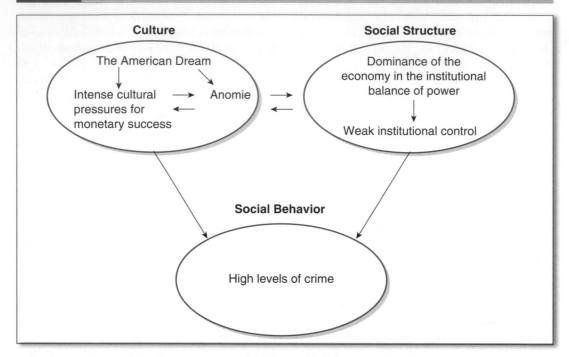

In principle, these other elements of culture could counterbalance the anomic pressures emanating from the American Dream. However, the very same institutional dynamics that contribute to the pressures to "innovate" in the pursuit of economic goals also make it less likely that the anomic pressures inherent in the American Dream will in fact be counterbalanced by other cultural forces. . . .

This generalized anomie ultimately explains, in our view, the unusually high levels of gun-related violence in the United States. In the final analysis, guns are very effective tools for enforcing compliance. The American penchant for owning guns and using them reflects, in other words, a more general anomic cultural orientation, a willingness to pursue goals by any means necessary. The basic social organization of the United States contributes to high levels of crime in another way. Institutions such as the family, schools, and the polity bear

responsibility not only for socialization, and hence the normative control associated with culture, but also for the more external type of social control associated with social structure. External control is achieved through the active involvement of individuals in institutional roles and through the dispensation of rewards and punishments by institutions. . . .

Weak institutions invite challenge. Under conditions of extreme *competitive* individualism, *people* actively resist institutional control. Not only do they fall from the insecure grasp of powerless institutions, sometimes they deliberately, even proudly, push themselves away. The problem of external control by major social institutions, then, is inseparable from the problem of the internal regulatory force of social norms, or anomie. Anomic societies will inevitably find it difficult and costly to exert social control over the behavior of people who feel free to use whatever means prove most effective in reaching personal goals. Hence, the very sociocultural dynamics that make American institutions weak also enable and entitle Americans to defy institutional controls. If Americans are exceptionally resistant to social control—and therefore exceptionally vulnerable to criminal temptations—the resistance occurs because they *live in* a society that enshrines the unfettered pursuit of individual material success above all other values. In the United States, anomie is considered a virtue. . . .

◼ Notes

1. In our analysis of the value foundations of the American Dream, we rely heavily on Marco Orru's (1990) excellent exegesis of Merton's theory. Characterization[s] of this cultural ethos that are very similar to ours can also be found in studies of the "success theme" in American literature. . . . See Hearn (1977) and Long (1985).

2. Authors' calculations from the 1998 GSS [General Social Survey].

3. For commentary on the notion of the "customer-driven classroom," see The Teaching Professor (1994). The references to "accountability" [and] "value-added" education are from Kozol (1992, p. 277). See also Kozol's (1991) discussion of the state of American public education, and Bellah et al.'s (1991, 170) critique of the idea of an "education industry."

4. For a particularly insightful discussion of the penetration of market-based norms and metaphors into noneconomic realms of social life, see Schwartz (1994a, 1994b).

◼ References

Bassis, M. S., Gelles, R. J., & Levine, A. (1991). *Sociology: An introduction* (4th ed.). New York: McGraw-Hill.

Bellah, R. N., Madsen, R., Sullivan, W. M., Swidler, A., & Tipton, S. M. (1991). *The good society.* New York: Knopf.

Currie, E. (1991). Crime in the market society: From bad to worse in the nineties. *Dissent* (Spring), 254–259.

Downes, D., & Rock, P. (1982). *Understanding deviance: A guide to the sociology of crime and rule breaking.* Oxford, UK: Clarendon.

Edsall, T. B. (1992, February). Willie Horton's message. *New York Review, 13,* 7–11.

Elkins, S. M. (1968). *Slavery: A problem in American institutional and intellectual life* (2nd ed.). Chicago: University of Chicago Press.

Hearn, C. R. (1977). *The American dream in the Great Depression.* Westport, CT: Greenwood.

Heilbroner, R. (1991, November). A pivotal question unanswered. *The World & I: A Chronicle of Our Changing Era,* 538–540.

Hochschild, J. (1995). *Facing up to the American Dream: Race, class, and the soul of the nation.* Princeton, NJ: Princeton University Press.

Kozol, J. (1991). *Savage inequalities: Children in America's schools.* New York: Crown.

Kozol, J. (1992, September). Whittle and the privateers. *The Nation, 21,* 272–278.

Long, E. (1985). *The American dream and the popular novel.* Boston: Routledge & Kegan Paul.

Merton, R. K. (1968). *Social theory and social structure* (enlarged ed.). New York: Free Press.

Orru, M. (1990). Merton's instrumental theory of anomie. In J. Clark, C. Modgil, & S. Modgil (Eds.), *Robert K. Merton: Consensus and controversy* (pp. 231–240). London: Falmer.

Parsons, T. (1951). *The social system.* New York: Free Press.

Passas, N. (1990). Anomie and corporate deviance. *Contemporary Crises, 14,* 157–178.

Polanyi, K. (1957). *The great transformation: The political and economic origins of our time.* Boston: Beacon. (Original work published 1944)

Schwartz, B. (1994a). *The costs of living: How market freedom erodes the best things in life.* New York: Norton.

Schwartz, B. (1994b). On morals and markets. *Criminal Justice Ethics, 13,* 61–69.

Stark, R., Kent, L., & Doyle, D. P. (1982). Religion and delinquency: The ecology of a lost relationship. *Journal of Research in Crime and Delinquency, 19,* 4–24.

Surette, R. (1992). *Media, crime, and criminal justice: Images and realities.* Pacific Grove, CA: Brooks/Cole.

Taylor, I., Walton, P., & Young, J. (1973). *The new criminology.* New York: Harper & Row.

REVIEW QUESTIONS

1. Which portions of Merton's strain theory do the authors emphasize most of all? How does this influence the overall perspective of this selection?

2. Which concepts do the authors propose that were not present in Merton's original proposal of strain theory? Which of these do you believe is most important?

3. Given the current state of the economy in the United States, do you think this perspective has more validity or less? Explain why you answered yes or no.

READING

In this selection, Robert Agnew explains how his extended model of general strain theory is an improvement over the traditional strain theory proposed by Merton. Specifically, general strain theory includes many elements that cause strain and frustration beyond simply the economic reasons that Merton emphasized. You will also see that Agnew's general strain theory places an emphasis on feelings of anger, as well as the primary causes of such feelings, especially the injustice that individuals tend to feel from not being treated fairly. The reader will see that the general strain theory framework is more elaborate, and hence more valid, than the traditional strain model by Merton, simply because it is far more versatile in terms of explaining what causes individuals to be strained or frustrated and thus more likely to commit criminal acts.

Building on the Foundation of General Strain Theory

Specifying the Types of Strain Most Likely to Lead to Crime and Delinquency

Robert Agnew

General strain theory (GST) argues that strains or stressors increase the likelihood of negative emotions like anger and frustration. These emotions create pressure for corrective action, and crime is one possible response (Agnew 1992). Crime may be a method for reducing strain (e.g., stealing the money you desire), seeking revenge, or alleviating negative

SOURCE: Robert Agnew, "Building on the Foundation of General Strain Theory: Specifying the Types of Strain Most Likely to Lead to Crime and Delinquency," *Journal of Research in Crime and Delinquency* 38, no. 4 (2001): 319–61. Copyright © 2001 Sage Publications, Inc. Used by permission of Sage Publications, Inc.

emotions (e.g., through illicit drug use). GST builds on previous strain theories in several ways: most notably, by pointing to several new categories of strain, including the loss of positive stimuli (e.g., loss of a romantic partner, death of a friend), the presentation of negative stimuli (e.g., physical assaults and verbal insults), and new categories of goal blockage (e.g., the failure to achieve justice goals). Recent research demonstrates that many of the specific strains falling under these categories are related to crime and delinquency (see Agnew 2001a for a summary; Aseltine, Gore, and Gordon 2000; Mazerolle et al. 2000; Piquero and Sealock 2000). The specification of these new categories of strain in GST's greatest strength.

This strength, however, is also GST's biggest weakness. GST is so broad that researchers have little guidance as to the specific types of strain to examine in their research. Hundreds of types of strain fall under the major categories of strain listed by GST, as reflected in recent inventories of stressful life events, chronic stressors, and daily life events or hassles (see Cohen, Kessler, and Gordon 1995; Herbert and Cohen 1996 for overviews). And even these inventories do not measure many of the strains described by GST. Furthermore, the broadness of GST makes it difficult to falsify. As Jensen (1995) stated, "If strain can be defined in so many different ways, then strain theory is virtually unfalsifiable. There is always a new measure that might salvage the theory" (p. 152).

It is therefore crucial that GST more precisely specify the types of strain most likely to lead to crime and delinquency. This article represents an attempt to do that. First, strain is defined. Although Agnew (1992) presented a general definition of strain, the term has nevertheless been used in different ways by researchers and it is important to clarify its meaning. Second, previous tests of GST are reviewed to determine what they say about the types of strain most likely to lead to crime. Third, the characteristics of those types of strain most likely to lead to crime are described. Briefly, such strains (1) are seen as unjust, (2) are seen as high in magnitude, (3) are associated with low social control, and (4) create some pressure or incentive to engage in crime. Fourth, these characteristics are then used to predict the likelihood that several types of strain will result in crime. Fifth, suggestions for empirical research are provided.

▨ What Is Strain?

Before discussing the types of strain most likely to lead to crime, it is first necessary to clarify what is meant by the term *strain*. Agnew (1992) stated that strain refers to "relationships in which others are not treating the individual as he or she would like to be treated" (p. 48). Even so, researchers use the term in different ways. Some refer to an objective event or condition (e.g., the infliction of physical abuse, the receipt of poor grades at school), some to the individual's evaluation of an event or condition (e.g., whether juveniles like the way their parents or teachers treat them), and some to the emotional reaction to an event or condition (e.g., whether respondents are angry at how others treat them). To help clarify the meaning of strain, the following definitions are proposed.

Objective strains refer to events or conditions that are disliked by most members of a given group. So, if we state that an individual is experiencing objective strain, we mean that he or she is experiencing an event or condition that is usually disliked by members of his or her group. Many events and conditions are disliked by most people, regardless of group membership (e.g., physical assault, lack of adequate food and shelter). The evaluation of other events and conditions varies with group characteristics, such as gender and age (e.g., Broidy and Agnew 1997; Elder, George, and Shanahan 1996). It is, of course, important for researchers to consider the possibility of such group differences when constructing measures of objective strain.

Empirically, it is possible to determine the objective strains for group members in several ways. Observational research is one method. Anderson (1999), for example, described many of the objective strains in a poor, inner-city, African American community. Surveying a representative sample of group members or people familiar with the group is another method, and both have been employed in the stress research (Turner and Wheaton 1995). In particular, respondents can be asked whether they (or group members) would dislike a range of events and conditions. It is important to present respondents with preestablished lists of events/conditions and to ask them to list events/conditions not on the list. This helps to ensure that a complete list of objective strains is developed.[1]

Subjective strains refer to events or conditions that are disliked by the people who are experiencing (or have experienced) them. So, if we state that individuals are experiencing subjective strain, we mean that they are experiencing an event or condition that *they* dislike. One of the key findings to emerge from the stress research is that individuals often differ in their subjective evaluation of the same objective strains. For example, people differ in how they subjectively evaluate such objective strains as divorce and the death of a family member. The subjective evaluation of an objective strain is a function of a range of factors, including individual traits (e.g., irritability), personal and social resources (e.g., self-esteem, self-efficacy, social support), goals/values/identities, and a range of life circumstances (for overviews, see Dohrenwend 1998; Kaplan 1996; Lazarus 1999). Wheaton (1990), for example, found that the quality of a prior marriage strongly influenced how people evaluated their divorce, with people in bad marriages evaluating their divorce is positive terms. It is also important to note that an individual's evaluation of an objective strain frequently changes over time as the individual copes with the strain. So, although there is a relationship between objective and subjective strain, it is far from perfect.

Most of the research on strain theory employs measures of objective strain (although see Agnew and White 1992). Researchers ask individuals whether they have experienced a certain event or condition (e.g., Did you fail any classes? Do your parents yell at you?); no effort is made to measure the individual's subjective evaluation of this event/condition. This may cause researchers to underestimate the support for strain theory because objective strains sometimes create little subjective strain. This does not mean, however, that researchers should simply employ subjective measures of strain. It is important to examine objective strains as well because this allows us to better distinguish external events from the subjective evaluation of such events. We can then examine individual and group differences in both the exposure to external events/conditions likely to cause strain and the subjective evaluation of those events/conditions. Furthermore, we can explore the factors that influence individual and group differences in the subjective evaluation of the same external events and conditions. This is critical if we are to fully explain individual and group differences in crime. As an illustration, Bernard (1990) argued that poor, inner-city residents have higher rates of violence not only because they experience more objective strains but also because they are more sensitive to such strains (also see Thoits 1995 on individual and group differences in the "vulnerability" to stressors).

The emotional response to an event or condition is closely linked to subjective strain. Subjective strain deals with the individual's evaluation of an event or condition. There are many definitions of emotion, but most state that a central component of an emotion is an evaluation of or an affective response to some object or behavior or idea. Most theorists, however, go on to state that emotions involve more than an evaluation or affective response. For example, they also involve changes in physiological or bodily sensations (see Berkowitz 1993; Smith-Lovin 1995; Thoits 1989). Building on this argument, I would contend that subjective strain is distinct from the full emotional reaction to strain.

Two individuals may evaluate an event/condition in the same way; that is, they may both dislike it an equal amount. So, they have the same level of subjective strain. One may become angry in response to the strain, however, whereas the other may become depressed. And they may differ in the degree to which they experience certain emotions, so one may become quite angry, whereas the other may experience only mild anger. So the same subjective strain may result in rather different emotional reactions. Again, a range of individual and environmental factors influences the emotional reaction to subjective strain. The potential utility of distinguishing between subjective strain and the emotional reaction to strain is highlighted by Broidy and Agnew (1997). They argued that males and females often differ in their emotional reaction to subjective strains. Although both males and females may experience anger, the anger of females is more likely to be accompanied by feelings of guilt, depression, and anxiety. These additional emotions are said to reduce the likelihood of other-directed crime, thereby helping us explain gender differences in such crime.

▨ Research on the Types of Strain Most Likely to Lead to Crime and Delinquency

Agnew (1992) described those types of events and conditions most likely to be classified as objective strains and to result in subjective strain. Such events/conditions involve goal blockage, the loss of positive stimuli, and/or the presentation of negative stimuli. They are also high in magnitude (degree), recent, and of long duration. But as indicated earlier, hundreds of events/conditions meet these criteria, and so there are potentially hundreds of objective and subjective strains. Agnew did *not* discuss whether certain of these strains are more likely to result in crime than others. Rather, he treated these strains as more or less equivalent in terms of their impact on crime. He argued that whether they result in crime is largely a function of the characteristics of the individuals experiencing the strain. In particular, strain is most likely to lead to crime when individuals lack the skills and resources to cope with their strain in legitimate manner, are low in conventional social support, are low in social control, blame their strain on others, and are disposed to crime. This article builds on Agnew by arguing that the effect of strain on crime is a function not only of individual characteristics but also of the type of strain experienced by the individual. Certain types of strain—either objective or subjective strain—are more likely to result in crime than other types.

Previous research on GST provides some information about the types of strain most likely to lead to crime, although much of this research suffers from two problems that severely limit its utility. First, most tests of GST examine only a small portion of the strains described by Agnew (1992). These tests tend to make use of existing data sets, which were not collected for the purpose of testing GST. As a consequence, many key strain measures are missing—particularly measures of the types of goal blockage described by Agnew and measures of certain types of negative treatment, like peer abuse and experiences with racial discrimination and prejudice. So we have little idea whether these types of strain are related to delinquency. Second, most tests of GST examine the effect of a single, cumulative strain measure on delinquency. In some cases, a measure of stressful life events is employed. Hoffmann and associates, for example, tested

GST using a 16- to 18-item measure that focuses on events like "death, illness, or accidents among family or friends; changes in school or residence; parental divorce or separation; and family financial problems" (Hoffmann and Cerbone 1999; Hoffmann and Miller 1998; Hoffmann and Su 1997; also see Aseltine et al. 2000). In other cases, the cumulative strain measure is a composite of several scales and/or items measuring a range of different types of strain, such as neighborhood problems, negative relations with adults, the failure to achieve educational and occupational goals, breaking up with a romantic partner or friend, and getting lower grades than you deserve (e.g., Mazerolle 1998; Mazerolle et al. 2000; Mazerolle and Piquero 1997). The use of such cumulative measures means that we lack information on the effect of the individual strain measures.

Researchers employ cumulative measures of strain because Agnew (1992) argued that it is not the effect of one specific strain or stressor that is important; rather, it is the cumulative effect of all the strains experienced by the individual. He recommended combining individual strain measures into a single scale so as to better estimate this cumulative effect (pp. 62–63). It is assumed that all or most of the individual strain measures in the cumulative scale make some contribution to crime. As will be argued below, there is good reason to question this assumption. Most cumulative measures encompass a wide range of strains, and it is likely that some contribute to crime and some do not. Given this fact, it is not surprising that most cumulative measures have only a moderate impact on crime. A consideration of different types of strain, however, might reveal that some have a strong impact on crime, whereas others have little or no impact.

Some tests of GST do examine the impact of different types of strain on crime among adolescents. Agnew and White (1992) examined the effect of eight strain measures on delinquency, including both general and specific measures. They found that negative life events, life hassles, negative relations with adults, and parental fighting are significantly associated with delinquency. Neighborhood problems, unpopularity with the opposite sex, occupational strain, and clothing strain are not associated with delinquency. Paternoster and Mazerolle (1994) examined the effect of five strain measures on delinquency. They found that neighborhood problems, negative life events, school/peer hassles, and negative relations with adults are significantly associated with subsequent delinquency, whereas a measure of educational and occupational expectations is not (see Mazerolle 1998 for information on gender differences in the effect of these strain measures). Aseltine et al. (2000) found that family and peer conflict (through anger) are related to selected types of delinquency. Agnew and Brezina (1997) found that poor relations with peers is related to delinquency, whereas unpopularity with peers is not. Piquero and Sealock (2000) found that physical and emotional abuse in the household (toward the juvenile and others) is related to delinquency (also see Brezina 1999). Tests of classic strain theory typically find that the failure to achieve educational and occupational goals is *not* related to delinquency (see Agnew 1995a). The failure to achieve economic goals, however, may be related to delinquency (Burton and Dunaway 1994).

Many other studies have not set out to test GST but have examined types of strain that fall under the theory. Several studies found that adolescent crime is significantly related to criminal victimization; parental abuse and neglect; parental rejection; disciplinary techniques that are excessive, very strict, erratic, and/or punitive (e.g., nagging, yelling, threats, insults, and/or hitting); family conflict; parental divorce/separation; and negative experiences at school (low grades, poor relations with teachers, and the perception that school is boring and a waste of time). Summaries of these studies are provided in Agnew (1992,

1995b, 1997, 2001a, 2001b). Studies of adults suggest that crime is related to marital problems, work in the secondary labor market, unemployment in certain cases, and possibly the failure to achieve economic goals (Agnew et al. 1996; Baron and Hartnagel 1997; Cernkovich, Giordano, and Rudolph 2000; Colvin 2000; Crutchfield and Pitchford 1997; Sampson and Laub 1993; Uggen 2000). There has not been enough good research on other types of strain to draw any firm conclusions about their relationship to crime.

The above studies, then, suggested that certain types of strain are related to crime whereas others are not. At this point, it seems safe to conclude that crime is related to verbal and physical assaults, including assaults by parents, spouse/partners, teachers, and probably peers. Crime is also related to parental rejection, poor school performance, and work problems, including work in the secondary labor market. Crime is not related to the expected failure to achieve educational/ occupational success or to unpopularity with peers. Beyond that, the relationship between various strains and crime is unclear.

These data pose a major problem for GST: Why is it that only some types of strain are related to crime? At present, GST offers little guidance in this area. GST, for example, does not allow us to explain why verbal and physical assaults are related to crime, but the failure to achieve educational/occupational goals and unpopularity with peers is not. All of these strains fall under the categories listed by Agnew (1992), and they are frequently high in magnitude (degree), recent, and of long duration.

Recent versions of GST do argue that certain types of strain are especially relevant to crime (Agnew and Brezina 1997; Broidy and Agnew 1997). Agnew (1997, 2001a, 2001b), for example, argued that although many types of goal blockage may lead to delinquency, the failure to achieve monetary, autonomy, and "masculinity" goals are of special importance. And he argued that although a range of negative or noxious stimuli may cause delinquency, physical and verbal assaults are of special importance. These suggestions, however, are not derived from theory. Rather, they represent ad hoc attempts to explain empirical findings or to incorporate other theoretical and empirical work into GST. Much theoretical and empirical work, for example, suggests that threats to one's status, particularly one's masculine status, contribute to crime in certain groups (Anderson 1999; Messerschmidt 1993). Likewise, some theoretical and empirical work suggests that the blockage of autonomy goals contributes to delinquency (Agnew 1984; Moffitt 1993; Tittle 1995).

And although empirical research is starting to point to those types of strain that are and are not related to delinquency, it is not wise to depend on such research to fully resolve this issue. There are hundreds of specific types of strain; it will take empirical researchers a long while to determine their relative importance (although observational research and open-ended, intensive interviews can be of some help here). Furthermore, we would still lack an explanation of why some types of strain have a greater effect on crime than other types. The lack of such an explanation might cause us to overlook certain important types of strain. It is therefore important for GST to better explain why some types of strain are more likely to lead to crime than other types.

Strain is most likely to lead to crime when it is seen as unjust, is seen as high in magnitude, is associated with low social control, and creates some pressure or incentive to engage in criminal coping. If these arguments are correct, types of strain that meet these conditions should be more strongly related to crime than types that do not (although the precise relationship between strain and crime is a function of the characteristics of *both* the strain and the people experiencing the strain). So at the most basic level, researchers should test the above arguments by classifying strains on the above characteristics and then examining the relative impact of these strains on crime. The classification of strains just presented can be used as a starting point for such research. Ideally, researchers should compare the criminal behavior of people who have experienced the above strains. As an alternative, researchers can present people with vignettes describing these types of strain and then ask how likely they or others would be or respond to them with crime (see Mazerolle and Piquero 1997 for a model).

The Cumulative Effect of Strain

This strategy for testing strain theory differs from the approach now taken by most researchers, who examine the impact of cumulative measures of strain on crime—with these cumulative measures often containing types of strain that differ widely on the above characteristics. Although researchers should not ignore the argument that strains may have a cumulative effect on crime, it is most important at this point to determine which types of strain are most strongly related to crime. Once this is determined, researchers can then explore the cumulative impact of strain on crime. Cumulative scales can be created by combining those types of strain that have a significant impact on crime—perhaps weighting them by their regression coefficients. A similar strategy has been successfully employed in the stress literature (Herbert and Cohen 1996; Turner and Wheaton 1995; Wheaton et al. 1997; also see Agnew and White 1992). Or researchers can determine whether strains interact with one another in their impact on crime through the creation of interaction terms (see Wheaton et al. 1997; note the argument that moderate levels of prior stress sometimes *reduce* the negative effects of current stressors).

Distinguishing Strain From Social Control and Social Learning

Researchers testing GST all confront a major problem: Many of the "strain" measures they use—like low grades or harsh parental discipline—can also be taken as social control or social-learning measures. Researchers usually deal with this problem by assigning some measures to the strain camp, some to the social control camp, and some to the social-learning camp. They then try to justify these assignments, although their arguments are often less than convincing. Agnew (1995c) explained why this is so, noting that most variables have implications for strain, social control, *and* social-learning theories. Harsh discipline, for example, is often classified as a type of strain, but some claim that it leads to crime by reducing attachment to parents or implicitly teaching the child that violence is acceptable under certain conditions (see Brezina 1998). It is therefore difficult to classify an independent variable as a purely strain, social control, or social-learning variable. This article makes the same argument: Most types of strain have implications for social control and the social learning of crime. Furthermore, it is argued that those types of strain most likely to lead to crime are those that are associated with low social control and the social learning of crime.

This argument raises a major problem: If those types of strain most strongly related to crime are associated with low control and the social learning of crime, how do we know whether these strains affect crime for reasons related to strain, social control, or social-learning theories? Agnew's (1995c) solution to this problem was to examine the intervening processes described by these theories. Although these theories have many of the same independent variables in common, they differ in terms of their specification of intervening processes. Strain theory argues that these variables increase crime through their effect on negative emotions, control theory argues that they lower the perceived costs of crime, and social-learning theory argues that they influence the perceived desirability of crime. A few studies have attempted to examine such intervening processes, and they typically find that the processes associated with all three theories are operative (see Agnew 1985; Brezina 1998).[2] Unfortunately, most existing data sets do not allow for the proper examination of these intervening processes (see Schieman 2000 and Stone 1995 for discussions of certain of the problems involved in measuring the key negative emotion of anger).

There is a second strategy that may be employed to determine if a strain measure affects crime for reasons related to strain, social control, or social-learning theory. Certain strain measures may affect crime because they reduce social control and/or foster the social learning of crime. As indicated, harsh discipline is said to reduce attachment to parents and foster beliefs conducive to violence. In such cases, we can examine the effect of the strain measure on crime while controlling for the relevant social control and social-learning variables. For example, we can examine the effect of harsh discipline on crime while controlling for parental attachment and beliefs conducive to violence. Or we can examine the effect of teacher conflicts while controlling for attachment to teachers, attachment to school, and grades. If the strain measure still affects crime after such controls, support for strain theory is increased. This strategy cannot be followed in all cases, however. Certain strain measures—like low grades—directly index the respondent's level of social control or social learning. Therefore, it is not possible to control for the relevant control or social-learning variables. Also, there is some risk in arguing that the *direct* effect of the strain measure on crime is best explained by strain theory. Researchers may have failed to control for or properly measure all relevant social control and social-learning variables. And it is possible that the strain measure affects crime for reasons other than those offered by strain, social control, and social-learning theories (e.g., genetic factors may influence both exposure to strain and levels of crime).

Finally, a third strategy sometimes allows us to determine whether strain variables affect crime for reasons distinct from those offered by social control theory. According to the logic of control theory, neutral relationships with other individuals and groups should have the same effect on crime as negative relationships. For example, a juvenile who does not care about her parents should be just as delinquent as a juvenile who dislikes or hates her parents. Both juveniles are equally free to engage in delinquency; that is, both have nothing to lose through delinquency. According to the logic of strain theory, however, the juvenile who hates her parents should be higher in delinquency than the juvenile who does not care about her parents. This is because the juvenile who hates her parents is under more strain. Her hatred likely stems from unpleasant relations with her parents, and it is stressful to live with people you hate. This prediction is easily tested with certain data sets, but researchers rarely compare juveniles who dislike/hate their parents with juveniles who neither like nor dislike their parents (see Nye 1958 for an exception). Similar analyses can be conducted in other areas. For example, researchers can compare the criminal behavior of individuals who hate their grades or jobs with those who do not care about their grades or jobs. If strain theory is correct, individuals who hate their grades or jobs should be higher in crime.

None of these strategies allows us to perfectly determine whether strain variables affect crime for reasons related to strain, social control, or social-learning theories, but taken together they can shed much light on this problem.

Measuring Strain

Many current measures of strain are quite simplistic; single-item measures of specific strains are often employed, with these measures providing little information about the magnitude, injustice, or other dimensions of the strain. A similar situation characterizes the stress literature, although stress researchers are starting to collect more detailed information on stressors to better estimate things like their magnitude. For example, some stress researchers have abandoned simple checklist measures and are employing intensive interviews with semi-structured probes (see Herbert and Cohen 1996; Wethington et al. 1995; Wheaton 1996). Such techniques were developed because respondents often report trivial stressors when checklist measures are used—even when such checklists attempt to focus on serious stressors (Dohrenwend 2000; Herbert and Cohen 1996; Wethington et al. 1995). Also, many stress researchers now recognize that the circumstances associated with the stressors have an important effect on its impact. It is difficult to employ intensive interviews in the large-scale surveys often conducted by criminologists, but criminologists can do a much better job of measuring strain in such surveys. As an illustration, one need only compare the measures of economic strain typically employed by criminologists with those commonly used in the family research. Economic strain is not simply measured in terms of low income or a two- or three-item index of socioeconomic status. Rather, family researchers examine such things as (1) family per capita income; (2) unstable work history, which includes changing to a worse job, demotions, and being fired or laid off; (3) debt-to-asset ratio; and (4) increases or decreases in family income in the past year. Furthermore, researchers recognize that these types of economic strain do not affect all families in the same way. So, more direct measures of economic strain are sometimes employed as well. For example, parents are asked about the extent to which the family has enough money for clothing, food, medical care, and bills. They are also asked about the changes they have had to make to cope with economic hardship, like moving, taking an additional job, canceling medical insurance, and obtaining government assistance(e.g., Conger et al. 1992; Fox and Chancey 1998; Voydanoff 1990; also see Agnew et al. 1996; Cernkovich et al. 2000).

⊠ Conclusion

GST is usually tested by examining the effect of selected types of strain on crime. Researchers, however, have little guidance when it comes to selecting among the many hundreds of types of strain that might be examined. And they have trouble explaining why only some of the strains they do examine are related to crime. This article builds on GST by describing the characteristics of strainful events and conditions that influence their relationship to crime. As indicated, strains are most likely to lead to crime when they (1) are seen as unjust, (2) are seen as high in magnitude, (3) are associated with low social control, and (4) create some pressure or incentive to engage in criminal coping. Based on these characteristics, it is argued that certain types of strain will be unrelated or only weakly related to crime. Such strains include the failure to achieve educational and occupational success, the types of strain that have dominated the research on strain theory. Such strains also include many of the types of strain found in stressful life events scales, which are commonly used to test GST. And it is argued that other types of strain will be more strongly related to crime, including types that have received much attention in the criminology literature (e.g., parental rejection; erratic, harsh parental discipline; child abuse and neglect; negative school experiences) and types that have received little attention (e.g., the inability to achieve selected goals, peer abuse, experiences with prejudice and discrimination).

The arguments presented in this article should have a fundamental impact on future efforts to test GST because they identify those types of strain that should and should not be related to crime. And in doing so, these arguments make it easier to falsify GST. Furthermore, these arguments help explain the contradictory results of past research on strain theory; for example, they help explain why the failure to achieve educational and occupational success is usually not related to crime, whereas verbal and physical assaults usually have a relatively strong relationship to crime.

These arguments also have important policy implications. Agnew (1992) argued that two major policy recommendations flow from GST: reduce the exposure of individuals to strain and reduce the likelihood that individuals will cope with strain through crime (by targeting those individual characteristics conducive to criminal coping). This article suggests a third recommendation: alter the characteristics of strains in ways that reduce the likelihood they will result in crime. Despite our best efforts, many individuals will be exposed to strain. For example, parents, teachers, and criminal justice officials will continue to sanction individuals in ways that are disliked. We can, however, alter the ways in which these sanctions are administered so as to reduce the likelihood that they will (1) be seen as unjust, (2) be seen as high in magnitude, (3) reduce social control, and (4) create some pressure or incentive to engage in crime. In fact, this is one of the central thrusts behind the restorative justice and related movements (see Bazemore 1998; Briathwaite 1989; Sherman 1993, 2000; Tyler 1990). These movements point to ways in which criminal justice officials can increase the perceived justice of sanctions, reduce the perceived magnitude of sanctions, sanction in ways that increase rather than reduce social control, and sanction in ways that create little pressure or incentive for crime. Recommendations in these areas include treating offenders with respect; making them aware of the harm they have caused; giving them some voice in determining sanctions; tempering the use of severe, punitive sanctions; and reintegrating offenders with conventional society through a variety of strategies—like reintegration ceremonies and the creation of positive roles for offenders. Certain parent-training and school-based programs are also structured in ways that reduce the likelihood that strains like disciplinary efforts will be administered in ways that increase the likelihood of criminal coping (see Agnew 1995d, 2001b).

This article, then, extends Agnew's (1992) GST in a way that substantially improves its ability to explain and control crime. Although Agnew (1992) argues that the reaction to strain is largely a function of individual characteristics, this article argues that the reaction to strain is a function of both individual characteristics and the characteristics of the strain that is being experienced. Strain is most likely to lead to crime when individuals possess characteristics conducive to criminal coping (as described in Agnew 1992) and they experience types of strain conducive to criminal coping (as described above). This extension of strain theory parallels recent developments in the stress literature. Like Agnew (1992), stress researchers argued that the impact of stressors on outcome variables was largely a function of individual characteristics like coping skills and social support. Stress researchers, however, have increasingly come to realize that stressors do not have comparable impacts on outcome variables. Certain stressors are significantly related to outcome variables—most often measures of mental and physical health—whereas others are not (e.g., Aseltine et al. 2000; Aseltine, Gore, and Colten 1998; Brown 1998; Wethington et al. 1995; Wheaton et al. 1997; Dohrenwend 1998). So we must consider both the nature of the stressor and the characteristics of the individual experiencing the stressor.

Like Agnew's (1992) original statement of GST, however, the arguments in this article are in need of further research and elaboration. The predictions regarding the impact of specific types of strain on crime are tentative. Researchers should use the methods described in this article to better determine the extent to which these and other types of strain are seen as unjust, are seen as high in magnitude, are associated with low social control, and create some pressure or incentive for crime. Such research should improve the accuracy of the predictions that are made. Furthermore, researchers should pay attention to the impact of group membership in such research. For example, it is likely that there are group differences in the extent to which certain strains are seen as unjust or high in magnitude.[3] In addition, researchers should examine whether particular strains have a greater impact on some types of crime than other types. For example, some research suggests that certain strains are more strongly related to aggression/violence than to other types of crime (e.g., Agnew 1990; Aseltine et al. 1998, 2000; Mazerolle et al. 2000; Mazerolle and Piquero 1997). (Likewise, the stress research reveals that some stressors are more strongly related to some types of negative outcomes than to others.) The arguments presented in this article, then, are still in need of much development, but that does not diminish their central thrust—some strains are more likely than others to result in crime.

Notes

1. Most of the research in criminology simply assumes that certain events or conditions are disliked by most of the people being studied. This is probably a reasonable assumption in most cases (e.g., criminal victimization), although it is a more questionable assumption in other cases (e.g., changing schools). A potentially more serious problem with the criminology research is that researchers rarely employ a complete or comprehensive list of objective strains. Researchers usually examine only a few types of objective strain—often overlooking many of the most important types. For example, interviews with adolescents suggest that peer conflict and abuse are among the most important types of objective strain in this group, but such conflict/abuse is rarely considered by researchers (although see Agnew 1997; Agnew and Brezina 1997; Ambert 1994; Aseltine, Gore, and Gordon 2000; Seiffge-Krenke 1995). Likewise, experiences with racial prejudice and discrimination are seldom considered by researchers, despite evidence that such experiences are a major type of objective strain among African Americans and others (Ambert 1994; Anderson 1999). Recent research suggests that the failure to examine the full range of stressors can lead

researchers to substantially underestimate the effect of stress or strain (Turner, Wheaton, and Lloyd 1995).

2. One should also take account of the possibility that anger may indirectly affect crime by reducing the perceived costs of crime and increasing the perceived desirability of crime, as indicated earlier in this article.

3. Explaining the origins of such differences is, of course, central to any effort to develop the macroside of GST (see the excellent discussions in Anderson 1999; Bernard 1990; Colvin 2000; and Messerschmidt 1993).

References

Agnew, Robert. 1984. "Autonomy and Delinquency." *Sociological Perspectives* 27: 219–36.

———. 1985. "A Revised Strain Theory of Delinquency." *Social Forces* 64: 151–67.

———. 1990. "The Origins of Delinquent Events: An Examination of Offender Accounts." *Journal of Research in Crime and Delinquency* 27: 267–94.

———. 1992. "Foundation for a General Strain Theory of Crime and Delinquency." *Criminology* 30: 47–87.

———. 1995a. "Strain and Subcultural Theories of Criminality." In *Criminology: A Contemporary Handbook,* edited by Joseph F. Sheley. Belmont, CA: Wadsworth.

———. 1995b. "The Contribution of Social-Psychological Strain Theory of the Explanation of Crime and Delinquency." In *The Legacy of Anomie Theory, Advances in Criminological Theory,* Vol. 6, edited by Freda Adler and William S. Laufer. New Brunswick, NJ: Transaction.

———. 1995c. "Testing the Leading Crime Theories: An Alternative Strategy Focusing on Motivational Processes." *Journal of Research in Crime and Delinquency* 32: 363–98.

———. 1995d. "Controlling Delinquency: Recommendations from General Strain Theory." In *Crime and Public Policy,* edited by Hugh D. Barlow. Boulder, CO: Westview.

———. 1997. "Stability and Change in Crime over the Life Course: A Strain Theory Explanation." In *Developmental Theories of Crime and Delinquency, Advances in Criminological Theory,* Vol. 7, edited by Terence P. Thornberry. New Brunswick, NJ: Transaction.

———. 2001a. "An Overview of General Strain Theory." In *Explaining Criminals and Crime,* edited by Raymond Paternoster and Ronet Bachman. Los Angeles: Roxbury.

———. 2001b. *Juvenile Delinquency: Causes and Control.* Los Angeles: Roxbury.

Agnew, Robert, and Timothy Brezina. 1997. "Relational Problems with Peers, Gender and Delinquency." *Youth and Society* 29: 84–111.

Agnew, Robert, Francis T. Cullen, Velmer S. Burton, Jr., T. David Evans, and R. Gregory Dunaway. 1996. "A New Test of Classic Strain Theory." *Justice Quarterly* 13: 681–704.

Agnew, Robert, and Helene Raskin White. 1992. "An Empirical Test of General Strain Theory." *Criminology* 30: 475–99.

Ambert, Ann-Marie. 1994. "A Qualitative Study of Peer Abuse and Its Effects: Theoretical and Empirical Implications." *Journal of Marriage and the Family* 56: 119–30.

Anderson, Elijah. 1999. *Code of the Street.* New York: Norton.

Aseltine, Robert H., Jr., Susan Gore, and Mary Ellen Colten. 1998. "The Co-occurrence of Depression and Substance Abuse in Late Adolescence." *Development and Psychopathology* 10: 549–70.

Aseltine, Robert H., Jr., Susan Gore, and Jennifer Gordon. 2000. "Life Stress, Anger and Anxiety, and Delinquency: An Empirical Test of General Strain Theory." *Journal of Health and Social Behavior* 41: 256–75.

Baron, Stephen W., and Timothy F. Hartnagel. 1997. "Attributions, Affect, and Crime: Street Youths' Reactions to Unemployment." *Criminology* 35: 409–34.

Bazemore, Gordon. 1998. "Restorative Justice and Earned Redemption." *American Behavioral Scientist* 41: 768–813.

Berkowitz, Leonard. 1993. *Aggression: Its Causes, Consequences, and Control.* New York: McGraw-Hill.

Bernard, Thomas J. 1990. "Angry Aggression among the 'Truly Disadvantaged.'" *Criminology* 28: 73–96.

Briathwaite, John. 1989. *Crime, Shame, and Reintegration.* Cambridge, UK: Cambridge University Press.

Brezina, Timothy. 1998. "Adolescent Maltreatment and Delinquency: The Question of Intervening Processes." *Journal of Research in Crime and Delinquency* 35: 71–99.

———. 1999. "Teenage Violence toward Parents as an Adaptation to Family Strain." *Youth and Society* 30: 416–44.

Broidy, Lisa, and Robert Agnew. 1997. "Gender and Crime: A General Strain Theory Perspective." *Journal of Research in Crime and Delinquency* 34: 275–306.

Brown, George W. 1998. "Loss and Depressive Disorders." In *Adversity, Stress, and Psychopathology,* edited by Bruce P. Dohrenwend. New York: Oxford University Press.

Burton, Velmer S., Jr., and R. Gregory Dunaway. 1994. "Strain, Relative Deprivation, and Middle-Class Delinquency." In *Varieties of Criminology,* edited by Greg Barak. Westport, CT: Praeger.

Cernkovich, Stephen A., Peggy C. Giordano, and Jennifer L. Rudolph. 2000. "Race, Crime, and the American Dream." *Journal of Research in Crime and Delinquency* 37: 131–70.

Cohen, Sheldon, Ronald C. Kessler, and Lynn Underwood Gordon. 1995. *Measuring Stress.* New York: Oxford University Press.

Colvin, Mark. 2000. *Crime and Coercion: An Integrated Theory of Chronic Criminality.* New York: St. Martin's.

Conger, Rand D., Katherine J. Conger, Glen H. Elder, Jr., Frederick O. Lorenz, Ronald L. Simons, and Lee B. Whitbeck. 1992. "A Family Process Model of Economic Hardship and Adjustment of Early Adolescent Boys." *Child Development* 63: 526–41.

Crutchfield, Robert D., and Susan R. Pitchford. 1997. "Work and Crime: The Effects of Labor Stratification." *Social Forces* 76:93–118.

Dohrenwend, Bruce P. 1998. *Adversity, Stress, and Psychopathology.* New York: Oxford University Press.

———. 2000. "The Role of Adversity and Stress in Psychopathology: Some Evidence and Its Implications for Theory and Research." *Journal of Health and Social Behavior* 41: 1–19.

Elder, Glen H., Jr., Linda K. George, and Michael J. Shanahan. 1996. "Psychosocial Stress over the Life Course." In *Psychosocial Stress,* edited by Howard B. Kaplan. San Diego, CA: Academic Press.

Fox, Greer Linton, and Dudley Chancey. 1998. "Sources of Economic Distress." *Journal of Family Issues* 19: 725–49.

Herbert, Tracy B., and Sheldon Cohen. 1996. "Measurement Issues in Research on Psychosocial Stress." In *Psychosocial Stress,* edited by Howard B. Kaplan. San Diego, CA: Academic Press.

Hoffmann, John P., and Felice Gray Cerbone. 1999. "Stressful Life Events and Delinquency Escalation in Early Adolescence." *Criminology* 37: 343–74.

Hoffmann, John P., and Alan S. Miller. 1998. "A Latent Variable Analysis of Strain Theory." *Journal of Quantitative Criminology* 14: 83–110.

Hoffmann, John P., and S. Susan Su. 1997. "The Conditional Effects of Stress on Delinquency and Drug Use: A Strain Theory Assessment of Sex Differences." *Journal of Research in Crime and Delinquency* 34: 46–78.

Jensen, Gary F. 1995. "Salvaging Structure through Strain: A Theoretical and Empirical Critique." In *The Legacy of Anomie Theory, Advances in Criminological Theory,* Vol. 6, edited by Freda Adler and William S. Laufer. New Brunswick, NJ: Transaction.

Kaplan, Howard B. 1996. "Psychosocial Stress from the Perspective of Self Theory." In *Psychosocial Stress,* edited by Howard B. Kaplan. San Diego, CA: Academic Press.

Lazarus, Richard S. 1999. *Stress and Emotion: A New Synthesis.* New York: Springer.

Mazerolle, Paul. 1998. "Gender, General Strain, and Delinquency: An Empirical Examination." *Justice Quarterly* 15: 65–91.

Mazerolle, Paul, Velmer S. Burton, Jr., Francis T. Cullen, T. David Evans, and Gary L. Payne. 2000. "Strain, Anger, and Delinquency Adaptations: Specifying General Strain Theory." *Journal of Criminal Justice* 28: 89–101.

Mazerolle, Paul, and Alex Piquero. 1997. "Violent Responses to Strain: An Examination of Conditioning Influences." *Violence and Victims* 12: 323–43.

Messerschmidt, James W. 1993. *Masculinities and Crime.* Lanham, MD: Rowman and Littlefield.

Moffitt, Terrie E. 1993. "'Life-Course Persistent' and 'Adolescent-Limited' Antisocial Behavior: A Developmental Taxonomy." *Psychological Review* 100: 674–701.

Nye, Ivan. 1958. *Family Relationships and Delinquent Behavior.* New York: John Wiley.

Paternoster, Raymond, and Paul Mazerolle. 1994. "General Strain Theory and Delinquency: A Replication and Extension." *Journal of Research in Crime and Delinquency* 31: 235–63.

Piquero, Nicole Leeper, and Miriam D. Sealock. 2000. "Generalizing General Strain Theory: An Examination of an Offending Population." *Justice Quarterly* 17: 449–84.

Sampson, Robert J., and John H. Laub. 1993. *Crime in the Making.* Cambridge, MA: Harvard University Press.

Schieman, Scott. 2000. "Education and the Activation, Course, and Management of Anger." *Journal of Health and Social Behavior* 41: 20–39.

Seiffge-Krenke, Inge. 1995. *Stress, Coping, and Relationships in Adolescence.* Mahwah, NJ: Lawrence Erlbaum.

Sherman, Lawrence W. 1993. "Defiance, Deterrence, and Irrelevance: A Theory of Criminal Sanctions." *Journal of Research in Crime and Delinquency* 30: 445–73.

———. 2000. "The Defiant Imagination: Consilience and the Science of Sanctions." Presented at the University of Pennsylvania, Philadelphia.

Smith-Lovin, Lynn. 1995. "The Sociology of Affect and Emotion." In *Sociological Perspectives on Social Psychology,* edited by Karen S. Cook, Gary Alan Fine, and James S. House. Needham Heights, NY: Allyn and Bacon.

Stone, Arthur A. 1995. "Measurement of Affective Response." In *Measuring Stress,* edited by Sheldon Cohen, Ronald C. Kessler, and Lynn Underwood Gordon. New York: Oxford University Press.

Sykes, Gresham M., and David Matza. 1957. "Techniques of Neutralization: A Theory of Delinquency." *American Sociological Review* 22: 664–70.

Thoits, Peggy A. 1989. "The Sociology of Emotions." *Annual Review of Sociology* 15: 317–42.

———. 1995. "Stress, Coping, and Social Support Processes: Where Are We? What Next?" *Journal of Health and Social Behavior* (Extra Issue): 53–79.

Tittle, Charles R. 1995. *Control Balance: Toward a General Theory of Deviance.* Boulder, CO: Westview.

Turner, R. Jay, and Blair Wheaton. 1995. "Checklist Measurement of Stressful Life Events." In *Measuring Stress,* edited by Sheldon Cohen, Ronald C. Kessler, and Lynn Underwood Gordon. New York: Oxford University Press.

Turner, R. Jay, Blair Wheaton, and Donald A. Lloyd. 1995. "The Epidemiology of Social Stress." *American Sociological Review* 60: 104–25.

Tyler, Tom. 1990. *Why People Obey the Law.* New Haven, CT: Yale University Press.

Uggen, Christopher. 2000. "Work as a Turning Point in the Life Course of Criminals: A Duration Model of Age. Employment, and Recidivism." *American Sociological Review* 67: 529–46.

Voydanoff, Patricia. 1990. "Economic Distress and Family Relations: A Review of the Eighties." *Journal of Marriage and the Family* 52: 1099–115.

Wethington, Elaine, George W. Brown, and Ronald C. Kessler. 1995. "Interview Measurement of Stressful Life Events." In *Measuring Stress,* edited by Sheldon Cohen, Ronald C. Kessler, and Lynn Underwood Gordon. New York: Oxford University Press.

Wheaton, Blair. 1990. "Life Transitions, Role Histories, and Mental Health." *American Sociological Review* 55: 209–24.

———. 1996. "The Domains and Boundaries of Stress Concepts." In *Psychosocial Stress,* edited by Howard B. Kaplan. San Diego. CA: Academic Press.

Wheaton, Blair, Patricia Roszell, and Kimberlee Hall. 1997. "The Impact of Twenty Childhood and Adult Traumatic Stressors on the Risk of Psychiatric Disorder." In *Stress and Adversity over the Life Course: Trajectories and Turning Points,* edited by Ian H. Gottlib. New York: Cambridge University Press.

REVIEW QUESTIONS

1. What types of strain does Agnew propose in this article that go beyond those presented by Merton's strain theory? Explain these new forms/categories of strain.

2. Which types of strain or characteristics of strain does Agnew claim are most important in causing criminal activity? What other types of theories do these characteristics of strain involve?

3. What types of policy implications does Agnew claim should be used given the findings of this study? Which of these do you believe is likely to be most effective and why?

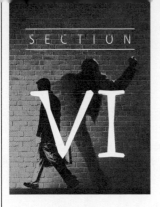

THE CHICAGO SCHOOL AND CULTURAL/ SUBCULTURAL THEORIES OF CRIME

The introduction to this section will examine the origin and evolution of the Chicago/Ecological School theory, otherwise known as the ecological perspective or the theory of social disorganization. We will also discuss modern research on this theory, which assumes that the environment people live in determines their behavior. Finally, the assumptions and dynamics of cultural/subcultural theory in society will be discussed, with an emphasis on differences in certain models emphasizing inner-city subcultures and other modern examples (e.g., street gangs). We will finish this section introduction by reviewing the policy implications that have been suggested by this perspective of crime.

◪ Introduction

This section examines the Chicago School of criminology, which is otherwise known as the ecological perspective or theory of social disorganization, for reasons that will become very clear. The Chicago School evolved there because the city at that time (late 19th and early 20th centuries) desperately needed answers for its exponentially growing problem of delinquency and crime. Thus, this became a primary focus in the city of Chicago, where total chaos prevailed at the time.

A significant portion of the Chicago perspective involved the transmission of cultural values to other peers, and even across generations, as the older youths relayed their antisocial values and techniques to the younger children. Thus, the cultural/subcultural perspective is also a key area of this theoretical model. This cultural aspect of the Chicago model is also examined in this section, as well as other subculture frameworks of offending behaviors.

◪ The School of Ecology and the Chicago School of Criminology

Despite its name specifying one city, the **Chicago School of criminology** represents one of the most valid and generalizable theories we will discuss in this book, in the sense that many of its propositions can be readily applied to the growth and evolution of virtually all cities around the world. The Chicago School, which is often referred to as the **Ecological School** or the theory of **social disorganization**, also represents one of the earliest examples of balancing theorizing with scientific analysis and at the same time guiding important programs and policy implementations that still thrive today. Perhaps most important, the Chicago School of criminology was the epitome of using theoretical development and scientific testing to help improve conditions in society when it was most needed, which can be appreciated only by understanding the degree of chaos and crime that existed in Chicago in the late 1800s and early 1900s.

Cultural Context: Chicago in the 1800s and Early 1900s

Experts have determined that 19th-century Chicago was the fastest-growing city in U.S. history.[1] Census data show the population went from about 5,000 in the early 1800s to more than 2 million by 1900; put another way, the population more than doubled every decade during the 19th century.[2] This massive rate of growth—much faster than that seen in other large U.S. cities such as Boston, Baltimore, New York, Philadelphia, and San Francisco—was due to Chicago's central geographic position. It was in many ways land-locked because although it sits on Lake Michigan, there was no water route to the city from the Atlantic Ocean until the Erie Canal opened in 1825, which provided access to the Great Lakes region for shipping and migration of people. Three years later came

---

[1]For an excellent discussion of the early history of Chicago, see Thomas J. Bernard, *The Cycle of Juvenile Justice* (New York: Oxford University Press, 1992).

[2]See discussion in George B. Vold, Thomas J. Bernard, and Jeffrey B. Snipes, Theoretical Criminology, 5th ed. (New York: Oxford University Press, 2002), 117–22.

the first U.S. passenger train, the Baltimore & Ohio railroad, with a route from a mid-Atlantic city to central areas. These two transportation advancements created a continuous stream migration to the Chicago area, increased again when the transcontinental railroad was completed in 1869, linking both coasts with the U.S. Midwest.[3]

It is important to keep in mind that in the early to mid-1800s, many large U.S. cities had virtually no formal social agencies to handle problems of urbanization: No social workers, building inspectors, garbage collectors, or even police officers. Once police agencies were introduced, their duties often included finding lost children and collecting the garbage, primarily because there weren't other agencies to perform these tasks. Therefore, communities were largely responsible for solving their own problems, including crime and delinquency. By the late 1800s, however, Chicago was largely made up of citizens who did not speak a common language and did not share each other's cultural values. This phenomenon is consistent with Census Bureau data from that era, which shows that 70% of Chicago residents were foreign born and another 20% were first-generation Americans. Thus, it was almost impossible for these citizens to organize themselves to solve community problems because in most cases, they could not even understand each other.

This resulted in the type of chaos and normlessness that Durkheim predicted would occur when urbanization and industrialization occurred too rapidly; in fact, Chicago represented the archetypal example of a society in an anomic state, with almost a complete breakdown in control. One of the most notable manifestations of this breakdown in social control was that children were running wild on the streets in gangs, with adults making little attempt to intervene. So delinquency was soaring, and it appeared that the gangs controlled the streets as much as any other group.

The leaders and people of Chicago needed theoretical guidance to develop solutions to their problems, particularly regarding the high rates of delinquency. This was a key factor in why the Department of Sociology at the University of Chicago became so important and dominant in the early 1900s. Essentially, modern sociology developed in Chicago because the city needed it the most to solve its social problems. Thus, Chicago became a type of laboratory for the sociological researchers, and they developed a number of theoretical models of crime and other social ills that are still shown to be empirically valid today.

Ecological Principles in City Growth and Concentric Circles

In the 1920s and 1930s, several new perspectives of human behavior and city growth were offered by sociologists at the University of Chicago. The first relevant model was proposed by Robert E. Park, who claimed that much of human behavior, especially the way cities grow, follow the basic principles of ecology that had been documented and applied to wildlife for many years at that point.[4] Ecology is essentially the study of the dynamics and processes through which plants and animals interact with the environment. In an application of Darwinian theory, Park proposed that the growth of cities follows a natural pattern and evolution.

---

[3]These dates were taken from *The World Almanac, 2000, Millennium Collector's Edition*, (Mahwah, NJ: Primedia Reference, 2000).

[4]Robert E. Park, "Human Ecology," *American Journal of Sociology* 42 (1936); Robert E. Park, *Human Communities* (Glencoe, IL: The Free Press, 1952).

Specifically, Park claimed that cities represent a type of complex organism with a sense of unity composed of the interrelations among the citizens and groups in the city. Park applied the ecological principle of symbiosis to explain the dependency of various citizens and units on each other: Everyone is better off working together as a whole. Furthermore, Park claimed that all cities would contain identifiable clusters, which he called **natural areas**, where the cluster had taken on a life or organic unity by itself. To clarify, many cities have neighborhoods that are made up of primarily one ethnic group or are distinguished by certain features. For example, New York City's Hell's Kitchen, Times Square, and Harlem represent areas of one city that have each taken on a unique identity; however, each of them contributes to the whole makeup and identity of the city. The same can be seen in other cities, such as Baltimore, which in a two-mile area has the Inner Harbor, Little Italy, and Fell's Point, with each area complementing the other zones. From Miami to San Francisco to New Orleans, all cities across America, and throughout the world for that matter, contain these identifiable natural areas.

Applying several other ecological principles, Park also noted that some areas (or species) may invade and dominate adjacent areas (species). The dominated area or species can recede, migrate to another location, or die off. In wildlife, an example is the incredible proliferation of a weed called kudzu, which grows at an amazing pace and has very large leaves. It grows on top of other plants, trees, fields, and even houses, seeking to cover everything in its path and steal all of the sunlight needed by other plants. Introduced to the United States in the 1800s at a world exposition, this weed was originally used to stop erosion but got out of control, especially in the southeastern region of the United States. Now the weed costs the government more than $350 million each year in destruction of crops and other flora. This is a good example of a species that invades, dominates, and causes the recession of other species in the area.

A similar example can be found in the introduction of bison on Santa Catalina Island off the Southern California coast in the 1930s. About three dozen buffalo were originally imported to the island for a movie shoot, and the producers decided not to spend the money to remove them after the shoot, so they have remained and multiplied. Had this occurred in many parts of the United States, it would not have caused a problem. However, the largest mammal native to the island before the bison was a four-pound fox. So the buffalo—now numbering in the hundreds, to the point where several hundred were recently shipped to their native western habitat—have destroyed much of the environment, driving to extinction some plants and animals unique to Catalina Island. Like the kudzu, the bison came to dominate the environment; in this case, other species couldn't move off the island and died off.

Park claimed that a similar process occurs in human cities as some areas invade other zones or areas, and the previously dominant area must relocate or die off. This is easy to see in modern times with the growth of what is known as *urban sprawl*. Geographers and urban planners have long acknowledged the detriment to traditionally stable residential areas when businesses move into an area. Some of the most recent examples involve the battles between long-time homeowners against the introduction of malls, businesses, and other industrial centers in a previously zoned residential district. The media have documented such fights, especially with the proliferation of such establishments as Wal-Mart or Super K-Marts, which residents perceive, and perhaps rightfully so, as an invasion. Such an invasion can create chaos in a previously stable residential community due to increased traffic, transient population, and perhaps most important, crime. Furthermore, some cities are granting power to such development through eminent domain, in which the local government can take land from the homeowners to rezone and import businesses.

When Park developed his theory of ecology, he observed the trend of businesses and factories invading the traditionally residential areas of Chicago, which caused major chaos and breakdown in stability in those areas. Readers, especially those who were raised in suburban or rural areas, can likely relate to this; going back to where they grew up, they can often see fast growth. Such development can devastate the informal controls (such as neighborhood networks or family ties, etc.) as a result of invasion by a highly transient group of consumers and residents who do not have strong ties to the area.

This leads to a psychological indifference toward the neighborhood, in which no one cares about protecting the community any longer. Those who can afford to leave the area do, and those who can't afford to get out remain until they can save enough money to do so. When Park presented his theory of ecology in the 1920s, factories moving into the neighborhood often meant having a lot of smoke billowing out of chimneys. No one wanted to live in such a place, particularly at a time when the effects of pollution were not understood and such smoke stacks had no filters. Certain parts of Chicago and other U.S. cities were perpetually covered by smog these factories created. In highly industrial areas, the constant and vast coverage of smoke and pollutants made it seem to be snowing or overcast most of the time. It is easy to see how such invasions can completely disrupt the previously dominant and stable residential areas of a community.

Park's ideas became even more valid and influential with the complementary perspective offered by Ernest W. Burgess.[5] He proposed a theory of city growth in which cities were seen as growing not simply on the edges but from the inside outward. It is easy to observe cities growing on the edges, as in the example of urban sprawl, but Burgess claimed that the source of the growth was in the center of the city. Growth of the inner city puts pressure on adjacent zones, which in turn begin to grow into the next adjacent zones, following the ecological principle of "succession" identified by Park. This type of development is referred to as radial growth, meaning beginning on the inside and rippling outward.[6]

An example of this can be seen by watching a drop of water fall into the center of a bucket filled with water. The waves from the impact will form circles that ripple outward. This is exactly how Burgess claimed that cities grow. Although the growth of cities is most visible on the edges, largely due to development of business and homes where only trees or barren land existed before, the reason for growth on the edges is due to pressure forming from the very heart of the city. Another good analogy is the "domino effect," because pressure from the center leads to pressure to grow on the next zone, which leads to pressure on the adjacent zones, and so forth.

Burgess also specified the primary zones—five pseudo-distinctive natural areas in a constant state of flux due to growth—that all cities appear to have. He depicted these zones as a set of concentric circles. The first innermost circle was called Zone I, or the central business district due to costs. This area of a city contains the large business buildings, modern skyscrapers that are home to banking, chambers of commerce, courthouses, and other essential business/political centers such as police headquarters, post offices, and so on.

---

[5]Ernest W. Burgess, "The Growth of the City," in *The City*, ed. Robert Park, Ernest W. Burgess, and Roderick D. McKenzie (Chicago: University of Chicago Press, 1928).

[6]See Vold et al., *Theoretical Criminology*, 118–21.

Just outside the business district is the unnumbered "factory zone." It is perhaps the most significant in terms of causing crime because it invaded the previously stable residential zones in Zone II, which was identified as the **zone in transition**. Zone II is appropriately named because it was truly in a state of transition from residential to industrial, primarily because this was the area of the city in which business and factories were invading residential areas. Zone II was the area that was most significantly subjected to the ecological principles Park suggested: invasion, domination, recession, and succession. Subsequent criminological theorists focused on this zone.

According to Burgess's theory of concentric circles, Zone III was the "workingmen's homes," relatively modest houses and apartment buildings; Zone IV consisted of higher-priced family dwellings and more expensive apartments, and Zone V was the suburban or commuter zone. These outer three zones Burgess identified were of less importance in terms of crime, primarily because as a general rule, the farther a family could move out of the city the better the neighborhood was in terms of social organization and the lower the rate of social ills (e.g., poverty, delinquency). The important point of this theory of concentric circles is that the growth of each inner zone puts pressure on the next zone to grow and push into the next adjacent zone.

It is easy for readers to see examples of **concentric circles theory**. Wherever you live in the United States, any major city provides real-life evidence of the validity of this perspective. For example, whether people drive on Interstate 95 through Baltimore or Interstate 5 through Los Angeles, they will see the same pattern of city structure. As they approach each of the cities, they see suburban wealth in the homes and buildings, often hidden by trees off the highway. Closer to the cities, they see homes and buildings deteriorate in terms of value. Because parts of the highway systems near Baltimore and Los Angeles are somewhat elevated above the ground, drivers entering Zone II can easily see the prevalence of factories and the highly deteriorated nature of the areas. Today, many 20th-century factories have been abandoned or are limited in use; these factory zones consist of rusted-out or demolished buildings. Zone II is also often the location of subsidized or public housing. Only the people who can't afford to live anywhere else are forced to live in these neighborhoods. Finally, as drivers enter the inner city of skyscrapers, the conditions improve dramatically because the major businesses have invested their money here. Compared to Zone II, this innermost area is a utopia.

This theory applies around the world, and we challenge readers to find any major city throughout the world that did not develop this way. Nowadays, some attempts have been made to plan their community development, and others have experienced the convergence of several patterns of concentric circles as central business districts (i.e., Zone Is) are developed in what was previously suburb (i.e., Zone Vs). However, for the most part, the theoretical framework of concentric circles still has a great deal of support. In fact, even cities found in Eastern cultures have evolved this way. Therefore, Park's application appears to be correct: Cities grow in a natural way across time and place, abiding by the natural principles of ecology.

Shaw and McKay's Theory of Social Disorganization

Clifford Shaw and Henry McKay drew heavily on their colleagues at the University of Chicago in devising their theory of social disorganization, which became known as the

Chicago School theory of criminology.[7] Shaw had been producing excellent case studies for years on the individual (i.e., micro) level before he took on theorizing on the macro (i.e., structural) level of crime rates.[8] However, once he began working with McKay, he devised perhaps the most enduring and valid model of why certain neighborhoods have more social problems, such as delinquency, than others.

In this model, Shaw and McKay proposed a framework that began with the assumption that certain neighborhoods in all cities have more crime than other parts of the city, most of them located in Burgess's Zone II, which is the zone in the transition from residential to industrial, due to the invasion of factories. According to Shaw and McKay, the neighborhoods that have the highest rates of crime typically have at least three common problems (see Figure 6.1): physical dilapidation, poverty, and heterogeneity (which is a fancy way of saying a high cultural mix). There were other common characteristics to these neighborhoods that Shaw and McKay noted, such as a highly transient population, meaning that people constantly move in and out of the area, as well as unemployment among the residents of the neighborhood.

As noted in Figure 6.1, other social ills are included as antecedent factors in the theoretical model. The antecedent social ills tend to lead to a breakdown in social organization, which is why this model is referred to as the theory of social disorganization. Specifically, it is predicted that the antecedent factors of poverty, heterogeneity, and physical dilapidation lead to a state of social disorganization, which in turn leads to crime and delinquency. This means that the residents of a neighborhood that fits the profile of having a high rate of poor, culturally mixed residents in a dilapidated area cannot come together to solve problems, such as delinquency among youth.

One of the most significant contributions of Shaw and McKay was that they demonstrated that the prevalence and frequency of various social ills—such as poverty, disease, and low birth weight—tend to overlap with higher delinquency rates. Regardless of what social problem is measured, higher rates are almost always clustered in the zone in transition. Shaw and McKay believed there was a breakdown of informal social controls in these areas and that children began to learn offending norms from their interactions with peers on the street, through what they called "play activities."[9] Thus, the breakdown in the conditions of the neighborhood leads to social disorganization, which in turn leads to delinquents learning criminal activities from older youth in the neighborhood. Ultimately, the failure of the neighborhood residents to organize themselves allows the older youth to govern the behavior of the younger children. Basically, the older youth in the area provide a system of organization where the neighborhood cannot, so younger children follow them.

---

[7]Clifford Shaw and Henry D. McKay, *Juvenile Delinquency and Urban Areas* (Chicago: University of Chicago Press, 1942); Clifford Shaw and Henry D. McKay, *Juvenile Delinquency and Urban Areas,* Rev. ed (Chicago: University of Chicago Press, 1969).

[8]Clifford Shaw, *Brothers in Crime* (Chicago: University of Chicago Press, 1938); Clifford Shaw, *The Jackroller* (Chicago: University of Chicago Press, 1930); Clifford Shaw, *The Natural History of a Delinquent Career* (Chicago: University of Chicago Press, 1931).

[9]Shaw, *Brothers in Crime,* 354–55.

| Figure 6.1 | Model of Shaw and McKay's Theory of Social Disorganization |

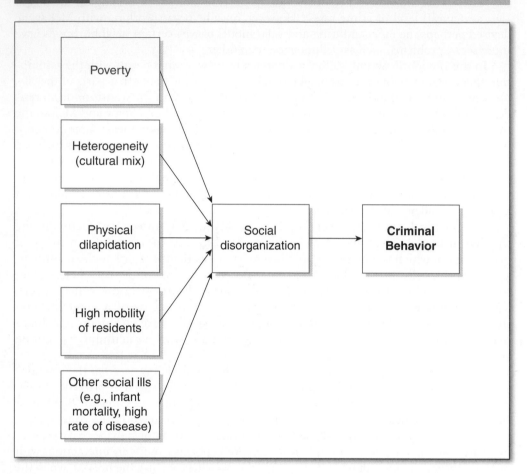

One of the best things about Shaw and McKay's theoretical model is that the researchers support their propositions with data from the U.S. census and city records showing that neighborhoods with high rates of poverty and physical dilapidation and high cultural mix also had the highest rates of delinquency and crime. Furthermore, the high rates of delinquency and other social problems were consistent with Burgess's framework of concentric circles, in that the highest rates were observed for the areas that were in Zone II, the zone in transition. There was an exception to the model: The Gold Coast area along the northern coast of Lake Michigan was notably absent from the high rates of social problems, particularly delinquency, even though it was geographically in Zone II according to the otherwise consistent model of concentric circles and neighborhood zones.

So the findings of Shaw and McKay were as predicted in the sense that high delinquency rates occurred in areas where factories were invading the residential district. Furthermore, Shaw and McKay's longitudinal data showed that it did not matter which

| Figure 6.2 | Zone Map of Male Delinquents in Chicago 1925–1933 |

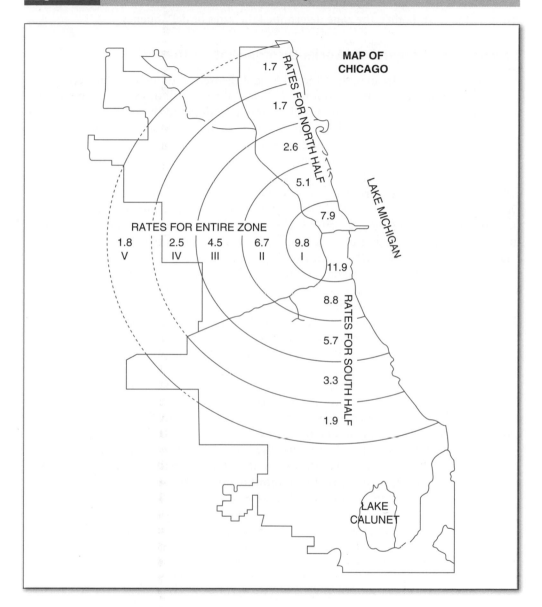

ethnic groups lived in Zone II (zone in transition); all groups (with the exception of Asians) that lived in that zone had high delinquency rates during their residency. On the other hand, once most of an ethnic group had moved out of Zone II, the delinquency rate among its youth decreased significantly.

This finding rejects the notion of **social Darwinism** because it is clearly not the culture that influences crime and delinquency but rather the criminogenic nature of the environment. If ethnicity or race made a difference, the delinquency rates in Zone II would fluctuate based on who lived there, but the rates continued to be high from one group to the next. Rather, the zone determined the rates of delinquency.

Reaction and Research on Social Disorganization Theory

Over the last few decades, the Chicago School theoretical framework has received an enormous amount of attention from researchers.[10] Virtually all of the research has supported Shaw and McKay's version of social disorganization and the resulting high crime rates in neighborhoods that exhibit such deprived conditions. Modern research has supported the theoretical model proposed by Shaw and McKay, specifically in terms of the high crime rates in disorganized neighborhoods. Also, virtually every city that has an elevated highway (such as Richmond, Virginia; Baltimore, Maryland; Los Angeles, California), visually supports Shaw and McKay's model of crime in concentric circles. Drivers entering those cities can see the pattern of dilapidated structures in the zone of transition surrounding the inner-city area. Before and after this layer of dilapidated structures, drivers encounter a layer of houses and residential areas that seem to increase in quality as the driver gets farther away from the inner-city area.

Some critics, however, have raised some valid concerns regarding the original model, arguing that Shaw and McKay's original research did not actually measure their primary construct: social disorganization. Although this criticism is accurate, recent research has shown that the model is valid even when valid measures of social disorganization are included in the model.[11] Such measures of social disorganization include simply asking members of the neighborhood how many neighbors they know by name or how often they observe unsupervised peer groups in the area.

Additional criticisms of Shaw and McKay's formulation of social disorganization focus on the emphasis that the theory places on the macro, or aggregate, level of analysis. Although their theory does a good job of predicting which neighborhoods have higher crime rates, the model does not even attempt to explain why most youths in the worst areas do not become offenders. Furthermore, their model does not attempt to explain why some youths—although a very small number—in the best neighborhoods (in Zone V) choose to commit crime. However, the previous case

---

[10]John Laub, "Urbanism, Race, and Crime," *Journal of Research in Crime and Delinquency* 20 (1983): 283–98; Robert Sampson, "Structural Sources of Variation in Race-Age-Specific Rate of Offending Across Major U.S. Cities," *Criminology* 23 (1985): 647–73; J. L. Heitgard and Robert J. Bursik, "Extracommunity Dynamics and the Ecology of Delinquency," *American Journal of Sociology* 92 (1987): 775–87; Robert Bursik, "Social Disorganization and Theories of Crime and Delinquency: Problems and Prospects," *Criminology* 26 (1988): 519–51; Ralph Taylor and Jeanette Covington, "Neighborhood Changes in Ecology and Violence," *Criminology* 26 (1988): 553–89; Robert Bursik, "Ecological Stability and the Dynamics of Delinquency," in *Crime and Community*, ed. Albert J. Reiss and Morris H. Tonry (Chicago: University of Chicago Press, 1986), 35–66; Robert Sampson, "Transcending Tradition: New Directions in Community Research, Chicago Style—The American Society of Criminology 2001 Sutherland Address," *Criminology* 40 (2002): 213–30; Robert Sampson, J. D. Morenoff, and T. Gannon-Rowley, "Assessing 'Neighborhood Effects': Social Processes and New Directions in Research," *Annual Review of Sociology* 28 (2002): 443–78; P. O. Wikstrom and Rolf Loeber, "Do Disadvantaged Neighborhoods Cause Well-Adjusted Children to Become Adolescent Delinquents?" *Criminology* 38 (2000): 1109–42.

[11]Robert Sampson and W. Byron Grove, "Community Structure and Crime: Testing Social Disorganization Theory," *American Journal of Sociology* 94 (1989): 774–802.

studies completed and published by Clifford Shaw attempted to address the individual (micro) level of offending.

Also, there was one notable exception to Shaw and McKay's proposition that all ethnic/racial groups have high rates of delinquency/crime while they live in Zone II. Evidence showed that when Japanese made up a large portion of residents in this zone in transition, they had very low rates of delinquency. Thus, as in most theoretical models in social science, there was an exception to the rule.

Perhaps the biggest criticism of Shaw and McKay's theory, one that has yet to be adequately addressed, deals with their blatant neglect in proposing ways to ameliorate the most problematic source of criminality in the Zone II, transitional zone neighborhoods. Although they clearly point to the invasion of factories and businesses into residential areas as a problem, they do not recommend how to slow such invasion. This is likely due to political and financial concerns: The owners of the factories and businesses partially financed their research and later funded their primary policy implementation. Furthermore, this neglect is represented in their failure to explain the exception of the Gold Coast in their results and conclusions.

Despite the criticisms and weaknesses of the Chicago School perspective of criminology, this theory resulted in one of the largest programs to date in attempting to reduce delinquency rates. Clifford Shaw was put in charge of establishing the Chicago Area Project (CAP), which created neighborhood centers in the most crime-ridden neighborhoods of Chicago. These centers offered activities for youth and tried to establish ties between parents and officials in the neighborhood. Although this program was never scientifically evaluated, it still exists, and many cities have implemented programs that are based on this model. For example, Boston implemented a very similar program, which was evaluated by Walter Miller.[12] This evaluation showed that although the project was effective in establishing relationships and interactions between local gangs and community groups and in providing more educational and vocational opportunities, it seemed to fail in reducing delinquent/criminal behavior. Thus, the overall conclusion made by experts was that the Boston project and other similar programs, like the CAP, typically fail to prevent criminal behavior.[13]

⊠ Cultural and Subcultural Theories of Crime

Cultural/subcultural theories of crime assume that there are unique groups in society that socialize their children to believe that certain activities that violate conventional law are good and positive ways to behave. Although it is rather difficult to find large groups of people or classes that fit this definition, it may be that some subcultures or isolated groups of individuals buy into a different set of norms than the conventional, middle-class set of values.

---

[12]Walter B. Miller, "The Impact of a 'Total-Community' Delinquency Control Project," *Social Problems* 10 (1962): 168–91.

[13]For a review, see Richard Lundman, *Prevention and Control of Juvenile Delinquency*, 2nd ed. (New York: Oxford University Press, 1993). Also see review by Vold et al., *Theoretical Criminology*, 125–26.

Early Theoretical Developments and Research in Cultural/Subcultural Theory

One of the key developments of cultural theory is the 1967 work of Ferracuti and Wolfgang, who examined the violent themes of a group of inner-city youth from Philadelphia.[14] Ferracuti and Wolfgang's primary conclusion was that violence is a culturally learned adaptation to deal with negative life circumstances and that learning such norms occurs in an environment that emphasizes violence over other options.[15] These researchers based their conclusion on an analysis of data that showed great differences in rates of homicide across racial groups. However, Ferracuti and Wolfgang were clear that their theory was based on subcultural norms. Specifically, they proposed that no subculture can be totally different from or totally in conflict with the society of which it is a part.[16] This brings the distinction of culture and subculture to a forefront.

A culture represents a distinct set of norms and values among an identifiable group of people, values that are summarily different from those of the mainstream culture. For example, communism is distinctly different from capitalism because it emphasizes equality over competition, and it values utopia (i.e., everyone gets to share all profits) over the idea that the best performer gets the most reward. So it can be said that communists tend to have a different culture than capitalists. There is also a substantial difference between a culture and a subculture, which is typically only a pocket of individuals who may have a set of norms that deviate from conventional values. Therefore, what Ferracuti and Wolfgang developed is not so much a cultural theory as much as a subcultural theory.

This is also seen in the most prominent (sub)culture theory, which was presented by Walter Miller.[17] Miller's theoretical model proposed that the entire lower class had its own cultural value system. In this model, virtually everyone in the lower class believed in and socialized the values of six **focal concerns**: fate, autonomy, trouble, toughness, excitement, and smartness. Fate means luck, or whatever life dealt you; it disregards responsibility and accountability for one's actions. Autonomy is the value of independence from authority. Trouble means staying out of legal problems, as well as getting into and out of personal difficulties (e.g., pregnancy). Toughness is maintaining your reputation on the street in many ways. Excitement is the engagement in activities, some illegal, that help liven up an otherwise mundane existence of being lower class. Smartness puts an emphasis on "street smarts" or the ability to con others. Miller thought that members of the lower class taught these six focal concerns as a culture or environment (or "milieu," as stated in the title of his work).

A more recent subculture model, proposed by Elijah Anderson, has received a lot of attention in the past few years.[18] This theory focuses on African Americans; because of the very deprived conditions in the inner cities, Black Americans feel a sense of hopelessness, isolation, and despair, Anderson asserts. He clearly notes that although many African

---

[14]Vold et al., *Theoretical Criminology*, 165–69; Frank Schmalleger, *Criminology Today*, 4th ed. (Upper Saddle River, NJ: Prentice Hall, 2006).

[15]Franco Ferracuti and Marvin Wolfgang, *The Subculture of Violence: Toward an Integrated Theory of Criminology* (London: Tavistock, 1967).

[16]As quoted from Schmalleger, *Criminology Today*, 230–31.

[17]Walter B. Miller, "Lower Class Culture as a Generating Milieu of Gang Delinquency," *Journal of Social Issues* 14 (1958): 5–19.

[18]Elijah Anderson, *Code of the Streets* (New York: W. W. Norton, 1999).

Americans believe in middle-class values, these values have no weight on the street, particularly among young urban males. According to Anderson, *The Code of the Streets*, which was the appropriate title of his book, is to maintain one's reputation and demand respect. For example, to be disrespected ("dissed") is considered grounds for a physical attack. Masculinity and control of one's immediate environment are treasured characteristics; this is perceived as the only thing people can control, given the harsh conditions in which they live (e.g., unemployment, poverty, etc.).

Criticisms of Cultural Theories of Crime

Studies on cultural theories of crime, at least in the United States, find no large groups that blatantly deny the middle-class norms of society. Miller's model of lower-class focal concerns is not consistent across the entire lower class. Studies show that most adults in the lower-class attempt to socialize their children to believe in conventional values, such as respect for authority, hard work, and delayed gratification, and not the focal concerns that Miller specified in his model.[19] Ferracuti and Wolfgang admitted that their research findings led them to conclude that their model was more of a subcultural perspective and a distinctly different culture. There may be small groups or gangs that have subcultural normative values, but that doesn't constitute a completely separate culture in society. Perhaps the best subculture theories are those presented by Cohen or Cloward and Ohlin (see Section V), in their variations of strain theory that emphasize the formations of gangs among lower-class male youth. If there are subcultural groups in U.S. society, they seem to make up a very small percentage of the population, which somewhat negates the cultural/subcultural perspective of criminality. But this type of perspective may be important regarding the criminality of some youth offenders.

▨ Policy Implications

Many of the policy implications that are suggested by the theoretical models proposed in this chapter are rather ironic. Regarding social disorganization, there is a paradox that exists, in the sense that the very neighborhoods most desperately in need of becoming organized to fight crime are the same inner-city ghetto areas where it is, by far, most difficult to cultivate such organization (e.g., neighborhood watch or "Block Watch" groups). Rather, the neighborhoods that have high levels of organization tend to be those that already have very low levels of crime because the residents naturally "police" their neighbors' well-being and property; they have a stake in the area remaining crime-free. Although there are some anecdotal examples of success of neighborhood watch programs in high-crime neighborhoods, most of the empirical evidence shows that this approach is "almost uniformly unsupportive" in its ability to reduce crime in such neighborhoods.[20] Furthermore, many studies of these neighborhood watch programs find that the groups actually increase the fear of crime in a number of neighborhoods, perhaps due to the heightened awareness regarding the crime issues in such areas.[21]

---

[19]Vold et al., *Theoretical Criminology*.

[20]John Worrall, *Crime Control in America: An Assessment of the Evidence* (Boston: Allyn & Bacon, 2006), 95.

[21]Ibid.

Perhaps the most notable program that resulted from the Chicago School/social disorganization model—the Chicago Area Project (CAP), and similar programs—have been dubbed failures in reducing crime rates among the participants. Still, there have been some advances in trying to get residents of high-crime areas to become organized in fighting crime. The more specific the goals regarding crime reduction—such as more careful monitoring of high-level offenders (more intensive supervised probation) or better lighting in dark places—the more effective the implementation will be.[22]

Regarding cultural/subcultural programs, some promising intervention and outreach programs have been suggested by such models. Many programs attempt to build pro-social attitudes among high-risk youth, often young children. For example, a recent evaluation showed that a program called *Peace Builders*, which focuses on children in early grades, was effective in producing gains in conflict resolution, development of pro-social values, and reductions in aggression; a follow-up showed that these attributes were maintained for a long period of time.[23] Another recent anti-aggression training program for foster-home boys showed positive effects in levels of empathy, self-efficacy, and attribution style among boys who had exhibited early-onset aggression.[24] Ultimately, there are effective programs out there that promote pro-social norms and culture. More efforts should be given to promoting such programs that will help negate the anti-social cultural norms of individuals, especially among high-risk youth.

⬚ Conclusion

In this section, we examined theoretical perspectives proposing that the lack of social organization in broken down and dilapidated neighborhoods is unable to contain delinquency and crime. Furthermore, we discussed how this model of crime was linked to processes derived from ecological principles. This type of approach has been tested numerous times, and virtually all studies show that the distribution of delinquents/crime activity is consistent with this model.

We then discussed the ability of cultural and subcultural theories to explain criminal activity. Empirical evidence shows that cultural values make a contribution to criminal behavior, but that the existence of an actual alternative culture in our society has not been found. However, some subcultural pockets, particularly regarding inner-city youth gangs, certainly exist and provide some validity for this perspective of crime. Furthermore, the Chicago School perspective plays a role because these subcultural groups tend to be found in zones of transition.

Finally, we examined some policy implications that are suggested by these theoretical models. Regarding social disorganization, we noted that neighborhood crime-fighting

---

[22]Lundman, *Prevention and Control*; Sampson and Grove, "Community Structure"; Vold et al., *Theoretical Criminology*.

[23]D . J. Flannery, M. I. Singer, and K. L. Wester, "Violence, Coping, and Mental Health in a Community Sample of Adolescence," *Violence and Victims* 18 (2003): 403–18; For a review of this research, see Stephen G. Tibbetts, "Perinatal and Developmental Determinants of Early Onset of Offending: A Biosocial Approach for Explaining the Two Peaks of Early Antisocial Behavior," in *The Development of Persistent Criminality*, ed. J. Savage (New York: Oxford University Press, 2009), 179–201.

[24]K. Weichold, "Evaluation of an Anti-aggressiveness Training with Antisocial Youth," *Gruppendynamik* 35 (2004): 83105; for a review, see Tibbetts, "Perinatal and Developmental."

groups are hardest to establish in high-crime neighborhoods and easiest to build in those neighborhoods with an already low rate of crime. Nevertheless, there have been some successes. We also looked at intervention and outreach programs based on the cultural and subcultural perspectives.

⊠ Section Summary

- ◆ We examined how principles of ecology were applied to the study of how cities grow, as well as to the study of crime, by researchers at University of Chicago, which became known as the Chicago (or Ecological) School of criminology.
- ◆ We reviewed the various zones of concentric circles theory, also a key contribution of the Chicago School of criminology, and explored which zones are most crime prone.
- ◆ We examined why the findings from the Chicago School of criminology showed that social Darwinism was not accurate in attributing varying crime rates to ethnicity or race.
- ◆ We reviewed much of the empirical evidence regarding the theory of social disorganization and examined the strengths and weaknesses of this theoretical model.
- ◆ We discussed the cultural/subcultural model presented by Ferracuti and Wolfgang, as well as the cultural model of inner-city urban youth presented by Anderson.
- ◆ We discussed Miller's theory of lower-class culture, particularly its six focal concerns.
- ◆ We reviewed the strengths and weaknesses of cultural and subcultural theories of crime, based on empirical evidence.

KEY TERMS

Chicago School of criminology	Ecological School/ perspective	Social Darwinism
Concentric circles theory	Focal concerns	Social disorganization
Cultural/subcultural theories	Natural areas	Zone in transition

DISCUSSION QUESTIONS

1. Identify and discuss an example of the ecological principles of invasion, domination, and succession among animals or plants that was not discussed in this introduction.

2. Can you see examples of the various zones that Shaw and McKay described in the town/city where you live (or the one nearest you)? Try obtaining a map or sketching a plot of this city or town, and then draw the various concentric circles where you think the zones are located.

3. What forms of organization and disorganization have you observed in your own neighborhood? Try to supply examples of both, if possible.

4. Can you provide modern-day examples of different cultures/subcultures in the United States? What regions, or parts of the country, would you say have different cultures that are more conducive to crime?

5. Do you know people who believe most or all of Miller's focal concerns? What is their social class? What are their other demographic features (age, gender, urban/rural, etc.)?

6. Do you know individuals who seem to fit either Ferracuti and Wolfgang's cultural theory or Anderson's model of inner-city youth's street code? Why do you believe they fit such a model?

WEB RESOURCES

Chicago School of Criminology

http://www.helium.com/items/865770-an-overview-of-the-chicago-school-theories-of-criminology

http://www.associatedcontent.com/article/771561/an_overview_of_the_chicago_school_theories.html

Subcultural Theories

http://www.umsl.edu/~keelr/200/subcult.html

READING

In this selection, Clifford Shaw and Henry McKay present a theoretical model of various characteristics of neighborhoods that contribute to higher crime and delinquency rates. Specifically, they examine physical, economic, and population factors that contribute to higher rates of delinquency in certain communities. Such observations of certain neighborhoods provide the basis for the theory of social disorganization (also known as the Chicago or Ecological School of criminology). While reading this selection, readers are encouraged to think about the places they have lived or visited that fit these characteristics; it is likely such neighborhoods have high crime rates. Because this theory fits virtually all cities around the world, and because of their methodology, Shaw and McKay are generally considered two of the most prominent criminologists of the 20th century.

SOURCE: Clifford R. Shaw and Henry D. McKay, *Juvenile Delinquency and Urban Areas,* 2nd ed. (Chicago: University of Chicago Press, 1969), 140–63. Copyright © 1969 University of Chicago Press. Reprinted by permission of The University of Chicago Press.

Delinquency Rates and Community Characteristics

Clifford R. Shaw and Henry D. McKay

The question has been asked many times: "What is it, in modern city life that produces delinquency?" Why do relatively large numbers of boys from the inner urban areas appear in court with such striking regularity, year after year, regardless of changing population structure or the ups and downs of the business cycle? [Elsewhere] a different series of male delinquents were presented which closely parallel one another in geographical distribution although widely separated in time, and the close resemblance of all these series to the distribution of truants and of adult criminals was shown. Moreover, many other community characteristics—median rentals, families on relief, infant mortality rates, and so on—reveal similar patterns of variation throughout the city. The next step would be to determine, if possible, the extent to which these two sets of data are related. How consistently do they vary together, if at all, and how high is the degree of association?

Where high zero-order correlations are found to exist uniformly between two variables, with a small probable error, it is possible and valid to consider either series as an approximate index, or indicator, of the other. This holds true for any two variables which are known to be associated or to vary concomitantly. The relationship, of course, may be either direct or inverse. In neither case, however, is there justification in assuming, on this basis alone, that the observed association is of a cause-and-effect nature; it may be, rather, that both variables are similarly affected by some third factor. Further analysis is needed. Controlled experimentation is often useful in establishing the degree to which a change in one variable "causes" or brings about a corresponding change in the other. In the social

field, however, experimentation is difficult. Instead, it is often necessary to rely upon refined statistical techniques, such as partial correlation, which, for certain types of data, enable the investigator to measure the effects of one factor while holding others relatively constant. By the method of successive redistribution, also, the influence of one or more variables may be held constant. Thus, it is possible to study the relationship between rates of delinquents and economic status for a single nationality group throughout the city or for various nationality groups in the same area or class of areas. This process may be extended indefinitely, subject only to the limitations of the available data. In the analysis to be presented, both of the latter methods have been used in an attempt to determine how much weight should be given to various more or less influential factors.

Several practical considerations prevent the neat and precise statistical analysis which would be desirable. The characteristics studied represent only a sampling of the myriad forms in which community life and social relationships find expression. The rate of delinquents must itself be thought of as an imperfect enumeration of the delinquents and an index of the larger number of boys engaging in officially proscribed activities. Not only will there be chance fluctuations in the amount of alleged delinquency from year to year, but the policy of the local police officer in referring boys to the Juvenile Court, the focusing of the public eye upon conditions in an area, and numerous other matters may bring about a change in the index without any essential change in the underlying delinquency-producing influences in the community or in the behavior resulting therefrom. If the infant mortality rates or the rates of families on relief are looked upon as indexes of economic status or of the social organization of a community, it is obvious that they can be considered only very crude indicators at best. The perturbing influence of other variables must always be considered.

Certain exceptional conditions are known to limit the value of other variables chosen as indicators of local community differentiation. Median rental has been used widely because of its popularity as an index of economic status, although in Chicago such an index is far from satisfactory when applied to areas of colored population. The Negro is forced to pay considerably higher rents than the whites for comparable housing—thus his economic level is made to appear higher on the basis of rental than it actually is. Similarly, rates of increase or decrease of population are modified in Negro areas by restrictions on free movement placed upon the Negro population. Thus, in certain areas the population is increasing where it normally would be expected to decrease if there were no such barriers. Likewise, the percentage of families owning homes is not entirely satisfactory as an economic index in large urban centers, where many of the well-to-do rent expensive apartments. It is, however, an indication of the relative stability of population in an area.

Correlation of series of rates based on geographical areas is further complicated by the fact that magnitude of the coefficient is influenced by the size of the area selected. This tendency has been noted by several writers, but no satisfactory solution of the problem has been offered. If it be borne in mind that a correlation of area data is an index of geographical association for a particular type of spatial division only, rather than a fixed measure of functional relationship, it will be apparent that a change in area size changes the meaning of the correlation. Thus, an r of .90 or above for two series of rates calculated by square-mile areas

indicates a high degree of association between the magnitudes of the two rates in most of the square miles but does not tell us the exact degree of covariance for smaller or larger areas.

With these limitations clearly in mind, a number of correlation coefficients and tables of covariance are presented. The statistical data characterizing and differentiating local urban areas may be grouped under three headings: (1) physical status, (2) economic status, and (3) population composition. These will be considered, in turn, in relation to rates of delinquents.

Indexes of Physical Status in Relation to Rates of Delinquents

The location of major industrial and commercial developments, the distribution of buildings condemned for demolition or repair, and the percentage increase or decrease in population by square-mile areas were presented [elsewhere] as indications of the physical differentiation of areas within the city. Quantitative measures of the first two are not available, but inspection of the distribution maps shows clearly that the highest rates of delinquents are most frequently found in, or adjacent to, areas of heavy industry and commerce. These same neighborhoods have the largest number of condemned buildings. The only notable exception to this generalization, for Chicago, appears in some of the areas south of the central business district.

There is, of course, little reason to postulate a direct relationship between living in proximity to industrial developments and becoming delinquent. While railroads and industrial properties may offer a field for delinquent behavior, they can hardly be regarded as a cause of such activities. Industrial invasion and physical deterioration do, however, make an area less desirable for residential purposes. As a consequence, in time there is found a movement from this area of those people able to afford more attractive surroundings. Further, the decrease in the number of buildings available for residential purposes leads to a decrease in the population of the area.

Population Increase or Decrease. Increase or decrease of population and rates of delinquents, by square-mile areas, do not exhibit a linear relationship. A relatively slight difference in rate of decrease of population, or of rate of increase for areas where the increase is slight, is generally associated with a considerable change in rates of delinquents; while for large differences in rates of increase of population, where increase is great, there is little or no consistent difference in rates of delinquents. Thus, areas increasing more than 70 per cent show no corresponding drop in rates of delinquents, although the relationship is clear up to this point. . . .

[T]here is a similarity between the pattern of distribution of delinquency and that of population growth or decline. The data do not establish a causal relationship between the two variables, however. The fact that the population of an area is decreasing does not impel a boy to become delinquent. It may be said, however, that decreasing population is usually related to industrial invasion of an area and contributes to the development of a general situation conducive to delinquency.

▧ Population Composition in Relation to Rates of Delinquency

In Chicago, as in other northern industrial cities, as has been said, it is the most recent arrivals—persons of foreign birth and those who have migrated from other sections of this country—who find it necessary to make their homes in neighborhoods of low economic level. Thus the newer European immigrants are found concentrated in certain areas, while Negroes from the rural South and Mexicans occupy others of comparable status. Neither of these population categories, considered separately, however, is suitable for correlation with rates of delinquents, since some areas of high rates have a predominantly immigrant population and others are entirely or largely Negro. Both categories, however, refer to groups of low economic status, making their adjustment to a complex urban environment. Foreign-born and Negro heads of families will therefore be considered together in order to study this segregation of the newer arrivals, on a city-wide scale.

Percentage of Foreign-Born and Negro Heads of Families.[1] When the rates of delinquents in the 1927–33 series are correlated with the percentage of foreign-born and Negro heads of families as of 1930, by 140 square-mile areas, the coefficient is found to be .60 ± .03. Similarly, when the 1917–23 delinquency data are correlated with percentages of foreign-born and Negro heads of families for 1920, by the 113 areas into which the city was divided for that series, the coefficient is .58 ± .04. When rates of delinquents are calculated for the classes of areas . . . wide variations are found between the rates in the classes where the percentage of foreign-born and Negro heads of families is high and in those where it is low. . . . Since the number of foreign-born heads of families in the population decreased and the number of Negroes increased between 1920 and 1930, the total proportions of foreign-born and Negro heads of families in each class do not correspond. The variation with rates of delinquents, however, remains unchanged.

While it is apparent from these data that the foreign born and the Negroes are concentrated in the areas of high rates of delinquents, the meaning of this association is not easily determined. One might be led to assume that the relatively large number of boys brought into court is due to the presence of certain racial or national groups, were it not for the fact that the population composition of many of these neighborhoods has changed completely, without appreciable change in their rank as to rates of delinquents. Clearly, one must beware of attaching causal significance to race or nativity. For, in the present social and economic system, it is the Negroes and the foreign born, or at least the newest immigrants, who have least access to the necessities of life and who are therefore least prepared for the competitive struggle. It is they who are forced to live in the worst slum areas and who are least able to organize against the effects of such living.

In Chicago three kinds of data are available for the study of nativity, nationality, and race in relation to rates of delinquents. These data concern (1) the succession of nationality groups in the high-rate areas over a period of years; (2) changes in the national and racial backgrounds of children appearing in the Juvenile Court; and (3) rates of delinquents for particular racial, nativity, or nationality groups in different types of areas at any given moment. In evaluating the significance of community characteristics found to be associated with high rates of delinquents, the relative weight of race, nativity, and nationality must be understood. . . .

It appears to be established, then, that each racial, nativity, and nationality group in Chicago displays widely varying rates of delinquents; that rates for immigrant groups in particular show a wide historical fluctuation; that diverse racial, nativity, and national groups possess relatively similar rates of delinquents in similar social areas; and that each of these groups displays the effect of disproportionate concentration in its respective areas at a given time. In the face of these facts it is difficult to sustain the contention that, by themselves, the factors of race, nativity, and nationality are vitally related to the problem of juvenile delinquency. It seems necessary to conclude, rather, that the significantly higher rates of delinquents

found among the children of Negroes, the foreign born, and more recent immigrants are closely related to existing differences in their respective patterns of geographical distribution within the city. If these groups were found in the same proportion in all local areas, existing differences in the relative number of boys brought into court from the various groups might be expected to be greatly reduced or to disappear entirely.

It may be that the correlation between rates of delinquents and foreign-born and Negro heads of families is incidental to relationships between rates of delinquents and apparently more basic social and economic characteristics of local communities. Evidence that this is the case is seen in two partial correlation coefficients computed. Selecting the relief rate as a fair measure of economic level, the problem is to determine the relative weight of this and other factors. The partial correlation coefficient between rate of delinquents and percentage of families on relief, holding constant the percentage of foreign-born and Negro heads of families, in the 140 areas, is .76 ± .02. However, the coefficient for rates of delinquents and percentage of foreign-born and Negro heads of families, when percentage of families on relief is held constant, is only .26 ± .05. It is clear from these coefficients, therefore, that the percentage of families on relief is related to rates of delinquents in a more significant way than is the percentage of foreign-born and Negro heads of families.

It should be emphasized that the high degree of association between rates of delinquents and other community characteristics . . . does not mean that these characteristics must be regarded as causes of delinquency, or vice versa. Within certain types of areas differentiated in city growth, these phenomena appear together with such regularity that their rates are highly correlated. Yet the nature of the relationship between types of conduct and given physical, economic, or demographic characteristics is not revealed by the magnitude either of zero-order or partial correlation coefficients, or of other measures of association.

A high degree of association may lead to the uncritical assumption that certain factors are causally related, whereas further analysis shows the existing association to be entirely adventitious. This is apparently the case with the data on nativity, nationality, and race. . . . That, on the whole, the proportion of foreign-born and Negro population is higher in areas with high rates of delinquents there can be little doubt; but the facts furnish ample basis for the further conclusion that the boys brought into court are not delinquent *because* their parents are foreign born or Negro but rather because of other aspects of the total situation in which they live. In the same way, the relationship between rates of delinquents and each associated variable should be explored, either by further analysis, by experimentation, or by the study of negative cases.

☒ Summary

It has been shown that, when rates of delinquents are calculated for classes of areas grouped according to rate of any one of a number of community characteristics studied, a distinct pattern appears—the two sets of rates in each case varying together. When values of these other community characteristics, in turn, are calculated for classes of areas grouped by rate of delinquents, the same consistent trends appear. . . . The data . . . indicate a high degree of association between rates of delinquents and other community characteristics when correlations are computed on the basis of values in square-mile areas or similar subdivisions, and a still closer general association by large zones or classes of areas. . . .

☒ Note

1. The categories "foreign born" and "Negro" are not compatible, since the former group is made up primarily of adults, while the latter includes all members of the race. The classification "heads of families" has been used; therefore, foreign-born and Negro family heads are entirely comparable groupings. The census classification "other races" has been included—a relatively small group, comprising Mexicans, Japanese, Chinese, Filipinos, etc.

REVIEW QUESTIONS

1. What do Shaw and McKay have to say about delinquency rates in neighborhoods located in or near heavy industrial areas? What types of physical indicators are present in such areas?

2. What do Shaw and McKay conclude regarding the crime rates of neighborhoods containing a population composition high in minorities or recent immigrants?

3. What do Shaw and McKay claim would happen to delinquency/crime rates if racial/ethnic groups, as well as recent immigrants, were equally distributed across all neighborhoods of a city (such as Chicago)?

❖

READING

In this selection, Christopher Lowenkamp, Frank Cullen, and Travis Pratt present an empirical study that seeks to replicate the findings of a well-respected previous test of the Chicago School model of social disorganization theory (as originally presented by Clifford Shaw and Henry McKay). Specifically, this well-respected previous test of this theoretical model was performed by Robert Sampson and W. Byron Groves (1989), using data from the 1982 British Crime Survey. It generally supported the social disorganization theoretical framework, especially regarding the mediating effects of informal social controls in explaining the impact of structural factors (e.g., poverty) on neighborhood crime rates. In this study, Lowenkamp et al. find that a replication of this earlier study by Sampson and Groves (1989), which utilizes more recently collected data, also revealed a high level of empirical support for the social disorganization theoretical framework. The few differences between this and the earlier test (by Sampson and Groves) actually revealed more support for the Chicago School model. While reviewing this study, readers are encouraged to consider the places they have lived or visited that fit these characteristics.

Replicating Sampson and Groves's Test of Social Disorganization Theory

Revisiting a Criminological Classic

Christopher T. Lowenkamp, Francis T. Cullen, and Travis C. Pratt

Although Shaw and McKay's work linking social disorganization to community crime rates exerted considerable influence on criminological theory during the 1950s, by the mid-1980s, their perspective came to be considered by many as "little more than an interesting footnote in the history of community-related research" (Bursik 1986:36; see also, Pfohl 1985:158). Since that time, however, the social disorganization perspective has experienced a dramatic revitalization, reemerging from the dustbin of spent criminological paradigms to challenge for the status as a preeminent macro-level theory.

No single work did more to polish the tarnished image of social disorganization theory than Sampson and Groves's essay, published in 1989, "Community Structure and Crime: Testing Social-Disorganization Theory." Their research ostensibly showed that consistent

SOURCE: Adapted from Christopher T. Lowenkamp, Francis T. Cullen, and Travis C. Pratt, "Replicating Sampson and Groves's Test of Social Disorganization Theory: Revisiting a Criminological Classic." *Journal of Research in Crime and Delinquency* 40, no. 4 (2003): 351–73. Copyright © 2003 Sage Publications, Inc. Used by permission of Sage Publications, Inc.

with the core predictions of the social disorganization perspective, the impact of structural
factors on crime rates was mediated in important ways by direct measures of informal social
controls (Taylor 2001). It is noteworthy that
contemporary publications now cite Sampson
and Groves's (1989) work as providing "a convincing test of Shaw and McKay's social disorganization thesis" (Bellair 1997:679) and as
"one of the more important studies in criminological literature over the past decade"
(Veysey and Messner 1999: 156). It is among
the most cited criminological works and has
become a staple in criminological textbooks
published after 1989 (see, e.g., Akers 1999;
Siegel 2000; Sykes and Cullen 1992; Tittle and
Paternoster 2000).[1] In short, "Community
Structure and Crime: Testing Social-Disorganization Theory" has emerged as a classic in
the field of criminology.

As with other classic empirical works,
however, the question arises as to whether
Sampson and Groves's (1989) research offers
virtually unshakable support for social disorganization theory or merely produced an idiosyncratic finding that was unique to a certain
time and place. To gain insight on this issue,
subsequent studies ideally would have replicated Sampson and Groves's work in different
contexts and with refined measures of the
study's core variables. But precise replications
of their empirical test have been hampered by
the failure of existing macro-level data sets to
contain intervening, direct measures of informal social controls, including the full set of
variables specified in the original study. A limited number of studies have moved beyond
structural proxies to assess components of
social disorganization or similar processes
(Bellair 1997; Morcnoff ct al. 2001; Roundtree
and Land 1996; Roundtree and Warner 1999;
Sampson et al. 1997; Sampson et al. 1999; Velez
2001; Warner and Roundtree 1997). Yet, again,
none has attempted to provide a *systematic* test
of Sampson and Groves's complete model.
Accordingly, we have a classic study—cited

repeatedly in the literature—that remains, in
effect, nonreplicated.

In this context, we suggest an alternative
strategy for replicating Sampson and Groves's
(1989) apparent empirical confirmation of
social disorganization theory. We propose to
revisit the British Crime Survey (BCS) (Home
Office Crime and Criminal Justice Unit 1994)
a decade later to investigate whether the findings reported in the original analysis will
remain stable as social disorganization theory
would predict. If the findings change in meaningful ways, it would suggest that the credence
accorded Sampson and Groves's "classic" test
is not warranted and that social disorganization theory cannot account for crime rates in
a later time period, even though virtually the
same data set and measures are being used.
Phrased differently, it would call into question
whether the social disorganization perspective
is a "general theory" capable of accounting for
ecological variations in crime rates across
time and place.

If, however, their findings remain largely
constant, it would suggest that the initial
analysis by Sampson and Groves (1989) was
not idiosyncratic but rather captured an ongoing empirical reality. Much as in other scientific fields, it would show that their conclusions
could be reproduced under "laboratory conditions" (so to speak) that were similar to, but
independent of, the original investigation.
Admittedly, these findings would not necessarily mean that results from Britain would generalize, without specification, to the United
States and other social contexts. Nonetheless,
these results would bolster social disorganization theory's claim to be a perspective that
warrants its place as a vital macro-level theory
of crime and as deserving of further empirical
investigation.

We begin by revisiting Sampson and
Groves's (1989) classic test in more detail. This
discussion serves as a prelude for a replication of
their study using 1994 data from the BCS. This
year of the BCS was chosen because it allows for

the closest replication of Sampson and Groves's study. As we shall see, in important ways, their findings are largely replicated by our analysis, suggesting that social disorganization theory is potentially useful in accounting for variation in macro-level crime rates across communities.

Finally, we note the growing recognition within criminology—and social science generally—of the value of systematic replication (Gendreau in press). Breaking new theoretical and/or empirical ground is perhaps of prime importance, for such efforts potentially push the field to new horizons. Nevertheless, the emphasis on innovation over systematic replication has a cost: the inability to build cumulative knowledge and to be able to state with confidence what is, and is not, known. Even when theories have been subjected to tens of empirical tests, it is difficult to assess the perspectives' merits because of the failure of the research to build in an organized way on one another (e.g., measure all core constructs, include similar measures) (Kempf 1993). In this regard, the current study should be seen as an initial, but much needed, attempt to contribute to the cumulative knowledge base on Sampson and Groves's (1989) analysis of social disorganization theory.

◼ Sampson and Groves's Test of Social Disorganization Theory

Measuring Social Disorganization: Intervening Variables

As Sampson and Groves (1989) noted, most ecological research inspired by Shaw and McKay (1972) examined the effects of structural antecedents (heterogeneity, socioeconomic status [SES], and mobility) on crime rates (see also Bursik 1986, 1988; Byrne and Sampson 1986). Although suggestive, these theoretical tests failed to provide direct measures of whether social disorganization

prevailed within the ecological units under investigation (Veysey and Messner 1999). In essence, the lack of measures of the theorized mediators/intervening variables have created a "black box" phenomenon in which little is known about what exists between structural variables on the left side of a causal model and crimes rates on the other side of the model. Because there are no measures of the intervening variables, assumptions and inferences—plausible as they may be—must be invoked to explain the processes that might specify the relationships between the structural and dependent variables. Notably, this dearth of attention to measures of mediating processes plagues not only social disorganization theory, but also ecological theories in general (Byrne and Sampson 1986).

To test social disorganization theory with direct measures of intervening variables, Sampson and Groves (1989) used data from the 1982 BCS.[2] Their goal was to examine the mediating effects of intervening variables on the structural antecedents hypothesized to be causal influences on social disorganization and, subsequently, criminality within a community. Building on Shaw and McKay's (1972) original model of social disorganization, Sampson and Groves (1989) included three measures of community-level variables thought to cause social disorganization: low economic status, ethnic heterogeneity, residential mobility. In extending Shaw and McKay's (1972) model, Sampson and Groves specified variables for two additional sources of social disorganization: urbanization and family disruption (a variable drawn from Sampson's [1987] earlier work).

The uniqueness of Sampson and Groves's (1989) work stems from the authors' construction of direct measures of social disorganization. These were treated as intervening variables in their main analyses. These measures include sparse local friendship networks, low organizational participation, and unsupervised teenage peer groups. These measures were largely developed based on works by Kasarda and Janowitz (1974), Krohn (1986) and on the relation of these works to Shaw and McKay's (1972) theoretical framework. Each measure was based on aggregated data from individual respondents in 238 communities.

Empirical Support for Social Disorganization Theory

To determine the impact of structural antecedents on social disorganization and on crime, Sampson and Groves (1989) developed and analyzed several weighted least squares (WLS) regression models. The results of their analyses indicated that each of the structural factors significantly predicted the degree to which respondents perceived unsupervised peer groups to be a problem in their community (R^2=.30). The model predicting the level of local friendship networks indicated that only residential stability and urbanization had significant relationships with the dependent measure ($R^2 = .26$). Only one structural variable, SES, had a significant relationship with the level of organizational participation ($R^2 = .07$).

Turning to the results regarding the dependent variables, crime and delinquency, eight models were presented by Sampson and Groves (1989). Models predicting rates of mugging ($R^2 = .61$), stranger violence ($R^2 = .15$), total victimization ($R^2 = .42$), burglary ($R^2 = .39$), auto theft ($R^2 = .30$), and vandalism ($R^2 = .21$) indicated, with few exceptions, that the intervening variables thought to measure social disorganization had significant relationships with the dependent variables and mediated some of the direct effects of the structural variables. Regarding rates of self-reported offending, unsupervised peer groups mediated much of the effects of the structural variables when predicting violent offending ($R^2 = .06$) Along with local friendship networks, unsupervised peer groups also mediated the effects

of the structural variables when predicting self-reported property offenses ($R^2 = .09$).

Replicating Sampson and Groves's Test

Although Sampson and Groves's (1989) study has not been systematically replicated outside their article,[3] their original data have recently been reassessed by Veysey and Messner (1999) through the use of LISREL instead of WLS regression analysis. The findings from the reanalysis were largely consistent with those from Sampson and Groves (1989): (1) the social disorganization variables mediated, although not completely, the effect of structural characteristics on rates of community victimization, and (2) the variable of unsupervised peer groups had the largest impact on the outcome measures. We should also note, however, that Veysey and Messner's (1999) LISREL analysis revealed that the measures of structural antecedents had various direct effects on rates of crime (as measured by victimization data). This finding, they suggest, supports a position that is conceptually inconsistent with Sampson and Groves's (1989) version of social disorganization theory because the disorganization perspective would argue that the effects of structural factors should be completely mediated by measures of each community's level of disorganization.

Nevertheless, in their article, Sampson and Groves (1989) did not contend that the intervening variables in their model mediated fully the effects of the five structural factors. Their original path model showed direct effects of structural factors, although, admittedly, they highlighted the ability of their measures of neighborhood disorganization and informal social control to mediate the impact of these factors on levels of crime. Thus, Veysey and Messner's (1999) findings that the social disorganization variables did not completely mediate the relationships between the structural variables and rates of crime is perhaps best seen as an important clarification of, rather than a failure to replicate, Sampson and Groves's (1989) analysis. In any case, we also replicate Veysey and Messner's (1999) analysis and discuss its implications for social disorganization theory.

⬚ Method

As with many surveys conducted repeated times, additions and deletions have been made to the content of information gathered in the BCS. After reviewing the codebooks for the BCSs conducted after 1982, we determined that the data from the 1994 survey contain those questions and information most similar to the data used by Sampson and Groves (1989). More recent BCSs no longer contain the same or, in some cases, similar questions employed by Sampson and Groves in the construction of their measures. In any event, the 1994 survey provided us with the data—collected a decade after the original study—to construct, in most instances, measures identical to those used by Sampson and Groves. In instances where identical measures could not be developed, very similar measures were constructed—each of which is noted below. The use of similar measures is significant because it means that differences in measurement cannot be employed to explain any divergent findings that might arise between the two studies.

Sample

The data for this study come from the 1994 BCS. The sampling method used is a multistage random probability design where 600 postal code sectors are selected by a systematic sampling method from a stratified list of postal code sectors. The probability for selection of any particular postal code sector was proportional to the total number of delivery points within that sector. Once 600 sectors were selected, 36 addresses were randomly selected from each

sector yielding a total of 21,600 addresses. After removing all ineligible addresses and nonresponding households, the sample size was reduced to 14,617. This averages approximately 25 respondents per sector. The BCS collects data on the characteristics of adults and children that reside within the household as well as information pertaining to residential history, perceptions of crime, problems in the area, social habits, type of community, community involvement, social cohesion, criminal victimization, and experience with the police.

This sampling method differs from that used in 1982, where 60 addresses from 238 electoral wards were selected. After removing ineligible addresses and nonresponding households, the final sample size for 1982 was 10,905. Checks on the differences between these two sampling frames indicate no appreciable differences in crime rate trends due to the sampling frame (Lynn 1997).

Measures of Intervening Variables

The intervening variables in this model are very similar to those used by Sampson and Groves (1989). We have included measures of unsupervised teenage peer groups, local friendship networks, and organizational participation. Sampson and Groves measured *local friendship networks* based on responses to the question that asked the number of friends that lived in the local community. A mean was calculated for these responses and served as the measure of local friendship networks. Our local friendship indicator was derived from responses to the question, "Thinking of the people who live in this area, by this area, I mean a 10–15 minute walk from here, how many would you regard as friends or acquaintances?" We also calculated a mean response and used this as our measure.

As measured by Sampson and Groves (1989), *organizational participation* was the percentage of respondents that reported in meetings of clubs or committees the week before the interview. Our measure of organizational participation is the percentage of people that reported participating in a meeting of a club or committee the last time they were out in the evening.

The final intervening variable, *unsupervised peer groups,* was measured by Sampson and Groves (1989) as the percentage of respondents that reported groups of teenagers making a nuisance of themselves as a very common problem in their neighborhood. Our measure differed slightly in that the question relating to teenage groups asked respondents how big a problem teenage groups hanging around on the streets was in their neighborhood. We calculated the percentage of people that reported teenage groups to be a very big problem. Although this measure differs from the 1982 BCS, for the purpose at hand, it is most likely a preferable measure. As Sampson and Groves (1989) note, by removing the mention of "nuisance" behavior, the concern that respondents are confounding crime rates with "nuisance" behavior of teens is obviated.

Measures of Structural Variables

Consistent with Sampson and Groves (1989), we constructed measures of SES, residential stability, ethnic heterogeneity, family disruption, and urbanization. We followed identical processes in constructing these measures (e.g., formulas, summed z scores percentages, and means). Wherever possible, we used the same questions and responses.

SES is a composite of measure of z scores tapping education (the percentage of people with a higher degree or teaching certification), occupation (percentage in professional or managerial occupations using the BCS social class ratings), and income (percentage reporting incomes in the top two income brackets).

Residential stability, as measured by Sampson and Groves (1989), was the percentage of people

that were brought up in the area within a 15-minute walk from home. Due to changes in the questions on the BCS main questionnaire, our measure of residential stability is the percentage of people that reported living in the area (within a 15 minute walk) for 10 or more years.

Ethnic heterogeneity is calculated using the same formula as Sampson and Groves (1989) $(1 - \Sigma pi^2)$ where *pi* is the fraction of people in a particular group. Our measure of ethnic heterogeneity differed from Sampson and Groves's in that the 1994 BCS used seven rather than five categories of ethnicity.

Family disruption was the sum of z scores for the percentage of people surveyed that were either separated or divorced and the percentage of people that were separated, divorced, or single and had children in their household. This measure is identical to that used by Sampson and Groves (1989).

Urbanization, finally, was coded as either 0 or 1 according to the BCS data, where 1 represents an urban area. In 1982, the Planning and Research Applications Group's (PRAG) inner-city designations were used to determine which areas were inner-city areas. In 1984, the method for identifying inner-city areas was changed to where inner-city status is determined by the population density, percentage of owner occupied housing, and/or the social class of the areas based on standard occupational classification. This is the same procedure used in subsequent surveys.

Measure of Crime

Consistent with Sampson and Groves's (1989) final analysis (as well as the reanalysis of the 1982 BCS data by Veysey and Messner 1999), our measure of crime is the *total victimization rate,* which reflects the rate of personal and household victimization. This variable consists of the sum of crimes that can be classified as personal or household victimization and then standardizing it by the number of respondents in the area.

Results

The results of the replication of Sampson and Groves's (1989) analysis using the 1994 BCS data are presented here in two sections. First, we report the results of the social disorganization models using the 1994 BCS data to determine—on their own—whether they support the major propositions specified by social disorganization theory according to Sampson and Groves. Second, we present a series of statistical comparisons between the standardized parameter estimates from the 1982 wave of the BCS to determine whether any of the results from the present replication study are significantly different from those of the original study.

Social Disorganization Models From the 1994 BCS

As with Sampson and Groves's (1989) analysis, our replication using the 1994 BCS data proceeded in two stages. First, we examined the effect of the five structural characteristics on the three measures of social disorganization outlined by Sampson and Groves. Second, the effects of the social disorganization measures as "intervening" variables (i.e., as mediators of the effects of the structural characteristics) predicting neighborhood crime rates were assessed.

Structural Characteristics and Social Disorganization

Table 1 contains the results of the WLS regression models testing the effects of the structural characteristics on the three dimensions of social disorganization using the 1994 BCS. For comparative purposes (which will be explicitly addressed below), the WLS coefficients from Sampson and Groves's (1989) analysis of the 1982 BCS data are reprinted in Table 1 as well. Consistent with the propositions set forth by Sampson and Groves, the first model indicates that SES, ethnic heterogeneity, and urbanization all exert a statistically significant inverse effect on local friendship networks. Family disruption was unrelated to local friendship networks, yet residential stability was positively

Table 1   Weighted Least Squares Regression Standardized Parameter Estimates of the Effects of Structural Variables on Measures of Social Disorganization ($n = 600$)

Structural Variable	Local Friendship Networks		Unsupervised Peer Groups		Organizational Participation	
	1982	1994	1982	1994	1982	1994
Socioeconomic status	−.06	−.11**	−.34***	−.23***	.17**	.07
Ethnic heterogeneity	.02	−.27***	.13**	.09**	−.06	−.18***
Residential stability	.42***	.26***	.12*	.04	−.09	.16***
Family disruption	−.03	−.02	.22***	.08*	−.02	−.09**
Urbanization	−.27***	−.11**	.15**	.29***	−.10	.03
R-Square	.26***	.22***	.30***	.28***	.07***	.10***

*$p < .10.$ **$p < .05.$ ***$p < .01.$

and significantly related to this indicator of social disorganization.

The WLS model predicting unsupervised peer groups follows a similar pattern. Levels of ethnic heterogeneity, family disruption, and urbanization were all positively and significantly related to the existence of unsupervised peer groups. Unlike the previous model, however, residential stability was not a statistically significant predictor of this dimension of social disorganization in this model. Nevertheless, also consistent with social disorganization theory, the effect of SES on unsupervised peer groups was inverse and statistically significant.

The WLS model predicting the social disorganization measure of organizational participation is not quite as robust as the previous two models. Compared to the R-square values of the models predicting local friendship networks and unsupervised peer groups (.22 and .28, respectively), the R-square for the model predicting organizational participation (.10) is considerably smaller. Even so, consistent with social disorganization theory, ethnic heterogeneity, residential stability, and family disruption were all significantly related to organizational participation in the theoretically expected direction. Inconsistent with the propositions made by Sampson and Groves (1989), however, SES and urbanization failed to significantly predict levels of organizational participation.

Social Disorganization and Total Victimization

Table 2 contains the results of the WLS model using the structural characteristics and the social disorganization measures to predict total criminal victimization. Two issues are important when examining this model: (1) whether the social disorganization measures significantly predict total criminal victimization, and (2) whether the social disorganization

| Table 2 | Weighted Least Squares Regression Standardized Parameter Estimates of the Effects of Structural Variables and Measures of Social Disorganization on Total Victimization Rates ($n = 600$) |

Independent Variable	Total Victimization Rate	
	1982	1994
Socioeconomic status	−.03	.06
Ethnic heterogeneity	.08	.06
Residential stability	.03	−.06
Family disruption	.20***	.17***
Urbanization	.21***	.03
Local friendship networks	−.12**	−.14***
Unsupervised peer groups	.34***	.40***
Organizational participation	−.11**	−.06
R-square	.42***	.28***

**p < .05. ***p < .01.

measures mediate the effects of the structural characteristics on total victimization.

With regard to the first issue, all three of the social disorganization measures were significantly related to total criminal victimization in the theoretically expected direction. The effect is particularly strong for the relationship between unsupervised peer groups and total victimization. These results are generally supportive of the social disorganization model articulated by Sampson and Groves (1989). The issue of the mediating effects of the social disorganization variables, on the other hand, offers less support for the social disorganization paradigm. In particular, after controlling for the measures of social disorganization, the structural characteristics of residential stability and family disruption still maintained statistically significant relationships with total victimization. Alternatively, on controlling for the social disorganization measures, SES, ethnic heterogeneity, and urbanization failed to significantly predict total victimization.

Summary of 1994 BCS Replication

Our test of social disorganization theory using the 1994 BCS data is, on balance, supportive of the major propositions made by the theory in three respects. First, each of the five structural characteristics significantly predicted at least two of the three social disorganization measures specified by Sampson and Groves (1989). Second, each of the social disorganization measures was significantly related to rates of total criminal victimization. Finally, although the predicted relationships did not always occur, in the majority of cases, the social disorganization measures did mediate the effects of the structural characteristics on levels of total criminal victimization. While this analysis does provide considerable support for the social disorganization perspective, it still remains to be seen whether the results found here are more or less consistent with those found in Sampson and Groves's analysis. In short, it is necessary to empirically explore the degree to which the results generated by the analysis of the 1982 BCS data differ from our replication study using the 1994 BCS data.

Statistical Comparisons Between the 1982 and 1994 Parameter Estimates

Using the results from the 1994 BCS analysis, 95 percent confidence intervals were constructed around each of the standardized parameter estimates from the WLS regression models. Once the upper and lower boundaries of the confidence intervals were established, we could then determine whether the parameter estimates from our replication were significantly different from Sampson and Groves's (1989) analyses of the 1982 BCS data (i.e., the coefficients presented in Table 1 and Table 2).[5] In essence, this series of significance tests contained two main objectives. First, they were

intended to reveal whether the results of Sampson and Groves's (1989) original study would, for the most part, be reproduced using a different data set (i.e., to determine whether the results would be consistent across data sources). Second, if significant differences did emerge, these tests were also designed to uncover whether such differences do, or do not, support social disorganization theory (i.e., do the significant differences between coefficients indicate more or less support for the theory?).

Structural Characteristics and Social Disorganization

Table 3 contains the results of the significance tests between the 1982 and 1994 BCS coefficients. Less than one-third of the coefficient comparisons (7 of 23) indicated a statistically significant difference. For the most part, these differences were found in the effects of the structural characteristics on the social disorganization variables. It is important to note, however, that none of these significant differences are inconsistent with social disorganization theory.

The effects of residential stability and urbanization on local friendship networks were smaller in the 1994 sample (yet still in the expected directions), but the effect of ethnic heterogeneity is significantly stronger in the 1994 analysis. The structural predictors of unsupervised peer groups followed a similar trend: the effect of family disruption was about one-third as strong in the 1994 sample (yet still in the expected direction), but the influence of urbanization was nearly twice as strong in the 1994 analysis. Only the effect of residential stability was significantly different for the structural predictors of organizational participation. This relationship was nonsignificant in the 1982 analysis (and the direction of this coefficient was negative), yet the 1994 data show that residential

	Table 3	Comparison of Standardized Weighted Least Squares Regression Coefficients Across 1982 and 1994 British Crime Survey Samples			

Variable	Friendship Networks	Unsupervised Peer Groups	Organized Participation	Total Victimization
Structural characteristics				
Socioeconomic status	NS	NS	NS	NS
Ethnic heterogeneity	Sig[a]	NS	NS	NS
Residential stability	Sig[a]	NS	Sig[a]	NS
Family disruption	NS	Sig[a]	NS	NS
Urbanization	Sig[a]	Sig[a]	NS	Sig[a]
Social disorganization measures				
Local friendship networks	—	—	—	NS
Unsupervised peer groups	—	—	—	NS
Organizational participation	—	—	—	NS

a. Standardized parameter estimates differ significantly, yet the estimate generated by the 1994 British Crime Survey sample is consistent with social disorganization theory.

stability has a positive and statistically significant relationship with organizational participation.

Social Disorganization and Total Victimization

As for the full model predicting total criminal victimization, only the effect of urbanization was significantly different from the 1982 to the 1994 BCS data. In Sampson and Groves's (1989) analysis, the direct effect of urbanization on total victimization was not fully mediated by the social disorganization variables. The magnitude of the urbanization coefficient in the 1994 analysis, however, is about one-seventh of the size of the parameter estimate from the 1982 sample. This finding from the 1994 BCS data indicates greater support for social disorganization theory since the direct effect of this structural characteristic is even more dampened by the presence of the social disorganization variables.

As an added check on the direct and indirect effects of the structural characteristics on the total victimization rate, consistent with Veysey and Messner's (1999) analysis of the 1982 BCS data we reestimated the full model in LISREL. As can be seen in Table 4, these results largely conform to the propositions set forth by Sampson and Groves (1989). To be sure, the overall model provides a relatively good fit to the data, with a goodness-of-fit index (GFI) of .95 and an adjusted goodness-of-fit index (AGFI) of .84—both of which are consistent with Veysey and Messner's (1999) results when they allowed the error terms of the five structural characteristics to correlate (i.e., their analytic improvement over the path analysis techniques employed by Sampson and Groves).[6] Even so, it is noteworthy that much of the total effects of the structural characteristics on the total victimization rate are indirect through the social disorganization variables.

Table 4	Direct and Indirect Effects of Structural Characteristics and Social Disorganization Variables on the Total Victimization Rate Using LISREL			
Variable	Direct Effect	Indirect Through	Indirect Effect	Total
Socioeconomic status	.060	Organizational participation	−.001	−.081
		Unsupervised teen groups	−1.40	
Ethnic heterogeneity	.060	Unsupervised teen groups	.048	.108
Urbanization	.030	Organizational participation	.005	.057
		Local friendship networks	.022	
Residential stability	−.070	Local friendship networks	−.052	−.122
Family disruption	.170	Unsupervised teen groups	.056	.226
Local friendship networks	−.140	Unsupervised teen groups	−.012	−.152
Organizational participation	−.060	Unsupervised teen groups	−.016	−.076
Unsupervised teen groups	.400	—	—	.400

GFI = .95. AFGI = .84.

Summary on 1982 and 1994 BCS Comparisons

Our replication of the test of social disorganization theory conducted by Sampson and Groves (1989), using the 1994 BCS data, generally mirrors the results that were found using the 1982 BCS data. Like Sampson and Groves's study, our analysis reveals a relatively high level of empirical support for the social disorganization perspective. Furthermore, the magnitudes of the parameter estimates across the two samples are generally similar. Indeed, significant differences between the coefficients across the two data sets appeared in fewer than one-third of the comparisons. Notably, in each instance where the parameter estimates from the 1994 sample did differ significantly from the 1982 data source, the results generated from the more recent sample indicated a greater degree of support for social disorganization theory than was previously revealed. On balance, therefore, our analysis provides both empirical support for the social disorganization perspective and support for the conclusion that Sampson and Groves's results were not idiosyncratic to the 1982 BCS data.

⊠ Discussion

Criminologists often embrace particular theoretical explanations for reasons that have little to do with a given theory's empirical validity (Cole 1975; Gould 1996; Hagan 1973; Lilly, Cullen, and Ball 2002; Pfohl 1985; Pratt and Cullen 2000). In this regard, social disorganization theory's "second wind" could be due, at least in part, to the emergence of growing public concern over the patterns of "incivility" and "disorder" that characterize certain portions of high-crime urban areas (Beckett 1997; DiIulio 1997; Skogan 1992; Wilson and Kelling 1982). Nevertheless, the empirical validation that Sampson and Groves's (1989) study accorded the social disorganization perspective clearly was integral to the perspective's revitalization.

Although subsequent studies have also revealed a certain measure of support for the major tenets of social disorganization theory (Bellair 1997; and Bergen and Herman 1998; Bursik and Grasmick 1993; Krivo and Peterson 1996; Messner and Sampson 1991; Miethe, Hughes, and McDowall 1991; Morenoff and Sampson 1997; Sampson and Raudenbush 1999; Sampson et al. 1997; Warner and Pierce

1993; Warner and Roundtree 1997), none have attempted to replicate Sampson and Groves's (1989) analysis systematically and with a different data set. This gap in the research literature is potentially problematic. One does not need to look terribly far to see the risks inherent in accepting a theoretical premise as valid—and constructing a set of social policies consistent with the theory—based on the results of a single study that may have produced results that were unique to a particular point in time and space.[7] More generally, the reproducibility of findings is a key feature of science—an activity that lacks the luster of discovering a new idea but is the linchpin to establishing whether initial claims should be accepted as "knowledge" (Gendreau in press; Hall 2000). In this context, and given the extant void in the criminological literature, our analysis was intended to assess whether the results of Sampson and Groves's (1989) study reflected time-specific phenomena or an enduring social reality. Upon comparing the findings of our analyses of the 1994 BCS data to those of Sampson and Groves's models assessing the 1982 wave of the BCS, four conclusions can be drawn.

First, independent of any comparisons to Sampson and Groves's (1989) work, our analysis can be viewed on its own as having revealed a high level of support for social disorganization theory. Our measures of local friendship networks, unsupervised peer groups, and organizational participation effectively mediated—to a large extent—the relationships between certain structural characteristics of neighborhoods and rates of criminal victimization. These measures of social disorganization were, in turn, significantly related to the total victimization rate (with the exception of organizational participation). Thus, the major propositions specified by social disorganization theory—that certain structural characteristics of communities affect the ability of residents to impose social control mechanisms over their members, and that the loss of such control mechanisms affects rates of crime—are supported here.

Second, the results of our replication are generally consistent with those found in the original study by Sampson and Groves (1989). Indeed, the major findings related to the direct, indirect, and mediating effects of community-level structural characteristics and social disorganization variables are generally reproduced here with a different data set. Our replication, therefore, indicates a pattern of consistency among the relationships specified by Sampson and Groves across the 1982 and 1994 BCS data sets.

Third, not all of the relationships revealed in the original work of Sampson and Groves (1989) were identically replicated using the 1994 BCS data. In particular, nearly a third of the parameter estimates for the relationships specified and tested in the original study and our replication differed significantly. Even so, it is important to note that when parameter estimates did differ significantly, they all diverged in a direction that was consistent with social disorganization theory (i.e., stronger mediating effects of the social disorganization variables on the relationships between the structural characteristics and total victimization).

Fourth, when taken together, these factors indicate that the theoretical attention that has been accorded to Sampson and Groves's (1989) "Community Structure and Crime: Testing Social-Disorganization Theory" is warranted. Although we cannot specify the degree to which these findings would generalize to social settings outside of Britain, our results suggest that Sampson and Groves's study appears to have captured an ongoing empirical reality. In short, our replication reinforces the view that Sampson and Groves's essay deserves the title of a "criminological classic."

In concluding that the analysis provides added support for social disorganization theory, however, we are not arguing for theoretical stagnation. Thus, compared to the measures used to assess the intervening or mediating variables in our study, additional or more finely calibrated measures of social disorganization might account more extensively for the direct effects of structural antecedents. But the alternative possibility is that the direct effects reflect causal processes that other theories are better suited to explain. Furthermore, as others have pointed out (Veysey and Messner 1999), the effects mediating disorganization factors in Sampson and Groves's model may themselves be open to alternative theoretical interpretation. Sampson and Groves chose to interpret social disorganization theory as a perspective that linked crime rates to the breakdown of informal social control. Other scholars, however, have argued that Shaw and McKay (1972) offered a "mixed model" of delinquency, which tied crime both to the weakening of control *and* to the transmission and learning of criminal values (Kornhauser 1978). Mediating variables, such as unsupervised peers, thus might capture not only the attenuation of control but also the presence of criminal subcultures and social learning (Veysey and Messner 1999). Further research is needed to reveal more precisely why variables have the effects they do and the implications of these finding for competing theories of crime.

Equally salient, the social disorganization perspective itself is evolving. Perhaps the most noteworthy advance is Sampson et al's (1997) introduction of the concept of "collective efficacy." At times, it appears that the concept of collective efficacy is the "opposite" of social disorganization (Taylor 2001:128). However, the theoretical advance comes in at least three ways. First, the concept of efficacy involves not only informal control but also "trust" or "social cohesion" (Sampson et al. 1997:920)—or what might be called "social support" (Cullen 1994). That is, control and trust are interrelated and mutually reinforcing. Second, the notion of "efficacy" implies a resource that can be activated. The willingness to apply controls is found in social disorganization theory as well,

but in Sampson et al.'s (1997) work, the choice of the word efficacy implies more explicitly that collectives differ in their ability to actively fight against neighborhood conditions that are criminogenic. Efficacy is not just a static state of being but capital that can be "called into play" when needed (e.g., to clear drug pushers from a community playground).Third, the collective efficacy model moves beyond social disorganization theory's traditional neglect of structural inequality (Pfohl 1985), and sees antecedent variables not as separate entities but as interrelated. In particular, they link collective efficacy to the *concentration* of disadvantage and advantage (Sampson et al. 1997, 1999).

In many ways, these theoretical extensions are to be welcomed. Nonetheless, three observations are warranted. First, collective efficacy encapsulates the core elements of social disorganization, with the model including informal social control in its measures and conceptualization of efficacy. Accordingly, studies providing empirical confirmation for social disorganization theory—such as our analyses and Sampson and Groves's (1989) original work—lend indirect support for collective efficacy theory. Second, it has yet to be shown that collective efficacy theory is empirically superior to social disorganization theory. Unlike tests of individual-level theories (Hirschi 1969), macro-level theories only rarely are tested against one another. Social disorganization theory—especially if tested in all its complexity as a "mixed model"—might rival collective efficacy in its ability to explain variation in crime rates. Third, if social disorganization theory is the product of 1920s Chicago (Lilly et al. 2002; Pfohl 1985), collective efficacy theory is the product of its time—current-day Chicago (Sampson et al. 1997). It remains to be seen whether this perspective will have more explanatory power than social disorganization theory when applied to other contexts, including, for example, urban areas located elsewhere in the United States, in Europe, and in rapidly growing developing nations.

✎ Notes

1. According to the Social Sciences Citation Index, Sampson and Groves's 1989 publication has been cited 232 times, including 96 times since 1999 and 32 times in the past 14 months. In a recent article, Wright, Malia, and Johnson (1999) reviewed the most cited criminology and criminal justice works in 107 articles from the leading sociology journals. Sampson and Groves's publication ranked third, higher than a number of noteworthy works (e.g., Black 1976; Blau and Blau 1982; Cohen and Felson 1979; Gottfredson and Hirschi 1990; Kornhauser 1978; Merton 1968).

2. Sampson and Groves (1989) focused primarily on the 1982 BCS data. They supplemented their analysis with the 1984 BCS data, and the results were largely consistent with those from the 1982 sample. Thus, consistent with Sampson and Groves's original focus, in the present study, we compare our results to their analyses of the 1982 BCS data.

3. We should note that in their original article, Sampson and Groves (1989) attempted to confirm their 1982 analysis with a partial replication of data drawn from the 1984 BCS (e.g., there was no measure of organizational participation). Their results were consistent with the findings drawn from the 1982 analysis and lent added support to social disorganization theory. Our replication is patterned after the more complete 1982 study. In either case, however, our independent replication is conducted on BCS data collected a full decade after the surveys analyzed by Sampson and Groves.

5. Traditional methods for comparing the equality of coefficients (see, e.g., Brame et al. 1998; Clogg, Petkova, and Haritou 1995; Cohen 1983) require the use of both the metric parameter estimates and the standard errors from each model. Since we only have such information (specifically, the standard errors) for our 1994 analysis, the use of confidence intervals for this purpose is the next methodologically rigorous method at our disposal.

6. Veysey and Messner (1999) did, however, arrive at a model with a Goodness-of-Fit Index of .988 and an Adjusted Goodness-of-Fit Index of .939 when they freed the parameters to estimate direct effects for socioeconomic status, ethnic heterogeneity, residential stability, and family disruption on the total victimization rate.

7. For example, in their classic Minneapolis experiment, Sherman and Berk (1984) found that arrest had a deterrent effect on domestic violence recidivism. Subsequent replications, however, produced inconsistent results and the specification that arrest primarily had effects on offenders with social bonds (see, e.g., Dunford, Huizinga, and Elliott 1990; Sherman and Berk 1984; Sherman et al. 1991, 1992).

✎ References

Akers, Ronald L. 1999. *Criminological Theories: Introduction and Evaluation*. 2nd ed. Chicago: Fitzroy Dearborn.

Beckett, Katherine. 1997. *Making Crime Pay: Law and Order in Contemporary American Politics*. New York: Oxford University Press.

Bellair, Paul E. 1997. "Social interaction and Community Crime: Examining the Importance of Neighborhood Networks." *Criminology* 35: 677–703.

Bergen, Albert, and Max Herman. 1998. "Immigration, Race, and Riot: The 1992 Los Angeles Uprising." *American Sociological Review* 63: 39–54.

Black, Donald. 1976. *The Behavior of Law*. New York: Academic Press.

Blau, Judith R., and Peter M. Blau. 1982. "The Cost of Inequality: Metropolitan Structure and Violent Crime." *American Sociological Review* 47: 114–29.

Brame, Robert, Raymond Paternoster, Paul Mazerolle, and Alex Piquero. 1998. "Testing for the Equality of Maximum-Likelihood Regression Coefficients Between Two Independent Equations." *Journal of Quantitative Criminology* 14: 245–61.

Bursik, Robert J. 1986. "Delinquency Rates as Sources of Ecological Change." Pp. 63–74 in *The Social Ecology of Crime*, edited by James M. Byrne and Robert J. Sampson. New York: Springer-Verlag.

———. 1988. "Social Disorganization and Theories of Crime and Delinquency: Problems and Prospects." *Criminology* 26: 519–51.

Bursik, Robert J., and Harold G. Grasmick. 1993. "Economic Deprivation and Neighborhood Crime Rates, 1960–1980." *Law and Society Review* 27: 263–83.

Byrne, James M., and Robert J. Sampson. 1986. "Key Issues in the Social Ecology of Crime." Pp. 1–22 in *The Social Ecology of Crime*, edited by James M. Byrne and Robert J. Sampson. New York: Springer-Verlag.

Clogg, Clifford C., Eva Petkova, and Adamantios Haritou. 1995. "Statistical Methods for Comparing Regression Coefficients Between Models." *American Journal of Sociology* 100: 1261–93.

Cohen, Ayala. 1983. "Comparing Regression Coefficients Across Subsamples: A Study of the Statistical Test." *Sociological Methods and Research* 12: 77–94.

Cohen, Lawrence E., and Marcus Felson. 1979. "Social Change and Crime Rate Trends: A Routine Activity Approach." *American Sociological Review* 44: 588–608.

Cole, Stephen. 1975. "The Growth of Scientific Knowledge: Theories of Deviance as a Case Study." Pp. 175–200 in *The Idea of Social Structure: Papers in Honor of Robert K. Merlon,* edited by Lewis A. Coser. New York: Harcourt Brace Jovanovich.

Cullen, Francis T. 1994. "Social Support as an Organizing Concept for Criminology: Presidential Address to the Academy of Criminal Justice Sciences." *Justice Quarterly* 11: 527–59.

DiIulio, John J. 1997. "Are Voters Fools? Crime, Public Opinion, and Democracy." *Corrections Management Quarterly* 1 (3): 1–5.

Dunford, Franklyn, David Huizinga, and Delbert S. Elliott 1990. "The Role of Arrest in Domestic Assault: The Omaha Experiment." *Criminology* 28: 183–206.

Gendreau, Paul. In press. "We Must Do a Better Job of Cumulating Knowledge." *Canadian Psychology.*

Gottfredson, Michael R., and Travis Hirschi. 1990. *A General Theory of Crime.* Stanford, CA: Stanford University Press.

Gould, Stephen Jay. 1996. *The Mismeasure of Man.* Rev. ed. New York: W. W. Norton.

Hagan, John. 1973. "Labeling and Deviance: A Case Study in the 'Sociology of the Interesting.'" *Social Problems* 20: 447–58.

Hall Nancy S. 2000. "The Key Role of Replication in Science." *The Chronicle of Higher Education,* November 10, B14.

Hirschi, Travis, 1969. *Causes of Delinquency.* Berkeley: University of California Press.

Home Office Crime and Criminal Justice Unit Office of Population Census and Social Surveys Division. 1994. *British Crime Survey* (computer file). Colchester, Essex, UK: The Data Archive.

Kasarda, John D., and Morris Janowitz. 1974. "Community Attachment in Mass Society." *American Sociological Review* 39: 328–39.

Kempf, Kimberly L. 1993. "The Empirical Status of Hirschi's Control Theory." Pp. 143–85 in *New Directions in Criminological Theory: Advances in Criminological Theory,* Vol. 4, edited by Freda Adler and William S. Laufer. New Brunswick, NJ: Transaction.

Kornhauser, Ruth R. 1978. *Social Sources of Delinquency: An Appraisal of Analytic Models.* Chicago: University of Chicago Press.

Krivo, Lauren J., and Ruth D. Peterson. 1996. "Extremely Disadvantaged Neighborhoods and Urban Crime." *Social Forces* 75: 619–50.

Krohn, Marvin D. 1986. "The Web of Conformity: A Network Approach to the Explanation of Delinquent Behavior." *Social Problems* 33: 81–93.

Lilly, Robert J., Francis T. Cullen, and Richard A. Ball. 2002. *Criminological Theory: Context and Consequences.* 3rd ed. Thousand Oaks, CA: Sage.

Lynn, Peter. 1997. "Sampling Frame Effects on the British Crime Survey." *Journal of the Royal Statistical Society Series* 160 (2): 253–69.

Merton, Robert K. 1968. *Social Theory and Social Structure.* New York: Free Press.

Messner, Steven F., and Robert J. Sampson. 1991. "The Sex Ratio, Family Disruption, and Rates of Violent Crime: The Paradox of Demographic Structure." *Social Forces* 69: 693–713.

Miethe, Terance D., Michael Hughes, and David McDowall. 1991. "Social Change and Crime Rates: An Evaluation of Alternative Theoretical Approaches." *Social Forces* 70: 165–85.

Morenoff, Jeffrey D., and Robert J. Sampson. 1997. "Violent Crime and the Spatial Dynamics of Neighborhood Transition: Chicago, 1970-1990." *Social Forces* 76: 31–64.

Morenoff, Jeffrey D., Robert J. Sampson, and Stephen W. Raudenbush, 2001. "Neighborhood Inequality, Collective Efficacy, and the Spatial Dynamics of Urban Violence." *Criminology* 39: 517–59.

Pfohl, Stephen J. 1985. *Images of Deviance and Social Control: A Sociological History.* New York: McGraw-Hill.

Pratt, Travis C., and Francis T. Cullen. 2000. "The Empirical Status of Gottfredson and Hirschi's General Theory of Crime: A Meta-Analysis." *Criminology* 38: 931–64.

Roundtree, Pamela Wilcox, and Kenneth C. Land. 1996. "Burglary Victimization, Perceptions of Crime Risk, and Routine Activities: A Multilevel Analysis Across Seattle Neighborhoods and

Census Tracts." *Journal of Research in Crime and Delinquency* 33: 147–80.

Roundtree, Pamela Wilcox, and Barbara D. Warner. 1999. "Social Ties and Crime: Is the Relationship Gendered?" *Criminology* 37: 789–812.

Sampson, Robert J. 1987. "Urban Black Violence: The Effect of Male Joblessness and Family Disruption." *American Journal of Sociology* 93: 348–82.

Sampson, Robert J., and W. Byron Groves. 1989. "Community Structure and Crime: Testing Social-Disorganization Theory." *American Journal of Sociology* 94: 774–802.

Sampson, Robert J., Jeffery D. Morenoff, and Felton Earls. 1999. "Beyond Social Capital: Spatial Dynamics of Collective Efficacy for Children." *American Sociological Review* 64: 633–60.

Sampson, Robert J., and Stephen W. Raudenbush. 1999. "Systematic Social Observation of Public Spaces: A New Look at Disorder in Urban Neighborhoods." *American Journal of Sociology* 105: 603–51.

Sampson, Robert J., Stephen W. Raudenbush, and Felton Earls. 1997. "Neighborhoods and Violent Crime: A Multilevel Study of Collective Efficacy." *Science* 277: 918–24.

Shaw, Clifford R., and Henry D. McKay. 1972. *Juvenile Delinquency and Urban Areas.* Chicago: University of Chicago Press.

Sherman, Lawrence W., and Richard A. Berk. 1984. "The Specific Deterrent Effects of Arrest for Domestic Assault." *American Sociological Review* 49: 261–72.

Sherman, Lawrence W., Janell D. Schmidt, Dennis P. Rogan, Patrick R. Gartin, Ellen G. Cohn, Dean J. Collins, and Anthony R. Bacich. 1991. "From Initial Deterrence to Long-Term Escalation: Short-Custody Arrest for Poverty Ghetto Domestic Violence." *Criminology* 29: 821–50.

Sherman, Lawrence W., Douglas A. Smith, Janell D. Schmidt, and Dennis P. Rogan. 1992. "Crime, Punishment, and Stake in Conformity; Legal and Informal Control of Domestic Violence." *American Sociological Review* 57: 680–90.

Siegel, Larry. 2000. *Criminology.* 7th ed. Belmont, CA: Wadsworth.

Skogan, Wesley G. 1992. *Disorder and Decline: Crime and the Spiral of Decay in American Neighborhoods.* Berkeley: University of California Press.

Sykes, Gresham M., and Francis T. Cullen. 1992. *Criminology.* 2nd ed. New York: Harcourt Brace Jovanovich.

Taylor, Ralph B. 2001. "The Ecology of Crime, Fear, and Delinquency: Social Disorganization Versus Social Efficacy. In *Explaining Criminals and Crime: Essays in Contemporary Criminological Theory,* edited by Raymond Paternoster and Ronet Bachman. Los Angeles: Roxbury.

Tittle, Charles R., and Raymond Paternoster. 2000. *Social Deviance and Crime: An Organizational and Theoretical Approach.* Los Angeles: Roxbury.

Velez, Maria B. 2001. "The Role of Public Social Control in Urban Neighborhoods: A Multi-Level Analysis of Victimization Risk." *Criminology* 39: 837–64.

Veysey, Bonita M., and Steven F. Messner. 1999. "Further Testing of Social Disorganization Theory: An Elaboration of Sampson and Groves's 'Community Structure and Crime'." *Journal of Research in Crime and Delinquency* 36: 156–74.

Warner, Barbara D., and Glenn L. Pierce. 1993. "Reexamining Social Disorganization Theory Using Calls to the Police as a Measure of Crime." *Criminology* 31: 493–517.

Warner, Barbara D., and Pamela Wilcox Roundtree. 1997. "Local Ties in a Community and Crime Model: Questioning the Systemic Nature of Informal Social Control." *Social Problems* 44: 520–36.

Wilson, James Q., and George L. Kelling. 1982. "Broken Windows: The Police and Neighborhood Safety." *Atlantic Monthly,* March, 29–38.

Wright, Richard A., Michael Malia, and C. Wayne Johnson. 1999. "Invisible Influence: A Citation Analysis of Crime and Justice Articles Published in Leading Sociology Journal." *Journal of Crime and Justice* 22: 147–65.

REVIEW QUESTIONS

1. Identify and discuss the three types of variables used by Lowenkamp et al. (as well as Sampson and Grove) to measure the key intervening variables in the model of social disorganization. Which of these three do you feel is the best measure of community (or informal) controls/organization, and why? Which of these do you feel is the least applicable, and why?

2. Identify and discuss the five types of structural variables of neighborhoods that Lowenkamp et al. (as well as Sampson and Grove) claim are mediated by the intervening variables (which you discussed in Review Question #1). Which of these structural variables do you think is most vital in predicting crime rates of neighborhoods, and why? Which structural variable do you think is least vital, and why?

3. If you were asked to make a conclusion as a social scientist regarding the current state of empirical evidence or support for Shaw and McKay's proposed model of social disorganization, what would you conclude? What key weaknesses do you see with the model or the way it has been tested?

4. To what extent do the findings presented by Lowenkamp et al. (or Samson and Grove) fit what you have observed in your own local communities?

❖

READING

In this selection, Marvin Wolfgang and Franco Ferracuti propose a theoretical model that explains how an entire other culture (e.g., subculture) can exist, even within an existing normative framework by a majority of individuals in a society. The authors propose a "subculture of violence," which they note is predominately held by young males. They argue that violence is a theme in the "values that make up the life-style, the socialization process, the interpersonal relationships of individuals living in similar conditions." This selection reveals the core elements of this theory, including the social learning and cultural context of this subculture, as well as conclusions that can be made regarding the existence of such subcultures. While reading this selection, readers are encouraged to think about people they know or grew up with and consider if they can relate adherence to subcultural norms regarding violence to anyone in their past.

The Subculture of Violence

Marvin E. Wolfgang and Franco Ferracuti

⬛ The Thesis of a Subculture of Violence

This section examines the proposition that there is a subculture of violence . . . to extend a theoretical formulation regarding the existence of subcultures in general if we are to hypothesize a particular subculture of violence. It would be difficult to support an argument that a subculture exists in relation to a single cultural interest, and the thesis of a subculture of violence does not suggest a monolithic character.

It should be remembered that the term itself—subculture—presupposes an already existing complex of norms, values, attitudes, material traits, etc. What the subculture-of-violence formulation further suggests is simply that there is a potent theme of violence current in the cluster of values that make up the life-style, the socialization process, the interpersonal relationships of individuals living in similar conditions.

The analysis of violent aggressive behavior has been the focus of interest of many social and biological researchers, and psychology has attempted to build several theories to explain its phenomenology, ranging from the death-aggression instinct of the psychoanalytic school to the frustration-aggression hypothesis. The present discussion is the result of joint explorations in theory and research in psychology and in sociology, using the concept of subculture as a learning environment. Our major area of study has been assaultive behavior, with special attention to criminal homicide . . . [S]ome of the main trends in criminological thinking related to this topic must now be anticipated for the proper focus of the present discussion.

Isolated sectional studies of homicide behavior are extremely numerous, and it is not our intention to examine them in this chapter. There are basically two kinds of criminal homicide: (1) premeditated, felonious, intentional murder; (2) slaying in the heat of passion, or killing as a result of intent to do harm but without intent to kill. A slaying committed by one recognized as psychotic or legally insane, or by a psychiatrically designated abnormal subject involves clinical deviates who are generally not held responsible for their behavior, and who, therefore, are not considered culpable. We are eliminating these cases from our present discussion, although subcultural elements are not irrelevant to the analysis of their psychopathological phenomenology.

Probably fewer than five per cent of all known homicides are premeditated, planned intentional killings, and the individuals who commit them are most likely to be episodic offenders who have never had prior contact with the criminal law. Because they are rare crimes often planned by rationally functioning individuals, perhaps they are more likely to remain undetected. We believe that a type of analysis different from that presented here might be applicable to these cases. Our major concern is with the bulk of homicides—the passion crimes, the violent slayings—that are not premeditated and are not psychotic manifestations. Like Cohen, who was concerned principally with most delinquency that arises from the "working-class" ethic, so we are focusing [on] the preponderant kind of homicide, although our analysis . . . will include much of the available data on homicide [in] general.

Social Learning and Conditioning

Studies of child-rearing practices in relation to the development of aggressive traits lead us to consider briefly one important theoretical development from the field of general and social psychology which, in our opinion, provides the theoretical bridge between an individual's violent behavior and his subcultural value allegiance. We are referring to what is generally included under the heading of "social learning." Issues which arise in any analysis of the structure and phenomenology of subcultures are the process of transmitting the subculture values, the extent of individual differences in the strength of allegiance to those values, and the fact that not all the individuals with ecological propinquity share value and motive identity with the surrounding culture. The process of social learning, through a number of mechanisms ranging from repetitive contacts to the subtler forms of imitation and identification, involves the acquisition of value systems in early childhood and their integration in the complex personality trait-value-motive system, which makes up the adult global individuality.

A recent paper by Jeffery (1965) summarizes a number of theoretical statements accepting the general principle that criminal behavior can be explained as learned behavior if conceptualized as operant behavior and reinforced through reward and immediate gratification. However, the complexity of learning theory and the serious uncertainties that still plague its core concepts have thus far produced an heuristic deficiency in transferring from theory and experimental laboratory research to field applications. The same may be said about transferring to diagnostically oriented studies of the differential psychology of violent offenders. Admittedly, the transposition from laboratory and animal experimentation to the street corner and the prison is not easy, is somewhat speculative, and may prove impossible until the social-learning approach can produce measurable, economical, and valid diagnostic instruments.

An interesting beginning towards such a development in criminology has been made by Eysenck and his collaborators. A general restatement of the theory can be found in Trasler, and earlier statements appear in Mowrer, for example. Bandura's rich production follows a social-learning approach, and his recent books with Walters provide a detailed discussion of mechanisms, patterns, and implications for the application of a behavioristic learning approach to analysis of personality development.

Indoctrination into a subculture can take place through early-infancy learning processes. However, not only does this indoctrination prove difficult to reconstruct in an individual diagnostic process, or impossible to demonstrate in the laboratory, but it is confused with individual differences. These differentials in the imitation and identification processes beg the central question of why equally exposed individuals terminally behave differently and exhibit values and norms that resist attempts to classify them into discrete, yet uniform categories. Eysenck has approached this problem through introduction of the concept of individual differences in conditioning, including, by extension, social conditioning. This approach assumes that, whereas introverts are easier to condition and therefore more readily absorb socialized values, extroverts are resistant to conditioning and dominated by anti-social impulsive reactions. The conceptualization can be extended to include social learning of whole antisocial value systems. These notions, if logically followed, would postulate two types of violent offender: (1) the introversive, who are socialized into a subculture of violence through conditioning, and are frequent in specific ecological settings; (2) the extroversive, impulsive, unsocialized types, who cut across social, cultural, and subcultural strata. Both types can exhibit violent behavior, but the etiology and the probability of such behavior vary along with basic psychological make-up, i.e., a set of inherited characteristics which, in Eysenck's terms, dichotomize individuals according to biological determinants that place them in a given position on the introversion-extroversion continuum. Only modest confirmation is so far available for this far-reaching conceptualization. An extension of the behavioristic learning theory into therapy has been proposed by Bandura and Walters and analyzed by several others. The advisability of granting scientific status to an approach which is still highly experimental has been seriously questioned.

Although the social-learning approach still awaits confirmation, it does provide a conceptually useful bridge between the sociological, the psychological, and clinical constructs which we have discussed in the preceding pages. It also furnishes us with the possibility of utilizing two other personality theories which have a definite place in the transposition of the concept of subculture from sociology to psychology. Dissonance theory, as one of these, constitutes an elegant, if unproved, link between the cognitive aspects of subcultural allegiance, the psychoanalytic mechanism of projection, and the internal consistency (with consequent reduction of the anxiety level) which constitutes the differential characteristics of members of the subculture. No dissonance is experienced so long as the value system of the individual is not confronted by different or certainly conflicting values. The treatment implications of the concept of cognitive dissonance in relation to subculture allegiance are obvious, and point to the need to fragment and rearrange antisocial group alliances. The utilization of cognitive dissonance in this way in the prevention of international conflict has been advocated, for example, by Stagner and Osgood. Stagner has, however, carefully analyzed the importance of perceptual personality theory to individual and group aggression. A subculture allegiance entails an organization or reorganization of the process of personality formation as a process of learning to perceive objects, persons, and situations as attractive or threatening, in accordance with subcultural positive and negative valences.

The general psychological contributions from conditioning, learning theory, cognitive dissonance, perceptual personality theory, are indeed far from providing a total theoretical system as a counterpart to the sociological notions about subcultures. However, we are convinced that these behavioral constructs of social learning not only are the most directly related to subculture theory but also are capable of generating an integrated theory in criminology.

The Cultural Context

Like all human behavior, homicide and other violent assaultive crimes must be viewed in terms of the cultural context from which they spring. De Champneuf, Guerry, Quetelet early in the nineteenth century, and Durkheim later, led the way toward emphasizing the necessity to examine the *physique sociale,* or social phenomena characterized by "externality," if the scientist is to understand or interpret crime, suicide, prostitution, and other deviant behavior. Without promulgating a sociological fatalism,

analysis of broad macroscopic correlates in this way may obscure the dynamic elements of the phenomenon and result in the empirical hiatus and fallacious association to which Selvin refers. Yet, because of wide individual variations, the clinical, idiosyncratic approach does not necessarily aid in arriving at Weber's *Verstehen,* or meaningful adequate understanding of regularities, uniformities, or patterns of interaction. And it is this kind of understanding we seek when we examine either deviation from, or conformity to, a normative social system.

Sociological contributions have made almost commonplace, since Durkheim, the fact that deviant conduct is not evenly distributed throughout the social structure. There is much empirical evidence that class position, ethnicity, occupational status, and other social variables are effective indicators for predicting rates of different kinds of deviance. Studies in ecology perform a valuable service for examining the phenomenology and distribution of aggression, but only inferentially point to the importance of the system of norms. Anomie, whether defined as the absence of norms (which is a doubtful conceptualization) or the conflict of norms (either normative goals or means), or whether redefined by Powell as "meaninglessness," does not coincide with most empirical evidence on homicide. Acceptance of the concept of anomie would imply that marginal individuals who harbor psychic anomie that reflects (or causes) social anomie have the highest rates of homicides. Available data seem to reject this contention. Anomie as culture conflict, or conflict of norms, suggests, as we have in the last section, that there is one segment (the prevailing middle-class value system) of a given culture whose value system is the antithesis of, or in conflict with, another, smaller segment of the same culture. This conceptualism of anomie is a useful tool for referring to subcultures as ideal types, or mental constructs. But to transfer this norm-conflict approach from the social to the individual level, theoretically making the individual a repository of culture conflict, again does not conform to the patterns of known psychological and sociological data. This latter approach would be forced to hypothesize that socially mobile individuals and families would be most frequently involved in homicide, or that persons moving from a formerly embraced subvalue system to the predominant communal value system would commit this form of violent deviation in the greatest numbers. There are no homicide data that show high rates of homicides among persons manifesting higher social aspirations in terms of mobility. It should also be mentioned that anomie, as a concept, does not easily lend itself to psychological study.

That there is a conflict of value systems, we agree. That is, there is a conflict between a prevailing culture value and some subcultural entity. But commission of homicide by actors from the subculture at variance with the prevailing culture cannot be adequately explained in terms of frustration due to failure to attain normative-goals of the latter, in terms of inability to succeed with normative-procedures (means) for attaining those goals, nor in terms of an individual psychological condition of anomie. Homicide is most prevalent, or the highest rates of homicide occur, among a relatively homogeneous subcultural group in any large urban community. Similar prevalent rates can be found in some rural areas. The value system of this group, we are contending, constitutes a subculture of violence. From a psychological viewpoint, we might hypothesize that the greater the degree of integration of the individual into this subculture, the higher the probability that his behavior will be violent in a variety of situations. From the sociological side, there should be a direct relationship between rates of homicide and the extent to which the subculture of violence represents a cluster of values around the theme of violence.

Except for war, probably the most highly reportable, socially visible, and serious form of violence is expressed in criminal homicide.

Data show that in the United States rates are highest among males, non-whites, and the young adult ages. Rates for most serious crimes, particularly against the person, are highest in these same groups. In a Philadelphia study of 588 criminal homicides, for example, non-white males aged 20–24 had a rate of 92 per 100,000 compared with 3:4 [3.4] for white males of the same ages. Females consistently had lower rates than males in their respective race groups (non-white females, 9:3 [9.3]; white females, 0:4 [0.4], in the same study), although it should be . . . that non-white females have higher rates than white males.

It is possible to multiply these specific findings in any variety of ways; and although a subcultural affinity to violence appears to be principally present in large urban communities and increasingly in the adolescent population, some typical evidence of this phenomenon can be found, for example, in rural areas and among other adult groups. For example, a particular, very structured, subculture of this kind can be found in Sardinia, in the central mountain area of the island. Pigliaru has conducted a brilliant analysis of the people from this area and of their criminal behavior, commonly known as the *vendetta barbaricina*. . . .

. . . . We suggest that, by identifying the groups with the highest rates of homicide, we should find in the most intense degree a subculture of violence; and, having focused on these groups, we should subsequently examine the value system of their subculture, the importance of human life in the scale of values, the kinds of expected reaction to certain types of stimulus, perceptual differences in the evaluation of stimuli, and the general personality structure of the subcultural actors. In the Philadelphia study it was pointed out that:

the significance of a jostle, a slightly derogatory remark, or the appearance of a weapon in the hands of an adversary are stimuli differentially perceived and interpreted by Negroes and whites, males and females. Social expectations of response in particular types of social interaction result in differential "definitions of the situation." A male is usually expected to defend the name and honor of his mother, the virtue of womanhood . . . and to accept no derogation about his race (even from a member of his own race), his age, or his masculinity. Quick resort to physical combat as a measure of daring, courage, or defense of status appears to be a cultural expression, especially for lower socio-economic class males of both races. When such a culture norm response is elicited from an individual engaged in social interplay with others who harbor the same response mechanism, physical assaults, altercations, and violent domestic quarrels that result in homicide are likely to be common. The upper-middle and upper social class value system defines subcultural mores, and considers many of the social and personal stimuli that evoke a combative reaction in the lower classes as "trivial." Thus, there exists a cultural antipathy between many folk rationalizations of the lower class, and of males of both races, on the one hand, and the middle-class legal norms under which they live, on the other.

This kind of analysis, combined with other data about delinquency, lower-class social structure, its value system, and its emphasis on aggression, suggest the thesis of a violent subculture, or, by pushing the normative aspects a little further, a *subculture of* violence. Among many juvenile gangs, as has repeatedly been pointed out, there are violent feuds, meetings, territorial fights, and the use of violence to prove "heart," to maintain or to acquire "rep."

Physical aggression is often seen as a demonstration of masculinity and toughness. We might argue that this emphasis on showing masculinity through aggression is not always supported by data. If homicide is any index at all of physical aggression, we must remember that in the Philadelphia data non-white females have [homicide] rates often two to four times higher than the rates of white males. Violent behavior appears more dependent on cultural differences than on sex differences, traditionally considered of paramount importance in the expression of aggression. . . .

. . . . It appears valid to suggest that there are, in a heterogeneous population, differences in ideas and attitudes toward the use of violence and that these differences can be observed through variables related to social class and possibly through psychological correlates. There is evidence that modes of control of expressions of aggression in children vary among the social classes. Lower-class boys, for example, appear more likely to be oriented toward direct expression of aggression than are middle-class boys. The type of punishment meted out by parents to misbehaving children is related to this class orientation toward aggression. Lower-class mothers report that they or their husbands are likely to strike their children or threaten to strike them, whereas middle-class mothers report that their type of punishment is psychological rather than physical; and boys who are physically [punished] express aggression more directly than those who are punished psychologically. As Martin has suggested, the middle-class child is more likely to turn his aggression inward; in the extreme and as an adult he will commit suicide. But the lower-class child is more accustomed to a parent-child relationship which during punishment is for the moment that of attacker and attacked. The target for aggression, then, is external; aggression is directed toward others. //

The existence of a subculture of violence is partly demonstrated by examination of the social groups and individuals who experience the highest rates of manifest violence. This examination need not be confined to the study of one national or ethnic group. On the contrary, the existence of a subculture of violence could perhaps receive even cross-cultural confirmation. Criminal homicide is the most acute and highly reportable example of this type of violence, but some circularity of thought is obvious in the effort to specify the dependent variable (homicide), and also to infer the independent variable (the existence of a subculture of violence). The highest rates of rape, aggravated assaults, persistency in arrests for assaults (recidivism) among these groups with high rates of homicide are, however, empirical addenda to the postulation of a subculture of violence. Residential propinquity of these same groups reinforces the socio-psychological impact which the integration of this subculture engenders. Sutherland's thesis of "differential association," or a psychological reformulation of the same theory in terms of learning process, could effectively be employed to describe more fully this impact in its intensity, duration, repetition, and frequency. The more thoroughly integrated the individual is into this subculture, the more intensely he embraces its prescriptions of behavior, its conduct norms, and integrates them into his personality structure. The degree of integration may be measured partly and crudely by public records of contact with law, so high arrest rates, particularly high rates of assault crimes and rates of recidivism for assault crimes among groups that form the culture of violence, may indicate allegiance to the values of violence.

We have said that overt physical violence often becomes a common subculturally expected response to certain stimuli. However, it is not merely rigid conformity to the demands and expectations of other persons, as Henry and Short seem to suggest, that results in the high probability of homicide. Excessive, compulsive, or apathetic conformity [of] middle-class individuals to the value system of their social group is a widely recognized cultural malady. Our concern is with the value elements of violence as an integral component of the subculture which experiences high rates of homicide. It is conformity to *this* set of values, and not rigid conformity *per se*, that gives important meaning to the subculture of violence. . . .

It is not far-fetched to suggest that a whole culture may accept a value set dependent upon violence, demand or encourage adherence to violence, and penalize deviation. . . . Homicide, it appears, is often a situation not unlike that of confrontations in wartime combat, in which two individuals committed to the value of violence come together, and in which chance, prowess, or possession of a particular weapon dictates the identity of the slayer and of the slain. The peaceful non-combatant in both sets of circumstances is penalized, because of the allelomimetic behavior of the group supporting violence, by his being ostracized as an out-group member, and he is thereby segregated (imprisoned, in wartime, as a conscientious objector) from his original group. If he is not segregated, but continues to interact with his original group in the public street or on the front line that represents the culture of violence, he may fall victim to the shot or stab from one of the group who still embraces the value of violence. . . .

. . . . We have said that overt use of force or violence, either in interpersonal relationships or in group interaction, is generally viewed as a reflection of basic values that stand apart from the dominant, the central, or the parent culture. Our hypothesis is that this overt . . . (and often illicit) expression of violence (of which homicide is only the most extreme) is part of a subcultural normative system, and that this system is reflected in the psychological traits of the subculture participants. In the light of our discussion of the caution to be exercised in interpretative analysis, in order to tighten the logic of this analysis, and to support the thesis of a subculture of violence, we offer the following corollary propositions:

1. *No subculture can be totally different from or totally in conflict with the society of which it is a part.* A subculture of violence is not entirely an expression of violence, for there must be interlocking value elements shared with the dominant culture. . . .

2. *To establish the existence of a subculture of violence does not require that the actors sharing in these basic value elements should express violence in all situations.* The normative system designates that in some types of social interaction a violent and physically aggressive response is either expected or required of all members sharing in that system of values. . . .

3. *The potential resort or willingness to resort to violence in a variety of situations emphasizes the penetrating and diffusive character of this culture theme.* The number and kinds of situations in which an individual uses violence may be viewed as an index of the extent to which he has assimilated the values associated with violence. . . .

4. *The subcultural ethos of violence may be shared by all ages in a sub-society, but this ethos is most prominent in a limited age group, ranging from late adolescence to middle age.* . . . [T]he known empirical distribution of conduct, which expresses the sharing of this violence theme, shows greatest localization, incidence, and frequency in limited sub-groups and reflects differences in learning about violence as a problem-solving mechanism.

5. *The counter-norm is nonviolence.* Violation of expected and required violence is most likely to result in ostracism from the group. . . .

6. The development of favorable attitudes toward, and the use of, violence in a subculture usually involve learned behavior and a process of differential learning, association, or identification. Not all persons exposed—even equally exposed—to the presence of a subculture of violence absorb and share in the values in equal portions. Differential personality variables must be considered in an integrated social-psychological approach to an understanding of the subcultural aspects of violence. . . .

7. The use of violence in a subculture is not necessarily viewed as illicit conduct and the users therefore do not have to deal with feelings of guilt about their aggression. Violence can become a part of the life style, the theme of solving difficult problems or problem situations. . . . [W]hen the attacked see their assaulters as agents of the same kind of aggression they themselves represent, violent retaliation is readily legitimized by a situationally specific rationale, as well as by the generally normative supports for violence. . . .

REVIEW QUESTIONS

1. What do Wolfgang and Ferracuti claim regarding two categories of homicide? What do they say regarding these two types of homicide (e.g., what percentage of homicides does each category make up of in the total), and why?

2. What do the authors conclude regarding the social learning and conditioning of homicides in the subculture of violence?

3. What do Wolfgang and Ferracuti claim regarding the cultural context of this subculture of violence?

4. The authors provide seven conclusions in this selection. Which conclusion do you most agree with and why? Which conclusion do you most disagree with and why?

READING

In this selection, Anderson presents a cultural/subcultural theory of the inner-city "ghetto poor." He presents a model that emphasizes the interpersonal aggression of such individuals, in which the main factor of the code is respect. An individual must maintain this respect, at all costs, even if it means resorting to lethal violence to maintain such status, especially if the individual has been "dissed" in public. Anderson goes further in this selection by identifying both "decent" and "street" families, which provides more insight into the environment and culture of inner-city youths. Anderson also provides some insightful examples regarding the way that individuals try (or campaign) for respect in their communities. Ultimately, it is about being seen as a "man" among their peers, and acquiring such "manhood" is seen as vitally important among this group.

The Code of the Streets

Elijah Anderson

Of all the problems besetting the poor inner-city black community, none is more pressing than that of interpersonal violence and aggression. It wreaks havoc daily with the lives of community residents and increasingly spills over into downtown and residential middle-class areas. Muggings, burglaries, carjackings, and drug-related shootings, all of which may leave their victims or innocent bystanders dead, are now common enough to concern all urban and many suburban residents. The inclination to violence springs from the circumstances of life among the ghetto poor—the lack of jobs that pay a living wage, the stigma of race, the fallout from rampant drug use and drug trafficking, and the resulting alienation and lack of hope for the future.

Simply living in such an environment places young people at special risk of falling victim to aggressive behavior. Although there are often forces in the community which can counteract the negative influences, by far the most powerful being a strong, loving, "decent" (as inner-city residents put it) family committed to middle-class values, the despair is pervasive enough to have spawned an oppositional culture, that of "the streets," whose norms are often consciously opposed to those of mainstream society. These two orientations—decent and street—socially organize the community, and their coexistence has important consequences for residents, particularly children growing up in the inner city. Above all, this environment means that even youngsters whose home lives reflect mainstream values—and the majority of homes in the community do—must be able to handle themselves in a street-oriented environment.

This is because the street culture has evolved what may be called a code of the streets, which amounts to a set of informal rules governing interpersonal public behavior, including violence. The rules prescribe both a proper comportment and a proper way to respond if challenged. They regulate the use of violence and so allow those who are inclined to aggression to precipitate violent encounters in an approved way. The rules have been established and are enforced mainly by the street-oriented, but on the streets the distinction between street and decent is often irrelevant; everybody knows that if the rules are violated, there are penalties. Knowledge of the code is thus largely defensive; it is literally necessary for operating in public. Therefore, even though families with a decency orientation are usually opposed to the values of the code, they often reluctantly encourage their children's familiarity with it to enable them to negotiate the inner-city environment.

At the heart of the code is the issue of respect—loosely defined as being treated "right," or granted the deference one deserves. However, in the troublesome public environment of the inner city, as people increasingly feel buffeted by forces beyond their control, what one deserves in the way of respect becomes more and more problematic and uncertain. This in turn further opens the issue of respect to sometimes intense interpersonal negotiation. In the street culture, especially among young people, respect is viewed as almost an external entity that is hard-won but easily lost, and so must constantly be guarded. The rules of the code in fact provide a framework for negotiating respect. The person whose very appearance—including his clothing, demeanor, and way of moving—deters transgressions feels that he possesses, and may be considered by others to possess, a measure of respect. With the right amount of respect, for instance, he can avoid "being bothered" in public. If he is bothered, not only may he be in physical danger but he has been disgraced or "dissed" (disrespected). Many of the forms that dissing can take might seem petty to middle-class people (maintaining eye contact for too long, for

example), but to those invested in the street code, these actions become serious indications of the other person's intentions. Consequently, such people become very sensitive to advances and slights, which could well serve as warnings of imminent physical confrontation.

This hard reality can be traced to the profound sense of alienation from mainstream society and its institutions felt by many poor inner-city black people, particularly the young. The code of the streets is actually a cultural adaptation to a profound lack of faith in the police and the judicial system. The police are most often seen as representing the dominant white society and not caring to protect inner-city residents. When called, they may not respond, which is one reason many residents feel they must be prepared to take extraordinary measures to defend themselves and their loved ones against those who are inclined to aggression. Lack of police accountability has in fact been incorporated into the status system: the person who is believed capable of "taking care of himself" is accorded a certain deference, which translates into a sense of physical and psychological control. Thus the street code emerges where the influence of the police ends and personal responsibility for one's safety is felt to begin. Exacerbated by the proliferation of drugs and easy access to guns, this volatile situation results in the ability of the street-oriented minority (or those who effectively "go for bad") to dominate the public spaces.

▧ Decent and Street Families

Although almost everyone in poor inner-city neighborhoods is struggling financially and therefore feels a certain distance from the rest of America, the decent and the street family in a real sense represent two poles of value orientation, two contrasting conceptual categories. The labels "decent" and "street," which the residents themselves use, amount to evaluative judgments that confer status on local residents.

The labeling is often the result of a social contest among individuals and families of the neighborhood. Individuals of the two orientations often coexist in the same extended family. Decent residents judge themselves to be so while judging others to be of the street, and street individuals often present themselves as decent, drawing distinctions between themselves and other people. In addition, there is quite a bit of circumstantial behavior—that is, one person may at different times exhibit both decent and street orientations, depending on the circumstances. Although these designations result from so much social jockeying, there do exist concrete features that define each conceptual category.

Generally, so-called decent families tend to accept mainstream values more fully and attempt to instill them in their children. Whether married couples with children or single-parent (usually female) households, they are generally "working poor" and so tend to be better off financially than their street-oriented neighbors. They value hard work and self-reliance and are willing to sacrifice for their children. Because they have a certain amount of faith in mainstream society, they harbor hopes for a better future for their children, if not for themselves. Many of them go to church and take a strong interest in their children's schooling. Rather than dwelling on the real hardships and inequities facing them, many such decent people, particularly the increasing number of grandmothers raising grandchildren, see their difficult situation as a test from God and derive great support from their faith and from the church community.

Extremely aware of the problematic and often dangerous environment in which they reside, decent parents tend to be strict in their child-rearing practices, encouraging children to respect authority and walk a straight moral line. They have an almost obsessive concern about trouble of any kind and remind their children to be on the lookout for people and situations that might lead to it. At the same time, they are themselves polite and considerate of others, and teach their children to be the same way. At home, at work, and in church, they strive hard to maintain a positive mental attitude and a spirit of cooperation.

So-called street parents, in contrast, often show a lack of consideration for other people and have a rather superficial sense of family and community. Though they may love their children, many of them are unable to cope with the physical and emotional demands of parenthood, and find it difficult to reconcile their needs with those of their children. These families, who are more fully invested in the code of the streets than the decent people are, may aggressively socialize their children into it in a normative way. They believe in the code and judge themselves and others according to its values.

In fact the overwhelming majority of families in the inner-city community try to approximate the decent-family model, but there are many others who clearly represent the worst fears of the decent family. Not only are their financial resources extremely limited, but what little they have may easily be misused. The lives of the street-oriented are often marked by disorganization. In the most desperate circumstances people frequently have a limited understanding of priorities and consequences, and so frustrations mount over bills, food, and, at times, drink, cigarettes, and drugs. Some tend toward self-destructive behavior; many street-oriented women are crack-addicted ("on the pipe"), alcoholic, or involved in complicated relationships with men who abuse them. In addition, the seeming intractability of their situation, caused in large part by the lack of well-paying jobs and the persistence of racial discrimination, has engendered deep-seated bitterness and anger in many of the most desperate and poorest blacks, especially young people. The need both to exercise a measure of control and to lash out at somebody is often reflected in the adults' relations with their children. At the least, the frustrations of persistent poverty shorten the fuse in such people—contributing to a lack of patience with anyone, child or adult, who irritates them.

In these circumstances a woman—or a man, although men are less consistently present in children's lives—can be quite aggressive with children, yelling at and striking them for the least little infraction of the rules she has set down. Often little if any serious explanation follows the verbal and physical punishment. This response teaches children a particular lesson. They learn that to solve any kind of interpersonal problem one must quickly resort to hitting or other violent behavior. Actual peace and quiet, and also the appearance of calm, respectful children conveyed to her neighbors and friends, are often what the young mother most desires, but at times she will be very aggressive in trying to get them. Thus she may be quick to beat her children, especially if they defy her law, not because she hates them but because this is the way she knows to control them. In fact, many street-oriented women love their children dearly. Many mothers in the community subscribe to the notion that there is a "devil in the boy" that must be beaten out of him or that socially "fast girls need to be whupped." Thus, much of what borders on child abuse in the view of social authorities is acceptable parental punishment in the view of these mothers.

Many street-oriented women are sporadic mothers whose children learn to fend for themselves when necessary, foraging for food and money any way they can get it. The children are sometimes employed by drug dealers or become addicted themselves. These children of the street, growing up with little supervision, are said to "come up hard." They often learn to fight at an early age, sometimes using short-tempered adults around them as role models. The street-oriented home may be fraught with anger, verbal disputes, physical aggression, and even mayhem. The children observe these goings-on, learning the lesson that might makes right. They quickly learn to hit those who cross them, and the dog-eat-dog

mentality prevails. In order to survive, to protect oneself, it is necessary to marshal inner resources and be ready to deal with adversity in a hands-on way. In these circumstances physical prowess takes on great significance.

In some of the most desperate cases, a street-oriented mother may simply leave her young children alone and unattended while she goes out. The most irresponsible women can be found at local bars and crack houses, getting high and socializing with other adults. Sometimes a troubled woman will leave very young children alone for days at a time. Reports of crack addicts abandoning their children have become common in drug-infested inner-city communities. Neighbors or relatives discover the abandoned children, often hungry and distraught over the absence of their mother. After repeated absences, a friend or relative, particularly a grandmother, will often step in to care for the young children, sometimes petitioning the authorities to send her, as guardian of the children, the mother's welfare check, if the mother gets one. By this time, however, the children may well have learned the first lesson of the streets: survival itself, let alone respect, cannot be taken for granted; you have to fight for your place in the world.

⊠ Campaigning for Respect

These realities of inner-city life are largely absorbed on the streets. At an early age, often even before they start school, children from street-oriented homes gravitate to the streets, where they "hang"—socialize with their peers. Children from these generally permissive homes have a great deal of latitude and are allowed to "rip and run" up and down the street. They often come home from school, put their books down, and go right back out the door. On school nights eight- and nine-year-olds remain out until nine or ten o'clock (and teenagers typically come in whenever they want to). On the streets they play in groups that often become the source of their primary social bonds. Children from decent homes tend to be more carefully supervised and are thus likely to have curfews and to be taught how to stay out of trouble.

When decent and street kids come together, a kind of social shuffle occurs in which children have a chance to go either way. Tension builds as a child comes to realize that he must choose an orientation. The kind of home he comes from influences but does not determine the way he will ultimately turn out—although it is unlikely that a child from a thoroughly street-oriented family will easily absorb decent values on the streets. Youths who emerge from street-oriented families but develop a decency orientation almost always learn those values in another setting—in school, in a youth group, in church. Often it is the result of their involvement with a caring "old head" (adult role model).

In the street, through their play, children pour their individual life experiences into a common knowledge pool, affirming, confirming, and elaborating on what they have observed in the home and matching their skills against those of others. And they learn to fight. Even small children test one another, pushing and shoving, and are ready to hit other children over circumstances not to their liking. In turn, they are readily hit by other children, and the child who is toughest prevails. Thus the violent resolution of disputes, the hitting and cursing, gains social reinforcement. The child in effect is initiated into a system that is really a way of campaigning for respect.

In addition, younger children witness the disputes of older children, which are often resolved through cursing and abusive talk, if not aggression or outright violence. They see that one child succumbs to the greater physical and mental abilities of the other. They are also alert and attentive witnesses to the verbal and physical fights of adults, after which they compare notes and share their interpretations of the event. In almost every case the victor is the person who physically won the altercation, and this person often enjoys the esteem and respect of onlookers. These experiences reinforce the lessons the children have learned at home: might makes right, and toughness is a virtue, while humility is not. In effect they learn the social meaning of fighting. When it is left virtually unchallenged, this understanding becomes an ever more important part of the child's working conception of the world. Over time the code of the streets becomes refined.

Those street-oriented adults with whom children come in contact—including mothers, fathers, brothers, sisters, boyfriends, cousins, neighbors, and friends—help them along in forming this understanding by verbalizing the messages they are getting through experience: "Watch your back." "Protect yourself." "Don't punk out." "If somebody messes with you, you got to pay them back." "If someone disses you, you got to straighten them out." Many parents actually impose sanctions if a child is not sufficiently aggressive. For example, if a child loses a fight and comes home upset, the parent might respond, "Don't you come in here crying that somebody beat you up; you better get back out there and whup his ass. I didn't raise no punks! Get back out there and whup his ass. If you don't whup his ass, I'll whup your ass when you come home." Thus, the child obtains reinforcement for being tough and showing nerve.

While fighting, some children cry as though they are doing something they are ambivalent about. The fight may be against their wishes, yet they may feel constrained to fight or face the consequences—not just from peers but also from caretakers or parents, who may administer another beating if they back down. Some adults recall receiving such lessons from their own parents and justify repeating them to their children as a way to toughen them up. Looking capable of taking care of oneself as a form of self defense is a dominant theme among both street-oriented and decent adults who worry about the safety of their children. There is thus at times a convergence in their child-rearing practices, although the rationales behind them may differ.

▨ Self-Image Based on "Juice"

By the time they are teenagers, most youths have either internalized the code of the streets or at least learned the need to comport themselves in accordance with its rules, which chiefly have to do with interpersonal communication. The code revolves around the presentation of self. Its basic requirement is the display of a certain predisposition to violence. Accordingly, one's bearing must send the unmistakable, if sometimes subtle, message to "the next person" in public that one is capable of violence and mayhem when the situation requires it, that one can take care of oneself. The nature of this communication is largely determined by the demands of the circumstances but can include facial expressions, gait, and verbal expressions—all of which are geared mainly to deterring aggression. Physical appearance, including clothes, jewelry, and grooming, also plays an important part in how a person is viewed; to be respected, it is important to have the right look.

Even so, there are no guarantees against challenges, because there are always people around looking for a fight to increase their share of respect—or "juice," as it is sometimes called on the street. Moreover, if a person is assaulted, it is important, not only in the eyes of his opponent but also in the eyes of his "running buddies," for him to avenge himself. Otherwise he risks being "tried" (challenged) or "moved on" by any number of others. To maintain his honor he must show he is not someone to be "messed with" or "dissed." In general, the person must "keep himself straight" by managing his position of respect among others; this involves in part his self-image, which is shaped by what he thinks others are thinking of him in relation to his peers.

Objects play an important and complicated role in establishing self-image. Jackets, sneakers, gold jewelry reflect not just a person's taste, which tends to be tightly regulated among adolescents of all social classes, but also a willingness to possess things that may require defending. A boy wearing a fashionable, expensive jacket, for example, is vulnerable to attack by another who covets the jacket and either cannot afford to buy one or wants the added satisfaction of depriving someone else of his. However, if the boy forgoes the desirable jacket and wears one that isn't "hip," he runs the risk of being teased and possibly even assaulted as an unworthy person. To be allowed to hang with certain prestigious crowds, a boy must wear a different set of expensive clothes—sneakers and athletic suit—every day. Not to be able to do so might make him appear socially deficient. The youth comes to covet such items—especially when he sees easy prey wearing them.

In acquiring valued things, therefore, a person shores up his identity—but since it is an identity based on having things, it is highly precarious. This very precariousness gives a heightened sense of urgency to staying even with peers, with whom the person is actually competing. Young men and women who are able to command respect through their presentation of self—by allowing their possessions and their body language to speak for them—may not have to campaign for regard but may, rather, gain it by the force of their manner. Those who are unable to command respect in this way must actively campaign for it—and are thus particularly alive to slights.

One way of campaigning for status is by taking the possessions of others. In this context, seemingly ordinary objects can become trophies imbued with symbolic value that far exceeds their monetary worth. Possession of the trophy can symbolize the ability to violate somebody—to "get in his face," to take something of value from him, to "dis" him, and thus to enhance one's own worth by stealing someone else's. The trophy does not have to be something material. It can be another person's sense of honor, snatched away with a derogatory remark. It can be the outcome of a fight. It can be the imposition of a certain standard, such as a girl's getting herself recognized as the most beautiful. Material things, however, fit easily into the pattern. Sneakers, a pistol, even somebody else's girlfriend, can become a trophy. When a person can take something from another and then flaunt it, he gains a certain regard by being the owner, or the controller, of that thing. But this display of ownership can then provoke other people to challenge him. This game of who controls what is thus constantly being played out on inner-city streets, and the trophy—extrinsic or intrinsic, tangible or intangible—identifies the current winner.

An important aspect of this often violent give-and-take is its zero-sum quality. That is, the extent to which one person can raise himself up depends on his ability to put another person down. This underscores the alienation that permeates the inner-city ghetto community. There is a generalized sense that very little respect is to be had, and therefore everyone competes to get what affirmation he can of the little that is available. The craving for respect that results gives people thin skins. Shows of deference by others can be highly soothing, contributing to a sense of security, comfort, self-confidence, and self-respect. Transgressions by others which go unanswered diminish these feelings and are believed to encourage further transgressions. Hence one must be ever vigilant against the transgressions of others or even *appearing* as if transgressions will be tolerated. Among young people, whose sense of self-esteem is particularly vulnerable, there is an especially heightened concern with being disrespected. Many inner-city young men in particular crave respect to such a degree that they will risk their lives to attain and maintain it.

The issue of respect is thus closely tied to whether a person has an inclination to be violent, even as a victim. In the wider society people may not feel required to retaliate physically after an attack, even though they are aware that they have been degraded or taken advantage of. They may feel a great need to defend themselves *during* an attack, or to behave in such a way as to deter aggression (middle-class people certainly can and do become victims of street-oriented youths), but they are much more likely than street-oriented people to feel that they can walk away from a possible altercation with their self-esteem intact. Some people may even have the strength of character to flee, without any thought that their self-respect or esteem will be diminished.

In impoverished inner-city black communities, however, particularly among young males and perhaps increasingly among females, such flight would be extremely difficult. To run away would likely leave one's self-esteem in tatters. Hence people often feel constrained not only to stand up and at least attempt to resist during an assault but also to "pay back"—to seek revenge—after a successful assault on their person. This may include going to get a weapon or even getting relatives involved. Their very identity and self-respect, their honor, is often intricately tied up with the way they perform on the streets during and after such encounters. This outlook reflects the circumscribed opportunities of the inner-city poor. Generally people outside the ghetto have other ways of gaining status and regard, and thus do not feel so dependent on such physical displays.

By Trial of Manhood

On the street, among males these concerns about things and identity have come to be expressed in the concept of "manhood." Manhood in the inner city means taking the prerogatives of men with respect to strangers, other men, and women—being distinguished as a man. It implies physicality and a certain ruthlessness. Regard and respect are associated with this concept in large part because of its practical application: if others have little or no regard for a person's manhood, his very life and those of this loved ones could be in jeopardy. But there is a chicken-and-egg aspect to this situation: one's physical safety is more likely to be jeopardized in public because manhood is associated with respect. In other words, an existential link has been created between the idea of manhood and one's self-esteem, so that it has become hard to say which is primary. For many inner-city youths, manhood and respect are flip sides of the same coin; physical and psychological well-being are inseparable, and both require a sense of control, of being in charge.

The operating assumption is that a man, especially a real man, knows what other men know—the code of the streets. And if one is not a real man, one is somehow diminished as a person, and there are certain valued things one simply does not deserve. There is thus believed to be a certain justice to the code, since it is considered that everyone has the opportunity to know it. Implicit in this is that everybody is held responsible for being familiar with the code. If the victim of a mugging, for example, does not know the code and so responds "wrong," the perpetrator may feel justified even in killing him and may feel no remorse. He may think, "Too bad, but it's his fault. He should have known better."

So when a person ventures outside, he must adopt the code—a kind of shield, really—to prevent others from "messing with" him. In these circumstances it is easy for people to think they are being tried or tested by others even when this is not the case. For it is sensed that something extremely valuable is at stake in every interaction, and people are encouraged to rise to the occasion, particularly with strangers. For people who are unfamiliar with the code— generally people who live outside the inner city—the concern with respect in the most ordinary interactions can be frightening and incomprehensible. But for those who are invested in the code, the clear object of their demeanor is to discourage strangers from even thinking about testing their manhood. And the sense of power that attends the ability to deter others can be alluring even to those who know the code without being heavily invested in it—the decent inner-city youths. Thus a boy who has been leading a basically decent life can, in trying circumstances, suddenly resort to deadly force.

Central to the issue of manhood is the widespread belief that one of the most effective ways of gaining respect is to manifest "nerve." Nerve is shown when one takes another person's possessions (the more valuable the better), "messes with" someone's woman, throws the first punch, "gets in someone's face," or pulls a trigger. Its proper display helps on the spot to check others who would violate one's person and also helps to build a reputation that works to prevent future challenges. But since such a show of nerve is a forceful expression of disrespect toward the person on the receiving end, the victim may be greatly offended and seek to retaliate with equal or greater force. A display of nerve, therefore, can easily provoke a life-threatening response, and the background knowledge of that possibility has often been incorporated into the concept of nerve.

True nerve exposes a lack of fear of dying. Many feel that it is acceptable to risk dying over the principle of respect. In fact, among the hard-core street-oriented, the clear risk of violent death may be preferable to being "dissed" by another. The youths who have internalized this attitude and convincingly display it in their public bearing are among the most threatening people of all, for it is commonly assumed that they fear no man. As the people of the community say, "They are the baddest dudes on the street." They often lead an existential life that may acquire meaning only when they are faced with the possibility of imminent death. Not to be afraid to die is by implication to have few compunctions about taking another's life. Not to be afraid to die is the quid pro quo of being able to take somebody else's life—for the right reasons, if the situation demands it. When others believe this is one's position, it gives one a real sense of power on the streets. Such credibility is what many inner-city youths strive to achieve, whether they are decent or street-oriented, both because of its practical defensive value and because of the positive way it makes them feel about themselves. The difference between the decent and the street-oriented youth is often that the decent youth makes a conscious decision to appear tough and manly; in another setting—with teachers, say, or at his part-time job—he can be polite and deferential. The street-oriented youth, on the other hand, has made the concept of manhood a part of his very identity; he has difficulty manipulating it—it often controls him.

⌧ Girls and Boys

Increasingly, teenage girls are mimicking the boys and trying to have their own version of "manhood." Their goal is the same—to get respect, to be recognized as capable of setting or maintaining a certain standard. They try to achieve this end in the ways that have been established by the boys, including posturing, abusive language, and the use of violence to resolve disputes, but the issues for the girls are different. Although conflicts over turf and status exist among the girls, the majority of disputes seem rooted in assessments of beauty (which girl in a group is "the cutest"), competition over boyfriends, and attempts to regulate other people's knowledge of and opinions about a girl's behavior or that of someone close to her, especially her mother.

A major cause of conflicts among girls is "he say, she say." This practice begins in the early school years and continues through high school. It occurs when "people," particularly girls, talk about others, thus putting their

"business in the streets." Usually one girl will say something negative about another in the group, most often behind the person's back. The remark will then get back to the person talked about. She may retaliate or her friends may feel required to "take up for" her. In essence this is a form of group gossiping in which individuals are negatively assessed and evaluated. As with much gossip, the things said may or may not be true, but the point is that such imputations can cast aspersions on a person's good name. The accused is required to defend herself against the slander, which can result in arguments and fights, often over little of real substance. Here again is the problem of low self-esteem, which encourages youngsters to be highly sensitive to slights and to be vulnerable to feeling easily "dissed." To avenge the dissing, a fight is usually necessary.

Because boys are believed to control violence, girls tend to defer to them in situations of conflict. Often if a girl is attacked or feels slighted, she will get a brother, uncle, or cousin to do her fighting for her. Increasingly, however, girls are doing their own fighting and are even asking their male relatives to teach them how to fight. Some girls form groups that attack other girls or take things from them. A hard-core segment of inner-city girls inclined toward violence seems to be developing. As one thirteen-year-old girl in a detention center for youths who have committed violent acts told me, "To get people to leave you alone, you gotta fight. Talking don't always get you out of stuff." One major difference between girls and boys: girls rarely use guns. Their fights are therefore not life-or-death struggles. Girls are not often willing to put their lives on the line for "manhood." The ultimate form of respect on the male-dominated inner-city street is thus reserved for men.

⌧ "Going for Bad"

In the most fearsome youths, such a cavalier attitude toward death grows out of a very limited view of life. Many are uncertain about how long they are going to live and believe they could die violently at any time. They accept this fate; they live on the edge. Their manner conveys the message that nothing intimidates them; whatever turn the encounter takes, they maintain their attack—rather like a pit bull, whose spirit many such boys admire. The demonstration of such tenacity "shows heart" and earns their respect.

This fearlessness has implications for law enforcement. Many street-oriented boys are much more concerned about the threat of "justice" at the hands of a peer than at the hands of the police. Moreover, many feel not only that they have little to lose by going to prison but that they have something to gain. The toughening-up one experiences in prison can actually enhance one's reputation on the streets. Hence the system loses influence over the hard core who are without jobs, with little perceptible stake in the system. If mainstream society has done nothing *for* them, they counter by making sure it can do nothing *to* them.

At the same time, however, a competing view maintains that true nerve consists in backing down, walking away from a fight, and going on with one's business. One fights only in self-defense. This view emerges from the decent philosophy that life is precious, and it is an important part of the socialization process common in decent homes. It discourages violence as the primary means of resolving disputes and encourages youngsters to accept nonviolence and talk as confrontational strategies. But "if the deal goes down," self-defense is greatly encouraged. When there is enough positive support for this orientation, either in the home or among one's peers, then nonviolence has a chance to prevail. But it prevails at the cost of relinquishing a claim to being bad and tough, and therefore sets a young person up as, at the very least, alienated from street-oriented peers and quite possibly a target of derision or even violence.

Although the nonviolent orientation rarely overcomes the impulse to strike back in an encounter, it does introduce a certain confusion, and so can prompt a measure of soul-searching, or even profound ambivalence. Did the person back down with his respect intact or did he back down only to be judged a "punk"—a person lacking manhood? Should he or she have acted? Should he or she have hit the other person in the mouth? These questions beset many young men and women during public confrontations. What is the "right" thing to do? In the quest for honor, respect, and local status—which few young people are uninterested in—common sense most often prevails, which leads many to opt for the tough approach, enacting their own particular versions of the display of nerve. The presentation of oneself as rough and tough is very often quite acceptable until one is tested. And then that presentation may help the person pass the test, because it will cause fewer questions to be asked about what he did and why. It is hard for a person to explain why he lost the fight or why he backed down. Hence many will strive to appear to "go for bad," while hoping they will never be tested. But when they are tested, the outcome of the situation may quickly be out of their hands, as they become wrapped up in the circumstances of the moment.

▨ An Oppositional Culture

The attitudes of the wider society are deeply implicated in the code of the streets. Most people in inner-city communities are not totally invested in the code, but the significant minority of hard-core street youths who are have to maintain the code in order to establish reputations, because they have—or feel they have—few other ways to assert themselves. For these young people the standards of the street code are the only game in town. The extent to which some children—particularly those who through upbringing have become most alienated and those lacking in strong and conventional social support—experience, feel, and internalize racist rejection and contempt from mainstream society may strongly encourage them to express contempt for the more conventional society in turn. In dealing with this contempt and rejection, some youngsters will consciously invest themselves and their considerable mental resources in what amounts to an oppositional culture to preserve themselves and their self-respect. Once they do, any respect they might be able to garner in the wider system pales in comparison with the respect available in the local system; thus they often lose interest in even attempting to negotiate the mainstream system.

At the same time, many less alienated young blacks have assumed a street-oriented demeanor as a way of expressing their blackness while really embracing a much more moderate way of life; they, too, want a nonviolent setting in which to live and raise a family. These decent people are trying hard to be part of the mainstream culture, but the racism, real and perceived, that they encounter helps to legitimate the oppositional culture. And so on occasion they adopt street behavior. In fact, depending on the demands of the situation, many people in the community slip back and forth between decent and street behavior.

A vicious cycle has thus been formed. The hopelessness and alienation many young inner-city black men and women feel, largely as a result of endemic joblessness and persistent racism, fuels the violence they engage in. This violence serves to confirm the negative feelings many whites and some middle-class blacks harbor toward the ghetto poor, further legitimating the oppositional culture and the code of the streets in the eyes of many poor young blacks. Unless this cycle is broken, attitudes on both sides will become increasingly entrenched, and the violence, which claims victims, black and white, poor and affluent, will only escalate.

REVIEW QUESTIONS

1. What type of individuals is Anderson discussing? Do you personally know any individuals who fit this description?

2. Which of Anderson's explanations do you find most credible? Which portions do you find least credible?

3. Given that Anderson's description is valid for inner-city, urban youth, to what extent do you feel that this explains crime rates in the United States? Which significant portion do you feel that this model does not cover?

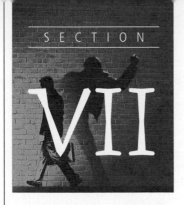

SOCIAL PROCESS/ LEARNING THEORIES OF CRIME

The introduction of this section will discuss Sutherland's development of differential association theory and how this evolved into Akers's work of differential reinforcement and other social learning theories, such as techniques of neutralization. Then the modern state of research on these theories will be discussed. The introduction will also discuss the evolution of control theories of crime, with an emphasis on social bonding and the scientific evidence found regarding the key constructs in Hirschi's control theories, two of the most highly regarded perspectives according to criminological experts and their studies.[1]

Most of the social process theories assume that criminal behavior is learned behavior, which means that criminal activity is actually learned from others through social interaction, much like riding a bike or playing basketball. However, other social process theories, namely control theories, assume that offending is the result of natural tendencies and thus must be controlled by social processes. This section will explore both of

---

[1]Anthony Walsh and Lee Ellis, "Political Ideology and American Criminologists' Explanations for Criminal Behavior," *The Criminologist* 24 (1999): 1, 14.

these theoretical frameworks and explain how social processes are vital to both perspectives in determining criminal behavior.

Introduction

Social process theories assume that individuals learn why and how to commit crime through a process of socialization. These theories claim that people learn crime like they learn any other behavior, such as riding a bike or playing basketball. Namely, they learn this from significant others, such as from family, peers, or coworkers.

This introduction explores the theories and empirical research regarding why it is useful to examine the role of socialization in criminal behavior. Social process theories examine how individuals interact with other individuals and groups and how the learning that takes place in these interactions leads to a propensity toward criminal activity. These theories focus on how such behavior is learned through social interaction and transmitted among individuals.

This introduction begins with social process theories known as **learning theories**. Such learning theories attempt to explain how and why individuals learn from significant others to engage in criminal rather than conventional behavior. Next, we discuss **control theories,** which emphasize personal or socialization factors that prevent individuals from engaging in selfish, antisocial behaviors.

Learning Theories

In this section, we review theories that emphasize how individuals learn criminal behavior through interacting with significant others, people with whom they typically associate. These learning theories assume that people are born with no tendency toward or away from committing crime. This concept is referred to as *tabula rasa,* or "blank slate," meaning that all individuals are completely malleable and will believe what they are told by their significant others and act accordingly. Thus, such theories of learning tend to explain how criminal behavior is learned through cultural norms. One of the main concepts in learning theories is the influence of peers and significant others on an individual's behavior.

In this section, three learning theories are discussed: (1) differential association theory, (2) differential identification theory, and (3) differential reinforcement theory; then it examines techniques of neutralization.

Differential Association Theory

Edwin Sutherland introduced his **differential association theory** in the late 1930s.[2] He proposed a theoretical framework that explained how criminal values could be culturally transmitted to individuals from their significant others. Sutherland proposed a theoretical model that included nine principles, all of them outlined in one of this section's readings.

---

[2]Edwin H. Sutherland, *Principles of Criminology*, 3rd ed. (Philadelphia: Lippincott, 1939).

To summarize, perhaps the most interesting principle is the first: that criminal behavior is learned. This was a radical departure from previous theories (e.g., Lombroso's born criminal theory, Goddard's feeblemindedness theory, Sheldon's body-type theory). Sutherland was one of the first to state that criminal behavior was the result of normal social processes, resulting when individuals associated with the wrong type of people, often by no fault on their part. By associating with crime-oriented people, whether they are parents or peers, an individual will inevitably choose to engage in criminal behavior because that is what he or she has learned, Sutherland thought.

Perhaps the most important of Sutherland's principles, and certainly the most revealing one, was No. 6 in his framework: "A person becomes delinquent because of an excess of definitions favorable to violation of law over definitions unfavorable to violation of law."[3] Sutherland noted that this principle represented the essence of differential association theory. It suggests that people can have associations that favor both criminal and noncriminal behavior patterns. If an

▲ **Photo 7.1** Edwin H. Sutherland (1883–1950), University of Chicago and Indiana University, author of Differential Association Theory

individual is receiving more information and values that are pro-crime than anti-crime, the individual will inevitably engage in such criminal activity. Also, Sutherland claimed that such learning could take place only in interactions with significant others, and not via television, movies, radio, or other media.

It is important to understand the cultural context at the time Sutherland was developing his theory. In the early 20th century, most academics, and society for that matter, believed that there was something abnormal or different about criminals. Sheldon's body-type theory was popular in the same time period, as was the use of IQ to pick out people who were of lower intelligence and supposedly predisposed to crime (both of these theories were covered in Sections III and IV). So the common assumption when Sutherland created the principles of differential association theory was that there was essentially something wrong with individuals who commit crime.

In light of this common assumption, Sutherland's proposal—that criminality is learned just like any conventional activity—was extremely profound. This suggests that any normal

---

[3]Edwin H. Sutherland and Donald R. Cressey, *Principles of Criminology*, 5th ed. (Chicago: Lippincott, 1950) p. 78.

person, when exposed to attitudes favorable toward crime, will learn criminal activity and that the processes and mechanisms of learning are the same for crime as for most legal, everyday behaviors: namely, social interaction with your family and friends, and not reading books or watching movies. How many of us learned to play basketball or other sports by reading a book about it? Virtually no one learns how to play sports this way. Rather, we learn the techniques (e.g., how to do a jump shot) and motivations for playing sports (e.g., it is fun and/or you might be able to earn a scholarship) through our friends, relatives, coaches, and other people close to us. According to Sutherland, crime is learned the same way; our close associates teach us both the techniques (e.g., how to steal a car) and the motivations (e.g., it is fun and/or you might be able to sell it or its parts). While most criminologists tend to take it for granted that criminal behavior is learned, the idea was rather unique and bold when Sutherland presented his theory of differential association.

It is important to keep in mind that differential association theory is just as positivistic as earlier biological and psychological theories. Sutherland clearly believed that if people were receiving more information that breaking the law was good, then they would inevitably commit crimes. There is virtually no allowance for free will and rational decision making in this model of offending. Rather, people's choices to commit crime are determined through their social interactions with those close to them; they do not actually make the decisions to engage (or not engage) in criminal activities. So differential association can be seen as a highly positive, deterministic theory, much like Lombroso's born criminal and Goddard's feeblemindedness (see Section III), except that instead of biological or psychological traits causing crime, it is social interaction and learning. Furthermore, Sutherland claimed that individual differences in biological and psychological functioning have little to do with criminality; however, this idea has been discounted by modern research, which shows such variations do in fact affect criminal behavior, largely because such biopsychological factors influence the learning processes of individuals, thereby directly impacting the basic principles of Sutherland's theory (see Sections III and IV).

Classical Conditioning: A Learning Theory With Limitations. Sutherland used the dominant psychological theory of learning of his era as the basis for his theory of differential association. This model was classical conditioning, which was primarily developed by Ivan Pavlov. **Classical conditioning** assumes that animals, as well as people, learn through associations between stimuli and responses.[4] Organisms, animals, or people are somewhat passive actors in this process, meaning that they simple receive and respond in natural ways to various forms of stimuli; over time, they learn to associate certain stimuli with certain responses.

For example, Pavlov showed that dogs, which naturally are afraid of loud noises such as bells, could be quickly conditioned not only to be less afraid of bells, but to actually desire and salivate at their sound. A dog naturally salivates when presented with meat, so when this presentation of an unconditioned stimuli (meat) is given, a dog will always salivate (unconditioned response) in anticipation of eating. Pavlov demonstrated through a series of experiments that if a bell (conditioned stimuli) is always rung at the same time the dog is presented with meat, then the dog will learn to associate what was previously a

---

[4]For a discussion, see George B. Vold, Thomas J. Bernard, and Jeffrey B. Snipes, *Theoretical Criminology*, 5th ed. (New York: Oxford University Press, 2002), 156–57.

negative stimulus with a positive stimulus (food). Thus, the dog will very quickly begin salivating at the ringing of a bell, even when meat is not presented. When this occurs, it is called a conditioned response, because it is not natural; however, it is a very powerful and effective means of learning, and it sometimes takes only a few occurrences of coupling the bell ringing with meat before the conditioned response takes place.

One modern use of this in humans is the administration of drugs that make people ill when they drink alcohol. Alcoholics are often prescribed drugs that make them very sick, often to the point of vomiting, if they ingest any alcohol. The idea is that they will learn to associate feelings of sickness with drinking and thus stop wanting to consume alcohol. One big problem with this strategy is that alcoholics often do not consistently take the drugs, so they quickly slip back into addiction. Still, if they were to maintain their regimen of drugs, it would likely work because people do tend to learn through association.

This type of learning model was used in the critically acclaimed 1964 novel (and later a motion picture), *A Clockwork Orange*. In this novel, the author, Anthony Burgess, tells the story of a juvenile murderer who is "rehabilitated" by doctors who force him to watch hour after hour of violent images while simultaneously giving him drugs that make him sick. In the novel, the protagonist is "cured" in only two weeks of this treatment, having learned to consistently associate violence with sickness. However, once he is let out, he lacks the ability to choose violence and other antisocial behavior, which is seen as losing his humanity. Therefore, the ethicists order a reversal treatment and make him back into his former self, a violent predator. Although a fictional piece, *A Clockwork Orange* is probably one of the best illustrations of the use of classical conditioning in relation to criminal offending and rehabilitation.

Another example of classical conditioning is the association we make with certain smells and sounds. For example, all of us can relate good times to smells that were present during those occasions. If a loved one or someone we dated wore a certain perfume or cologne, the scent of that at a later time can bring back memories. When our partners go out of town, we can smell their pillow, and it reminds us of them because we associate their smell with their being. Or perhaps the smell of a turkey cooking in an oven always reminds us of Thanksgiving or other holidays. Regarding associations of sounds, we all can remember one or more songs that remind us of both happy and sad times in our life. Often these songs will replay on the radio, and it takes us back to those occasions, whether they are good or bad. People with post-traumatic stress disorder (PTSD) also experience sound associations; war veterans, for example, may "hit the deck" when a car backfires. These are all clear examples of classical conditioning and associating stimuli with responses.

Since Sutherland's theory was published, many of the principles outlined in his model have come under scrutiny. Follow-up research has shown some flaws, as well as misinterpretations, of his work.[5] Specifically, Sutherland theorized that crime occurs when the ratio of associations favorable to violation of the law outweigh associations favorable to conforming to the law. However, measuring this type of ratio is all but impossible for social scientists.[6]

---

[5]Edwin H. Sutherland and Donald R. Cressey, *Criminology*, 9th ed. (Philadelphia: Lippincott, 1974).

[6]R. L. Matsueda and K. Heimer, "Race, Family Structure, and Delinquency: A Test of Differential Association and Social Control Theories," *American Sociological Review* 47 (1987): 489–504; Charles R. Tittle, M. J. Burke, and E. F. Jackson, "Modeling Sutherland's Theory of Differential Association: Toward an Empirical Clarification," *Social Forces* 65 (1986): 405–32.

Another topic of criticism involves Sutherland's claim that all criminals learn the behavior from others *before* they engage in such activity. However, many theorists have noted that an individual may engage in criminal activity without being taught such behavior, then seek out others with attitudes and behavior similar to their own.[7] So do the individuals learn to commit crime after they were taught by delinquent peers, or do they start associating with similar "delinquents" or "criminals" once they have initiated their offending career (i.e., birds of a feather flock together)? This exact debate was examined by researchers, and the most recent studies point to *both* causal processes occurring: Criminal associations cause more crime *and* committing crime causes more criminal associations. Both are key in the causal process, so Sutherland was missing half this equation.[8]

Another key criticism is that if all individuals are born with a blank slate and all criminal behavior is learned, then who committed crime in the first place? Who could expose the first criminal to the definitions favorable to violation of law? Furthermore, what factor(s) caused that individual to do it first if it was not learned? Obviously, if it were due to any other factor(s) than learning—and it must have been because there was no one to teach it—then it obviously was not explained by learning theories. This criticism cannot be addressed, so it is somewhat ignored in the scientific literature.

Despite the criticisms and flaws, much research supports Sutherland's theory. For example, researchers have found that older criminals teach younger delinquents.[9] In addition, delinquents often associate with criminal peers prior to engaging in criminal activity.[10] Furthermore, research has shown that the nexus of criminal friends, attitudes, and activity are highly associated.[11] Still, Sutherland's principles are quite vague and elusive in terms of measurement, which renders them difficult in terms of measuring by social scientists.[12] Related to these issues, perhaps one of the biggest problems with Sutherland's formulation of differential association is that he used primarily one type of learning model—classical conditioning—to formulate most of his principles, and thus he neglects other important ways that we learn attitudes and behavior from others. Ultimately, Sutherland's principles are hard to test; more current models of his framework have incorporated other learning models and thus are easier to test so that empirical validity can be demonstrated.

Glaser's Concept of Differential Identification. Another reaction to Sutherland's differential association dealt with the influence of movies and television, as well as other reference

---

[7]Tittle et al., "Modeling Sutherland's Theory."

[8]Terence Thornberry, "Toward an Interactional Theory of Delinquency," Criminology 25 (1987): 863–87.

[9]Kenneth Tunnell, "Inside the Drug Trade: Trafficking from the Dealer's Perspective," *Qualitative Sociology* 16 (1993): 361–81.

[10]Douglas Smith, Christy Visher, and G. Roger Jarjoura, "Dimensions of Delinquency: Exploring the Correlates of Participation, Frequency, and Persistence of Delinquent Behavior," *Journal of Research in Crime and Delinquency* 28 (1991): 6–32.

[11]Matthew Ploeger, "Youth Employment and Delinquency: Reconsidering a Problematic Relationship," *Criminology* 35 (1997): 659–75.

[12]Reed Adams, "The Adequacy of Differential Association Theory," *Journal of Research in Crime and Delinquency* 1 (1974): 1–8; James F. Short, "Differential Association as a Hypothesis: Problems of Empirical Testing," *Social Problems* 8 (1960).

groups outside of one's significant others. As stated above, Sutherland claimed that learning of criminal definitions could take place only through social interactions with significant others, as opposed to reading a book or watching movies. However, in 1956, Daniel Glaser proposed the idea of **differential identification theory**, which allows for learning to take place not only through people close to us but also through other reference groups, even distant ones such as sports heroes or movie stars with whom the individual has never actually met or corresponded.[13] Glaser claimed that it did not matter much whether the individual had a personal relationship with the reference group(s); in fact, he argued that the group could be imaginary, such as fictitious characters in a movie or book. The important thing, according to Glaser, was that the individual must identify with the person or character and thus behave in ways that fit the norm set of this reference group or person.

Glaser's proposition has been virtually ignored, with the exception of Dawes's study of delinquency in 1973, which found that identification with people other than parents was strong when youths perceived a greater degree of rejection from their parents.[14] Given the profound influence of movies, music, and television on today's youth culture, it is obvious that differential identification was an important addition to Sutherland's framework, and more research should examine the validity of Glaser's theory in contemporary society. Although Glaser and others modified differential association, the most valid and respected variation is differential reinforcement theory.

Differential Reinforcement Theory

In 1965, C. R. Jeffrey provided an extensive critique and reevaluation of Sutherland's differential association theory. He argued that the theory was incomplete without some attention to an updated social psychology of learning (e.g., operant conditioning and modeling theories of learning).[15] He wanted Sutherland to account for the fact that people can be conditioned into behaving certain ways, such as by being rewarded for conforming behavior. Then, in 1966, Robert Burgess and Ronald Akers criticized and responded to Jeffrey's criticism by proposing a new theory that incorporated some of these learning models into Sutherland's basic framework.[16] The result was what is now known as **differential reinforcement theory**. Ultimately, Burgess and Akers argued that by integrating Sutherland's work with contributions from the field of social psychology, criminal behavior could be more clearly understood.

In some ways, differential reinforcement theory may appear to be no different than rational choice theory (see Section II). To an extent this is true because both models focus on reinforcements and punishments that occur after an individual offends. However, differential reinforcement theory can be distinguished from the rational choice perspective.

---

[13]Daniel Glaser, "Criminality Theories and Behavioral Images," *American Journal of Sociology* 61 (1956): 433–44.

[14]K. J. Dawes, "Family Relationships, Reference Others, Differential Identification and Their Joint Impact on Juvenile Delinquency," PhD dissertation (Ann Arbor, MI: University Mircrofilms, 1973).

[15]C. R. Jeffery, "Criminal Behavior and Learning Theory, *The Journal of Criminal Law, Criminology, and Police Science* 56 (1965): 294–300.

[16]Robert Burgess and Ronald Akers, "A Differential Association-Reinforcement Theory of Criminal Behavior, *Social Problems* 14 (1966): 131.

The latter assumes that humans are born with the capacity for rational decision making, whereas the differential reinforcement perspective assumes people are born with a blank slate and, thus, must be socialized and taught how to behave through various forms of conditioning (e.g., operant and classical), as well as modeling.

Burgess and Akers developed seven propositions to summarize differential reinforcement theory, which largely represent efficient modifications of Sutherland's original nine principles of differential association.[17] The strong influence of social psychologists is illustrated in their first statement, as well as throughout the seven principles. Although differential reinforcement incorporates the elements of modeling and classical conditioning learning models in its framework, the first statement clearly states that the essential learning mechanism in social behavior is the operant conditioning, so it is important to understand what operant conditioning is and how it is evident throughout life.

Operant Conditioning. The idea of **operant conditioning** was primarily developed by B. F. Skinner,[18] who ironically was working just across campus from Edwin Sutherland when he was developing differential association theory at Indiana University. As in modern times, academia was too intradisciplinary and intradepartmental. Had Sutherland been aware of Skinner's studies and theoretical development, he likely would have included it in his original framework. In his defense, operant conditioning was not well known or researched at the time; as a result, Sutherland incorporated the then-dominant learning model, classical conditioning. Burgess and Akers went on to incorporate operant conditioning into Sutherland's framework.

Operant conditioning is concerned with how behavior is influenced by reinforcements and punishments. Furthermore, operant conditioning assumes that the animal or human being is a proactive player in seeking out rewards, and not just a passive entity that receives stimuli as classical conditioning assumes. Behavior is strengthened or encouraged through reward (**positive reinforcement**) and avoidance of punishment (**negative reinforcement**). For example, if someone is given a car for graduation from college, that would be a positive reinforcement. On the other hand, if a teenager who has been grounded is allowed to start going out again because he or she has better grades, this would be a negative reinforcement because they are now being rewarded via the avoidance of something negative. Like different types of reinforcement, punishment comes in two forms as well. Behavior is weakened, or discouraged, through adverse stimuli (**positive punishment**) or lack of reward (**negative punishment**). A good example of positive punishment would be a good old-fashioned spanking, because it is certainly a negative stimulus; anything that directly presents negative sensation or feelings is a positive punishment. On the other hand, if parents take away car privileges from a teenager who broke curfew, that would be an example of negative punishment because the parents are removing a positive aspect or reward.

Some notable examples of operant conditioning include teaching a rat to successfully run a maze. When rats take the right path and finish the maze quickly, they are either positively reinforced (e.g., rewarded with a piece of cheese) or negatively reinforced (e.g., not

---

[17]For a more recent version of these principles, see Ronald Akers, *Deviant Behavior: A Social Learning Approach,* 3rd ed. (Belmont, CA: Wadsworth, 1985); also see Ronald L. Akers and Christine S. Sellers, *Criminological Theories: Introduction, Evaluation, and Application,* 4th ed. (Los Angeles, CA: Roxbury, 2004).

[18]B. F. Skinner, *Science and Human Behavior* (New York: Macmillan, 1953).

zapped with electricity as they were when they chose the wrong path). On the other hand, when rats take wrong turns or do not complete the maze in adequate time, they are either positively punished (e.g., zapped with electricity) or negatively punished (e.g., not given the cheese they expect to receive). The rats, like humans, tend to learn the correct behavior very fast using such consistent implementation of reinforcements and punishments.

In humans, such principles of operant conditioning can be found even at very early ages. In fact, many of us have implemented such techniques (or been subjected to them) without really knowing they were called operant conditioning. For example, during toilet training, children learn to use the bathroom to do their natural duty, rather than doing it in their pants. To reinforce the act of going to the bathroom on a toilet, we encourage the correct behavior by presenting positive rewards, which can be as simple as applauding or giving a child a piece of candy for a successful job. While parents (we hope) rarely proactively use spanking in toilet training, there is an inherent positive punishment involved when children go in their pants; namely, they have to be in their dirty diaper for a while, not to mention the embarrassment that most children feel when they do this. Furthermore, negative punishments are present in such situations because they do not get the applause or candy, so the rewards have been removed.

Of course, this does not apply only to early behavior. An extensive amount of research has shown that humans learn attitudes and behavior best through a mix of reinforcements and punishments throughout life. In terms of criminal offending, studies have clearly shown that rehabilitative programs that appear to work most effectively in reducing recidivism in offenders are those that have opportunities for rewards as well as threats for punishments. Empirical research has combined the findings from hundreds of such studies of rehab programs, showing that the programs that are most successful in changing attitudes and behavior of previous offenders are those that offer at least four reward opportunities for every one punishment aspect of the program.[19] So whether it is training children to "go potty" correctly or altering criminals' thinking and behavior, operant conditioning is a well-established form of learning that makes differential reinforcement theory a more valid and specified model of offending than differential association.

Whether deviant or conforming behavior occurs and continues "depends on the past and present rewards or punishment for the behavior, and the rewards and punishment attached to alternative behavior."[20] This is in contrast to Sutherland's differential association model, which looked only at what happened before an act (i.e., classical conditioning), not at what happens after the act is completed (i.e., operant conditioning); Burgess and Akers's model looks at both. Criminal behavior is likely to occur, Burgess and Akers theorized, when its rewards outweigh the punishments.

Bandura's Model of Modeling/Imitation. Another learning model that Burgess and Akers emphasized in their formulation of differential reinforcement theory was the element of **imitation and modeling.** Although Sutherland's original formulation of differential association theory was somewhat inspired by Gabriel Tarde's concept of imitation,[21] the

---

[19]P. Van Voorhis, Michael Braswell, and David Lester. *Correctional Counseling*, 3rd ed. (Cincinnati, OH: Anderson, 2000).

[20]Ronald L. Akers, *Deviant Behavior: A Social Learning Approach*, 2nd ed. (Belmont, CA: Wadsworth Publishing, 1977), 57.

[21]Gabriel Tarde, *Penal Philosophy*, trans. Rapelje Howell (Boston: Little, Brown, 1912).

nine principles did not adequately emphasize the importance of modeling in the process of learning behavior. Sutherland's failure was likely due to the fact that Albert Bandura's primary work in this area had not occurred when Sutherland was formulating differential association theory.[22]

Through a series of experiments and theoretical development, Bandura demonstrated that a significant amount of learning takes place without any form of conditioning. Specifically, he claimed that individuals can learn even if they are not rewarded or punished for behavior (i.e., operant conditioning) and even if they have not been exposed to associations between stimuli and responses (i.e., classical conditioning). Rather, Bandura proposed that people learn much of their attitudes and behavior from simply observing the behavior of others, namely through mimicking what others do. This is often referred to as "monkey see, monkey do," but it is not just monkeys that do this. Like most animal species, humans are biologically hard-wired to observe and learn the behavior of others, especially elders, to see what behavior is essential for survival and success.

Bandura showed that simply observing the behavior of others, especially adults, can have profound learning effects on the behavior of children. Specifically, he performed experiments in which a randomized experimental group of children watched a video of adults acting aggressively toward Bo-Bo dolls (which are blow-up plastic dolls) and a control group of children that did not watch such a video. Both groups of children were then sent into a room containing Bo-Bo dolls, and the experimental group who had seen the adult behavior mimicked their elders by acting far more aggressively toward the dolls than the children in the control group. The experimental group had no previous associations of more aggressive behavior toward the dolls and good feelings or motivations, let alone rewards for such behavior. Rather, the children became more aggressive themselves simply because they were imitating what they had seen older people do.

Bandura's findings have important implications for the modeling behavior of adults (and peers) and for the influence of television, movies, video games, and other factors. Furthermore, the influences demonstrated by Bandura supported a phenomenon commonplace in everyday life. Mimicking is the source of fashion trends—wearing low-slung pants or baseball hats turned a certain way. Styles tend to ebb and flow based on how some respected person (often a celebrity) wears clothing. This can be seen very early in life; parents must be careful what they say and do because their children, as young as two years old, imitate what their parents do. This continues throughout life, especially in teenage years as young persons imitate the "cool" trends and styles, as well as behavior. Of course, sometimes this behavior is illegal, but individuals are often simply mimicking the way their friends or others are behaving, with little regard for potential rewards or punishments. Ultimately, Bandura's theory of modeling and imitation adds a great deal of explanation in a model of learning, and differential reinforcement theory included such influences, whereas Sutherland's model of differential association did not, largely because the psychological perspective had not yet been developed.

Burgess and Akers's theory of differential reinforcement has also been the target of criticism by theorists and researchers. Perhaps the most important criticism of differential

---

[22]See Albert Bandura, *Principles of Behavior Modification* (New York: Holt, Rinehart, and Winston, 1969); Albert Bandura, *Aggression: A Social Learning Analysis* (Englewood Cliffs, NJ: Prentice Hall, 1973); Albert Bandura, *Social Learning Theory* (Englewood Cliffs, NJ: Prentice Hall, 1977).

reinforcement theory is that it appears tautological, which means that the variables and measures used to test its validity are true by definition. To clarify, studies testing this theory have been divided into four groups of variables/factors: associations, reinforcements, definitions, and modeling.

Some critics have noted that if individuals who report that they associate with those who offend, are rewarded for offending, believe offending is good, and have seen many of their significant others offend, they will inevitably be more likely to offend. In other words, if your friends and/or family are doing it, there is little doubt that you will be doing it.[23] For example, critics would argue that a person who primarily hangs out with car thieves, knows he will be rewarded for stealing cars, believes stealing cars is good and not immoral, and has observed many respected others stealing cars will commit auto theft himself inevitably. However, it has been well argued that such criticisms of tautology are not valid because none of these factors necessarily make offending by the respondent true by definition.[24]

Differential reinforcement theory has also faced the same criticism that was addressed to Sutherland's theory: that delinquent associations may take place after criminal activity, rather than before. However, Burgess and Akers's model clearly has this area of criticism covered in the sense that differential reinforcement covers what comes after the activity, not just what happens before it. Specifically, it addresses the rewards or punishments that follow criminal activity, whether those rewards come from friends, parents, or other members/institutions of society.

It is arguable that differential reinforcement theory may have the most empirical validity of any contemporary (nonintegrated) model of criminal offending, especially considering that studies have examined a variety of behaviors, ranging from drug use to property crimes to violence. The theoretical model has also been tested in samples across the United States, as well as other cultures such as South Korea, with the evidence being quite supportive of the framework. Furthermore, a variety of age groups have been examined, ranging from teenagers to middle-aged adults to the elderly, with all studies providing support for the model.[25]

Specifically, researchers have found that the major variables of the theory had a significant effect in explaining marijuana and alcohol use among adolescents.[26] The researchers concluded that the "study demonstrates that central learning concepts are amenable to meaningful questionnaire measurement and that social learning theory can

---

[23]Mark Warr, "Parents, Peers, and Delinquency," *Social Forces* 72 (1993): 247–64; Mark Warr and Mark Stafford, "The Influence of Delinquent Peers: What They Think or What They Do?" *Criminology* 29 (1991): 851–66.

[24]See Akers and Sellers, *Criminological Theories,* 98–101.

[25]See studies including Ronald Akers and Gang Lee, "A Longitudinal Test of Social Learning Theory: Adolescent Smoking," *Journal of Drug Issues* 26 (1996): 317–43; Ronald Akers and Gang Lee, "Age, Social Learning, and Social Bonding in Adolescent Substance Abuse," *Deviant Behavior* 19 (1999): 1–25; Ronald Akers and Anthony J. La Greca, "Alcohol Use Among the Elderly: Social Learning, Community Context, and Life Events," in *Society, Culture, and Drinking Patterns Re-examined,* ed. David J. Pittman and Helene Raskin White (New Brunswick, NJ: Rutgers Center of Alcohol Studies, 1991), 242–62; Sunghyun Hwang, "Substance Use in a Sample of South Korean Adolescents: A Test of Alternative Theories," PhD dissertation (Ann Arbor, MI: University Microfilms, 2000); Sunghyun Hwang and Ronald Akers, "Adolescent Substance Use in South Korea: A Cross-Cultural Test of Three Theories," in *Social Learning Theory and the Explanation of Crime: A Guide for the New Century,* ed. Ronald Akers and Gary F. Jensen (New Brunswick, NJ: Transaction Publishers, 2003).

[26]Ronald Akers, Marvin D. Krohn, Lonn Lanza-Kaduce, and Marcia Radosevich, "Social Learning and Deviant Behavior: A Specific Test of a General Theory," *American Sociological Review* 44 (1979): 638.

be adequately tested with survey data."[27] Other studies have also supported the theory when attempting to understand delinquency, cigarette smoking, and drug use.[28] Therefore, the inclusion of three psychological learning models, namely classical conditioning, operant conditioning, and **modeling/imitation,** appears to have made differential reinforcement one of the most valid theories of human behavior, especially in regard to crime.

Neutralization Theory

Neutralization theory is associated with Gresham Sykes and David Matza's **techniques of neutralization**[29] and Matza's **drift theory.**[30] Like Sutherland, both Sykes and Matza thought that social learning influences delinquent behavior, but they also asserted that most criminals hold conventional beliefs and values. More specifically, Sykes and Matza argued that most criminals are still partially committed to the dominant social order. According to Sykes and Matza, youth are not immersed in a subculture that is committed to either extremes of complete conformity or complete nonconformity. Rather, these individuals vacillate, or *drift*, between these two extremes and are in a state of "transience".[31]

While remaining partially committed to conventional social order, youth can "drift" into criminal activity, Sykes and Matza claimed, and avoid feelings of guilt for these actions by justifying or rationalizing their behavior. This typically occurs in their teenage years, when social controls (e.g., parents, family, etc.) are at their weakest point, and peer pressures/associations are at their highest level. Why is this called neutralization theory? The answer is that people justify and rationalize behavior through "neutralizing" it or making it appear not so serious. They make up situational excuses for behavior that they know is wrong to alleviate the guilt they feel for doing such immoral acts. In many ways, this resembles Freud's defense mechanisms, which allow us to forgive ourselves for the bad things we do, even though we know they are wrong. The specific techniques of neutralization outlined by Sykes and Matza in 1957 are much like excuses for inappropriate behavior.

Techniques of Neutralization. Sykes and Matza identified methods or techniques of neutralization[32] that people use to justify their criminal behavior. These techniques allow people to neutralize or rationalize their criminal and delinquent acts by making them

---

[27]Akers et al., "Social Learning," 651.

[28]Richard Lawrence, "School Performance, Peers and Delinquency: Implications for Juvenile Justice," *Juvenile and Family Court Journal* 42 (1991): 59–69; Marvin Krohn, William Skinner, James Massey, and Ronald Akers, "Social Learning Theory and Adolescent Cigarette Smoking: A Longitudinal Study," *Social Problems* 32 (1985): 455–71; L. Thomas Winfree, Christine Sellers, and Dennis Clason, "Social Learning and Adolescent Deviance Abstention: Toward Understanding the Reasons for Initiating, Quitting, and Avoiding Drugs," *Journal of Quantitative Criminology* 9 (1993): 101–23; Ronald Akers and Gang Lee, "A Longitudinal Test of Social Learning Theory: Adolescent Smoking," *Journal of Drug Issues* 26 (1996): 317–43.

[29]Gresham M. Sykes and David Matza, "Techniques of Neutralization: A Theory of Delinquency," *American Sociological Review* 22 (1957): 664–70.

[30]David Matza, *Delinquency and Drift* (New York: John Wiley, 1964).

[31]Ibid., 28.

[32]Sykes and Matza, "Techniques of Neutralization."

look as though they are conforming to the rules of conventional society. If individuals can create such rationalizations, then they are free to engage in criminal activities without serious damage to their conscience or self-image. According to Sykes and Matza, there are five common techniques of neutralization:

1. *Denial of responsibility:* Individuals may claim they were influenced by forces outside themselves and that they are not responsible or accountable for their behavior. For example, many youths blame their peers for their own behavior.

2. *Denial of injury:* This is the rationalization that no one was actually hurt by their behavior. For instance, if someone steals from a store, they may rationalize this by saying that the store has insurance, so there is no direct victim.

3. *Denial of the victim:* Offenders see themselves as avengers and the victim as the wrongdoer. For example, some offenders believe that a person who disrespects or "disses" them deserves what he or she gets, even if it means serious injury.

4. *Condemnation of the condemners:* Offenders claim the condemners (usually the authorities who catch them) are hypocrites. For instance, one may claim that police speed on the highway all the time, so everyone else is entitled to drive higher than the speed limit.

5. *Appeal to higher loyalties:* Offenders often overlook the norms of conventional society in favor of the rules of a belief they have or a group to which they belong. For example, people who kill doctors who perform abortions tend to see their crimes as above the law because they are serving a higher power.

Although Sykes and Matza specifically labeled only five techniques of neutralization, it should be clear that there may be endless excuses people make up to rationalize behaviors they know are wrong. Techniques of neutralization have been applied to white-collar crime, for example. Several studies have examined the tendency to use such excuses to alleviate guilt for engaging in illegal corporate crime; they point out new types of excuses white-collar criminals use to justify their acts, techniques that were not discussed in Sykes and Matza's original formulation.[33]

Studies that have attempted to empirically test neutralization theory are, at best, inconsistent. For example, Agnew argued that there are essentially two general criticisms of studies that support neutralization theory.[34] First, theorists and researchers have noted that some neutralization techniques are very difficult to measure, as opposed to a commitment to unconventional attitudes or norms.[35] The second major criticism is the concern

---

[33]For a review and a study on this topic, see Nicole Piquero, Stephen Tibbetts, and Michael Blankenship, "Examining the Role of Differential Association and Techniques of Neutralization in Explaining Corporate Crime," *Deviant Behavior* 26 (2005): 159–88.

[34]Robert Agnew, "The Techniques of Neutralization and Violence," *Criminology* 32 (1994): 563–64.

[35]W. William Minor, "The Neutralization of Criminal Offense, *Criminology* 18 (1980): 116; W. William Minor, "Neutralization as a Hardening Process: Considerations in the Modeling of Change," *Social Forces* 62 (1984): 995–1019. See also Roy L. Austin, "Commitment, Neutralization, and Delinquency," in *Juvenile Delinquency: Little Brother Grows Up,* ed. Theordore N. Ferninand (Beverly Hills, CA: Sage); Quint C. Thurman, "Deviance and the Neutralization of Moral Commitment: An Empirical Analysis," *Deviant Behavior* 5 (1984): 291–304.

that criminals may not use techniques of neutralization *prior* to committing a criminal offense but rather only *after* committing a crime. As estimated by previous studies, this temporal ordering can be problematic in terms of causal implications when neutralization *follows* a criminal act.[36] This temporal ordering problem results from research conducted at a single point in time. Some would argue that the temporal ordering problem is not a major criticism because individuals may be predisposed to make up such rationalizations for their behavior regardless of whether they do it before or after the act of offending. Such a propensity may be related to low self-control theory, which we examine later in this section.

Summary of Learning Theories

Learning theories tend to emphasize the social processes of how and why individuals learn criminal behavior. These theories also focus on the impact of significant others involved in the socialization process, such as family, friends, and teachers. Ultimately, empirical research has shown that learning theories are key in our understanding of criminal behavior, particularly in terms of whether criminal behavior is rewarded or punished. In summary, if individuals are taught and rewarded for performing criminal acts by the people they interact with on a day-to-day basis, in all likelihood they will engage in illegal activity.

⬚ Control Theories

The learning theories discussed in the previous section assume that individuals are born with a conforming disposition. By contrast, control theories assume that all people would naturally commit crimes if it wasn't for restraints on their innate selfish tendencies. Social control perspectives of criminal behavior thus assume that there is some type of "basic human nature" and that all human beings exhibit "antisocial tendencies." Such theories are concerned with why individuals *don't* commit crime or deviant behaviors. Control theorists ask: What is it about society and human interaction that causes people *not* to act on their impulses?

The assumption that people have innate antisocial tendencies is a controversial one because it is nearly impossible to test. Nevertheless, some recent evidence supports the idea that human beings are inherently selfish and antisocial by nature. Specifically, researchers have found that most individuals are oriented toward selfish and aggressive behaviors at an early age, with such behaviors peaking at end of the second year (see Figure 7.1).[37]

An example of antisocial dispositions appearing early in life was reported by Tremblay and LeMarquand, who found that most young children's (particularly boys)

---

[36]Travis Hirschi, *Causes of Delinquency* (Berkeley: University of California Press, 1969): 207. See also Mark Pogrebin, Eric Poole, and Amos Martinez, "Accounts of Professional Misdeeds: The Sexual Exploitation of Clients by Psychotherapists," *Deviant Behavior* 13 (1992): 229–52; John Hamlin, "The Misplaced Concept of Rational Choice in Neutralization Theory," *Criminology* 26 (1988): 425–38.

[37]Michael Lewis, S. Alessandri, and M. Sullivan, "Expectancy, Loss of Control, and Anger in Young Infants," *Developmental Psychology* 25 (1990): 745–51; A. Restoin, D. Rodriguez, V. Ulmann, and H. Montagner, "New Data on the Development of Communication Behavior in the Young Child with his Peers," *Recherches de Psychologie Sociale* 5 (1985): 31–56; R. Tremblay, C. Japel, D. Perusse, P. McDuff, M. Boivin, M. Zoccolillo, and J. Montplaisir, "The Search for the Age of 'Onset' of Physical Aggression: Rousseau and Bandura revisited," *Criminal Behaviour and Mental Health* 9 (1999): 8–23.

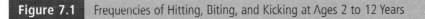

Figure 7.1 Frequencies of Hitting, Biting, and Kicking at Ages 2 to 12 Years

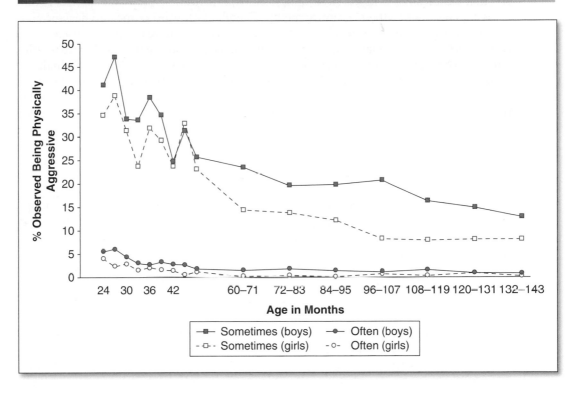

aggressive behaviors peaked at age 27 months. These behaviors included hitting, biting, and kicking others.[38] Their research is not isolated; virtually all developmental experts acknowledge that toddlers exhibit a tendency to show aggressive behaviors toward others. This line of research would seem to support the notion that people are predisposed toward antisocial, even criminal, behavior.

Control theorists do not necessarily assume that people are predisposed toward crime in a way that remains constant throughout life. On the contrary, research shows that most individuals begin to desist from such behaviors starting at around age two. This trend continues until approximately age five, with only the most aggressive individuals (i.e., chronic offenders) continuing into higher ages.

It is important to note that at the same time selfish and aggressive behaviors decline, self-consciousness is formed. In addition, social emotion—such as shame, guilt, empathy, and pride—begin to appear.[39] This observation is critical because it is what separates

---

[38]Richard Tremblay and D. LeMarquand, "Individual Risk and Protective Factors," in *Child Delinquents: Development, Intervention, and Service Needs*, ed. R. Loeber and D. Farrington (London: Sage, 2001): 137–64.

[39]For reviews of supporting studies, see June Price Tangney and K. W. Fischer, *Self-Conscious Emotions: The Psychology of Shame, Guilt, Embarrassment, and Pride* (New York: Guilford Press, 1995); Michael Lewis, *Shame: The Exposed Self* (New York: Macmillan, 1992).

control theories from the classical school of criminology and the predispositional theories that we already discussed. According to control theories, without appropriate socialization, people act on their "pre-programmed" tendency toward crime and deviance.

In short, control theories claim that all individuals have natural tendencies to commit selfish, antisocial, and even criminal behavior. So what is it that curbs this natural propensity? Many experts believe the best explanation is that individuals are socialized and controlled by social attachments and investments in conventional society. This assumption regarding the vital importance of early socialization is probably the primary reason why control theories are currently the most popular and accepted theories among criminologists.[40] We will now discuss several early examples of these control theories.

Early Control Theories of Human Behavior

Thomas Hobbes. Control theories are found in a variety of disciplines, including biology, psychology, and sociology. Perhaps the earliest significant use of social control in explaining deviant behavior is found in a perspective offered by the 17th-century Enlightenment philosopher, Thomas Hobbes (see Section I). Hobbes claimed that the natural state of humanity was one of selfishness and self-centeredness to the point of constant chaos, characterized by a state of warfare between individuals. He stated that all individuals are inherently disposed to take advantage of others in order to improve their own personal well-being.[41]

However, Hobbes also claimed that the constant fear created by such selfishness resulted in humans rationally coming together to create binding contracts that would keep individuals from violating others' rights. Even with such controlling arrangements, however, Hobbes was clear that the selfish tendencies people exhibit could never be extinguished. In fact, they explained why punishments were necessary to maintain an established social contract among people.

Durkheim's Idea of Awakened Reflection and Collective Conscience. Consistent with Hobbes's view of individuals as naturally selfish, Durkheim later proposed a theory of social control in the late 1800s that suggested humans have no internal mechanism to let them know when they are fulfilled.[42] To this end, Durkheim coined the terms "automatic spontaneity" and "awakened reflection." Automatic spontaneity can be understood with reference to animals' eating habits. Specifically, animals stop eating when they are full, and they are content until they are hungry again; they don't start hunting right after they have filled their stomach with food. In contrast, awakened reflection concerns the fact that humans do not have such an internal, regulatory mechanism. That is because people often acquire resources beyond what is immediately required. Durkheim went so far as to say that "our capacity for feeling is in itself an insatiable and bottomless abyss."[43] This is

---

[40]Walsh and Ellis, "Political Ideology."

[41]Thomas Hobbes, *Leviathan* (Cambridge: Cambridge University Press, 1651/1904).

[42]Emile Durkheim, *The Division of Labor in Society* (New York: Free Press, 1893/1965); Emile Durkheim, *Suicide* (New York: Free Press, 1897/1951).

[43]Emile Durkheim, *Suicide*, 246–47. Also, much of this discussion, is adapted from Raymond Paternoster and Ronet Bachman, *Explaining Criminals and Crime* (Los Angeles, CA : Roxbury, 2001).

one of the reasons that Durkheim believed crime and deviance are quite normal, even essential, in any society.

Durkheim's "awakened reflection" has become commonly known as greed. People tend to favor better conditions and additional fulfillment because they apparently have no biological or psychological mechanism to limit such tendencies. As Durkheim noted, the selfish desires of mankind "are unlimited so far as they depend on the individual alone . . . the more one has, the more one wants."[44] Thus, society must step in and provide the "regulative force" that keeps humans from acting too selfishly.

One of the primary elements of this regulative force is the *collective conscience*, which is the extent of similarities or likenesses that people share. For example, almost everyone can agree that homicide is a serious and harmful act that should be avoided in any civilized society. The notion of collective conscience can be seen as an early form of the idea of social bonding, which has become one of the dominant theories in criminology.[45]

According to Durkheim, the collective conscience serves many functions in society. One such function is the ability to establish rules that control individuals from following their natural tendencies toward selfish behavior. Durkheim also believed that crime allows people to unite together in opposition against deviants. In other words, crime and deviance allow conforming individuals to be "bonded" together in opposition against a common enemy, as can be seen in everyday life when groups come together to face opposition. This enemy consists of the deviants who have not internalized the code of the collective conscience.

Many of Durkheim's ideas hold true today. Just recall a traumatic incident you may have experienced with other strangers (e.g., being stuck in an elevator during a power outage, weathering a serious storm, or being involved in a traffic accident). Incidents such as this bring people together and permit a degree of bonding that would not take place in everyday life. Crime, Durkheim argued, serves a similar function.

How is all of this relevant today? Most control theorists claim that individuals commit crime and deviant acts not because they are lacking in any way, but because certain controls have been weakened in their development. This assumption is consistent with Durkheim's theory, which we discussed previously (also see Section V).

Freud's Concept of Id, Superego, and Ego. Although psychoanalytic theory would seem to have few similarities with a sociological positivistic theory, in this case, it is extremely complementary. One of Freud's most essential propositions is that all individuals are born with a tendency toward inherent drives and selfishness due to the **id** domain of the psyche (see Figure 7.2).[46] According to Freud, all people are born with equal amounts of Id drives (e.g., libido, food, etc.) and motivations toward selfishness and greed. Freud said this inherent selfish tendency must be countered by controls produced from the development of the **superego,** which is the subconscious domain of the psyche that contains our conscience. According to Freud, the superego is formed through the interactions between

---

[44]Durkheim, *Suicide*, 254.

[45]A good discussion of Durkheim's concepts, particularly that of the collective conscience, can be found in George B. Vold, Thomas J. Bernard, and Jeffrey B. Snipes, *Theoretical Criminology*, 4th ed. (New York: Oxford University Press, 1998), 124–139.

[46]Sigmund Freud, "The Ego and the Id," in *The Complete Psychological Works of Sigmund Freud*, Vol. 19, ed. J. Strachey (London: Hogarth Press, 1923/1959).

Figure 7.2   Freud's Model of the Three Domains of the Psyche

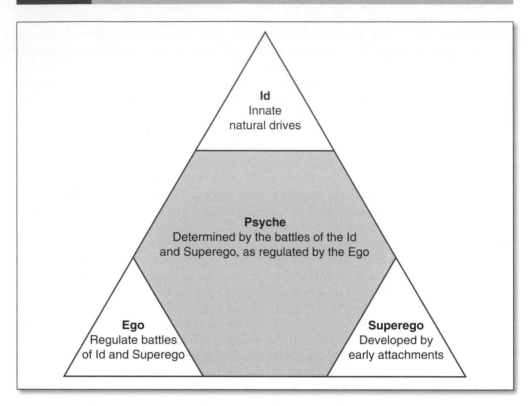

a young infant/child and his or her significant others. As you can see, the control perspective has a long history in many philosophical and scientific disciplines.

These two drives of the subconscious domains of the id and superego are regulated, Freud thought, by the only conscious domain of the psyche: the **ego**. This ego mediates the battles between our innate drives (id) and our socialized constraints (superego); it represents our personality. There have been a number of applications of Freud's theoretical model to criminality, such as having a deficient superego (due to a lack of early attachments) or a weak ego (which fails to properly regulate the battle between the id and superego). The main point is that Freud was an early control theorist and that his theoretical model was highly influential among psychologists in the early 1900s as they tried to determine why certain individuals committed criminal offenses.[47]

Early Control Theories of Crime

Throughout the 1950s and 1960s, criminologists borrowed and built on some of the ideas just discussed. Until that time, most research in the criminological literature was dominated by the learning theories discussed earlier in this section or social structure

---

[47]Ibid.

theories, such as the Chicago School or Merton's strain theory (see Section V). While early control theories may not be particularly popular in this day and age, they were vitally important in the sense that they laid the groundwork for future theoretical development.

Reiss's Control Theory. One of the first control theories of crime was proposed by Albert Reiss in 1951. Reiss claimed that delinquency was a consequence of weak ego or superego controls among juvenile probationers.[48] Reiss found no explicit motivation for delinquent activity. Rather, he thought it would occur in the absence of controls or restraints against such behavior.

Like Freud, Reiss believed that the family was the primary source through which deviant predispositions were discouraged. Furthermore, Reiss claimed that a sound family environment would provide for an individual's needs and the essential emotional bonds that are so important in socializing individuals. Another important factor in Reiss's model was close supervision, not only by the family but also by the community. He said that individuals must be closely monitored for delinquent behavior and adequately disciplined when they break the rules.

Personal factors, such as the ability to restrain one's impulses and delay gratification, were also important in Reiss' framework. These concepts are very similar to later, more modern concepts of control theory, which have been consistently supported by empirical research.[49] For this reason, Reiss was ahead of his time when he first proposed his control theory. Although the direct tests of Reiss's theory have provided only partial support for it, his influence is apparent in many contemporary criminological theories.[50]

Toby's Concept of "Stake in Conformity." Soon after Reiss's theory was presented, a similar theory was developed. In 1957, Jackson Toby proposed a theory of delinquency and gangs.[51] He claimed that individuals were more inclined to act on their natural inclinations when the controls on them were weak. Like most other control theorists, Toby claimed that such inclinations toward deviance were distributed equally across all individuals. Furthermore, he emphasized the concept of a **stake in conformity** that supposedly prevents most people from committing crime. The stake in conformity Toby was referring to is the extent to which individuals have investments in conventional society. In other words, how much is a person willing to risk when he or she violates the law?

Studies have shown that stake in conformity is one of the most influential factors in individuals' decisions to offend. People who have nothing to lose are much more likely to take risks and violate others' rights than those who have relatively more invested in social institutions.[52]

One distinguishing feature of Toby's theory is his emphasis on peer influences in terms of both motivating and inhibiting antisocial behavior depending on whether most

---

[48]Albert Reiss, "Delinquency as the Failure of Personal and Social Controls," *American Sociological Review* 16 (1951): 196–207.

[49]For a comprehensive review of studies of low self-control, see Travis Pratt and Frank Cullen, "The Empirical Status of Gottfredson and Hirschi's General Theory of Crime: A Meta-Analysis," *Criminology* 38 (2000): 931–64.

[50]See Vold et al. *Theoretical Criminology*, 4th ed., 202–203.

[51]Jackson Toby, "Social Disorganization and Stake in Conformity: Complementary Factors in the Predatory Behavior of Hoodlums," *Journal of Criminal Law, Criminology, and Police Science* 48 (1957): 12–17.

[52]Travis Hirschi, *Causes of Delinquency* (Berkeley, CA: University of California Press, 1969); Robert Sampson and John Laub, *Crime in the Making: Pathways and Turning Points in Life* (Cambridge: Harvard University Press, 1993).

peers have low or high stakes in conformity. Toby's stake in conformity has been used effectively in subsequent control theories of crime.

Nye's Control Theory. A year after Toby introduced the stake in conformity, F. Ivan Nye (1958) proposed a relatively comprehensive control theory that placed a strong focus on the family.[53] Following the assumptions of early control theorists, Nye claimed that there was no significant positive force that caused delinquency because such antisocial tendencies are universal and would be found in virtually everyone if not for certain controls usually found in the home.

Nye's theory consisted of three primary components of control. The first component was internal control, which is formed through social interaction. This socialization, he claimed, assists in the development of a conscience. Nye further claimed that if individuals are not given adequate resources and care, they will follow their natural tendencies toward doing what is necessary to protect their interests.

Nye's second component of control was direct control, which consists of a wide range of constraints on individual propensities to commit deviant acts. Direct control includes numerous types of sanctions, such as jail and ridicule, and the restriction of one's chances to commit criminal activity. Nye's third component of control was indirect control, which occurs when individuals are strongly attached to their early caregivers. For most children, it is through an intense and strong relationship with their parents or guardians that they establish an attachment to conventional society. However, Nye suggested that when the needs of an individual are not met by their caregivers, inappropriate behavior can result.

As shown in Figure 7.3, Nye predicted a U-shaped curve of parental controls in predicting delinquency. Specifically, he argued that either no controls (i.e., complete freedom)

Figure 7.3	Nye's Control Theory

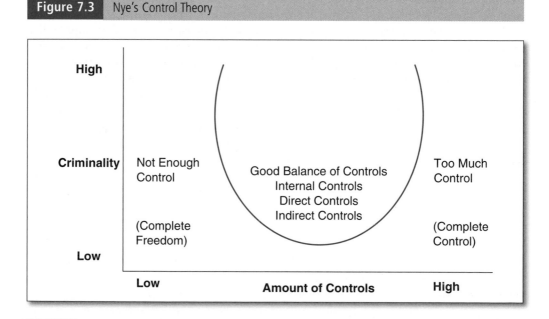

---

[53]F. Ivan Nye, *Family Relationships and Delinquent Behavior* (New York: Wiley, 1958).

or too much control (i.e., no freedom at all) would predict the most chronic delinquency. He believed that a healthy balance of freedom and parental control was the best strategy for inhibiting criminal activity. Some recent research supports Nye's prediction.[54] Contemporary control theories, such as Tittle's control-balance theory, draw heavily on Nye's idea of having a healthy balance of controls and freedom.[55]

Reckless's Containment Theory. Another control theory, known as **containment theory**, has been proposed by Walter Reckless.[56] This theory emphasizes both inner containment and outer containment, which can be viewed as internal and external controls. Reckless broke from traditional assumptions of social control theories by identifying predictive factors that *push* and/or *pull* individuals toward antisocial behavior. However, the focus of his theory remained on the controlling elements, which can be seen in the emphasis placed on containment in the theory's name.

Reckless claimed that individuals can be *pushed* into delinquency by their social environment, such as by a lack of opportunities for education or employment. Furthermore, he pointed out some individual factors, such as brain disorders or risk-taking personalities, could push some people to commit criminal behavior. Reckless also noted that some individuals could be *pulled* into criminal activity by hanging out with delinquent peers, watching too much violence on television, and so on. All told, Reckless went beyond the typical control theory assumption of inborn tendencies. In addition to these natural dispositions toward deviant behavior, containment theory proposes that extra pushes and pulls can motivate people to commit crime.

Reckless further claimed that the pushes and pulls toward criminal behavior could be enough to force individuals into criminal activity unless they are sufficiently *contained* or controlled. Reckless claimed that such containment should be both internal and external. By "internal containment" he meant building a person's sense of self, which helps the person resist the temptations of criminal activity. According to Reckless, other forms of inner containment include the ability to internalize societal norms. With respect to external containment, Reckless claimed that social organizations, such as school, church, and other institutions, are essential in building bonds that inhibit individuals from being pushed or pulled into criminal activity.

Reckless offered a visual image of containment theory, which we present in Figure 7.4. The outer circle (Circle 1) in the figure represents the social realm of pressures and pulls (e.g., peer pressure, etc.), whereas the innermost circle (Circle 4) symbolizes a person's individual-level pushes to commit crime, such as predispositions or personality traits that are linked to crime. In between these two circles are the two layers of controls, external containment (Circle 2) and internal containment (Circle 3). The structure of Figure 7.4 and the examples included in each circle are those specifically noted by Reckless.[57]

---

[54]Ruth Seydlitz, "Complexity in the Relationships among Direct and Indirect Parental Controls and Delinquency," *Youth and Society* 24 (1993): 243–75.

[55]Charles Tittle, *Control Balance: Toward a General Theory of Deviance* (Boulder, CO: Westview, 1995).

[56]Walter Reckless, *The Crime Problem*, 4th ed. (New York: Appleton-Century-Crofts, 1967).

[57]Ibid., 479.

Figure 7.4   Reckless's Containment Theory

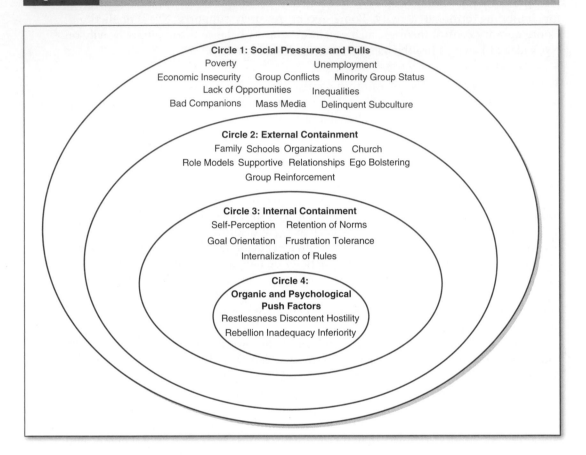

While some studies have shown general support for containment theory, others offer more support for some components, such as internalization of rules, than for other factors, such as self-perception, in accounting for variations in delinquency.[58] External factors may be more important than internal ones. Furthermore, some studies have noted weaker support for Reckless's theory among minorities and females, who may be more influenced by their peers or other influences. Thus, the model appears to be most valid for White males, at least according to empirical studies.[59]

One of the problems with containment theory is that it does not go far enough toward specifying the factors that are important in predicting criminality, especially

---

[58]Richard Lawrence, "School Performance, Containment Theory, and Delinquent Behavior," *Youth and Society* 7 (1985): 69–95; Richard A. Dodder and Janet R. Long, "Containment Theory Reevaluated: An Empirical Explication," *Criminal Justice Review* 5 (1980): 74–84.

[59]William E. Thompson and Richard A. Dodder, "Containment Theory and Juvenile Delinquency: A Reevaluation through Factor Analysis," *Adolescence* 21 (1986): 365–76.

regarding specific groups of individuals. For example, an infinite number of concepts could potentially be categorized as either a push or pull toward criminality, or as an inner or outer containment of criminality. Thus, the theory could be considered too broad or vague and not specific enough to be of practical value. To Reckless's credit, however, containment theory has increased the exposure of control theories of criminal behavior. And although support for containment theory has been mixed, there is no doubt that it has influenced other, more recent control theories.[60]

Modern Social Control Theories

As the previous sections attest, control theory has been around, in various forms, for some time. Modern social control theories build on these earlier versions of social control and add a level of depth and sophistication. Two modern social control theories are Matza's drift theory and Hirschi's social bonding theory.

Matza's Drift Theory. The theory of drift, or drift theory, presented by David Matza in 1964, claims that individuals offend at certain times in their life when social controls—such as parental supervision, employment, and family ties—are weakened.[61] In developing his theory, Matza criticized earlier theories and their tendency to predict too much crime. For example, the Chicago School would incorrectly predict that all individuals in bad neighborhoods will commit crime. Likewise, strain theory predicts that all poor individuals will commit crime. Obviously, this is not true. Thus, Matza claimed that there is a degree of determinism (i.e., Positive School) in human behavior but also a significant amount of free will (i.e., Classical School). He called this perspective "**soft determinism**," which is the gray area between free will and determinism. This is illustrated in Figure 7.5.

Returning to the basics of Matza's theory, he claimed that individuals offend at the time in life when social controls are most weakened. As is well known, social controls are most weakened for most individuals during the teenage years. At this time, parents and other caretakers stop having a constant supervisory role, and at the same time, teenagers generally do not have too many responsibilities—such as careers or children—that would inhibit them from experimenting with deviance. This is very consistent with the well-known age-crime relationship; most individuals who are arrested experience this in their teenage years.[62] Once sufficient ties are developed, people tend to mature out of criminal lifestyles.

Matza further claimed that when supervision is absent and ties are minimal, the majority of individuals are the most "free" to do what they want. Where, then, does the term *drift* come from? During the times when people have few ties and obligations, they will "drift" in and out of delinquency, Matza proposed. He pointed out that previous

---

[60]Ronald L. Akers, *Criminological Theories: Introduction, Evaluation, and Application,* 3rd ed. (Los Angeles, CA: Roxbury, 2000), 103–4.

[61]Matza, *Delinquency and Drift.*

[62]OJJDP Statistical Briefing Book. Arrest data for 1980–1997 from unpublished data from the Federal Bureau of Investigation and for 1998, 1999, and 2000 from *Crime in the United States* reports (Washington, DC: Government Printing Office, 1999, 2000, and 2001, respectively).

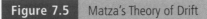

Figure 7.5   Matza's Theory of Drift

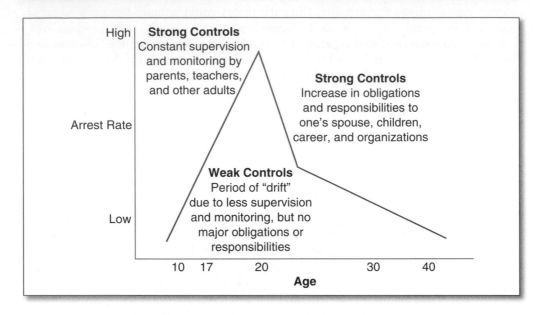

theories are unsuccessful in explaining this age-crime relationship: "Most theories of delinquency take no account of maturational reform; those that do often do so at the expense of violating their own assumptions regarding the constrained delinquent."[63]

Matza insisted that drifting is not the same as a commitment to a life of crime. Instead, it is experimenting with questionable behavior and then rationalizing it. The way youth rationalize behavior that they know to be wrong is through the learning of techniques of neutralization discussed earlier.

Drift theory goes on to say that individuals do not reject the conventional normative structure. On the contrary, much offending is based on neutralizing or adhering to **subterranean values,** which young people have been socialized to use as a means of circumventing conventional values. This is basically the same as asserting one's independence, which tends to occur with a vengeance during the teenage years.

Subterranean values are quite prevalent and underlie many aspects of our culture, which is why Matza's drift theory is also classified as a learning theory, as discussed earlier in this section. For example, while it is conventional to believe that violence is wrong, boxing matches and injury-prone sports are some of the most popular spectator activities. Such phenomena create an atmosphere that readily allows neutralization or rationalization of criminal activity.

We will see other forms of subterranean values when we discuss risk-taking and low self-control later in this section. In many contexts, such as business, risk-taking and aggressiveness are seen as desirable characteristics, so many individuals are influenced by

---

[63]"Age-arrest rate curve" is loosely based on data provided by Federal Bureau of Investigation, *Crime in the United States Report* (Washington, DC: Government Printing Office, 1997).

such subterranean values. This, according to Matza, adds to individuals' likelihood of drifting into crime and delinquency.

Matza's theory of drift seems sensible on its face, but empirical research examining the theory has shown mixed results.[64] One of the primary criticisms of Matza's theory, which even he acknowledged, is that it does not explain the most chronic offenders, the people who are responsible for the vast majority of serious, violent crimes. Chronic offenders often offend long before and well past their teenage years, which clearly limits the predictive value of Matza's theory.

Despite its shortcomings, Matza's drift theory appears to explain why many people offend exclusively during their teenage and young adult years, but then grow out of it. Also, the theory is highly consistent with several of the ideas presented by control theorists, including the assumption that (1) selfish tendencies are universal, (2) these tendencies are inhibited by socialization and social controls, and (3) the selfish tendencies appear at times when controls are weakest. The theory goes beyond the previous control theories by adding the concepts of soft determinism, neutralization, and subterranean values, as well as the idea that in many contexts, selfish and aggressive behaviors are not wrong but actually desirable.

Hirschi's Social Bonding Theory. Perhaps the most influential social control theory was presented by Travis Hirschi in 1969.[65] Hirschi's model of **social bonding theory** takes an assumption from Durkheim that "we are all animals, and thus naturally capable of committing criminal acts."[66] However, as Hirschi acknowledged, most humans can be adequately socialized to become tightly bonded to conventional entities, such as families, schools, communities, and the like. Hirschi said that the stronger a person is bonded to conventional society, the less prone to engaging in crime he or she will be. More specifically, the stronger the *social bond*, the less likely it is that an individual will commit criminal offenses.

As shown in Figure 7.6, Hirschi's social bond is made up of four elements: (1) attachment, (2) commitment, (3) involvement, and (4) moral belief. The stronger or more developed a person is in each of the four elements, the less likely he or she will be to commit crime. Let us now consider each element in detail.

The most important factor in the social bond is *attachments*, which consist of affectionate bonds between an individual and his or her significant others. Attachment is vitally important for the internalization of conventional values. Hirschi said, "The essence of internalization of norms, conscience, or superego thus lies in the attachment of the individual to others."[67] Hirschi made it clear, as did Freud, that strong early attachments are the most important factor in developing a social bond. The other constructs in the social bond—commitment, involvement, and belief—are contingent on adequate attachment to others, he argued. That is, without healthy attachments, especially early in life, the probability of acting inappropriately increases.

Commitment, the second element of Hirschi's social bond, is the investment a person has in conventional society. This has been explained as one's "stake in conformity," or

---

[64]See Vold et al., *Theoretical Criminology*, 4th ed., 205–7.

[65]Travis Hirschi, *Causes of Delinquency* (Berkeley: University of California Press, 1969).

[66]Ibid., 31, in which Hirschi cites Durkheim.

[67]Ibid., 18.

Figure 7.6   Hirschi's Social Bonding Theory

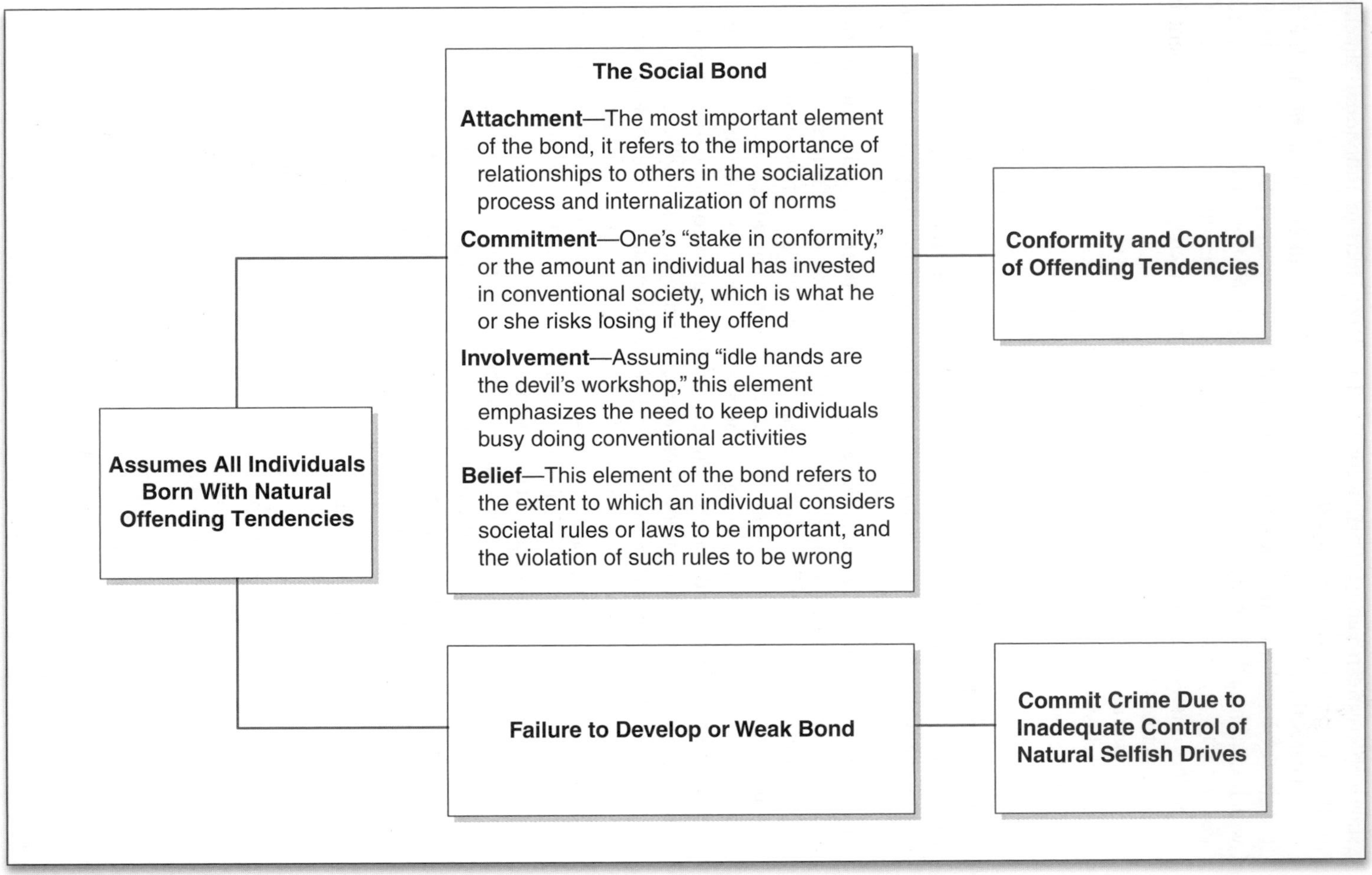

Assumes All Individuals Born With Natural Offending Tendencies

The Social Bond

Attachment—The most important element of the bond, it refers to the importance of relationships to others in the socialization process and internalization of norms

Commitment—One's "stake in conformity," or the amount an individual has invested in conventional society, which is what he or she risks losing if they offend

Involvement—Assuming "idle hands are the devil's workshop," this element emphasizes the need to keep individuals busy doing conventional activities

Belief—This element of the bond refers to the extent to which an individual considers societal rules or laws to be important, and the violation of such rules to be wrong

Conformity and Control of Offending Tendencies

Failure to Develop or Weak Bond

Commit Crime Due to Inadequate Control of Natural Selfish Drives

what is at risk of being lost if one gets caught committing crime. If people feel they have much to lose by committing crime, they will probably not do so. In contrast, if someone has nothing to lose, what is to prevent that person from doing something he or she may be punished for? The answer is, of course, not much. And this, some theorists claim, is why it is difficult to control so-called "chronic offenders." Trying to instill a commitment to conventional society in such individuals is extremely difficult.

Another element of the social bond is *involvement*, which is the time spent in conventional activities. The assumption is that time spent in constructive activities will reduce time devoted to illegal behaviors. This element of the bond goes back to the old adage that "idle hands are the devil's workshop."[68] Hirschi claimed that taking an active role in all forms of conventional activities can inhibit delinquent and criminal activity.

The last element of the social bond is *beliefs*, which have generally been interpreted as moral beliefs concerning the laws and rules of society. This is one of the most examined, and consistently supported, aspects of the social bond. Basically, individuals who feel that a course of action is against their moral beliefs are much less likely to pursue it than individuals who don't see a breach of morality in such behavior. For example, we all probably know some people who see drunk driving as a very serious offense because of the injury and death it can cause. However, we also probably know individuals who don't see a problem with such behavior. The same can be said about speeding in a car, shoplifting from a store, or using marijuana; people differ in their beliefs about most forms of criminal activity.

Hirschi's theory has been tested by numerous researchers and has, for the most part, been supported.[69] However, one criticism is that the components of the social bond may predict criminality only if they are defined in a certain way. For example, with respect to the involvement element of the bond, studies have shown that not all conventional activities are equal when it comes to preventing delinquency. Only academic or religious activities seem to have consistent effects on inhibiting delinquency. In contrast, many studies show that teenagers who date or play sports actually have an increased risk of committing crime.[70]

Another major criticism of Hirschi's theory is that the effect of attachments on crime depends on to whom one is attached. Studies have clearly and consistently shown that attachment to delinquent peers is a strong predictor of criminal activity.

Finally, some evidence indicates that social bonding theory may better explain why individuals start offending than why they continue or escalate in their offending. One reason for this is that Hirschi's theory does not elaborate on what occurs after an individual commits criminal activity. This is likely the primary reason why some of the more complex integrated theories of crime, often attribute the initiation of delinquency to a breakdown in the social bond. However, they typically see other theories (such as differential reinforcement) as being better predictors of what happens after the initial stages of the criminal career.[71]

---

[68]Ibid., 22.

[69]For a review, see Akers, *Criminological Theories*, 105–110.

[70]Ibid.

[71]Delbert Elliott, David Huizinga, and Suzanne Ageton, *Explaining Delinquency and Drug Use* (Beverly Hills: Sage, 1985).

Despite the criticism it has received, Hirschi's social bonding theory is still one of the most accepted theories of criminal behavior.[72] It is a relatively convincing explanation for criminality because of the consistent support that it has found among samples of people taken from all over the world.[73]

Integrated Social Control Theories

Although we will review integrated theories in detail in Section IX, it is worthwhile to briefly discuss the two integrated models that most incorporate the control perspective into their framework. These two integrated models are control-balance theory and power-control theory. Both have received considerable attention in the criminological literature. Other integrated theories that incorporate control theory to a lesser extent include Braithwaite's shaming theory and Sampson and Laub's life-course theory. These will be covered in more detail in Sections IX and X.

Tittle's Control-Balance Theory. Presented by Charles Tittle in 1995, **control-balance theory** proposes that (1) the amount of control to which one is *subjected* and (2) the amount of control one can *exercise* determine the probability of deviance occurring. The "balance" between these two types of control, he argued, can even predict the *type* of behavior that is likely to be committed.[74]

Tittle argued that a person is least likely to offend when he or she has a balance of controlling and being controlled. Furthermore, the likelihood of offending increases when these become unbalanced. If individuals are more controlled (Tittle calls this "control deficit"), then the theory predicts they will commit predatory or defiant acts. In contrast, if an individual possesses an excessive level of control (Tittle calls this control surplus), then he or she will be more likely to commit acts of exploitation or decadence. Note that excessive control is not the same as excessive self-control. Tittle argues that people who are controlling, that is, who have excessive control over others, will be predisposed toward inappropriate activities.

Initial empirical tests of control-balance theory have reported mixed results, with both surpluses and deficits predicting the same types of deviance.[75] In addition, researchers have uncovered differing effects of the control-balance ratio on two types of deviance that are contingent on gender. This finding is consistent with the gender-specific support found for Reckless's containment theory described earlier in this section.[76]

---

[72]Walsh and Ellis, "Political Ideology."

[73]D. Wong, "Pathways to Delinquency in Hong Kong and Guangzhou, South China,"

International Journal of Adolescence and Youth 10 (2001): 91–115; A. Vazsonyi and M. Killias, "Immigration and Crime Among Youth in Switzerland," *Criminal Justice and Behavior* 28 (2001): 329–66; M. Eisner, "Crime, Problem Drinking, and Drug Use: Patterns of Problem Behavior in Cross-National Perspective," *Annals of the American Academy of Political and Social Science* 580 (2002): 201–25.

[74]Charles Tittle, *Control Balance: Toward a General Theory of Deviance* (Boulder, CO: Westview, 1995).

[75]Alex Piquero and Matthew Hickman, "An Empirical Test of Tittle's Control Balance Theory," *Criminology* 37 (1999): 319–42; Matthew Hickman and Alex Piquero, "Exploring the Relationships Between Gender, Control Balance, and Deviance," *Deviant Behavior* 22 (2001): 323–51.

[76]Matthew Hickman and Alex Piquero, "Exploring the Relationships Between Gender, Control Balance, and Deviance," *Deviant Behavior* 22 (2001): 323–51.

Hagan's Power-Control Theory. **Power-control theory** is an integrated theory that was proposed by John Hagan and his colleagues.[77] The primary focus of this theory is on the level of control and patriarchal attitudes, as well as structure in the household, which are influenced by parental positions in the workforce. Power-control theory assumes that in households where the mother and father have relatively similar levels of power at work (i.e., balanced households), mothers will be less likely to exert control on their daughters. These balanced households will be less likely to experience gender differences in the criminal offending of the children. However, households in which mothers and fathers have dissimilar levels of power in the workplace (i.e., unbalanced households) are more likely to suppress criminal activity in daughters. In addition, assertiveness and risky activity among the males in the house will be encouraged. This assertiveness and risky activity may be a precursor to crime.

Most empirical tests of power-control have provided moderate support for the theory, while more recent studies have further specified the validity of the theory in different contexts.[78] For example, one recent study reported that the influence of mothers, not fathers, on sons had the greatest impact on reducing the delinquency of young males.[79] Another researcher has found that differences in perceived threats of embarrassment and formal sanctions varied between more patriarchal and less patriarchal households.[80] Finally, studies have also started measuring the effect of patriarchal attitudes on crime and delinquency.[81] Power-control theory is a good example of a social control theory in that it is consistent with the idea that individuals must be socialized and that the gender differences in such socialization make a difference in how people will act throughout life.

A General Theory of Crime: Low Self-Control

In 1990, Travis Hirschi, along with his colleague Michael Gottfredson, proposed a general theory of low self-control, which is often referred to as the general theory of

---

[77]John Hagan, *Structural Criminology* (Newark, NJ: Rutgers University Press, 1989); John Hagan, A. Gillis, and J. Simpson, "The Class Structure of Gender and Delinquency: Toward a Power-Control Theory of Common Delinquent Behavior," *American Journal of Sociology* 90 (1985): 1151–78; John Hagan, A. Gillis, and J. Simpson, "Clarifying and Extending Power-Control Theory," *American Journal of Sociology* 95 (1990): 1024–37; John Hagan, J. Simpson, and A. Gillis, "Class in the Household: A Power-Control Theory of Gender and Delinquency," *American Journal of Sociology* 92 (1987): 788–816.

[78]John Hagan, J. Simpson, and A. Gillis, "Class in the Household: A Power-Control Theory of Gender and Delinquency," *American Journal of Sociology* 92 (1987): 788–816; B. McCarthy and John Hagan, "Gender, Delinquency, and the Great Depression: A Test of Power-Control Theory," *Canadian Review of Sociology and Anthropology* 24 (1987): 153–77; Merry Morash and Meda Chesney-Lind, "A Reformulation and Partial Test of the Power-Control Theory of Delinquency," *Justice Quarterly* 8 (1991): 347–77; S. Singer and M. Levine, "Power-Control Theory, Gender, and Delinquency: A Partial Replication with Additional Evidence on the Effects of Peers," *Criminology* 26 (1988): 627–47.

[79]B. McCarthy, John Hagan, and T. Woodward, "In the Company of Women: Structure and Agency in a Revised Power-Control Theory of Gender and Delinquency," *Criminology,* 37 (1999): 761–88.

[80]Brenda Sims Blackwell, "Perceived Sanction Threats, Gender, and Crime: A Test and Elaboration of Power-Control Theory," *Criminology* 38 (2000): 439–88.

[81]Brenda Sims Blackwell, "Perceived Sanction Threats, Gender, and Crime: A Test and Elaboration of Power-Control Theory," *Criminology* 38 (2000): 439–88; Kristin Bates and Chris Bader, "Family Structure, Power-Control Theory, and Deviance: Extending Power-Control Theory to Include Alternate Family Forms," *Western Criminology Review* 4 (2003).

crime.[82] This theory has led to a significant amount of debate and research in the field since its appearance, more than any other contemporary theory of crime. Like the previous control theories of crime, this theory assumes that individuals are born predisposed toward selfish, self-centered activities and that only effective child-rearing and socialization can create self-control. Without such adequate socialization (i.e., social controls) and reduction of criminal opportunities, individuals will follow their natural tendencies to become selfish predators. Furthermore, the general theory of crime assumes that self-control must be established by age 10. If it has not formed by that time, then, according to the theory, individuals will forever exhibit low self-control.

Although Gottfredson and Hirschi still attribute the formation of controls as coming from the socialization processes, the distinguishing characteristic of this theory is its emphasis on the individual's ability to control himself or herself. That is, the general theory of crime assumes that people can take a degree of control over their own decisions and, within certain limitations, "control" themselves.

The general theory of crime is accepted as one of the most valid theories of crime.[83] This is probably because it identifies only one primary factor that causes criminality—low self-control. But **low self-control theory** may actually implicate a series of personality traits and behavior, including risk-taking, impulsiveness, self-centeredness, short-term orientation, and quick temper. For example, recent research has supported the idea that inadequate child-rearing practices tend to result in lower levels of self-control among children and that these low levels produce various risky behaviors, including criminal activity.[84] Such propensities toward low self-control can manifest in varying forms across an individual's life. For example, teenagers with low self-control will likely hit or steal from peers, and as they grow older, will be more likely to gamble or cheat on taxes.

Psychological Aspects of Low Self-Control. Criminologists have recently claimed that low self-control may be due to the emotional disposition of individuals. For example, one study showed that the effects of low self-control on intentions to commit drunk driving and shoplifting were tied to individuals' perceptions of pleasure and shame. More specifically, the findings of this study showed that individuals who had low self-control had significantly lower levels of anticipated shame but significantly higher levels of perceived pleasure in committing both drunk driving and shoplifting.[85] These results suggest that individuals who lack self-control will be oriented toward gaining pleasure

---

[82]Michael Gottfredson and Travis Hirschi, *A General Theory of Crime* (Palo Alto, CA: Stanford University Press, 1990).

[83]For an excellent review of studies regarding low self-control theory, see Travis Pratt and Frank Cullen, "The Empirical Status of Gottfredson and Hirschi's General Theory of Crime: A Meta-Analysis," *Criminology* 38 (2000): 931–64. For critiques of this theory, see Ronald Akers, "Self-Control as a General Theory of Crime," *Journal of Quantitative Criminology* 7 (1991): 201–11. For a study that demonstrates the high popularity of the theory, see Walsh and Ellis, "Political Ideology."

[84]Carter Hay, "Parenting, Self-Control, and Delinquency: A Test of Self-Control Theory," *Criminology* 39 (2001): 707–36; K. Hayslett-McCall and T. Bernard, "Attachment, Masculinity, and Self-Control: A Theory of Male Crime Rates," *Theoretical Criminology* 6 (2002): 5–33.

[85]Alex Piquero and Stephen Tibbetts, "Specifying the Direct and Indirect Effects of Low Self-Control and Situational Factors in Offenders' Decision Making: Toward a More Complete Model of Rational Offending," *Justice Quarterly* 13 (1996): 481–510.

and taking advantage of resources and toward avoiding negative emotional feelings (e.g., shame) that are primarily induced through socialization.

Physiological Aspects of Low Self-Control. Low self-control can also be tied to physiological factors. Interestingly, research has shown that chronic offenders show greater arousal toward danger and risk taking than the possibility of punishment.[86] This arousal has been measured by monitoring brain activity in response to certain stimuli. The research suggests that individuals are *encouraged* to commit risky behavior due to physiological mechanisms that reward their risk-taking activities by releasing "pleasure" chemicals in their brain.[87]

In a similar vein, recent studies show that chronic gamblers tend to get a physiological "high" (such as a sudden, intense release of brain chemicals similar to that following a small dose of cocaine) from the activity of betting, particularly when they are gambling with their own money and risking a personal loss.[88] Undoubtedly, a minority of individuals thrive off of risk-taking behaviors significantly more than others. This suggests that physiological as well as psychological differences may explain why certain individuals favor risky behaviors.

Researchers have also found that criminal offenders generally perceive a significantly lower level of internal sanctions (e.g., shame, guilt, embarrassment) than do non-offenders.[89] So, in summary, a select group of individuals appear to derive physiological and psychological pleasure from engaging in risky behaviors, while simultaneously being less likely to be inhibited by internal emotional sanctions. Such a combination, Gottfredson and Hirschi claimed, is very dangerous and helps explain why some impulsive individuals often end up in prison.

Finally, the psychological and physiological aspects of low self-control may help explain the gender differences observed between males and females. Specifically, studies show that females are significantly more likely to experience internal emotional sanctioning for offenses they have committed than are males.[90] In other words, there appears to be something innately different about males and females that helps explain the differing levels of self-control each possesses.

Summary of Control Theories

Control perspectives are among the oldest and most respected explanations of criminal activity. The fundamental assumption that humans have an inborn, selfish disposition that must be controlled through socialization distinguishes control theories from

---

[86]Adrian Raine, *The Psychopathology of Crime* (San Diego, CA: Academic Press, 1993).

[87]Anthony Walsh, *Biosocial Criminology: Introduction and Integration* (Cincinnati, OH: Anderson, 2002).

[88]Christopher D. Fiorillo, Phillippe N. Tobler, and Wolfram Schultz, "Discrete Coding of Reward Probability and Uncertainty by Dopamine Neurons," *Science* 299 (2003): 1898–1902.

[89]Harold Grasmick and Robert Bursik, "Conscience, Significant Others, and Rational Choice: Extending the Deterrence Model," *Law and Society Review* 24 (1990): 837–861; Stephen Tibbetts, "Shame and Rational Choice in Offending Decisions," *Criminal Justice and Behavior* 24 (1997): 234–55.

[90]Stephen Tibbetts and Denise Herz, "Gender Differences in Factors of Social Control and Rational Choice," *Deviant Behavior* 17 (1996): 183–208.

other theories of crime. The control perspective's longevity as one of the most popular criminological theories demonstrates its legitimacy as an explanation of behavior. This is likely due to the dedication and efforts of criminologists who are constantly developing new and improved versions of control theory, many of which we have discussed here.

Policy Implications

Numerous policy implications can be taken from the various types of social learning and control theories presented here. We will concentrate on those that are likely to be most effective and pragmatic in helping to reduce criminal behavior.

A number of policy implications can be drawn from the various learning models. Perhaps their most important suggestion is to supply many opportunities for positive reinforcements, or rewards, for good behavior. Such reinforcements have been found to be far more effective than punishments, especially among criminal offenders.[91] Furthermore, studies show that the most effective rehabilitation programs for offenders should be based on a cognitive-behavioral approach, which teaches individuals to "think before they act."[92] Furthermore, evaluation studies have shown that simply grouping offenders together for counseling or peer-therapy sessions is not an effective strategy; rather, it appears that such programs often show no effect, or actually increase offending among participants, perhaps because they tend to learn more antisocial attitudes from such sessions.[93] Ultimately, the offender programs that emphasize positive reinforcements and are based on a cognitive-behavioral approach are those that show the greatest success.

Regarding the policy implications of control theories, we will focus on the early social bonding that must take place and the need for more parental supervision to help an individual develop/learn self-control and create healthy, strong bonds to conventional society. Most control theories assume that individuals are predisposed to criminal behavior, so the primary focus of programs should be to reduce this propensity toward such behavior. According to most control perspectives, the most important factor in preventing/controlling this predisposition involves early parenting and building attachments or ties to pro-social aspects of the individuals' environment.

Thus, perhaps the most important policy recommendation is to increase the ties between early caregivers/parents and their children. A variety of programs try to increase the relationship and bonding that takes place between infants/young children and their parents, as well as to monitor the supervision that takes place in this dynamic. Such programs have consistently been shown to be effective in preventing/reducing criminality in high-risk children, especially when such programs involve home visitations by health care (e.g., nurses) and social care experts (e.g., social workers).[94] By visiting the homes of high-risk

---

[91]Van Voorhis et al., *Correctional Counseling.*

[92]Ibid.

[93]Ibid. Also see reviews of such programs in Richard J. Lundman, *Prevention and Control of Juvenile Delinquency* (New York: Oxford University Press, 1993); Akers and Sellers, *Criminological Theories,* 101–8.

[94]Joycen Carmouche and Joretta Jones, "Her Children, Their Future," *Federal Prisons Journal* 1(1989): 23, 26–27.

children, workers can provide more direct, personal attention in helping aid and counsel parents about how best to nurture, monitor, and discipline their young children.[95] These types of programs may lead to more control over individuals' behavior, while building stronger bonds to society and developing self-control among these high-risk individuals.

▧ Conclusion

In this section's introduction, we have discussed a wide range of theories that may appear to be quite different. However, all of the criminological theories in this section share an emphasis on social processes as the primary reason for why individuals commit crime. This is true regarding the learning theories, which propose that people are taught to commit crime, as well as the control theories, which claim people offend naturally and rather must be taught *not* to commit crime. Despite their seemingly opposite assumptions of human behavior, the fact is that learning and control theories both identify socialization, or the lack thereof, as the key cause of criminal behavior.

We also examined some of the key policy recommendations that have been suggested by both of these theoretical perspectives. Specifically, we noted that programs that simply group offenders together only seem to reinforce their tendency to offend, whereas programs that take a cognitive-behavioral approach and use many reward opportunities appear to have some effect in reducing recidivism. Also, we concluded that programs that involve home-visitations by experts (e.g., nurses, counselors) tend to aid in developing more effective parenting and building of social bonds among young individuals, which helps build strong attachments to society and develop self-control.

▧ Section Summary

- First, we discussed what distinguishes learning theories of crime from other perspectives.
- We then discussed Sutherland's differential association theory and how this framework was improved by Akers's differential reinforcement theory.
- We examined in depth the psychological learning model of classical conditioning, as well as its limitations.
- We then explored two other learning models that were the basis of differential reinforcement theory, namely operant conditioning and learning according to modeling/imitation.
- We reviewed the theory of neutralization, including the five original techniques of neutralization presented by Sykes and Matza.
- We also reviewed several early forms of social control theory, such as Hobbes, Freud, and Durkheim.

---

[95]For a review, see Stephen G. Tibbetts, "Perinatal and Developmental Determinants of Early Onset of Offending: A Two-Peak Model Approach," in *The Development of Persistent Criminality*, ed. Johanna Savage (New York: Oxford University Press, 2009), 179–201.

◆ Then we examined the early social control theories of crime, presented by Reiss, Toby, and Nye, and Reckless's containment theory.

◆ We then examined more modern social control theories, such as Hirschi's social bonding theory.

◆ Integrated social control theories were briefly examined, including Tittle's control-balance theory and Hagan's power-control theory.

◆ Finally, we reviewed low self-control theory, from both psychological and physiological perspectives.

KEY TERMS

Classical conditioning

Containment theory

Control-balance theory

Control theories

Differential association theory

Differential identification theory

Differential reinforcement theory

Drift theory

Ego

Id

Imitation and modeling

Learning theories

Low self-control theory

Modeling/imitation

Negative punishment

Negative reinforcement

Neutralization theory

Operant conditioning

Positive punishment

Positive reinforcement

Power-control theory

Social bonding theory

Soft determinism

Stake in conformity

Subterranean values

Superego

Tabula rasa

DISCUSSION QUESTIONS

1. What distinguishes learning theories from other criminological theories?

2. What distinguishes differential association from differential reinforcement theory?

3. What did differential identification add to learning theories?

4. Which technique of neutralization do you use/relate to the most? Why?

5. Which technique of neutralization do you find least valid? Why?

6. Which element of Hirschi's social bond do you find you have highest levels of?

7. Which element of Hirschi's social bond do you find you have lowest levels of?

8. Can you identify someone you know who fits the profile of a person with low self-control?

9. Which aspects of the low self-control personality do you think you fit?

10. Regarding Matza's theory of drift, do you think this relates to when you or your friends committed crime in life? Studies show that most people commit crimes when they are in their teens or 20s, or at least know people who do (e.g., drinking under age 21, speeding).

WEB RESOURCES

Differential Association Theory

http://www.criminology.fsu.edu/crimtheory/sutherland.html

Differential Reinforcement Theory

http://www.as.wvu.edu/~sbb/comm221/chapters/rf.htm

Social/Self-Control Theory

http://www.apsu.edu/oconnort/crim/crimtheory13.htm

http://www.criminology.fsu.edu/crimtheory/hirschi.htm

Techniques of Neutralization

http://www.hewett.norfolk.sch.uk/CURRIC/soc/crime/sykes_ma.htm

READING

This selection may be one of the most important entries in the book, primarily because it is written by Edwin H. Sutherland, who is generally regarded as the most important criminologist of the 20th century. One of the key reasons that Sutherland is held in such high regard is because of his proposed theoretical model of criminal behavior, which is known as differential association theory. This theory contains nine propositions, which are the focus of this selection. Readers should keep in mind that when these nine propositions were presented in the early to mid-1900s, the primary emphasis of criminological theory and research was on how offenders are physically (e.g., body-type theories) or psychologically (e.g., IQ) different from nonoffenders. Although Sutherland's emphasis on the social learning of criminal behavior is often assumed by modern criminologists, at that time it was seen as a major break from widely accepted theoretical frameworks.

Specifically, Sutherland claims that offenders are no different, physically or psychologically, from nonoffenders. Rather, he concludes that individuals engage in offending because they are exposed to significant others (e.g., family, friends, etc.) who teach them the norms and techniques beneficial for committing crime. It is also important to keep in mind that Sutherland's theory is just as deterministic as previous theories (e.g., Lombroso's theory, Goddard's feeblemindedness theory, Sheldon's body-type theory) in the sense that Sutherland claimed that people do not have free choice or free will in determining their actions; rather, it all comes down to whom they associate with the most, and their attitudes regarding the violation of law.

SOURCE: Edwin H. Sutherland and Donald Cressey, *Principles of Criminology*, 5th ed., ed. E. H. Sutherland and D. Cressey (Chicago: J. B. Lippincott, 1955), 77–80. Copyright © 1955 by J.B. Lippincott Company. Reprinted by permission of the Estate of Donald Cressey, via The Copyright Clearance Center.

It is also important to note that Sutherland based his theory on classical conditioning, which was the dominant psychological model of learning when Sutherland developed differential association theory. Classical conditioning is based on Pavlov's concept of learning through association between stimuli and responses. Good examples can be easily found in everyday life, such as when people hear a certain song on the radio or smell a certain scent that reminds them of a particular event from their past. Although there is no doubt that classical conditioning, or learning via association, is certainly a valid way that individuals learn, we will see later in this section that Sutherland's reliance on only this one model of learning somewhat limited his theory because it does not take other learning models into account.

A Sociological Theory of Criminal Behavior

Edwin H. Sutherland

⬛ Explanation of Criminal Behavior

The following statement refers to the process by which a particular person comes to engage in criminal behavior.

1. *Criminal behavior is learned.* Negatively, this means that criminal[ity] is not inherited, as such; also, the person who is not already trained in crime does not invent criminal behavior, just as a person does not make mechanical inventions unless he has had training in mechanics.

2. *Criminal behavior is learned in interaction with other persons in a process of communication.* This communication is verbal in many respects but includes also "the communication of gestures."

3. *The principal part of the learning of criminal behavior occurs within intimate personal groups.* Negatively, this means that the impersonal agencies of communication, such as movies and newspapers, play a relatively unimportant part in the genesis of criminal behavior.

4. When criminal behavior is learned, the learning includes (a) techniques of committing the crime, which are sometimes very complicated, sometimes very simple; (b) the specific direction of motives, drives, rationalizations, and attitudes.

5. *The specific direction of motives and drives is learned from definitions of the legal codes as favorable or unfavorable.* In some societies an individual is surrounded by persons who invariably define the legal codes as rules to be observed, while in others he is surrounded by persons whose definitions are favorable to the violation of the legal codes. In our American society these definitions are almost always mixed, with the consequences that we have culture conflict in relation to the legal codes.

6. A person becomes delinquent because of an excess of definitions favorable to violation of law over definitions unfavorable to violation of law. This is the principle of differential association. It refers to both criminal and anti-criminal associations and has to do with counteracting forces. When persons become criminal, they do

so because of contacts with criminal patterns and also because of isolation from anti-criminal patterns. Any person inevitably assimilates the surrounding culture unless other patterns are in conflict; a Southerner does not pronounce "r" because other Southerners do not pronounce "r." Negatively, this proposition of differential association means that associations which are neutral so far as crime is concerned have little or no effect on the genesis of criminal behavior. Much of the experience of a person is neutral in this sense, e.g., learning to brush one's teeth. This behavior has no negative or positive effect on criminal behavior except as it may be related to associations which are concerned with the legal codes. This neutral behavior is important especially as an occupier of the time of a child so that he is not in contact with criminal behavior during the time he is so engaged in the neutral behavior.

7. *Differential associations may vary in frequency, duration, priority, and intensity.* This means that associations with criminal behavior and also associations with anti-criminal behavior vary in those respects. "Frequency" and "duration" as modalities of associations are obvious and need no explanation. "Priority" is assumed to be important in the sense that lawful behavior developed in early childhood may persist throughout life, and also that delinquent behavior developed in early childhood may persist throughout life. This tendency, however, has not been adequately demonstrated, and priority seems to be important principally through its selective influence. "Intensity" is not precisely defined but it has to do with such things as the prestige of the source of a criminal or anti-criminal pattern and with emotional reactions related to the associations. In a precise

description of the criminal behavior of a person these modalities would be stated in quantitative form and a mathematical ratio be reached. A formula in this sense has not been developed, and the development of such a formula would be extremely difficult.

8. The process of learning criminal behavior by association with criminal and anti-criminal patterns involves all of the mechanisms that are involved in any other learning. Negatively, this means that the learning of criminal behavior is not restricted to the process of imitation. A person who is seduced, for instance, learns criminal behavior by association, but this process would not ordinarily be described as imitation.

9. While criminal behavior is an expression of general needs and values, it is not explained by those general needs and values since non-criminal behavior is an expression of the same needs and values. Thieves generally steal in order to secure money, but likewise honest laborers work in order to secure money. The attempts by many scholars to explain criminal behavior by general drives and values, such as, the happiness principle, striving for social status, the money motive, or frustration, have been and must continue to be futile since they explain lawful behavior as completely as they explain criminal behavior. They are similar to respiration, which is necessary for any behavior but which does not differentiate criminal from non-criminal behavior.

It is not necessary, at this level of explanation, to explain why a person has the associations which he has; this certainly involves a complex of many things. In an area where the delinquency rate is high a boy who is sociable, gregarious, active, and athletic is very likely to come in contact with the other boys in the neighborhood, learn delinquent behavior from them, and become a gangster; in the same neighborhood the psychopathic boy who is isolated, introvert, and inert may remain at home, not become acquainted with the other boys in the neighborhood, and not become delinquent. In another situation, the sociable, athletic, aggressive boy may become a member of a scout troop and not become involved in delinquent behavior. The person's associations are determined in a general context of social organization. A child is ordinarily reared in a family; the place of residence of the family is determined largely by family income; and the delinquency rate is, in many respects, related to the rental value of the houses. Many other factors enter into this social organization, including many of the small personal group relationships.

The preceding explanation of criminal behavior is stated from the point of view of the person who engages in criminal behavior. As indicated earlier, it is possible, also, to state sociological theories of criminal behavior from the point of view of the community, nation, or other group. The problem, when thus stated, is generally concerned with crime rates and involves a comparison of the crime rates of various groups or the crime rates of a particular group at different times. The explanation of a crime rate must be consistent with the explanation of the criminal behavior of the person, since the crime rate is a summary statement of the number of persons in the group who commit crimes and the frequency with which they commit crimes. One of the best explanations of crime rates from this point of view is that a high crime rate is due to social disorganization. The term "social disorganization" is not entirely satisfactory and it seems preferable to substitute for it the term "differential social organization." The postulate on which this theory is based, regardless of the name, is that crime is rooted in the social organization and is an expression of that social organization. A group may be organized for criminal behavior or organized against criminal behavior. Most communities are organized both for criminal and anticriminal behavior and in that sense the crime rate is an expression of the differential group organization. Differential group organization as an explanation of variations in crime rates is consistent with the differential association theory of the processes by which persons become criminals. . . .

REVIEW QUESTIONS

1. Using Sutherland's first three propositions, how does he claim criminal behavior is learned? What does he have to say about the effect of movies and media on the learning of criminal behavior?

2. What two types of learning does Sutherland claim takes place in his fourth proposition? Provide an example of each of these two types.

3. Sutherland's sixth proposition is often considered the best summary of his theory. Explain what this proposition means as if you were trying to tell a person who knows nothing about criminology. Also, do you agree with it or not?

4. What four types of associations are identified by Sutherland in his seventh proposition? Explain each, and provide your opinion on which of the four types you think are most important in determining criminal activity.

READING

In this selection, Ron Akers presents a theoretical model known as differential reinforcement theory. Although some, including Akers himself, often refer to this theory as social learning theory, this label is a bit confusing because there are many social learning theories. We prefer *differential reinforcement theory* as the name of this framework because when you say this, criminologists know exactly which theoretical model you are referring to and also because the term *reinforcement* specifies the key concept that distinguishes this model from all other social learning theories.

Differential reinforcement theory builds on the framework provided by Sutherland's differential association theory but adds two important models of social learning that were not included in Sutherland's theory. Specifically, Sutherland based his differential association model on only one type of social learning: classical conditioning. While including classical conditioning in its framework, Akers's theory of differential reinforcement adds two additional learning models, operant conditioning and modeling and imitation. Operant conditioning is based on B. F. Skinner's work in psychology, which emphasizes whether punishments or reinforcements (i.e., rewards) occur after a given activity. The other learning model is imitation/modeling, which is largely based on Bandura's psychological model of learning through observation, such as children watching what adults do and then imitating their behavior, in other words "monkey see, monkey do." Differential reinforcement theory is an improvement over Sutherland's differential association theory because it takes all three learning processes—classical conditioning, operant conditioning, and imitation/modeling—into account in the theoretical model. While reading this selection, readers are encouraged to consider how the primary concepts and propositions of the theory occur in their life, not necessarily in terms of criminal behavior, but the learning of any behavior. Another important aspect of Sutherland's theory that Akers's differential reinforcement theory adopts is that the processes involved in the learning of criminal behavior are the same as the processes for learning all types of conventional behavior.

A Social Learning Theory of Crime

Ronald L. Akers

⬛ Concise Statement of the Theory

The basic assumption in social learning theory is that the same learning process, operating in a context of social structure, interaction, and situation, produces both conforming and deviant behavior. The difference lies in the direction of the process in which these mechanisms operate. In both, it is seldom an either-or, all-or-nothing process; what is involved, rather, is the balance of influences on behavior. That balance usually exhibits some stability over time, but it can become unstable and change with time or circumstances. Conforming and deviant behavior is learned by all of the mechanisms in this process, but the theory proposes that the principal mechanisms are in that part of the process in which differential reinforcement (instrumental learning through rewards and punishers) and imitation (observational learning) produce both overt behavior and cognitive definitions that function as discriminative (cue) stimuli for the behavior. Always implied, and sometimes made explicit when these concepts are called upon to account for deviant/conforming behavior, is the understanding that the behavioral processes in operant and classical conditioning are in operation (see below). However, social learning theory focuses on four major concepts—differential association, differential reinforcement, imitation, and definitions. The central proposition of the social learning theory of criminal and deviant behavior can be stated as a long sentence proposing that criminal and deviant behavior is more likely when, on balance, the combined effects of these four main sets of variables instigate and strengthen nonconforming over conforming acts:

> The probability that persons will engage in criminal and deviant behavior is increased and the probability of their conforming to the norm is decreased when they differentially associate with others who commit criminal behavior and espouse definitions favorable to it, are relatively more exposed in-person or symbolically to salient criminal/deviant models, define it as desirable or justified in a situation discriminative for the behavior, and have received in the past and anticipate in the current or future situation relatively greater reward than punishment for the behavior.

The probability of conforming behavior is increased and the probability of deviant behavior is decreased when the balance of these variables moves in the reverse direction.

Each of the four main components of this statement can be presented as a separate testable hypothesis. The individual is more likely to commit violations when:

1. He or she differentially associates with other[s] who commit, model, and support violations of social and legal norms.

2. The violative behavior is differentially reinforced over behavior in conformity to the norm.

3. He or she is more exposed to and observes more deviant than conforming models.

4. His or her own learned definitions are favorable toward committing the deviant acts.

⊠ General Principles of Social Learning Theory

Since it is a general explanation of crime and deviance of all kinds, social learning is not simply a theory about how novel criminal behavior is learned or a theory only of the positive causes of that behavior. It embraces variables that operate to both motivate and control delinquent and criminal behavior, to both promote and undermine conformity. It answers the questions of why people do and do not violate norms. The probability of criminal or conforming behavior occurring is a function of the variables operating in the underlying social learning process. The main concepts/variables and their respective empirical indicators have been identified and measured, but they can be viewed as indicators of a general latent construct, for which additional indicators can be devised (Akers & La Greca, 1991; Akers & Lee, 1996).

Social learning accounts for the individual becoming prone to deviant or criminal behavior and for stability or change in that propensity. Therefore, the theory is capable of accounting for the development of stable individual differences, as well as changes in the individual's behavioral patterns or tendencies to commit deviant and criminal acts, over time and in different situations. . . . The social learning process operates in each individual's learning history and in the immediate situation in which the opportunity for a crime occurs.

Deviant and criminal behavior is learned and modified (acquired, performed, repeated, maintained, and changed) through all of the same cognitive and behavioral mechanisms as conforming behavior. They differ in the direction, content, and outcome of the behavior learned. Therefore, it is inaccurate to state, for instance, that peer influence does not explain adolescent deviant behavior since conforming behavior is also peer influenced in adolescence. The theory expects peer influences to be implicated in both; it is the content and direction of the influence that is the key.

The primary learning mechanisms are differential reinforcement (instrumental conditioning), in which behavior is a function of the frequency, amount, and probability of experienced and perceived contingent rewards and punishments, and imitation, in which the behavior of others and its consequences are observed and modeled. The process of stimulus discrimination/generalization is another important mechanism; here, overt and covert stimuli, verbal and cognitive, act as cues or signals for behavior to occur. As I point out below, there are other behavioral mechanisms in the learning process, but these are not as important and are usually left implied rather than explicated in the theory.

The content of the learning achieved by these mechanisms includes the simple and complex behavioral sequences and the definitions (beliefs, attitudes, justifications, orientations) that in turn become discriminative for engaging in deviant and criminal behavior. The probability that conforming or norm-violative behavior is learned and performed, and the frequency with which it is committed, are a function of the past, present, and anticipated differential reinforcement for the behavior and the deviant or nondeviant direction of the learned definitions and other discriminative stimuli present in a given situation.

These learning mechanisms operate in a process of differential association—direct and indirect, verbal and nonverbal communication, interaction, and identification with others. The relative frequency, intensity, duration, and priority of associations affect the relative amount, frequency, and probability of reinforcement of conforming or deviant behavior and exposure of individuals to deviant or conforming norms and behavioral models. To the extent that the individual can control with whom she or he associates, the frequency, intensity, and duration of those associations are themselves affected by how rewarding or aversive they are. The principal learning is through differential association with those persons and groups (primary, secondary, reference, and symbolic) that comprise or control the individual's major sources of reinforcement, most salient behavioral models, and most effective definitions and other discriminative stimuli for committing and repeating behavior. The reinforcement and discriminative stimuli are mainly social (such as socially valued rewards and punishers contingent on the behavior), but they are also nonsocial (such as unconditioned physiological reactions to environmental stimuli and physical effects of ingested substances and the physical environment).

Sequence and Reciprocal Effects in the Social Learning Process

Behavioral feedback effects are built into the concept of differential reinforcement—actual or perceived changes in the environment produced by the behavior feed back on that behavior to affect its repetition or extinction, and both prior and anticipated rewards and punishments influence present behavior. Reciprocal effects between the individual's behavior and definitions or differential association are also reflected in the social learning process. This process is one in which the probability of both the initiation and the repetition of a deviant or criminal act (or the initiation and repetition of conforming acts) is a function of the learning history of the individual and the set of reinforcement contingencies and discriminative stimuli in a given situation. The typical process of initiation, continuation, progression, and desistance is hypothesized to be as follows:

1. The balance of past and current associations, definitions, and imitation of deviant models, and the anticipated balance of reinforcement in particular situations, produces or inhibits the initial delinquent or deviant acts.

2. The effects of these variables continue in the repetition of acts, although imitation becomes less important than it was in the first commission of the act.

3. After initiation, the actual social and nonsocial reinforcers and punishers affect the probability that the acts will be or will not be repeated and at what level of frequency.

4. Not only the overt behavior, but also the definitions favorable or unfavorable to it, are affected by the positive and negative consequences of the initial acts. To the extent that they are more rewarded than alternative behavior, the favorable definitions will be strengthened and the unfavorable definitions will be weakened, and it becomes more likely that the deviant behavior will be repeated under similar circumstances.

5. Progression into more frequent or sustained patterns, rather than cessation or reduction, of criminal and deviant behavior is promoted to the extent that reinforcement, exposure to deviant models, and norm-violating definitions are not offset by negative formal and informal sanctions and norm abiding definitions.

The theory does not hypothesize that definitions favorable to law violation always precede and are unaffected by the commission of criminal acts. Although the probability of a criminal act increases in the presence of favorable definitions, acts in violation of the law do occur (through imitation and reinforcement) in the absence of any thought given to whether the acts are right or wrong. Furthermore, the individual may apply neutralizing definitions retroactively to excuse or justify an act without having contemplated them beforehand. To the extent that such excuses become associated with successfully mitigating others' negative sanctions or one's self-punishment, however, they become cues for the repetition of deviant acts. Such definitions, therefore, precede committing the same acts again or committing similar acts in the future.

Differential association with conforming and nonconforming others typically precedes the individual's committing crimes and delinquent acts. This sequence of events is sometimes disputed in the literature because it is mistakenly believed to apply only to differential peer association in general or to participation in delinquent gangs in particular without reference to family and other group associations. It is true that the theory recognizes peer associations as very important in adolescent deviance and that differential association is most often measured in research by peer associations. But the theory also hypothesizes that the family is a very important primary group in the differential association process, and it plainly stipulates that other primary and secondary groups besides peers are involved (see Sutherland, 1947, pp. 164–65). Accordingly, it is a mistake to interpret differential association as referring only to peer associations. The theoretical stipulation that differential association is causally prior to the commission of delinquent and criminal acts is not confined to the balance of peer associations; rather, it is the balance (as determined by the modalities) of family, peer, and other associations. According to the priority principle, association, reinforcement, modeling, and exposure to conforming and deviant definitions occurring within the family during childhood, and such antisocial conduct as aggressiveness, lying, and cheating learned in early childhood, occur prior to and have both direct and selective effects on later delinquent and criminal behavior and associations. . . .

The socializing behavior of parents, guardians, or caretakers is certainly reciprocally influenced by the deviant and unacceptable behavior of the child. However, it can never be true that the onset of delinquency precedes and initiates interaction in a particular family (except in the unlikely case of the late-stage adoption of a child who is already delinquent or who is drawn to and chosen by deviant parents). Thus, interaction in the family or family surrogate always precedes delinquency.

But this is not true for adolescent peer associations. One may choose to associate with peers based on similarity in deviant behavior that already exists. Some major portion of this behavioral similarity results from previous association with other delinquent peers or from anticipatory socialization undertaken to make one's behavior match more closely that of the deviant associates to whom one is attracted. For some adolescents, gravitation toward delinquent peers occurs after and as a result of the individual's involvement in delinquent behavior. However, peer associations are most often formed initially around interests, friendships, and such circumstances as neighborhood proximity, family similarities, values, beliefs, age, school attended, grade in school,

and mutually attractive behavioral patterns that have little to do directly with co-involvement or similarity in specifically law-violating or serious deviant behavior. Many of these factors in peer association are not under the adolescents' control, and some are simply happenstance. The theory does not, contrary to the Gluecks' distorted characterization, propose that "accidental differential association of non-delinquents with delinquents is the basic cause of crime" (Glueck & Glueck, 1950, p. 164). Interaction and socialization in the family precedes and affects choices of both conforming and deviant peer associations.

Those peer associations will affect the nature of models, definitions, and rewards/punishers to which the person is exposed. After the associations have been established, their reinforcing or punishing consequences as well as direct and vicarious consequences of the deviant behavior will affect both the continuation of old and the seeking of new associations (those over which one has any choice). One may choose further interaction with others based on whether they too are involved in deviant or criminal behavior; in such cases, the outcomes of that interaction are more rewarding than aversive and it is anticipated that the associates will more likely approve or be permissive toward one's own deviant behavior. Further interaction with delinquent peers, over which the individual has no choice, may also result from being apprehended and confined by the juvenile or criminal-justice system.

These reciprocal effects would predict that one's own deviant or conforming behavioral patterns can have effects on choice of friends; these are weaker in the earlier years, but should become stronger as one moves through adolescence and gains more control over friendship choices. The typical sequence outlined above would predict that deviant associations precede the onset of delinquent behavior more frequently than the sequence of events in which the delinquent associations begin only after the peers involved have already separately and individually established similar patterns of delinquent behavior. Further, these behavioral tendencies that develop prior to peer association will themselves be the result of previous associations, models, and reinforcement, primarily in the family. Regardless of the sequence in which onset occurs, and whatever the level of the individual's delinquent involvement, its frequency and seriousness will increase after the deviant associations have begun and decrease as the associations are reduced. That is, whatever the temporal ordering, differential association with deviant peers will have a causal effect on one's own delinquent behavior (just as his actions will have an effect on his peers).

Therefore, both "selection," or "flocking" (tendency for persons to choose interaction with others with behavioral similarities), and "socialization," or "feathering" (tendency for persons who interact to have mutual influence on one another's behavior), are part of the same overall social learning process and are explained by the same variables. A peer "socialization" process and a peer "selection" process in deviant behavior are not mutually exclusive, but are simply the social learning process operating at different times. Arguments that social learning posits only the latter, that any evidence of selective mechanisms in deviant interaction run counter to social learning theory (Strictland, 1982; Stafford & Ekland-Olson, 1982), or that social learning theory recognizes only a recursive, one-way causal effect of peers on delinquent behavior (Thornberry et al., 1994; Catalano et al., 1996) are wrong.

Behavioral and Cognitive Mechanisms in Social Learning

The first statement in Sutherland's theory was a simple declarative sentence maintaining that criminal behavior is learned, and the eighth statement declared that this involved all the mechanisms involved in any learning. What little Sutherland added in his (1947, p. 7)

commentary downplayed imitation as a possible learning mechanism in criminal behavior. He mentioned "seduction" of a person into criminal behavior as something that is not covered by the concept of imitation. He defined neither imitation nor seduction and offered no further discussion of mechanisms of learning in any of his papers or publications. Recall that filling this major lacuna in Sutherland's theory was the principal goal of the 1966b Burgess-Akers reformulation. To this end we combined Sutherland's first and eighth statements into one: "Criminal behavior is learned according to the principles of operant conditioning." The phrase "principles of operant conditioning" was meant as a shorthand reference to all of the behavioral mechanisms of learning in operant theory that had been empirically validated.

Burgess and I delineated, as much as space allowed, what these specific learning mechanisms were: (1) operant conditioning, differential reinforcement of voluntary behavior through positive and negative reinforcement and punishment; (2) respondent (involuntary reflexes), or "classical," conditioning; (3) unconditioned (primary) and conditioned (secondary) reinforcers and punishers; (4) shaping and response differentiation; (5) stimulus discrimination and generalization, the environmental and internal stimuli that provide cues or signals indicating differences and similarities across situations that help elicit, but do not directly reinforce, behavior; (6) types of reinforcement schedules, the rate and ratio in which rewards and punishers follow behavior; (7) stimulus-response constellations; and (8) stimulus satiation and deprivation. We also reported research showing the applicability of these mechanisms of learning to both conforming and deviant behavior.

Burgess and I used the term "operant conditioning" to emphasize that differential reinforcement (the balance of reward and punishment contingent upon behavioral responses) is the basic mechanism around which the others revolve and by which learning most relevant to conformity or violation of social and legal norms is produced. This was reflected in other statements in the theory in which the only learning mechanisms listed were differential reinforcement and stimulus discrimination.

We also subsumed imitation, or modeling, under these principles and argued that imitation "may be analyzed quite parsimoniously with the principles of modern behavior theory," namely, that it is simply a sub-class of behavioral shaping through operant conditioning (Burgess and Akers, 1966b, p. 138). For this reason we made no specific mention of imitation in any of the seven statements. Later, I became persuaded that the operant principle of gradual shaping of responses through "successive approximations" only incompletely and awkwardly incorporated the processes of observational learning and vicarious reinforcement that Bandura and Walters (1963) had identified. Therefore, without dismissing successive approximation as a way in which some imitative behavior could be shaped, I came to accept Bandura's conceptualization of imitation. That is, imitation is a separate learning mechanism characterized by modeling one's own actions on the observed behavior of others and on the consequences of that behavior (vicarious reinforcement) prior to performing the behavior and experiencing its consequences directly. Whether the observed acts will be performed and repeated depends less on the continuing presence of models and more on the actual or anticipated rewarding or aversive consequences of the behavior. I became satisfied that the principle of "observational learning" could account for the acquisition, and to some extent the performance, of behavior by a process that did not depend on operant conditioning or "instrumental learning." Therefore, in later discussions of the theory, while continuing to posit differential reinforcement as the core behavior-shaping mechanism, I included imitation as another primary mechanism in acquiring behavior.

Where appropriate, discriminative stimuli were also specifically invoked as affecting behavior, while I made only general reference to other learning mechanisms.

Note that the term "operant conditioning" in the opening sentence of the Burgess-Akers revision reflected our great reliance on the orthodox behaviorism that assumed the empirical irrelevance of cognitive variables. Social behaviorism, on the other hand, recognizes "cognitive" as well as "behavioral" mechanisms (see Bandura, 1969; 1977a; 1977b; 1986; 1989; Grusec, 1992; Staats, 1975). My social learning theory of criminal behavior retains a strong element of the symbolic interactionism found in Sutherland's theory (Akers, 1985, pp. 39–70). As a result, it is closer to cognitive learning theories, such as Albert Bandura's, than to the radical operant behaviorism of B. F. Skinner with which Burgess and I began. It is for this reason, and the reliance on such concepts as imitation, anticipated reinforcement, and self-reinforcement, that I have described social learning theory as "soft behaviorism" (Akers, 1985, p. 65).

The unmodified term "learning" implies to many that the theory only explains the acquisition of novel behavior by the individual, in contrast to behavior that is committed at a given time and place or the maintenance of behavior over time (Cornish and Clarke, 1986). It has also been interpreted to mean only "positive" learning of novel behavior, with no relevance for inhibition of behavior or of learning failures (Gottfredson and Hirschi, 1990). As I have made clear above, neither of these interpretations is accurate. The phrase that Burgess and I used, "effective and available reinforcers and the existing reinforcement contingencies," and the discussion of reinforcement occurring under given situations (Burgess & Akers, 1966b, pp. 141, 134) make it obvious that we were not proposing a theory only of past reinforcement in the acquisition of a behavioral repertoire with no regard for the reward/cost balance obtaining at a given time and place. There is nothing in the learning principles that restrict[s] them to prior socialization or past history of learning. Social learning encompasses both the acquisition and the performance of the behavior, both facilitation and inhibition of behavior, and both learning successes and learning failures. The learning mechanisms account not only for the initiation of behavior but also for repetition, maintenance and desistance of behavior. They rely not only on prior behavioral processes but also on those operating at a given time in a given situation. . . .

⬚ Definitions and Discriminative Stimuli

[In] *The Concept of Definitions*, Sutherland asserted that learning criminal behavior includes "techniques of committing the crime which are sometimes very complicated, sometimes very simple" and the "specific direction of motives, drives, rationalizations and attitudes" (1947, p. 6). I have retained both definitions and techniques in social learning theory, with clarified and modified conceptual meanings and with hypothesized relationships to criminal behavior. The qualification that "techniques" may be simple or complex shows plainly that Sutherland did not mean to include only crime-specific skills learned in order to break the law successfully. Techniques also clearly include ordinary, everyday abilities. This same notion is retained in social learning theory.

By definition, a person must be capable of performing the necessary sequence of actions before he or she can carry out either criminal or conforming behavior—inability to perform the behavior precludes committing the crime. Since many of the behavioral techniques for both conforming and criminal acts are the same, not only the simple but even some of the complex skills involved in carrying out crime are not novel to most or many of us. The required component parts of the complete skill are acquired in essentially conforming or neutral contexts to which we have been exposed—driving a car, shooting a gun, fighting with fists, signing checks, using a computer, and so on. In most white-collar crime, the same skills needed to carry out a job legitimately are put to illegitimate use. Other skills are specific to given deviant acts—safe cracking, counterfeiting, pocket picking, jimmying doors and picking locks, bringing off a con game, and so on. Without tutelage in these crime-specific techniques, most people would not be able to perform them, or at least would be initially very inept.

Sutherland took the concept of "definitions" in his theory from W. I. Thomas's "definition of the situation" (Thomas and Thomas, 1928) and generalized it to orienting attitudes toward different behavior. It is true that "Sutherland did not identify what constitutes a definition 'favorable to' or 'unfavorable to' the violation of law" (Cressey, 1960, p. 53). Nevertheless . . . there is little doubt that "rationalizations" and "attitudes" are subsumed under the general concept of definitions—normative attitudes or evaluative meanings attached to given behavior. Exposure to others' shared definitions is a key (but not the only) part of the process by which the individual acquires or internalizes his or her own definitions. They are orientations, rationalizations, definitions of the situation, and other attitudes that label the commission of an act as right or wrong, good or bad, desirable or undesirable, justified or unjustified.

In social learning theory, these definitions are both general and specific. General beliefs include religious, moral, and other conventional values and norms that are favorable to conforming behavior and unfavorable to committing any of a range of deviant or criminal acts. Specific definitions orient the person to particular acts or series of acts. Thus, there are people who believe that it is morally wrong to steal and that laws against theft should be obeyed, but at the same time see little wrong with smoking marijuana and rationalize that it is all right to violate laws against drug possession. The greater the extent to which one holds attitudes that disapprove of certain acts, the less likely one is to engage in them. Conventional beliefs are negative toward criminal behavior. The more strongly one has learned and personally believes in the ideals of honesty, integrity, civility, kindness, and other general standards of morality that condemn lying, cheating, stealing, and harming others, the less likely he or she is to commit acts that violate social and legal norms. Conversely, the more one's own attitudes approve of, or fail to condemn, a behavior, the greater the chances are that he or she will engage in it. For obvious reasons, the theory would predict that definitions in the form of general beliefs will have less effect than specific definitions on the commission of specific criminal acts.

Definitions that favor criminal or deviant behavior are basically positive or neutralizing. Positive definitions are beliefs or attitudes that make the behavior morally desirable or wholly permissible. They are most likely to be learned through positive reinforcement in a deviant group or subculture that carries values conflicting with those of conventional society. Some of these positive verbalizations may be part of a full-blown ideology of politically dissident, criminal, or deviant groups. Although such ideologies and groups can be identified, the theory does not rest only on this type of definition favorable to deviance; indeed, it proposes that such positive definitions occur less frequently than neutralizing ones.

Neutralizing definitions favor violating the law or other norms not because they take the acts to be positively desirable but because they justify or excuse them. Even those who commit deviant acts are aware that others condemn the behavior and may themselves define the behavior as bad. The neutralizing definitions view the act as something that is probably undesirable but, given the situation, is nonetheless justified, excusable, necessary, all right, or not really bad after all. The process of

acquiring neutralizing definitions is more likely to involve negative reinforcement; that is, they are verbalizations that accompany escape or avoidance of negative consequences like disapproval by one's self or by society.

While these definitions may become part of a deviant or criminal subculture, acquiring them does not require participation in such subcultures. They are learned from carriers of conventional culture, including many of those in social control and treatment agencies. The notions of techniques of neutralization and subterranean values (Sykes and Matza, 1957; Matza and Sykes, 1961; Matza, 1964) come from the observation that for nearly every social norm there is a norm of evasion. That is, there are recognized exceptions or ways of getting around the moral imperatives in the norms and the reproach expected for violating them. Thus, the general prohibition "Thou shalt not kill" is accompanied by such implicit or explicit exceptions as "unless in time of war," "unless the victim is the enemy," "unless in self-defense," "unless in the line of duty," "unless to protect others"! The moral injunctions against physical violence are suspended if the victim can be defined as the initial aggressor or is guilty of some transgression and therefore deserves to be attacked.

The concept of neutralizing definitions in social learning theory incorporates not only notions of verbalizations and rationalizations (Cressey, 1953) and techniques of neutralization (Sykes & Matza, 1957) but also conceptually similar if not equivalent notions of "accounts" (Lyman & Scott, 1970), "disclaimers" (Hewitt & Stokes, 1975), and "moral disengagement" (Bandura, 1976, 1990). Neutralizing attitudes include such beliefs as "Everybody has a racket"; "I can't help myself, I was born this way"; "It's not my fault"; "I am not responsible"; "I was drunk and didn't know what I was doing"; "I just blew my top"; "They can afford it"; "He deserved it." Some neutralizations (e.g., nonresponsibility) can be generalized to a wide range of disapproved and criminal behavior. These and other excuses and justifications for committing deviant acts and victimizing others are definitions favorable to criminal and deviant behavior.

Exposure to these rationalizations and excuses may be through after-the-fact justifications for one's own or others' norm violations that help to deflect or lessen punishment that would be expected to follow. The individual then learns the excuses either directly or through imitation and uses them to lessen self-reproach and social disapproval. Therefore, the definitions are themselves behavior that can be imitated and reinforced and then in turn serve as discriminative stimuli accompanying reinforcement of overt behavior. Deviant and criminal acts do occur without being accompanied by positive or neutralizing definitions, but the acts are more likely to occur and recur in situations the same as or similar to those in which the definitions have already been learned and applied. The extent to which one adheres to or rejects the definitions favorable to crime is itself affected by the rewarding or punishing consequences that follow the act.

☒ References

Akers, R. L. (1985). *Deviant behavior: A social learning approach.* Belmont, CA: Wadsworth.

Akers, R. L., & La Greca, A. J. (1991). Alcohol use among the elderly: Social learning, community context, and life events. In D. J. Pittman & H. R. White (Eds.), *Society, culture, and drinking patterns re-examined* (pp. 242–262). New Brunswick, NJ: Rutgers Center of Alcohol Studies.

Akers, R. L., & Lee, G. (1996). A longitudinal test of social learning theory: Adolescent smoking. *Journal of Drug Issues, 26,* 317–343.

Bandura, A. (1969). *Principles of behavior modification.* New York: Holt, Rinehart & Winston.

Bandura, A. (1976). *Analysis of delinquency and aggression.* New York: Lawrence Erlbaum.

Bandura, A. (1977a). Self-efficacy: Toward a unifying theory of behavioral change. *Psychological Review, 84,* 191–215.

Bandura, A. (1977b). *Social learning theory.* Englewood Cliffs, NJ: Prentice Hall.

Bandura, A. (1986). *Social foundations of thought and action: A social cognitive theory.* Englewood Cliffs, NJ: Prentice Hall.

Bandura, A. (1989). Human agency and social cognitive theory. *American Psychologist, 44,* 1175–1184.

Bandura, A. (1990). Selective activation and disengagement of moral control. *Journal of Social Issues, 46,* 27–46.

Bandura, A., & Walters, R. H. (1963). *Social learning and personality development.* New York: Holt, Rinehart & Winston.

Burgess, R. L., & Akers, R. L. (1966b). A different association-reinforcement theory of criminal behavior. *Social Problems, 14,* 128–147.

Catalano, R. F., Kosterman, R., Hawkins, J. D., Newcomb, M. D., & Abbott, R. D. (1996). Modeling the etiology of adolescent substance use: A test of the social development model. *Journal of Drug Issues, 26,* 429–455.

Cornish, D. B., & Clarke, R. V. (1986). *The reasoning criminal: Rational choice perspectives on offending.* New York: Springer-Verlag.

Cressey, D. R. (1953). *Other people's money.* Glencoe, IL: Free Press.

Cressey, D. (1960). Epidemiology and individual conduct. *Pacific Sociological Review, 3,* 47–58.

Glueck, S., & Glueck, E. (1950). *Unraveling juvenile delinquency.* Cambridge, MA: Harvard University Press.

Gottfredson, M., & Hirschi, T. (1990). *A general theory of crime.* Stanford, CA: Stanford University Press.

Grusec, J. E. (1992). Social learning theory and developmental psychology: The legacies of Robert Sears and A. Bandura. *Developmental Psychology, 28,* 776–786.

Hewitt, J. P., & Stokes, R. (1975). Disclaimers. *American Sociological Review, 40,* 1–11.

Lyman, S. M., & Scott, M. B. (1970). *A sociology of the absurd.* New York: Appleton-Century-Crofts.

Matza, D. (1964). *Delinquency and drift.* New York: Wiley.

Matza, D., & Sykes, G. M. (1961). Juvenile delinquency and subterranean values. *American Sociological Review, 26,* 712–719.

Staats, A. (1975). *Social behaviorism.* Homewood, IL: Dorsey.

Stafford, M. C., & Ekland-Olson, S. (1982). On social learning and deviant behavior: A reappraisal of the findings. *American Sociological Review, 47,* 167–169.

Strictland, D. E. (1982). Social learning and deviant behavior: A specific test of a general theory: A comment and critique. *American Sociological Review, 47,* 162–167.

Sutherland, E. H. (1947). *Criminology.* Philadelphia: J. B. Lippincott.

Sykes, G. M., & Matza, D. (1957). Techniques of neutralization: A theory of delinquency. *American Sociological Review, 22,* 664–670.

Thomas, W. I., & Thomas, D. S. (1928). *The child in America: Behavior problems and programs.* New York: Knopf.

Thornberry, T. P., Lizotte, A. J., Krohn, M. D., Farnworth, M., & Jang, S. J. (1994). Delinquent peers, beliefs, and delinquent behavior: A longitudinal test of interactional theory. *Criminology, 32,* 47–83.

REVIEW QUESTIONS

1. What are the four main components of the theory of differential reinforcement according to Akers? Which of these propositions is primarily based on Sutherland's differential association theory? Which of these components is based on operant conditioning? Which of these are based on imitation/modeling? Explain your reasons in each case.

2. Akers seems to emphasize the importance of behavioral feedback and/or reciprocal effects? Explain what these concepts mean, and why they are important.

3. What does Akers mean when he talks about the concept of definitions? Provide a thorough explanation and some examples.

❖

READING

In this selection, Nicole Piquero, Stephen Tibbetts and Michael Blankenship present a scientific test of differential association theory and techniques of neutralization in explaining unethical business decisions among graduate business students (many of whom had years of experience in the business world). In this study, a sample of 133 masters of business administration (MBA) students (with 21% being Executive MBA students) were presented with a hypothetical scenario that involved the possible distribution of a drug that would probably lead to injury and death among users, and then they were asked what their decisions would be if they were the CEO of the company that makes that drug.

We think, and hope, readers will be surprised by the high number of respondents who claimed that they would continue to market and/or distribute the drug, even while knowing that the drug is likely to cause harm to the users. Furthermore, the findings of the study appear to generally support many of the propositions offered by the theoretical

SOURCE: Nicole L. Piquero, Stephen G. Tibbetts, and Michael Blankenship, "Examining the Role of Differential Association and Techniques of Neutralization in Explaining Corporate Crime," *Deviant Behavior* 26, no. 2 (1996):159–88. Copyright © 1996 by Taylor & Francis Informa UK Ltd, Journals. Reproduced with permission of Taylor & Francis Informa UK Ltd via Copyright Clearance Center.

frameworks of differential association and techniques of neutralization. While reading this selection, readers are encouraged to ask themselves what they would do if they (as a CEO) were to make such a decision of whether to market and/or distribute such a potentially damaging drug, as well as what factors would influence their decision.

Examining the Role of Differential Association and Techniques of Neutralization in Explaining Corporate Crime

Nicole Leeper Piquero, Stephen G. Tibbetts, and Michael B. Blankenship

Researchers have applied several theories to account for white-collar and corporate crime including deterrence/rational choice (Braithwaite and Makkai 1991; Makkai and Braithwaite 1994; Paternoster and Simpson 1993, 1996; Simpson and Koper 1992; Simpson et al. 1998), self-control (Barlow 1990; Benson and Moore 1992; Geis 2000; Herbert et al. 1998; Hirschi and Gottfredson 1987, 1989; Reed and Yeager 1996; Simpson and Piquero 2002; Steffensmeier 1989; Yeager and Reed 1998), and organizational strain (Finney and Lesieur 1982; Simpson and Koper 1997; Vaughan 1983) theories. In particular, differential association theory and techniques of neutralization have seldom been applied to explain occurrence of white-collar crime (see Benson 1985 and Hollinger 1991).

For the most part, differential association theories tend to explain white-collar crime as a result of the learned definitions and experiences that occur within the workplace (Sutherland 1949). As workers learn the drives, motivations, and rationales to commit white-collar crime, and when the definitions favorable toward committing white-collar crime outweigh the definitions against law violations,

criminal behavior is more likely to occur. Several studies have been applied in this vein, including Cressey's (1953) application to embezzlement, Geis's (1967) study of price-fixing in the heavy electrical equipment industry, and, most recently, Vaughan's (1996) incorporation of differential association theory into a cultural argument underlying the NASA space shuttle disaster. More recent studies have applied elements of this theoretical model, with the addition of punishment/reinforcement components to a variety of crimes across various age groups (for a review, see Akers 2004).

Although useful, the previous applications of differential association and techniques of neutralization are limited in their ability to realize their full explanatory potential for both white-collar and corporate crimes. Most studies of differential association and techniques of neutralization are anecdotal accounts of criminal events after they have occurred. In this paper, we move beyond these efforts with an original data collection designed to assess the role of definitions, drives, motives, and neutralizations in explaining a particular type of corporate crime, dangerous pharmaceutical drug distribution.

▧ Theoretical Framework

When Sutherland (1940) first introduced the concept of white-collar crime he was responding to the traditional explanations of crime and criminality that explained criminal behavior as the result of poverty or social class status. In addition to sensitizing the field to the concept of white-collar crime, Sutherland also introduced a new approach to explaining the general process of all criminality, differential association theory (Sutherland 1947). He argued that criminal behavior is learned much like any behavior is learned, namely through exposure of differential associations, especially those of primary intimate groups. It is through the associations with those who approve of illegal behavior and those who do not that the process of learning occurs. The actual learning process involves acquiring not only the mechanical techniques (i.e., how to break into a car or how to price fix) of a crime but also the intellectual techniques (e.g., motives, drives, rationales) that allow the individual to use the mechanical behaviors they have learned (Hamlin 1988). Crime results when the weight of definitions favorable to law violation exceeds the weight of unfavorable definitions.

Sutherland (1947) highlighted the importance of differential associations by suggesting that the frequency, duration, priority, and intensity of associations will vary. In other words, those associations that "are exposed first (priority), more frequently, for a longer time (duration), and with greater intensity (importance)" will have a greater impact upon the person (Akers 2000:73). Intimate personal groups, such as family and friends, are often regarded as the most important or influential associations to the individual. However, corporate crime scholars and organizational theorists have argued that an equally important group is the one that forms in the workplace (Jackall 1988; Reed and Yeager 1996; Vaughan 1996, 1998; Yeager and Reed 1998).

Since individuals can and often do have many differential associations (or influential groups), a particularly interesting quagmire that has yet to be reconciled by learning theorists is what happens when two or more of these groups have conflicting views. One approach to solving this dilemma offered by corporate criminologists is to conceptualize motivation as a social construction that may vary across different social contexts (Yeager and Reed 1998). Vaughan (2002), for example, refers to the situated nature of action by suggesting that an individual's action (referring to both the meaning given to an act as well as the individual's choice) is situated within an immediate social setting that is vulnerable to influences from larger institutional, structural, and cultural arrangements. Therefore, the influences upon an individual's decision-making process can and will vary across different social settings. In other words, the choice an individual makes in one setting, such as the decision to violate the law in the workplace to meet company deadlines, may not be the same choice that would be made in a different situation, such as violating traffic laws in order to get their kid to tennis practice on time.

The influence of the social context or environment is particularly salient when dealing with decision-making in a corporate or business setting. Although the socialization into the corporate culture may or may not contradict the rules and expectations of other realms of life (e.g., family and peers), it undoubtedly has an important influence on how workplace decisions are made, particularly when individuals exhibit strong loyalties to their firms (Reed and Yeager 1996; Yeager and Reed 1998). Therefore, while the influence of the corporate environment may not be the first association made, it easily could be regarded as an association that is frequent and of long duration (occurring daily as the individual goes to work at the same job) as well as important (to continue to earn an income). Thus, the corporate culture becomes an important intimate personal group that influences, in part, an individual's decision-making process.

Since the publication of Sutherland's (1947) original statement of differential association, the theory has seen its share of criticisms (Tittle et al. 1986), extensions into social learning theory (Akers 1973; Burgess and Akers 1966), and empirical assessments (Matsueda 1988). Furthermore, Akers (2004) has extended the theory of differential association to include concepts of reinforcement, which include whether an individual will be rewarded or punished by their actions. Another particular clarification of the theory—techniques of neutralization—focuses on explaining how the specific drives and motives associated with the criminal behavior are developed (Hamlin 1988).

Sykes and Matza (1957) proposed a theory of juvenile delinquency that contends that individuals are not committed to criminal values and beliefs but rather adhere mostly to conventional values, attitudes, and beliefs. Individuals learn ways to neutralize or justify their criminal actions; these rationalizations are called techniques of neutralization. Therefore, individuals become delinquent by learning techniques of neutralization rather than learning criminal definitions. These learned techniques allow the individual to justify or neutralize their behavior, thereby allowing him or her to keep a non-criminal image and the ability to drift back and forth between law-abiding and law-breaking behavior (Matza 1964). Sykes and Matza (1957) identified five major types of techniques of neutralization: denial of responsibility, denial of injury, denial of the victim, condemnation of the condemners, and appeal to higher loyalties. Other researchers proposed two additional techniques, defense of necessity (Minor 1981) and metaphor of ledger (Klockars 1974).

Defense of necessity implies that "if an act is perceived as necessary, then one need not feel guilty about its commission, even if it's considered morally wrong in the abstract" (Minor 1981:298). A violator, therefore, can lessen the amount of guilt by focusing on the fact that the act was necessary and the isolates had no choice in the matter. Metaphor of the ledger works the notion of balancing good and bad acts. When it is believed that an individual has accumulated a surplus of "good" acts, then a "bad" act can be mitigated because of the surplus of good that has accumulated (Klockars 1974). In this case, the individual does not focus on a deviant/criminal personality but rather focuses on the deviant act, in light of the good they have done.

While Sykes and Matza (1957) specifically focused on explaining juvenile delinquency (see also Agnew 1994), scholars have not been precluded from applying the theory to other realms of offending (Cromwell and Thurman 2003). White-collar criminals, like most individuals, are believed to have not committed themselves to a criminal way of life. Instead, they maintain a crime-free self-image (Benson 1985). Thus, techniques of neutralization theory may be a unique and important theoretical lens within which to view white-collar offending (Green 1997).

◪ Current Research

The current research builds upon previous work on learning theories and white-collar crime in three ways. First, by utilizing a vignette research design, we are able to circumvent some of the issues raised by temporal ordering concerns. Temporal ordering is important when studying techniques of neutralization in order to differentiate between neutralizing an act and rationalizing an act. Neutralization occurs *prior* to the act whereas rationalizations occur *after* the act (Green 1997). Studies utilizing cross-sectional data in order to test the assumptions of neutralization theory have come under attack for not appropriately differentiating the causal mechanism inherent in the theory (Hollinger 1991). Two plausible research methodologies can be utilized to overcome this criticism. First, longitudinal data collected from a minimum of two points in time could help to clearly establish the temporal ordering of events. However, because of the dynamic nature of the neutralization process, the time lag between Time 1 and Time 2 events may have changed. Thus, a second approach utilizing vignette designs, such as used in the current study, may be a promising approach, since they allow for the instantaneous manipulation of both conditions and variables.

Second, we investigate techniques of neutralization used by white-collar criminals for a particular type of corporate crime, which has not been explored in extant research. Finally, we present an empirical assessment of differential association and techniques of neutralization to account for a type of corporate crime at the individual level. This is important as most of this line of research has employed the rational choice framework (Paternoster and Simpson 1993, 1996; Simpson et al. 1998). We believe that insights from both a differential association and techniques of neutralization approach may serve as a helpful complement to the existing studies of corporate crime decision-making (Akers 1990).

The purpose of this paper is to explore how differential associations and techniques of neutralization relate to intentions to engage in a type of corporate crime, the distribution (or recall) of a pharmaceutical drug known to be harmful to consumers. Specifically, we test four hypotheses. First, we assess how respondent's associations with others (coworkers, friends, business professors, and boards of directors) related to their own decision-making process. Second, we assess how respondents employ a variety of techniques of neutralization in their decision-making process. Third, we examine whether these effects hold after controlling for

several other variables, including deterrence considerations. Finally, we examine whether age moderates how techniques of neutralization related to offending decisions.

Method

Data were collected from 133 students enrolled in an MBA program at one university. Respondents were enrolled in either the traditional MBA program or the executive MBA program that is specifically designed for those who have considerable business experience or are working full-time. All students were informed prior to the distribution of the survey that they were part of a study, and all participants gave voluntary consent. These students were asked to complete a questionnaire during class that presented a scenario regarding the promotion and sales of a hypothetical drug—"Panalba"—that was banned by the Food and Drug Administration (FDA). This scenario was based on prior work by Armstrong (1977), and respondents were asked to report their decision on the avenue that should be taken, as well as to estimate their perceptions of others regarding their decision. Students also were asked to respond to a variety of questions that dealt with their attitudes toward various aspects (e.g., employing organization's attitude) of such a situation. Finally, participants were asked to provide demographic/characteristic information about themselves.[1]

Dependent Variable

Solution/Intentions to Commit Corporate Crime

The dependent measure is respondents' self-reported level of the extent to which they would further or inhibit the distribution of "Panalba," which (given the scenario) is known to harm persons (see Appendix).[2] This measure was coded on a scale ranging from 1 (recall "Panalba" immediately and destroy all existing inventories), 2 (stop production of "Panalba" immediately but allow existing inventories to be sold), 3 (stop all advertising and promotion of "Panalba" but continue distribution to those physicians who request it), 4 (continue efforts to effectively market "Panalba" until its sale is actually banned), 5 (continue efforts to effectively market "Panalba" while taking legal, political, and other actions to prevent its banning), or 6 (continue efforts to market "Panalba" in other countries after the FDA bans the drug in the U.S.). The mean score on this item was 2.57, which can be interpreted as the average respondent chose to either stop production but allow existing inventories to be sold or to stop all advertising and promotion but continue distribution to physicians who request it. Descriptive information for all variables is provided in Table 1.

Table 1	Descriptive Statistics for All Variables (N = 133)			
Variable	Minimum	Maximum	M	SD
Intentions	1.00	6.00	2.57	1.67
Recall immediately		n = 60		
Stop production—sell existing inventories		n = 28		
Stop advertising—distribute on request		n = 30		
Continue production and market until ban		n = 8		
Impede the ban while producing and marketing		n = 17		
Market outside U.S. even after FDA ban		n = 13		
Coworkers agree	1.00	6.00	4.58	1.12
Friends agree	2.00	6.00	3.80	1.16
Business professors agree	1.00	6.00	4.40	1.27
Board of directors agree	1.00	6.00	4.07	1.44
Government exaggerates	1.00	5.00	2.82	1.09
Regulations impede	1.00	5.00	2.55	1.01
Profit most important	1.00	5.00	2.90	1.24
Caveat emptor motto	1.00	5.00	2.30	1.07
Anything to make profit	1.00	5.00	2.42	1.11
Age	22.00	55.00	30.43	7.36
Gender	1.00	2.00	1.33	0.47
Race	1.00	2.00	0.76	0.46
Religious	0.00	1.00	0.86	0.35
Protestant	0.00	1.00	0.62	0.49
Political orientation	1.00	5.00	3.28	1.04
Executive MBA	1.00	2.00	1.21	0.41
Employer worried about legal suits	1.00	5.00	3.74	1.14

Independent Variables

Favorable Definitions

Differential association/reinforcement theory suggests that criminal behavior results when there is an excess of definitions favorable to the violation of law. A major criticism of differential association is that measuring the presence of favorable definitions is very difficult to accomplish (Tittle et al. 1986). In order to gain an understanding of the respondents' perceptions of favorable or supportive definitions, we asked them a series of questions regarding the extent to which some of those closest (coworkers, friends, business professors, boards of directors) to them would support the decision they made. Responses ranged from 1 (strongly disagree) to 6 (strongly agree). Given established theoretical frameworks (Akers 1973; Sutherland 1947, 1949) and previous studies (for a review, see Akers 2004), all of the four differential association/reinforcement measures were predicted to be positively correlated with intentions to commit corporate crime.

Techniques of Neutralization

In order to maintain a crime-free self-image, individuals may employ any number of techniques of neutralization. Due to the fact that the theory was originally designed to explain juvenile delinquency, we have operationalized our techniques of neutralization around the same general concepts put forth by Sykes and Matza (1957) with modifications made to apply the concepts into the workplace. All responses ranged from 1 (strongly disagree) to 5 (strongly agree) with 3 being the midpoint (neutral or no opinion). Higher scores on all five items indicated stronger levels of neutralization, and consistent with theoretical frameworks (Sutherland 1949) and previous studies (Hollinger 1991) were predicted to be positively associated with higher

levels of intentions to commit corporate crime. Furthermore, these dispositions toward more neutralization techniques were expected to interact with age, such that older respondents were predicted to use more neutralization in decisions to commit corporate crime.

Government Exaggerates Dangers to Consumers

Denial of injury was operationalized by asking respondents to give the response that best expressed their belief for the following statement: "The government exaggerates the danger to consumers from most products." This statement allows the individual to deny the injury of the act by claiming that the government is overly cautious in assessing the danger to the public. This belief also relates to denial of responsibility in the sense that it implies that there is minimal danger in the use of marketed products, so the companies that produce such items should not be held responsible if injuries do happen to occur from usage.

Regulations Impede

In order to tap into the condemnation of the condemners technique, respondents were asked to give the response that best expressed their belief for the following statement: "Government regulations impede business." This statement allows the individual to apply the blame of the act onto the strict regulations placed on business by government.

Profit Most Important

Respondents were asked to give the response that best expressed their belief for the following statement: "Profit is emphasized above everything else at my place of work" in order to serve as a proxy for the technique of appeal to higher loyalty. This statement allows the individual to alleviate feelings associated with the act by placing the blame onto the goals of the organization or company for which he or she works.

Caveat Emptor Motto

In order to tap into the denial of victim technique, respondents were asked to give the response that best expressed their belief for the following statement: "*Caveat emptor* (let the buyer beware) is the motto of my employer." This statement shifts the blame of the act onto the victim by allowing the respondent to claim that the victim should have known better.

Anything to Make a Profit

Denial of responsibility was operationalized by asking respondents to give the response that best expressed their belief for the following statement: "Where I work, it is all right to do anything to make a profit unless it is against the law." This statement allows the individual to deny responsibility of the event by placing the responsibility of the act onto the organization in which he or she is employed. In other words, the individual dodges blame by allowing the company, by means of rules or culture, to be responsible for the actions taken.

Control Variables

Seven variables were measured to control for respondents' demographic characteristics. Age of the respondent was coded as reported and ranged from 22–55, with a mean of 30.43. Based on the results of previous studies, age was expected to be positively related to intentions to commit corporate crime. Gender was coded as 1 (male) or 2 (female), with 33% of the sample consisting of females. Based on previous studies, females were expected to be less likely to commit corporate crime than males. Race was coded as 1 (non-white) or 2 (white), with 24% of the sample being non-white; due to the lack of studies involving minorities, no predictions were made regarding the association of this variable with intentions to commit corporate crime. Executive MBA program was coded a 1 (regular program) or 2 (executive

program), with 21% of the sample attending the executive program. Based on previous studies, respondents who were enrolled in the Executive MBA program were predicted to have higher intentions to commit corporate crime than the respondents in the Regular MBA program. Religious was coded as 0 (other) or 1 (religious), with 14% of the sample reporting no religious affiliation; respondents who reported being religious were predicted to have lower intentions to commit corporate crime than those who reported not being religious. Protestant was coded as 0 (non-Protestant) or 1 (Protestant), with 38% of the sample reporting a non-Protestant affiliation. Based on prior studies, respondents who reported being Protestant were predicted to have lower scores on intentions to commit corporate crime than those who did not report Protestant affiliation. Finally, political orientation was coded as 1 (very liberal) to 3 (neutral) to 5 (very conservative), with the mean of the sample reporting themselves as 3.28, or relatively conservative. Although there has been little empirical research on this factor in disposing one toward corporate deviance, we predicted that the respondents who reported being more conservative would be more likely to have higher scores on intentions to commit corporate crime.

Lastly, we included a variable that taps into the theoretical component of perceptual deterrence. As previously noted, the majority of studies examining intentions to engage in white-collar/corporate misconduct are drawn from the rational choice framework. Thus, we included a variable to take into account the possibility of a deterrent effect. Respondents were asked to give the response that best expressed their belief for the following statement: "My employer worries about penalties that government regulators might impose because of violations." Responses ranged from 1 (strongly disagree) to 5 (strongly agree) with 3 being the midpoint (neutral or no opinion).

Consistent with previous studies in the area of deterrence and rational choice research, it was predicted that the respondents who had high levels of worries regarding penalties that government regulators might impose due to violations would have lower scores on intentions to commit corporate crime.

We estimated a series of multivariate equations in which the effects of other variables are controlled. The first of these models included only the four differential association variables in predicting intentions to commit corporate crime. Consistent with the bivariate analysis, one differential association variable was, as predicted, positively associated with intentions to commit corporate crime: board of directors agree.

Interestingly, the degree to which coworkers agree with the decision did not have a significant impact on determining the intentions to offend. It was expected that coworkers would have perhaps the biggest influence on intentions to offend, due to their proximity and importance in everyday operations, and especially the development of mores in the office environment. Even more interesting is the significant inverse effect that two differential association variables—closest friends and business professor agree—have on intentions to commit corporate crime. This finding implies that employees not only disregard the norms and feelings of those outside their corporate workplace, but they actually defy those beliefs. Overall the model of the differential association variables accounted for approximately 27% of the variation in intentions to commit corporate crime, which is relatively good considering the model only included four independent variables.

A second multivariate equation was estimated that included only the five techniques of neutralization variables in predicting intentions to commit corporate crime. Two of the five neutralization measures were found, as predicted, to have positive effects on intentions

to commit corporate crime. These two variables were the perception that the government exaggerates dangers to consumers and the belief that profit is most important. The other three neutralization measures were not significantly related to intentions to commit corporate crime and, thus, it is not surprising that this model did not explain much of the variation in intentions to commit corporate crime, explaining approximately 14% of the variation in such intentions.

A fully-specified model was estimated (see Table 2), which included all differential association variables, techniques of neutralization variables, and all control variables. This model showed positive effects for closest coworkers agree and board of directors agree, which were consistent with predictions that differential association measures would be positively associated with intentions to commit corporate crime. Predicted effects also were observed for two techniques of neutralization measures: government exaggerates dangers to consumers and profit is most important, which supported previous analyses. Results of the model that were counter to expectations were the inverse effects observed for closest friends agreeing and business professors agreeing with intentions to commit crime. Interestingly, the fully specified model accounted for approximately 48% of the variation in intentions to commit corporate crime, and the theoretical model accounted for virtually all of the influence of demographic variables (e.g., female, white, religious, political orientation, executive program, and age).

Finally, consideration was given to the possible interaction between age and techniques of neutralization due to previous studies of such non-linear relationships (Hollinger 1991). An estimate of the effect of "age-neutralization" interaction on intentions to commit corporate crime was estimated and the coefficient was significant ($p < .01$), so separate models of techniques of neutralization

Table 2	OLS Regression Coefficients of Independent Variables Predicting Intentions to Commit Corporate Offending (N = 133)			
Variable	b	SE	B	t
Coworkers agree	.399	.136	.243	2.939**
Friends agree	−.540	.113	−.359	−4.756**
Business professors agree	−.277	.104	−.202	−2.655**
Board of directors agree	.202	.095	.169	2.112*
Government exaggerates	.433	.124	.277	3.501**
Regulations impede	.244	.140	.144	1.740
Profit most important	.384	.114	.278	3.368**
Caveat emptor motto	.200	.130	.127	1.542
Anything to make profit	.104	.131	.067	.795
Gender	−.517	.292	−.144	−1.774
Race	−.302	.324	−.074	−934
Executive MBA	−.600	.389	−.145	−1.542
Religious	−.308	.448	−.060	−.687
Protestant	.208	.291	.059	.714
Political orientation	.071	.137	.042	.518
Age	.032	.020	.148	1.673
Employer worried about suits	−.160	.118	−.102	−1.356
Constant	6.530	1.284		5.087**

$R^2 = .485$. *$p < .05$, two-tailed; **$p < .01$, two-tailed.

(with control variables) were estimated for respondents who were aged under 35 and for respondents who were 35 or older (see Table 3). Respondents were split into these groups because other recent studies examining inter-action have separated cases based on a 75%/25% criteria; furthermore, we performed the analysis with other percentage breakdowns (50%/50%) and they resulted in substantially similar results. Additional models also were estimated that incorporated the differential association variables into these interaction equations, but the results were substantively similar (i.e., no changes in direction or signifi-cance of the neutralization effects); thus, only the models with the neutralization and control variables are included due to the problems of

the limited sample size of the older (over 35 years old) group.

These models showed that respondents aged 35 or older were significantly more influ-enced by techniques of neutralization than were the younger respondents. Specifically, the younger respondents were significantly influenced by only one of the neutralization measures—government exaggerates dangers to consumers—whereas the older respondents were significantly influenced by this neutraliza-tion, as well as two additional forms of neutral-ization: profit most important and the belief that anything is justified to make a profit.

Ultimately, the amount of variation explained in these two models was drastically different. The model for respondents under

Table 3	Differential Effects Across Age—Influence of Techniques of Neutralization on Intentions to Commit Corporate Offending							
	Age 35 or Older (*n* = 36)				Age Under 35 (*n* = 97)			
Variable	b	SE	B	t	b	SE	B	t
Government exaggerates	.789	.266	.530	2.965**	.548	.169	.351	3.237**
Regulations impede	.299	.310	.161	.965	.260	.205	.160	1.269
Profit most important	1.042	.286	.745	3.642**	.039	.150	.029	.258
Caveat emptor motto	.290	.248	.180	1.166	.007	.186	.004	.036
Anything to make profit	.737	.353	.426	2.090*	.008	.172	.006	.046
Gender	−1.312	.734	−.304	−1.788	−.628	.389	−.189	−.1.64
Race	−.130	.668	−.027	−.194	−.250	.472	−.065	−.529
Executive MBA	−1.152	.584	−.306	−1.973	.063	.706	.011	.090
Religious	−.418	.886	−.077	−.471	.244	.620	−.051	−.394
Protestant	−.299	.544	−.080	−.549	−.366	.416	.110	.880
Political orientation	.124	.289	.069	.429	.036	.190	.022	.188
Age	.053	.063	.122	.831	−.014	.057	−.030	−.247
Constant	9.321	3.907		2.386*	6.466	1.980		3.265**
	R^2 .648; Adjusted R^2 .464				R^2 .164; Adjusted R^2 .046			

*$p < .05$, two-tailed; **$p < .01$, two-tailed.

the age of 35 only explained approximately 16% of variation in offending intentions, whereas the model for respondents 35 and older accounted for 65% of variation in intentions to commit corporate crime. Thus, the results strongly support the findings of previous studies (Hollinger 1991) that found techniques of neutralization were employed more by older respondents in their decision to commit corporate crime as compared to their younger counterparts.

▧ Conclusions

Recognizing that few theoretical models have been applied to understanding unethical corporate decision-making, we obtained data from 133 MBA students and executives in training in an effort to understand how differential association

and techniques of neutralization theories accounted for decisions to market and produce an illegal drug. The findings of this study support the application of differential association and techniques of neutralization theoretical frameworks in understanding decisions to commit corporate crime. Furthermore, subsequent analyses supported the predicted and observed interaction between age and neutralization techniques in such decisions regarding corporate crime. Specifically, older respondents compared to their younger counterparts were more likely to employ techniques of neutralization in decisions to commit obvious violations regarding corporate activities.

The results of this study suggest that an employee's corporate climate, namely attitudes of their closest coworkers and the perceived attitudes of the board of directors had a positive and significant influence on what they would do when they had to decide whether to engage in acts that were unethical. On the other hand, the perceived attitudes or reactions of those outside of this corporate workplace—their closest friends and business professors—did not have much influence at all on what they would decide; in fact, the estimates indicate that respondents were actually inclined to go against what they believed their friends and business teachers felt was the right decision. These results suggest the powerful effects of the corporate climate and the relative insignificance of influences external to the workplace environment (see Vaughan 1996). It is disappointing and somewhat disturbing to acknowledge the implications of our findings. After all, many of the respondents, particularly the older and more experienced participants, essentially reported that they would take actions that would lead to injuring or killing innocent people, while knowingly disregarding the ethical responsibilities they learned from their professors and defying moral values of their closest friends. Thus, this may represent a specific form of differential association theory to be applied to corporate crime, whereby an individual's corporate environment can subvert the other associations and normative values that the individual has learned. Perhaps this is due to the extensive amounts of hours spent at work and immersed in the corporate climate, as well as the fact that survival instincts are likely to kick in when the actions they take at work decide what type of life they will be providing for themselves and their families. Regardless of the reasons, it is clear that business employees such as those in our sample are far more likely to be affected by their corporate environments in their corporate decisions than by ethical standards they have learned outside of the workplace.

Findings from this study also support the use of techniques of neutralization in that several of the neutralization measures that were included had a significant effect on decisions to commit corporate crime. Specifically, the idea that the government exaggerates dangers to the consumer reeks of the traditional "denial of injury" as identified by Sykes and Matza (1957), in the sense that it says that no real harm is actually caused or inflicted by corporate misbehavior. It also points to the idea of "denial of responsibility" that suggests products on the market are generally safe so the corporate producers should not take blame for the few injuries that occur. Obviously, neither of these beliefs are true, especially regarding the further marketing and production of a dangerous drug (such as "Panalba"). Apparently however, this does not appear to stop individuals, particularly older and more experienced professionals, from adopting such attitudes. Perhaps more disconcerting is the attitude of profit being most important, which was observed for the older respondents. For many experienced individuals, the bottom-line is all that matters, regardless of any other concerns. This reflects Sykes and Matza's (1957) principle of "appeal to higher loyalty," and the company's profit is their primary (if not only) loyalty. This is precisely the attitude that should be targeted for altering by business (and other) professors in college courses. If nothing else, we believe this study shows that Sykes and Matza's original framework for techniques of neutralization, while specifically designed to explain delinquent behavior, can be applied to other forms of criminality such as corporate criminality.

Finally, the findings of this study support those of other empirical researchers, namely Hollinger (1991) who found that older individuals are more likely to employ neutralization techniques in corporate environments than their younger counterparts. This result is somewhat surprising given that techniques of neutralization were originally designed by Sykes and Matza (1957) to explain juvenile delinquency. However, Sykes and Matza never claimed that such a model would not apply to other forms of offending, and furthermore they likely did not anticipate the extent to which corporate crime, and other modern forms of crime, would evolve that appear to fit their model so well. We believe that not only is the Sykes and Matza model salient for decisions to commit corporate crime, as our findings suggest, but that it is even more applicable for such crimes and for older persons than it is for delinquency.

Another notable finding of this study is that individuals did not appear to be deterred by legal sanctions. The measure of respondents' perceptions of their employer being concerned about legal suits did not have a significant effect when other factors were accounted for. This does not support deterrence models of compliance and reiterates the need for more informal measures, such as education, in curbing corporate crime as well as other forms of criminal activity. After all, formal deterrence is not likely to work in environments where the social climate is so strong as to suggest that it (crime) is part of the everyday functioning of the company/agency. As Benson (1985) noted, antitrust violators were most likely to argue the everyday character and historical continuity of the offense. Their criminal activity was seen as part of everyday business. Consistent with our findings, such attitudes almost always come from the top (e.g., board of directors) and are typically endemic throughout the company (e.g., coworkers/peers) (Reed and Yeager 1996). It is obvious that different measures must be explored for curbing such practices, and this implicates education and socialization prior to employment, as well as after. Consistent with this idea is recent empirical research that has shown a relationship between academic cheating in college and participation in unethical business practices (Sims, 1993). Individuals are likely to learn cheating practices quite early on and carry them when they move on to the corporate world. Despite our null findings regarding the influence of professors, much more must be done in this area to develop more effective means for instilling and maintaining ethical values and responsibility.

In the end, our modest effort has shown that differential associations and techniques of neutralization are related to corporate illegalities, even in the face of strong deterrent messages. Moreover, our research also points to important age-moderating effects that indicate older individuals are more responsive to some techniques of neutralization than are younger individuals. Delineating the reasons for this (and related) findings is an important task for future research.

⊠ Notes

1. We recognize and appreciate the concerns associated with student samples. However, two aspects of this type of sample are appealing. First, corporate crime decision-making data do not exist in the community or at the national level. Second, this sample is comprised of people who will be the future leaders of business and are therefore in a unique position to provide valuable formation. The average age of the sample is 30 and many respondents have professional business experience.

2. Much of the decision-making research uses hypothetical scenarios to solicit offending intentions. Extant research shows a strong correlation between projected and actual behavior Green 1989; Kim and Hunter 1993, and Pogarsky's (2004) recent research details this relationship.

⊠ References

Agnew, Robert. 1994. "The Techniques of Neutralization and Violence." *Criminology* 32: 555–80.

Akers, Ronald L. 1973. *Deviant Behavior: A Social Learning Approach.* Belmont, CA: Wadsworth.

———. 1990. "Rational Choice, Deterrence, and Social Learning Theories: The Path Not Taken." *Journal of Criminal Law and Criminology* 81: 653–76.

———. 2000. *Criminological Theories: Introduction, Evaluation, and Application.* Los Angeles: Roxbury.

———. 2004. *Criminological Theories: Introduction, Evaluation, and Application,* 4th ed. Los Angeles: Roxbury.

Armstrong, J. S. 1977. "Social Irresponsibility in Management." *Journal of Business Research* 5:185–213.

Barlow, Hugh D. 1990. "Explaining Crime and Analogous Acts, or the Unrestrained Will Grab at Pleasure Whenever They Can." *Journal of Criminal Law and Criminology* 82:229–42.

Benson, Michael L. 1985. "Denying the Guilty Mind: Accounting for Involvement in White-Collar Crime." *Criminology* 23:583–607.

Benson, Michael L., and Elizabeth Moore. 1992. "Are White-Collar and Common Offenders the Same? An Empirical and Theoretical Critique of a Recently Proposed General Theory of Crime." *Journal of Research in Crime and Delinquency* 29:251–72.

Braithwaite, John, and Toni Makkai. 1991. "Testing an Expected Utility Model of Corporate Deterrence." *Law and Society Review* 25:7–39.

Burgess, Robert L., and Ronald L. Akers. 1966. "A Differential Association-Reinforcement Theory of Criminal Behavior." *Social Problems* 14:128–47.

Cressey, Donald. 1953. *Other People's Money.* New York: The Free Press.

Cromwell, Paul, and Quint Thurman. 2003. "The Devil Made Me Do It: Use of Neutralizations by Shoplifters." *Deviant Behavior* 24:535–50.

Finney, Henry C., and Henry R. Lesieur. 1982. "A Contingency Theory of Organizational Crime." Pp. 255–99 in *Research in the Sociology of Organizations,* Vol, 1, edited by S. B. Bacharach. Greenwich, CT: JAI Press.

Geis, Gilbert. 1967. "White Collar Crime: The Heavy Electrical Equipment Antitrust Case of 1961." Pp. 139–51 in *Criminal Behavior Systems: A Typology,* edited by M. B. Clinard and R. Quinney. New York: Holt, Rinehart, and Winston, Inc.

———. 2000. "On the Absence of Self-Control as the Basis for a General Theory of Crime: A Critique." *Theoretical Criminology* 4: 35–53.

Green, D. 1989. "Measures of Illegal Behavior in Individual-Level Deterrence Research." *Journal of Research in Crime and Delinquency* 26: 253–75.

Green, Gary S. 1997. *Occupational Crime.* Chicago: Nelson-Hall.

Hamlin, John E. 1988. "The Misplaced Role of Rational Choice in Neutralization Theory." *Criminology* 26: 425–38.

Herbert, Carey, Gary S. Green, and Victor Larragoite. 1998. "Clarifying the Reach of a General Theory of Crime for Organizational Offending: A Comment on Reed and Yeager." *Criminology* 36: 867–83.

Hirschi, Travis, and Michael Gottfredson. 1987. "Causes of White-Collar Crime." *Criminology* 25: 949–74.

———. 1989. "The Significance of White-Collar Crime for a General Theory of Crime." *Criminology* 27: 359–71.

Hollinger, Richard C. 1991. "Neutralizing in the Workplace: An Empirical Analysis of Property Theft and Production Deviance." *Deviant Behavior* 12:169–202.

Jackall, Robert. 1988. *Moral Mazes: The World of Corporate Managers.* New York: Oxford University Press.

Kim, Min-Sun, and John E. Hunter. 1993. "Relationships Among Attitudes, Behavioral Intentions, and Behavior: A Meta-Analysis of Past Research, Part 2." *Communications Research* 20: 331–64.

Klockars, Carl B. 1974. *The Professional Fence.* New York: The Free Press.

Makkai, Toni, and John Braithwaite. 1994. "The Dialects of Corporate Deterrence." *Journal of Research in Crime and Delinquency* 31: 347–73.

Matsueda, Ross L. 1988. "The Current State of Differential Association Theory." *Crime and Delinquency* 34: 277–306.

Matza, David. 1964. *Delinquency and Drift.* New York: John Wiley.

Minor, W. William. 1981. "Techniques of Neutralization: A Reconceptualization and Empirical Examination." *Journal of Research in Crime and Delinquency* 18: 295–318.

Paternoster, Raymond, and Sally Simpson. 1993. "A Rational Choice Theory of Corporate Crime." Pp. 37–58 in *Routine Activities and Rational Choice: Advances in Criminological Theory,* Vol, 5, edited by R. V. Clarke and M. Felson. New Brunswick, NJ: Transaction Publishers.

———. 1996. "Sanction Threats and Appeals to Morality; Testing a Rational Choice Model of Corporate Crime." *Law and Society. Review* 30: 549–83.

Pogarsky, Greg. 2004. "Projected Offending and Contemporaneous Rule Violation: Implications for Heterotypic Continuity." *Criminology* 42: 111–35.

Reed, Gary E., and Peter Cleary Yeager. 1996. "Organizational and Neonclassical Criminology: Challenging the Reach of a General Theory of Crime." *Criminology* 34: 357–82.

Simpson, Sally S., and Christopher S. Koper. 1992. "Deterring Corporate Crime." *Criminology* 30: 347–75.

———. 1997. "The Changing of the Guard: Top Management Characteristics, Organizational

Strain, and Antitrust Offending." *Journal of Quantitative Criminology* 13: 373–404.

Simpson, Sally S., Raymond Paternoster, and Nicole Leeper Piquero. 1998. "Exploring the Micro-Macro Link in Corporate Crime Research." Pp. 35–68 in *Research in the Sociology of Organizations,* Vol.15, edited by P. A. Bamberger and W. J. Sonnenstuhl. Stamford, CT: JAI Press.

Simpson. Sally S., and Nicole Leeper Piquero. 2002. "Low Self-Control, Organizational Theory, and Corporate Crime." *Law and Society Review* 36: 509–47.

Sims, Randi L. 1993. "The Relationship Between Academic Dishonesty and Unethical Business Practices." *Journal of Education for Business* 69: 207–11.

Steffensmeier, Darrell. 1989. "On the Causes of "White-Collar" Crime: An Assessment of Hirschi and Gottfredson's Claims." *Criminology* 27: 325–58.

Sutherland, Edwin H. 1940. "White Collar Criminality." *American Sociological Review* 10: 132–39.

———. 1947. *Principles of Criminology:* Philadelphia: J. B. Lincott.

———. 1949. *White-Collar Crime.* New York: Holt, Rinehart, Winston.

Sykes, Gresham M., and David Matza. 1957. "Techniques of Neutralization: A Theory of Delinquency." *American Sociological Review* 22: 664–70.

Tittle, C. R., Burke. M. J., & Jackson, E. F. (1986), "Modeling Sutherland's Theory of Differential Association: Toward an Empirical Clarification". *Social Forces* 65: 405–32.

Vaughan, Diane. 1983. *Controlling Unlawful Organizational Behavior.* Chicago: University of Chicago Press.

———. 1996. *The Challenger Launch Decision.* Chicago: University of Chicago Press.

———. 1998. "Rational Choice, Situated Action, and the Social Control of Organizations." *Law and Society Review* 32: 23–61.

———. 2002. "Criminology and the Sociology of Organizations." *Crime, Law, and Social Change* 37: 117–36.

Yeager, Peter Geary, and Gary E. Reed. 1998. "Of Corporate Persons and Straw Men: A Reply to Herbert, Green, and Larragoite." *Criminology* 36: 885–97.

⊠ Appendix

Vignette

As chief executive officer of a Fortune 500 pharmaceutical company, you have just received information about a product known as "Panalba."

Panalba, a "fixed-ratio" prescription antibiotic containing a combination of drugs, has been on the market for over 13 years and now accounts for 50 million dollars in yearly sales, which is 15% of the corporation's gross income in the U.S. Panalba is marketed under a different name in foreign markets, and profits are roughly comparable to those in the U.S.

Over the past 20 years, numerous medical scientists have objected to the sale of most fixed-ratio drugs, arguing that: (1) no evidence proves these fixed-ratio drugs are superior to single drugs; and (2) detrimental side effects, including death, are at least doubled for fixed-ratio drugs. For example, estimates indicate that Panalba causes 14 to 22 deaths per year that could be prevented by using a single drug manufactured by a competitor. Despite these recommendations to remove fixed-ratio drugs from the market, physicians have continued to prescribe them because they offer a "shotgun" approach to treatment for the physician who is unsure of his or her diagnosis.

Recently a National Academy of Science–National Research Council panel conducted extensive research on fixed-ratio drugs and recommended that the Food and Drug Administration (FDA) ban the sale of Panalba. One of the panel members stated to the press that, "There are few instances in medicine when so many experts have agreed unanimously and without reservation" about banning the sale of Panalba. This view was typical of comments made by other members of the panel and of comments made about fixed-ratio drugs over the past 20 years. The panel concluded that while all drugs have potential negative side effects, those associated with Panalba far exceeded the benefits derived from the drug.

The corporation just received FDA notification of its intention to ban the distribution of Panalba in the U.S. Should the ban become effective, the corporation will have to stop distribution of the drug in the U.S. and remove existing inventories from the market. Such action could also threaten the overseas market for Panalba. The corporation has no substitute for Panalba, so consumers will have to switch to a competitor's product. Some substitutes offer benefits equivalent to those of Panalba, but without known serious side effects. The retail price of the substitutes is approximately the same as for Panalba. The repercussions from the loss of Panalba will reduce shareholders' profits, jeopardize employment, and threaten the future viability of the corporation.

On the other hand, research suggests that it is extremely unlikely that bad publicity from this case would have any significant effect on the long-term profits of the corporation. Therefore, there are several alternatives to the FDA's proposed action.

REVIEW QUESTIONS

1. What two "new" types of techniques of neutralization are discussed in this study that were not among the original five presented by Sykes and Matza? Do you feel these are important forms of neutralization techniques? How so?

2. If you were the CEO of the company that manufactured this drug, what would you decide given the choices of the seven-point scale provided in the dependent measure? Provide a justification for your decision.

3. What do you think is the reason why older (and more experienced) respondents are more likely to justify their decisions to injure or kill users of the drug? How does this relate to differential association theory? Can you think of other reasons or theoretical models for such decisions?

❖

READING

In this selection, David May estimates the influence of fear on carrying firearms to school among approximately 8,000 public high school students. He bases this test on two theories, the first being a variation of Sutherland's differential association theory, which we have examined in previous selections in this section. The second theory is that of Travis Hirschi's social bonding theory, which we have not examined in previous selections, so we will now briefly review it here, and a more elaborate explanation can be found in the section introduction.

Source: David C. May, "Scared Kids, Unattached Kids, or Peer Pressure: Why Do Students Carry Firearms to School?" *Youth and Society* 31, no. 1 (1999): 100–27. Copyright © 1999 Sage Publications, Inc. Used by permission of Sage Publications, Inc.

Hirschi's social bonding theory, a well-known control theory, is actually quite straightforward. First, like all other control theories, Hirschi's theory assumes that individuals are born selfish, greedy, and so on; thus, it assumes that criminal offending is a natural state of human beings, which is consistent with Sigmund Freud's concept of being born with an "id" that constantly seeks pleasure even if it means violating others. So instead of explaining why people offend, the challenge is to explain why human beings actually conform or do not violate the law. Hirschi's social bonding theory claims that what inhibits individuals from engaging in offending behavior is the strength of one's "social bond" to conventional society. According to Hirschi, this social bond is made up of four components, which we will now briefly review.

Specifically, the social bond consists of four elements or components: attachments to others, commitment to conventional society, involvement in conventional activities, and moral beliefs that violation of law is wrong. The stronger or higher the levels on each of these components of the social bond, the less likely individuals are to conform to rules of society and the less likely they are to follow their natural tendencies to commit crimes against others.

While reading this selection, readers are encouraged to put themselves in the shoes of public high school students and the fear they may have of being victimized, as well as whether such fear could justify carrying a firearm to school. In addition, readers should consider the validity of the two theoretical models that were used by the author to explain this phenomenon and whether you feel these two models were good choices. Would you have chosen other theoretical frameworks if you had performed this study?

Scared Kids, Unattached Kids, or Peer Pressure

Why Do Students Carry Firearms to School?

David C. May

Firearm homicides are the second leading cause of death for youngsters 15 to 19 years old and are the leading cause of death for Black males aged 10 to 34 years (Fingerhut, 1993). Between 1980 and 1990, there was a 79% rise in the number of juveniles committing murder with guns (Senate Hearings, 1993) and in 1990, 82% of all murder victims aged 15 to 19 years and 76% of victims aged 20 to 24 years were killed by firearms (Roth, 1994).

The problem of violent crime among adolescents is particularly acute in places where youth spend much of their time. Arguably, there is no single place where youth consistently congregate more than at school. Although deaths at school as a result of violence are rare events (76 violent student deaths in the 1992–1994 academic years), the general perception is that school-associated violence is on the increase (Kachur et al., 1996). Although Kachur et al. indicate that this does not appear to be the case, school violence, and particularly gun violence at school, is a problem that is not easily ignored. Some have argued that approximately 100,000 students take guns to school every day (Senate Hearings, 1993; Wilson & Zirkel, 1994). Estimates of the percentage of students who take guns to school vary widely, however; the percentage has been determined to be anywhere from .5% (Chandler, Chapman, Rand, & Taylor, 1998) to approximately 9% who carry a gun to school at least "now and then" (Sheley & Wright, 1993, p. 5). Furthermore, a majority of students say it would be little or no trouble to get a gun if they wanted one (Sheley & Wright, 1993). It is obvious that firearms in our schools are a problem.

Although numerous studies have examined prevalence of firearms possession among school-aged adolescents (Callahan & Rivara, 1992; Center to Prevent Handgun Violence, 1993; Chandler et al. 1998; Hechinger, 1992; Roth, 1994; Wilson & Zirkel, 1994), the study of causes of firearm possession at school has all but been neglected. Of those researchers who have attempted to examine determinants of adolescent firearm possession, many argue that protection is an often cited reason for weapon possession and provide various explanations for this finding (Asmussen, 1992; Bergstein, Hemenway, Kennedy, Quaday, & Ander, 1996; Blumstein, 1995; Hemenway, Prothrow-Stith, Bergstein, Ander, & Kennedy, 1996; Sheley, 1994; Sheley & Wright, 1993); in fact, Sheley and Wright (1995) determined that 89% of the gun carriers in their sample of youth in correctional facilities and inner-city youth felt that self-protection was a *very important* reason for owning a handgun. Only one study (Sheley & Brewer, 1995), however, examines the association between fear and firearm possession at school (see below for detailed discussion).

There have been many studies that have attempted to determine characteristics of adult firearms owners as well as research that seeks to determine individuals' motivation for owning guns. One explanation that has surfaced to account for why adults own firearms is the "fear and loathing hypothesis" (Wright, Rossi, & Daly, 1983, p. 49). The fear and loathing hypothesis suggests that people buy guns in response to their fear of crime and other incivilities present in our society. According to the fear and loathing hypothesis, individuals, fearful of elements of the larger society (e.g., crime and violence), go through a mental process in which they begin to deplore crime, criminals, and the like and purchase firearms for protection. Recently, many researchers have diminished the loathing aspect of the hypothesis and have examined what would more accurately be called the fear of criminal victimization hypothesis.

Several studies have attempted to test this hypothesis, with mixed results (Arthur, 1992; Lizotte, Bordua, & White, 1981; Smith & Uchida, 1988; Wright & Marston, 1975). These studies, however, are limited in that they use only adult populations; excluding the study conducted by Sheley and Brewer (1995), the fear of criminal victimization hypothesis has never been tested using a sample of adolescents.

Sheley and Brewer (1995), in an examination of suburban high school students in Louisiana, were the first to specifically test the fear of criminal victimization hypothesis among adolescents by investigating the effect of fear on adolescent gun possession at school. They determined that fear had a nonsignificant association with carrying firearms to school, particularly for males. Their study was limited, however, by the fact that they used a single-item indicator of fear of criminal victimization, a method often criticized in the fear of crime literature (see Ferraro, 1995, for a review).

Although there may be no reason to treat gun ownership among adults as a deviant behavior, gun possession among juveniles should be treated as such. Despite the fact that legislation allows adults to carry firearms in many areas, it is illegal in all areas for juveniles to carry firearms to school. This act is not only a violation of the law, it also undermines the authority of schoolteachers and other officials. For these reasons, this act must be considered delinquent.

Furthermore, it is expected that the correlates of delinquency should be the same as the correlates of juvenile gun ownership. Although many of the aforementioned evaluations offer various explanations as to why youths possess firearms, often arguing that the possession is due to structural or personality factors, none attempt to test the effects of fear of criminal victimization on adolescent firearm possession, while controlling for competing explanations of delinquent behavior These studies, although accepting that carrying a weapon for protection may be due to fear of criminal victimization, fail to examine the specific relationship between fear of criminal victimization and firearm possession. This study attempts to fill that void.

The study makes two major contributions to the literature concerning adolescent firearm possession. First, I test the effect of the fear of criminal victimization hypothesis on juvenile firearms possession at school using a cumulative index to represent fear of criminal victimization, an effort heretofore unexplored. Second, this analysis enhances the relevant scholarship by controlling for variables in Hirschi's (1969) social control theory and a derivative of Sutherland's (1939/1947) differential association theory, a much-needed improvement in this area (see Benda & Whiteside, 1995, for a discussion).

Following social control theory, youth whose bond to society is weakest will be more likely to carry guns to school; advocates of differential association would argue that delinquent tendencies of a juvenile's associates would induce the adolescent to carry firearms to school. On the other hand, following the fear of criminal victimization hypothesis, those youth who are most fearful would be most likely to carry guns to school. Thus, this study seeks to determine if fear of criminal victimization is associated with juvenile gun possession at school, controlling for two acknowledged explanations of delinquency.

⊠ Fear of Criminal Victimization Hypothesis

The fear of criminal victimization hypothesis argues that firearm ownership and possession among some people has been motivated by fear of criminal victimization (Newton & Zimring, 1969). Consequently, according to the fear of criminal victimization hypothesis, a group of gun owners purchase guns for protection from crime and criminals; consequently, those who are most fearful of criminal victimization will be those most likely to own a firearm.

According to the fear of crime literature, females, the elderly, those with lower income and education, urban residents, and African Americans are more fearful than their counterparts (Baumer, 1985; Belyea & Zingraff, 1988; Box, Hale, & Andrews, 1988; Braungart, Braungart, & Hoyer, 1980; Clemente & Kleiman, 1976, 1977; Garofalo, 1979; Kennedy & Krahn, 1984; Kennedy & Silverman, 1984; Larson, 1982; Lawton & Yaffee, 1980; Parker, 1988; Sharp & Dodder, 1985). It follows that these groups would also be the most likely to purchase and carry guns.

As alluded to earlier, tests of this hypothesis have produced inconclusive results. Some authors find that those who are most fearful are most likely to own guns for protection (Lizotte et al., 1981; Smith & Uchida, 1988); other researchers have found little evidence that fear of crime influences carrying a gun for protection (Adams, 1996; Arthur, 1992; Bankston & Thompson, 1989; Bankston, Thompson, Jenkins, & Forsyth, 1990; Williams & McGrath, 1976; Wright et al., 1983).

Although a tremendous volume of research has been produced in the area of fear of criminal victimization and its effect on adult gun ownership and use, two glaring inadequacies remain. The fear of criminal victimization hypothesis has not been tested on adolescents; by the same token, it has not been tested either against, or along with, any other theory of juvenile delinquency or as an explanation for delinquent behavior. The purpose of this study is to fill this void within the literature. This study will attempt to determine the impact of the fear of criminal victimization on juvenile firearms possession. This relationship then will be tested against the relationship between the control that social instructions (such as family and school) have and the influence that peers have on adolescents' firearm possession, to determine the association that these factors have on adolescent gun possession.

The fear of criminal victimization hypothesis, social control theory, and differential association all offer explanations for juvenile firearm possession; the substance of each theory's explanation, however, varies greatly. According to the fear of criminal victimization hypothesis, adolescents who are more fearful would be more likely to carry guns; advocates of control theory would argue that adolescents with weaker bonds to society would be more likely to carry guns; and finally, proponents of differential association theory would argue that youth whose peers were most approving of engagement in delinquency would be most likely to carry guns. Thus, if, after controlling for two known explanations of deviance, the relationship between fear of criminal victimization and adolescent firearm possession still persists, support for the fear of criminal victimization hypothesis will increase.

✺ Method

Sample

Data for this study were obtained from a study of Mississippi high school students conducted in the spring of 1992. As the South has the

highest rate of gun ownership by region (Kleck, 1997), and Mississippi was the site of the first of several highly publicized schoolyard shootings in the academic year of 1997–1998, gun-carrying behavior of Mississippi adolescents is of particular interest in this field of inquiry.

Dependent Variable

The dependent variable is juvenile firearm possession at school and was operationalized through the questions, "How many times have you carried a gun to school?" Responses were collapsed into two categories, with those who answered *never* coded (0) and those who responded *one or more* times coded as (1). Only 637 students (8.1%) indicated that they had carried a firearm to school one or more times.

Independent Variables

The independent variables include an index representing fear of criminal victimization, an index representing the strength of the adolescent's bond to society (social control), and a similar index representing deviant attitudes of the adolescent's peers (differential association). Due to their significant associations with delinquent behavior found in the literature, race, gender, residence, number of parental figures in the household, socioeconomic status, and gang membership are used as control variables. Finally, as neighborhood incivility has been determined to be a strong predictor of fear of crime among adults (Bursik & Grasmik, 1993; Will & McGrath, 1995), an index of perceived neighborhood incivility is included as a control variable as well. The questions used to measure the independent variables are in the appendix.

Perceived Neighborhood Incivility Index

The exogenous variable, perceived neighborhood incivility, was obtained by constructing an index using responses to statements concerning how the respondent viewed his or her neighborhood, such as "There are drug dealers in my neighborhood." The indicators used to construct the perceived neighborhood incivility scale are in the appendix. Item analysis was conducted on the index and Cronbach's alpha was used to determine its reliability. The construct demonstrated an internal reliability of .753, indicating that the index is a reliable measure of perceived neighborhood incivility.

Fear of Criminal Victimization Index

The fear of criminal victimization hypothesis states that those who are more fearful will be more likely to carry guns (Wright et al., 1983). This hypothesis does not postulate that excessive fear leads to crime commission. In contrast, its premise is that the fearful will carry firearms to protect themselves from crime.

Fear of crime among the adolescents was measured through construction of a fear of criminal victimization index composed of nine items. The items used to measure the respondent's fear are also in the appendix. The Cronbach's alpha for the scale was .896. It is hypothesized that those respondents scoring higher on the fear scale will be more likely to have carried a gun to school.

Social Control Index

The research reviewed earlier suggests that those adolescents with stronger bonds betweens themselves and family and school institutions will be less likely to commit delinquent acts (Hirschi, 1969; Jensen, 1972; Reiss, 1951; Wiatrowski et al., 1981). The adolescent's involvement with social institutions was measured by creating a social control index. The index consists of 11 questions or statements, which are included in the appendix. The Cronbach's alpha for the scale was .734. It is hypothesized that those adolescents scoring at the lower end of the scale will be more likely to

carry a firearm to school than those on the higher end of the scale.

Differential Association Index

As noted from the research reviewed earlier, those adolescents whose peers have the greatest amount of definitions favorable to violation of the law and engage in the most deviant acts are more likely to engage in delinquent acts (see Williams & McShane, 1999, for a review). Following differential association theory, an index was created to represent the deviant definitions and activities of one's peers.

Cronbach's alpha was used to measure the reliability of the index. The α for the scale was .614. It is hypothesized that those adolescents scoring at the higher end of the scale will be more likely to carry a firearm to school than those on the lower end of the scale.

Demographic Characteristics

It is feasible to suggest that as adolescents age and become familiar with their school surroundings and peers, they will be more able and thus more likely to obtain a firearm if they desire. Thus, it is hypothesized that older youth will be more likely to bring firearms to school than their younger counterparts.

Race will also be used as a control variable. As the overwhelming majority of Mississippi residents fall into one of two racial categories (White and African American), responses to the race question were dichotomized into two categories: White and Black. Several researchers have determined that Whites are less delinquent than Blacks (Cernkovich & Giordano, 1987; Rosen, 1985) and Whites are less fearful than non-Whites; subsequently, Blacks will be more likely to carry guns to school.

Regardless of the theory employed when predicting delinquency, males are more likely than females to be delinquent (Cernkovich & Giordano, 1987, 1992; Hirschi & Gottfredson,

1994). However, the fear of criminal victimization hypothesis would predict that females, the more fearful group, would be more likely to carry firearms. Sheley and Wright (1995) further determined that protective gun possession was more common among females than males in their sample. The question used to assess firearm possession in this study, however, does not allow the researcher to gauge reasons for possession. Thus, even though females may be more likely to possess firearms for protection, it is hypothesized that males will be much more likely to carry guns to school than will their female counterparts, a finding confirmed in numerous studies (e.g. Arria, Wood, & Anthony, 1995; Sheley & Wright, 1995).

Several researchers have identified an inverse relationship between socioeconomic status and delinquency (Reiss, 1951; Rosen, 1985; Wiatrowski et al., 1981). People with lower incomes also are more fearful. Thus, those youth from households with lower household income will be more likely to carry firearms to school.

Although some have argued that family structure does not affect delinquency, particularly violent delinquency (Salts, Lindholm, Goddard, & Duncan, 1995), many have indicated that juveniles from disrupted households (those with one parent) will be more likely than those from intact households (two or more adults in the household) to commit delinquent acts (Gove & Crutchfield, 1982; Hirschi & Gottfredson, 1994; Matsueda & Heimer, 1987; Reiss, 1951). Such findings support Hirschi's control theory. According to Hirschi (1969), two parents will have more control over the juvenile and his or her whereabouts than will one, especially if one of the two parents is not employed. It is reasonable to assume that the relationship will be the same for firearm possession at school as well. Following control theory, it is expected that those juveniles from disrupted family settings will be more likely to carry a gun to school than will their counterparts from households with two parents.

Finally, gang members have been demonstrated to engage in disproportionate amounts of delinquency, including carrying guns (Block & Block, 1993; Blumstein, 1995; Dukes, Martinez, & Stein, 1997; Knox, 1991; Spergel, 1990). Thus, it is hypothesized that gang members will be more likely to carry guns as well.

⊠ Results

Characteristics of the sample are presented in Table 1. Table 1 classifies the sample into gun carriers and nongun carriers. There were several statistically significant differences between the two groups. First, the mean score was significantly higher for the gun carriers than for the nongun carriers on the fear scale (17.101 vs. 14.271, $p < .001$) and the perceived incivility scale (7.139 vs. 5.842, $p < .001$). Gun carriers also scored significantly lower on the social control index (27.250) than did the nongun carriers (30.568, $p < .001$) and significantly higher on the differential association index (6.932 vs. 5.394, $p < .001$).

The percentage distributions in Table 1 also reveal important differences between gun and nongun carriers. As expected, the percentage of gun carriers who were male (83.7%) is much larger than the percentage of nongun carriers who were male (41.5%), as was the percentage of gun carriers who were gang members (61.5% vs. 4.6%). The percentage of gun carriers who were Black (55.4%) was also larger than the percentage of nongun carriers who were Black (46.2%). A lower proportion of the gun carriers came from two-parent households (50.7%) than did nongun carriers (58.0%). Furthermore, a larger percentage of youth from the 17 to 18 years and 19 to 20 years age groups (61.1% and 11.8%, respectively) were found among the gun carriers than among the nongun carriers (54.6% and 5.2%, respectively). Finally, there was little difference between gun carriers and nongun carriers across household income or place of residence.

Table 1	Characteristics of Gun Carriers and Nongun Carriers				

	Gun Carriers		Nongun Carriers	
Variable	**M**	**SD**	**M**	**SD**
Social control scale	27.250***	4.915	30.568	4.599
Differential association scale	6.932***	1.946	5.394	1.598
Fear scale	17.101***	6.215	14.271	4.812
Perceived incivility	7.139***	2.302	5.842	2.048
	n	**%**	**n**	**%**
Gender				
Male	533	83.7	3,011	41.5
Female	104	16.3	4,238	58.5
Total n	637		7,249	
Race				
White	284	44.6	3,901	53.8
Black	353	55.4	3,348	46.2
Total n	637		7,249	
Age (years)				
13–14	6	0.9	50	0.7
15–16	166	26.1	2,866	39.6
17–18	389	61.1	3,954	54.6
19–20	75	11.8	376	5.2
Total n	636		7,246	
Parental arrangement				
2 or more parents (intact) household	323	50.7	4,205	58.0
Single-parent (disrupted) household	313	49.1	3,036	41.9
Total n	636		7,241	
Family income				
Less than $10,000	8	13.8	922	12.7
$10,000–$19,999	101	15.9	1,368	18.9
$20,000–$29,999	140	22.0	1,546	21.3
$30,000–$39,999	120	18.8	1,503	20.7
More than $40,000	159	25.0	1,652	22.8
Total n	608		6,991	
Gang membership				
Yes	286	61.5	315	4.6
No	179	38.5	6,523	95.4
Total n	465		6,838	

(Continued)

Table 1	(Continued)			
	n	%	*n*	%
Place of residence				
Rural farm	72	11.3	579	8.0
Rural nonfarm	101	15.9	1,580	21.8
Outside suburbs	90	14.1	1,020	14.1
In suburbs	169	26.5	1,831	25.3
Near center of city	202	31.7	2,212	30.5
Total *n*	634		7,222	

***$p < .001$

The results presented in Table 1 indicate that significant associations exist between adolescent firearm possession and the independent variables included in this study. To determine the effect of the factors reflected in the perceived incivility index, the fear index, the social control index, and the differential association index on adolescent firearm possession, controlling for the demographic variables, logistic regression will be used.

The results of regressing gun possession at school on the variables in this study are presented in Table 2. The first logistic regression model presented in Table 2 regresses adolescent gun possession on those demographic variables that have been determined to have statistically significant associations with firearm ownership among adults, namely, gender, race, age, household income, place of residence, and gang membership. In addition, due to the adolescent nature of the sample, a variable representing parental presence in the household is included as well. In the second model, perceived neighborhood incivility and the index representing fear of crime are added and adolescent firearm possession is regressed on these variables and the set of demographic variables. The final model includes the index representing social control and differential association, with the previous demographic variables, the fear of crime index, and the

index representing perceived neighborhood incivility. As there was an inordinate amount of missing data for the variables included in the social control, differential association, and perceived neighborhood incivility indices (which, analysis reveals, was due to respondent attrition because of their placement at the end of the questionnaire), race-sex subgroup mean substitution is used throughout the analysis.

The results obtained from regressing adolescent firearm possession on the demographic variables are presented in the first model in Table 2. The results indicate that, as expected, gender, race, age, parental structure, income, and gang membership had statistically significant associations with adolescent firearm possession. The regression coefficients indicate that males were six times as likely as females to possess a firearm at school whereas Blacks were 1.4 times as likely as Whites to carry a firearm to school. Older adolescents were also more likely to carry guns to school than were their younger counterparts. The association between parental structure and adolescent firearm possession was also significant, as those youth from households with other than two adults were significantly more likely than their counterparts from two-parent homes to carry a firearm to school. Those students from families with higher incomes were more likely than students from families with lower

Table 2 Logistic Regression Results of Adolescent Gun Carrying on Demographic Variables, Incivility, Fear of Criminal Victimization, Social Control Index, and Differential Association Index

Variable	Model 1 β/SE	Model 1 Exp (B)	Model 1 Wald	Model 2 β/SE	Model 2 Exp(B)	Model 2 Wald	Model 3 β/SE	Model 3 Exp (B)	Model 3 Wald
Male	1.815/.132***	6.142	189.404	1.804/.133***	6.073	173.511	1.482/.137***	4.402	116.693
Black	.353/.117**	1.423	9.157	.208/.122	1.232	2.925	.487/1.28****	1.627	14.503
Age	.460/.086***	1.585	28.715	.470/.087***	1.601	29.481	.524/.089***	1.689	34.497
Two-parent home	−.301/.110**	.740	7.445	−.291/.112**	.748	6.696	−.264/.115*	.768	5.238
Family income	.084/.143*	1.088	3.879	.133/.044**	1.142	9.239	.120/.045**	1.128	7.214
Urban	−.028/.040	.973	.485	−.051/.041	.950	1.598	−.051/.042	.950	1.494
Gang member	2.198/.119***	9.003	338.761	2.004/.124***	7.415	262.080	1.665/.130***	5.283	165.247
Perceived incivility				.162/.023***	1.176	48.024	.105/.025***	1.111	18.186
Fear index				.069/.009***	1.072	59.212	.070/.009***	1.072	55.702
Social control index							−.079/.012***	.924	43.573
Differential association index							.277/.031***	1.319	78.637
Constant	−5.044/.287***		264.682	−7.759/.346***		358.552	−6.737/.593***		129.117
Listwise n	7,303			7,303			7,303		
−2 Log Likelihood	2,855.631			2,740.868			2,574.519		
Goodness of Fit	6,619.220			6,608.242			6,074.746		
Model chi-square	759.356****			874.119***			1,040.468***		

*p < .05. **p < .01. ***p < .001.

incomes to carry guns to school. Finally, gang members were nine times as likely to carry firearms to school as adolescents who were not in gangs. Place of residence had no statistically significant effect on adolescent firearm possession at school.

The second model in Table 2 presents the results of regressing adolescent firearm possession on the variables included in the first model with the addition of the perceived neighborhood incivility index and the fear of criminal victimization index. Males were again six times as likely as females to indicate that they had carried a firearm to school. Interestingly, however, with the addition of perceived neighborhood incivility and the fear of crime index, the effect of race became nonsignificant. Youth from two-parent homes were again significantly less likely than youth from other than two-parent homes to carry a firearm to school, whereas older youth continued to be significantly more likely than younger youth to carry a firearm to school. Gang members remained significantly more likely to carry a firearm to school than youth who were not in gangs. The association between household income and adolescent firearm possession remained statistically significant, with those youth from households with greater income again significantly more likely to carry a gun to school than their counterparts from households with lower incomes. The perceived incivility index and the fear of criminal victimization index had statistically significant effects on adolescent firearm possession at school, indicating that those youth who perceived their neighborhood to be most disorderly and those youth who were most fearful of criminal victimization were significantly more likely to carry a gun to school than their counterparts, who perceived their neighborhood to be less disorderly and who were less fearful, respectively. Thus, following the fear of criminal victimization hypothesis, those youth who are most fearful and who perceive their neighborhoods as most disorderly are most likely to carry a gun to school.

The third model in Table 2 presents the results of regressing adolescent firearm possession on the variables included in the second model with the addition of the social control index and the differential association index. Males, Blacks, older respondents, youth from other than two-parent homes, gang members, and respondents with higher levels of household income continued to be significantly more likely than their counterparts to carry a firearm to school, whereas those adolescents who perceived their neighborhoods as disorderly and those adolescents who were most fearful were again significantly more likely than their counterparts to indicate that they had carried a firearm to school. Moreover, those who scored lower on the social control index and those who scored higher on the differential association index respectively) were significantly more likely to indicate that they had carried a firearm to school. Thus, as expected, those youth with weaker bonds to society and those youth with more deviant friends were more likely to carry a firearm to school.

⊠ Discussion

The results of this study generally support the findings of previous studies dealing with social control, differential association, and delinquent acts. Those children who scored lower on the social control scale, exhibiting lower parental and familial attachment, were significantly more likely to possess firearms at school, as were those youth who scored higher on the index measuring differential association. Both of these findings are consistent with the literature reviewed earlier.

The findings concerning family structure also conform with those from other studies; namely, those children from single-parent homes were significantly more likely to carry guns to school. The evaluation also is consistent with previous studies in demonstrating that males, Blacks, and gang members were more likely to commit the delinquent act in question. As expected, age was positively associated with gun carrying; the older youth were more likely to carry guns to school than were their younger counterparts.

Household income also had a statistically significant positive association with carrying firearms to school; those youth who were from families with higher household income were more likely to carry firearms to school. As this finding is contrary to the literature concerning social control, differential association, and the fear of criminal victimization hypothesis, this finding deserves some explanation. There are two divergent explanations for this finding. It could be that those youth from homes with higher household income had greater financial resources to purchase firearms, thus allowing them to carry a gun to school, whereas other youth might choose some less expensive weapon. This study is limited in that it is unable to examine that relationship.

A second, and perhaps more likely, explanation is that the relationship between household income and adolescent firearm possession may be due to the methodology of the study. Adolescents were asked to estimate their household income; it is quite probable that many of the youth did not have this knowledge and may have exaggerated their household income. Finally, the size of the sample also leads one to question the substantive value of this finding. In a sample this large, associations may achieve statistical significance ($p < .05$) but not substantive significance. Whatever the explanation, this finding should be viewed with extreme caution.

Place of residence had no statistically significant association with adolescent firearm possession. This contradicted the hypothesis concerning the relationship between these variables. One explanation for this might be that the adolescents are unaware of the actual size of their residence; furthermore, inasmuch as the sample consists of adolescents from Mississippi, a primarily rural state, even those who classified themselves as living at the center of the city probably live in areas that could not be characterized as traditional, urban residents.

The most interesting finding from this study concerns the relationship between perceived neighborhood incivility, fear of criminal victimization, and adolescent firearm possession. Adolescents who perceived their neighborhoods as most disorderly and who were most fearful were significantly more likely to carry a gun to school. These findings support the fear of criminal victimization hypothesis: at least in part, students carry guns to school because of the fear of classmates and the perceived criminogenic conditions of their neighborhoods that they experience. It is evident that fear of criminal victimization in juveniles creates similar reactions to fear of criminal victimization in adults.

On further review, the association might not be as peculiar as it sounds. With the prevalence of guns in our society, guns are readily available for juveniles. As mentioned earlier, a majority of students know where to get a gun if they desire. Those students who are most fearful might take action to get a gun and take that gun to school for protection. Another plausible explanation might be that those juveniles who are most delinquent are also most fearful of other delinquents with whom they may interact. If this is the case, their fear might be a result instead of a cause of delinquency. It could be that the delinquents develop their fear after they have committed delinquent acts.

It also could be that adolescents' fear might be explained by the lifestyles approach, which states that those individuals who put themselves in situations in which crimes will occur are more likely to commit crimes and to be victimized by them (Cohen & Felson, 1979; Hindelang, Gottfredson, & Garofalo, 1978). These adolescents could very well develop the fear because they see so many delinquent acts being committed by their peers (as this study shows, those adolescents whose peers were most delinquent, as well as those youth who were gang members, were also more likely to carry firearms) and thus become fearful that one day their delinquent friends, or adversaries of their friends, might harm them. It is beyond the nature of this study to test this temporal relationship. However, further research in this area would be beneficial. A longitudinal study might be used to determine which occurs first; the fear of the firearm-related delinquency.

⬚ Implications

The purpose of this study was to examine the association between the fear of criminal victimization hypothesis and adolescent firearm possession, while controlling for more traditional explanations of delinquency—namely, differential association and social control theory. Analysis of the results indicate that the fear of criminal victimization hypothesis, social control theory, and differential association are all statistically significant indicators of adolescent firearm possession.

The implications of this study are twofold. First, social control theory, differential association theory, and the fear of criminal victimization hypothesis have a significant association with firearms possession. Second, the fear of criminal victimization hypothesis seems to apply to gun ownership and possession at least as well among juveniles as among adults.

As the findings from this study indicate, at least part of the reason that youth carry firearms to school is because of a fear of criminal victimization at school and a perception that their neighborhoods are dens of criminal activity. The findings from this study explicate a phenomenon known to social science researchers for quite some time but only recently identified and prioritized by policy makers. To reduce the number of firearms in school, the circumstances that cause youth to bring firearms to school must be reduced. Although taking steps to alleviate the problem of fear on the school grounds is important, it is equally important that these steps be taken

in the neighborhoods where these youth live as well. It is imperative that we find measures to reduce the fear of crime of students at school, as well as their fear of crime in their neighborhoods.

There are a number of steps that could be implemented in this regard that are already in place with adult populations. First, as Kenney and Watson (1998) suggest, adolescents should be empowered to make use of school and police resources to reduce fear and disorder problems at school and at home. By implementing problem-solving methods widely used by police and Neighborhood Watch programs throughout the country, adolescents can be empowered to identify sources of fear of crime in their environment and identify and implement steps aimed at alleviating those conditions.

Second, it is well demonstrated that much of the public perception of crime in the United States is strongly influenced by the media. Thus, it is imperative that more accurate information about the problem of guns and violence in schools be presented. Universities throughout the country regularly publish reports identifying the crime that goes on at their campus. It is quite possible that if junior high and high schools throughout the country were to implement a similar reporting system, the perception that crime, particularly crime with guns, is rampant at schools could be alleviated and thus reduced.

Finally, an important step in this effort to curb violence and fear of violence among adolescents might be to ensure that adolescents do not carry firearms to school. Although various measures have been implemented in an attempt to curb this activity (e.g., metal detectors, banning book bags, increased use of locker searches), each needs to be evaluated to determine its effectiveness in combating crime at school. Those measures that are most effective subsequently can be implemented in schools throughout the country. When these measures are institutionalized, the first step will be taken in reducing firearm possession at school and concomitantly reducing school violence.

⊠ References

Adams, K. (1996). Guns and gun control. In T. J. Flanagan & D. R. Longtime (Eds.), *Americans view crime and justice: A national public opinion survey* (pp. 109-123). Thousand Oaks, CA: Sage.

Arria, A. M., Wood, N. P., & Anthony, J. (1995). Prevalence of carrying a weapon and related behaviors in urban schoolchildren, 1989 to 1993. *Archives of Pediatric and Adolescent Medicine, 149,* 1345–1350.

Arthur, J. A. (1992). Criminal victimization, fear of crime, and handgun ownership among Blacks: Evidence from national survey data. *American Journal of Criminal Justice, 16*(2), 121–141.

Asmussen, K. J. (1992). Weapon possession in public high schools. *School Safety, 28,* 28–30.

Bankston, W., & Thompson, C. (1989). Carrying firearms for protection: A causal model. *Sociological Inquiry, 59*(1), 75–87.

Bankston, W., Thompson, C., Jenkins, Q., & Forsyth, C. (1990). The influence of fear of crime, gender, and Southern culture on carrying firearms for protection. *Sociological Quarterly, 31*(2), 287–305.

Baumer, T. (1985). Testing a general model of fear of crime: Data from a national sample. *Journal of Research in Crime and Delinquency, 22*(3), 239–255.

Belyea, M. J., & Zingraff, M. T. (1988). Fear of crime and residential location. *Rural Sociology, 53*(4), 473–486.

Benda, B. B., & Whiteside, L. (1995). Testing an integrated model of delinquency using LISREL. *Journal of Social Service Research, 21*(2), 1–32.

Bergstein, J. M., Hemenway, D., Kennedy, B., Quaday, S., & Ander, R. (1996). Guns in young hands: A survey of urban teenagers' attitudes and behaviors related to handgun violence. *Journal of Trauma, Injury, Infection, and Critical Care, 41*(5), 794–798.

Block, C. R., & Block, R. (1993). *Street gang crime in Chicago,* Washington, DC: National Institute of Justice.

Blumstein, A. (1995). *Violence by young people: Why the deadly nexus?* Washington, DC: National Institute of Justice.

Box, S., Hale, C., & Andrews, G. (1988). Explaining fear of crime. *British Journal of Criminology, 28*(3), 340–356.

Braungart, M., Braungart, R., & Hoyer, W. (1980). Age, sex, and social factors in fear of crime. *Sociological Focus, 13*(1), 55–66.

Bursik, R. J., Jr., & Grasmick, H. G. (1993) *Neighborhoods and crime: The dimensions of effective community control.* Lexington, MA: Lexington Books.

Callahan, C. M., & Rivara, F. P. (1992). Urban high school youth and handguns: A school-based survey. *Journal of the American Medical Association, 267*(22), 3038–3042.

Center to Prevent Handgun Violence. (1993). *Kids carrying guns.* Washington, DC: Government Printing Office.

Cernkovich, S. A., & Giordano, P. C. (1987). Family relationships and delinquency. *Criminology, 25*(2), 295–321.

Cernkovich, S. A., & Giordano, P. C. (1992). School bonding, race, and delinquency. *Criminology, 30*(2), 261–289.

Chandler, K. A., Chapman, C. D., Rand, M. R., & Taylor, B. M. (1998). *Students' reports of school crime: 1989 and 1995.* Washington, DC: Department of Education and Justice.

Clemente, F., & Kleiman, M. (1976). Fear of crime among the aged. *The Gerontologist, 16*(3), 207–210.

Clemente, F., & Kleiman, M. (1977, December). Fear of crime in the United States: A multivariate analysis. *Social Forces, 56,* 519–531.

Cohen, L. E., & Felson, M. (1979). Social change and crime rate trends: A routine activities approach. *American Sociological Review, 44,* 588–608.

Dukes, R. L., Martinez, R. O., & Stein, J. A. (1997). Precursors and consequences of membership in youth gangs. *Youth and Society, 29*(2), 139–165.

Ferraro, K. F. (1995). *Fear of crime: Interpreting victimization risk.* Albany: State University of New York Press.

Fingerhut, L. (1993, March 23). *Firearm mortality among children, youth, and young adults 1-34 years of age, trends and current status: United States, 1985-90: Advance Data.* Hyattsville, MD: National Center for Health Statistics.

Garofalo, J. (1979). Victimization and the fear of crime. *Journal of Research on Crime and Delinquency, 16,* 80–97.

Gove, W. R., & Crutchfield, R. D. (1982). The family and juvenile delinquency. *Sociological Quarterly, 23*(3), 301–319.

Hechinger, F. M. (1992). *Fateful choice: Healthy youth for the 21st century.* New York: Hill and Wang.

Hemenway, D., Prothrow-Stith, D., Bergstein, J. M., Ander, R., & Kennedy, B. P. (1996). Gun carrying among adolescents. *Law and Contemporary Problems, 59*(1), 39–53.

Hindelang, M., Gottfredson, M., & Garofalo, J. (1978). *Victims of personal crime: An empirical foundation for a theory of personal victimization.* Cambridge, MA: Ballinger.

Hirschi, T. (1969). *Causes of delinquency.* Berkeley: University of California Press.

Hirschi, T, & Gottfredson, M. (1994). *The generality of deviance.* New Brunswick, NJ: Transaction Publishers.

Jensen, G. F. (1972). Parents, peers, and delinquent action: A test of the differential association perspective. *American Journal of Sociology, 78*(3), 562–575.

Kachur, S. P., Stennies, G. M., Powell, K. E., Modzeleski, W., Stephens, R., Murphy, R., Kresnow, M., Sleet, D., & Lowry, R. (1996). School-associated violent deaths in the United States, 1992 to 1994. *Journal of the American Medical Association, 275*(22), 1729–1733.

Kennedy, L. W., & Krahn, H. (1984). Rural-urban origin and fear of crime: The case for rural baggage. *Rural Sociology, 49*(2), 247–260.

Kennedy, L. W., & Silverman, R. A. (1984). Significant others and fear of crime among the elderly. *International Journal of Aging and Human Development, 20*(4), 241–256.

Kenney, D. J., & Watson, T. S. (1998). *Crime in the schools: Reducing fear and disorder with student problem solving.* Washington, DC: Police Executive Research Forum.

Kleck, G. (1997). *Targeting guns: Firearms and their control.* Hawthorne, NY: Aldine.

Knox, G. W. (1991). *An introduction to gangs.* Berrien Springs, MI: Vande Vere.

Larson, C. J. (1982). City size, fear, and victimization. *Free Inquiry in Creative Sociology, 10*(1), 13–22.

Lawton, M. P., & Yaffee, S. (1980). Victimization and fear of crime in elderly housing tenants. *Journal of Gerontology, 35,* 768–79.

Lizotte, A. J., Bordua, D. J., & White, C. S. (1981). Firearms ownership for sport and protection: Two not so divergent models. *American Sociological Review, 46,* 499–503.

Matsueda, R. L., & Heimer, K. (1987). Race, family structure, and delinquency: A test of differential association and social control theories. *American Sociological Review, 52*(6) 826–840.

Newton, G. D., & Zimring, F. E. (1969). *Firearms and violence in American life: A staff report to the national commission on the causes and prevention of violence.* Washington, DC: Government Printing Office.

Parker, K. D. (1988). Black-White differences in perceptions of fear of crime. *Journal of Social Psychology, 128*(4), 487–494.

Reiss, A. J. Jr. (1951). Delinquency as the failure of personal and social controls. *American Sociological Review, 16*(2), 196–206.

Rosen, L. (1985). Family and delinquency: Structure or function? *Criminology, 23*(3), 553–573.

Roth, J. A. (1994). *Firearms and violence.* Washington, DC: National Institute of Justice.

Salts, C. J., Lindholm, B. W., Goddard, H. W., & Duncan, S. (1995). Predictive variables of violent behavior in adolescent males. *Youth and society, 26*(3), 377–399.

Senate Hearings. (1993). *Children and gun violence.* Washington, DC: Government Printing Office.

Sharp, P. M., & Dodder, R. A., (1985). Victimization and the fear of crime: Some consequences by age and sex. *International Journal of Contemporary Sociology, 22*(1, 2), 149–161.

Sheley, J. F. (1994). Drug activity and firearms possession and use by juveniles. *The Journal of Drug Issue, 24*(3), 363–382.

Sheley, J. F., & Brewer, V. E. (1995). Possession and carrying of firearms among suburban youth. *Public Health Reports, 110,* 18–26.

Sheley, J. F., & Wright, J. D. (1993). *Gun acquisition and possession in selected juvenile samples.* Washington, DC: National Institute of Justice.

Sheley, J. F., & Wright, J. D. (1995). *In the line of fire: Youth, guns, and violence in urban America.* Hawthorne, NY: Aldine.

Smith, D. A., & Uchida, C. D. (1988). The social organization of self-help: A study of defensive weapon ownership. *American Sociological Review, 53,* 94–102.

Spergel, I. A. (1990). Youth gangs: Continuity and change. In M. Tonry & N. Morris (Eds.), *Crime and justice: A review of research* (Vol. 12, pp. 171–275). Chicago: University of Chicago Press.

Sutherland, E. H. (1947). *Principles of Criminology* (4th ed.). Philadelphia: J. B. Lippincott. (Original work published 1939)

U. S. Bureau of the Census. (1990). *1990 Census of population, social, and economic characteristics, United States.* Washington, DC: Author.

Wiatrowski, M. D., Griswold, D. B., & Roberts, M. K. (1981). Social control theory and delinquency. *American Sociological Review, 46,* 525–541.

Will, J. A., & McGrath, J. H. (1995). Crime, neighborhood perceptions, and the underclass: The relationship between fear of crime and class position. *Journal of Criminal Justice, 23*(2), 163–176.

Williams, F. P., III, & McShane, M. D. (1999). *Criminological theory* (3d ed.). Englewood Cliffs, NJ: Prentice Hall.

Williams, J. S., McGrath, J. H., III. (1976). Why people own guns. *Journal of Communication, 26,* 22–30.

Wilson, J. M., & Zirkel, P. A. (1994). When guns come to school. *American School Board Journal, 181,* 32–34.

Wright, J. D., & Marston, L. (1975). The ownership of the means of destruction: Weapons in the United States. *Social Problems, 23,* 92–107

Wright, J. D., Rossi, P., & Daly, K. (1983). *Under the gun: Weapons, crime, and violence in America.* Hawthorne, NY: Aldine.

REVIEW QUESTIONS

1. What does the author have to say about the impact of firearms in causing injury/death among juveniles and the existing research regarding this phenomenon? Do you find these statistics and existing findings convincing that this is a major problem in U.S. society? State your reasons.

2. How important do you feel the fear of victimization is among public high school students? How important do you feel it is for other individuals in U.S. society? Do you feel it justifies carrying weapons, even if it is illegal to do so?

3. How does fear of crime affect your daily life? For example, do you lock your doors/windows at home when you leave, or do you lock your car doors, or do you avoid going out to certain places/certain times? If not, why don't you take any of these precautions against crime?

4. According to the findings of the study, to what extent do the two theories (Sutherland's and Hirschi's) appear to have importance in explaining the reasons why high school students carry firearms to school out of fear? What does the author propose as policy implications? Do you agree with these recommendations, and explain your reasoning?

❖

READING

In this selection, Douglas Longshore, Eunice Chang, Shih-chao Hsieh, and Nena Messina examine a test of approximately 1,000 male drug offenders. The authors base this study on two control theoretical frameworks: social bonding theory and low self-control theory, which were both presented by Travis Hirschi (with the latter being co-authored by Michael Gottfredson).

SOURCE: Douglas Longshore, Eunice Chang, Shih-chao Hsieh, and Nena Messina, "Self-Control and Social Bonds: A Combined Control Perspective on Deviance," *Crime and Delinquency* 50, no. 4 (2004): 542–64. Copyright © 2004 Sage Publications, Inc. Used by permission of Sage Publications, Inc.

Although previous selections in this section have examined the former, low self-control theory has not been thoroughly examined in this section, so we will briefly review it now.

In 1990, Michael Gottfredson and Travis Hirschi presented a theory of criminality in their book, *A General Theory of Crime*. Like other control theories, it assumes that human beings are born with tendencies toward selfishness, greed, and so on and thus are naturally programmed to commit crime. Like other control theories, they claim that there are ways that individuals are inhibited or controlled from committing crimes against others, and encouraged to conform to the rules of society.

Gottfredson and Hirschi claim individuals are inhibited from committing crime by developing self-control, which is the result of consistent discipline, monitoring, and lack of abuse/neglect among caregivers of children. Unfortunately, as the theory of low self-control notes, numerous children do not have such consistent monitoring or discipline and, therefore, never develop adequate levels of self-control. If such adequate levels of self-control are not learned or acquired by the approximate age of 10, then those individuals will have problems with self-control, and thus criminality, for the rest of their lives, the authors believe. Like Hirschi's previous control theory of social bonding, this more recent theory of low self-control can be seen as related to Sigmund Freud's concept of the early formation of the psyche, in the sense that what occurs early in life (in this case, prior to the age of 10), is unlikely to be changed regardless of what occurs afterward. Thus, low self-control theory claims that if people have not developed self-control by the age of 10 or before, it will never be acquired.

Gottfredson and Hirschi propose at least six factors that individuals with low self-control would exhibit. These factors include: temperamental, risk-taking propensities, impulsive behavior (with little regard for long-term consequences), propensity for physical tasks (over intellectual activities), self-centeredness, and orientation toward simple tasks (as opposed to long-term commitments, such as long-term employment or education). Gottfredson and Hirschi claim that people who exhibit such qualities are far more likely to engage in criminal activity.

While examining this selection, readers are encouraged to think about people they have known among their own families, peers, or fellow students who seem to fit the personality of low self-control. In addition, readers should consider the extent to which they believe bonds to conventional society have limited their deviant or criminal behavior and which types of bonds have been strongest or weakest.

Self-Control and Social Bonds

A Combined Control Perspective on Deviance

Douglas Longshore, Eunice Chang, Shih-chao Hsieh, and Nena Messina

According to the general theory of crime (Gottfredson & Hirschi, 1990), variation in the propensity to engage in crime and other deviance is mainly a function of individual differences in self-control. The general theory of crime, featuring self-control as the central explanatory factor, contrasts with Hirschi's (1969) social bonding theory, in which deviance is a result of weak social bonds such as poor attachment to others and low involvement in conventional activities. It is not clear whether or how these two perspectives in control theory can be reconciled. One possibility, explored in the current study, is that social bonds mediate the relationship between self-control and deviance. It has also been suggested that association with deviant peers may mediate the influence of social bonds on deviance (Krohn, Massey, Skinner, & Lauer, 1983; Marcos, Bahr, & Johnson, 1986); that is, people whose peers expose them to and reinforce deviance are more likely to engage in deviance themselves and to have weak bonds to conventional peers (Akers, 1994). In addition, association with deviant peers may be characteristic of people with low self-control and may mediate the effect of Low Self-Control on deviance. Prior research has not tested these possibilities.

Theory integration can help to resolve disparate conceptual approaches in the field of criminology (Bernard & Snipes, 1996; Messner, Krohn, & Liska, 1989). Through identification of dominant themes, premises, hypotheses, and findings common across different disciplines or causal propositions, theory integration may lead to an "intellectual account" (Tittle, 2000) that offers more conceptual richness and greater predictive power than any one theory individually. For reasons explained below, we believe that the self-control and social bonding perspectives might be combined into one explanatory model in which social bonds and deviant peer association are treated as processes through which Low Self-Control exerts some of its influence on deviance. Using longitudinal data from a sample of 1,036 adult male drug offenders, we examined relationships between self-control, social bonds, deviant peer association, and drug use. We also tested the degree to which social bonds and peer association mediate the relationship between self-control and drug use.

Method

Sample

The article is based on data collected between 1991 and 1995 for an evaluation of Treatment Alternatives to Street Crime (TASC) programs in five U.S. cities. TASC programs assess the drug treatment needs of offenders in local criminal justice systems, refer drug-involved offenders to treatment and other services, and monitor their status. Treatment may be in lieu of, or an adjunct to, routine criminal justice processing. Evaluation results are reported in Anglin, Longshore, and Turner (1999) and Turner and Longshore (1998).

Data required for this analysis were complete for a sample of 1,036 adult male offenders. The ethnic breakdown of this sample was 59% African Americans; 35% non-Hispanic Whites; and 5% others, mostly Hispanics. Ages ranged from 18 to 64 years ($M = 30.8$) and offenders had completed 10.3 years of school on average. Of the sample, 25% said they were currently married or living with someone, and 66% were not employed at the time of the baseline interview. Most offenders had extensive criminal histories, two thirds (66%) had at least two prior felony convictions, and just more than one third (36%) had been younger than 15 years when first arrested. Most (77%) had been incarcerated at least once, and all offenders were on probation at the time of the baseline interview. Involvement in drug use was extensive, as would be expected in a sample of adult drug offenders referred to treatment by criminal justice. Lifetime use of marijuana was reported by 91% of the sample, crack cocaine by 47%, powder cocaine by 54%, and heroin by 28%.

Measures

Items intended to capture the constructs of interest were factor analyzed in SAS by means of maximum likelihood estimation and direct quartimin rotation. Items that formed reliable and distinct factors corresponding to the intended constructs were retained in confirmatory factor analyses and employed as factor indicators in subsequent analyses. Where appropriate, items were scored in reverse. Factor loadings from the confirmatory factor analysis are shown in Table 1. Low self-control, social bonds, deviant peer association, and baseline drug use were measured in an initial interview. Subsequent drug use was measured in a follow-up interview 6 months later. More than 80% of the sample were located and completed the follow-up interview.

Low self-control. Low self-control was measured with three multi-item indicators: impulsivity, based on four self-report items (e.g., "you act on the spur of the moment without stopping to think"); self-centeredness, based on four self-report items (e.g., "you look out for yourself first, even if it makes things hard on other people"); and volatile temper, based on four self-report items (e.g., "you lose your temper pretty easily"). Response options were *never, rarely, sometimes, often,* and *almost always.* When necessary, item scores were reversed so that higher values represent lower self-control. The three indicators are among the constituent elements of low self-control defined by Gottfredson and Hirschi (1990); see also Grasmick et al. (1993) Psychometric properties of these indicators are acceptable (Longshore et al., 1996, 1998).

Distinctive relationships between deviant conduct and some of self-control's constituent elements have been found (Arneklev et al., 1993; Longshore et al., 1996). However, self-control can defensibly be analyzed as a unidimensional construct (Arneklev et al., 1993; Evans et al., 1997; Grasmick et al., 1993; Piquero & Rosay, 1998)—at least among men (Longshore et al., 1996; Longshore et al., 1998), and a unitary measure of self-control is appropriate in an analysis testing hypotheses derived from a theory in which self-control is viewed as a unitary construct (Nagin & Paternoster, 1993).

Table 1	Confirmatory Factor Analysis
Factor	**Standardized Factor Loadings**
Low self-control	
Impulsivity	0.55
Self-centeredness	0.61
Volatile temper	0.64
Attachment	
Cooperation	0.81
Enjoyment	0.83
Listening/helping	0.78
Involvement	
Currently married/living with	0.64
Ever married/lived with	0.70
Duration of current/last employment	0.37
Beliefs	
Children should obey	0.27
Things called crime don't hurt	0.42
Okay to sneak into game/movie	0.59
Okay to sell alcohol to minors	0.52
Religious commitment	
Religion important	0.70
Religious preference	0.66
Born again	0.52
Drug/alcohol peers	
Friends like to drink	0.62
Friends use drugs	0.97
Previous drug use	
Number of drugs used	0.71
Frequency of use (logged)	0.99
Days of use (logged)	0.98
Follow-up drug use	
Number of drugs used	0.72
Frequency of use (logged)	0.99
Days of use (logged)	0.98

Attachment. Attachment was measured by three indicators of affective ties among family members "when you are around other members of your family." Questions asked how often there is (a) "a feeling of cooperation," (b) "enjoyment in being together," and (c) "an interest in listening and helping each other." The recall period for these questions was the past 6 months.

Response options were *never, sometimes, about half the time, usually,* and *always.* Higher values indicate stronger attachment.

These questions, derived from Marcos et al. (1986) and Krohn et al. (1983), are typical of Attachment indicators employed in tests of social bonding theory (Akers, 1994). Because our focus was on bonds constraining adult behavior, questions pertained to the current family, not the family of origin. All men in the sample reported interaction with their families during the baseline recall period (past 6 months); thus, coding and interpretation of responses was not complicated by lack of recent interaction with family members.

Involvement. Involvement in a conventional lifestyle is the temporal aspect of social bonding. Involvement indicators in our data set pertained to the person's history and stability of intimate relationships; specifically, (a) whether the person is currently married or living with someone, (b) whether he has ever been married or lived with someone, and (c) duration of current or most recent employment. Higher values indicate greater involvement. These indicators are similar to those in Sampson and Laub (1990) and Burton, Cullen, Evans, and Dunaway (1994).

Moral belief. This bond represents adherence to a general belief that the rules of conventional society are binding. Our belief factor was based on endorsement of four items: (a) "many things called crime do not really hurt anyone": (b) "when parents set down a rule, children should obey"; (c) "it is okay to sneak into a ballgame or movie without paying"; and (d) "even though it is against the law, it is okay to sell alcohol to

minors." Response options were: *strongly disagree, disagree, undecided, agree,* and *strongly agree.* Scoring on items (1), (3), and (4) was reversed so that higher values represent stronger endorsement of conventional moral belief. These items were employed in Marcos et al. (1986) and Massey and Krohn (1986).

Religious commitment. The fourth bonding factor is stake in conformity, or devotion to conventional lines of action (Nagin & Paternoster, 1994). Commitment is typically measured as educational or job aspirations, time spent on homework or other conventional activities, and/or religiosity (Akers, 1994). However, a key problem in bonding studies is that Commitment, so measured, is difficult to distinguish from Involvement, especially when the commitment indicators implicitly or explicitly ask about time spent in conventional activities or time invested in conventional goals (Conger, 1976; Krohn et al., 1983; Massey & Krohn, 1986).

In accord with other research (e.g., Akers, 1994; Burkett & Warren, 1987; Krohn et al., 1983), we used indicators of religiosity to capture the constraining effect of Commitment. The three indicators were (a) "how important is religion in your life" (*not important, a little important, important,* or *very important*); (b) "how would you describe your current religious preference" (*none* versus *Catholic, Protestant, Jewish,* or *other*); and (c) "do you consider yourself a born-again Christian" (no or yes). Higher values indicate stronger religious commitment.

Association with substance-using peers. For a deviant peer association measure specific to the deviant conduct at issue (drug use), we used two indicators: (a) "how many of your friends like to drink" and (b) "how many of your friends use illegal drugs." (A third possible indicator, how many of your friends are involved in crime, did not load on this factor.) Response options were *none, some, about half, most,* or *all.* Higher values indicate greater association with substance-using peers.

Notably, this measure is not based on a simple count of peers who used drugs or alcohol. By measuring substance-using peers as a (non-numerical) proportion of total peers, we accounted for the fact that social networks can include conventional as well as deviant others in varying proportions and that ties to deviant others may not contribute to misconduct if such ties are outweighed by conventional ones (Marcos et al., 1986).

Drug use. At baseline and follow-up, offenders were asked to report their drug-use patterns over the prior 6 months. Respondents reported drug use for the 6-month period on a month-by-month basis moving backward. This procedure followed previous interview techniques shown to produce good recall data on alcohol and drug use (Fals-Stewart, O'Farrell, Rutigliano, Freitas, & McFarlin, 2000; O'Farrell, Fals-Stewart, & Murphey, 2003). Indicators for this factor were number of drugs used, frequency of drug use (log transformed), and number of days of drug use (log transformed) in the prior 6 months.

Analysis of Data

As a first step in the analysis, we adjusted for possible effects of assignment to TASC and demographic variables (race and age). Following the procedure employed by Newcomb and Bentler (1988), we partialed all three variables out of each relationship in the correlation matrix, thus removing their influence from the entire system of theory-relevant factors. In subsequent causal modeling, we were therefore able to maintain focus on predictors drawn from the two control perspectives. This procedure reduced the possibility of misspecification of the relationships of theoretical interest and, at the same time, avoided adding unduly to the complexity of the analytic model. Race and age indicators were

based on self-report. TASC assignment was measured as a dummy variable (TASC group = 1; comparison group = 0). We considered adjusting for additional background characteristics such as employment history, educational background, criminal history, and treatment experience. However, none of these was significantly correlated with the drug-use outcome. Accordingly, it was not necessary to include them in the partialing procedure.

We used confirmatory factor analysis to test the adequacy of the proposed measurement model and relationships among the latent factors. Each hypothesized factor predicted its proposed indicators, and factors were allowed to intercorrelate. Next we tested a structural equation model in which (a) low self-control predicted all four social bonds, (b) these five factors predicted substance-using peer association, and (c) all six factors predicted drug use at follow-up. Baseline drug use was employed as an additional predictor so that scores on the follow-up measure would reflect greater involvement in drug use after adjustment for baseline use. Paths were dropped from the initial model if they were not significant. The significance of possible indirect effects of low self-control on drug use was also examined.

The closeness of our hypothetical model to the empirical data was evaluated through goodness-of-fit indexes, one of which is the chi-square/degrees of freedom ratio. A chi-square value no more than twice the degrees of freedom in the model generally indicates a plausible, well-fitting model inasmuch as large sample sizes make it difficult to obtain nonsignificant chi-squares. In addition, the Comparative Fit Index (CFI), which ranges from 0 to 1, indicates the improvement in fit of the hypothesized model compared to a model of complete independence among the measured variables (Bentler, 1995). Values of 0.9 and higher are desirable and indicate that at least 90% of the covariation in the data was reproduced by the hypothesized model

(Bentler & Stein, 1992). Inasmuch as multivariate kurtosis was large (normalized Mardia's coefficient = 36.86), we relied on the Satorra-Bentler chi-square and robust CFI as the appropriate fit statistics, taking non-normality into account (Bentler & Dudgeon, 1996; Byrne, 1994).

Our measures of low self-control, social bonds, peer association, and baseline drug use were coterminous. The outcome measure was drug use during the 6-month follow-up period; after adjustment for baseline use, this measure reflected change in degree of drug-use involvement over that period. Thus, the temporal order is clear from predictors to outcome measure but not among the predictors themselves. We do not see the latter as a major problem. In the general theory of crime, self-control is said to be established early in life and to remain stable thereafter (Gottfredson & Hirschi, 1990). In his analysis of data from the Cambridge delinquent development study, Polakowski (1994) found that self-control had indeed remained "moderately stable" across a 4-year span (see also Arneklev, Cochran, & Gainey, 1996; Moffitt, Caspi, Silva, & Stouthamer-Loeber, 1995). Thus it is logical to use a coterminous self-control measure as an exogenous factor in an analysis in which the endogenous factors, social bonds and deviant peer association, are based on data also collected at baseline.

⚅ Results

We examined bivariate relationships between Low Self-Control and other factors to be included in the model. As shown in Table 2, Low Self-Control was strongly and inversely related to all four factors indicating strength of conventional social bonds. Offenders with low self-control also reported that a greater proportion of their peers were involved in substance use. Finally, measures of drug use were higher among persons with low self-control.

The four bonding factors were related consistently, though not always significantly, to drug use. The direction of these relationships was as hypothesized. Persons reporting more drug use at follow-up appeared to have weaker conventional bonds. Moral Belief was the bond most strongly linked to subsequent drug use. The other bonding factors were modestly related to subsequent drug use. The substance-using peer factor was associated positively with subsequent drug use.

The final structural equation model, with nonsignificant paths deleted, is shown in Figure 1 (parameter estimates changed only slightly with removal of nonsignificant paths).

Table 2	Correlation Matrix								
	M	**SD**	**1**	**2**	**3**	**4**	**5**	**6**	**7**
1. Low Self-Control	19.61	6.58							
2. Religious Commitment	2.95	1.64	−.30						
3. Attachment	7.5	3.7	−.38	.27					
4. Involvement	3.44	5.10	−.06[ns]	.14	.06[ns]				
5. Belief	15.62	2.33	−.42	.22	.15	.12			
6. Drug/Alcohol Peers	3.16	2.69	.26	−.13	−.09	−.03[ns]	−.22		
7. Follow-up Drug Use	3.71	3.49	.17	−.08	−.04[ns]	−.01[ns]	−.19	.27	

NOTE: All correlations except those marked ns are significant, $p < .05$.

Figure 1 Final Path Model

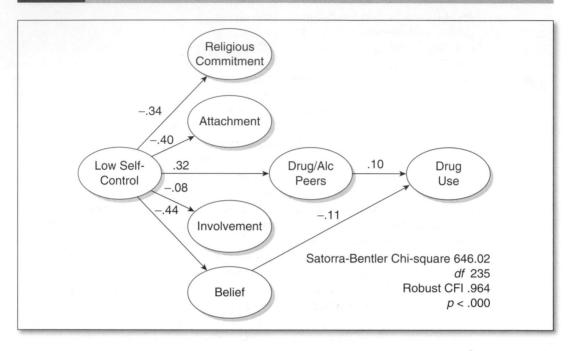

Satorra-Bentler Chi-square 646.02
df 235
Robust CFI .964
p < .000

Fit statistics for the model were highly favorable. The robust CFI = .964; Satorra-Bentler chi-square (*df* = 235, *n* = 1,036) = 646.02; *p* < .000. Substance-using peer association and one bonding factor, Moral Belief, had significant direct paths to subsequent drug use. Neither Low Self-Control nor any of the three other bonding factors directly predicted drug use. Thus, the bivariate relationship between low self-control and drug use was fully mediated by substance-using peer association and moral belief. About 19% of the variance in drug use (R^2 = .185) was explained by the model.

✖ Discussion

In the general theory of crime (Gottfredson & Hirschi, 1990), the propensity to engage in crime and other deviance is determined mainly by individual differences in self-control. This proposition contrasts with Hirschi's

(1969) own earlier view that deviance is mainly a result of weak social bonds. We tested the possibility that these two control perspectives might be integrated by positing social bonds, along with Deviant Peer Association, as outcomes of Low Self-Control and as mediators of the relationship between Low Self-Control and drug use. Moreover, this test was based on a data set that addressed important issues in prior research: data were longitudinal rather than cross-sectional; the sample was composed of adult men rather than juveniles; and deviant involvement, specifically illegal drug use, was substantial among these men.

All four bonding factors and the peer association factor were related strongly, in the expected direction, with low self-control. These results are consistent with the proposition that people with low self-control will also lack close emotional ties to conventional others, spend less time in conventional activities, evince a weaker commitment to conventional lifestyles, reject the view that prevailing moral values are binding on the individual, and associate with others involved in deviant conduct (Evans et al., 1997; Gottfredson & Hirschi, 1990; Hirschi, 1996; Nagin & Paternoster, 1994; Short, 1997; Wright et al., 2001). Bonding factors were also related as expected to drug use, although the relationship for three of the factors was weak. These results are consistent with social-bonds research using samples drawn from nonoffender or juvenile populations. That research has confirmed hypotheses in social bonding theory but has typically found that bonding measures explain only a modest portion of the variance in deviant outcomes (Akers, 1994; Kempf, 1993).

Conventional moral belief was the only social bond that mediated the relationship between low self-control and drug use. This result too is consistent with research using nonoffender and juvenile samples—research in which Moral Belief has emerged as the social bond most consistently correlated with deviance (Kempf, 1993). For example, in Elliott and Menard (1996), Moral Belief was the sole bonding factor associated with juvenile delinquency in general and with drug use in particular. Burkett and Warren (1987) found an association between moral belief and adolescent marijuana use (see also Krohn et al., 1983; Marcos et al., 1986). In Williams (1985) and Williams and Hawkins (1989), crime and marijuana use by adults were associated with Moral Belief (see also Burton et al., 1994). Finally, in the study by Evans et al., (1997), reported above, moral belief ("internal criminal values") appeared to mediate partially the relationship between low self-control and adult crime.

Gottfredson and Hirschi (1990) argued that the apparent influence of weak social bonding and deviant peer association on deviance is spurious because all these characteristics can be traced to low self-control, a trait emerging early in life and remaining stable. Our findings argue against viewing social bonds and deviant peer association as causally

trivial or irrelevant. Instead, an integrated explanatory model served to identify the more proximal processes through which low self-control may exert at least some of its influence on deviance. More specifically, and in accord with the implications of findings in Evans et al. (1997), we were able to identify Moral Belief and Deviant Peer Association as the mediating processes at work.

Although results support an integrated theoretical control perspective, only about 19% of the variance in drug use was explained by our model. The measures available to us may not have captured the underlying constructs adequately. However, other longitudinal research on low self-control has explained, at best, about the same portion of the variance in crime and other outcomes (e.g., Grasmick et al., 1993; Longshore, 1998b; Longshore et al., 1996). Similarly, longitudinal research on social bonds has explained only a limited portion of the variance in the outcomes tested (Akers, 1994). For example, bonding factors explained about 15% of the variance in adolescent smoking and about 2% of the variance in delinquency in longitudinal analyses reported by, respectively, Krohn et al. (1983) and Agnew (1985). One possible reason is that samples used in many prior studies were not involved in serious deviance; variability in the outcome measures may accordingly have been low. Unlike those samples, ours was composed of adult male offenders with extensive criminal histories, and the dependent variable in our analysis captured serious deviant involvement (frequent use of illegal drugs, chiefly marijuana and cocaine). Still, most of the variance in that outcome was left unexplained. Below we suggest alternatives by which explained variance might be improved.

The only bonding factor mediating the relationship between low self-control and drug use was conventional moral belief. This finding suggests that the mediating role of social bonding occurs mainly in the realm of internal constraint (Moral Belief) rather than the realm of affective ties (Attachment), investment in conventional lifestyle (religious commitment), or time spent in conventional activity (Involvement); that is, if low self-control influences deviance via social bonding, its effect may operate through internalization of deviant values and/or neutralization of conventional values (Kempf, 1993). In addition, conventional moral belief has been found to predict help seeking for drug problems and unfavorable attitudes toward drug use (Longshore, 1998a; Longshore, Grills, Annon, & Anglin, 1997; Longshore & Sanders-Phillips, 2000), and the desire to regain moral standing as a member of conventional society seems central to the recovery process (Biernacki, 1986; Waldorf, Reinarman, & Murphy, 1991). The mediating role of Moral Belief suggests that low self-control influences drug use partly by weakening the person's stake in conformity or, conversely, by elevating the person's feelings of social exclusion or stigma.

The predictive strength of substance-using peers may reflect differential association or social learning; that is, deviant conduct may be determined in part by normative and interpersonal influences, differential reinforcement of deviance, and modeling effects of substance-using peers (Akers, 1994). Such effects are not anticipated in theories of self-control or social bonding. However, the path from substance-using peers to deviant conduct can be read as consistent with control perspectives. It may, first, represent a sorting effect; persons with low self-control tend to flock together and share a taste for risk (Hirschi, 1969; Hirschi & Gottfredson, 1995). Substance-using peers may, second, represent exposure to greater opportunity for drug use (Evans et al., 1997; Kaplan, 1995).

We suggest three avenues for control-theory research attempting to increase the amount of variance explained in deviant conduct. First, the causal link from low self-control to bonds has been conceptualized in a simple one-way model in which low self-control is established early in life, remains stable, and

later has adverse effects on social bonds. The causal processes may be more dynamic, however. Weak social ties early in life may undermine the development of adequate self-control and sensitivity to others, thus setting in motion a vicious cycle in which weak social bonds and low self-control reinforce each other (Short, 1997). Greater variance might be explained in data sets designed to model these more complex causal processes.

Second, control theorists may gain insight by examining the role of low self-control and social bonds within more comprehensive integrated models, such as modified strain (Agnew, 1985, 1992), problem behavior (Elliott, Huizinga, & Menard, 1989), or control balance (Tittle, 1995). Causal factors in the problem behavior model, for example, are deviant as well as conventional bonds, early socialization, strain, and social disorganization. Low self-control might be folded into that model as an outcome of early socialization with effects on subsequent strain and on both types of bonding. In control balance theory, self-control might be handled as a constituent element of constraint (Tittle, 1995) or as a factor influencing one's ability to balance control effectively and thus reducing the likelihood that the person will use deviance to try to resolve a control imbalance (Tittle, 1997). Integrated theories can explain a healthy proportion of variance (50% or better) in crime and delinquency (e.g., Le Blanc, Ouimet, & Tremblay, 1988: Elliott et al., 1985; Matsueda & Heimer, 1987) and may be especially applicable to more serious and persistent deviance (Cohen & Vila, 1996). In short, research using an integrated theory approach might illuminate the conditions under which control factors exert strong effects on deviant behavior and might serve to locate these factors within an overall causal nexus.

A third approach to improving the variance explained by control factors is to identify moderators, that is, contingencies under which self-control and/or social bonds exert more influence on deviance. Among the set of potential moderating factors are aptitudes and skills for crime, motivations to commit crime, competing motivations that might divert people from crime despite low self-control or weak bonds, and rational choice variables (Tittle, 1995). Another possibility is that self-control is most closely linked to crime in early-onset than in late-onset offenders. Among early-onset offenders, behavioral problems indicative of low self-control manifest themselves early in childhood and result in weakened bonds to parents and others. Crime and other deviance are more serious and persistent among early-onset cases (Blackson, Tarter, & Mezzich, 1996; Jeglum-Bartusch, Lynam, Moffitt, & Silva, 1997; Lynam, 1996; Moffitt, 1993; Paternoster & Brame, 1997). Thus the causal processes in which control factors are pivotal in the production of later deviance may be stronger and easier to model among early-onset cases.

Finally, apart from social bonds, what other mediators might help to explain the effects of low self-control on deviance? As noted above, people with low self-control may experience greater strain, which may, in turn, lead to more deviant involvement (Elliott et al., 1989). Deterrence or rational-choice factors such as perceived pleasure of offending and perceived consequences of offending may play a mediating role as well (Nagin & Paternoster, 1993; Piquero & Tibbetts, 1996) if people with lower self-control derive more pleasure from offending, fail to foresee negative consequences, and discount such consequences more heavily.

In summary, the combination of self-control and social control perspectives shed some light on the causal processes by which low self-control may influence later deviance. However, more conceptual clarity may be gained by testing low self-control within broader integrated theories that account for factors outside the control tradition; identifying personal traits or social circumstances under which low self-control has more predictive value; and exploring the processes or mediating factors, including but not limited to social bonds, that explain the effects of low self-control on deviance.

◪ References

Agnew, R. (1985). Social control theory and delinquency: A longitudinal test. *Criminology, 23,* 47–62.

Agnew, R. (1992). Foundation for a general strain theory of crime and delinquency. *Criminology, 30,* 47–87.

Akers, R. L. (1994). *Criminological theories: Introduction and evaluation.* Los Angeles: Roxbury.

Anglin, M. D., Longshore, D., & Turner, S. (1999). Treatment alternatives to street crime: An evaluation of five programs. *Criminal Justice and Behavior, 26,* 168–195.

Arneklev, B J., Cochran, J. K., & Gainey, R. R. (1996, November). *Assessing the stability of low self-control.* Paper presented at the Annual Meeting of the American Society of Criminology, Chicago.

Arneklev, B. J., Grasmick, H. G., Tittle, C. R., & Bursik, R. J. (1993). Low self-control and imprudent behavior. *Journal of Quantitative Criminology, 9,* 225–247.

Bentler, P. M. (1995). *EQS structural equations program manual.* Encino, CA: Multivariate Software.

Bentler, P. M., & Dudgeon, P. (1996). Covariance structure analysis: Statistical practice, theory, and directions. *Annual Review of Psychology, 47,* 563–592.

Bentler, P. M., & Stein, J. A. (1992). Structural equation modeling in medical research. Statisti*cal Methods in Medical Research, 1,* 159–181.

Bernard, T. J., & Snipes, J. N. (1996)). Theoretical integration in criminology. In M. Tonry (Ed.), *Crime and justice: A review of restart* (Vol. 20, pp. 301–348). Chicago: University of Chicago Press.

Biernacki, P. (1986). *Pathways from heroin addiction: Recovery without treatment.* Philadelphia: Temple University Press.

Blackson, T. C., Tarter, R. E., & Mezzich, A. C. (1996). Interaction between childhood temperament and parental discipline practices on behavioral adjustment in preadolescent sons of substance abuse and normal fathers. *American Journal of Drug and Alcohol Abuse,* 22(3), 335–348.

Burkett, S. R., & Warren, B. O. (1987). Religiosity, peer associations, and adolescent marijuana use: A panel study of underlying caused structures. *Criminology, 25*(1), 109–131.

Burton, V. S., Jr., Cullen, F. T., Evans, T. D., & Dunaway, R. G. (1994). Reconsidering strain theory: Operationalization, rival theories, and adult criminality. *Journal of Quantitative Criminology, 10,* 213–239.

Byrne, B. M. (1994). *Structural equation modeling with EQS and EQS/Windows.* Thousand Oaks, CA: Sage.

Cohen, L. E., & Vila, B. J. (1996). Self-control and social control: An exposition of the Gottfredson-Hirschi/Sampson-Laub debate. *Studies on Crime and Crime Prevention, 5,* 125–150.

Conger, R. D. (1976). Social control and social learning models of delinquent behavior: A synthesis. *Criminology, 14,* 17–39.

Elliott, D., & Menard, S. (1996). Delinquent friends and delinquent behavior: Temporal and developmental patterns. In J. D. Hawkins (Ed.), *Delinquency and crime: Current theories* (pp. 28–67). New York: Cambridge University Press.

Elliott, D., Huizinga, D., & Ageton, S. S. (1985). *Explaining delinquency and drug use.* Beverly Hills, CA: Sage.

Elliott, D., Huizinga, D., & Menard, S. (1989). *Multiple problem youth delinquency, substance use, and mental health problems.* New York: Springer-Verlag.

Evans, T. D., Cullen, F. T., Burton, V. S., Jr., Dunaway, R. G., & Benson, M. L. (1997). The social consequences of self-control: Testing the general theory of crime. *Criminology, 35*(3), 475–504.

Fals-Stewart, W., O'Farrell. T. J., Rutigliano, P., Freitas, T., & McFarlin, S. K. (2000). The timeline followback reports of psychoactive substance use: Psychometric properties. *Journal of Consulting and Clinical Psychology, 68,* 134–144.

Gottfredson, M. R., & Hirschi, T. (1990). *A general theory of crime.* Stanford, CA: Stanford University Press.

Grasmick, H. G., Tittle, C. R., Bursik, R. J., & Arneklev, B. J. (1993). Testing the core empirical implications of Gottfredson and Hirschi's general theory of crime. *Journal of Research in Crime and Delinquency, 30,* 5–29.

Hirschi, T. (1969). *Causes of delinquency.* Berkeley: University of California Press.

Hirschi, T. (1996, November). *Control theory and the stability assumption: Inherent or imposed?* Paper presented at the Annual Meeting of the American Society of Criminology, Chicago.

Hirschi, T., & Gottfredson, M. R. (1995). Control theory and the life-course perspective. *Studies on Crime and Crime Prevention, 4*(2), 131–142.

Jeglum-Bartuseh, D. R., Lynam, D. R., Moffitt, T. E., & Silva, P. A. (1997). Is age important? Testing a general versus a developmental theory of antisocial behavior. *Criminology, 35*(1), 13–48.

Kaplan, H. B. (1995). *Drugs, crime and other deviant adaptations: Longitudinal studies.* New York: Plenum.

Kempf, K. L. (1993). The empirical status of Hirschi's control theory. In F. Adler & W. S. Laufer (Eds.), *New directions in criminological theory* (Vol. 4, pp. 143–185). New Brunswick, NJ: Transaction Publishers.

Krohn, M. D., Massey. J. L., Skinner, W. F., & Lauer, R. M. (1983). Social bonding theory and adolescent cigarette smoking: A longitudinal analysis. *Journal of Health and Social Behavior. 24*(4), 337–349.

Le Blanc, M., Ouimet, M., & Tremblay, R. E. (1988). An integrative control theory of delinquent behavior: A validation, 1976–1985. *Psychiatry, 51,* 164–176.

Longshore, D. (1998a). Drug problem recognition among Mexican American drug users. *Hispanic Journal of Behavioral Sciences, 20*(2), 270–275.

Longshore, D. (1998b). Self-control and criminal opportunity: A prospective test of the general theory of crime. *Social Problems, 45*(1), 103–114.

Longshore, D., Grills, C., Annon, K., & Anglin, M. D. (1997). Desire for help among African American drug users. *Journal of Drug Issues, 27*(4), 755–770.

Longshore, D., & Sanders-Phillips, K. (2000). Moral belief and drug problem recognition in three ethnic groups. In J. A. Levy, R. C. Stephens, & D. C. McBride (Eds.), *Emergent issues in the field of drug abuse* (pp. 177–191). Stamford, CT: Jai.

Longshore, D., Stein, J., & Turner, S. (1998). Reliability and validity of a self-control measure: Rejoinder *Criminology, 36*(1), 175–182.

Longshore, D., Turner, S., & Stein, J. A. (1996). Self-control in a criminal sample: An examination of construct validity. *Criminology, 34*(2), 209–228.

Lynam, D, R. (1996). Early identification of chronic offenders: Who is the fledgling psychopath? *Psychological Bulletin. 120*(2), 209–234.

Marcos, A. C., Bahr, S. J., & Johnson, R. E. (1986). Test of a bonding/association theory of adolescent drug use. *Social Forces, 65*(1), 135–161.

Massey, J. L., & Krohn, M. D. (1986). A longitudinal examination of an integrated social process model of deviant behavior. *Social Forces, 65*(1), 106–134.

Matsueda, R. L., & Heimer, K. (1987). Race, family structure, and delinquency: A test of differential association and social control theories. *American Sociological Review, 52*, 826–840.

Messner, S. F., Krohn, M. D., & Liska, A. E. (1989). *Theoretical integration in the study of deviance and crime.* Albany: State University of New York Press.

Moffitt, T. E. (1993). Adolescence-limited and life-course-persistent antisocial behavior: A developmental taxonomy. *Psychological Review, 100*(4), 674–701.

Moffitt, T. E., Caspi, A., Silva, P. A., & Stouthamer-Loeber, M. (1995). Individual differences in personality and intelligence are linked to crime: Cross-context evidence from nations, neighborhoods, genders, races and age-cohort s. *Current Perspectives on Aging and the Life Cycle, 4*, 1–34.

Nagin, D. S., & Paternoster, R. (1993). Enduring individual differences and rational choice theories of crime. *Law and Society Review, 27*, 467–496.

Nagin, D. S., & Paternoster, R. (1994). Personal capital and social control: The difference implications of a theory of individual differences in criminal offending. *Criminology, 32*(4), 581–606.

Newcomb, M. D., & Bentler, P. M. (1988). Impact of adolescent drug use and social support on problems of young adults: A longitudinal study. *Journal of Abnormal Psychology, 97*, 64–75.

O'Farrell, T. J., Fals-Stewart, W., & Murphey, M. (2003). Concurrent validity of a brief self-report drug use frequency measure. *Addictive Behaviors, 28*, 327–337.

Paternoster, R., & Brame, R. (1997). Multiple routes to delinquency? A test of developmental and general theories of crime. *Criminology, 35*(1), 49–84.

Piquero, A., & Rosay, A. B. (1998). The reliability and validity of Grasmick et al.'s self-control scale: A comment on Longshore et al. *Criminology. 36*, 157–173.

Piquero, A., & Tibbetts, S. (1996). Specifying the direct and indirect effects of low self-control and situational factors in offenders' decision making: Toward a more complete model of rational offending. *Justice Quarterly, 13*(3), 481–510.

Polakowski, M. (1994). Linking self- and social-control deviance: Illuminating the structure underlying a general theory of crime and its relation to deviant activity. *Journal of Quantitative Criminology, 10*, 41–78.

Sampson, R. J., & Laub, J. H. (1990). Crime and deviance over the life course: The salience of adult social bonds. *American Sociological Review, 55*. 609–627.

Short, J. R., Jr. (1997). *Poverty, ethnicity, and violent crime.* Boulder, CO: Westview.

Tittle, C. (1995). *Control balance: Toward a general theory of deviance.* Boulder, CO: Westview.

Tittle, C. (1997, November). *The limits of theoretical integration.* Paper presented at the Annual Meeting of the American Society of Criminology. San Diego, CA.

Tittle, C. (2000). Theoretical developments in criminology. In *The nature of crime: Continuity and change* (Vol. 1, pp. 51–101). Washington, DC: U.S. Department of Justice, Office of Justice Programs.

Turner, S., & Longshore, D. (1998). Evaluating the Treatment Alternatives to Street Crime (TASC) program. In J. Petersilia (Ed.), *Community corrections: Probation, parole, and intermediate sanctions* (pp. 134–141). New York: Oxford University Press.

Waldorf, D., Reinarman, C., & Murphy, S. (1991). *Cocaine changes: The experience of using and quitting.* Philadelphia: Temple University Press.

Williams, F. P., III. (1985). Deterrence and social control: Rethinking the relationship. *Journal of Criminal Justice, 13*, 141–151.

Williams, K. R., & Hawkins, R. (1989). Controlling male aggression in intimate relationships. *Law and Society Review, 23*(4), 591–612.

Wright, B. R., Caspi, A., Moffitt, T. E., & Silva, P. A. (2001). The effects of social ties on crime vary by criminal propensity: A life-course model of interdependence. *Criminology, 39*(2), 321–351.

1. What component of the social bond was found to have direct effects on drug use by the participants in this study?

2. Are you surprised that other elements of the social bond did not have direct effects on drug use? Which component of the social bond were you most surprised did not have an effect on drug use, and why?

3. Explain the finding that low self-control had an indirect effect on drug use. Which variable intervened in this relationship, and is this consistent with your experience or with the experience with others you know who use drugs?

❖

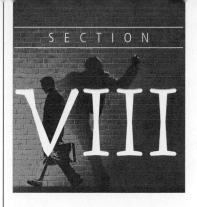

VIII

SOCIAL REACTION, CRITICAL, AND FEMINIST MODELS OF CRIME

The introduction of this section will discuss the evolution of social reaction/labeling theory, reviewing contributions made by early theorists, as well as the modern developments in this area. The introduction will then discuss social conflict/reaction models of criminal behavior, with an emphasis on the foundational assumptions and principles of Marx, as well as the more criminological applications of Marxist and conflict theory by Bonger, Turk, Vold, and others. This introduction will also review modern applications of various forms of feminism theory.

⊠ Introduction

During the 1960s and early 1970s, social reaction/labeling theories, as well as the various critical/conflict and feminist theories, became popular. At the time, society was looking for theories that placed the blame for criminal offending on government authorities, either the police or societal institutions like the economic structure or class structure. In this section introduction, we explore these various theories with a special emphasis on how they radically altered the way that crime and law were viewed, as well as how these perspectives highly represented the overall climate in the United States at that time. Specifically, many groups of people—particularly the lower-class, minorities, and women—were fighting for their rights during this period, and this manifested itself in criminological theory and research.

✎ Labeling/Social Reaction Theory

Social reaction theory, otherwise referred to as **labeling theory**, is primarily concerned with how an individual's personal identity is highly influenced by the way that society or authority tends to categorize them as offenders. Such categorization or labeling as an offender becomes a "self-fulfilling prophecy," in this perspective, and results in individuals confirming their status as criminals or delinquents by increasing the frequency or seriousness of their illegal activity. Furthermore, this perspective assumes that there is a tendency to put negative labels on lower-class individuals or minorities as offenders significantly more often than on middle- or upper-class White people.[1]

This perspective assumes that people who are labeled offenders have virtually no choice but to conform to the role that they have been "assigned" by society. Thus, social reaction theory claims that recidivism can be reduced by limiting stigmatization by authorities (e.g., law enforcement) and society. This is referred to as the "hands off" policy, and it became very popular in the 1960s and early 1970s.[2] Policies that became popular during this period were diversion, decriminalization, and deinstitutionalization (known as the "Ds") (see Section XI for a discussion of these policies). All of these attempted to get youthful or first-time offenders out of the formal justice system as soon as possible to avoid stigmatizing or labeling them as offenders. Today, these very policies have led critics to dismiss labeling theory by claiming that it promotes lenient and ineffective policy in sentencing practices.[3]

Labeling theory was based on seminal work by George Mead and Charles Cooley, which emphasized the importance of the extreme ways that individuals react to and are influenced by the social reaction to their roles and behavior. George Herbert Mead, who was a member of the Chicago School (see Section VI), said that a person's sense of self is constantly constructed and reconstructed through the various social interactions a person has on a daily basis.[4] Every person is constantly aware of how he or she is judged by others through social interactions.

Readers can probably relate to this in the sense that they have experienced how differently they are treated in stores or restaurants if they are dressed nicely, as opposed to

---

[1]George B. Vold, Thomas J. Bernard, and Jeffrey B. Snipes, *Theoretical Criminology*, 5th ed. (New York: Oxford University Press, 2002).

[2]Edwin Schur, *Radical Nonintervention: Rethinking the delinquency problem* (Englewood Cliffs, NJ: Prentice Hall, 1973).

[3]Ronald Akers and Christine Sellers, *Criminological Theories*, 4th ed. (Los Angeles: Roxbury, 2004).

[4]George H. Mead, *Mind, Self, and Society* (Chicago: University of Chicago, 1934).

being less well dressed; as you have observed, there is a significant difference in the way one is treated. Also, when growing up, you probably heard your parents or guardians warn that you should not hang out with "Johnny" or "Sally" because they are "bad kids." Or perhaps you were a "Johnny" or "Sally" at some point. Either way, you can see how certain individuals can be labeled by authorities or society and then ostracized by mainstream groups. This can lead to isolation and typically results in a person having only other "bad" kids or adults to hang out with. This results in a type of feedback system, in which the person begins associating with others who will only increase their propensity toward illegal activity.[5]

Many strain theorists claim that certain demographic factors, such as the offender's social class or the neighborhood where a certain offense took place, may make it more likely that the offender will be caught and labeled by authorities. This claim is quite likely to be true, especially given recent policing strategies that target areas/neighborhoods that have a high rate of crime. This is the side of social reaction/labeling theory that deals with the disproportionate rate at which members of the lower class and minorities are labeled as offenders.

Some of the earliest labeling theorists laid the groundwork for this perspective, long before it became popular in the 1960s. For example, in the 1930s, Frank Tannenbaum noted the "**dramatization of evil**" that occurred when youth were arrested and charged with their first offense.[6] Later, other theorists such as Edwin Lemert contributed a highly important causal sequence to how labeling affects criminality among those who are labeled. Lemert said that individuals, typically youth, commit **primary deviance**, which is not serious (i.e., it is nonviolent) and not frequent, but they happen to be caught by police and are subsequently labeled.[7] The stigma of the label makes them think of themselves as offenders and forces them to associate only with other "offenders." This results in what Lemert referred to as **secondary deviance**, in which offending is more serious (often violent) and far more frequent. Thus, the causal model that Lemert describes is illustrated as:

Primary Deviance → Caught and Labeled → Secondary Deviance

According to Lemert's model, if the label/stigma was not placed on a young or first-time offender, then the more serious and more frequent offending of secondary deviance would not take place. Therefore, Lemert's model is highly consistent with the labeling approach's hands-off policies such as diversion, decriminalization, and deinstitutionalization. If you ignore such behavior, Lemert reasons, it will tend to go away. However, since the mid-1970s, the "get-tough" approach has become highly dominant, so such policies are not emphasized by society or policymakers today.

Research on labeling theory suffered a significant blow in the 1970s and 1980s when empirical findings showed consistently that formal arrests/sanctions did not tend to result in ways that supported traditional labeling theory.[8] In fact, most people who are arrested

[5]Howard S. Becker, *Outsiders: Studies in the Sociology of Deviance* (New York: Free Press, 1963).

[6]Frank Tannenbaum, *Crime and the Community* (Boston: Ginn, 1938).

[7]Edwin Lemert, *Human Deviance, Social Problems, and Social Control,* 2nd ed. (Englewood Cliffs, NJ: Prentice Hall, 1972).

[8]Marvin Wolfgang, Robert Figlio, and Thorsten Sellin, *Delinquency in a Birth Cohort* (Chicago: University of Chicago Press, 1972).

once are never arrested again, which tends to support deterrence theory and does not support labeling theory. In addition, some experts have concluded that "the preponderance of research finds no or very weak evidence of [formal] labeling effects. . . . The soundest conclusion is that official sanctions by themselves have neither strong deterrent nor a substantial labeling effect."[9] Furthermore, some theorists have questioned the basic assumptions of labeling theory, pointing out that the label does not cause the initial (or "primary") offending and that labeling theorists largely ignore the issue of what is causing individuals to engage in illegal activity in the first place. Also, labeling theorists do not recognize the fact that offenders who are caught tend to be the ones who are committing more crimes than those who are not caught; in fact, there tends to be a strong relationship between being caught and committing multiple offenses.[10]

However, more contemporary research and theorizing have emphasized more informal forms of labeling, such as labeling by the community, parents, or friends. Studies have shown more support for the influence of this informal labeling on individuals' behavior

▲ **Photo 8.1** Karl Marx

(see Zhang's study later in this section). After all, it only makes sense that informal labeling by people with whom you interact on a daily basis (i.e., parents, friends, neighbors, employers, etc.) will have more impact in terms of how you feel about yourself than labeling by police or other authorities, which tends to be temporary or situational.

✑ Marxist Theories of Crime

Based on the writings of Karl Marx, Marxist theories of crime focus on the fact that people from the lower classes (i.e., poor) are arrested and charged with crimes at a disproportionate rate. As in conflict criminology, Marxist theories emphasize the effects of a capitalistic society on how justice is administered, describing how society is divided by money and power.[11] Marx said the law is the tool by which the *bourgeoisie* (the ruling class in a country without a ruling aristocracy; e.g., industrialists and financiers

---

[9]Akers and Sellers, *Criminological Theories*, 142.

[10]Charles Wellford, "Labeling Theory and Criminology," *Social Problems* 22 (1975): 313–32.

[11]Karl Marx, *Selected Works of Karl Marx and Frederick Engels* (Moscow: Foreign Languages Publishing House, 1962).

in Western industrialized countries) controls the lower classes (the ***proletariat*** and the lowest group, the *lumpenproletariat*) and keeps them in a disadvantaged position. In other words, the law is used as a mechanism by which the middle or upper class maintains their dominance over the lower classes. More specifically, Marx claimed that law is used as a tool to protect the economic interests and holdings of the bourgeoisie, as well as to prevent the lower classes from gaining access to financial resources.[12] Thus, Marxist theories propose that economic power can be translated into legal or political power and substantially accounts for the general disempowerment of the majority.

Willem Bonger. One early key theorist who applied Marxist theory to crime was Willem Bonger, who emphasized the relationship between the economy and crime but did not believe simply being poor would cause criminal activity. Rather, he thought crime resulted because capitalism caused a difference in the way individuals felt about society and their place in it. In the early 1900s, Bonger said that the contemporary economic structure, particularly capitalism, was the cause of crime in the sense that it promoted a system based on selfishness and greed.[13] Such selfishness manifests itself in competition among individuals, which is obvious in interactions and dealings for goods and resources. This competition and selfishness lead to more isolation, individualism, and "egoistic tendencies," which promote a strong focus on self-interest at the expense of communitarianism and societal well-being. Bonger believed that this strong focus on the individual leads to criminal behavior.[14] He also stressed the association between social conditions (largely the result of economic systems) and criminal offending; because of cultural differences, crime can be a normal, adaptive response to social and economic problems, he argued. The poor often develop a strong feeling of injustice, which also contributes to their entering into illegal activity.[15]

Richard Quinney. Although Bonger's theory did not become popular in the early 1900s, when his book was originally published, his ideas received a lot of attention when a neo-Marxist period began in the early 1970s. This renewed interest in Marxist theory was coupled with harsh criticisms leveled at the existing theoretical frameworks, which is why these neo-Marxist theories are often referred to as "critical theories." This time, Marxist theories of crime became quite popular, largely because the social climate desired such perspectives. Whereas notable European theorists in this vein include Ian Taylor, Paul Walton, and Jock Young,[16] one of the key figures in this neo-Marxist perspective in the United States was Richard Quinney.[17]

---

[12]Vold et al., *Theoretical Criminology.*

[13]Willem Bonger, *Criminality and Economic Conditions* (Bloomington: Indiana University Press,1969 [originally published in 1905]).

[14]Vold et al., *Theoretical Criminology*; Akers and Sellers, *Criminological Theories.*

[15]Bonger, *Economic Conditions.*

[16]Ian Taylor, Paul Walton, and Jock Young, *The New Criminology: For a Social Theory of Deviance* (New York: Harper & Row, 1973).

[17]Richard Quinney, *Critique of Legal Order: Crime Control in Capitalist Society* (Boston: Little, Brown and Company, 1974).

Like Bonger, Quinney claimed that crime was caused by the capitalistic economic structure and the emphasis on materials that this system produced. One way that Quinney's theory goes beyond Bonger is that Quinney further proposed that even the crimes committed by the upper classes were caused by capitalism. To clarify, Quinney claimed that such acts were crimes of "domination and repression" committed by the elite to keep the lower classes down or to protect their property, wealth, and power.[18] A good example is white-collar crimes, which almost always involve raising profits or income, whether for individual or company advantage; such crimes often result in losses to the relatively lower-income clients or customers.

Evidence Regarding Marxist Theories of Crime

Many critics noted that these seminal Marxist theories of crime were too simplistic, as well as somewhat naïve in the sense that they seemed to claim that the capitalistic economic system was the only reason for crime and that socialism/communism was the only sure way to reduce crime in the United States.[19] Now, even most Marxist theorists reject this proposition. Thus, more modern frameworks have been presented that place more emphasis on factors that stem from capitalism. For example, Colvin and Pauly presented a theoretical model that claims delinquency and crime are the result of problematic parenting, which results from the degrading and manipulative treatment that parents of lower-class children get in the workplace.[20] However, the empirical tests of this more modern Marxist theory have demonstrated rather weak effects regarding the importance of capitalism (or parenting practices resulting from employment positions/social class).[21] Thus, there does not seem to be much empirical support for Marxist or neo-Marxist theories of crime, which is perhaps why this theoretical framework is not one of the primary models currently accepted by most criminologists.[22]

⊠ Conflict Theories of Crime

Conflict theories of crime assume that all societies are in a process of constant change and that this dynamic process inevitably creates conflicts among various groups.[23] Much of the conflict among these groups is due to the competition to have a group's interests promoted, protected, and often put into law. If all groups were equally powerful and had the same amount of resources, such battles would involve much negotiation and compromise;

---

[18]Quinney, *Legal Order*; also see Akers and Sellers, *Criminological Theories*, 224–26, for a discussion.

[19]For a discussion, see Akers and Sellers, *Criminological Theories*, 226–31.

[20]Mark Colvin and John Pauly, "A Critique of Criminology: Toward an Integrated Structural Marxist Theory of Delinquency Production," *American Journal of Sociology* 89 (1983): 513–51.

[21]See Sally Simpson and Lori Elis, "Is Gender Subordinate to Class? An Empirical Assessment of Colvin and Pauly's Structural Marxist Theory of Delinquency," *Journal of Criminal Law and Criminology* 85 (1994): 453–80; see further discussion in Akers and Sellers, *Criminological Theories*.

[22]Lee Ellis and Anthony Walsh, "Criminologists' Opinions about the Causes and Theories of Crime and Delinquency," *The Criminologist* 24 (1999): 1–4.

[23]William Chambliss and Robert Seidman, *Law, Order, and Power* (Reading, MA: Addison-Wesley, 1971).

however, groups tend to differ significantly in the amount of power or resources. Thus, laws can be created and enforced such that powerful groups can exert dominance over the weaker groups. So like Marxist theories, law is seen as a tool by which some groups became and maintain dominance over less powerful groups. Furthermore, this state of inequality and resulting oppression creates a sense of injustice and unfairness among members of the less powerful groups, and such feelings are a primary cause of crime.[24]

There are several types of **conflict theory**, and fittingly for this framework (as well as inherently supportive of the model), the theorists from the varying types often give scathing reviews of the other types of conflict theory. Marxist theories are one example. Critics have noted that many communistic countries (e.g., Cuba, Soviet Union, etc.) have high rates of crime, whereas some countries that have a capitalistic economic structure have very low crime rates, such as Sweden.

Another type of conflict theory is referred to as pluralistic; it argues that instead of one or a few groups holding power over all the other groups, a multitude of groups all must compete on a relatively fair playing field.[25] However, this latter type of conflict theory is not one of the more popular versions among critical theorists because it is sometimes seen as rather naïve and idealistic.[26] Some of the key theorists in the **pluralistic (conflict) perspective** are Thorsten Sellin, George Vold, and Austin Turk.

Thorsten Sellin. Thorsten Sellin applied Marxist and conflict perspectives, as well as numerous other types of models, to studying the state of cultural diversity in industrial societies. Sellin claimed that separate cultures will diverge from a unitary, mainstream set of norms and values held by the dominant group in society.[27] Thus, these minority groups that break off from the mainstream will establish their own norms. Furthermore, when laws are enacted, they will reflect only the values and interests of the dominant group, which causes what Sellin referred to as "border culture conflict." This conflict of values that manifests itself when different cultures interact can cause a backlash by the weaker groups, which tend to react defiantly or defensively. According to Sellin, the more unequal the balance of power, the worse the conflict tends to be.[28]

George Vold. Another key conflict theorist was George Vold, who presented his model in his widely used textbook, *Theoretical Criminology*.[29] Vold claimed that people are naturally social and inevitably form groups out of shared needs, values, and interests. Because various groups compete with each other for power and to promote their values and interests, each group competes for control of political processes, including the power to create and enforce laws that can suppress the other groups. Some critics have argued that Vold put too much emphasis on the battle for creation of laws, as opposed to the enforcement of laws.[30]

---

[24]Vold et al., *Theoretical Criminology.*

[25]Akers and Sellers, *Criminological Theories.*

[26]Quinney, *Legal Order.*

[27]Thorsten Sellin, *Culture Conflict and Crime* (New York: Social Science Research Council, 1938).

[28]Sellin, *Culture Conflict.*

[29]George Vold, *Theoretical Criminology* (New York: Oxford University Press, 1958).

[30]Akers and Sellers, *Criminological Theories.*

Austin Turk. Like the other conflict theorists, Austin Turk assumed that the competition for power among various groups in society is the primary cause of crime.[31] Turk emphasized the idea that a certain level of conflict among groups can be very beneficial because it reminds citizens to consider whether the status quo or conventional standards can be improved. This type of idea is very similar to Durkheim's proposition that a certain level of crime is healthy for society because it defines moral boundaries and sometimes leads to progress (see Section V). Another aspect of Turk's theorizing that separates him from other conflict theorists is that he saw conflict among the various components of the criminal justice system. For example, the police often are at odds with the courts and district attorney's office. Turk and other conflict theorists will be discussed in much more depth in the selected readings later in this section.

Evidence Regarding Conflict Theories of Crime

Empirical tests of conflict theories are rare, likely because of the nature of the framework, which lends itself to a global view of societal structure and a perhaps infinite number of interest groups who are constantly in play for power.[32]

However, one notable study found evidence of a relationship between U.S. states that had a large number of interest groups and violent crime, but not property crime.[33] The authors concluded that these findings demonstrated the need for more discussion about how competitiveness in the United States affects criminal behavior, but no other studies have examined the influence of political interest groups on criminal behavior. It is rather difficult to test conflict theory in other ways. Perhaps conflict theory researchers should build an agenda of more rigorous ways to test the propositions of their theoretical perspective; as it stands, it remains quite vague. The few studies do not seem rigorous enough to persuade other criminologists or readers toward accepting the validity of this model.[34]

Despite the lack of empirical research supporting the conflict (and Marxist/critical) theories of crime, there is little doubt that such perspectives have contributed much to the theorizing and empirical studies of criminologists regarding this framework. In fact, the American Society of Criminology (ASC)—which is probably the largest and best known professional society in the discipline—has a special division made up of experts devoted to this area of study. So it is likely that the theorizing and empirical research will be greatly enhanced in the near future. Furthermore, it is clear that criminologists have acknowledged the need to explore the various issues presented by the conflict and Marxist perspectives in research in criminal justice and offending.

⊠ Feminist Theories of Crime

About the same time that Marxist theories of crime were becoming popular in the early-1970s, the feminist perspective began to receive attention; this was a key period in the

---

[31]Austin Turk, *Criminality and Legal Order* (Chicago: Rand McNally, 1969).

[32]See discussion by Akers and Sellers, *Criminological Theories*, 210–12.

[33]Gregory G. Brunk and Laura A. Wilson, "Interest Groups and Criminal Behavior," *Journal of Research in Crime and Delinquency* 28 (1991): 157–73.

[34]Ellis and Walsh, "Criminologists' Opinions."

women's rights movement. The Feminist School of criminology began largely as a reaction to the lack of rational theorizing about why females commit crime and why they tend to be treated far differently by the criminal justice system.[35] Prior to the 1970s, theories of why girls and women engage in illegal activities were primarily based on false stereotypes.

Much of the attention of theorists in this area can be broken into two categories: the "gender ratio" issue and the "generalizability" issue. The gender ratio issue refers to theorizing and research that examines why females so often commit less serious, less violent offenses than males. Some experts feel that this does not matter; however, if we understood why females commit far less violence, then perhaps we could apply such knowledge to reducing male offending.[36] The generalizability argument is consistent with the idea some make about the gender ratio issue; specifically, many of the same critics argue that theorists should simply take the findings found for male offending and generalize them to females. However, given the numerous differences found among males and females in what predicts their offending patterns, simply generalizing across gender is not a wise thing to do.[37]

Another important issue in feminist research on crime is that women today have more freedom and rights than those in past generations. Seminal theories of female crime in the 1970s predicted that this would result in far higher offending rates for women.[38] However, this has not been seen in serious violent crimes; the increases in property and

▲ **Photo 8.2** Female Gang Members

[35]Meda Chesney-Lind and Lisa Pasko, *The Female Offender: Girls, Women, and Crime* (Thousand Oaks, CA: Sage, 2004).

[36]For a discussion, see Stephen Tibbetts and Denise Herz, "Gender Differences in Factors of Social Control and Rational Choice," *Deviant Behavior* 17 (1996): 183–208.

[37]Tibbetts and Herz, "Gender Differences."

[38]Freda Adler, *Sisters in Crime: The Rise of the New Female Criminal* (New York: McGraw Hill, 1975).

public order crimes are typically committed by girls or women who have not benefited from such freedom and rights, for example, those who do not have much education or strong employment records.

Also, there are numerous forms of feminism, and thus many types of feminist theories of crime, as pointed out by Daly and Chesney-Lind.[39] One of the earliest was **liberal feminism**, which assumes that differences between males and females in offending were due to the lack of female opportunities in education and employment and that as more females were given such opportunities, they would come to resemble males in terms of offending.

Another major feminist perspective of crime is **critical/radical feminism**, which emphasizes the idea that many societies (such as the United States) are based on a structure of patriarchy, in which males dominate virtually every aspect of society, such as politics, family structure, and economy. It is hard to contest the primary assumption of this theory. Despite the fact that more women than ever hold professional, white-collar jobs, men still get paid a significant amount more than women for the same position, on average. Furthermore, the U.S. Senate and House of Representatives—and other high political offices such as president/vice president, cabinet, U.S. Supreme Court—are still made up primarily (or exclusively) by men. So the United States, like most other countries in the world, appears to be based on patriarchy.

The extent to which this model explains female criminality, however, remains to be seen. Regarding serious crimes, it is not clear why this perspective would expect higher or lower rates of female criminal behavior. Regarding some delinquent offenses, it may partially explain the greater tendency to arrest females for certain offenses. For example, virtually every self-report study ever conducted shows that males run away far more than females; however, FBI data show that female juveniles are arrested for running away far more often than males. This model may provide the best explanation for this difference. Females are more "protected"—that is, reported and arrested for running away—because they are considered to be more like property in our "patriarchal" society. This is just one explanation, but it appears to be somewhat valid.

Similar to critical or radical feminism is **Marxist feminism**, which emphasizes men's ownership and control of the means of economic production, thus focusing solely on the economic structure. Marxist feminists point out that men control economic success in our country, as well as in virtually every country in the world, and that this flows from capitalism. One of the primary assumptions of capitalism is a "survival of the fittest" or "the best person for the job," which would seem to favor women. Studies have found that women in the United States do far better, despite our capitalistic system, than those in most other countries. Furthermore, women in countries based on Marxism have a less favorable lifestyle and are no better off economically than those in the United States. Whether or not one believes in a Marxist economic structure, it does not readily explain female criminality.

Another feminist perspective is that of **socialist feminism,** which moved away from economic structure (e.g., Marxism) as the primary detriment for females and placed a focus on control of reproductive systems. This model believes that women should take control of their own bodies and their reproductive functions in order to control their criminality. It is not entirely clear how females' taking charge of their reproductive

---

[39]Kathleen Daly and Meda Chesney-Lind, "Feminism and Criminology," *Justice Quarterly* 5 (1988): 536–58.

destinies can increase or reduce their crime rates. Although no one can deny that data show females who reproduce frequently, especially in inner-city, poor environments, tend to offend more often than other females, it appears that other factors mediate these effects. Women who want a good future tend to take more precautions against becoming pregnant; on the other hand, the very females who most need to take precautions against getting pregnant are the least likely to do so, despite the availability of numerous forms of contraception. This is one of the many paradoxes in our field. It is unclear how much socialist feminism has contributed to an understanding of female criminality.

An additional perspective of feminist criminology is that of **postmodern feminism**, which holds that an understanding of women as a group, even by other women, is impossible because every person's experience is unique. If this is true, we should give up discussing female criminal theory and theories of criminality in general—along with all studies of medicine, astronomy, psychology, and so on because every person interprets each observation subjectively. According to postmodern feminists, there is no point in measuring anything. Thus, this model is based on "anti-science" and has contributed nothing to the study of female criminality.

In all these variations of the feminist perspectives, little emphasis is placed on parental differences in how children are disciplined and raised. Studies have clearly shown that parents, often without realizing it, tend to globally reward young boys for completing a task (e.g., "You are such a good boy"), whereas they tend to tell a young girl that she did a good "job." On the other hand, when young boys do not successfully complete a task, most parents tend to excuse the failure (e.g., "It was a hard thing to do, don't worry"), whereas for young girls the parents will often "globally" evaluate them for the task (e.g., "Why couldn't you do it?"). Although numerous psychological studies have found this tendency, it has yet to make it into the mainstream criminological theories of crime.

Evidence Regarding Feminist Theories of Crime

As discussed above, there is no doubt that female offenders were highly neglected by traditional models of criminological theory, and given that they make up at least 50% of the population of the world, it is important they be covered and explained by such theories. Furthermore, we also discussed the fact that if we know why females everywhere commit far less violence than men, it would likely go a long way toward policy implications for reducing the extremely higher rates of violence among males. However, in other ways, the feminist theories of crime have not been supported.

For instance, as noted above, the seminal feminist crime theories specifically proposed that as women became liberated, their rates of crime would become consistent with the rates of male offending.[40] Not only did this fail to occur, but the evidence actually supports the opposite trend; specifically, the females who were given the most opportunities (e.g., education, employment, status, etc.) were the *least* likely to offend, whereas the women who were not liberated or given such opportunities were the *most* likely to engage in criminal behavior.[41]

---

[40]Adler, *Sisters in Crime.*

[41]For a discussion, see Akers and Sellers, *Criminological Theories*; also see discussion in Vold et al., *Theoretical Criminology.*

On the other hand, one major strength of feminist theories of crime is that they have led to a number of studies showing that the factors causing crime in males are different than those for females. For example, females appear to be far more influenced by internal, emotional factors; they are more inhibited by moral emotions, such as shame, guilt, and embarrassment.[42] Ultimately, there is no doubt that feminist theories of crime have contributed much to the discourse and empirical research regarding why females (as well as males) commit crime. In fact, some highly respected criminology/criminal justice journals have been created to deal with that very subject. So in that sense, the field has recognized and accepted the need to explore the various issues involved in examining feminist theorizing and research in offending and the justice system.

⊠ Policy Implications

A variety of policy implications have come from the theoretical perspectives in this section. Regarding social reaction/labeling theory, several policies have evolved, known as the "D"s: diversion, decriminalization, and deinstitutionalization. **Diversion,** which is now commonly used, involves trying to get cases out of the formal justice system as soon as possible. Courts try to get many juvenile cases and, in most recent times, drug possession cases diverted to a less formal, more administrative process (e.g., drug courts, youth accountability boards/teams, etc.). Such diversion programs appear to have saved many billions of dollars for offenders who would have been otherwise incarcerated, while providing a way for first-time or relatively non-serious offenders not to experience the stigmatizing effects of being incarcerated. Although empirical evaluations of such diversion programs are mixed and suffer from methodological problems (i.e., the individuals who volunteer or qualify for such programs are likely the "better" cases among the sample population), some studies have shown promise for such programs.[43]

Regarding **decriminalization,** there have been numerous examples of reducing the criminality of certain illegal activities. A good example is the legal approach to marijuana possession in California. Unlike other jurisdictions, California does not incarcerate individuals for possessing less than an ounce of marijuana; rather, they receive a citation, similar to a parking ticket. There are many other forms of decriminalization, which is to be distinguished from legalization, which makes an act completely legal and not subject to legal sanction. The purpose of decriminalization is to reduce emphasis and resources placed on offenders who pose less danger to society, but in terms of social reaction/labeling theory, decriminalization also reduces the stigmatization of individuals who are relatively minor offenders, who would likely become more serious offenders if they were incarcerated with more chronic offenders.

Another policy implication of this section is **deinstitutionalization.** In the early 1970s, federal laws were passed that ordered all youth status offenders to be removed from incarceration facilities. This has not been accomplished; some are still being placed in such facilities. However, the number and rate of status offenders being placed in incarceration

---

[42]For a review, see Tibbetts and Herz, "Gender Differences."

[43]For a review, see John Worrall, *Crime Control in America: An Assessment of the Evidence* (Boston, MA: Allyn & Bacon, 2006): 228–29.

facilities has declined, avoiding any further stigmatization and integration into further criminality. Overall, this deinstitutionalization has kept relatively minor, often first-time offenders from experiencing the ordeals of incarceration.

Additional policy implications that can be inferred from this section involve providing more economic and employment opportunities to those who do not typically have access to such options. From a historical perspective, such as the New Deal during the Great Depression of the 1930s, providing more employment opportunities can greatly enhance well-being of the population, and in that period, there was a very significant decrease in the crime/homicide rate (see Introduction). Today, perhaps nothing could be more important in our nation than creating jobs; the ability to do this will largely determine the future crime rates.

Finally, there are numerous policy implications regarding feminist theory and perspectives of crime. Of primary importance is to include such perspectives in future research and theoretical developments. Furthermore, it is important to realize that females offend far less than males; if criminologists could figure out why, this could be a landmark finding in reducing crime significantly. Ultimately, the policy implications from this section indicate that the more attention and resources given to disenfranchised groups (e.g., youth, minorities, women, etc.), the less likelihood that they will offend, and the better we will understand the reasons why they offend.

▧ Conclusion

This section introduction examined the theories that place responsibility for individuals' criminal behavior on societal factors. Specifically, this section introduction discussed theories of social reaction or labeling, critical perspectives, and feminist theories of criminal behavior. All of these theories have a common theme: They emphasize the use of the legal or criminal justice systems to target or label certain groups of people (poor, women, etc.) as criminals, while protecting the interests of those who have power (i.e., White males). Given the evidence discussed in this section, readers will have enough evidence to make an objective conclusion about whether or not this perspective is valid, as well as which aspects of these theories are more supported by empirical research and which are more in question.

▧ Section Summary

- ◆ First, we explored the basic assumptions of social reaction/labeling theory.
- ◆ We then reviewed the primary theoretical concepts and propositions of labeling theory, especially the importance of distinguishing primary deviance from secondary deviance.
- ◆ We discussed the current state of labeling theory, which emphasizes informal sources of social reaction, not just the formal sources as in traditional models.
- ◆ Then we examined Marxist theories of crime, as well as numerous subsequent versions of this theory, such as those developed by Bonger, Chambliss, and Quinney.
- ◆ We also examined conflict theories, including theories by Sellin, Vold, and Turk.
- ◆ Finally, we examined the basic assumptions of feminist perspectives of crime, as well as some of the more notable implications that can be derived from this perspective.

KEY TERMS

Bourgeoisie

Conflict theory

Critical/radical feminism

Decriminalization

Deinstitutionalization

Diversion

Dramatization of evil

Labeling theory

Liberal feminism

Marxist feminism

Pluralist (conflict) perspective

Postmodern feminism

Primary deviance

Proletariat

Secondary deviance

Socialist feminism

Symbolic interaction

DISCUSSION QUESTIONS

1. What are the major assumptions of labeling/social reaction theory, and how does this differ significantly from other traditional theories of crime?

2. According to Lemert, what is the difference between primary deviance and secondary deviance?

3. How can you relate personally to being labeled, even if not for offending?

4. How do Marx's ideas relate to the study of crime? Provide some examples.

5. Which conflict theory do you buy into the most? The least? Why?

6. What are the key assumptions and features of the various feminist perspectives?

7. Which type of feminist theory do you believe is the most helpful for explaining crime?

WEB RESOURCES

Labeling Theory

http://www.apsu.edu/oconnort/crim/crimtheory14.htm

Conflict Theories

http://www.criminology.fsu.edu/crimtheory/conflict.htm

Marxist Theory

www.sociology.org.uk/pcdevmx.doc

http://law.jrank.org/pages/819/Crime-Causation-Sociological-Theories-Critical-theories.html

Feminist Theories of Crime

http://www.keltawebconcepts.com.au/efemcrim1.htm

READING

In this selection, Lening Zhang provides a brief review of the history of labeling theory, discussing its prominence in the 1960s and the declaration in 1985 that labeling theory was "dead," largely because the tests of the theory up to that time were based solely on formal labeling, such as that by law enforcement, courts, or corrections, which showed little to no effect for the labeling perspective. Then Zhang reviews the various studies and perspectives that revitalized labeling theory in the 1990s, which largely consisted of introducing the informal labeling process by significant others (e.g., peers, parents, employers, etc.), in other words, the labeling that takes place during interactions with people or agencies who are not part of the formal justice system (i.e., police, courts, and corrections).

Zhang then presents a test of this informal labeling process. Specifically, Zhang uses a national sample of youths, called the National Youth Survey (NYS), to test his predictions that delinquency among about 1,700 youths produces informal labeling; that such informal labeling by parents produces social isolation among youth who have been labeled by such informal sources; and that this social isolation increases the likelihood of recidivism or subsequent delinquency.

While reading this selection, readers are encouraged to think about when they were a teenager and their parents may have told them to stay away or not hang out with certain youths in their neighborhood or school. Readers are also encouraged to consider what types of effects this had on such people, who were likely socially isolated and in many cases had only other "bad" kids to hang out with, which likely increased their delinquent activity. Perhaps they have siblings or cousins who experienced such informal labeling, or maybe the reader actually experienced this type of labeling personally.

Informal Reactions and Delinquency

Lening Zhang

In formulating a **symbolic interaction** theory of delinquency, Matsueda (1992) recently developed a model of reflected appraisals and behavior based on Felson's (1985, 1989) and Kinch's (1963) work. A reflected appraisal is how one perceives the way others see one. Matsueda's model predicts that actual delinquent acts affect both actual and reflected appraisals by significant others. In turn, both actual and reflected appraisals influence

SOURCE: Lening Zhang, "Informal Reactions and Delinquency." *Criminal Justice and Behavior* 24, no. 1 (1997): 129–50. Copyright © 1997 Sage Publications, Inc. Used by permission of Sage Publications, Inc.

subsequent delinquent behavior. Also, actual appraisals by significant others have an effect on reflected appraisals of others, and prior delinquent behavior directly affects subsequent delinquency.[1] Drawing on labeling theory, Matsueda also argued that these predictions derived from the model implied the role of informal labeling in accounting for subsequent delinquency. Youths who have engaged in delinquent behavior should be more likely to be labeled delinquent by significant others. Significant others' labeling increases the probability of further delinquency. Although his study shed light on the relationship between informal labeling processes and subsequent life and behavioral adjustments, Matsueda did not fully address the issue because the focus of his study was not on this issue. Using Matsueda's basic framework, the present research specified a comprehensive theoretical model of the informal labeling process and tested this model with data from the National Youth Survey (NYS; Elliott & Ageton 1980), a longitudinal study of delinquency and drug use.

The labeling perspective on deviance has undergone an uneven development. During the 1960s, labeling, or reaction, theory emerged as a new and dominant perspective in criminology. In the mid-1970s, the perspective was subjected to serious critiques (Gibbs, 1966, 1972; Gove, 1980; Hagan, 1974; Tittle, 1975, 1980a; Wellford, 1975) and by 1985 was pronounced "dead" (Paternoster & Iovanni, 1989, p. 359). After its unpopular position in the study of crime for several years, the perspective has been revitalized since the late 1980s and the early 1990s (Berk, Campell, Klap, & Western, 1992; Farrell, 1989; Gove & Hughes, 1989; Hagan & Palloni, 1990; Heimer & Matsueda, 1994; Link, 1982, 1987; Link & Cullen, 1983; Link, Cullen, Frank, & Wozniak, 1987; Link, Cullen, Struening, Shrout, & Dohrenwend, 1989; Matsueda, 1992; Palamara, Cullen, & Gersten, 1986; Pate & Hamilton, 1992; Paternoster & Iovanni, 1989; Sampson, 1986; Sherman & Smith, 1992; Tittle 1988; Tittle & Curran, 1988; Triplett & Jarjoura, 1994).

This revitalization reflects new theoretical and research interests in labeling theory, which indicate the potential power and capacity of the theory for explaining deviance and crime. However, the new developments do not represent a simple return to the traditional version of labeling theory. They entail new attempts to elaborate, to specify, and to expand the theory in a new context of studies in criminology. As Paternoster and Iovanni (1989) pointed out, the above-mentioned efforts suggested some of the components that might constitute a *neo-labeling theory*.

Consistent with these new research interests in the labeling perspective, the present study addressed an important but relatively neglected issue—the informal labeling process and delinquency.

Despite this relative neglect, a few scattered early studies (Alvarez, 1968; Black, 1970; Black & Reiss, 1970; DeLamater, 1968; Orcutt, 1973; Swigert & Farrell, 1978; Tittle, 1975) involved attempts to explore the issues of informal reactions. A notable example was Orcutt's (1973) research, in which he differentiated between formal and informal reactions on the basis of the labeling perspective and noted the underemphasis on informal reactions. He argued that studies of the labeling perspective must pay greater attention to informal reactions.

According to Swigert and Farrell (1978), the evaluations and views of social audiences, such as parents and close friends, might have significant effects on the self-evaluations of labeled deviants. Hence informal groups may be crucial for self-identity and behavioral adjustment of labeled individuals. A similar argument was offered by Tittle (1975). He suggested that cultural patterns in different communities might serve as important variables that could interact with official formal reactions and affect the outcome of formal reactions.

However, Tittle observed, these effects of cultural patterns had not yet been addressed theoretically and empirically.

In addition, several early studies dealt with family reactions to drinking problems and mental disorders (Jackson, 1954, 1962; Sampson, Messinger, & Towne, 1962; Yarrow, Schartz, Murphy, & Deasy, 1955). These studies indicated how family reactions change in response to drinking and mental problems from "inclusive" to "exclusive" reactions (Orcutt, 1973). Although these early studies appealed for attention to informal reactions to deviance, studies of labeling phenomena have focused primarily on formal official reactions.

Recently, in the resurgence of labeling theory, some scholars (Braithwaite, 1989; Heimer & Matsueda, 1994; Matsueda, 1992; Pate & Hamilton, 1992; Paternoster & Iovanni, 1989; Tittle, 1988; Triplett & Jarjoura, 1994) again have called attention to informal reactions to deviance. An interesting formulation of this issue was Braithwaite's (1989) specification of informal reactions associated with "reintegrative shaming." He argued that

> if the labeling perspective is to be the stimulus to testable propositions with any hope of consistent empirical support, then a strategy is required for predicting the circumstances where labeling will be counterproductive and where it will actually reduce crime. (p. 20)

To meet this challenge, he identified two types of shaming: reintegrative and disintegrative (stigmatization) shaming. Reintegrative shaming refers to expressions of community disapproval with gestures of reacceptance into the community of law-abiding citizens. Such informal labeling reduces crime. In contrast, disintegrative shaming divides the community by creating a class of outcasts. It is conceived of as criminogenic labeling.

In discussing the secondary deviance hypothesis of labeling theory, Paternoster and Iovanni (1989) similarly called for more attention to informal rather than official reactions, whereas Triplett and Jarjoura (1994) formulated an integrated model of the labeling perspective and social learning theory. In their model, (a) parents' labeling of their child and (b) the youth's interpretation that parents are labeling him or her as delinquent are treated as key variables that may predict parental attachment, school attachment, and subsequent delinquency. Their findings indicate that both objective and subjective labels are factors in accounting for a child's relationship with major socialization sources, such as school and friends and subsequent delinquency. Thus they concluded that the labeling perspective could contribute more to criminological theory and research than it did in the past.

In the deterrence literature, informal sanctions have been demonstrated to have much stronger effects on deviance than do formal sanctions, and they significantly mediate the relationship between formal sanctions and deviance. As Tittle (1980b) observed,

> Social control as a general process seems to be rooted almost completely in informal sanctioning. Perceptions of formal sanction probabilities or severities do not appear to have much of an effect, and those effects that are evident turn out to be dependent upon perceptions of informal sanctions. (p. 241)

For a review of the deterrence literature, see Braithwaite (1989) and Williams and Hawkins (1986). Recently, on the basis of the Dade County spouse assault experiment, Pate and Hamilton (1992) examined the interaction effects between formal and informal deterrents on domestic violence. They found that formal arrest had no independent effect on the occurrence of a subsequent domestic assault.

Its effect was contingent on employment status, with formal arrest more likely to exhibit a significant deterrent effect for employed suspects than for unemployed suspects. These observations in the deterrence literature also may be valuable and useful in exploring the role of informal reactions in the labeling process.

All of these studies suggest that informal reactions play an important role in explaining deviant behavior. Following this general line of studies, the present study attempted to further clarify some conceptual issues by specifying informal reactions as a dependent variable, an independent variable, and an intervening variable in the labeling process.

Treating informal reactions as a dependent variable is concerned with explaining why some people come to be labeled deviant by significant others, such as parents, friends, and neighbors or the public (for a general discussion of societal reaction as a dependent variable, see Gove, 1980; Orcutt, 1973). The explanation involves three analytic dimensions: (a) Informal labeling is caused by formal labeling such as police arrest and court hearings; (b) actual deviant behavior leads to informal reactions; and (c) status characteristics of individuals, including those of both labelee and labeler (e.g., race, sex, social status) are relevant to informal reactions.

Viewing informal reaction as an independent variable focuses on the consequences that follow if a person is labeled deviant by significant others or the public. Four possible consequences can be specified. First, informal reactions may lead to formal (official) reactions. For instance, neighbors' complaints of someone's behavior may lead to formal actions by official agencies (Black, 1970; Black & Reiss, 1970). Second, informal reactions may push the labeled person to commit further deviance. Third, informal reactions may have negative consequences in other areas of the labeled person's life, such as interpersonal relationships. Finally, as deterrence theory predicts, informal reactions may be important factors in deterring further deviance.

In considering informal reactions as an intervening variable, the major concerns are with the mediating role of informal reactions in explaining the effects of formal reactions and actual deviant behavior on further deviance and other areas of a person's life. The role can be specified in two possibly opposite directions underlying efforts at social control. If informal labeling following and accompanying formal labeling or actual deviant behavior comes with reintegrative shaming, it may play an important and positive role in rehabilitating a rule breaker. This is a positive direction expected by a society. For example, Braithwaite (1989) emphasized the contributions of the informal mechanisms to low crime rate in Japanese society. In contrast, informal reactions may reinforce the stigma imposed by official agencies and thereby become a significant factor leading to further deviance.

These conceptual specifications suggest a framework for examining the relationship of informal labeling processes to delinquency. On the basis of available data from the NYS, the author developed a specific model focused on this relationship (see Figure 1). The model uses Matsueda's (1992) symbolic interactionist framework and extends his study to reflect the conceptual specifications proposed in the present study. The key extension is that the present study introduces an important variable—social isolation—into the model. First, in the model, social isolation from significant others is treated as a dependent variable caused by informal labeling. As previously specified, informal reactions may have negative consequences for other areas of the labeled person's life, one of which is negative change in interpersonal relations. Recently, an important dimension of the trend to revitalize labeling theory has involved efforts to focus on the effect of formal reactions on other areas of a person's life rather than just on subsequent deviance. Specifically, in the research areas of labeling theory and mental illness, Link and his colleagues (Link, 1982, 1987; Link et al.,

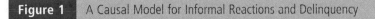

Figure 1 A Causal Model for Informal Reactions and Delinquency

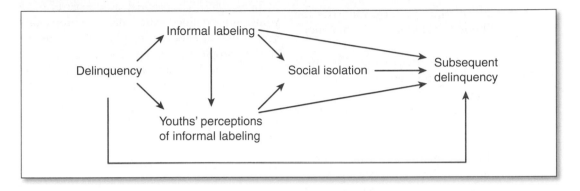

NOTE: Controls for gender, race, age, parent education, and family income are included in all regressions but are not shown in the diagram.

1987, 1989) noted the previous overemphasis on the effect of formal labeling on subsequent mental disorders in comparison with the neglect of specifying the direct effect of formal labeling on problems of life adjustment, such as marriage and job problems, experienced by ex-mental patients. Their studies provided evidence for the relationship between formal labeling and life adjustment. Similarly, it is reasonable to speculate that informal labeling not only increases the possibility of further delinquency involvement but also has negative consequences for other areas of the labeled person's life, such as interpersonal relationships.

Second, social isolation is also treated as an intervening variable between informal reactions and subsequent delinquency. The theoretical hypothesis is that informal reactions result in social isolation from significant others, and, in turn, such isolation increases the probability of further delinquency. It is possible that a negative label, through a series of reinforcing conditions, such as rejection by significant others, increases the probability of further deviance. Link (1982) pointed out that if this were demonstrated by further research, the eventual results of such a program of

research offered the genuine possibility of developing a theory that assigned a partial role to labeling in the etiology of deviance. Similarly, Liska (1987) proposed a clear diagram regarding the association of labeling with further deviance. His diagram assumed that the effect of societal reaction on further deviance was through its effect on self-concept, interpersonal networks, and structured opportunities. Therefore, the present model posits that social isolation has a mediating role for the effect of informal labeling on subsequent delinquency involvement, even though this mediating role may be only partial.

In summary, the predictions from this model are as follows:

1. Delinquent acts are a significant predictor of informal labeling.

2. Both delinquent acts and informal labeling positively affect youths' perceptions of informal labeling.

3. Both informal labeling and youths' perceptions of informal labeling lead to two possible consequences: social isolation from significant others experienced

by the labeled youths and subsequent delinquency.

4. Social isolation, in turn, increases the likelihood of subsequent delinquency.

Finally, as previous studies (see Gibbons & Krohn, 1991, for a review) have indicated, stable delinquent patterns exist. Therefore, a path included in the model predicts that prior delinquency is a significant predictor of subsequent delinquency.

An additional dimension of efforts to revitalize labeling theory involves attempts to specify a variety of contingencies that potentially influence the effect of official labeling on subsequent deviance (Berk et al., 1992; Palamara et al., 1986; Ray & Downs, 1986; Sherman & Smith, 1992). Studies have demonstrated that official reactions do affect the likelihood of subsequent deviance but that the effects are contingent on specific circumstances, such as gender, marital status, and employment status. The effect of official punishment on subsequent deviance is not uniform across different individual characteristics and different social contexts. In this vein, the present study, hypothesizing that women are more vulnerable to informal reactions than are men, assessed gender differences with the model. In addition, the present study also speculated that youth's perceptions of different kinds of informal labeling by different significant others (e.g. parents, friends, and teachers) may have differential associations with preceding variables such as delinquent behavior and subsequent variables such as social isolation. Therefore, the present study assessed the model with respect to youths' perceptions of different kind of informal labeling.

⊠ Method

Sample

The data used for the present study came from the NYS, conducted by Elliott and his colleagues (Elliott & Ageton, 1980, Elliot, Huizinga, & Ageton, 1985; Elliot, Huizinga, & Menard, 1989). The original sample size of 1,725 youths was obtained from a national probability sample of households by employing a multistage cluster sample frame. The age range of these youths at the first wave was 11 to 17 years.

The data from the first two waves of the NYS were used for the present study. The first wave survey was conducted in early 1977 and the second in early 1978.

Measurement

The NYS used personal interview to collect self-reports of delinquent acts; parents' labeling of their child; the child's perceptions of parents', friends', and teachers' labeling; and the child's perceptions of social isolation from family, from friends, and at school. The variables included in the model presented above were measured as follows:

Informal labeling. Because only data concerning parents' attitudes toward their child were available in the NYS, the informal labeling was reflected in parents' labeling. The parents' labeling consisted of a set of four measures, including "bad kid," "gets into trouble," "breaks rules," and "does things against the law." These measures reflected the extent to which parents labeled their child as deviant, with a 5-point scale ranging from 1 (*strongly disagree*) to 5 (*strongly agree*).

Child's perceptions of informal labeling. Youth respondents were asked to indicate the extent to which their parents, their friends, and their teachers labeled them as deviant. The items and scoring for youth responses were the same as those for parents' labeling.

Social isolation. Youth respondents reported their perceptions of interpersonal relationships with their families, their friends, and at school. Each type of relationship was assessed by five items. A 5-point scale ranging from 1 (*strongly disagree*) to 5 (*strongly agree*) was used for each item. The scoring for these items was rearranged by the present investigator so that high scores represented high isolation. For isolation related to family, the items were as follows: "outsider with family," "feel lonely with family," "family not interested in problems," "family listens to problems," and "feel close to family." For isolation from friends, the five items were the following: "don't fit with friends," "friends don't take interest," "feel lonely with friends," "feel close to friends," and "friends listen to problems." Similarly, five items reflected isolation at school: 'Teachers don't call on me," "nobody at school cares," "don't belong at school," "feel lonely at school," and "teachers don't ask me to work on projects."

Delinquent acts. The NYS included a delinquency inventory to represent the entire range of delinquent acts. Each of the self-reported delinquent acts was coded in two parts: (a) absolute frequency and (b) categorical responses, ranging from 1 (*never*) to 9 (*two to three times per day*). Following Elliott et al. (1985) and Matsueda (1992), the present study used a 24-item scale of general delinquency with the categorical responses because the categorical responses have less skewed distributions. The delinquency items included auto theft, $5 theft, $5 to $50 theft, buying stolen goods, runaway, concealed weapon, aggravated assault, prostitution, sexual intercourse, gang fights, sold marijuana, hit parents, hit teachers, hit students, disorderly conduct, sold drugs, joyriding, sexual assault, strong-armed students, strong-armed teachers, strong-armed others, breaking and entering, panhandled.

The present study created several additive indexes to represent these variables in the model (i.e., indexes for parents' labeling; youths' perceptions of parents', friends', and teachers' labeling; social isolation from family, from friends, and at school; and delinquency)

by using the above items. The standardized reliability coefficient for each index ranged from .65 to .75.

In addition to these key variables in the model, the present study included the following control variables that may be related to the informal labeling process: race; age; family income, measured by a 10-point scale ranging from 1 (*$6,000 or less*) to 10 (*$38,001 or more*); and parental education, measured by a 7-point scale ranging from 1 (*some grade school*) to 7 (*graduate degree*).

Based on time and logical order, self-reported delinquency in the first wave of the NYS consisted of delinquent acts prior to parents' labeling and youths' perceptions of informal labeling. In the first wave, the adolescent respondents reported their delinquency during the previous year. Also, in the first wave, parents were interviewed to report their current attitudes toward their child, and youths reported their current perceptions of parents', friends', and teacher's labeling and their perceptions of isolation from family, from friends, and at school. Thus parents' labeling, youths' perceptions of informal labeling, and social isolation were measured in the first wave. In addition, race, age, family income, and parental education were measured in the first wave. The measure of subsequent self-reported delinquency occurred in the second wave.

Ordinary least squares regression (OLS) was employed to assess the model. In addition, the present study examined the interactions between gender and each of the primary independent variables to assess the possible role of gender differences in the informal labeling process.

☒ Results

The present study begins its analysis with the effect of delinquency on parents' labeling (see Table 1). Delinquency significantly and positively affected parents' labeling of their child ($\beta = 23$). Also, there were significant associations

Table 1	Regression for Effects of Delinquency on Parents' Labeling, With Control Variables Included

Independent Variable	β	t
Delinquency	.23	9.54*
Gender	.04	1.72
Race	−.13	−4.96*
Age	−.07	−2.93*
Parent education	−.13	−4.83*
Family income	−.03	−1.21

NOTE: Beta (β) = Standardized regression coefficient. R^2 = .12.

*p <.05.

and parental education and parents' labeling, which indicated that younger and non-White youths and youths whose parents' education was lower were more likely to be labeled as deviant by their parents. These results supported the predictions of labeling theory that individual and social disadvantages are related to the labeling process.

Table 2 presents the effects of delinquency and parents' labeling on youths' perceptions of parents', friends', and teachers' labeling. Three equations are included in Table 2 for the three dependent variables—perceived parents', friends', and teachers' labeling. Both parents' labeling and delinquency had significant effects on youths' perceptions of parents' labeling β =.29 and .25, respectively), friends' labeling (β =.22 and .35, respectively), and teachers' labeling (β =.23 and .36, respectively).

In addition to the direct effect of delinquency on youths' perceptions of significant others' labeling, delinquency had indirect effects on these youths' perceptions through parents' labeling. However, these indirect effects were fairly small. On the basis of the direction of these indirect effects, they increased the amount of total positive effects of delinquency on youths' perceptions of informal labeling.

Furthermore, both gender and parent education were significantly associated with youths' perceptions of parents', friends', and teachers' labeling, whereas family income

Table 2	Regression for Effects of Delinquency and Parents' Labeling on Youths' Perceptions of Informal Labeling by Parents, Friends, and Teachers, With Control Variables Included

	Youths' Perceptions of Informal Labeling					
	Parents		Friends		Teachers	
Independent Variable	β	t	β	t	β	t
Parents' labeling	.29	12.00*	.22	9.28*	.23	10.04*
Delinquency	.25	10.57*	.35	15.19*	.36	15.41*
Gender	.05	2.19*	.11	4.68*	.09	4.05*
Race	.02	0.02	−.01	−0.71	−.01	−0.31
Age	−.04	−1.86	.03	1.53	−.01	−0.62
Parent education	−.06	−2.54*	−.08	−3.17*	−.08	−3.32*
Family income	−.05	−1.97*	.06	2.37*	.03	1.18

NOTE: Beta (β) = Standardized regression coefficient. R^2 = .22 for parents, .25 for friends, and .26 for teachers.

*p < .05.

exhibited significant effects on youths' perceptions of parents' and friends' labeling. Men and youths whose parental education was lower were more likely to perceive labeling by parents, friends, and teachers. Youths who came from families with lower income were more likely to perceive labeling by parents and friends.

Table 3 presents the results of regressions of social isolation on parents' labeling and youths' perceptions of significant others' labeling. Similar to Table 2, Table 3 includes three equations for the three dependent variables—isolation from family, from friends, and at school. First, youths' perceptions of parents' labeling were a significant predictor of isolation from family (β = .31), from friends (β = .20), and at school (β = .24). The greater the youths' perceptions of parents' labeling, the greater were their perceptions of isolation from all three kinds of significant others.

Second, youths' perceptions of friends' labeling had a positive effect only on isolation from friends (β = .18), and youths' perceptions of teachers' labeling had a positive effect only on isolation at school (β = .13). Third, parents' labeling significantly affected isolation from family (β = .05). Furthermore, on the basis of the causal order, parents' labeling had indirect effects on isolation from significant others through youths' perceptions of significant others' labeling. All of these findings were consistent with propositions of labeling theory. Informal labeling appears to lead to negative changes in interpersonal relationships, as assessed by degree of isolation.

Finally, the results in Table 3 also show that delinquency was negatively related to youths' perceptions of friends' isolation (β = –11). This may imply that delinquency is an important factor leading to closer group association. In

Table 3	Regression for Effects of Delinquency; Parents' Labeling; Youths' Perceptions of Informal Labeling by Parents, Friends, and Teachers on Youths' Perceptions of Social Isolation from Family, Friends, and at School, With Control Variables Included

	Youths' Perceptions of Informal Labeling					
	Family		**Friends**		**School**	
Independent Variable	β	t	β	t	β	t
Parents' labeling	.05	2.10*	−.01	−0.33	.01	0.43
Delinquency	.04	1.38	−.11	−3.81*	.01	0.25
Youths' perceptions of parents' labeling	.31	9.43*	.20	5.84*	.24	7.05*
Youths' perceptions of friends' labeling	.05	1.30	.18	4.32*	.03	0.77
Youths' perceptions of teachers' labeling	.06	1.60	−.06	−1.34	.13	3.10*
Gender	−.12	−4.97*	.07	2.81*	−.03	−1.41
Race	−.07	−2.77*	−.12	−4.51*	−.05	−2.04*
Age	.13	5.69*	−.07	−2.84*	−.04	−1.86
Parent education	−.04	−0.02	.02	0.89	−.04	−1.42
Family income	−.02	−0.56	−.06	−2.24*	−04	−1.31

NOTE: Beta (β) = Standardized regression coefficient. R^2 = .22 for isolation from family, .13 for isolation from friends, and .16 for isolation at school.

*$p < .05$.

addition, non-White youths were more likely to perceive isolation from family, from friends, and at school. Older and female youths were more likely to perceive isolation from family, whereas younger and male youths were more likely to perceive isolation from friends. Family income was significantly and negatively related to isolation from friends.

The results of a full regression equation predicting subsequent delinquency are reported in Table 4. Consistent with labeling theory, parents' labeling and youths' perceptions of teachers' labeling yielded significantly positive coefficients ($\beta = .08$ and .08, respectively). They predicted an increased possibility of subsequent delinquency. Also, parents' labeling had an indirect effect on subsequent delinquency via perceived teachers' labeling. Inconsistent with labeling theory, social isolation from family,

Table 4	Regression for Effects Hypothesized Predictors on Subsequent Delinquency	
Independent Variable	**β**	**t**
Parents' labeling	.08	3.66*
Prior delinquency	.50	21.17*
Youths' perceptions of parents' labeling	−.01	−0.26
Youths' perceptions of friends' labeling	.03	0.93
Youths' perceptions of teachers' labeling	.08	2.41*
Family isolation	−.02	−0.94
Friends' isolation	−.01	−0.06
School isolation	−.02	−0.81
Gender	.09	4.18*
Race	.01	0.52
Age	.09	4.04*
Parent education	−.01	−0.11
Family income	−.03	−1.10

NOTE: Beta (β) = Standardized regression coefficient. $R^2 = .39$.
*$p < .05$.

from friends, and at school evidenced no significant positive effects on subsequent delinquency. Thus social isolation had no mediating role for the relationship between informal labeling and subsequent delinquency.[2]

The results in Table 4 also revealed that, as predicted, prior delinquency was significantly and positively related to subsequent delinquency ($\beta = .50$). Furthermore, consistent with labeling theory, part of the total effect of prior delinquency on subsequent delinquency was positively mediated by parents' labeling and youths' perceptions of teachers' labeling. Finally, the significant coefficients for gender ($\beta = .09$) and age ($\beta = .09$) indicated that older and male youths were more likely to be involved in delinquency.

The present study also examined possible interactions between gender and the primary variables. The results revealed three kinds of significant interaction effects. First, there was a significant interaction of delinquency and gender on parents' labeling ($\beta = -.56$), which indicated that delinquency exerted a greater effect on parents' labeling for women than for men. This was consistent with the hypothesis that women are more likely to suffer informal labeling. Second, parents' labeling and gender had an interaction effect on subsequent delinquency ($\beta = .20$), which indicated that parents' labeling was more likely to increase the probability of delinquency involvement for men than for women. This was at odds with the hypothesis proposed in the present study. Third, there was a significant interaction effect between prior delinquency and gender on subsequent delinquency ($\beta = .57$). Consistent with previous research, prior delinquency exhibited a greater effect on further delinquency for men than for women. That is, a stable delinquent pattern was more likely to exist among men than women.

◤ Discussion

There were several noteworthy findings in the present study. First, delinquent behavior significantly increased the probability of parents' labeling of their child, with the probability greater for women than men. Also, as Matsueda (1992) reported, some demographic variables exhibited significant effects on parents' labeling of their children as deviant. Non-White, younger youths and youths whose parents had lower education were more likely to be labeled as deviant by their parents. These findings were consistent with the propositions of labeling theory, which predict that individual and social disadvantages are related to the labeling process.

Second, delinquent behavior and parents' labeling exerted significant effects on youths' perceptions of labeling by significant others, including parents, friends, and teachers. This implies that delinquent behavior and informal labeling may damage youths' previous conventional self-identity through their self-perceptions of their own delinquent acts and significant others' labeling, thereby increasing the probability of their self-degradation and self-labeling. This finding was also consistent with the labeling perspective.

Third, youths' perceptions of parents' labeling were significantly and positively related to their perceptions of isolation from family, from friends, and at school. Parents' labeling was more likely to increase the probability of family isolation, and youths' perceptions of friends' and teachers' labeling exerted, respectively, positive and significant effects on isolation from friends and at school. Thus labeling by significant others and perceptions of the labeling were more likely to lead to feelings of social rejection. Although youths' perceptions of parents' labeling were a source of feelings of isolation from all three kinds of significant others (i.e., family, friends, and school); specificity of effects also existed, in that (a) parents' labeling led to feelings of isolation from family, (b) youths' perceptions of friends' labeling led to feelings of isolation from friends, and (c) youths' perceptions of teachers' labeling led to feelings of isolation

from school. These findings were consistent with the hypothesis that informal labeling has direct and negative consequences for other areas of a person's life.

Fourth, prior delinquency significantly affected subsequent delinquency, with the effect greater for male youths than female youths. This was consistent with previous findings of stable delinquent patterns among youths and among male youths in particular. Furthermore, parents' labeling and youths' perceptions of teachers' labeling were significantly and positively related to subsequent delinquency. Also, a gender difference was uncovered, indicating that parents' labeling was more likely to be a negative factor leading to subsequent delinquency for male youths than for female youths. Again, these results were in agreement with the prediction of labeling theory that deviant labels increase the likelihood of further delinquency involvement, and they reaffirmed the importance of recent efforts to specify the variety of contingencies that may influence the effect of a deviant label on further deviant behavior.

Fifth, youths' perceptions of social isolation from significant others had no direct and significant effect on subsequent delinquency and thus had no mediating role for the effect of informal labeling on subsequent delinquency. This was at odds with the hypothesis derived from labeling theory.

Finally, some indirect effects were found in the present study. For instance, part of the total effect of delinquency on social isolation from significant others was mediated by parents' labeling and youths' perceptions of parents', friends', and teachers' labeling. These findings were consistent with recent studies of labeling theory, which have adopted a "softer" rather than a "harder" stance by specifying a variety of intervening variables that may, in part, account for the association of labeling with delinquency. As Paternoster and Iovanni (1989) pointed out, this softer stance, compared with a deterministic one, can represent more plausibly the classic implication of labeling theory.

In addition to these major results related to the predictions derived from the model, an unanticipated but important finding was the negative relationship between delinquency and youths' perceptions of isolation from their friends. This finding may have two implications. First, delinquency may be a medium or vehicle that ties youths together, even though their relationships are not close and intimate. Second, according to self-derogation theory (Kaplan, 1975, 1980), adolescents are motivated to commit delinquency in order to enhance their self-esteem, and, thus, appreciative companionship is necessary to satisfy the motivation. Therefore, delinquency and peer companionship may be positively correlated.

The informal labeling process appears to play an important role in explaining youths' life and behavioral adjustments. Elaborations and specifications of this informal labeling process should be valuable in the development of "neolabeling" theory. Second, in some important aspects, such as parents' labeling, the informal labeling process is not uniform across gender. This reaffirms the trend of revitalizing labeling theory by specifying a variety of contingencies. Such specifications of contingencies should be an important requirement for any attempts to develop a neolabeling theory. Third, social isolation from significant others appears to have no significant effect on further delinquency. This challenges the proposition derived from labeling theory that social rejection caused by deviant labels necessarily results in further deviance. It may be that any such relationship, if it exists, depends on the nature and type of social rejection and deviant behavior. Future research is needed to address these important issues. Such research would contribute to the further development of neolabeling theory.

⊠ Notes

1. More recent work by Heimer and Matsueda (1994) has extended Matsueda's study by adding some variables of social control, such as attachment to family and friends. This social control model, based on Matsueda's (1992) model of reflected appraisals, has been tested by Heimer and Matsueda.

2. Tests for multicollinearity among youths' perceptions of parents', friends', and teachers' labeling and among youths' perceptions of isolation from family, from friends, and at school indicated no multicollinearity problems in each equation.

⊠ References

Alvarez, R. (1968). Informal reactions to deviance in simulated work organizations: A laboratory experiment. *American Sociological Review, 33,* 895–912.

Berk, R. A., Campell, A., Klap. R., & Western, B. (1992). The deterrent effect of arrest: A Bayesian analysis of four field experiments. *American Sociological Review, 57,* 689–708.

Black, D. J. (1970). Production of crime rates. *American Sociological Review, 35,* 733–748.

Black, D. J., & Reiss. A. J. (1970). Police control of juveniles. *American Sociological Review, 35,* 63–47.

Braithwaite, J. (1989). *Crime, shame, and reintegration.* Cambridge. England: Cambridge University Press.

DeLamater, J. (1968). On the nature of deviance. *Social Forces, 46,* 445–455.

Elliott, D. S., & Ageton, S. S. (1980). Reconciling race and class differences in self-reported and official estimates of delinquency. *America Sociological Review, 40,* 95–110.

Elliott, D. S., Huizinga, D., & Ageton, S. S. (1985). *Explaining delinquency and drug use.* Beverly Hills. CA: Sage.

Elliott, D. S., Huizinga. D., & Menard, S. (1989). *Multiple problem youth: Delinquency, substance use, and mental health problems.* New York: Springer-Verlag.

Farrell, R. A. (1989). Cognitive consistency in deviance causation: A psychological elaboration of an integrated system model. In S. F. Messner, M. D. Krohn, & A. E. Liska (Eds.), *Theoretical integration in the study of deviance and crime: Problems and prospects* (pp. 77–92). Albany: State University of New York Press.

Felson, R. B. (1985). Reflected appraisal and the development of self. *Social Psychology Quarterly, 48,* 71–77.

Felson, R. B. (1989). Parents and reflected appraisal process: A longitudinal analysis. *Journal of Personality and Social Psychology, 56,* 965–971.

Gibbons, D. C., & Krohn, M. D. (1991). *Delinquent behavior* (5th ed.). Eaglewood Cliffs, NJ: Prentice Hall.

Gibbs, J. P. (1966). Conception of deviant behavior: The old and the new. *Pacific Sociological Review, 9,* 9–14.

Gibbs, J. P. (1972). Issues in defining deviant behavior. In R. A. Scott & J. D. Douglas (Eds.), *Theoretical perspectives on deviance* (pp. 39–68). New York: Basis Books.

Gove, W. R. (1980). The labeling perspective: An overview. In W. R. Gove (Ed.), *The labeling of deviance: Evaluating a perspective* (2nd ed., pp. 9–33). Beverly Hills, CA: Sage.

Gove, W. R., & Hughes, M. (1989). A theory of mental illness: An attempted integration of biological, psychological, and social variables. In S. F. Messner, M. D. Krohn, & A. E. Liska (Eds.), *Theoretical integration in the study of deviance and crime: Problems and prospects* (pp 61–76). Albany: State University of New York Press.

Hagan, J. (1974). Extra-legal attitudes and criminal sanctioning: An assessment and a sociological view. *Law and Society Review, 8,* 357–383.

Hagan, J., & Palloni, A. (1990). The social reproduction of a criminal class in working-class London. *American Journal of Sociology, 96,* 265–299.

Heimer, K., & Matsueda, R. L. (1994). Role-taking, role commitment, and delinquency: A theory of differential social control. *American Sociological Review, 59,* 356–390.

Jackson, J. K. (1954). The adjustment of the family to the crisis of alcoholism. *Quarterly Journal of Studies on Alcohol, 15,* 564–586.

Jackson, J. K. (1962). Alcoholism and the family. In D. J. Pittman & C. R. Snyder (Eds.), *Society, culture, and drinking patterns* (pp. 472–479). New York: Wiley.

Kaplan, H. B. (1975). Increase in self-rejection as an antecedent of deviant responses. *Journal of Youth and Adolescence, 4,* 438–458.

Kaplan, H. B. (1980). *Deviant behavior in defense of self.* New York: Academic Press.

Kinch, J. W. (1963). A formalized theory of the self-concept. *American Journal of Sociology, 68,* 481–486.

Link, B, (1982). Mental patient status, and income: An examination of the effects of a psychiatric label. *American Sociological Review, 47,* 456–478.

Link, B. (1987). Understanding labeling effects in the area of mental disorders: An assessment of the effects of expectations of rejection. *American Sociological Review, 52,* 1004–1081.

Link, B., & Cullen, F. T. (1983). Reconsidering the social rejection of external patients: Levels of attitudinal response. *American Journal of Community Psychology, 11,* 261–273.

Link, B., Cullen, F. T., Frank, J., & Wozniak, J. F. (1987). The social rejection of former mental patients: Understanding why labels matter. *American Journal of Sociology, 92,* 1461–1500.

Link, B., Cullen, F. T., Struening, E., Shrout, P. E., & Dohrenwend, B. P. (1989). A modified labeling theory approach to mental disorders: An empirical assessment. *American Sociological Review, 54,* 400–423.

Liska, A. E. (1987). *Perspectives on deviance.* Englewood Cliffs, NJ: Prentice Hall.

Matsueda, R. L. (1992). Reflected appraisals, parental labeling, and delinquency: Specifying a symbolic interactionist theory. *American Journal of Sociology, 97,* 1577–1611.

Orcutt, J. D. (1973). Social reaction and the response to deviation in small groups. *Social Forces, 52,* 259–267.

Palamara, F., Cullen, F. T., & Gersten, J. C. (1986). The effect of police and mental health intervention on juvenile delinquency: Specifying contingencies in the impact of formal reaction. *Journal of Health and Social Behavior, 27,* 90–105.

Pate, A. M., & Hamilton, E. E. (1992). Formal and informal deterrents to domestic violence: The Dade County spouse assault experiment. *American Sociological Review, 57,* 691–697.

Paternoster, R., & Iovanni, L. (1989). The labeling perspective and delinquency: An elaboration of the theory and an assessment of the evidence. *Justice Quarterly, 6,* 395–394.

Ray, M. C., & Downs, W. R. (1986). An empirical test of labeling theory using longitudinal data. *Journal of Research in Crime and Delinquency, 23,* 169–194.

Sampson, H., Messinger, S. L., & Towne, R. D. (1962). Family processes and becoming a mental patient. *American Journal of Sociology, 68,* 88–96.

Sampson, R. J. (1986). Effects of socioeconomic context on social reaction to juvenile delinquency. *American Sociological Review, 51,* 876–885.

Sherman, L. W., & Smith, D. A. (1992). Crime, punishment, and stake in community: Legal and informal control of domestic violence. *American Sociological Review, 57,* 680–690.

Swigert, L. V., & Farrell, R. A. (1978). Referent others and deviance causation; A neglected dimension in labeling research. In M. D. Krohn & R. L. Akers (Eds.), *Crime, law, and sanctions; theoretical perspectives* (pp. 59–72). Beverly Hills, CA: Sage.

Tittle, C. R. (1975). Deterrence or labeling? *Social Forces, 53,* 399–410.

Tittle, C. R. (1980a). Labeling and crime: An empirical evaluation. In W. R. Gove (Ed.), *The labeling of deviance: Evaluating a perspective* (2nd ed., pp. 241–263). Beverly Hills, CA: Sage.

Tittle, C. R. (1980b). *Sanctions and social deviance: The question of deterrence.* New York: Praeger.

Tittle, C. R. (1988). Two empirical regularities (maybe) in search of an explanation: Commentary on the age/crime debate. *Criminology, 26,* 75–86.

Tittle, C. R., & Curran, D. A. (1988). Contingencies for dispositional disparities in juvenile justice. *Social Forces, 67,* 23–58.

Triplett, R. A., &. Jarjoura, G. R. (1994). Theoretical and empirical specification of a model of informal labeling. *Journal of Quantitative Criminology, 10,* 241–276.

Wellford, G. F. (1975). Labeling theory and criminology: An assessment. *Social Problems, 22,* 332–345.

Williams, K. R., & Hawkins, R. (1986). Perceptual research on general deterrence: A critical review. *Law and Society Review, 20,* 545–547.

Yarrow, M. R., Schartz. C. G., Murphy, H. S., & Deasy, L. C. (1955). The psychological meaning of mental illness in the family. *Journal of Social Issues, 11,* 12–24.

REVIEW QUESTIONS

1. How does Zhang measure informal labeling, the child's perceptions of informal labeling, and social isolation? Do you agree that such measures are valid? How would you improve such measures?

2. What did findings from this study show regarding parents' labeling? Which types of the youths' perceptions of labeling had the greatest effect on their subsequent delinquency? Explain these findings.

3. Given the findings of this study, do you believe that informal labeling has an important impact on recidivism or reoffending by youths who have been labeled as a "bad" kid? What types of policies would you advise based on these findings?

READING

In this selection, Ian Taylor, Paul Walton, and Jock Young examine an original approach for explaining criminality, namely a perspective based on the context of a capitalist political economy. Such a radical perspective says crime is defined by society's economically powerful people, who also govern its enforcement. The authors base their assumptions and propositions on the work of Karl Marx and other critical theorists, who claim that the law and the enforcement of such laws are determined by those in power in a given society.

SOURCE: Ian Taylor, Paul Walton, and Jock Young, "Marx, Engels, and Bonger on Crime and Social Control," in *The New Criminology: For a Social Theory of Deviance,* ed. I. Taylor, P. Walton, and J. Young (London: Routledge & Kegan Paul, 1973), 219–35. Copyright © 1973 Routledge. Reproduced by permission of Taylor & Francis Books UK.

Willem Bonger, in the early 1900s, was one of the first to use this model for explaining criminal behavior in society. He argued that the capitalistic system of the economy led to greed and egoism, which promoted more selfish and greedy behavior among individuals. Bonger concluded that crime could be reduced, and even eliminated, in a given society, if an economic system was based on the progress of the entire society and if the legal system removed all bias for the wealthy.

While reading this section, readers are encouraged to consider whether they believe that complete fairness can ever be reached in any justice system, particularly the system currently in place in the United States. Also, readers are encouraged to consider the crime levels in capitalistic societies, such as the United States, Canada, and England, comparing crime levels to other nations that are not based on capitalism, such as Cuba and China. Specifically, would readers feel safer in such socialistic societies, as compared to capitalistic societies?

Marx, Engels, and Bonger on Crime and Social Control

Ian Taylor, Paul Walton, and Jock Young

⊠ Willem Bonger and Formal Marxism

In the study of crime and deviance, the work of Willem Bonger (1876–1940) . . . has assumed the mantle of the Marxist orthodoxy—if only because (with the exception of untranslated writers inside the Soviet bloc) no other self-proclaimed Marxist has devoted time to a full-scale study of the area. Bonger's criminology is an attempt to utilize some of the formal concepts of Marxism in the understanding of the crime rates of European capitalism in the late nineteenth and early twentieth centuries. Importantly, however, Bonger's efforts appear, for us, not so much the application of a fully fledged Marxist theory as they are a recitation of a "Marxist catechism" in an area which Marx had left largely untouched—a recitation prompted by the growth not of the theory itself, but by the growth of a sociological pragmatism.

Bonger must, therefore, be evaluated in his own terms, in terms of the competence of his extension of the formal concepts of Marxism to the subject-matter, rather than in terms of any claim that might be made for him as the Marxist criminologist.

In at least two respects, Bonger's analysis of crime differs in substance from that of Marx. On the one hand, Bonger is clearly very much more seriously concerned than Marx with the causal chain linking crime with the precipitating economic and social conditions. On the other, he does not confine his explanations to working-class crime, extending his discussions to the criminal activity of the industrial bourgeoisie as defined by the criminal laws of his time. Whilst differing from Marx in these respects, however, Bonger is at one with his mentor in attributing the activity itself to demoralized individuals, products of a dominant capitalism.

Indeed, in both Marx and Bonger, one is aware of a curious contradiction between the "image of man" advanced as the anthropological underpinning of "orthodox" Marxism and the questions asked about men who deviate. . . . The criminal thought, which runs through the bulk of Bonger's analysis of crime, is seen as the product of the tendency in industrial capitalism to create "egoism" rather than "altruism" in the structure of social life. It is apparent that the notion performs two different notions for Bonger, in that he is able to argue, at different points, that, first, "the criminal thought" is engendered by the conditions of misery forced on sections of the working class under capitalism and that, second, it is also the product of the greed encouraged when capitalism thrives. In other words, as an intermediary notion, it enabled Bonger to circumvent the knotty problem of the relationship between general economic conditions and the propensity to economic crime.

Now, whilst the ambiguity in the notion may help Bonger's analysis, it does not stem directly from his awareness of dual problems. For Bonger, it does appear as an autonomous psychic and behavioural quality which is to be deplored and feared; "the criminal thought," and its associated "egoism," are products of the brutishness of capitalism, but at the same time they do appear to "take over" individuals and independently direct their actions.

The Marxist perspective, of course, has always emphasized the impact that the dominant mode of production has had on social relationships in the wider society, and, in particular, has spelt out the ways in which a capitalist means of production will tend to "individuate" the nature of social life. But to understand that "egoism" and "individuation" are products of particular sets of social arrangements is to understand that egoism and individuation have no force or influence independently of their social context. For Bonger, the "criminal thought"—albeit a product of the egoistic structure of capitalism—assumes an independent status as an intrinsic and behavioural quality of certain (criminal) individuals. It is enormously paradoxical that a writer who lays claim to be writing as a sociologist and a Marxist should begin his analysis with an assumed individual quality (which he deplores) and proceed only later to the social conditions and relationships sustaining and obstructing the acting-out quality.

In the first place, the emphasis in Bonger on "the criminal thought" as an independent factor for analysis is equivalent to the biological, physiological, and sociological (or environmental) factors accorded an independent and causative place in the writings of the positivist theorists of crime. The limitations of this approach have been pointed out, amongst others, by Austin T. Turk (1964, pp. 454–55):

> students of crime have been preoccupied with the search for an explanation of distinguishing characteristics of criminality, almost universally assuming that the implied task is to develop scientific explanations of the behaviour of persons who deviate from "legal norms." The quest has not been very successful . . . the cumulative impact of efforts to specify and explain differences between "criminal" and "non-criminal" cultural and behaviour patterns is to force serious consideration of the possibility that there may be no significant differences between the overwhelming majority of legally identified criminals and the relevant general population, i.e., that population whose concerns and expectations impinge directly and routinely upon the individuals so identified.

More succinctly: "the working assumption has been that crime and non-crime are classes of behaviour instead of simply labels associated with the processes by which individuals come to occupy ascribed . . . statuses of criminal and non-criminal."

It is a comment on the nature of Bonger's Marxism that the actor is accorded such an idealistic independence; when to have started with a model of a society within which there are conflicting interests and a differential distribution of power would have revealed the utility of the criminal law and the "criminal" label (with a legitimating ideology derived from academia) to the powerful elites of capitalist society. In fact, of course, a criminology which proceeds in recognition of competing social interests has two interrelated tasks of explanation. Certainly it has the task of explaining the causes for an individual's involvement in "criminal" behaviour, but, prior to that, it has the task of explaining the derivation of the "criminal" label (whose content, function and applicability we have argued will vary across time, across cultures, and internally within a social structure).

One cannot entirely avoid the conclusion that Bonger's analysis, irrespective of the extent to which it is guided by a reading and acceptance of Marxist [precepts], is motivated (and confused) by a fear of those with "criminal thoughts." For Bonger "criminal thought" is by and large a product of the lack of moral training in the population. Moral training has been denied the proletariat, in particular, because it is not the essential training for work in an industrializing society. The spread of "moral training" is the antidote to "criminal thoughts," but, since such an education is unlikely under the brutish capitalism of the imperialist period, capitalism—or more precisely, the economic conditions (of inequality and accumulation)—are indeed a cause of crime.

In so far as Bonger displays any concern for the determinant nature of social relationships of production, he does so in order to illustrate the tendencies of different social arrangements to encourage... "criminal thoughts" in the population at large. As against the ameliorarist school, which saw an inevitable advance of man from conditions of primitive and brutish living to societies in which altruistic relationships

would predominate, Bonger, in fundamental agreement with the value placed on altruism and liberalism, identified the advent of capitalism with the break in the process of civilizing social relationships.... Bonger comments: "The fact that the duty of altruism is so much insisted upon is the most convincing proof that it is not generally practised."

The demise of egoism, and the creation of social conditions favourable to the "criminal thought" parallels, for Bonger, the development of social arrangements of production as described by Marx.... Under capitalism, the transformation of work from its value for use to its value for exchange (as fully described by Marx) is responsible for the "cupidity and ambition," the lack of sensitivity between men, and the declining influence of men's ambitions on the actions of their fellows.... Capitalism, in short, "has developed egoism at the expense of altruism."

"Egoism" constitutes a favourable climate for the commission of criminal acts, and this, for Bonger, is an indication that an environment in which men's social instincts are encouraged has been replaced by one which confers legitimacy on asocial or "immoral" acts of deviance. The commission of these acts, as Bonger explicitly states in *Introduction to Criminology* (1935), has a demoralizing effect on the whole of the body politic.

Bonger's substantive analysis of types of crime, covering a range of "economic crimes," "sexual crimes," "crimes from vengeance and other motives," "political crimes" and "pathological crimes," is taken up with a demonstration of the ways in which these crimes are causatively linked with an environment encouraging egoistic action. Even involvement of persons born with "psychic defects" in criminal activity can be explained in terms of these enabling conditions (Bonger, 1916, p. 354):

These persons adapt themselves to their environment only with difficulty...
[and] have a smaller chance than others

to succeed in our present society, where the fundamental principle is the warfare of all against all. Hence they are more likely to seek for means that others do not employ (prostitution, for example).

The whole of Bonger's analysis, however much it is altered or qualified at particular points in his discussion, rests on the environmental determinism of his "general considerations." In a social structure encouraging of egoism, the obstacles and deterrents to the emergence of the presumably ever-present "criminal thought" are weakened and/or removed; whereas, for example, under primitive communism, the communality was constructed around, and dependent upon, an interpersonal altruism. Capitalism is responsible for the free play granted to the pathological will, the "criminal thought" possessed by certain individuals.

The bulk of Bonger's work, indeed, so far from being an example of dialectical procedure, is a kind of positivism in itself, or at least an eclecticism reminiscent of "inter-disciplinary" positivism. Where the general theory appears not to encompass all the facts (facts produced by positivist endeavor), mediations of various kinds are introduced. In Bonger, it is possible to find examples of the elements of anomie theory, differential opportunity theory and, at times, the frameworks of structural-functionalism (much of it well in advance of its time). In his discussion of economic crime, for example, Bonger (1969, p. 108) approached a Mertonian stance on larceny:

> Modern industry manufactures enormous quantities of goods without the outlet for them being known. The desire to buy must, then, be excited in the public. Beautiful displays, dazzling illuminations, and many other means are used to attain the desired end. The perfection of this system is reached in

the great modern retail store, where persons may enter freely, and see and handle everything—where, in short, the public is drawn as a moth to a flame. The result of these tactics is that the cupidity of the crowd is highly excited.

And Bonger is not unaware of the general, or the more limited, theories of criminality and deviance produced by the classical thinkers of his time and earlier. Where appropriate, Bonger attempts to incorporate elements of these competing theorists, though always in a way which subordinates their positions to his own "general considerations." On Gabriel Tarde's "law of imitation," for example, which purports to explain criminality as a function of association with "criminal types," Bonger (1969, p. 85) writes:

> In our present society, with its pronounced egoistic tendencies, imitation strengthens these, as it would strengthen the altruistic tendencies produced by another form of society. . . . It is only as a consequence of the predominance of egoism in our present society that the error is made of supposing the effect of imitation to be necessarily evil.

Our concern here is not to dispute particular arguments in Bonger for their own sakes, but rather to point to the way in which a single-factor environmentalism is given predominance, with secondary considerations derived from the body of existing literature being introduced eclectically. That is, Bonger's method, though resting on an environmentalism explicitly derived from Marx, appears in the final analysis as a method reminiscent of the eclectism practised by positivist sociologists operating with formal concepts lacking a grounding in history and structure.

This eclectic approach is accompanied by a crudely statistical technique of verification and elaboration. We are presented, amongst other things, with statistical demonstrations of the relationship between levels of educational attainment and violent crime, declines in business and "bourgeois" crime (fraud, etc.), degrees of poverty and involvement in sexual crime (especially prostitution), crimes of "vengeance" and the season of the year and many more. Consistently, the objective is to demonstrate the underlying motivation as being bound up with an egoism induced and sustained by the environment of capitalism. . . . And, lest we should think that egoism is directly a product of poverty and subordination, as opposed to being a central element of a general moral climate, Bonger is able to offer explanations of crime among the bourgeoisie. These crimes he sees to be motivated by need, in cases of business decline and collapse, or by cupidity. In the latter case, "what [men] get by honest business is not enough for them, they wish to become richer." In either case, Bonger's case is contingent on the moral climate engendered by the economic system (Bonger, 1969, p. 138):

It is only under special circumstances that this desire for wealth arises, and . . . it is unknown under others. It will be necessary only to point out that although cupidity is a strong motive with all classes of our present society, it is especially so among the bourgeoisie, as a consequence of their position in the economic life.

Now, Bonger's formal Marxism does enable him to make an insightful series of comments about the nature of the deprivations experienced under capitalism. Judged in Bonger's own terms—that is, in terms of the social positivism of his time—his work surpasses much that was, and is, available. Notably, Bonger's discussion[s] of the effects of the subordination of women (and its contribution to the aetiology of female criminality) and of "militarism" (in sustaining an egoistic and competitive moral climate) seem far ahead of their time.

Writing of the criminality of women, for example, Bonger (1969, p. 58) asserts that:

The great power of a man over his wife, as a consequence of his economic preponderance, may equally be a demoralizing cause. It is certain that there will always be abuse of power on the part of a number of those whom social circumstances have clothed with a certain authority. How many women there are now who have to endure the coarseness and bad treatment of their husbands, but would not hesitate to leave them if their economic dependence and the law did not prevent. . . .

The contemporary ring of these comments is paralleled in Bonger's comments, made, it should be remembered, at the time when the "Marxist" parties of Europe found their members rushing to the "national defence" in the "Great War." . . . Thus, whilst much of Bonger's formal Marxism appears as a form of abstracted and eclectic positivism when viewed across its canvas, he still derives a considerable benefit and understanding from the Marxist perspective in his sensitivity to the demoralizing and destructive consequences of the forms of domination characteristic of a capitalist society. Paradoxically, however, this sensitivity does not extend to an understanding of the nature of domination and social control in defining and delineating the field of interest itself, namely what passes for crime and deviance in societies where "law" is the law determined by powerful interests and classes in the population at large . . . .

Bonger asserts that "there are instances where an action stamped as criminal is not felt to be immoral by anybody." But these statements, and others like them, are made in passing and do not constitute the basis for the thoroughgoing analysis of the structure of laws and interests. And Bonger is ambivalent throughout on the role of social control in the creation of crime. He seems aware only in certain cases, of "societal reaction" in determining degrees of apprehension. So, for example: "the offences of which women are most often guilty are also those which it is most difficult to discover, namely those committed without violence. Then, those who have been injured are less likely to bring a complaint against a woman than against a man."

But later, in dealing with sexual crimes in general, Bonger uncritically accepts the official statistics of apprehension as an indication of "the class of the population that commits these crimes." In fact, Bonger's position is that the law (and its enforcement) whilst certainly the creation of a dominant class—is a genuine reflexion of some universal social and moral sentiment. . . . The manifest explanation for the inclusion within the criminal law of sanctions controlling behaviour which is not directly harmful to the class interests of the powerful is that the working classes themselves are not without power. That is, one supposes, it is in the interests of the powerful to operate a system of general social control in the interests of order (within which individual and corporate enterprise can proceed unimpeded). However, there is more than a suspicion that Bonger's equation of social control with a universal moral sentiment is based on a belief he shares with the bourgeoisie in order for its own sake. Socialism is preferable to capitalism because it is more orderly. . . .

Bonger's formal Marxism, therefore, tells us that the solution to the problems of criminality is not so much to challenge the labels and the processing of capitalist law as it is to wage a responsible and orderly political battle for the reform of a divisive social structure. Even in the case of political opposition, a crucial distinction is to be drawn between responsible activity (the acts of a noble man) and the irresponsible and pathological activity—especially that of the anarchist movement (characterized, argues Bonger, by "extreme individualism," "great vanity," "pronounced altruistic tendencies" "coupled with a lack of intellectual development"). . . .

. . . . For us, the outstanding feature of Bonger's essentially correctional perspective is that, quite aside from the premises on which it operates (the contingency of criminality on an egoistic moral climate), it does not reveal a consistent social psychology, or, by the same token, a systematic social theory. At one moment, the actor under consideration is seen to be inextricably caught up in a determined and identifiable set of circumstances (or, more properly, a set of economic relationships); at another, he appears as the victim of an assumed personal quality (the "criminal thought") sustained and (often) apparently developed by the moral-climate of industrial capitalism.

In so far as a social theory reveals itself in Bonger, the central assumptions on which it is built appear to be Durkheimian in nature rather than to derive from the avowedly Marxist theory of its author. Criminal man is consistently depicted not so much as a man produced by a matrix of unequal social relationships, nor indeed as a man attempting to resolve those inequalities of wealth, power and life-chances; rather, criminal man is viewed as being in need of social control. "Socialism," in this perspective, is an alternative and desirable set of social institutions, which carry with them a set of Durkheimian norms and controls. "Socialism" thus expressed is the resort of an idealist, wishing for the substitution of a competitive and egoistic moral climate by a context in which the co-operativeness of men is encouraged. Socialism is preferable to capitalism, most of all, because it will control the baser instincts of man. Bonger does not assert that the "egoistic" man will "wither away" under socialism: it is only that the social relationships of socialism will not reward the endeavours of an egoist. . . .

⊠ References

Bonger, W. (1916). *Criminality and economic conditions.* Boston: Little, Brown.

Bonger, W. (1935). *An introduction to criminology.* London: Methuen.

Bonger, W. (1969). *criminality and economic conditions.* Bloomington, IN: Indiana University Press.

Turk, Austin T. (1964). Prospects for theories of criminal behavior. *Journal of Criminal Law, Criminology and Police Science,* 55(December), 454–461.

REVIEW QUESTIONS

1. What portions of Marx's theory seem to make the most sense? Which do you agree least with?

2. Which of Bonger's concepts or propositions do you agree with most? Which of Bonger's concepts or propositions do you disagree with most?

3. Do you think U.S. society would be better if it followed the propositions presented by Marx and/or Bonger? If so, which country should the United States use as a model of economical structure? If not, what problems do you see in such models?

❖

READING

In this selection, Meda Chesney-Lind discusses feminist criminology as an outgrowth of the second wave of feminism and argues for combining racial factors with the feminist perspective as a dual focus for research. The author promotes this combined approach as the best method to advance the emphasis on both of these factors for future criminological research. She asserts that this combination of factors will lead to more research attention and that examinations of both race and feminism will be more likely to foster an understanding of criminality and criminal justice processing for both.

While reading this selection, readers are encouraged to consider the patriarchal nature of virtually every society in the world, but particularly the United States. There is likely no society or justice system in the world that views females as equal to males, despite the fact that often females are given more lenient sentences for most offenses than males in the United States, with a notable exception of status offenses by juveniles (e.g., running away). The author points out that the criminal justice system does not adequately account for the issues women, especially those of other races, face when they are incarcerated (e.g., the negative impact on their children), and the fact that many are from the lower class or disadvantaged groups makes this fact even more overwhelming for such women.

SOURCE: Meda Chesney-Lind, "Patriarchy, Crime, and Justice: Feminist Criminology in an Era of Backlash," *Feminist Criminology* 1, no. 1 (2006): 6–26. Copyright © 2006 Sage Publications, Inc. Used by permission of Sage Publications, Inc.

While examining this selection, readers should put themselves in the place of a police officer, prosecutor, or judge and consider if they would treat female suspects the same way as
male suspects if they held these positions. Furthermore, taking the role of a criminological
researcher/theorist, each reader should consider the value of understanding female offending. Specifically, because females offend far less often than males for virtually all violent
offenses, it may be useful to gather information about female offenders and then apply such
knowledge to policies that attempt to reduce higher rates of offending among males. Finally,
readers should consider such issues in terms of the theory of patriarchy that exists in all
countries, and particularly in our nation, as well in terms of racial/ethnic aspects.

Patriarchy, Crime, and Justice

Feminist Criminology in an Era of Backlash

Meda Chesney-Lind

A product of the second wave of the women's
movement, feminist criminology has been in
existence now for more than three decades.
Although any starting point is arbitrary, certainly
one could point to the publication of key journal
issues and books in the 1970s,[1] and it is clear that
a signal event was the founding of the Women
and Crime Division of the American Society of
Criminology in 1982 (Rafter, 2000, p. 9). Since
that time, the field has grown exponentially,
which makes it increasingly impossible to do
justice to all its dimensions in the space of an
article. This article, instead, focuses on the
challenges facing our important field as we enter
a millennium characterized by a deepening
backlash against feminism and other progressive
movements and perspectives.

⊠ Feminist Criminology in the 20th Century: Looking Backward, Looking Forward

The feminist criminology of the 20th century
clearly challenged the overall masculinist

nature of theories of crime, deviance, and
social control by calling attention to the
repeated omission and misrepresentation of
women in criminological theory and research
(Belknap, 2001; Cain, 1990; Daly & Chesney-
Lind, 1988). Turning back the clock, one can
recall that prior to path-breaking feminist
works on sexual assault, sexual harassment, and
wife abuse, these forms of gender violence were
ignored, minimized, and trivialized. Likewise,
girls and women in conflict with the law were
overlooked or excluded in mainstream works
while demonized, masculinized, and sexualized
in the marginalized literature that brooded on
their venality. Stunning gender discrimination,
such as the failure of most law schools to admit
women, the routine exclusion of women from
juries, and the practice of giving male and
female "offenders" different sentences for the
same crimes went largely unchallenged (see
Rafter, 2000, for a good overview of the history
in each of these areas).

The enormity of girls' and women's victimization meant that the silence on the role of
violence in women's lives was the first to
attract the attention of feminist activists and

scholars. Because of this, excellent work exists on the problem of women's victimization—especially in the areas of sexual assault, sexual harassment, sexual abuse, and wife battery (see, e.g., Buzawa & Buzawa, 1990; Dziech & Weiner, 1984; Estrich, 1987; D. Martin, 1977; Rush, 1980; Russell, 1986; Schechter, 1982; Scully, 1990).

In retrospect, the naming of the types and dimensions of female victimization had a significant impact on public policy, and it is arguably the most tangible accomplishment of both feminist criminology and grassroots feminists concerned about gender, crime, and justice. The impact on the field of criminology and particularly criminological theory was mixed, however, in part because these offenses did not initially seem to challenge androcentric criminology per se. Instead, the concepts of "domestic violence" and "victimology," although pivotal in the development of feminist criminology, also supplied mainstream criminologists and some criminal justice practitioners with a new area in which to publish, "new" crimes to study (and opportunities to secure funding), and new men to jail (particularly men of color and other marginalized men). More recently, the field of domestic violence has even been home to a number of scholars who have argued that women are as violent as men (for critical reviews, see DeKeseredy, Sanders, Schwartz, & Alvi, 1997; DeKeseredy & Schwartz, 1998; S. Miller, 2005). In part because of these trends, the focus on girls' and women's victimization has produced a range of challenges for feminist criminology and for feminist activists that have become even more urgent as we move into the new century.

Compared to the wealth of literature on women's victimization, interest in girls and women who are labeled, tried, and jailed as "delinquent" or "criminal" was slower to fully develop[2] in part because scholars of "criminalized" women and girls had to contend early on with the masculinization (or "emancipation") hypothesis of women's crime, which argues in part that "in the same way that women are demanding equal opportunity in the fields of legitimate endeavor, a similar number of determined women are forcing their way into the world of major crimes" (Adler, 1975, p. 3; see also Simon, 1975). Feminist criminologists, as well as mainstream criminologists, debated the nature of that relationship for the next decade and ultimately concluded it was not correct (Chesney-Lind, 1989; Steffensmeier, 1980; Weis, 1976), but this was a costly intellectual detour (and also a harbinger of things to come, as it turned out).

The 1980s and 1990s, however, would see breakthrough research on the lives of criminalized girls and women. Rich documentation of girls' participation in gangs, as an example, challenges earlier gang research that focuses almost exclusively on boys (see Chesney-Lind & Hagedorn, 1999; Moore, 1991). Important work on the role of sexual and physical victimization in girls' and women's pathways into women's crime (see Arnold, 1995; Chesney-Lind & Rodriguez, 1983; Chesney-Lind & Shelden, 1992; Gilfus, 1992) began to appear, along with work that suggests unique ways in which gender and race create unique pathways for girl and women offenders into criminal behavior, particularly in communities ravaged by drugs and overincarceration (Bourgois & Dunlap, 1993; Joe, 1995; Maher & Curtis, 1992; Richie, 1996). Needless to say, the focus on girls' and women's gender also highlights the fact that masculinity and crime need to be both theorized and researched (Bowker, 1998; Messerschmidt, 2000).

Instead of the "add women and stir" (Chesney-Lind, 1988) approach to crime theorizing of the past century (which often introduces gender solely as a "variable" if at all), new important work on the gender/crime nexus *theorizes gender*. This means, for example, drawing extensively on sociological notions of "doing gender" (West & Zimmerman, 1987) and examining the role of "gender regimes" (Williams, 2002) in the production of girls' and women's behavior. Contemporary approaches to gender and crime (see Messerschmidt, 2000; J. Miller, 2001) tend to avoid the problems of reductionism and determinism that characterize early discussions of gender and gender relations, stressing instead the complexity, tentativeness, and variability with which individuals, particularly youth, negotiate (and resist) gender identity (see Kelly, 1993; Thorne, 1993). J. Miller and Mullins (2005), in particular, have argued for the crafting of "theories of the middle range" that recognize that although society and social life are patterned on the basis of gender, it is also the case that the *gender order* (Connell, 2002) is "complex and shifting" (J. Miller & Mullins, 2005, p. 7).

⚒ Feminist Criminology and the Backlash

Feminist criminology in the 21st century, particularly in the United States, finds itself in a political and social milieu that is heavily affected by the backlash politics of a sophisticated and energized right wing—a context quite different from the field's early years when the initial intellectual agenda of the field evolved. Political backlash eras have long been a fixture of American public life, from reconstruction after the Civil War to the McCarthy era of the 1950s. Most of these have certain common characteristics, including a repression of dissent, imperialistic adventure, a grim record of racism, and "resistance to extending full rights to women" (Hardisty, 2000, p. 10).

The current backlash era, however, uses crime and criminal justice policies as central rather than facilitating elements of political agenda—a pattern clearly of relevance to feminist criminology. The right-wing intent to use the "crime problem" became evident very early. Consider Barry Goldwater's 1964 unsuccessful presidential campaign where he repeatedly used phrases such as *civil disorder* and *violence in the streets* in a "covertly racist campaign" to attack the civil rights movement (Chambliss, 1999, p. 14). Both Richard Nixon and Ronald Reagan refined the approach as the crime problem became a centerpiece of the Republican Party's efforts to wrest electoral control of southern states away from the Democratic Party. Nixon's emphasis on *law and order* and Reagan's *war on drugs* were both built on "white fear of black street crime" (Chambliss, 1999, p. 19). With time, crime would come to be understood as a code word for race in U.S. political life, and it became a staple in the Republican attacks on Democratic rivals. When Reagan's former vice president, George Bush Sr., ran for office, he successfully used the Willie Horton incident (where an African American on a prison furlough raped and murdered a woman) to derail the candidacy of Michael Dukakis in 1988 (Chambliss, 1999).

His son, George W. Bush, would gain the presidency as a direct result of backlash criminal justice policies, because felony disenfranchisement of largely African American voters in Florida was crucial to his political strategy in that state (Lantigua, 2001). In Bush's second election campaign, however, another feature would be added to the Republican mix: an appeal to "moral values." Included in the moral values agenda, designed to appeal to right-wing Christians, is the rolling back of the gains of the women's movement of the past century, including the recriminalization of abortion and the denial of civil rights to gay and lesbian Americans. Bush's nominee to the Supreme Court, John Roberts, has even questioned "whether encouraging homemakers to become lawyers contributes to the public good" (Goldstein, Smith, & Becker, 2005).

The centrality of both crime and gender in the current backlash politics means that feminist criminology is uniquely positioned to challenge right-wing initiatives. To do this effectively, however, the field must put an even greater priority on *theorizing patriarchy and crime*, which means focusing on the ways in which the definition of the crime problem and criminal justice practices support patriarchal practices and worldviews.

To briefly review, patriarchy is a sex/gender system[3] in which men dominate women and what is considered masculine is more highly valued than what is considered feminine. Patriarchy is a system of social stratification, which means that it uses a wide array of social control policies and practices to ratify male power and to keep girls and women subordinate to men (Renzetti & Curran, 1999, p. 3). Often, the systems of control that women experience are explicitly or implicitly focused on controlling female sexuality (such as the sexual double standard; Renzetti & Curran, 1999, p. 3). Not infrequently, patriarchal interests overlap with systems that also reinforce class and race privilege, hence, the unique need for feminist criminology to maintain the focus on intersectionality that characterizes recent research and theorizing on gender and race in particular (see Crenshaw, 1994).

Again, in this era of backlash, the formal system of social control (the law and criminal justice policies) play key roles in eroding the rights of both women and people of color, particularly African Americans but increasingly, other ethnic groups as well. Feminist criminology is, again, uniquely positioned to both document and respond to these efforts. To theorize patriarchy effectively means that we have done cutting-edge research on the interface between patriarchal and criminal justice systems of control and that we are strategic about how to get our findings out to the widest audience possible, issues to which this article now turns.

▧ Race, Gender, and Crime

If feminist criminology is to fully understand the interface between patriarchal control mechanisms and criminal justice practices in the United States, we must center our analysis on the race/gender/punishment nexus. Specifically, America's long and sordid history of racism and its equally disturbing enthusiasm for imprisonment must be understood as intertwined, and both of these have had a dramatic effect on African American women in particular (Bush-Baskette, 1998; Horton & Horton, 2005; Johnson, 2003; Mauer, 1999).

More than a century ago, W. E. B. Du Bois saw the linkage between the criminal justice system and race-based systems of social control very clearly. Commenting on the dismal failure of "reconstruction," he concluded,

> Despite compromise, war, and struggle, the Negro is not free. In well-nigh the whole rural South the black farmers are peons, bound by law and custom to an economic slavery from which the only escape is death or the penitentiary. (as quoted in Johnson, 2003, p. 284)

Although the role of race and penal policy has received increased attention in recent years, virtually all of the public discussion of the issues has focused on African American males (see, e.g., Human Rights Watch, 2000). More recently, the significant impact of mass incarceration on African American and Hispanic women has received the attention it deserves. Current data show that African American women account for "almost half (48 percent)" of all the women we incarcerate (Johnson, 2003, p. 34). Mauer and Huling's (1995) earlier research adds an important perspective here; they noted that the imprisonment of African American women grew by more than 828% between 1986 and 1991, whereas that of White women grew by 241% and of Black men by 429% (see also Bush-Baskette, 1998; Gilbert, 2001). Something is going on, and it is not just about race or gender; it is about both—a sinister synergy that clearly needs to be carefully documented and challenged.

Feminist criminologist Paula Johnson (2003), among others, advocated a "Black Feminist analysis of the criminal justice system." An examination of Black women's history from slavery through the Civil War and the postwar period certainly justifies a clear focus on the role that the criminal justice system played in the oppression of African American women and the role of prison in that system (Rafter, 1990). And the focus is certainly still relevant because although women sometimes appear to be the unintended victims of the war on drugs, this "war" is so heavily racialized that the result can hardly be viewed as accidental. African American women have always been seen through the "distorted lens of otherness," constructed as "subservient, inept, oversexed and undeserving" (Johnson, 2003, pp. 9–10), in short, just the "sort" of women that belong in jail and prison. Hence, any good work on criminalized women must also examine the ways in which misogyny and racism have long been intertwined themes in the control of women of color (as well as other women) in the United States, as the next section demonstrates.

◼ Media Demonization and the Masculinization of Female Offenders

As noted earlier, the second wave of feminism had, by the 1980s, triggered an array of conservative political, policy, and media responses. In her book *Backlash: The Undeclared War Against American Women*, Susan Faludi (1991), a journalist, was quick to see that the media in particular were central, not peripheral, to the process of discrediting and dismissing feminism and feminist gains. She focused specific attention on mainstream journalism's efforts to locate and publicize those "female trends" of the 1980s that would undermine and indict the feminist agenda. Stories about "the failure to get husbands, get pregnant, or properly bond

with their children" were suddenly everywhere, as were the very first stories on "bad girls"; Faludi noted that "NBC, for instance, devoted an entire evening news special to the pseudo trend of 'bad girls' yet ignored the real trends of bad boys: the crime rate among boys was climbing twice as fast as for girls" (p. 80).

Faludi's (1991) recognition of the media fascination with bad girls was prescient. The 1990s would produce a steady stream of media stories about violent and bad girls that continues unabated in the new millennium. Although the focus would shift from the "gansta girl," to the "violent girl," to the "mean girl" (Chesney-Lind & Irwin, 2004), the message is the same: Girls are bad in ways that they never used to be. As an example, the Scelfo (2005) article titled "Bad Girls Go Wild," published in the June 13, 2005, issue of *Newsweek*, describes "the significant rise in violent behavior among girls" as a "burgeoning national crisis" (p. 1).

Media-driven constructions such as these generally rely on commonsense notions that girls are becoming more like boys on both the soccer field and the killing fields.[4] Implicit in what might be called the "masculinization" theory (Chesney-Lind & Eliason, in press; Pollock, 1999) of women's violence is the idea that contemporary theories of violence (and crime more broadly) need not attend to gender but can, again, simply add women and stir. The theory assumes that the same forces that propel men into violence will increasingly produce violence in girls and women once they are freed from the constraints of their gender. The masculinization framework also lays the foundation for simplistic notions of "good" and "bad" femininity, standards that will permit the demonization of some girls and women if they stray from the path of "true" (passive, controlled, and constrained) womanhood.

Ever since the first wave of feminism, there has been no shortage of scholars and political commentators issuing dire warnings that women's demand for equality would result in a dramatic change in the character and frequency of women's crime (Pollak, 1950; Pollock, 1999; Smart, 1976). As noted earlier, in the 1970s, the notion that the women's movement was causing changes in women's crime was the subject of extensive media and scholarly attention (Adler, 1975; Chesney-Lind, 1989; Simon, 1975). Again, although this perspective was definitely refuted by the feminist criminology of the era (see Gora, 1982; Steffensmeier & Steffensmeier, 1980; Weis, 1976), media enthusiasm about the idea that feminism encourages women to become more like men and, hence, their "equals" in crime, remains undiminished (see Chesney-Lind & Eliason, in press).

As examples, *Boston Globe Magazine* proclaimed in an article as well as on the issue's cover, over the words *BAD GIRLS* in huge red letters, that "girls are moving into the world of violence that once belonged to boys" (Ford, 1998). And from *San Jose Mercury News* came a story titled "In a New Twist on Equality, Girls' Crimes Resemble Boys'" that features an opening paragraph arguing that

> juvenile crime experts have spotted a disturbing nationwide pattern of teenage girls becoming more sophisticated and independent criminals. In the past, girls would almost always commit crimes with boys or men. But now, more than ever, they're calling the shots. (Guido, 1998, p. 1B)

In virtually all the stories on this topic (including the Scelfo, 2005, article appearing in *Newsweek*), the issue is framed as follows. A specific and egregious example of female violence is described, usually with considerable, graphic detail about the injury suffered by the victim—a pattern that has been dubbed "forensic journalism" (Websdale & Alvarez, 1997, p. 123). In the *Mercury News* article, for example, the reader hears how a 17-year-old girl, Linna Adams, "lured" the victim into a car where her boyfriend "pointed a .357 magnum revolver at him, and the gun went off.

Rodrigues was shot in the cheek, and according to coroner's reports, the bullet exited the back of his head" (Guido, 1998, p. 1B). Websdale and Alvarez (1997) noted that this narrative style, while compelling and even lurid reading, actually gives the reader "more and more information about less and less" and stresses "individualistic explanations that ignore or de-emphasize the importance of wider social structural patterns of disadvantage" (p. 125).

These forensic details are then followed by a quick review of the Federal Bureau of Investigation's arrest statistics showing what appear to be large increases in the number of girls arrested for violent offenses. Finally, there are quotes from "experts," usually police officers, teachers, or other social service workers, but occasionally criminologists, interpreting the narrative in ways consistent with the desired outcome: to stress that girls, particularly African American and Hispanic girls whose pictures often illustrate these stories, are getting more and more like their already demonized male counterparts and, hence, becoming more violent (Chesney-Lind & Irwin, 2005).

There are two problems with this now familiar frame: One, there are considerable reasons to suspect that it is demonstrably wrong (i.e., that girls' violence is not increasing) and two, it has created a "self-fulfilling prophecy" that has had dramatic and racialized effects on girls' arrests, detentions, and referrals to juvenile courts across our country.

Although arrest data consistently show dramatic increases in girls' arrests for "violent" crimes (e.g., arrests of girls for assault climbed an astonishing 40.9%, whereas boys' arrests climbed by only 4.3% in the past decade; Federal Bureau of Investigation, 2004), other data sets, particularly those relying on self-reported delinquency, show no such trend (indeed they show a decline; Chesney-Lind, 2004; Chesney-Lind & Belknap, 2004; Steffensmeier, Schwartz, Zhong, & Ackerman, 2005). It seems increasingly clear that forces other than changes in girls' behavior have caused shifts in girls' arrests (including such forces as zero-tolerance policies in schools and mandatory arrests for domestic violence; Chesney-Lind & Belknap, 2004). There are also indications that although the hype about bad girls seems to encompass all girls, the effects of enforcement policies aimed at reducing "youth violence" weigh heaviest on girls of color whose families lack the resources to challenge such policies (Chesney-Lind & Irwin, 2005).

Take juvenile detention, a focus of three decades of deinstitutionalization efforts. Between 1989 and 1998, girls' detentions increased by 56% compared to a 20% increase seen in boy's detentions, and the "large increase was tied to the growth in the number of delinquency cases involving females charged with person offenses (157%)" (Harms, 2002, p. 1). At least one study of girls in detention suggests that "nearly half" the girls in detention are African American girls, and Latinas constitute 13%; Caucasian girls, who constitute 65% of the girl population, account for only 35% of those in detention (American Bar Association & National Bar Association, 2001, pp. 20–21).

It is clear that two decades of the media demonization of girls, complete with often racialized images of girls seemingly embracing the violent street culture of their male counterparts (see Chesney-Lind & Irwin, 2004), coupled with increased concerns about youth violence and images of "girls gone wild," have entered the self-fulfilling prophecy stage. It is essential that feminist criminology understand that in a world governed by those who self-consciously manipulate corporate media for their own purposes, newspapers and television may have moved from simply covering the police beat to constructing crime "stories" that serve as a "nonconspiratorial source of dominant ideology" (Websdale & Alvarez, 1997, p. 125). Feminist criminology's agenda must consciously challenge these backlash media narratives, as well as engage in "newsmaking criminology" (Barak, 1988), particularly with regard to constructions

of girl and women offenders. The question of how to do this is one that must also engage the field. As a start on such a discussion, consider the advice of Bertold Brecht (1966):

> One must have the courage to write the truth when the truth is everywhere opposed; the keenness to recognize it, and although it is everywhere concealed; the skill to manipulate it as a weapon; the judgment to select in whose hands it will be effective, and the cunning to spread the truth among such persons. (p. 133)

The advocacy work coupled with excellent research that one sees with reports issued by The Sentencing Project and the Center for Juvenile and Criminal Justice provide models of how this work might be done. It certainly requires that we work more closely with progressive journalists than many academics are used to, but given the success of these agencies in doing just that, feminist criminologists should consider this a priority and use our national and regional meetings, as a start, to develop strategies toward this end.

◼ Criminalizing Victimization

Many feminist criminologists have approached the issue of mandatory arrest in incidents of domestic assault with considerable ambivalence (see Ferraro, 2001). On one hand, as noted earlier, the criminalization of sexual assault and domestic violence was in one sense a huge symbolic victory for feminist activists and criminologists alike. After centuries of ignoring the private victimizations of women, police and courts were called to account by those who founded rape crisis centers and shelters for battered women and those whose path-breaking research laid the foundation for major policy and legal changes in the area of violence against women (see Schechter, 1982).

On the other hand, the insistence that violence against women be handled as a criminal matter threw victim advocates into an uneasy alliance with police and prosecutors—professions that feminists had long distrusted and with good reason (see Buzawa & Buzawa, 1990; Heidensohn, 1995; S. Martin, 1980). The criminal justice approach, however, was bolstered in the mid-1980s by what appeared to be overwhelming evidence that arrest decreased violence against women (Sherman & Berk, 1984). Although subsequent research would find that arrest was far less effective than originally thought (see Ferraro, 2001; Maxwell, Garner, & Fagan, 2002), for the policy world, the dramatic early research results seemed to ratify the wisdom of a law enforcement-centered approach to the problem of domestic violence. Ultimately, the combined effects of the early scientific evidence; political pressure from the attorney general of the United States, the American Bar Association, and others; and the threat of lawsuits against departments who failed to protect women from batterers "produced nearly unanimous agreement that arrest was the best policy for domestic violence" (Ferraro, 2001, p. 146).

As the academic debate about the effectiveness of arrest in domestic violence situations continued unabated, the policy of "mandatory arrest" became routinized into normal policing and quite quickly, other unanticipated effects began to emerge. When arrests of adult women for assault increased by 30.8% in the past decade (1994 to 2003), whereas male arrests for this offense fell by about 5.8% (Federal Bureau of Investigation, 2004, p. 275), just about everybody from the research community to the general public began to wonder what was happening. Although some, such as criminologist Kenneth Land, quoted in a story titled "Women Making Gains in Dubious Profession: Crime," attributed the increase to "role change over the past decades" that presumably created more females as "motivated offenders" (Anderson, 2003, p. 1), others were

not so sure. Even the Bureau of Justice Statistics looked at a similar trend (increasing numbers of women convicted in state courts for "aggravated assault") and suggested the numbers might be "reflecting increased prosecution of women for domestic violence" (Greenfeld & Snell, 1999, pp. 5–6).

Much like the increases seen in girls' assaults, this trend requires critical review, a process that takes the reader through the looking glass to a place that the feminists who worked hard to force the criminal justice system and the general public to take wife battery seriously could never have imagined. In this world, as in California recently, the female share of domestic violence arrests tripled (from 5% in 1987 to 17% in 1999; S. Miller, 2005, p. 21); and as it turned out, it was not just a California phenomenon.

Despite the power of the stereotypical scenario of the violent husband and the victimized wife, the reality of mandatory arrest practices has always been more complicated. Early on, the problem of "mutual" arrests—the practice of arresting both the man and the woman in a domestic violence incident if it is not clear who is the "primary" aggressor—surfaced as a concern (Buzawa & Buzawa, 1990). Nor has the problem gone away, despite efforts to clarify procedures (Bible, 1998; Brown, 1997); indeed, many jurisdictions report similar figures. In Wichita, Kansas, for example, women were 27% of those arrested for domestic violence in 2001 (Wichita Police Department, 2002). Prince William County, Maryland, saw the number of women arrested for domestic violence triple in a 3-year period, with women going from 12.9% of those arrested in 1992 to 21% in 1996 (Smith, 1996). In Sacramento, California, even greater increases were observed; there the number of women arrested for domestic violence rose by 91% between 1991 and 1996, whereas arrests of men fell 7% (Brown, 1997).

A Canadian study (Comack, Chopyk, & Wood, 2000) provides an even closer look at the impact of mandatory arrest on arrests of women for crimes of violence. Examining the gender dynamics in a random sample of 1,002 cases (501 men and 501 women) involving charges filed by the Winnipeg, Manitoba, police services for violent crimes during the period 1991 through 1995, the researchers found that the "zero-tolerance" policy implemented by the police force in 1993 had a dramatic effect on women's arrest patterns. Although the policy resulted in more arrests of both men and women for domestic violence, the impact on women's arrests was most dramatic. In 1991, domestic violence charges represented 23% of all charges of violence against women; by 1995, 58% of all violent crime charges against women were for partner violence (Comack et al., 2000, p. ii). Most significant, the researchers found that in 35% of the domestic violence cases involving women, the accused woman had actually called the police for help (only 5% of male cases showed this pattern; Comack et al., 2000, p. 15).

Susan Miller's (2005) study of mandatory arrest practices in the state of Delaware adds an important dimension to this discussion. Based on data from police ride-alongs, interviews with criminal justice practitioners, and observations of groups run for women who were arrested as offenders in domestic violence situations, Miller's study comes to some important conclusions about the effects of mandatory arrest on women.

According to beat officers S. Miller (2005) and her students rode with, in Delaware, they do not have a "pro-arrest policy, we have a pro-paper policy" (p. 100) developed in large part to avoid lawsuits. What initially surfaces as a seemingly minor, and familiar, lament begins to take on far more meaning. It emerges that at least in Delaware (but one suspects elsewhere), police departments, often in response to threatened or real lawsuits, have developed an "expansive definition" (S. Miller, 2005, p. 89) of domestic violence, including a wide range of family disturbances. As a consequence,

although the officers "did not believe there was an increase in women's use of violence" (S. Miller, 2005, p. 105), "her fighting back now gets attention too" (S. Miller, 2005, p. 107) because of this sort of broad interpretation of what constitutes domestic violence.

Another significant theme in police comments reflects male batterers' increased skill in deploying the criminal justice system to further intimidate and control their wives. Officers reported that men are now more willing to report violence committed by their wives and more willing to use "cross-filings" in securing protective orders against their wives and girlfriends. Police particularly resent what they regard as "bogus" violations of protective orders that are actually just harassment (S. Miller, 2005, p. 90).

None of the social service providers and criminal justice professionals S. Miller (2005) spoke with felt women had become more aggressively violent; instead, they routinely called the women "victims." They noted that at least in Delaware, as the "legislation aged," the name of the game began to be "get to the phone first" (S. Miller, 2005, p. 127). Social service workers noted that male batterers tended to use their knowledge of the criminal justice system and process as a way to threaten their wives with the loss of the children, particularly if they had managed to get the woman placed on probation for abuse. Workers echoed the police complaints about paperwork, noting it takes 8 hours to do the paperwork if an arrest is made, but then they made a crucial link to the arrest of women, noting that police, weary of being told they were the problem, have channeled at least some of their resentment into making arrests of women who act out violently (regardless of context or injury) because "according to police policy," they have to make an arrest.

Essentially, it appears that many mandatory arrest policies have been interpreted on the grounds of making an arrest if any violent "incident" occurs, rather than considering the context within which the incident occurs

(Bible, 1998). Like problematic measures of violence that simply count violent events without providing information on the meaning and motivation, this definition of *domestic violence* fails to distinguish between aggressive and instigating violence from self-defensive and retaliatory violence. According to S. Miller (2005) and other critics of this approach, these methods tend to produce results showing "intimate violence is committed by women at an equal or higher rate than by men" (p. 35). Although these findings ignited a firestorm of media attention about the "problem" of "battered men" in the United States (Ferraro, 2001, p. 137), the larger question of how to define *domestic violence* in the context of patriarchy is vital. Specifically, much feminist research of the sort showcased here is needed on routine police and justice practices concerning girls' and women's "violence." In particularly short supply are studies of girls' arrests, particularly those of girls of color (who are often detained for "assault"), although indirect evidence certainly suggests this is happening (see Chesney-Lind & Irwin, 2005). The evidence to date suggests the distinct possibility that in addition to the well-documented race and class problems, with draconian criminal justice approaches to domestic violence (S. Miller, 1989; Richie, 1996), we have a gender issue: Are these policies criminalizing women's (and girls') attempts to protect themselves?

▧ Women's Imprisonment and the Emergence of Vengeful Equity

When the United States embarked on a policy that might well be described as mass incarceration (Mauer & Chesney-Lind, 2002), few considered the impact that this correctional course change would have on women. Yet the number of women in jail and prison continues to soar (outstripping male increases for most of the past decade), completely untethered from women's crime rate, which has not increased by nearly the same amount (Bloom & Chesney-Lind, 2003). The dimensions of this shift are staggering: For most of the 20th century, we imprisoned about 5,000 to 10,000 women. At the turn of the new century, we now have more than 100,000 women doing time in U.S. prisons (Harrison & Beck, 2004, p. 1). Women's incarceration in the United States not only grew during the past century but also increased tenfold; and virtually all of that increase occurred in the final two decades of the century.

The number of women sentenced to jail and prison began to soar at precisely the same time that prison systems in the United States moved into an era that abandoned any pretense of rehabilitation in favor of punishment. As noted earlier, decades of efforts by conservative politicians to fashion a crime policy that would challenge the gains of the civil rights movement as well as other progressive movements in the 1960s and 1970s had, by the 1980s, borne fruit (Chambliss, 1999). Exploiting the public fear of crime, particularly crime committed by "the poor, mostly nonwhite, young, male inner-city dwellers" (Irwin, 2005, p. 8), all manner of mean-spirited crime policies were adopted. The end of the past century saw the war on drugs and a host of other "get tough" sentencing policies, all of which fueled mass imprisonment (see Mauer, 1999). The period also saw the development of what Irwin (2005) has called "warehouse prisons," a correctional regime that focuses on a physical plant designed to control (not reform), rigid enforcement of extensive rules, and easy transfer of unruly prisoners to even more draconian settings.

Although feminist legal scholars can and do debate whether equality under the law is necessarily good for women (see Chesney-Lind & Pollock-Byrne, 1995), a careful look at what has happened to women in U.S. prisons might serve as a disturbing case study of how correctional equity is implemented in practice. Such a critical review is particularly vital in an era where decontextualized notions of gender and race "discrimination" are increasingly and successfully deployed against the achievements of both the civil rights and women's movements (Pincus, 2001/2002).

Consider the account of Martha Sierra's experience of childbirth. As she

> writhed in pain at a Riverside hospital, laboring to push her baby into the world, Sierra faced a challenge not covered in the childbirth books: her wrists were shackled to the bed. Unable to roll on her side or even sit straight up, Sierra managed as best she could. The reward was fleeting . . . she watched as her daughter, hollering and flapping her arms, was taken from the room. (Warren, 2005, p. A1)

As difficult as it was to talk about giving birth while serving time in prison, Sierra was particularly "distressed and puzzled" by her medical treatment: "Did they think I was going to get up and run away?" asked the 28-year-old California prisoner (Warren, 2005, p. A1).

Sierra's story is unfortunately all too familiar to anyone who examines gender themes in modern correctional responses to women inmates. In fact, her experience is less horrific than the case of Michelle T., a former prisoner from Michigan who told Human Rights Watch (1996) that she was accompanied by two male correctional officers into the delivery room:

> According to Michelle T., the officers handcuffed her to the bed while she was in labor and positioned themselves where they could view her genital area while giving birth. She told [Human Rights Watch] they made derogatory comments about her throughout the delivery. (p. 249)

Basically, male prisoners have long used visits to hospitals as opportunities to escape, so correctional regimes have generated extensive security precautions to assure that escapes do not occur, including shackling prisoners to hospital beds (Amnesty International, 1999, p. 63). This is the dark side of the equity or parity model of justice—one which emphasizes treating women offenders as though they were men, particularly when the outcome is punitive, in the name of equal justice—a pattern that could be called vengeful equity.

Vengeful equity could have no better spokesperson than Sheriff Joe Arpaio who, when he defended his controversial chain gang for women in Maricopa County, Arizona, proclaimed himself an "equal opportunity incarcerator" and went on to explain his controversial move by saying,

> If women can fight for their country, and bless them for that, if they can walk a beat, if they can protect the people and arrest violators of the law, then they should have no problem with picking up trash in 120 degrees. (Kim, 1996, p. A1)

Other examples of vengeful equity can be found in the creation of women's boot camps, often modeled on the gender regimes found in military basic training. These regimes, complete with uniforms, shorn hair, humiliation, exhausting physical training, and harsh punishment for even minor misconduct have been traditionally devised to "make men out of boys." As such, feminist researchers who have examined them contended, they "have more to do with the rites of manhood" than the needs of the typical woman in prison (Morash & Rucker, 1990).

Although these examples might be seen as extreme, legal readings by correctional administrators and others that define any attention to legitimate gender differences as "illegal" have clearly produced troubling outcomes. It is obviously misguided to treat women as if they were men with reference to cross-gender supervision, strip searches, and other correctional regimes while ignoring the ways in which women's imprisonment has unique features (such as pregnancy and vulnerability to sexual assault). Recently, this approach has been correctly identified by Human Rights Watch (1996) as a major contributing factor to the sexual abuse of women inmates.

Reviewing the situation of women incarcerated in five states (California, Georgia, Michigan, Illinois, and New York) and the District of Columbia, Human Rights Watch (1996) concluded,

> Our findings indicate that being a woman prisoner in U.S. state prisons can be a terrifying experience. If you are sexually abused, you cannot escape from your abuser. Grievance or investigatory procedures, where they exist, are often ineffectual, and correctional employees continue to engage in abuse because they believe they will rarely be held accountable, administratively or criminally. Few people outside the prison walls know what is going on or care if they do know. Fewer still do anything to address the problem. (p. 1)

Human Rights Watch also noted that their investigators were "concerned that states' adherence to U.S. anti-discrimination laws, in the absence of strong safeguards against custodial sexual misconduct, has often come at the fundamental rights of prisoners" (p. 2).

Institutional subcultures in women's prisons, which encourage correctional officers to "cover" for each other, coupled with inadequate protection accorded women who file complaints, make it unlikely that many women prisoners will formally complain about abuse. In addition, the public stereotype of women in prison as bad makes it difficult for a woman inmate to support her case against a correctional officer in court. Finally, what little progress has been made is now threatened by recent legislation that curtails the ability of prisoners and advocates to commence a legal action concerning prison conditions (Stein, 1996, p. 24).

Finally, it appears that women in prison today are also recipients of some of the worst of the more traditional, separate-spheres approach to women offenders (which tends to emphasize gender difference and the need to focus on "saving" women by policing even minor behaviors, particularly sexual behaviors; Rafter, 1990). Correctional officers often count on the fact that women prisoners will complain, not riot, and as a result, often punish women inmates for offenses that would be ignored in male prisons. McClellan (1994) found this pattern quite clearly in her examined disciplinary practices in Texas prisons. Following up two cohorts of prisoners (one male and one female), she found most men in her sample (63.5%) but only a handful of women (17.1%) had no citation or only one citation for a rule violation. McClellan found that women prisoners not only received numerous citations but also were charged with different infractions than men. Most frequently, women were cited for "violating posted rules," whereas males were cited most often for "refusing to work" (McClellan, 1994, p. 77). Women were more likely than men to receive the most severe sanctions.

McClellan (1994) noted that the wardens of the women's prisons in her study stated quite frankly that they demand total compliance with every rule on the books and punish violations through official mechanisms. McClellan concluded by observing that there exists

> two distinct institutional forms of surveillance and control operating at the male and female facilities. . . . This policy not only imposes extreme constraints on adult women but also costs the people of the State of Texas a great deal of money. (p. 87)

Much good, early feminist criminology focuses on the conditions of girls and women in training schools, jails, and prisons (see Burkhart, 1976; Carlen, 1983; Faith, 1993; Freedman, 1981). Unfortunately, that work is now made much harder by a savvy correctional system that is extremely reluctant to admit researchers, unless the focus of the research is clearly the woman prisoner and not the institution. That said, there is much more need for this sort of criminology in the era of mass punishment, and the work that is being done in this vein (see Bloom, 2003; Owen, 1998) points to the need for much of the same. Huge numbers of imprisoned girls and women are targeted by male-based systems of "risk" and "classification" (Hannah-Moffat & Shaw, 2003) and then subjected to male-based interventions such as "cognitive behaviorism" to address their "criminal" thinking as though they were men (Kendall & Pollack, 2003). Good work has also been done on the overuse of "chemical restraints" with women offenders (Auerhahn & Leonard, 2000; Leonard, 2002). In short, as difficult as it might be to do, in this era of mass imprisonment, feminist criminology needs to find creative ways to continue to engage core issues in girls' and women's carceral control as a central part of our intellectual and activist agenda. As Adrian Howe (1994) put it, "Academics must not let 'theoretical rectitude' deter them from committing themselves as *academics* and *feminists* to campaigns on behalf of women lawbreakers" (p. 214).

◈ Theorizing Patriarchy: Concluding Thoughts

In 1899, Jane Addams was asked to address the American Academy of Political and Social Science. She took the occasion to reflect on the role of the social science of her day:

> As the college changed from teaching theology to teaching secular knowledge

the test of its success should have shifted from the power to save men's souls to the power to adjust them in healthful relations to nature and their fellow men. But the college failed to do this, and made the test of its success the mere collecting and dissemination of knowledge, elevating the means unto an end and falling in love with its own achievement. (Addams, 1899, pp. 339–340)

Perhaps Addams's use of the generic *he* to describe the universities of her day was more than a simple convention. Recall that when Addams worked in Chicago, criminology as a discipline was taking shape at the University of Chicago, whose researchers often relied heavily on contacts made at Hull House while systematically excluding women from its faculty ranks and distancing themselves from the female-dominated field of social work (see Deegan, 1988).

How do we avoid the pitfalls Addams (1899) observed in the male-dominated criminology of her day? This article argues that although feminist criminology has made a clear contribution to what might be described as the criminological project, it is positioned to play an even more central role in the era of political backlash. Certainly, we, as feminist scholars, shoulder many burdens, but perhaps the most daunting is the one articulated by Liz Kelly (as quoted in Heidensohn, 1995): "Feminist research investigates aspects of women's oppression while seeking at the same time to be a part of the struggle against it" (p. 71).

For feminist criminology to remain true to its progressive origins in very difficult times, we must seek ways to blend activism with our scholarship (and senior scholars, in particular, need to make the academy safe for their junior colleagues to do just that by redefining tenure criteria to make this work a part of "scholarship"). We must discuss the many tensions and difficulties with this work, again in an era of backlash when the right is actively patrolling faculty behavior (Horowitz & Collier, 1994), and be honest about the many challenges ahead. We must create venues for feminist criminology, including peer-reviewed journals (such as *Critical Criminology, Feminist Criminology*, and *Women & Criminal Justice*) while also writing for broader audiences, particularly practitioners and policy makers (see *Women, Girls & Criminal Justice*). We must engage in continued activism on the part of girls and women who are the victims of crimes and whose very experiences are being trivialized by well-funded backlash research (Hoff Sommers, 1994) while also documenting the problems those same girls and women have when they take their cases to court (see Estrich, 1987; Matoesian, 1993). It means close attention and continued vigilance about the situations of women working in various aspects of the criminal justice system, particularly as the right wing cynically appropriates concepts such as discrimination. Finally, and most important, it means activism on behalf of criminalized girls and women, the least powerful and most marginalized of all those we study.

Again, given the focus of the backlash, this article argues that feminist criminology is uniquely positioned to do important work to challenge the current political backlash. To do so effectively, however, it is vital that in addition to documenting that gender matters in the lives of criminalized women, we engage in exploration of the interface between systems of oppression based on gender, race, and class. This work will allow us to make sense of current crime-control practices, particularly in an era of mass incarceration, so that we can explain the consequences to a society that might well be ready to hear other perspectives on crime control if given them (consider the success of drug courts and some initiatives that encourage alternatives to incarceration; Mauer, 2002). Researching as well as theorizing both patriarchy and gender is crucial to feminist criminology so that we can craft work, as the

right wing does so effectively, that speaks to backlash initiatives in smart, media-savvy ways. To do this well means foregrounding the role of race and class in our work on gender and crime, as the work showcased here makes clear. There is simply no other way to make sense of key trends in both the media construction of women offenders and the criminal justice response that increasingly awaits them, particularly once they arrive in prison.

Finally, we must also do work that will document and challenge the policy and research backlash aimed at the hard fought and vitally important feminist and civil rights victories of the past century. To do any less would be unthinkable to those who fought so long to get us where we are today, and so it must be for us.

◾ Notes

1. One might cite the appearance of the classic special issue on women and crime of *Issues in Criminology*, edited by Dorie Klein and June Kress (1973); two important books on the topic of women and crime by Rita Simon (1975) and Freda Adler (1975); and the publication of Del Martin's (1977) *Battered Wives* and Carol Smart's (1976) *Women, Crime and Criminology*.

2. Early but important exceptions to this generalization are Klein and Kress (1973), Smart (1976), Crites (1976), Bowker, Chesney-Lind, and Pollock (1978), Chapman (1980), and Jones (1980). There has also been an encouraging outpouring of more recent work on women offenders in the past decade. See Belknap (2001), Chesney-Lind and Pasko (2004), and DeKeseredy (1999) for reviews of this recent work.

3. Sex-gender systems include the following elements: (a) the social construction of gender categories on the basis of biological sex, (b) a sexual division of labor in which specific tasks are allocated on the basis of sex, and (c) the social regulation of sexuality, in which particular forms of sexual expression are positively and negatively sanctioned (Renzetti & Curran, 1999, p. 3).

4. I owe this analogy to Frank Zimring who, in response to a question from a reporter, quipped, "Women's liberation didn't turn girls into boys—violence

is still particularly male. There has been much more diversification of gender roles on the soccer field than the killing field" (Ryan, 2003, p. 2).

References

Addams, J. (1899). A function of the social settlement. *Annals of the Academy of Political and Social Science, 13,* 323–345.

Adler, F. (1975). *Sisters in crime.* New York: McGraw-Hill.

American Bar Association and the National Bar Association. (2001, May 1). *Justice by gender: The lack of appropriate prevention, diversion and treatment alternatives for girls in the justice system.* Retrieved from http://www.abanet.org/crimjust/juvjus/justicebygenderweb.pdf

Amnesty International. (1999). *Not part of my sentence: Violations of the human rights of women in custody.* Washington, DC: Author.

Anderson, C. (2003, October 28). Women making gains in dubious profession: Crime. *Arizona Star,* p. A1.

Arnold, R. (1995). Processes of criminalization: From girlhood to womanhood. In M. B. Zinn & B. T. Dill (Eds.), *Women of color in American society* (pp. 136–146). Philadelphia: Temple University Press.

Auerhahn, K., & Leonard, E. (2000). Docile bodies? Chemical restraints and the female inmate. *The Journal of Criminal Law and Criminology, 90*(2), 599–634.

Barak, G. (1988). Newsmaking criminology: Reflections on the media, intellectuals, and crime. *Justice Quarterly, 5*(4), 565–587.

Belknap, J. (2001). *The invisible woman: Gender, crime and justice* (2nd ed.). Belmont, CA: Wadsworth.

Bible, A. (1998). When battered women are charged with assault. *Double-Time, 6*(1/2), 8–10.

Bloom, B. (Ed.). (2003). *Gendered justice.* Durham, NC: Carolina Academic Press.

Bloom, B., & Chesney-Lind, M. (2003). Women in prison: Vengeful equity. In R. Muraskin (Ed.), *It's a crime: Women and the criminal justice system* (pp. 175–195). New Jersey: Prentice Hall.

Bourgois, P., & Dunlap, E. (1993). Exorcising sex–for crack: An ethnographic perspective from Harlem. In M. Ratner (Ed.), *The crack pipe as pimp* (pp. 97–132). New York: Lexington Books.

Bowker, L. (Ed.). (1998). *Masculinities and violence.* Thousand Oaks, CA: Sage.

Bowker, L., Chesney-Lind, M., & Pollock, J. (1978). *Women, crime, and the criminal justice system.* Lexington, MA: D. C. Heath.

Brecht, B. (1966). *Galileo* (E. Bentley, Ed., C. Laughton, Trans.). New York: Grove.

Brown, M. (1997, December 7). Arrests of women soar in domestic assault cases. *Sacramento Bee.* Retrieved July 31, 2005, from http://www.sacbee.com/static/archive/news/projects/violence/part12.html

Burkhart, K. W. (1976). *Women in prison.* New York: Popular Library.

Bush-Baskette, S. (1998). The war on drugs as a war against Black women. In S. L. Miller (Ed.), *Crime control and women* (pp. 113–129). Thousand Oaks, CA: Sage.

Buzawa, E., & Buzawa, C. G. (1990). *Domestic violence: The criminal justice response.* Newbury Park, CA: Sage.

Cain, M. (1990). Realist philosophy and standpoint epistemologies or feminist criminology as a successor science. In L. Gelsthorpe & A. Morris (Eds.), *Feminist perspectives in criminology* (pp. 124–140). Buckingham, UK: Open University Press.

Carlen, P. (1983). *Women's imprisonment: A study in social control.* London: Routledge.

Chambliss, W. (1999). *Power, politics, and crime.* Boulder, CO: Westview.

Chapman, J. R. (1980). *Economic realities and the female offender.* Lexington: Lexington Books.

Chesney-Lind, M. (1988, July-August). Doing feminist criminology. *The Criminologist, 13,* 16–17.

Chesney-Lind, M. (1989). Girls' crime and woman's place: Toward a feminist model of female delinquency. *Crime & Delinquency, 35*(10), 5–29.

Chesney-Lind, M. (2004, August). Girls and violence: Is the gender gap closing? *National Electronic Network on Violence Against Women.* Retrieved from http://www.vawnet.org/DomesticViolence/Research/VAWnetDocs/ARGirlsViolence.php

Chesney-Lind, M., & Belknap, J. (2004). Trends in delinquent girls' aggression and violent behavior: A review of the evidence. In M. Putallaz & P. Bierman (Eds.), *Aggression, antisocial behavior and violence among girls: A development perspective* (pp. 203–222). New York: Guilford.

Chesney-Lind, M., & Eliason, M. (in press). From invisible to incorrigible: The demonization of marginalized women and girls. *Crime, Media, Culture: An International Journal.*

Chesney-Lind, M., & Hagedorn, J. M. (Eds.). (1999). *Female gangs in America: Essays on gender and gangs.* Chicago: Lakeview Press.

Chesney-Lind, M., & Irwin, K. (2004). From badness to meanness: Popular constructions of contemporary girlhood. In A. Harris (Ed.), *All about the girl: Culture, power, and identity* (pp. 45–56). New York: Routledge.

Chesney-Lind, M., & Irwin, K. (2005). Still "the best place to conquer girls": Gender and juvenile justice. In J. Pollock-Byrne & A. Merlo (Eds.), *Women, law, and social control* (pp. 271–291). Boston: Allyn & Bacon.

Chesney-Lind, M., & Pasko, L. (2004). *The female offender.* Thousand Oaks, CA: Sage.

Chesney-Lind, M., & Pollock-Byrne, J. (1995). Women's prisons: Equality with a vengeance. In J. Pollock-Byrne & A. Merlo (Eds.), *Women, law, and social control* (pp. 155–176). Boston: Allyn & Bacon.

Chesney-Lind, M., & Rodriguez, N. (1983). Women under lock and key. *The Prison Journal, 63,* 47–65.

Chesney-Lind, M., & Shelden, R. G. (1992). *Girls, delinquency and juvenile justice.* Belmont, CA: Wadsworth.

Comack, E., Chopyk, V., & Wood, L. (2000). *Mean streets? The social locations, gender dynamics, and patterns of violent crime in Winnipeg.* Ottawa, Ontario: Canadian Centre for Policy Alternatives.

Connell, R. W. (2002). *Gender.* Cambridge, UK: Polity.

Crenshaw, H. (1994). Mapping the margins: Intersectionality, identity politics, and violence against women of color. In M. A. Fineman & R. Mykitiuk (Eds.), *The public nature of private violence* (pp. 93–118). New York: Routledge.

Crites, L. (Ed.). (1976). *The female offender.* Lexington, MA: Lexington Books.

Daly, K., & Chesney-Lind, M. (1988). Feminism and criminology. *Justice Quarterly, 5*(4), 497–538.

Deegan, M. J. (1988). *Jane Addams and the men of the Chicago School, 1892-1918.* New Brunswick, NJ: Transaction Books

DeKeseredy, W. (1999). *Women, crime, and the Canadian criminal justice system.* Cincinnati, OH: Anderson.

DeKeseredy, W., Sanders, D., Schwartz, M., & Alvi, S. (1997). The meanings and motives for women's use of violence in Canadian college dating relationships. *Sociological Spectrum, 17,* 199–222.

DeKeseredy, W., & Schwartz, M. (1998, February). *Measuring the extent of woman abuse in intimate heterosexual relationships: A critique of the conflict tactics scales.* Retrieved from VAWnet Web site: http://www.vawnet.org/DomesticViolence/Research/VAWnetDocs/AR_ctscrit.php

Dziech, B. W., & Weiner, L. (1984). *The lecherous professor.* Boston: Beacon.

Estrich, S. (1987). *Real rape.* Cambridge, MA: Harvard University Press.

Faith, K. (1993). *Unruly women: The politics of confinement & resistance.* Vancouver, British Columbia, Canada: Press Gang.

Faludi, S. (1991). *Backlash: The undeclared war against American women.* New York: Anchor Doubleday.

Federal Bureau of Investigation. (2004). *Crime in the United States, 2003.* Washington, DC: Government Printing Office.

Ferraro, K. (2001). Women battering: More than a family problem. In C. Renzetti & L. Goodstein (Eds.), *Women, crime and criminal justice* (pp. 135–153). Los Angeles: Roxbury.

Ford, R. (1998, May 24). The razor's edge. *Boston Globe Magazine,* pp. 3, 22–28.

Freedman, E. (1981). *Their sisters' keepers: Women and prison reform, 1830–1930.* Ann Arbor: University of Michigan Press.

Gilbert, E. (2001). Women, race, and criminal justice processing. In C. Renzetti & L. Goodstein (Eds.), *Women, crime and criminal justice* (pp. 222–231). Los Angeles: Roxbury.

Gilfus, M. (1992). From victims to survivors to offenders: Women's routes of entry into street crime. *Women & Criminal Justice, 4*(1), 63–89.

Goldstein, A., Smith, J., & Becker, J. (2005, August 19). Roberts resisted women's rights. *Washington Post,* p. A1.

Gora, J. (1982). *The new female criminal: Empirical reality or social myth.* New York: Praeger.

Greenfeld, A., & Snell, T. (1999). *Women offenders: Bureau of Justice Statistics, special report.* Washington, DC: U.S. Department of Justice.

Guido, M. (1998, June 4). In a new twist on equality, girls' crimes resemble boys'. *San Jose Mercury News,* p. 1B–4B.

Hannah-Moffat, K., & Shaw, M. (2003). The meaning of "risk" in women's prisons: A critique. In B. Bloom (Ed.), *Gendered justice* (pp. 25–44). Durham, NC: Carolina Academic Press.

Hardisty, J. V. (2000). *Mobilizing resentment.* Boston: Beacon.

Harms, P. (2002, January). *Detention in delinquency cases, 1989-1998* (OJJDP Fact Sheet No. 1). Washington, DC: U.S. Department of Justice.

Harrison, P. M., & Beck, A. J. (2004). *Prisoners in 2003.* Washington, DC: U.S. Department of Justice, Bureau of Justice Statistics.

Heidensohn, F. (1995). Feminist perspectives and their impact on criminology and criminal justice in Britain. In N. H. Rafter & F. Heidensohn (Eds.), *International feminist perspectives in criminology* (pp. 63–85). Buckingham, UK: Open University Press.

Hoff Sommers, C. (1994). *Who stole feminism? How women have betrayed women.* New York: Simon & Schuster.

Horowitz, D., & Collier, P. (1994). *The heterodoxy handbook: How to survive the PC campus.* Lanham, MD: National Book Network.

Horton, J. O., & Horton, L. (2005). *Slavery and the making of America.* Oxford, UK: Oxford University Press.

Howe, A. (1994). *Punish and critique: Towards a feminist analysis of penality.* London: Routledge.

Human Rights Watch. (1996). *All too familiar: Sexual abuse of women in U.S. state prisons.* New York: Author.

Human Rights Watch. (2000, May). Punishment and prejudice: Racial disparities in the war on drugs. *Human Rights Watch Reports, 12*(2). Retrieved from http://www.hrw.org/reports/2000/usa/

Irwin, J. (2005). *The warehouse prison.* Los Angeles: Roxbury.

Joe, K. (1995). Ice is strong enough for a man but made for a woman: A social cultural analysis of methamphetamine use among Asian Pacific Americans. *Crime, Law and Social Change, 22,* 269–289.

Johnson, P. (2003). *Inner lives: Voices of African American women in prison.* New York: New York University Press.

Jones, A. (1980). *Women who kill.* New York: Fawcett Columbine.

Kelly, D. M. (1993). *Last chance high: How girls and boys drop in and out of alternative schools.* New Haven, CT: Yale University Press.

Kendall, K., & Pollack, S. (2003). Cognitive behaviorism in women's prisons. In B. Bloom (Ed.), *Gendered justice* (pp. 69–96). Durham, NC: Carolina Academic Press.

Kim, E.-K. (1996, August 26). Sheriff says he'll have chain gangs for women. *Tuscaloosa News,* p. A1.

Klein, D., & Kress, J. (Eds.). (1973). Women, crime and criminology [Special issue]. *Issues in Criminology, 8*(3).

Lantigua, J. (2001, April 30). How the GOP gamed the system in Florida. *The Nation,* pp. 1–8.

Leonard, E. (2002). *Convicted survivors: The imprisonment of battered women.* New York: New York University Press.

Maher, L., & Curtis, R. (1992). Women on the edge: Crack cocaine and the changing contexts of street-level sex work in New York City. *Crime, Law and Social Change, 18,* 221–258.

Martin, D. (1977). *Battered wives.* New York: Pocket Books.

Martin, S. (1980). *Breaking and entering: Police women on patrol.* Berkeley: University of California Press.

Matoesian, G. (1993). *Reproducing rape domination through talk in the courtroom.* Chicago: University of Chicago Press.

Mauer, M. (1999). *Race to incarcerate.* New York: New Press.

Mauer, M. (2002). State sentencing reforms: Is the "get tough" era coming to a close. *Federal Sentencing Reporter, 15,* 50–52.

Mauer, M., & Chesney-Lind, M. (Eds.). (2002). *Invisible punishment: The collateral consequences of mass imprisonment.* New York: New Press.

Mauer, M., & Huling, T. (1995). *Young Black Americans and the criminal justice system: Five years later.* Washington, DC: The Sentencing Project.

Maxwell, C., Garner, J. H., & Fagan, J. A. (2002). The preventive effects of arrest on intimate partner violence: Research, policy and theory. *Criminology & Public Policy, 2*(1), 51–80.

McClellan, D. S. (1994). Disparity in the discipline of male and female inmates in Texas prisons. *Women & Criminal Justice, 5*(20), 71–97.

Messerschmidt, J. W. (2000). *Nine lives: Adolescent masculinities, the body, and violence.* Boulder, CO: Westview.

Miller, J. (2001). *One of the guys: Girls, gangs, and gender.* New York: Oxford University Press.

Miller, J., & Mullins, C. (2005). *Taking stock: The status of feminist theories in criminology.* Unpublished manuscript.

Miller, S. (1989). Unintended side effects of pro-arrest policies and their race and class implications for battered women: A cautionary note. *Criminal Justice Policy Review, 3,* 299–317.

Miller, S. (2005). *Victims as offenders: Women's use of violence in relationships.* New Brunswick, NJ: Rutgers University Press.

Moore, J. (1991). *Going down to the barrio: Homeboys and homegirls in change.* Philadelphia: Temple University Press.

Morash, M., & Rucker, L. (1990). A critical look at the idea of boot camp as a correctional reform. *Crime & Delinquency, 36*(2), 204–222.

Owen, B. (1998). *"In the mix": Struggle and survival in a women's prison.* Albany: State University of New York Press.

Pincus, F. (2001/2002). The social construction of reverse discrimination: The impact of affirmative action on Whites. *Journal of Inter-Group Relations, 38*(4), 33–44.

Pollak, O. (1950). *The criminality of women.* Philadelphia: University of Pennsylvania Press.

Pollock, J. (1999). *Criminal women.* Cincinnati, OH: Anderson.

Rafter, N. H. (1990). *Partial justice: Women, prisons and social control.* New Brunswick, NJ: Transaction Books.

Rafter, N. H. (Ed.). (2000). *Encyclopedia of women and crime.* Phoenix, AZ: Oryx Press.

Renzetti, C., & Curran, D. J. (1999). *Women, men and society.* Boston: Allyn & Bacon.

Richie, B. (1996). *Compelled to crime: The gender entrapment of battered Black women.* New York: Routledge.

Rush, F. (1980). *The best kept secret: Sexual abuse of children.* New York: McGraw-Hill.

Russell, D. (1986). *The secret trauma.* New York: Basic Books.

Ryan, J. (2003, September 5). Girl gang stirs up false gender issue: Data show no surge in female violence. *San Francisco Chronicle,* p. 2.

Scelfo, J. (2005, June 13). Bad girls go wild. *Newsweek.* Retrieved July 31, 2005, from http://www.msnbcnsn.com/id.8101517/site/newsweek/page/2/

Schechter, S. (1982). *Women and male violence: The visions and struggles of the battered women's movement.* Boston: South End.

Scully, D. (1990). *Understanding sexual violence.* Boston: Unwin Hyman.

Sherman, L. W., & Berk, R. A. (1984). The specific deterrent effects of arrest for domestic assault. *American Sociological Review, 49*(1), 261–272.

Simon, R. (1975). *Women and crime.* Lexington, MA: Lexington Books.

Smart, C. (1976). *Women, crime and criminology: A feminist critique.* London: Routledge Kegan Paul.

Smith, L. (1996, November 18). Increasingly, abuse shows a female side: More women accused of domestic violence. *Washington Post,* p. B1.

Steffensmeier, D. J. (1980). Sex differences in patterns of adult crime, 1965–1977. *Social Forces, 58,* 1080–1108.

Steffensmeier, D. J., Schwartz, J., Zhong, H., & Ackerman, J. (2005). An assessment of recent trends in girls' violence using diverse longitudinal sources. *Criminology, 43,* 355–406.

Steffensmeier, D. J., & Steffensmeier, R. H. (1980). Trends in female delinquency: An examination of arrest, juvenile court, self-report, and field data. *Criminology, 18,* 62–85.

Stein, B. (1996, July). Life in prison: Sexual abuse. *The Progressive,* 23–24.

Thorne, B. (1993). *Gender play.* New Brunswick, NJ: Rutgers University Press.

Warren, J. (2005, June 19). Rethinking treatment of female prisoners. *Los Angeles Times,* p. A1.

Websdale, N., & Alvarez, A. (1997). Forensic journalism as patriarchal ideology: The newspaper construction of homicide-suicide. In D. Hale & F. Bailey (Eds.), *Popular culture, crime and justice* (pp. 123–141). Belmont, CA: Wadsworth.

Weis, J. G. (1976). "Liberation and crime": The invention of the new female criminal. *Crime and Social Justice, 6,* 17–27.

West, C., & Zimmerman, D. H. (1987). Doing gender. *Gender & Society, 1,* 125–151.

Wichita Police Department. (2002). *Domestic violence statistics: 2001.* Retrieved from http://wichitapolice.com/DV/DV_statistics.htm

Williams, L. S. (2002). Trying on gender, gender regimes, and the process of becoming women. *Gender & Society, 16,* 29–52.

REVIEW QUESTIONS

1. What does Chesney-Lind mean by the contemporary "backlash" against feminist criminology?

2. What are the primary reasons why the author believes that combining gender with race will lead to a better understanding of criminality, as well as handling by the criminal justice system?

3. Regarding the criminal justice system's handling of females, do you believe equal treatment is a rational goal? Why or why not? Discuss the same questions regarding individuals' race/ethnicity.

4. What does the author mean by "vengeful equity"? Explain this concept and give an example.

READING

In this selection, Amanda Burgess-Proctor presents a theoretical piece that examines the interplay among the various concepts that are so vital in conflict perspectives of criminality, namely race/ethnicity, socioeconomic status, and gender. She reveals the established and potential interactions between these various concepts, particularly their importance in terms of these various aspects regarding reasons why they commit crime, as well as the handling by such persons in the justice system. Although the author places a high emphasis on feminism, this selection involves important discussions about the most vital aspects of demographic characteristics that are key in virtually all conflict perspectives of crime and formal handling by the system.

While reading the first portion of Burgess-Proctor's article, readers should pay attention to the idea that there does not exist a single, unitary perspective of feminist criminology. Rather, there are various forms of feminist perspectives of crime that exist under the feminist grouping of theories in the discipline. Specifically, such categories included in her review are: liberal feminism, radical feminism, Marxist feminism, socialist feminism, and postmodern feminism. Although not exhaustive, these categories are currently the most common forms of feminist criminology and are largely based on differential emphases regarding what is most vital to their specific frameworks. For example, while Marxist feminists claim that women's economic oppression in capitalistic societies is the primary issue, liberal feminists claim that the differential way that males and females are socialized toward gender-specific roles in society is the primary problem facing women. This is just one example of the way feminist criminology is far more complex than most people, including many criminologists, realize.

Also while reading this selection, readers are encouraged to consider the importance of each of these demographic characteristics—race/ethnicity, social class, and gender—regarding their impact on crime and how such cases are treated in the criminal justice system. More than half of the readers of this text/selection may themselves be underrepresented minorities, women, or members of the working class, and the other readers likely know from experience some offenders who fit into these groups. Thus, all readers are encouraged to examine if they can relate to this perspective, either for themselves or others they know well. Ultimately, Burgess-Proctor presents an important framework in her presentation of multi-class and multi-racial feminism, and the issues she proposes regarding the interactions among these concepts represent the current state of understanding and recommendations of future research regarding race, class and feminism.

SOURCE: Amanda Burgess-Proctor, "Intersections of Race, Class, Gender, and Crime," *Feminist Criminology* 1, no. 1 (2006): 27–47. Copyright © 2006 Sage Publications, Inc. Used by permission of Sage Publications, Inc.

Intersections of Race, Class, Gender, and Crime

Future Directions for Feminist Criminology

Amanda Burgess-Proctor

Feminist criminology has survived the growing pains of its development during the 1970s to emerge as a mature theoretical orientation. Thanks to the pioneering generation of feminist criminologists who insisted that women's deviance was worthy of academic inquiry, as well as to the contemporary generation of feminist criminologists who have contributed immensely to our understanding of women as victims, offenders, and practitioners of the criminal justice system, feminist criminology now is routinely recognized by the broader discipline as a legitimate theoretical perspective (or more accurately, set of perspectives). More than 30 years after the first scholarship of its kind was produced, feminist studies of crime are more commonplace than ever before.

Two recent milestone events remind us of just how powerful an influence feminists have had in criminology during the past three decades. First, 2004 marked the 20th anniversary of the Division on Women and Crime, the unit of the American Society of Criminology whose members are dedicated to feminist criminology and to the study of issues related to women, gender, and crime. Second, the creation of this journal, the official publication of the Division on Women and Crime, serves as a testament to the demand that exists for feminist criminological scholarship. Together, these two important achievements provide the perfect opportunity to reflect on what lies ahead for feminist criminology. Now is the time to ask ourselves, In what direction is contemporary feminism heading, and how will developments in the broader feminist movement influence the future of feminist criminology in particular? How will the work of feminist criminologists be defined in the 21st century, and what opportunities exist for the advancement of feminist criminology in the coming years?

Throughout this article, I argue that the future of feminist criminology lies in our willingness to embrace a theoretical framework that recognizes multiple, intersecting inequalities. Contemporary feminist criminologists bear the responsibility of advancing an inclusive feminism, one that simultaneously attends to issues of race, class, gender, sexuality, age, nationality, religion, physical ability, and other locations of inequality as they relate to crime and deviance. Put simply, to advance an understanding of gender, crime, and justice that achieves universal relevance and is free from the shortcomings of past ways of thinking, feminist criminologists must examine linkages between inequality and crime using an intersectional theoretical framework that is informed by multiracial feminism.

To present a persuasive argument for using an intersectional approach in feminist studies of crime, I have divided this article into four sections. First, I start by tracing the development of feminist criminology from its inception in the early 1970s. This section begins with a brief discussion of various feminist perspectives and then outlines the development of feminist criminology within the context of the broader feminist movement. Second, I describe the evolution of feminist approaches to gender that occurred during second-wave feminism of the 1970s and 1980s. Specifically, this section considers the transition from "sameness" and "difference" models, which are informed by liberal feminism, to a "dominance" model informed by radical feminism. Third, I discuss the emergence during contemporary third-wave feminism of an alternative approach to gender: an intersectional model informed by multiracial feminism. Fourth, I underscore the importance of this intersectional model for feminist criminology by highlighting its theoretical, methodological, and praxis-related relevance, and I suggest future directions for feminist criminology with respect to intersections of race, class, gender, and crime.

Feminist Perspectives and the Development of Feminist Criminology

Overview of Feminist Perspectives

Before summarizing the development of feminist criminology, it is first necessary to point out that feminism does not refer to a unitary theory. Rather, there are multiple perspectives that fall under the rubric of feminism, each of which involves different assumptions about the source of gender inequality and women's oppression (Barak, Flavin, & Leighton, 2001; Daly & Chesney-Lind, 1988; Price & Sokoloff, 2004). Accordingly, feminist theory traditionally is divided into five major perspectives.[1]

First, liberal feminism regards gender role socialization as the primary source of women's oppression. In other words, men's social roles (e.g., competitive and aggressive) are afforded more social status and power than women's roles (e.g., nurturing and passive). Consequently, liberal feminists emphasize political, social, legal, and economic equality between women and men. Within criminology, liberal feminists view women's offending as a function of gender role

socialization as well; that is, women offend at a lower rate than men because their socialization provides them with fewer opportunities to engage in deviance.

Second, radical feminism identifies patriarchy, or male dominance, as the root cause of women's oppression. In other words, women experience discrimination because social relations and social interactions are shaped by male power and privilege. Within criminology, radical feminists often focus on manifestations of patriarchy in crimes against women, such as domestic violence, rape, sexual harassment, and pornography, and recognize that women's offending often is preceded by victimization, typically at the hands of men.[2]

Third, Marxist feminism attributes women's oppression to their subordinate class status within capitalist societies. In other words, the capitalist mode of production shapes class and gender relations that ultimately disadvantage women because women occupy the working class instead of the ruling class. Within criminology, Marxist feminists theorize that women's subordinate class status may compel them to commit crime as a means of supporting themselves economically.

Fourth, socialist feminism combines radical and Marxist perspectives to conclude that women's oppression results from concomitant sex- and class-based inequalities. In other words, class and gender work in tandem to structure society, and socialist feminists call for an examination of the ways in which gender relations are shaped by class and vice versa. Within criminology, socialist feminists examine causes of crime within the context of interacting gender- and class-based systems of power.

Fifth, postmodern feminism departs from the other feminist perspectives by questioning the existence of any one "truth," including women's oppression. In other words, postmodern feminists reject fixed categories and universal concepts in favor of multiple truths, and as such examine the effects of discourse and symbolic representation on claims about knowledge. Within criminology, postmodern feminists interrogate the social construction of concepts such as "crime," "justice," and "deviance" and challenge accepted criminological truths.

Although these are the five most commonly identified feminist perspectives, other perspectives are equally important to feminist theory. Black feminism and critical race feminism are centered on the experiences of Black women and women of color, and as such view women's oppression in terms of simultaneous gender-and race-based disadvantage (Collins, 2000; Crenshaw, 1991). As I discuss more fully in the third section of this article, this focus on intersecting systems of race and class makes Black and critical race feminism (along with socialist feminism) the precursors to multiracial feminism. Within criminology, Black and critical race feminists call attention to the discriminatory treatment of non-White women in the criminal justice system, such as judicial instructions to juries in rape trials to forgo the assumption that Black women are chaste (Crenshaw, 1991).

Alternatively, lesbian feminism links women's oppression to heterosexism and men's control of women's social spaces by "[taking] the radical feminist pessimistic view of men to its logical conclusion," where Third World feminism views women's oppression as a function of the economic exploitation of women in developing nations (Lorber, 2001, p. 99; see also Belknap, 2001). Examples of these latter two perspectives are not as common in criminology as examples of the previous perspectives, although this appears to be changing. For example, the past decade has seen an increase in domestic violence research that examines the experiences of lesbian (Renzetti, 1992), immigrant (Bui, 2004), and Muslim (Hajjar, 2004) battered women.

It is clear that each of these feminist perspectives represents a unique way of theorizing about women's oppression and about the linkages between inequality and crime. However, as I argue throughout this article, it is not any

of these perspectives but the perspective of multiracial feminism that is most relevant to feminist criminology in the 21st century.

▧ The Emergence of Feminist Criminology

Having outlined these feminist perspectives, it is now possible to trace the emergence of feminist criminology. Historically, the feminist movement is divided into three eras or waves. The first wave of feminism began in the United States with the birth of the abolitionist and women's suffrage movements in the mid- to late 1800s. Criminology itself was still developing at this time, as scholars such as Lombroso and Durkheim (in Europe) and Kellor (in the United States) began theorizing about crime and deviance (Beirne & Messerschmidt, 2000). Some 100 years later, the women's liberation and civil rights movements of the 1960s and 1970s marked the genesis of second-wave feminism. It is during this era that feminism made its appearance in criminology.

Theoretically speaking, feminist criminology developed because (primarily liberal) feminist scholars objected to the exclusion of gender from criminological analyses, an omission that seemed particularly glaring given that gender is such a strong predictor of offending (Blumstein, Cohen, Roth, & Visher, 1986; Steffensmeier & Allan, 1996), arrest (Stolzenberg & D'Alessio, 2004) and sentencing outcomes (Daly, 1994; Daly & Tonry, 1997). Feminist scholars were dissatisfied with the failure of mainstream criminology to recognize issues of gender inequality at all, as well as with the failure of critical and radical criminology to consider the relationship between inequality and crime outside of the narrow context of economic disparities, under which were subsumed issues of race and gender (see Beirne & Messerschmidt, 2000). In particular, feminist criminologists protested the exclusion of women's experiences in emerging "general" theories of crime, which were being developed by mainstream criminologists using almost exclusively male samples to predict patterns of male delinquency (Barak, 1998; Belknap, 2001; Chesney-Lind & Pasko, 2004; Daly & Chesney-Lind, 1988; Flavin, 2004; Miller, 2004; Milovanovic & Schwartz, 1996; Morash, 1999). These early feminist criminologists demanded that analyses of crime include consideration of gender in ways that had not occurred before.

It is important to note here that feminist criminology was born during a crucial juncture of the feminist movement. Shortly after the beginning of the second wave in the 1970s, feminists of minority-group status found that their experiences were underrepresented in mainstream feminism and subsequently levied sharp criticism toward their majority-group counterparts whose voices were purported to speak on behalf of all women. Feminists of color, lesbian feminists, Third World feminists, and feminists from other marginalized groups condemned the hegemony of White, middle-class, heterosexual experience that characterized mainstream second-wave feminism at that time. As I discuss more fully in the following section, it is against this backdrop that feminist criminology emerged, and claims of essentialism and reductionism soon plagued the broader feminist movement as well as feminist criminology in particular.

Finally, second-wave feminism gave way to third-wave feminism during the 1980s and 1990s. A defining feature of third-wave feminism is its focus on multiplicities or the belief that there exist multiple genders, races, and sexualities. With this idea in mind and echoing earlier criticisms, many third-wave feminists expressed dissatisfaction with the insufficient treatment of race, class, sexuality, and other locations of inequality in mainstream feminist scholarship. As a result, it is during contemporary third-wave feminism that intersectionality first appeared (Price & Sokoloff, 2004).[3] Intersectionality recognizes that systems of power such as race, class, and gender do not act alone to shape our experiences but rather, are multiplicative, inextricably linked, and simultaneously experienced. Feminist criminologists writing at this time also recognized the need for intersectionality (Daly, 1993; Daly & Stephens, 1995). For example, Daly and Stephens (1995) observed that an intersectional approach to studying crime explores

> how class, gender, and race (and age and sexuality) construct the normal and deviant . . . how these inequalities put some societal members at risk to be rendered deviant or to engage in law-breaking, and . . . how law and state institutions both challenge and reproduce these inequalities. (p. 193)

This timeline of events allows for a more thorough understanding of the development of feminist thought during second-wave and early third-wave feminism. In the following section, I describe in detail how feminist approaches to studying gender have evolved in the past 30 years, largely in response to the events outlined above.

⊠ The Evolution of Feminist Approaches to Gender: Sameness, Difference, and Dominance

Shortly after the inception of second-wave feminism, one question divided feminists more sharply than perhaps any other: Are women essentially similar to men such that the sexes should be treated equally (i.e., the sameness approach) or do women have distinctive characteristics that require special treatment to overcome their gender-based discrimination (i.e., the difference approach)? This question loomed large among feminists, with proponents of each side lamenting the threat posed by the opposing viewpoint to the advancement of women's rights. For example, opponents of the difference approach argued that difference often is nothing more than a euphemism for discrimination, as well as that championing women's differences ultimately leads to their exclusion from certain roles, particularly within the workplace (MacKinnon, 1991; Nagel & Johnson, 2004; Williams, 1991). Conversely, opponents of the sameness approach believed in a gender dichotomy and claimed that women suffer from an equal-treatment model because, under the guise of gender neutrality, women's status ultimately is measured against a dominant male norm (Daly & Chesney-Lind, 1988; MacKinnon, 1991; Nagel & Johnson, 2004).

The sameness/difference debate that surfaced in the broader feminist movement also appeared in feminist criminology. For example, advocates of the sameness approach supported gender-neutral, "equal treatment under the law" measures, such as symmetrical correctional programming meant to guarantee male and female inmates access to the same vocational and educational resources (Barak et al., 2001). Furthermore, supporters of this equal-treatment model worried about the hidden dangers of the difference approach, wherein women's difference from men translates into their need for greater legal protection, as in the case of statutory rape laws:

> Statutory rape is, in criminal law terms, a clear instance of a victimless crime, since all parties are, by definition, voluntary participants. In what sense, then, can [Supreme Court Justice] Rehnquist assert [in a 1981 decision upholding California's statutory rape law] that the woman is victim and the man offender? One begins to get an inkling when, later, the Justice explains that the statutory rape law is "protective" legislation. . . . The notion that men are frequently the sexual aggressors and that the law ought to be able to take that reality

into account in very concrete ways is hardly one that feminists could reject out of hand . . . it is therefore an area . . . in which we need to pay special attention to our impulses lest we inadvertently support and give credence to the very social constructs and behaviors we so earnestly mean to oppose. (Williams, 1991, pp. 20–21)

However, critics of the sameness approach argued that this model actually harms women because the law is not gender neutral but in fact assumes a male standard (MacKinnon, 1991; Williams, 1991). Given this male standard, women's legal claims have the potential to be regarded as requiring special or preferential treatment (Daly & Chesney-Lind, 1988). For example, considering parenthood as a mitigating factor in sentencing decisions, although applied "equally" to all defendants, disproportionately benefits female defendants with children (or "familied" women) and unintentionally punishes childless women (Barak et al., 2001; Daly, 1989; Daly & Tonry, 1997). Furthermore, critics recognized that the desire to standardize sentencing practices in the name of "equal justice" may carry unintended consequence for women:

> A major problem is that [equal-treatment] sentencing reforms are designed to reduce race- and class-based disparities in sentencing men. Their application to female offenders may yield equality with a vengeance: a higher rate of incarceration and for longer periods of time than in the past. Like reforms in divorce . . . and in child custody . . . devised with liberal feminist definitions of equality, sentencing reform also may prove unjust and may work ultimately against women. (Daly & Chesney-Lind, 1988, p. 525)

Despite the stark contrast in the orientation of the sameness and difference models, both approaches share a commonality: They are rooted in simplistic notions of women's and men's equality that are characteristic of liberal feminism. Recall that liberal feminism emphasizes political, social, legal, and economic equality between the sexes. Therefore, theoretical paradigms such as the sameness and difference models that fail to place gender relations in the context of patriarchy fit squarely within the perspective of liberal feminism. Furthermore, as many feminists have observed, the sameness and difference models also share the same fundamental flaw: They essentially ignore issues of power and privilege (Barak et al., 2001; MacKinnon, 1991; Sokoloff, Price, & Flavin, 2004). Put differently, both approaches fail to acknowledge disparities in power between the sexes; consequently, sameness and/or difference cannot be considered meaningfully without regard for women's subordinate status in a patriarchal society. As MacKinnon (1991) noted, although "men's differences from women are equal to women's differences from men . . . the sexes are not socially equal" (p. 85).

In an effort to improve upon previous models that were informed by liberal feminism and that offered elementary comparisons of men and women without regard for the effects of patriarchy, radical feminists writing at the beginning of the third wave argued for the adoption of a dominance approach to studying gender. The dominance approach is informed by radical feminism because it recognizes how patriarchy shapes gender relations and considers gender difference within the context of power and oppression (Barak et al., 2001; MacKinnon, 1991). "For women to affirm difference, when difference means dominance, as it does with gender, means to affirm the qualities and characteristics of powerlessness" (MacKinnon, 1991, p. 86). Moreover, supporters argued that the dominance approach was the only truly feminist paradigm. According to radical feminists, the sameness and difference approaches are masculinist insomuch as they use a male referent, whereas the dominance approach, "in that it sees the inequalities of the social world from the standpoint of the subordination of women to men, is feminist" (MacKinnon, 1991, p. 86).

Feminist criminologists also recognized the value of using the dominance approach. For example, Bark et al. (2001) maintain that the utility of the dominance model for feminist criminologists rests in its attention to power: "For example, proponents of the dominance approach have been instrumental in pressuring the legal system to abandon its 'hands-off' attitude toward domestic violence and to define wife battering and marital rape as crimes" (p. 154). Furthermore, the dominance approach has value for feminist criminologists because its emphasis on power and privilege dovetails with related issues of inequality in the criminal justice system.

However, despite its appeal, the dominance approach was criticized by many early third-wave feminists—primarily women of color, lesbian women, and women from other marginalized groups—for essentializing women (Andersen & Collins, 2004; Baca Zinn, Hondagneu-Sotello, & Messner, 2000; Barak et al., 2001; Belknap, 2001; Daly & Chesney-Lind, 1988). "Essentialism occurs when a voice—typically a white, heterosexual, and socioeconomically privileged voice—claims to speak for everyone" (Sokoloff et al., 2004, p. 12). By asserting that women universally suffer the effects of patriarchy, the dominance approach rests on the dubious assumption that all women, by virtue of their shared gender, have a common "experience" in the first place. In short, the dominance approach is reductionist because it assumes that all women are oppressed by all men in exactly the same ways or that there is one unified experience of dominance experienced by women. Soon after these criticisms first emerged in the broader feminist movement, feminist criminologists began contemplating their own acts of essentialism. In the words of Daly and Chesney-Lind (1988), "One of the many challenges for feminism in general and feminist criminology in particular is the paradox of acknowledging diversity among women while claiming women's unity in experiences of oppression and sexism" (p.502).

◪ Beyond Dominance: Multiracial Feminism and Intersections of Race, Class, and Gender

As discussed in the previous section, the promise of the dominance approach was not shared by women who felt marginalized by "the hegemony of feminisms constructed primarily around the lives of white-middle class women" (Baca Zinn & Thornton Dill, 1996, p. 321).[4] Moreover, the dominance approach, for all of its positive attributes, is unidimensional in the sense that it merely examines women's experiences vis-à-vis men's experiences. In contrast, contemporary feminists now face a more multidimensional question: How do we move away from the "false universalism embedded in the concept 'woman'" toward an examination of gender in the context of other locations of inequality (Baca Zinn & Thornton Dill, 1996, p. 322)?

This question is not easily answered, and previous attempts by feminist scholars to acknowledge systems of power other than gender resulted in a rather benign emphasis on "diversity." As Baca Zinn and Thornton Dill (1996) comment,

> Despite the much heralded diversity trend within feminist studies, difference is often reduced to mere pluralism; a "live and let live" approach where principles or relativism generate a long list of diversities which

begin with gender, class, and race and continue through a range of social structural as well as personal characteristics. . . . The major limitation of these approaches is the failure to attend to the power relations that accompany difference. Moreover, these approaches ignore the inequalities that cause some characteristics to be seen as "normal" while others are seen as "different" and thus, deviant. (p. 323)

For example, as "outsiders within" the feminist movement, women of color protested the complicity of "unitary theories of gender" in mainstream feminism (Baca Zinn & Thornton Dill, 1996, p. 321; see also Collins, 2000). Building on the criticisms wielded by feminists of color, a new paradigm has emerged in contemporary third-wave feminism that advances feminist thought beyond issues of sameness, difference, or dominance.[5] The "intersectional" model, informed by multiracial feminism, has succeeded in examining gender through the lens of difference while at the same time acknowledging the instrumental role of power in shaping gender relations. That is, guided by the perspective of multiracial feminism, the intersectional approach successfully attends to issues of power and dominance while achieving a universal relevance that has eluded previous approaches to studying gender.

It is sometimes difficult to disentangle concepts embedded in multiracial feminism from concepts embedded in intersections of race, class, and gender. In an attempt to elucidate these ideas, I have divided the following section in half. The first half describes key conceptual features of multiracial feminism. The second half discusses the advantages of adopting an intersectional (or "race-class-gender") approach to studying gender that is informed by multiracial feminism. Together, both halves of the following section are intended to highlight the importance of intersectionality for feminist criminology.

Multiracial Feminism

Multiracial feminism was pioneered by women of color who recognized the need to construct approaches to studying gender that attended to issues of power and difference in ways that previous models had not (Baca Zinn & Thornton Dill, 1996). Although this perspective is known by a variety of names, including intersectionality theory and multicultural feminism, the term multiracial feminism is preferred because it emphasizes "race as a power system that interacts with other structured inequalities to shape genders" (Baca Zinn & Thornton Dill, 1996, p. 324). Still, the main focus is on interlocking and multiple inequalities (Baca Zinn & Thornton Dill, 1996; Thompson, 2002).

Multiracial feminism has several key conceptual features that distinguish it from other feminist perspectives and that make it ideal for promoting theoretical advancement for feminist criminology. First, multiracial feminism proposes that gender relations do not occur in a vacuum but, instead, that men and women also are characterized by their race, class, sexuality, age, physical ability, and other locations of inequality (Baca Zinn & Thornton Dill, 1996). Put differently, this perspective emphasizes that a power hierarchy—what Collins (2000) calls the "matrix of domination"—exists in which people are socially situated according to their differences from one another. Feminists who operate within this perspective interpret gender as being socially constructed through interlocking systems of race, class, gender, and other sources of inequality (Baca Zinn & Thornton Dill, 1996). In turn,

> this structural pattern affects individual consciousness, group interaction, and group access to institutional power and privileges . . . within this framework, [the focus is] less on the similarities and differences among race, class, and gender than on patterns

of connection that join them. (Andersen & Collins, 2004, p. 7)

In a related manner, multiracial feminism proposes that these intersections occur simultaneously and, therefore, create a distinct social location for each individual (Baca Zinn & Thornton Dill, 1996). In other words, the various axes of the matrix of domination intersect to create a particular status within the broader social structure, which constitutes one's social location (Andersen & Collins, 2004). "A key element to class-race-gender is that social relations are viewed in multiple and interactive terms—not as additive" (Daly, 1993, p. 56). Thus, the main point here is that these elements work multiplicatively to shape one's social location. Although at a given time race, class, or gender might feel more relevant, "they are overlapping and cumulative in their effect on people's experience" (Andersen & Collins, 2004, p. 7).

Second, multiracial feminism calls attention to the ways in which intersecting systems of power act on all social-structural levels:

> Class, race, gender, and sexuality are components of *both* social structure and social interaction. Women and men are differently embedded in locations created by these cross-cutting hierarchies. As a result, women and men throughout the social order experience different forms of privilege and subordination. (Baca Zinn & Thornton Dill, 1996, p. 327, italics added)[6]

One important concept of multiracial feminism that underscores this point is "both/and" (e.g., see Collins, 2000). That is, all people simultaneously experience both oppression and privilege; no individual or group can be entirely privileged or entirely oppressed. In other words, the theoretical starting point [of this perspective] is that there are multiple and cross-cutting relations of class, race-ethnicity, gender, sexuality and age. This produces a matrix of domination taking a "both/and" form . . . not a simple additive model of structural subordinate relations. (Daly & Stephens, 1995, pp. 206–207)

Third, multiracial feminism is centered on the concept of *relationality*; that is, it assumes that groups of people are socially situated in relation to other groups of people based on their differences. "This means that women's differences are *connected* in systematic ways" (Baca Zinn & Thornton Dill, 1996, p. 327, emphasis in original). More important, relationality should not be thought of as a unilateral concept. To illustrate, multiracial feminism assumes that the experiences of women are structurally linked, such that some women benefit from the oppression of other women who occupy a lower social position, even when (or perhaps especially when) the former are not cognizant of the benefits that their privileged status provides. In this way, multiracial feminism retains an emphasis on power and privilege but avoids essentializing women's experiences. As Baca Zinn and Thornton Dill (1996) observe, "Multiracial feminism highlights the relational nature of dominance and subordination. Power is the cornerstone of women's differences" (p. 327).

Last, other key conceptual features of multiracial feminism include a focus on the interaction of social structure and women's agency, a reliance on a variety of methodological approaches, and an emphasis on understandings grounded in the lived experiences of dynamic groups of women (Baca Zinn & Thornton Dill, 1996). Together, these features distinguish multiracial feminism from other feminist perspectives. Although some feminist perspectives share certain of these conceptual features (e.g., postmodern feminism's emphasis on the social construction of reality; socialist, Black, and critical race feminism's attention to interacting systems of power), taken in total these characteristics make multiracial feminism truly unique among feminist perspectives.

◪ The Intersectional Approach to Gender: Using a Race-Class-Gender Framework

Feminists who operate from the perspective of multiracial feminism advocate for an intersectional approach to studying gender. The intersectional approach recognizes that race, class, gender, sexuality, and other locations of inequality are dynamic, historically grounded, socially constructed power relationships that simultaneously operate at both the microstructural and macro-structural levels (Andersen & Collins, 2004; Weber, 2001; Weber & Parra-Medina, 2003).

In many ways, the development of this intersectional approach to studying gender may be viewed as a natural progression of feminist thought. Recall that the sameness and difference approaches, which are informed by liberal feminism, were criticized by radical feminists who favored the dominance approach because it attends to issues of power and privilege. Likewise, feminists dissatisfied with the primacy of gender in the dominance approach have advocated for the adoption of the intersectional approach, which is informed by multiracial feminism and which does not prioritize gender over other systems of power. Thus, applying a race-class-gender framework to the study of gender may be regarded as the next step in the evolution of feminist thought.

An example from the sentencing literature helps to clarify this point. According to Daly and Tonry (1997), criminologists and legal scholars have adopted three modes of conceptualizing race and gender in criminal law and criminal justice practice: (a) law and practices as racist/sexist, (b) law and practices as White/male, and (c) law and practices as racialized/gendered. The first mode (racist/sexist) parallels the sameness approach because "differential treatment is seen as synonymous with discrimination" and because liberal feminists would "be most comfortable" with this mode (p. 236). The second mode (White/male) is equivalent to the dominance approach because it is concerned with challenging the use of a White, male referent. As Daly and Tonry note,

> Virtually all empirical work on race and gender disparities (and our assessment of it) is framed within a racist/sexist perspective in that the research centers on whether sanctions are applied differently across varied racial-ethnic, gender groups. However, a new generation of feminists and critical race scholars has raised questions about the limits this conceptualization imposes on theory, research and policy . . . [and] are more likely to embrace the latter two modes. (p. 235)

The third mode (racialized/gendered) corresponds to the intersectional approach because it "assumes that race and gender relations structure criminal law and justice system practices" in important ways (p. 237). Thus, the third (intersectional) mode of conceptualizing race and gender has evolved in response to the inadequacies of previous paradigms.

For these reasons, the intersectional approach that is informed by multiracial feminism offers feminists the broadest, richest, and most complete theoretical framework for studying gender. Because all social relations are racialized, including those that appear not to be, multiracial feminism achieves a universal relevance that remains elusive to other feminist perspectives. In Daly and Stephens's (1995) words,

Although many claim that black women are at the intersection of class, race, and gender, that statement is misleading. Black women are marked at the intersection as being on the subordinate side of these three relations, but all social groups (including middle-class white men) are at the intersection. (p. 205)

A race-class-gender framework is applicable to the lives of all people, regardless of their social location:

> At the same time that structures of race, class, and gender create disadvantages for women of color, they provide unacknowledged benefits for those who are at the top of these hierarchies—Whites, members of upper classes, and males. Therefore, multiracial feminism applies not only to racial ethnic women but also to women and men of all races, classes, and genders. (Baca Zinn & Thornton Dill, 1996, p. 327)

It is clear, then, that multiracial feminism and the intersectional framework through which it operates hold great promise for contemporary feminist scholarship.

Multiracial Feminism and Criminology: Intersections of Race, Class, Gender, and Crime

For several reasons, multiracial feminism and the intersectional approach can make important contributions to feminist criminology. Although still an emerging body of scholarship, some feminists (as well as some nonfeminists) have already begun urging the adoption of an intersectional approach to studying crime (e.g., Barak et al., 2001; Belknap, 2001; Britton, 2004; Daly, 1993,1997; Daly & Chesney-Lind, 1988: Daly & Stephens, 1995; Flavin, 2004; Lynch, 1996; Milovanovic & Schwartz, 1996; Price & Sokoloff, 2004; Sokoloff & Dupont, 2005; Zatz, 2000). However, despite this support for a race-class-gender framework, to date few feminist criminologists have embraced the intersectional approach. Below I outline several reasons why contemporary third-wave feminist criminologists should adopt an intersectional framework.

The Theoretical Relevance of Intersections

Criminologists have begun to recognize the importance of developing integrated criminological theories. As Barak (1998) argues, criminology stands to benefit from the integration of criminological perspectives. Although a thorough discussion of integrating criminologies is beyond the scope of this article, Barak's overall point helps justify the use of the intersectional approach and multiracial feminism. For example, Barak observes that race, class, and gender have been "autonomously" applied to the study of crime, whereas in the past decade feminist criminologists have begun to focus on the interaction between two or more of these variables as they relate to crime (p. 251). According to Barak, an integrated theoretical perspective "incorporates an appreciation of differences in the patterns of crime attributed to socialization, opportunities, and bias in the context that everyone's life is framed by inequalities of race, class, and gender" (p. 251).

In a similar manner, multiracial feminism emphasizes that intersecting system of race, class, and gender act as "structuring forces" affecting how people act, the opportunities that are available to them, and the way in which their behavior is socially defined (Lynch, 1996, p. 4). For criminologists, this includes examining how the legal system responds to individual offenders based on their social

locations (Barak et al., 2001). For example, Steffensmeier, Ulmer, and Kramer (1998) studied the interaction of race, gender, and age in sentencing decisions and discovered that outcomes are most punitive for defendants whose social locations place them at the margins of race, age, and gender systems—in other words, young Black men. The findings of Steffensmeier et al. "demonstrate the importance of considering the joint effects of race, gender, and age on sentencing, and of using interactive rather than additive models" (p. 763). In a similar manner, Spohn and Holleran (2000) discovered that punishment penalties also are paid by defendants in other marginalized social locations: young, unemployed Black and Hispanic men.

Therefore, concepts and propositions found in multiracial feminism may be integrated with concepts and propositions found in existing criminological theories to achieve the type of integrated perspective that Barak (1998) described. To illustrate, the preceding examples from the sentencing literature represent conceptual theoretical integration. That is, the concept of social location from multiracial feminism and the concept (or "focal concern") of offender blameworthiness from sentencing theory overlap in a meaningful, theoretical way to explain how defendants' demographic characteristics influence judicial decision making (Steffensmeier et al., 1998).

Moreover, initial attempts at theorizing about inequality and crime using an intersectional framework have already occurred. For example, Sampson and Wilson (1995) developed their theory of race, crime, and urban inequality by examining intersecting race and class inequalities. Taking this approach one step further, Lynch (1996) has attempted to develop a theory of race, class, gender inequality, and crime in four ways by (a) "linking race, class, and critical criminology to life course or life history research"; (b) "connecting race, class, and gender to the types of choices that are structured into people's lives"; (c) "demonstrating life course and structured choice effects by reviewing data on income, wealth, and power disparities that arise from race, gender and class inequality"; and (d) "examining how race, class and gender intersect to affect the production of crime" (p. 3).

Finally, several feminist criminologists have used intersectionality to theorize about the relationship between inequality and crime. For example, Richie (1996) shows how intersecting systems of race, class, and gender can lead battered Black women to commit criminal offenses. In her analysis, Richie uses the concept of gender entrapment to explain how "some women are forced or coerced into crime by their culturally expected gender roles, the violence in their intimate relationships, and their social position in the broader society" (p. 133). In a similar manner, Maher (1997) explores intersections of race, class, and gender in the lives of women who participate in the street-level drug economy. In her study, Maher describes how explanations of women's involvement with the drug economy have shifted from "primarily class-based explanations towards acknowledgement of a more complex set of cross-cutting influences— race/ethnicity, sex/gender, age, immigrant status, and other social relations" (p. 169). Most recently, Sokoloff and Dupont (2005) argue for the use of an intersectional framework in developing theories of domestic violence (see also Sokoloff & Pratt, 2005). Sokoloff and Dupont observe that intersectional approaches to domestic violence

> question the monolithic nature of woman battering, call for a greater emphasis on the structural causes of woman battering, caution against disempowering representations of marginalized battered women, and explore the complex role of culture in understanding woman abuse and our responses to it. (p. 40)

As these examples illustrate, an intersectional theoretical framework that is informed by multiracial feminism can be instrumental to the advancement of criminological theory.

⊠ Conclusion

For the reasons outlined in this article, feminist criminologists wishing to advance inclusive analyses of race, class, gender, and crime would do well to adopt an intersectional framework that is informed by multiracial feminism. In one of the few existing criminological texts advocating the use of the intersectional approach, Barak et al. (2001) noted several promising developments for intersectionality in criminology. Of these, perhaps most relevant to feminist criminology is "scholarship that shifts the emphasis [of criminology] on identifying systems of privilege that support existing systems of oppression but are rarely acknowledged by those who reap the benefits" (p. 234).

Indeed, feminist criminologists have long been critical of scholarship that remains blind to issues of power and privilege. For contemporary third-wave feminist criminologists, the time has come to build on the foundation that has been laid for us by our predecessors and to advance a feminist criminology that embraces all sources of oppression without prioritizing gender. After all, as multiracial feminism reveals, power, privilege, and oppression are multiplicative and intersecting according to race, class, gender, sexuality, nationality, age, and other defining social characteristics. As we take stock of our field and look ahead to the future, the words of Daly and Chesney-Lind (1988) are instructive:

> Turning to the future, we wonder what will happen as increasing numbers of white women, as well as men and women of color, enter the discipline

and try to find their place in it. One cannot expect that the first generation of new scholars will be confident or surefooted after centuries of exclusion from the academy. One might expect, however, that we will ask different questions or pursue problems which our discipline has ignored. (p. 506)

In the coming years, successful examination of the ways in which women (and men) in the criminal justice system experience oppression by virtue of their race, class, and gender characteristics must be grounded in an intersectional framework that is informed by multiracial feminism:

Research and theorizing must continue to reject the essentialism inherent in treating women as a unitary category. . . . We already know much about the ways in which race, class, and sexual inequality interweave with women's experiences as victims, offenders, and workers. The challenge for feminist criminology in the years to come lies in formulating theory and carrying out empirical studies that prioritize all of these dimensions, rather than relegating one or more of them to the background for the sake of methodological convenience. (Britton, 2004, p. 71)

In time, such advances will no doubt constitute the very core of feminist criminology in the 21st century.

Notes

1. See Daly and Chesney-Lind (1988) for a thoughtful comparison of these feminist perspectives.

2. For example, recent feminist research identifies the concept of "blurred boundaries" between women's victimization and offending experiences (e.g., Daly & Maher, 1998: Gaarder & Belknap, 2002; Moe, 2004).

3. According to Thompson (2002), intersectionality actually emerged during the 1970s. That is, at the beginning of the second wave of the feminist movement, women of color (as well as White antiracist women and others who felt marginalized by mainstream feminism) began calling for scholarship that simultaneously attended to issues of race, class, gender, and sexuality. However, the concept of intersecting inequalities first appeared in criminology during the 1980s, corresponding to feminism's third wave. Therefore, for the purposes of this article, intersectionality is presented as a product of third-wave feminism.

4. In fact, the mainstream second-wave feminist movement was labeled "hegemonic feminism" by some women of color and White antiracist women for its exclusive use of a White, middle-class, heterosexual female standard (Thompson, 2002).

5. Again, see Thompson (2002) for a discussion of the timing of the emergence of multiracial feminism.

6. However, Sokoloff and Dupont (2005) have argued that social structure is "rarely" the focus of analyses rooted in multiracial feminism.

References

Andersen, M., & Collins. P. H. (2004). *Race, class, and gender* (5th ed.). Belmont, CA: Wadsworth.

Baca Zinn, M., Hondagneu-Sotello, P., & Messner, M. A. (Eds.). (2000). *Gender through the prism of difference* (2nd ed.). Boston: Allyn & Bacon.

Baca Zinn, M., & Thornton Dill, B. (1996). Theorizing difference from multiracial feminism. *Feminist Studies, 22,* 321–331.

Barak, G. (1998). *Integrating criminologies.* Boston: Allyn & Bacon.

Barak, G., Flavin, J., & Leighton, P. S. (2001). *Class, race, gender, and crime: Social realities of justice in America.* Los Angeles: Roxbury.

Beirne, P., & Messerschmidt, J. (2000). *Criminology* (3rd ed.). Boulder, CO: Westview.

Belknap, J. (2001). *The invisible woman: Gender, crime, and justice* (2nd ed.). Belmont, CA: Wadsworth.

Blumstein, A., Cohen, J., Roth, J. A., & Visher, C. (1986). Introduction: Studying criminal careers. In A. Blumstein, J. Cohen. J. A. Roth. & C. Visher (Eds.), *Criminal careers and "career criminals"* (pp. 12–30). Washington, DC: National Academy Press.

Britton, D. M. (2004). Feminism in criminology: Engendering the outlaw. In P. J. Schram & B. Koons-Witt (Eds.), *Gendered (in)justice: Theory and practice in feminist criminology* (pp. 49–67). Long Grove. IL: Waveland Press.

Bui, H. (2004). *In the adopted land: Abused immigrant women and the criminal justice system.* Westport, CT: Praeger.

Chesney-Lind, M., & Pasko, L. (Eds.). (2004). *Girls, women, and crime.* Thousand Oaks, CA: Sage.

Collins. P. H. (2000). *Black feminist thought: Knowledge, consciousness, and the politics of empowerment* (2nd ed.). New York: Routledge.

Crenshaw, K. (1991). Demarginalizing the intersection of race and sex: A Black feminist critique of antidiscrimination doctrine, feminist theory, and antiracist politics. In K. Bartlett & R. Kennedy (Eds.), *Feminist legal theory* (pp. 57–80). Boulder, CO: Westview.

Daly, K. (1989). Rethinking judicial paternalism: Gender, work-family relations, and sentencing. *Gender & Society, 3,* 9–36.

Daly, K. (1993). Class-race-gender: Sloganeering in search of meaning. *Social Justice, 20,* 56–71.

Daly, K. (1994). Gender and punishment disparity. In M. Myers & G. Bridges (Eds.), *Inequality, crime, and social control* (pp. 117–133). Boulder, CO: Westview.

Daly, K. (1997). Different ways of conceptualizing sex/gender in feminist theory and their implications for criminology. *Theoretical Criminology, 1,* 25–51.

Daly, K., & Chesney-Lind, M. (1988). Feminism and criminology. *Justice Quarterly, 5,* 497–538.

Daly, K., & Maher. L. (1998). Crossroads and intersections: Building from feminist critique. In K. Daly & L. Maher (Eds.), *Criminology at the crossroads: Feminist readings in crime and justice* (pp. 1–17). New York: Oxford University Press.

Daly, K., & Stephens, D. J. (1995). The "dark figure" of criminology: Towards a Black and multi-ethnic feminist agenda for theory and research. In N. Hahn Rafter & F. Heidensohn (Eds.), *International feminist perspectives in criminology: Engendering a discipline* (pp. 189–215). Philadelphia: Open University Press.

Daly, K., & Tonry, M. (1997). Gender, race, and sentencing. In M. Tonry (Ed.), *Crime and justice: An annual review of research* (Vol. 22, pp. 201–252). Chicago: University of Chicago.

Flavin. J. (2004). Feminism for the mainstream criminologist: An invitation. In P. J. Schram & B. Koons-Witt (Eds.), *Gendered (in)justice: Theory and practice in feminist criminology* (pp. 68–92). Long Grove, IL: Waveland Press.

Gaarder, E., & Belknap, J. (2002). Tenuous borders: Girls transferred to adult court. *Criminology, 40,* 481–518.

Hajjar, L. (2004). Religion, state power, and domestic violence in Muslim societies: A framework for comparative analyses. *Law & Social Inquiry, 29,* 1–38.

Lorber, J. (Ed.). (2001). *Gender inequality: Feminist theories and politics* (2nd ed.). Los Angeles: Roxbury.

Lynch, M. J. (1996). Class, race, gender, and criminology: Structured choices and the life course. In D. Milovanovic & M. D. Schwartz (Eds.), *Race, gender, and class in criminology: The intersections* (pp. 3–28). New York: Garland.

MacKinnon, C. A. (1991). Difference and dominance: On sex discrimination. In K. T. Bartlett & R. Kennedy (Eds.), *Feminist legal theory* (pp. 81–94). Boulder, CO: Westview.

Maher, L. (1997). *Sexed work: Gender, race, and resistance in a Brooklyn drug market.* Oxford, UK: Oxford University Press.

Miller, J. (2004). Feminist theories of women's crime: Robbery as a case study. In B. R. Price & N. Sokoloff (Eds.), *The criminal justice system and women* (3rd ed., pp. 51–67). New York: McGraw-Hill.

Milovanovic, D., & Schwartz, M. D. (Eds.). (1996). *Race, gender, and class in criminology: The intersections.* New York: Garland.

Moe, A. (2004). Blurring the boundaries: Women's criminality in the context of abuse. *Women's Studies Quantity, 32,* 116–138.

Morash, M. (1999). A consideration of gender in relation to social learning and social structure: A general theory of crime and deviance. *Theoretical Criminology, 3,* 451–462.

Nagel, I., & Johnson, B. (2004). The role of gender in a structured sentencing system: Equal treatment, policy choices, and the sentencing of female offenders. In P. J. Schram & B. Koons-Witt (Eds.), *Gendered (in)justice: Theory and practice in feminist criminology* (pp. 198–235). Long Grove, IL: Waveland Press.

Price, B. R., & Sokoloff, N. (2004). *The criminal justice system and women* (3rd ed.). New York: McGraw-Hill.

Renzetti, C. (1992). *Violent betrayal: Partner abuse in lesbian relationships.* Newbury Park, CA: Sage.

Richie, B. (1996). *Compelled to crime: The gender entrapment of Black battered women.* New York: Routledge.

Sampson, R., & Wilson, W, J. (1995). Toward a theory of race, crime, and urban inequality. In J. Hagan & R. D. Peterson (Eds.), *Crime and inequality* (pp. 37–54). Stanford, CA: Stanford University Press.

Sokoloff, N., & Dupont, I. (2005). Domestic violence at the intersections of race, class, and gender. *Violence Against Women, 11,* 38–64.

Sokoloff, N., & Pratt, C. (2005). *Domestic violence at the margins: Readings on race, class, gender, and culture.* New Brunswick, NJ: Rutgers University Press.

Sokoloff, N., Price, B. R., & Flavin, J. (2004). The criminal law and women. In B. R. Price & N. Sokoloff (Eds.), *The criminal justice system and women* (3rd ed., pp. 11–29). New York: McGraw-Hill.

Spohn, C., & Holleran, D. (2000). The imprisonment penalty paid by young, unemployed Black and Hispanic male offenders. *Criminology, 38,* 281–306.

Steffensmeier, D., & Allan, E. (1996). Gender and crime: Toward a gendered theory of female offending. *Annual Review of Sociology, 22,* 459–487.

Steffensmeier, D., Ulmer, J., & Kramer, J. (1998). The interaction of race, gender, and age in criminal sentencing: The punishment cost of being young, Black, and male. *Criminology, 36,* 763–793.

Stolzenberg, L., & D'Alessio, S. (2004). Sex differences in the likelihood of arrest. *Journal of Criminal Justice, 32,* 443–454.

Thompson, B. (2002). Multiracial feminism: Recasting the chronology of second wave feminism. *Feminist Studies, 28,* 337–360.

Weber, L. (2001). *Understanding race, class, gender, and sexuality: A conceptual framework.* Boston: McGraw-Hill.

Weber, L., & Parra-Medina, D. (2003). Intersectionality and women's health: Charting a path to eliminating health disparities. In M. Texler Segal, V. Demos, & J. Jacobs Kronenfeld (Eds.), *Gender perspectives on health and medicine: Key themes* (pp. 181–229). Oxford, UK: Elsevier.

Williams, W. (1991). The equality crisis: Some reflections on culture, courts, and feminism. In K. T. Bartlett & R. Kennedy (Eds.), *Feminist legal theory* (pp. 15–34). Boulder, CO: Westview.

Zatz, M. (2000). Convergence of race, ethnicity, and class on court decision-making; Looking toward the 21st century. In J. Horney (Ed.), *Policies, process, & decisions of the criminal justice system: Criminal justice 2000* (Vol. 3, pp. 503–552). Washington, DC: U.S. Department of Justice.

REVIEW QUESTIONS

1. Burgess-Proctor asserts that all three factors—race/ethnicity, social class, and gender—are equally important when it comes to engaging in criminal behavior or processing by the criminal justice system. From your own experience, do you agree? If so, why? If not, which characteristic of these do you think is most important, and why?

2. Studies have consistently shown that most victims of homicide by rate are lower-class, Black males, and that the offenders by rate are lower-class, Black males. How do you think these findings relate to Burgess-Proctor's discussion? Is it consistent or not?

3. How do you think an upper-class White male would be treated by the criminal justice system, as compared to a lower-class Black female? What advice would you offer to the latter defendant?

4. Which of the five types of feminist criminological perspectives do you agree with the most? Which of these five types do you least agree with? Do your selections vary whether you are considering the reasons why females commit crime versus how females are handled by the criminal justice system? If so, how?

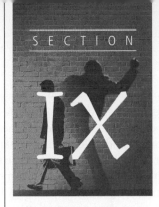
IX

LIFE-COURSE PERSPECTIVES OF CRIMINALITY

The introduction of this section will discuss the development of the life-course perspective in the late 1970s and its influence on modern research on criminal trajectories. The introduction will focus on explaining the various concepts in the life-course perspective, such as onset, desistence, and frequency, as well as the arguments against this perspective. Finally, the introduction of this section will review the current state of research regarding this perspective.

◪ Introduction

This introduction will present one of the most current and progressive approaches toward explaining why individuals engage in criminal activity, namely developmental theories of criminal behavior. **Developmental theories** are explanatory models of criminal behavior that follow individuals throughout their life course of offending, thus explaining the development of offending over time. Such developmental theories represent a break with past traditions of theoretical frameworks, which typically focused on the contemporaneous effects of constructs and variables on behavior at a given point in time. Virtually no theories attempted to explain the various stages (e.g., onset, desistence) of individuals' criminal careers, and certainly no models differentiated the varying factors that are important at each stage. Developmental theories have been prominent in modern times, and we believe that readers will agree that developmental theories have added a great deal to our understanding and thinking about why people commit criminal behavior.

⊠ Developmental Theories

Developmental theories, which are also to some extent integrated, are distinguished by their emphasis on the evolution of individuals' criminality over time. Developmental theories tend to look at the individual as the unit of analysis, and such models focus on the various aspects of the onset, frequency, intensity, duration, desistence, and other aspects of the individual's criminal career. The onset of offending is when the offender first begins offending, and desistence is when an individual stops committing crime. Frequency refers to how often the individual offends, whereas intensity is the degree of seriousness of the offenses he or she commits. Finally, duration is the length of an individual's criminal career.

Experts have long debated and examined these various aspects of the development of criminal behavior. For example, virtually all studies show an escalation from minor status offending (e.g., truancy, underage drinking, smoking tobacco) to petty crimes (e.g., shoplifting, smoking marijuana) to far more serious criminal activity, such as robbery and aggravated assault, and then murder and rape. This development of criminality is shown across every study that has ever been performed and demonstrates that, with very few exceptions, people begin with relatively minor offending and progress toward more serious, violent offenses.

Although this trend is undisputed, other issues are not yet resolved. For example, studies have not yet determined when a police contact or arrest becomes "early" onset. Most empirical studies draw the line at age 14, so that any arrest or contact prior to this time is considered early onset.[1] However, other experts would disagree and say that this line should be drawn earlier (say 12 years old) or even later (such as 16 years old). Still, however it is defined, early onset is one of the most important predictors of future criminality and chronic offending of any of the measures we have in determining who is most at risk for developing serious, violent offending behavior.

Perhaps the most discussed and researched aspect of developmental theory is that of offender frequency, which has been referred to as "lambda." Estimates of lambda, or average frequency of offending by criminals over a year period, vary greatly.[2] Some estimates of lambda are in the high single digits and some are in the triple digits. Given this large range, it does not do much good in estimating what the frequency of most offenders are. Rather, the frequency depends on many, many variables, such as what type of offenses the individual commits. Perhaps if we were studying only drug users or rapists, it would make sense to determine the average frequency of offending (or lambda), but given the general nature of most examinations of crime, such estimates are not useful. Still, the frequency of offending even within crime type varies so widely across individuals that we question its use in understanding criminal careers.

Before we discuss the dominant models of developmental theory, it is important to discuss the opposing viewpoint, which is that of complete stability in offending. Such counterpoint views assume that the developmental approach is a waste of time because

---

[1]For more discussion, see Stephen Tibbetts and Alex Piquero, "The Influence of Gender, Low Birth Weight, and Disadvantaged Environment in Predicting Early Onset of Offending: A Test of Moffitt's Interactional Hypothesis," *Criminology* 37 (1999): 843–78; Chris L. Gibson and Stephen Tibbetts, "A Biosocial Interaction in Predicting Early Onset of Offending," *Psychological Reports* 86 (2000): 509–18.

[2]For a review, see Samuel Walker, *Sense and Nonsense about Crime and Drugs: A Policy Guide*, 5th ed. (Belmont, CA: Wadsworth, 2001).

the same individuals who show antisocial behavior at early ages (before age 10) are those who will exhibit the most criminality in the teenage years, their 20s, their 30s, 40s, and so on. This framework is most notably represented by the theoretical perspective proposed by Gottfredson and Hirschi in their model of low self-control.

Antidevelopmental Theory: Low Self-Control Theory

In 1990, Travis Hirschi, along with his colleague Michael Gottfredson, proposed a general theory of low self-control as the primary cause of all crime and deviance; this is often referred to as the general theory of crime.[3] This theory has led to a significant amount of debate and research in the field since its appearance, more than any other contemporary theory of crime.

Like other control theories of crime, this theory assumes that individuals are born predisposed toward selfish, self-centered activities and that only effective child-rearing and socialization can create self-control. Without such adequate socialization (i.e., social controls) and reduction of criminal opportunities, individuals will follow their natural tendencies to become selfish predators. The general theory of crime assumes that self-control must be established by age 10. If it has not formed by that time, then, according to the theory, individuals will forever exhibit low self-control. This assumption of the formation of low self-control by age 10 or before is the oppositional feature of this theory to the developmental perspective. Once low self-control is set by age 10, there is no way to develop self-control afterward, the authors assert. In contrast, developmental theory assumes that people can indeed change over time.

Like others, Gottfredson and Hirschi attribute the formation of controls to socialization processes in the first years of life; the distinguishing characteristic of this theory is its emphasis on the individual's ability to control himself or herself, which is formed before age 10. That is, the general theory of crime assumes that people can take a degree of control over their own decisions and, within certain limitations, "control" themselves. The general theory of crime is accepted as one of the most valid theories of crime.[4] This is probably due to the parsimony, or simplicity, of the theory because it identifies only one primary factor that causes criminality—low self-control. But low self-control may actually consist of a series of personality traits, including risk-taking, impulsiveness, self-centeredness, short-term orientation, and quick temper. Recent research has supported the idea that inadequate child-rearing practices tend to result in lower levels of self-control among children and that these low levels produce various risky behaviors, including criminal activity.[5] It is important to note that this theory has a developmental component,

---

[3]Michael Gottfredson and Travis Hirschi, *A General Theory of Crime* (Palo Alto, CA: Stanford University Press, 1990).

[4]For an excellent review of studies regarding low self-control theory, see Travis Pratt and Frank Cullen, "The Empirical Status of Gottfredson and Hirschi's General Theory of Crime: A Meta-Analysis," *Criminology* 38 (2000): 931–64. For critiques of this theory, see Ronald Akers, "Self-Control as a General Theory of Crime," *Journal of Quantitative Criminology* 7 (1991): 201–11. For a study that demonstrates the high popularity of the theory, see Anthony Walsh and Lee Ellis, "Political Ideology and American Criminologists' Explanations for Criminal Behavior," *The Criminologist* 24 (1999): 1, 14.

[5]Carter Hay, "Parenting, Self-Control, and Delinquency: A Test of Self-Control Theory," *Criminology* 39 (2001): 707–36; K. Hayslett-McCall and T. Bernard, "Attachment, Masculinity, and Self-Control: A Theory of Male Crime Rates," *Theoretical Criminology* 6 (2002): 5–33.

Figure 9.1 Gottfredson and Hirschi's Theory of Low Self-Control

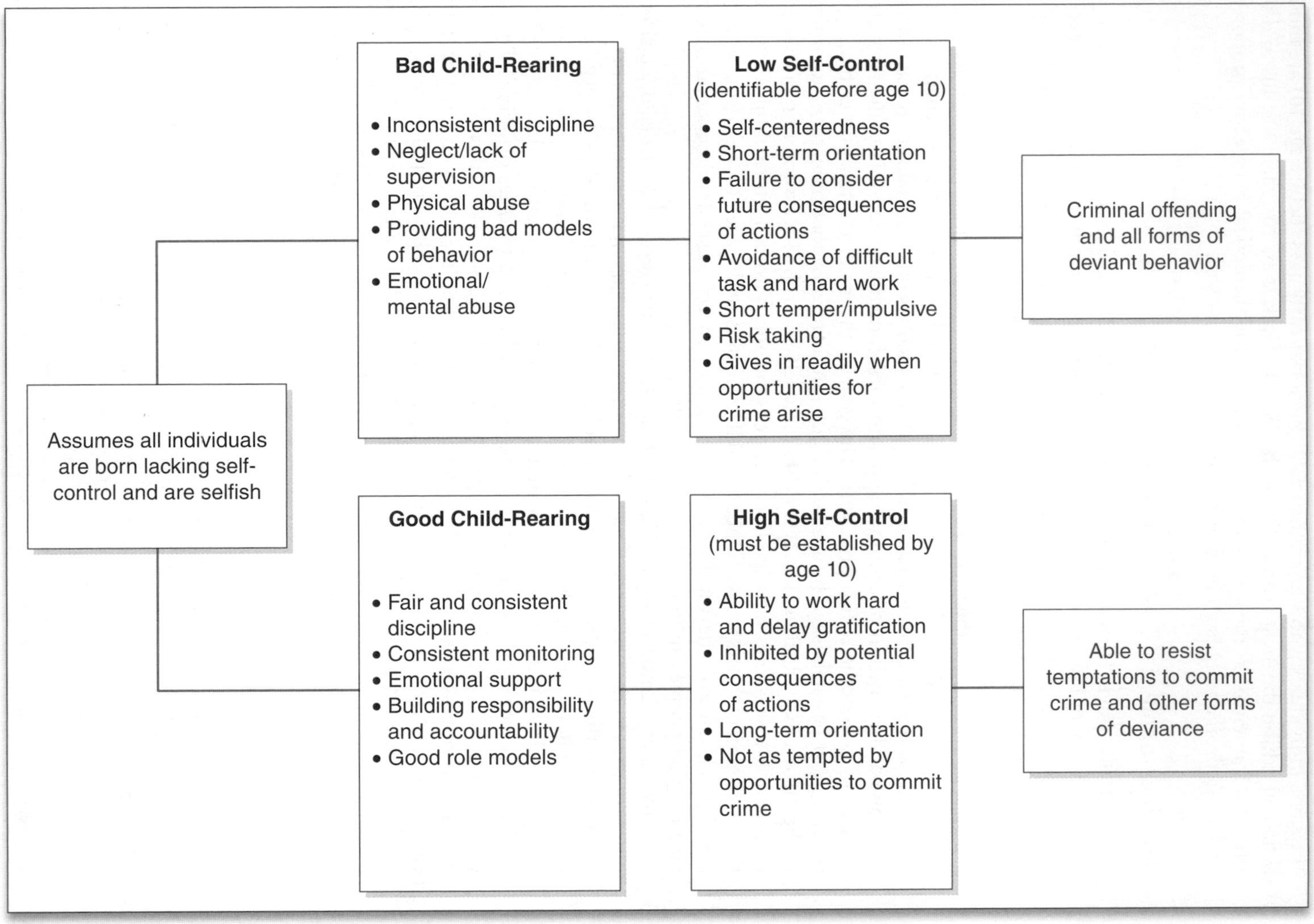

Bad Child-Rearing

- Inconsistent discipline
- Neglect/lack of supervision
- Physical abuse
- Providing bad models of behavior
- Emotional/mental abuse

Low Self-Control
(identifiable before age 10)

- Self-centeredness
- Short-term orientation
- Failure to consider future consequences of actions
- Avoidance of difficult task and hard work
- Short temper/impulsive
- Risk taking
- Gives in readily when opportunities for crime arise

Criminal offending and all forms of deviant behavior

Assumes all individuals are born lacking self-control and are selfish

Good Child-Rearing

- Fair and consistent discipline
- Consistent monitoring
- Emotional support
- Building responsibility and accountability
- Good role models

High Self-Control
(must be established by age 10)

- Ability to work hard and delay gratification
- Inhibited by potential consequences of actions
- Long-term orientation
- Not as tempted by opportunities to commit crime

Able to resist temptations to commit crime and other forms of deviance in the sense that it proposes that self-control develops during early years from parenting practices; thus, even this most notable antidevelopment theory actually includes a strong developmental aspect.

In contrast to Gottfredson and Hirschi's model, one of the most dominant and researched frameworks of the last 20 years, another sound theoretical model shows that individuals can change their life trajectories in terms of crime. Research shows that events or realizations can occur that lead people to alter their frequency or incidence of offending, sometimes to zero. To account for such extreme transitions, we must turn to the dominant life-course model of offending, which is that of Sampson and Laub's developmental model of offending.

Sampson and Laub's Developmental Model

Perhaps the best known and researched developmental theoretical model to date is that of Robert Sampson and John Laub.[6] Sampson and Laub have proposed a developmental framework that is largely based on a reanalysis of original data collected by Sheldon and Eleanor Glueck in the 1940s. As a prototype developmental model, individual stability and change is the primary focus of their theoretical perspective.

Most important, Sampson and Laub emphasized the importance of certain events and life changes, which can alter an individual's decisions to commit (or not commit) criminal activity. Although based on a social control framework, this model contains elements of other theoretical perspectives. First, Sampson and Laub's model assumes, like other developmental perspectives, that early antisocial tendencies among individuals, regardless of social variables, are often linked to later adult criminal offending. Furthermore,

▲ **Photo 9.1** Life-course theories attempt to uncover points of desistance, critical events that may change a person's life path from deviant behavior back to law-abiding status. Getting married is considered one of those events. Once one is married, free time to hang out with one's friends tends to disappear.

[6]Robert Sampson and John Laub, "Crime and Deviance Over the Life Course: The Salience of Adult Social Bonds," *American Sociological Review* 55 (1990): 609–27; Robert Sampson and John Laub, "Turning Points in the Life Course: Why Change Matters to the Study of Crime," *Criminology* 31 (1993): 301–26; Robert Sampson and John Laub, *Crime in the Making: Pathways and Turning Points Through Life* (Cambridge, MA: Harvard University Press, 1993).

some social structure factors (e.g., family structure, poverty, etc.) also tend to lead to problems in social and/or educational development, which then leads to crime. Another key factor in this development of criminality is the influence of delinquent peers or siblings, which further increases an individual's likelihood for delinquency.

However, Sampson and Laub also strongly emphasize the importance of **transitions**, or events that are important in altering trajectories toward or against crime, such as marriage, employment, or military service, drastically changing a person's criminal career. Sampson and Laub show sound evidence that many individuals who were once on a **trajectory** or path toward a consistent form of behavior, in this case, serious violent crime, suddenly (or gradually) halted their path due to such a transition or series of such transitions. In some ways, this model is a more specified form of David Matza's theory of drift, which we discussed in Section VII, in which individuals tend to grow out of crime and deviance due to the social controls imposed by marriage, employment, and so on. Still, Sampson and Laub's framework contributed much to the knowledge of criminal offending by providing a more specified and grounded framework that identified the ability of individuals to change their criminal trajectories via life-altering transitions, such as the effect that marriage can have on a man or woman, which is quite profound. In fact, recent research has consistently shown that marriage and full-time employment significantly reduced the recidivism of California parolees, and other recent studies have shown similar results from employment in later years.[7]

Moffitt's Developmental Taxonomy

Another primary developmental model that has had a profound effect on the current state of criminological thought and theorizing is Terrie **Moffitt's developmental theory/ taxonomy**, proposed in 1993.[8] Moffitt's framework distinguishes two types of offenders: **adolescence-limited** and **life-course persistent** offenders. Adolescence-limited offenders make up most of the general public and include all persons who committed offenses when they were teenagers and/or young adults. Their offending was largely caused by association with peers and a desire to engage in activities that were exhibited by the adults that they are trying to be. Such activities are a type of rite of passage and quite normal among all people who have normal social interactions with their peers in teenage or young adult years. It should be noted that a very small percentage (about 1 to 3%) of the population are nonoffenders, who quite frankly do not have normal relations with their peers and therefore do not offend at all, even in adolescence.

On the other hand, there exists a small group of offenders, referred to in this model as life-course persistent offenders. This small group, estimated to be 4 to 8% of offenders—albeit the most violent and chronic—commit the vast majority of the serious violent offenses in any society, such as murder, rape, and armed robbery. In contrast to the adolescence-limited offenders, the disposition of life-course persistent offenders toward offending is caused by an entirely different model: an interaction between neurological problems and the disadvantaged/criminological environments in which they are raised.

---

[7]Alex Piquero, Robert Brame, Paul Mazzerole, and Rudy Haapanen, "Crime in Emerging Adulthood," *Criminology* 40 (2002): 137–70; Chris Uggen, "Work as a Turning Point in the Life Course of Criminals: A Duration Model of Age, Employment, and Recidivism," *American Sociological Review* 65 (2000): 529–46.

[8]Terrie Moffitt, "Adolescence Limited and Life Course Persistent Antisocial Behavioral: A Developmental Taxonomy," *Psychological Review* 100 (1993): 674–701.

For example, if an individual had only neurological problems or only a poor, disadvantaged environment, then that individual would be unlikely to develop a life-course persistent trajectory toward crime. However, if a person has both neurological problems and a disadvantaged environment, then that individual would have a very high likelihood of becoming a chronic, serious, violent offender. This proposition, which has been supported by empirical studies,[9] suggests that it is important to pay attention to what happens early in life. Because illegal behaviors are normal among teenagers or young adults, more insight can be gained by looking at the years prior to age 12 to determine who is most likely to become chronic, violent offenders. Life-course persistent offenders begin offending very early in life and continue to commit crime far into adulthood, even middle age, whereas adolescence-limited offenders tend to engage in criminal activity only during teenage and young adult years. Moffitt's model suggests that more than one type of development explains criminality. Furthermore, this framework shows that certain types of offenders commit crime due to entirely different causes and factors.

▨ Policy Implications

There are many, perhaps an infinite number, of policy implications that can be derived from developmental theories of criminality. Thus, we will focus on the most important, which is that of prenatal and perinatal stages of life because the most significant and effective interventions can occur during this time. If policymakers hope to reduce early risk factors for criminality, they must insist on universal health care for pregnant women, as well as their newborn infants through the first few years of life. The United States is one of the few developed nations that does not guarantee this type of maternal and infant medical care and supervision. Doing so would go a long way to avoiding the costly damages (in many ways) of criminal behavior among youth at risk.[10]

Furthermore, there should be legally mandated interventions for pregnant women who are addicted to drugs or alcohol. Although this is a highly controversial topic, it appears to be a "no-brainer" that women who suffer from such addictions may become highly toxic to the child(ren) they carry and should receive closer supervision and more health care. There may be no policy implementation that would have as much influence on reducing future criminality in children as making sure their mothers do not take toxic substances while they are pregnant.[11]

Other policy implications include assigning special caseworkers for high-risk pregnancies, such as those involving low birth weight or low Apgar scores. Another advised intervention would be to have a centralized medical system that provides a "flag" for high-risk infants who have numerous birth or delivery complications, so that the doctors who are seeing them for the first time are aware of their vulnerabilities.[12] Finally, universal preschool should be funded and provided to all young children; studies have shown this leads to better performance once they enter school, both academically and socially.[13]

---

[9]See Tibbetts and Piquero, "Influence of Gender."

[10]John P. Wright, Stephen G. Tibbetts, and Leah Daigle, *Criminals in the Making: Criminality Across the Life Course* (Thousand Oaks, CA: Sage), chapter 11.

[11]Ibid.

[12]Ibid.

[13]Ibid.

Ultimately, as the many developmental theories have shown, there are many concepts and stages of life that can have a profound effect on the criminological trajectories that lives can take. However, virtually all of these models propose that the earlier stages of life are likely the most important in determining whether an individual will engage in criminal activity throughout life or not. Therefore, policymakers should focus their efforts on providing care and interventions in this time period.

Conclusion

This introduction presented a brief discussion of the importance of developmental, life-course theories of criminal behavior. This perspective is relatively new, becoming popular in the late-1970s, compared to other traditional theories explored in this text. Ultimately, this is one of the most "cutting-edge" areas of theoretical development, and life-course theories are likely to be the most important frameworks in the future of the field of criminological theory.

Then we examined the policy implications of this developmental approach, which emphasized the need to provide universal care for pregnant mothers, as well as their newborn children. Other policy implications included legally mandated interventions for mothers who are addicted to toxic substances (e.g., alcohol, drugs) and assignment of caseworkers for high-risk infants/children, such as those with birth or delivery complications. Such interventions would go a long way toward saving society the many problems (e.g., financial, victimization, etc.) that will persist without such interventions. Ultimately, a focus on the earliest stages of intervention will pay off the most and will provide the "biggest bang for the buck."

Section Summary

- Developmental or life-course theory focuses on the individual and following such individuals throughout life to examine their offending careers. In-depth consideration of such changes during the life course is of the highest concern, especially regarding general conclusions that can be made about the factors that tend to increase or decrease the risk that they will continue offending.
- Life-course perspectives emphasize such concepts as onset, frequency of offending, duration of offending, seriousness of offending, desistence of offending, and other factors that play a key role in when individuals offended and why the did so—or didn't do so—at certain times of their lives.
- There are many critics of the developmental/life-course perspective, particularly those who buy into the low self-control model, which is antidevelopmental in the sense that this model assumes that propensities for crime do *not* change over time but rather remain unchanged across life.
- One of the developmental models that has received the most attention is that by Sampson and Laub, which emphasizes transitions in life (e.g., marriage, military service, employment, etc.) that alter trajectories either toward or away from crime.

◆ Moffitt's developmental theory of chronic offenders (which she labeled life-course persistent offenders) versus more normal offenders (which she labeled adolescence-limited offenders) is the developmental model that has received the most attention over the last decade, and much of this research is supportive of the interactive effects of biology and environment combining to create chronic, habitual offenders.

KEY TERMS

Adolescence-limited offenders

Developmental theories

Life-course persistent offenders

Moffitt's developmental theory/taxonomy

Trajectory

Transitions

DISCUSSION QUESTIONS

1. What characteristic distinguishes developmental theories from traditional theoretical frameworks?

2. What aspects of a criminal career do experts consider important in such a model? Describe all of the aspects they look at in a person's criminal career.

3. Discuss the primary criticisms regarding the developmental perspective, particularly that presented by Gottfredson and Hirschi. Which theoretical paradigm do you consider the most valid? Why?

4. What transitions or trajectories have you seen in your life or your friends' lives that support Sampson and Laub's developmental model? What events encouraged offending or inhibited it?

5. Given Moffitt's dichotomy of life-course persistent offenders and adolescence-limited, which of these should be given more attention by research? Why do you feel this way?

WEB RESOURCES

Developmental Theories of Crime

http://www.apsu.edu/oconnort/crim/crimtheory06.htm

Terrie Moffitt's Theory

www.wpic.pitt.edu/research/famhist/PDF_Articles/APA/BF16.pdf

Sampson and Laub's Model

http://harvardmagazine.com/2004/03/twigs-bent-trees-go-stra.html

READING

In this selection, Stephen Tibbetts and Alex Piquero present one of the first empirical tests of Moffitt's interactional hypothesis, which predicts life-course persistent offending, as measured by one of Moffitt's key predictors, namely early onset of offending (a well-known predictor of chronic offending by all longitudinal studies). Using an inner-city sample of close to 1,000 youths, it was found that early onset of offending was predicted by the interaction between biosocial variables (in this case, low birth weight) and early social factors (in this case, weak family structure and lower income levels).

As readers examine this selection, they should focus on the propositions and concepts presented by Moffitt's theory, as well as the concepts and variables that are used to test her theoretical model. Furthermore, readers should concentrate on the interaction between the biosocial variable and the social/developmental variables that are used in this study, particularly the interaction effects that occur between biological factors and social/developmental aspects of their environment.

The Influence of Gender, Low Birth Weight, and Disadvantaged Environment in Predicting Early Onset of Offending

A Test of Moffitt's Interactional Hypothesis

Stephen G. Tibbetts and Alex R. Piquero

Similar to the observed strong and positive relationship between past and future offending (Gottfredson and Hirschi, 1990; Nagin and Paternoster, 1991; Paternoster et al., 1997; Robins, 1966, 1978; Wilson and Herrnstein, 1985), the age at which a first offense occurs (i.e., onset) is an important factor in predicting the future offending of individuals. Studies have consistently found that early onset is one of the best predictors of serious, high-rate offending in adolescence and adulthood (Blumstein et al., 1986; Dunford and Elliott, 1984; Elliott et al., 1984; Farrington, 1986; Farrington et al., 1990; LeBlanc and Fréchette, 1989; Loeber and LeBlanc, 1990; Nagin and Farrington, 1992a, 1992b; Patterson et al., 1992; Reiss and Roth, 1993; Sampson and Laub, 1993; Tolan, 1987; Wolfgang, 1983; Wolfgang et al., 1987). Furthermore, the association between early onset and persistent

SOURCE: Adapted from Stephen G. Tibbetts and Alex R. Piquero, "The influence of Gender, Low Birth Weight, and Disadvantaged Environment in Predicting Early Onset of Offending: A Test of Moffitt's Interactional Hypothesis," *Criminology* 37, no. 4 (1999): 843–78. Reprinted by permission of the American Society of Criminology.

offending appears across historical periods, geographical areas (e.g., London, Philadelphia, Boston, Racine, Montreal, and Dunedin), various forms of measuring deviant behavior (e.g., official, retrospective, prospective, etc.), and various methods of operationalizing onset (e.g., first offense, first arrest, first conviction, etc.).

Understanding early onset has profound implications for the study of juvenile, as well as adult, offending. By identifying the unique factors contributing to early onset, researchers and policy makers may better understand the offending patterns of some of the most serious offenders in society, as well as develop effective intervention strategies (Farrington and Hawkins, 1991; Loeber et al., 1991; Moffitt, 1993; Patterson et al., 1992).

Although understanding that early onset is important for the explanation of future offending, what determines early onset is an entirely different question. Knowledge of such determinants would enable researchers and policy makers to more accurately predict who is most at risk for early onset and persistent offending (Moffitt, 1993). Thus, the ability to identify determinants and processes in the development of early onset has profound implications for the prevention and control of a large portion of serious offending. Although few question these implications, Farrington et al.'s (1990) recent review of the literature surrounding early onset concluded that little is known about the causes of early onset.

In an attempt to fill this void, recent research has examined the influence of various factors on the development of early antisocial behavior, including neuropsychological problems (Gorenstein, 1990; Moffitt, 1990), antisocial predisposition (Patterson et al., 1992), poor psychomotor skills (Farrington and Hawkins, 1991), as well as family characteristics and economic deprivation (Farrington and Hawkins, 1991; Farrington et al., 1986; Moffitt, 1990).[1] Recently, some studies have examined biosocial interactions among some of the aforementioned psychological and

environmental determinants in concert with various biological and physiological characteristics, such as low birth weight, heart rate, and pre/perinatal complications in predicting early deviant behavior (Brennan et al., 1995; Moffitt, 1993; Raine et al., 1997b), as well as persistent and violent offending (Brennan et al., 1997; Denno, 1990; Kandel et al., 1990; Kandel and Mednick, 1991; Piquero and Tibbetts, 1999; Raine et al., 1994, 1997a; Reiss and Roth, 1993). The results of these studies imply that the development of early onset and serious offending is predicted by interactions among the determinants more so than the independent influence of each (e.g., Brennan et al., 1995:87). In a summary statement of this literature, Moffitt (1993:682) stated: "It is now widely acknowledged that personality and behavior are shaped in large measure by the interactions between the person and the environment." Although similar claims have been made by other biosocial theorists (e.g., Raine et al., 1994), the recently convened Panel on the Understanding and Control of Violent Behavior (Reiss and Roth, 1993:380) concluded that little is known about how and why individual and environmental factors influence behavior.

In an effort to shed light on this issue, Moffitt (1993) recently presented a developmental taxonomy to establish a theoretical framework that distinguishes the etiology of early onset (i.e., "life-course-persistent") from that of late onset (i.e., "adolescence-limited"). Life-course-persistent offenders begin their participation in antisocial behavior during early childhood and continue their participation in crime throughout the life course when most other offenders desist. The cause of antisocial behavior for the life-course persisters, according to Moffitt, is the result of an interaction between neuropsychological impairments and poor social environments. This "double hazard" of perinatal risk and social disadvantage increases the risk for deviant behavioral outcomes (Brennan and Mednick, 1997:272).

For Moffitt, a defining characteristic of life-course-persistent offenders is their early onset of offending.[2]

Adolescence-limited offenders, on the other hand, become antisocial for the first time during adolescence, but desist by young adulthood. For these offenders, the causes of crime coincide with the physical changes occurring around puberty, which to these youngsters signal that they should be entitled to adult social status. In essence, their delinquency is caused by the "social mimicry" of the antisocial style of life-course-persistent youths (Moffitt, 1993:686).[3]

The present study seeks to test Moffitt's (1993) interactional hypothesis for the development of early onset. First, we examine if the risk of neuropsychological deficit-disadvantaged environment interaction predicts early onset, as anticipated by Moffitt. Second, we examine Moffitt's (1993, 1994) prediction that life-course-persistent patterns of offending are limited almost exclusively to males (e.g., Moffitt et al., 1994:283, 296). Although Moffitt (1994) suggests the causal process leading to membership in the life-course-persistent group should be the same for both sexes, fewer early onset girls than early onset boys should exist because of differences in the prevalence or degree of exposure to the predictors.

⊠ Moffitt's Interactional Hypothesis

The Causes of Early Onset

Moffitt (1993:679) claims early offending is caused by an increased vulnerability to neuropsychological problems in early childhood interacting with disadvantaged environments. In the outline of her theory, Moffitt (1993:680) theorizes that perinatal complications, such as low birth weight, produce neuropsychological deficits in the infant's central nervous system. These deficits manifest themselves in a variety of ways, including temperament difficulties,

cognitive deficits, and poor test scores. Often, these children find themselves in deficient social environments. They are theorized to be at highest risk for persistent antisocial behavior.

This hypothesis is consistent with recent findings that show support for biosocial interactions in predicting offending patterns, in general, and violent offending, in particular (Brennan et al., 1997; Kandel et al., 1990; Kandel and Mednick, 1991; Lewis et al., 1979; McGauhey et al., 1991; Moffitt, 1990; Piquero and Tibbetts, 1999; Raine et al., 1994, 1997b; Ross et al., 1990). In fact, to date, there is no evidence that contradicts this hypothesis (Brennan and Mednick, 1997:276). Early results from Moffitt's (1990) ongoing New Zealand study showed that young boys with both low neuropsychological test scores and adverse home environments had a mean aggression score more than four times greater than that of boys with either neuropsychological problems or adverse homes alone (see also Moffitt et al., 1994). In the only longitudinal test of her theory, in which prospective measures of neuropsychological status were used to predict antisocial outcomes, Moffitt et al. (1994) found that poor neuropsychological scores were associated with early onset of delinquency for males, and predicted high levels of offending thereafter. However, poor neuropsychological scores were not related to late (i.e., adolescence) onset of offending. In a related piece, Moffitt et al. (1996) found that life-course-persistent path males differed from adolescence-limited path males in convictions or violent crime, personality profiles, school leaving, and bonds to family. For each of these markers, the life-course-persistent path males faired worse (i.e., the life-course-persistent males had more convictions for violent crime, had worse personality profiles, left school early, and had lower bonds to family than the adolescence-limited path males).

To reemphasize, for Moffitt, it is the interaction between a child's vulnerabilities to neuropsychological disorders and poor social environments that produces early onset, and not necessarily the independent influence of these determinants. The strongest predictors of early onset are measures of individual and family characteristics: cognitive abilities, school achievement, mental disorders, familial factors, socioeconomic status, temperament, lack of control, and so on (e.g., Henry et al., 1996). Vulnerability to neuropsychological problems is likely caused by such factors as heritability, low birth weight, early brain injury, complications at birth, etc. (Coren, 1993; Denno, 1990; Moffitt, 1993; Raine et al., 1994). Vulnerabilities to neuropsychological disorders are often perpetuated by the poor social environments in which many of these individuals reside (Chomitz et al., 1995; Hughes and Simpson, 1995; Moffitt, 1993, 1997).

Moffitt et al. (1994) claim that problem behavior begins early in childhood because neuropsychological dysfunctions disrupt normal development, and these deficits increase vulnerability to the criminogenic aspects of disadvantaged rearing environments. According to Moffitt (1993:682), these children "evoke a challenge to even the most loving and patient families."[4] This assertion is consistent with studies showing that low birth weight infants negatively influence the behavior of their caretakers (Tinsley and Parke, 1983).

With regard to low birth weight children, they are often clustered in lower social classes and in families with dismantled social structures, both of which increase their risk of experiencing higher rates of physical health problems, mental disorders, and depression (Hack et al., 1995; National Research Council, 1993:42). As a result, the parents of these children often lack the resources, time, cognitive abilities, and adequate disciplining practices "high-risk" children require (Moffitt, 1993). Although Moffitt (1993:681) uses the term "criminogenic environment" for most of her discussion, she implies that this is synonymous with a disadvantaged environment: "Vulnerable infants are disproportionately found in environments

that will not be ameliorative because many sources of neural maldevelopment co-occur with family disadvantage." In addition, Moffitt (1993:680–685) claims that important measures of this type of disadvantaged environment include socioeconomic status and family structure. In fact, studies predicting early onset or persistent, serious, and violent offending, or both, have used both socioeconomic status (Patterson et al., 1992) and poor familial environment (Raine et al., 1997b; Werner, 1987) as indicators for disadvantaged environments.

In the current investigation, low birth weight serves as a proxy for neuropsychological deficit. To be used in research assessing Moffitt's theory, low birth weight must first be associated with negative life outcomes, especially outcomes that take the form of criminal behavior. To the extent that this is true, an argument must be made for considering low birth weight an adequate proxy for risk of neuropsychological deficit. We now turn to a brief synopsis of this literature, particularly in its relation to Moffitt's theory, in general, and criminal/aggressive behavior, in particular.

Low Birth Weight

One strong and consistent indicator of increased risk for neuropsychological disorders is low birth weight. For many years, low birth weight has been considered a major obstetrical and pediatric problem (Taylor, 1976:108; see also Behrman e al., 1971; Paneth, 1995). Cigarette smoking by the mother, lack of prenatal care, drug/alcohol use by the mother during pregnancy, low socioeconomic status, poor diet, and mother's low educational level have all been found to be determinants of low birth weight (Coren, 1993; Denno, 1990; Moffitt, 1993; Shiono and Behrman, 1995). Low birth weight has been found to be associated with other birth complications, such as bleeding during pregnancy, placenta previa, abortion and still-birth history, etc. (Coren,

1993; Denno, 1990; Hardy et al., 1979; Niswander and Gordon, 1972).

The developmental sequelae for many low birth weight infants are varied. In fact, one review of the literature documented that low birth weight infants were three times more likely than controls to evidence neurological sequelae (McCormick, 1985). Adverse manifestations of low birth weight include problems in cognition, attention, neuromotor functioning, temperament, subnormal growth, illnesses, neurodevelopmental problems, learning difficulties hyperactivity, behavioral problems, visual defects, low intelligence, poor academic achievement, central nervous system damage, neurological abnormalities, psychiatric disorders, cerebral palsy, and so forth (Brand and Bignami, 1969; Brennan et al., 1997; Brennan and Mednick, 1997; Broman et al., 1975; Chess and Thomas, 1987; Coren, 1993; Denno, 1990; Fitzhardinge and Steven, 1972; Hack et al., 1994, 1995; Hardy et al., 1979; Hertzig, 1983; Kandel et al., 1990; Marlow et al., 1989; McCormick et al., 1990; McGauhey et al., 1991; Ross et al., 1990; Thomas and Chess, 1977; Werner and Smith, 1977; Whitaker et al., 1997). Many of the problems arising from a low birth weight condition have been implicated in the development of crime, in general, and violence, in particular (Moffitt, 1990, 1997; Raine et al., 1997b; Reiss and Roth, 1993:383; Szatmari et al., 1986).

In addition, low birth weight is commonly correlated with low socioeconomic status and level of parents' education (Denno, 1990; Hughes and Simpson, 1995; Penchaszadeh et al., 1972; Taylor, 1976). These two problems are overrepresented among minority, particularly African-American, populations (Chomitz et al., 1995), such that African-American women in the United States experience an especially high prevalence of low birth weight babies relative to other races and ethnicities in the United States and other parts of the world (Paneth, 1995; Shiono et al., 1986).

Although low birth weight has been linked with various adverse developmental implications, including neuropsychological problems, the link between low birth weight and criminal/aggressive behavior has not been extensively studied. However, the existing research on this relationship has documented that low birth weight, often in concert with inadequate social environments, is related to criminal and aggressive behavior.

Even though direct measures of neuropsychological deficits are difficult to come by, particularly with secondary data, there appears to be ample evidence that the association between low birth weight and behavioral problems/criminal behavior may be linked to early central nervous system dysfunction or development, neurological abnormalities, and neurodevelopmental problems and deficits (Brennan and Mednick, 1997; Breslau et al., 1988; Denno, 1985:719; Drillien et al., 1980; Lewis et al., 1979; Szatmari et al., 1990). Although the current research has no direct measure of neuropsychological deficit, we follow a strategy similar to the one recently employed by Brennan and Mednick (1997 p. 273; see also Brennan et al., 1997), in which an assumption was made that neuropsychological deficit/central nervous system dysfunction is likely a mediating factor between pre/perinatal complications (i.e., low birth weight) and adverse outcomes (i.e., criminal behavior).

> It is assumed that perinatal factors influence antisocial behavior through the mediating factor of neurological or central nervous system dysfunction. It is important to note that this assumption is never directly tested in the studies linking perinatal factors and antisocial outcome. . . . This research provides support for the theory that perinatal factors influence antisocial outcomes through the mediating factor of central nervous system dysfunction.

This assumption is consistent with previously documented research, as well as Moffitt's

(1993) hypothesis that medical and congenital risk factors, such as low birth weight, produce central nervous system damage (i.e., neuropsychological deficits), which, in turn, has been shown to increase the risk of criminal offending (see also Moffitt et al., 1994:282). Given that low birth weight is strongly associated with early and long-term neuropsychological development (Brennan and Mednick, 1997; Denno, 1990:156; Hack et al., 1995; McCormick, 1985), low birth weight appears to be a reasonable proxy for increased risk of neuropsychological disorders, brain injury, and early central nervous system dysfunction (Moffitt, 1995; see also Brennan et al., 1997:164; Brennan and Mednick, 1997:273; Denno, 1985:719; Pasamanick et al., 1956:613).

What Should the Early Onset Group Look Like? According to Moffitt (1993), less than 10% of the population are early onset offenders, but this group is large enough to be distinguished from other offenders. Although they represent a relatively small amount of the overall population, early onset offenders commit a relatively large portion of the more serious offenses in society (Moffitt, 1993). This finding is consistent with longitudinal studies reporting that persistent, serious offenders comprise less than 10% of birth cohorts (Farrington et al., 1990; Shannon, 1980; Wolfgang, 1983; Wolfgang et al., 1987).

Given that prior research has shown that life-course-persistent offenders are very rarely female (Denno, 1990; Moffitt, 1993, 1994; Moffitt et al., 1994; Wolfgang, 1983), it is no surprise that Moffitt (1993:674) concentrates on boys for the early onset group (i.e., life-course persisters): "chronic offending is . . . found among a relatively small number of males whose behavior problems are quite extreme." Toward this end, little effort has been made to examine the impact of the neuropsychological-environmental interaction hypothesis for girls. In one of their most recent investigations, Moffitt et al. (1994:283) failed to find evidence for the presence of girls in the life-course-persistent group.

To account for this, they (1994:283) offered three reasons: (1) neuropsychological disorders are rarer among females than among males, (2) childhood-onset conduct problems are very rare in girls, and (3) female delinquency lacks stability in relation to male delinquency (e.g., Denno, 1990:17; Reinish et al., 1979; Singer et al., 1968).

In sum, Moffitt is quite clear about the determinants of early onset. Aside from the interaction between neuropsychological and environmental characteristics, Moffitt (1994:39) states that much of the gender difference in the life-course-persistent group is attributable to sex differences in the exposure to risk factors for life-course-persistent antisocial behavior. For Moffitt, the causal process does not necessarily differ between males and females; rather, the difference in the prevalence or degree of exposure to predictors differs.

▧ Hypotheses

From the preceding discussions, we develop the following hypotheses:

> H1: The risk of neuropsychological deficit-disadvantaged environment interaction will predict early onset.
>
> H2: Although boys and girls follow the same path toward the development of early onset of offending, or membership in the life-course-persistent group of offenders, this neurological deficit-disadvantaged environment interaction will be magnified for boys.

The first hypothesis is grounded in Moffitt's (1993) work concerning the etiological correlates of the life-course-persistent offender. The second hypothesis is grounded in Moffitt's (1994:39) claim that girls are less likely than boys to encounter all of the putative links in the causal chain for the development of life-course-persistent behavior, and Moffitt et al.'s

(1994:292,296) finding that neuropsychological effects were limited to males.

As a result, although the process of cumulative continuity can apply to girls, it does so at a lower rate than it does for boys. Also, if the analysis uncovers that fewer girls than boys are at risk, but they develop the same early onset outcome, this finding would not be inconsistent with Moffitt's theory. Such a finding would replicate research completed by other scholars who found that high levels of perinatal risk were related to serious delinquency among females (Shanok and Lewis, 1981; Werner and Smith, 1992). On the other hand, if the analysis finds that there are girls at risk, but they do not develop the early onset outcome, this would be something not necessarily expected by Moffitt's theory, but interesting and important nevertheless (e.g., Stanton et al., 1991:958). Regardless of the outcome, little is known about the development of female risk, in general, and the correlates of an early onset of offending for females, in particular. While problems associated with a low base rate of offending and/or early onset are common with female samples (Brennan et al., 1997:164), information on this front is sorely needed for a better understanding of female crime.

◤ Data and Methods

The data used in this study were collected for the Longitudinal Study of Biosocial Factors Related to Crime and Delinquency in Pennsylvania (Denno, 1990). Data were collected from three sources. The first data source was the Collaborative Perinatal Project (CPP). Initiated in Philadelphia in 1959, the CPP was, at the time, one of the most ambitious and costly medical projects undertaken. Designed to obtain baseline data on birth defects, it includes a vast array of measures, including risk of neurological conditions, birth complications associated with cognitive abilities, familial conditions and socioeconomic status,

all of which are key constructs in Moffitt's (1993) theory. Furthermore, the CPP measures follow the subjects from birth to age 7, allowing researchers to trace the development of the individuals over time. The second and third sources of the data collected by Denno were from the Philadelphia public schools and the Philadelphia Police Department. School and police measures applied to the children from ages 7 to 14 and 7 to 18, respectively.

Detailed data have been organized and analyzed on the subsample of 987 youths who constitute the subjects selected by Denno (1990:29) from a larger sample of 2,958 black mothers who participated in the first four cohorts (1959–1962) of the CPP. Denno (1982; 1990:30) reports on comparisons made between the final sample and the excluded sample of subjects, the results of which showed no significant difference on a number of key variables.[6]

These data are appealing for the current study because the CPP was designed to measure the effects of family background and developmental variables on the offending of a cohort of black, inner-city youths followed from birth to late adolescence (Denno, 1990).

The 987 subjects used for Denno's study were infants of mothers who participated in the CPP and, therefore, reflect the characteristics of families who would be interested in receiving inexpensive maternity care provided by a public clinic at Pennsylvania Hospital between 1959 and 1962 (Denno, 1990). The 987 cases of the original sample were made up of children who: (1) had a black mother, (2) were located in a Philadelphia public school, (3) stayed in Philadelphia from ages 10 to 17, (4) were not among sibling members excluded from the sample to prevent possible biases in multiple family membership, and (5) received selected intelligence tests at age 7 and 14.

Given that inner-city, black youth are a high-risk population for violent and persistent offending (National Research Council, 1993; Reiss and Roth 1993; Wolfgang et al., 1987), Moffitt (1994: 38–39) has observed:

life-course-persistent antisocials may be anticipated at elevated rates among Black Americans because the putative root causes of this type are elevated by institutionalized prejudice and by poverty. . . . Moreover, among poor blacks, prenatal care is less available, infant nutrition is poorer, and the incidence of fetal exposure to toxic and infectious agents is greater, placing infants at high risk for the nervous system problems that research has shown to interfere with prosocial child development.

Relatedly, research has shown that African-American women are more likely to give birth to low birth weight babies relative to other races and minorities (Paneth, 1995:19). These two points, taken with the recent assertion by Moffitt (1997:126) that these problems are magnified in inner-cities, suggest that the Denno data appear valuable, and perhaps unique to testing Moffitt's interactional hypothesis.

From the 987 subjects in the publicly available data, a subsample was extracted that included the 220 subjects who had at least one recorded offense by age 18. In the present study, it is reasonable to study only offenders because Moffitt's hypothesis specifies an etiological distinction within the population of offenders. Because of missing data, 13 of these cases were removed. The resulting sample consisted of 207 offenders (144 boys and 63 girls) for which sufficient data were available.

▧ Measurement of Variables

Dependent Variable

Early Onset. The dependent variable for this study—early onset—was measured by the age at first offense as recorded by the Philadelphia Police Department. Age at first offense was the age an offender first experienced a police contact that resulted in an official arrest or a remedial disposition. In the current sample, age at first offense ranged from 8 to 18. Although there is no clear consensus for determining age of onset (Dean et al., 1996; Mazerolle, 1997; Nagin and Farrington, 1992a, 1992b; Paternoster et al., 1997; Simons et al., 1994), we use age 14 in determining early/late onset for several reasons. First, findings from research report the highest hazard rates for onset of offending before age 14 (Blumstein et al., 1986; Patterson et al., 1992). Second, this cutoff age was used in one of the most recent tests of Moffitt's theory (Moffitt et al., 1994) as well as Patterson's developmental theory (Simons et al., 1994).[8] In the present analysis, onset was recoded as 0 (age at first offense was 14 or older) or 1 (age at first offense was 13 or younger). The latter group of cases (N = 70) represents the early onset group, while the former group of cases (N = 137) represents the late onset groups.[9] It is also important to note that the number of offenders (70) who fit the criterion for early onset make up about 7.1% of the original sample of 987. This is consistent with Moffitt's (1993) claim that early offenders make up about 5% to 10% of the population.

Independent Variables

Low Birth Weight. Low birth weight was measured immediately upon delivery by the hospital staff in Philadelphia. Responses originally ranged from 3 to 12 pounds, but the variable was recoded as 1 (less than 6 pounds) or 0 (6 pounds or more).[10] In the present analysis, 67 (32%) of the 207 offenders were classified as having low birth weight.

Disadvantaged Environment. In this paper, we utilize two different measures of disadvantaged environment. The first is concerned with a weak familial structure; the second deals with a poor socioeconomic status. We examined these two measures of disadvantaged environment for two reasons. First, we are interested in performing sensitivity analysis to determine if the conclusions we draw from our operationa-lization of disadvantaged environment are sensitive to varying measurement strategies. Second, Moffitt's theory of the life-course-persistent offender does not prefer one of the disadvantaged environment measures over the other. In fact, in various places, Moffitt (1993:684, 1997:153) has argued or shown that both weak familial structure and poor socio-economic status are likely to exacerbate neuro-psychological or neurological abnormalities.

For example, in the initial presentation of her theory, Moffitt (1993:682) notes: "Vulnerable children are often subject to adverse homes and neighborhoods because their parents are vulnerable to problems too." Notice that in this particular passage, both adverse homes (weak family structure) and adverse neighborhoods (weak socioeconomic status) are discussed. In one of her most recent statements, Moffitt (1997:126; see also Moffitt, 1994:38) states:

> Neuropsychological vulnerable chil-dren might be found at elevated rates in inner-city neighborhoods because the sources of poor brain health are linked with institutionalized prejudice and poverty. . . . Among the inner-city poor, prenatal care is less avail-able, low birth weight is more common, infant nutrition is poorer, and the incidence of fetal exposure to toxic and infectious agents is greater, placing infants at high risk for the nervous system problems that research has shown to interfere with optimal child development.[11]

Socioeconomic Status. SES is believed to be a powerful influence on the development of early offenders (Moffitt, 1993; Patterson et al., 1992; Patterson and Yoerger, 1993). Moffitt (1993:682) claims that one of the best measures of a disadvantaged environment is SES. In their review of familial predictors of juvenile offending, Henry et al. (1993) found that low SES is widely associated with stable and pervasive antisocial behavior in late childhood. In the current research, SES was measured by a single-item, general SES score that is a composite measure comprising three indicators collected at age 7 for each child: education of head of household, income of head of household, and the occupation of the head of household. Higher scores are indicative of a higher SES.[12]

We should bear in mind, however, that this measure of SES should be viewed cautiously. As indicated by Denno (1990:30–31), the present sample exerts very little variation in SES because they were predominantly of lower SES when the CPP began. As such, "high" scores on the SES composite may still reflect high SES, but since the sample is relatively homogenous with regard to SES a "high" SES may give a somewhat false impression of a true high SES. We address this issue in further detail later in the text.

Family Structure. Research shows that the family structure of a child is an important factor in the development of early onset and serious offending (Denno. 1990; Henry et al., 1996; Kandel et al., 1990; Kolvin et al., 1988; Moffitt, 1993; Patterson et al., 1992; Smith and Jarjoura, 1988). Moffitt (1993) claims that a weak family structure is one of the primary features of a disadvantaged environment. Weak family structure was measured by summing the standardized scores of three indicators collected at age 7: the number of changes in mother's marital status, with whom the child lives, and whether the husband/father of the child was present in the household.[13] Higher scores on the scale indicated a weak, family structure. The reliability coefficient (Cronbach's α = .83) estimated for these indicators revealed that the family structure index was internally consistent.

Low Birth Weight × Disadvantaged Environment. In this study, we place particular emphasis on the biosocial interaction between low birth weight and the two measures of disadvantaged environment. Two interaction terms were created: low birth weight × family structure and low birth weight × SES. The raw scores of the component factors were mean-centered in order to rid the measures of nonessential ill-conditioning (the multicollinearity between the component variables produced by noncentered-ness, which would inevitably cause multicollinearity between the component variables and their product terms (Aiken and West, 1991; Jaccard et al., 1990). When we examined the zero-order correlations, multicollinearity did not appear to be a problem.

Table 1 contains descriptive statistics for two sets of groups: the original cohort and the offender-only group. Within the original cohort, we report descriptive information for the nonoffenders (N = 734) and offenders (N = 207). Within the offender-only group, we report descriptive statistics for females, coded 2 (N = 63), and males, coded 1 (N = 144). By presenting the descriptive data in such a format, one can compare not only the distribution of male offenders versus female offenders, but also the study sample vs. the original

cohort. Compared to the original cohort, offenders are significantly more likely to be male, to come from lower SES, and to have a weaker family structure. In the offender-only group, the only significant difference to emerge across gender concerns SES: males were more likely to come from higher SES. Interestingly, a similar percentage of early onset offenders were identified among boys (34%) and girls (33%). Of the 70 offenders who fit the study's criteria for early onset, 49 were boys and 21 were girls. Low birth weight did not vary significantly across the four groups.[14] We now turn to the results of our analysis.

▧ Results

As stated earlier, Moffitt's hypothesis specifies an etiological distinction within the population of offenders; thus, it is appropriate to study only offenders. Table 2 presents the results of three models; a full model that contains all 207 offenders and two gender-specific models, one for females and one for males. In all three models, the dependent variable is early onset (0 = late onset, 1 = early onset).[15]

| Table 1 | Comparison of Means on All Measures for Non-Offenders Versus Offenders and Female Offenders Versus Male Offenders |

| | Original Cohort | | | | Offenders Only | | | |
| | Non-Offenders (N = 734)[a] | | Offenders (N = 207)[b] | | Females (N = 63) | | Males (N = 144) | |
Variable	M	SD	M	SD	M	SD	M	SD
Low birth weight	.34	.47	.32	.47	.33	.48	.32	.47
SES	39.65	16.87	36.38*	17.10	32.49	15.29	38.08*	17.61
Weak family structure	−.13	2.57	.35*	2.61	.71	2.47	.19	2.66
Gender	1.56	.50	1.30*	.46	—	—	—	—
Early onset	—	—	—	—	.33	.48	.34	.48

a. Because of removal of cases with missing data, 33 cases were discarded from the original group of 767 nonoffenders.

b. Because of removal of cases with missing data, 13 cases were discarded from the original group of 220 offenders.

*$p < .05$, one-tailed.

| Table 2 | Logistic Regression Coefficients Predicting Early Onset: Full Model and Gender-Specific Models | | | | | | | | | | | |

	Full Model				Gender-Specific Models							
	(n = 207)				Females (n = 63)				Males (n = 144)			
Variable	B	S.E	Wald	exp(b)	B	S.E.	Wald	exp(b)	B	S.E.	Wald	exp(b)
Low birth weight	.79	.34	5.57*	2.21	1.00	.69	2.15	2.73	.71	.40	3.17	2.03
SES	−.03	.01	6.05*	.97	−.05	.03	3.33	.95	.02	.01	3.18	.98
Weak family structure	.04	.07	.37	1.04	.16	.15	1.06	1.17	.00	.08	.00	1.00
Gender	−.27	.35	.62	.76	–	–	–	–	–	–	–	–
Low birth weight × SES	.05	.02	5.92*	1.06	.08	.06	2.07	1.09	.05	.02	3.93*	1.05
Low birth weight × Weak family structure	.44	.15	7.96*	1.55	.37	.35	1.14	1.45	.44	.17	6.41*	1.56
Constant	−.43	.48	.82		−.99	.35	8.23*		.73	.19	14.76*	
X^2			24.63				14.29				13.81	
Df			6				5				5	
P			.001				.013				.016	
Model prediction rate			68%				73%				67%	

* $p < .05$, one-tailed.

We also present the log odds of the coefficients, found in the column exp(B), in which a value greater than 1 indicates an increase in early onset, a value less than 1 indicates a decrease in early onset, and a value of 1 suggests no effect on early onset.

In examining the additive effects in the full sample model, it can be seen that individuals born at low birth weight were more likely to have an early onset of offending, and individuals from a higher SES were less likely to incur an early onset. Though in the correct direction, neither weak family structure nor gender had a significant effect on early onset.

Although an inspection of the additive effects is informative, the central focus of this study concerns the hypothesis outlined by Moffitt that risk of neuropsychological deficit

(indicated in the present study by low birth weight) interacts with disadvantaged home environment. Both interactions have a significant effect on early onset. Both low birth weight × SES and low birth weight × weak family structure have a significant and positive effect on early onset. Thus when low birth weight is met with a weak family structure and a higher SES, individuals are more likely to incur an early onset of offending.[16]

So far, some evidence has emerged in support of Moffitt's interactional hypothesis. However, the question remains whether this interaction operates equally across gender. Turning to the gender-specific estimations in Table 2, some curious results emerge. In the female model, only one of the five variables exerts a significant effect ($p = .06$) on early onset.

Females who come from a higher SES are less likely to incur an early onset of offending. In terms of the other additive effects, neither low birth weight nor weak family structure significantly predicted early onset. Notably, in terms of the interaction effects, neither interaction term significantly predicted early onset for females. This result is consistent with Hypothesis 2, which specified Moffitt's prediction that the neuropsychological-environmental interaction should not be important for females.

Turning to the male sample, two of the three additive effects are significant ($p < .07$). Low birth weight had a positive effect on early onset, and higher SES exerted a negative effect on early onset. Weak family structure failed to significantly predict early onset. In contrast to the results for females, however, both interaction terms were significant for males.[17] The first interaction, low birth weight × SES, had a positive effect on early onset.[18] Consistent with Moffitt, the second interaction term, low birth weight × weak family structure, had a positive effect on early onset, suggesting that individuals residing in a weak family structure *and* born at low birth weight are more likely to incur an early onset.

The counterintuitive result we observed for the low birth weight × SES interaction may have much to do with the nature of the sample under investigation. Recall that mothers interested in inexpensive maternal care for the first seven years of life selected themselves (and subsequently their children) in the study project. As Denno (1990:30–31) reports, the sample was predominantly low SES. Thus, individuals scoring "high" on the SES indicator may still be of low SES relative to the general population, but give the somewhat clouded impression that they are of higher SES.

One reviewer suggested that the neuropsychological deficit-poor SES environment interaction could be interpreted to suggest that among those with lower SES, in our case very low SES, it is only those with neuropsychological deficits, such as low birth weight, that have risk for an early onset (e.g., Ross et al., 1990). In other words, the reviewer was asserting that low birth weight is a risk factor conditional on low SES.

To further explore the moderating effect of SES on the relationship between low birth weight and early onset, we partialed the SES measure into two groups: low SES (those scoring at the 25th percentile or below, $N = 50$) and high SES (those scoring above the 25th percentile, $N = 157$).[19] In Table 3, we present the results of two separate logistic

Table 3	Logistic Regression Coefficients Predicting Early Onset by Level of SES							
	Low SES (N = 50)				**High SES (N = 157)**			
Variable	B	SE(B)	Wald	Exp(B)	B	SE(B)	Wald	Exp(B)
Low birth weight	1.64	.69	5.53*	5.13	.58	.38	2.32	1.78
SES	−.11	.07	2.43	.90	−.03	.01	5.51*	.97
Weak family structure	.23	.15	2.13	1.25	.02	.07	.06	1.02
Gender	.64	.65	.97	1.90	−.56	.41	1.79	.57
Constant	−.392	1.74	5.06*	1.14	.11	.55	.04	.57
X^2		11.26				10.19		
Df		4				4		
P		.02				.04		
Model prediction rate		72%				69%		

* $p < .05$, one-tailed.

regressions, one for the low SES group, and one for the high SES group which includes gender, SES, weak family structure, and low birth weight.

In the first column, we present the results for those individuals in the low SES group. The important coefficient for present purposes is the effect of low birth weight on early onset. The effect of this variable is positive and significant, suggesting that those individuals born at low birth weight are more likely to incur an early onset. Turning to the second column, among those high in SES, the effect of low birth weight on early onset was positive, but not statistically significant.[20] Figures 1 and 2 graphically present these results. Figure 1 shows the relationship between low birth weight and early onset for the low SES group, and Figure 2 shows the same relationship for the high SES group, When selecting on those low on SES (Figure 1), individuals born at low birth weight have a mean onset of .611 and those individuals not born at low birth weight have a mean onset of .281 (F = 5.599, p = .02). When selecting on high SES (Figure 2), those born at low birth weight had a mean onset of .388 while those not born at low birth weight had a mean onset of .287 (F = 1.571, p = .21).

Taken together, these results appear to suggest that the effect of low birth weight on early onset is conditional on (low) SES.[21] When we consider these results in the context of Moffitt's theory, the initially apparent contradictory result obtained earlier for the low birth weight × SES interaction, with the original continuous coding for SES, appears to be a function of the fact that the sample is relatively homogenous with regard to SES. Thus, we view the supplementary results as being consistent with Moffitt's interactional hypothesis.

| Figure 1 | Low SES |

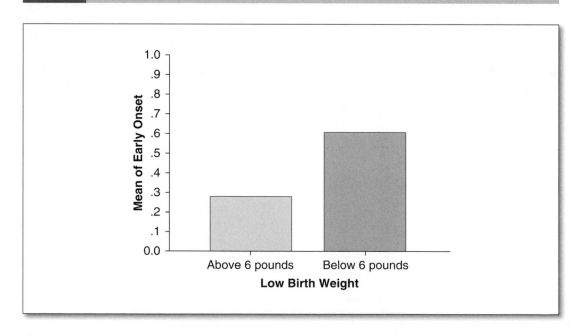

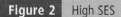

Figure 2 High SES

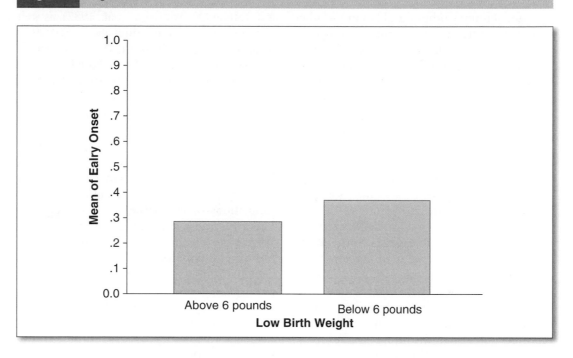

☒ Discussion and Conclusion

To the best of our knowledge, this study provides the first attempt to examine a prospective link between the interaction of neuropsychological risk and disadvantaged environment in explaining early onset in a longitudinal context, when the measurement of neuropsychological deficit is not composed of test scores, psychological assessments, or both (cf. Moffitt et al., 1994, 1996). This attempt is important because one of the neglected areas of biosocial research has been the examination of interactions between biological and environmental factors within a major, urban U.S. city (Brennan et al., 1997; Piquero and Tibbetts, 1999; Raine et al., 1997b).

The results of this study appear supportive of Moffitt's (1993, 1994) interactional hypothesis concerning the development of early onset of offending and further demonstrate that

influences of biosocial interaction are under way in early childhood (e.g., Brennan et al., 1995). Consistent with our first hypothesis, findings showed that (1) low birth weight interacts with a disadvantaged familial environment to predict early onset, and (2) low birth weight combined with low SES significantly increases the risk of an early onset. When gender-specific models were estimated, the results appeared to be in line with our second hypothesis that the neuropsychological risk-disadvantaged environment interactions were important for males, but not for females.[22] In terms of the additive effects, the results showed that across a variety of models, low SES poses a risk for early onset of offending, and the aggravating effect of low birth weight on early onset was more important for males than for females.

We should note that the data utilized in this study suffer from a few limitations. First, only one indicator of increased risk for neuropsychological deficit—low birth weight—was used. Future studies should examine other potential predictors of risk for neuropsychological deficits, such as birth complications, head injuries, mother's drug use/abuse, mother's health, and so on. This is important because we assumed: that low birth weight served as an adequate proxy for neuropsychological deficit. Although we are in the mainstream in this regard (Brennan and Mednick, 1997), we should point out that more direct measures of neuropsychological deficit are needed for a more direct test of the etiology of Moffitt's life-course-persistent offender.

Second, age at first officially recorded offense was used to measure early onset. Although many researchers have employed official measures for identifying early onset (Moffitt et al., 1994; Patterson et al., 1992; Simons et al., 1994), such indices would probably best be used in conjunction with other measures such as self-reported offending in future studies (e.g., Bartusch et al., 1997). Confidence in the results of the present study would be increased if the results of this study are confirmed, with different operationalizations of early onset.

Third, the sample comprised inner-city, black youths, and as a result, some generalizability problems may emerge. Inclusion in the sample was, in part, based on characteristics of the families who would be interested in receiving inexpensive maternity care provided by a public clinic (Denno, 1990:30). Moreover, given that low birth weight rates differ between African-Americans and whites, and that African-American babies usually weigh about 9 ounces less than white babies (Paneth, 1995:24), this subtle difference could have a demonstrable impact when infants are categorized into low and regular birth weight groups. However, our analysis focused on a single race, African-Americans. This potential limitation can also be viewed as a beneficial characteristic of the study because black youth are a high-risk population for violent and persistent offending (National Research Council, 1993; Reiss and Roth, 1993; Wolfgang et al., 1987), as well as membership in the life-course-persistent group of offenders (Moffitt, 1994). Nevertheless, because our analyses were based on a single ethnicity predominantly of lower SES, without other ethnic groups and/or African-Americans of a variety of different types of SES levels, the present analysis cannot articulate the role that (poor) ethnic background plays in the development of early onset. As a result, future studies should examine the validity of Moffitt's interactional hypothesis for other populations of youth (i.e., rural, white, etc.) and of varying levels of SES in order to further disentangle the interplay between individual and structural forces.

In addition to the future research directions just addressed, we also see potential for a few promising lines of inquiry with regard to birth weight. First, future studies may want to partial out the low birth weight categories into finer distinctions, such as very low birth weight and extremely low birth weight, because it is conceivable that infants in the lowest categories could suffer more damaging consequences compared with higher birth weight infants (e.g., Hack et al., 1994; McCormick et al., 1990, 1996; The Scottish Low Birthweight Study Group, 1992), and thus incur an early onset of offending.

Toward this end, we performed some preliminary, albeit crude, analyses that attempted to examine if the proportions of early onset were greater for those lowest in birth weight. Specifically, we created three groups of individuals: (1) those born over 6 pounds, (2) those born at low birth weight (between 5 and 6 pounds), and (3) those born at very low birth weight (between 3 and 4 pounds). We then calculated the mean of early onset across the three groups. Interestingly, the mean of early onset ($X^2 = .857$) was highest for the seven individuals who weighed 3 or 4 pounds at birth when compared to the 69 individuals who weighed between 5 and 6 pounds at birth (X^2 = .449) and the 144 individuals who weighed above 6 pounds at birth (X^2 = .292). The differences between the groups was statistically significant (F = 6.718, p = .001).

Measures of low birth weight could also be refined by incorporating information on gestational age. Although not directly expressed in her theory, it is possible that Moffitt's biosocial interaction may be magnified for those infants in the lowest of all possible birth weight groups. Also, additional measures of disadvantaged environment, such as parent-child interactions, could be incorporated into indices measuring this component of Moffitt's theory to create a more comprehensive estimate of disadvantaged environment (for a similar approach, see Raine et al., 1997b).

Our results also have some import for public policy. Although many avenues exist in terms of preventing low birth weight from occurring in the first place by targeting prenatal care (Alexander et al., 1995) and mother's lifestyle (Chomitz et al., 1995), the presence of a low birth weight infant should not be viewed as a dooming consequence. Recent evidence suggests that a substantial portion of the adverse outcomes of low birth weight reflects modifiable environmental factors (McCormick et al., 1996). Therefore, even with recent gains in medical technology that have allowed for the positive development of low birth weight infants, there is still room for work in the social, familial, and economic environment of the infant that could help alleviate the impact of low birth weight. Reviews of the literature surrounding the amelioration of perinatal and other related birth consequences suggest that supportive environments and early interventions stand a fighting chance at diminishing the consequences of birth-related difficulties (Brennan and Mednick, 1997), and such approaches may have an even more demonstrable impact on inner-city youths (Guerra et al., 1997).

▧ Notes

1. To be sure, the emphasis placed on early onset has not been wholeheartedly accepted by the criminological community. For some, early onset of offending is simply a behavioral manifestation of criminal propensity, such that individuals experiencing an early onset have higher criminal propensity than individuals experiencing onset later (i.e., persons having a lower criminal propensity) (Gottfredson and Hirschi, 1990; Hirschi and Gottfredson, 1983:574–77).

2. Some commonality exists with Moffitt's (1993) life-course-persistent offender and Gottfredson and Hirschi's (1990) low self-control offender. For both theories, early evidence of antisocial behavior, including an early age of onset, usually sets the stage for persistent offending throughout life. Where the theorists diverge, however, is in the attribution of the cause of this time-stable behavior. For Moffitt, it lies in the interaction of individual differences in neuropsychological development and poor environment. For Gottfredson and Hirschi, it is in low self-control and opportunity.

3. Some research appears to find support for the position that at least two (if not more) types of offenders exist (e.g., Bartusch et al., 1997; Caspi et al., 1994; D'Unger et al., 1998; Land et al., 1996; Land and Nagin, 1996; Moffitt et al., 1994: Nagin et al., 1995; Nagin and Land, 1993; Patterson et al., 1992; Simons et al., 1994).

4. Another recently proposed taxonomy (Patterson et al., 1992; Patterson and Yoerger, 1993; Simons et al., 1994) also claims that children with defiant orientations often overwhelm the efforts of authority figures to nurture or discipline them in their development.

6. Denno selected the 987 subjects from the larger sample of 2,958. Although she reports that there were no significant differences between the 987 subjects and the larger sample of 2,958, we do not have access to the larger sample. Our data on the 987 subjects were taken from the Inter-University Consortium for Political and Social Research (ICPSR) secondary data analysis archive. Nowhere in the publicly available data, nor in any of Denno's works (1984, 1985, 1990), is there information on the 2,958 subjects to which we can compare the 987 subjects. Thus, we have to assume that the publicly available data, on which the current analyses are based, are representative of the large sample of 2,958 subjects, as Denno indicates.

8. In preliminary analysis not shown here, we estimated the onset dependent variable in a continuous fashion. Both the continuous and dichotomies indicators were highly correlated (r = .75). When onset was specified in a contiguous fashion, the results were substantively similar to the dichotomous indicator of onset. As such, we only report the results for the dichotomous indicator of onset. Moreover, we point out that, for theoretical reasons, previous examinations of the determinants of early and late onset of offending have utilized age 14 as the marker of distinction (cf. Moffitt et al., 1994: Simons et al., 1994). Such an approach is consistent with using this age as a marker for the transition between childhood and adolescence (Bartusch et al., 1997; Tanner, 1978).

9. Although age at first recorded police contact may not be as accurate a measure of onset as self-reported offending, it has several advantages over other commonly used measures of onset of offending (e.g., age at first arrest, age at first conviction, etc.). While recent empirical examinations of early onset have used arrest and conviction data (e.g., Moffitt et al., 1994; Simons et al., 1994), age at first police contact provides a more sufficient measure of onset because many first contacts with police do not result in an arrest or conviction. The likelihood of a police contact to result in an arrest may be influenced by several factors including: police discretion, demeanor, demographic characteristics of offender/victim, characteristics of the community, etc (e.g., Smith, 1984).

10. This cutoff for low birth weight was chosen because most medical researchers use a cutoff of just below 6 pounds (i.e., low birth weight is actually identified in the literature as 5 pounds, 8 ounces) (Coren, 1993; Denno, 1990; Paneth, 1995:19; Shiono and Behrman. 1995:17). This cutoff was adopted by the World Health Organization (1950), who was interested in adopting a universal definition for low birth weight. Since the WHO's adoption of this cutoff, most studies have used the 6-pound marker for low birth weight. Although modern obstetric/pediatric research has improved the precision for the measurement of low birth weight (e.g., accounting for gestational age), our measure is limited to the medical technology of the 1950s, the time period within which the data were collected. There exists no separate cutoff for low birth weight between the sexes; thus, the 6 pound marker for low birth weight is used for both males and females.

One reviewer noted that this cutoff for low birth weight is rather liberal, and that it is at severe low birth weight that risk is really carried. Although this is certainly plausible, we offer two reasons as to why we do not explore this possibility in the present study. First, Moffitt's theoretical argument does not make any distinctions between the various types of low birth weight (i.e., low, very low, extremely low). For her theory, it is the presence of low birth weight per se that matters. Second, and more practically, our sample was made of only 207 cases. To partial birth weight into smaller categories would create problems associated with reduced sample sizes in birth weight categories that could have a detrimental impact on the power of the analysis, a problem that has plagued other researchers (McGauhey et al., 1991).

11. In her own work, Moffitt (1997:155–57) has begun to examine the joint influence of neuropsychological problems and poor neighborhood structure (i.e., in an inner-city context) in predicting delinquency and poor cognitive test outcomes. Using data from the Pittsburgh Youth Study, Moffitt identified two neighborhoods. "Bad" or disadvantaged neighborhoods were those in the worst 25% in measures of median household income, unemployment, families below the poverty level, density of males aged 10–14, female-headed households, or rate of separation/divorce, and "good" neighborhoods were those that were not extreme on any of these factors. Her preliminary results suggested that "good" neighborhoods seemed to protect boys from delinquency, but only if the boys were neuropsychologically healthy. Moreover, cognitive impulsivity scores were significantly worse, and delinquency scores significantly higher, among boys from disadvantaged or "bad" neighborhoods.

12. Although we view our usage of the age 7 measure of socioeconomic status as adequate, some may suggest that early (i.e., birth) SES indicators are more relevant to Moffitt's theory than those obtained later in childhood. We chose to retain the age 7 measure of socioeconomic status as the principle SES measure for several reasons. First, in her own work, Moffitt has utilized birth (Henry et al., 1996) and/or childhood/early adolescence (Moffitt et al., 1994; Moffitt, 1997) measures of SES and, thus, does not appear to prefer SES at one age over another. Second, because the family structure variables were obtained at age 7, we thought that keeping the environmental indicators measured concurrently would be more valid than measuring one of them early in life and the other later in life. Third, we also thought that the composite index of SES was a more internally valid indicator of SES

because it takes into consideration income, education, and employment rather than just income. Nevertheless, we also performed the interactional analyses with an indicator of SES measured solely as the mother's income in dollars while she was three months pregnant. The results (reported later) for the birth SES variable were substantively the same as those for the age 7 SES measure.

13. The number of changes in mother's marital status was coded 0 to 6 (number of changes in marital status) or 7 (seven or more changes in marital status). The living situation of the child was coded 0 (child lives with mother and father/stepmother and father/ mother and stepfather) or 1 (child lives with mother only/ father only/adoptive home/foster home/miscellaneous). Husband or father present in household was coded 0 (yes) or 1 (no). All responses were standardized before summation for the family structure scale.

14. Preliminary bivariate analysis showed that birth weight and early onset were significantly related. Of the 67 individuals born with low birth weight, 30 of them incurred an early onset ($\chi^2 = 5.316$, df = 1, p = 0.2). Of the 144 males, 46 were born with low birth weight. Of the 46 cases, 20 had an early onset, but this result was not statistically significant ($\chi^2 = 2.689$, df = 1, p =.10).Of the 63 females. 21 were born with low birth weight. Of these 21 cases, 10 had an early onset. Similar to males, this result was not statistically significant ($\chi^2 = 2.892$, df = 1, p = .09).

15. We also estimated the models with a continuous age of onset, as well as several other age cutoffs (e.g., age 13 and age 15). In all cases, the coefficient estimates were substantively the same.

16. The coding of the SES variable (i.e., higher values connote higher SES) makes the interpretation of this interaction term counterintuitive. We will return to this issue shortly.

17. A test (e.g., Clogg et al., 1995) of the difference of the interactions across gender failed to detect any significant difference. Results are available on request.

18. Similar to the full sample results, the interaction between low birth weight and SES is statistically significant, but the sign is in the opposite direction to that predicted.

19. We recognize that our usage of 25% as a cutoff for low SES is somewhat arbitrary. However, in some of her most recent work, Moffitt (1997) used a similar cutoff to signify poor environment. In related work, Escalona (1982) also utilized a 25th percentile

cutoff for low SES. To examine the robustness of our findings, we performed sensitivity analysis by defining the low SES group as those individuals at or below the 20th percentile, and the high SES group as those individuals above the 20th percentile. The results were substantively the same for both percentile cutoffs.

20. A coefficient comparison test failed to evidence a statistically significant difference between the low birth weight coefficients across SES groups ($z = 1.34$) at conventional levels of significance ($p < .05$). However, this result must be viewed in the context of a relatively small sample size, in general, and across SES groups, in particular.

21. Though not presented here, we also completed a similar analysis, but with a slightly different strategy. We created a dummy variable that was coded 1 when individuals were born at low birth weight and low in SES. When this variable was entered into a logistic regression equation predicting early onset, it evidenced a positive and statistically significant effect in predicting onset.

22. The null interaction effect observed for females may be caused by the small number of female offenders included in the analysis. As such, the difference in significance of the interaction terms across gender may be a spurious artifact of statistical power. It is entirely possible that the interaction proposed by Moffitt is salient for females; however, it may not have been different enough among boys and girls in the present sample to be detected at the relatively small sample size. We encourage future research endeavors to obtain a sample size larger than the one in the present study to give this hypothesis a better assessment.

◪ References

Aiken, Leona S., and Stephen G. West. 1991. *Multiple Regression: Testing and Interpreting Interactions.* London: Sage.

Alexander, Greg R., and Carol Korenbrot. 1995. The role of parental care in preventing low birth weight. *The Future of Children: Low Birth Weight* (Center for the Future of Children) 5:103–20.

Bartusch, Dawn R. J., Donald R. Lynam, Terrie E. Moffitt, and Phil A. Silva. 1997. Is age important? Testing a general versus a developmental theory of antisocial behavior. *Criminology* 35:13–48.

Berhman, Richard E., Gorham S. Babson, and Richard Lessel. 1971. Fetal and neonatal mortality in white middle class infants: Mortality risks by gestational age and weight. *American Journal of Diseases of Children* 121:486–89.

Blumstein, Alfred, Jacqueline Cohen, Jeffrey A. Roth, and Christy A. Visher. 1986. *Criminal Careers and "Career Criminals."* Washington, DC: National Academy Press.

Brand, M., and A. Bignami. 1969. The effects of chronic hypoxia on the neonatal and infantile brain. *Brain* 92:233–54.

Brennan, Patricia A., and Sarnoff A. Mednick. 1997. Medical histories of antisocial individuals. In *Handbook of Antisocial Behavior,* ed. D. M. Stoff, J. Breiling, and J. D. Maser. New York: John Wiley.

Brennan, Patricia A., Sarnoff A. Mednick, and Adrian Raine. 1997. Biosocial interactions and violence: A focus on perinatal factors. In *Biosocial Bases of Violence,* ed. Adrian Raine, Patricia A. Brennan, David P. Farrington, and Sarnoff A. Mednick. New York: Plenum Press.

Brennan, Patricia A., Sarnoff A. Mednick, and Jan Volavka. 1995. Biomedical factors in crime. In *Crime,* ed. James Q. Wilson and Joan Petersilia. San Francisco: ICS Press.

Breslau, Naomi, Nancy Klein, and Lida Allen. 1988. Very low birthweight: Behavioral sequelae at nine years of age. *Journal of the American Academy of Child and Adolescent Psychiatry* 27:605–12.

Broman, Sarah H., Paul L. Nichols, and Wallace A. Kennedy. 1975. *Preschool IQ: Prenatal and Early Developmental Correlates.* New York: John Wiley.

Caspi, Avshalom, Terrie E, Moffitt, Phil A. Silva, Magda Stouthamer-Loeber, Robert Krueger, and Pamela Schmutte. 1994. Are some people crime-prone? Replications of the personality-crime relationship across countries, genders, races, and methods. *Criminology* 32:163–96.

Chess, Stella and Alexander Thomas. 1987. *Origins and Evolution of Behavior Disorders: From Infancy to Early Adult Life.* Cambridge, MA: Harvard University Press.

Chomitz, Virginia Rail, Lilian W. Y. Cheung, and Elice Lieberman. 1995. The role of lifestyle in preventing low birth weight. *The Future of Children: Low Birth Weight* (Center for the Future of Children) 5:121–38.

Clogg, Clifford C., E. Petkova, and Admantios Haritou. 1995. Statistical methods for comparing regression coefficients between models. *American Journal of Sociology* 100:1261–93.

Coren, Stanley. 1993. *The Left-Hander Syndrome.* New York: Vintage.

Dean, Charles W., Robert Brame, and Alex R. Piquero. 1996. Criminal propensities discrete groups of offenders, and persistence in crime. *Criminology* 34:547–74.

Denno, Deborah. 1982. Sex differences in cognition and crime: early developmental, biological, and sociological correlates. PhD diss., University of Pennsylvania.

———. 1984. Neuropsychological and early environmental correlates of sex differences in crime. *International Journal of Neuroscience* 23:199–214.

———. 1985. Sociological and human developmental explanations of crime: Conflict or consensus? *Criminology* 23:711–41.

———. 1990. *Biology and Violence.* Cambridge, UK: Cambridge University Press.

Drillien, C. M., A. J. M. Thompson, and K. Burgoyne. 1980. Low birthweight children at early school age: A longitudinal study. *Developmental Medicine and Child Neurology* 22:26–47.

Dunford Franklyn, and Delbert Elliott. 1984. Identifying career offenders with self-reported data. *Journal of Research in Crime and Delinquency* 21:57–86.

D'Unger, Amy V., Kenneth C. Land, Patricia McCall, and Daniel S. Nagin. 1998. How many latent classes of delinquent/criminal careers? Results from mixed poisson regression analyses of the London, Philadelphia, and Racine cohort studies. *American Journal of Sociology* 103:1593–1630.

Elliott, Delbert, Suzanne Ageton, David Huizinga, Brian Knowles, and Rachelle Canter. 1984. *The Prevalence and Incidence of Delinquent Behavior: 1976-1980* (The National Youth Survey Report No. 26). Boulder, CO: Behavioral Research Institute.

Escalona, Sibylie K. 1982. Babies at double hazard: Early development of infants at biologic and social risk. *Pediatrics* 70:670–76.

Farrington, David P. 1986. Age and crime. In *Crime and Justice: An Annual Review of Research,* Vol. 7, ed. Michael Tonry and Norval Morris, 189–250. Chicago: University of Chicago Press.

Farrington, David P., B. Gallagher, Lynda Morley, Raymond Ledger, and Donald J. West. 1986. Unemployment, school leaving and crime. *British Journal of Criminology* 26:335–56.

Farrington. David P., and J. David Hawkins. 1991. Predicting participation, early onset, and later persistence in officially recorded offending. *Criminal Behavior and Mental Health* 1:1–33.

Farrington, David P., Rolf Loeber, Delbert S. Elliott, J. David Hawkins, Denise Kandel, Malcolm Klein, Joan McCord, David Rowe, and Richard Tremblay. 1990. Advancing knowledge about the onset of delinquency and crime. In *Advances in Clinical and Child Psychology,* Vol. 13, ed. B. Lahey and A. Kazdin. New York: Plenum.

Fitzhardinge, P., and E. Steven. 1972. The small-for-date infant: II. Neurological and intellectual sequelae. *Pediatrics* 50:50–57.

Gorenstein, E. 1990. Neuropsychology of juvenile delinquency. *Forensic Reports* 3:15–48.

Gottfredson, Michael, and Travis Hirschi. 1990. *A General Theory of Crime.* Stanford, CA: Stanford University Press.

Guerra, Nancy G., Beth Attar, and Roger P. Weissberg. 1997. Prevention of aggression and violence among inner-city youths. In *Handbook of Antisocial Behavior,* ed. D. M. Stoff, J. Breiling, and J. D. Maser. New York: John Wiley.

Hack, Maureen, H. Gerry Taylor, Nancy Klein, Robert Eiben, Christopher Schatschneider, and Nori Mercuri-Minich. 1994. School-age outcomes in children with birth weights under 750g. *New England Journal of Medicine* 331:757–59.

———. 1995. Long-term developmental outcomes of low birth weight infants. *The Future of Children: Low Birth Weight* (Center for the Future of Children) 5:176–96.

Hardy, Janet B., Joseph S. Drags, and Esther C. Jackson. 1979. *The First Year of Life.* Baltimore, Md.: Johns Hopkins University Press.

Henry, Bill, Avshalom Caspi, Terrie E. Moffitt, and Phil A. Silva. 1996. Temperamental and familial predictors of violent and nonviolent criminal convictions: Age 3 to age 18. *Developmental Psychology* 32:614–23.

Henry, Bill, Terrie E. Moffitt, Lee Robins, Felton Earls, and Phil A. Silva. 1993. Early familial predictors of child and adolescent antisocial behavior: Who are the mothers of delinquents? *Criminal Behavior and Mental Health* 3:97–118.

Hertzig, Margaret E. 1983. Temperamental and neurological status. In *Developmental Neuropsychiatry,* ed. Michael Rutter. New York: Guilford Press.

Hirschi, Travis, and Michael Gottfredson. 1983. Age and the explanation of crime. *American Journal of Sociology* 89:552–84.

Hughes, Dana, and Lisa Simpson. 1995. The role of social change in preventing low birth weight. *The Future of Children: Low Birth Weight* (Center for the Future of Children) 5:87–102.

Jaccard, James, Robert Turrisi, and Choi K. Wan. 1990. *Interaction Effects in Multiple Regression.* Newbury Park, CA: Sage.

Kandel, Elizabeth, Patricia A. Brennan, and Sarnoff A. Mednick. 1990. Minor physical anomalies and parental modeling of physical aggression predict adult violent offending. Unpublished manuscript, University of Southern California.

Kandel, Elizabeth, and Sarnoff A. Mednick. 1991. Perinatal complications predict violent offending. *Criminology* 29:519–30.

Kolvin, I., F. Miller, M. Fleeting, and P. Kolvin. 1988. Social and parenting factors affecting criminal-offense rates. *British Journal of Psychiatry* 152:80–90.

Land, Kenneth C., Patricia McCall, and Daniel S. Nagin. 1996. A comparison of Poisson, negative binomial, and semiparametric mixed poisson regression models with empirical applications to criminal careers data. *Sociological Methods and Research* 25:387–442.

Land, Kenneth C. and Daniel S. Nagin. 1996. Micromodels of criminal careers: A synthesis of the criminal careers and life course approaches via semiparametric mixed poisson regression models, with empirical applications. *Journal of Quantitative Criminology* 12:163–91.

LeBlanc, Marc, and Marcel Fréchette. 1989. *Male Criminal Activity From Childhood Through Youth.* New York: Springer-Verlag.

Lewis, Dorothy D., Shelley S. Shanock, and David A. Balla. 1979. Perinatal difficulties, head and face trauma, and child abuse in the medical histories of seriously delinquent children. *American Journal of Psychiatry* 136:419–23.

Loeber, Rolf, and Marc LeBlanc. 1990. Toward a developmental criminology. In *Crime and Justice: An Annual Review of Research,* Vol. 12, ed. Michael Tonry and Norval Morris. Chicago: University of Chicago Press.

Loeber, Rolf, Magda Stouthamer-Loeber, Welmoet Van Kammen, and David P. Farrington. 1991. Initiation, escalation, and desistance in juvenile offending and their correlates. *Journal of Criminal Law and Criminology* 82:36–82.

Marlow, N., B. L. Roberts, and R. W. I. Cooke. 1989. Motor skills in extremely low birthweight children at the age of 6 years. *Archives of Disease in Childhood* 64:839–47.

Mazerolle, Paul. 1997. Delinquent definitions and participation age: Assessing the invariance hypothesis. *Studies on Crime and Crime Prevention* 6:151–67.

McCormick, Marie C. 1985. The contribution of low birth weight to infant mortality and childhood morbidity. *New England Journal of Medicine* 312:82–90.

McCormick, Marie C., Steven L. Gortmaker, and Arthur M. Sobol. 1990. Very low birth weight children: Behavior problems and school difficulty in a national sample. *Journal of Pediatrics* 117:687–93.

McCormick, Marie C., Kathryn Workman-Daniels, and Jeanne Brooks-Gunn. 1996. The behavioral and emotional well-being of school-age children with different birthweights. *Pediatrics* 97:18–25.

McGauhey, Peggy J., Barbara Starfield, Cheryl Alexander, and Margaret E. Ensminger. 1991. Social environment and vulnerability of low birth weight children: A social-epidemiological perspective. *Pediatrics* 88:943–55.

Moffitt, Terrie E. 1990. The neuropsychology of delinquency: A critical review of theory and research. In *Crime and Justice: An Annual Review of Research,* Vol. 12, ed. Michael Tonry and Norval Morris, 99–169. Chicago: University of Chicago Press.

———. 1993. Adolescence-limited and life-course persistent antisocial behavior: A developmental taxonomy. *Psychological Review* 100:674–701.

———. 1994. Natural histories of delinquency. In *Cross-National Longitudinal Research on Human Development and Criminal Behavior,* ed. E. Weitekamp and H. Kerner. Amsterdam, The Netherlands: Kluwer Academic.

———. 1995. Personal communication. June 1995.

———. 1997. Neuropsychology, antisocial behavior, and neighborhood context. In *Violence and Childhood in the Inner City,* ed. Joan McCord. New York: Cambridge University Press.

Moffitt, Terrie E., Avshalom Caspi, Nigel Dickson, Phil A. Silva, and Warren Stanton. 1996. Childhood-onset versus adolescent-onset antisocial conduct problems in males: Natural history from ages 3 to 18 years. *Development and Psychopathology* 8:399–424.

Moffitt Terrie E., Donald Lynam, and Phil A. Silva. 1994. Neuropsychological tests predicting persistent male delinquency. *Criminology* 32:277–300.

Nagin, Daniel S., and David P. Farrington. 1992a The onset and persistence of offending. *Criminology* 30:501–23.

———. 1992b. The stability of criminal potential from childhood to adulthood. *Criminology* 30:235–60.

Nagin, Daniel S., David P. Farrington, and Terrie E. Moffitt. 1995. Life-course trajectories of different types of offenders. *Criminology* 33:111–39.

Nagin, Daniel S., and Kenneth C. Land. 1993. Age, criminal careers, and population heterogeneity: Specification and estimation of a nonparametric, mixed poisson model. *Criminology* 31:327–62.

Nagin, Daniel S., and Raymond Paternoster. 1991. On the relationship of past to future delinquency. *Criminology* 29:163–189.

National Research Council. 1993. *Losing Generations: Adolescents in High-Risk Settings.* Washington, DC: National Research Council.

Niswander, Kenneth R., and Myron Gordon. 1972. *The Women and Their Pregnancies.* Washington, DC: U.S. Department of Health, Education, and Welfare.

Paneth, Nigel S. 1995. The problem of low birth weight. *The Future of Children: Low Birth Weight* (Center for the Future of Children) 5:19–34.

Pasamanick, Benjamin, Martha E. Rogers, and Abraham M. Lilenfeld. 1956. Pregnancy experience and the development of behavior disorder in children. *American Journal of Psychiatry* 112:613–18.

Paternoster, Raymond, Charles W. Dean, Alex R. Piquero, Paul Mazerolle, and Rober Brame. 1997. Continuity and change in offending careers. *Journal of Quantitative Criminology* 13:231–66.

Patterson, Gerald, L. Crosby, and S. Vuchinich. 1992. Predicting risk for early police arrest. *Journal of Quantitative Criminology* 8:335–55.

Patterson, Gerald, and Karen Yoerger. 1993. A model for early onset of delinquent behavior. In *Crime and Mental Disorder,* ed. S. Hodgins. Newbury Park, CA: Sage.

Penchaszadeh, Victor B., Janet B. Hardy, E. David Mellits, Bernice Cohen, and Victor A. McKusick. 1972. Growth and development in an "inner city" population: An assessment of possible biological and environmental influences: I. Intra-uterine Growth. *Johns Hopkins Medical Journal* 131:384–97.

Piquero, Alex R., and Stephen G. Tibbetts. 1999. The impact of pre/perinatal disturbances and disadvantaged familial environments in predicting criminal offending. *Studies on Crime and Crime Prevention* 8(1):52–70.

Raine, Adrian, Patricia A. Brennan, David P. Farrington, and Sarnoff A. Mednick. 1997a. *Biosocial Bases of Violence.* New York: Plenum Press.

Raine, Adrian, Patricia A. Brennan, and Sarnoff A. Mednick. 1994. Birth complications combined with early maternal rejection at age 1 year predispose to violent crime at age 18 years. *Archives of General Psychiatry* 51:984–88.

———. 1997b. Interaction between birth complications and early maternal rejection in predisposing individuals to adult violence: Specificity to serious, early-onset violence. *American Journal of Psychiatry* 154:1265–71.

Reinish, Juan M., Ronald Gandelman, and Frances S. Spiegel. 1979. Prenatal influences on cognitive abilities: Data from experimental animals and human genetic and endocrine syndromes. In *Sex-Related Differences in Cognitive Functioning: Developmental Issues,* ed. Michelle Wittig and Anne Petersen. New York: Academic.

Reiss, Albert J., Jr., and Jeffrey Roth. 1993. *Understanding and Preventing Violence.* Washington, DC: National Academy Press.

Robins, Lee. 1966. *Deviant Children Grown Up.* Baltimore, MD: Williams and Wilkins.

———. 1978. Sturdy childhood predictors of adult antisocial behavior: Replications from longitudinal studies. *Psychological Medicine* 8:611–22.

Ross, Gail, Evelyn G. Lipper, and Peter A. Auld. 1990. Social competence and behavior problems in premature children at school age. *Pediatrics* 86:391–97.

Sampson, Robert J., and John H. Laub. 1993. *Crime in the Making: Pathways and Turning Points Through Life.* Cambridge, MA: Harvard University Press.

The Scottish Low Birthweight Study Group. 1992. The Scottish Low Birthweight Study: II. Language and attainment, cognitive status, and behavioral problems. *Archives of Disease in Childhood* 67:682–86.

Shannon, Lyle. 1980. Assessing the reliability of adult criminal careers to juvenile careers. In *Problems in American Social Policy,* ed. C. Abt. Cambridge: Abt Books.

Shanok, Shelley S., and Dorothy O. Lewis. 1981. Medical histories of female delinquents. *Archives of General Psychiatry* 38:211–13.

Shiono, Patricia H., and Richard E. Behrman. Low birth weight: Analysis and recommendations. *The Future of Children: Low Birth Weight* (Center for the Future of Children) 5:14–18.

Shiono, Patricia H., Mark A. Klebanoff, Barry I. Graubard. Heinz W. Bersendes, and George G. Rhoads. 1986. Birth weight among women of different ethnic groups. *Journal of the American Medical Association* 255:48–52.

Simons, Ronald L., Chyi-In Wu, Rand D. Conger, and Frederick D. Lorenz. 1994. Two routes to delinquency: Differences between early and late starters in the impact of parenting and deviant peers. *Criminology* 32:247–76.

Singer, Judith E., Milton Westphal, and Kenneth R. Niswander. 1968. Sex differences in the incidence of neonatal abnormalities and abnormal predictors in early childhood. *Child Development* 39:103–12.

Smith, Douglas A. 1984. The organizational context of legal control. *Criminology* 22:19–38.

Smith, Douglas A., and G. Roger Jarjoura. 1988. Social structure and criminal victimization. *Journal of Research in Crime and Delinquency* 25:27–52.

Stanton, Warren, Rob McGee, and Phil A. Silva. 1991. Indices of perinatal complications, family background, child rearing, and health as predictors of early cognitive and motor development. *Pediatrics* 88:954–59.

Szatmari, Peter, Marge Reitsma-Street, and David R. Offord. 1986. Pregnancy and birth complications in antisocial adolescents and their siblings. *Canadian Journal of Psychiatry* 31:513–16.

Szatmari, Peter, Saroj Saigal, Pete Rosenbaum, Dugal Campbell, and Susanne King. 1990. Psychiatric disorders at five years among children with birthweights < 1000g: A regional perspective. *Developmental Medicine and Child Neurology* 32:954–62.

Tanner, J. M. 1978. *Fetus into Man: Physical Growth from Conception to Maturity.* Cambridge, MA: Harvard University Press.

Taylor, E. 1976. *Beck's Obstetrical Practice and Fetal Medicine.* 10th ed. Baltimore, MD: Williams and Wilkins.

Thomas, Alexander, and Stella Chess. 1977. *Temperament and Development.* New York: Brunner/ Mazel.

Tinsley, Barbara R., and Ross D. Parke. 1983. The person-environment relationship: Lessons from families with preterm infants. In *Human Development: An Interactional Perspective,* ed. David

Magnuson and Vernon R. Allen. San Diego, CA: Academic Press.

Tolan, Patrick. 1987. Implications of age of onset for delinquency risk. *Journal of Abnormal Child Psychology* 15:47–65.

Werner, Emmy E. 1987. Vulnerability and resiliency in children at risk for delinquency: A longitudinal study from birth to young adulthood. In *Prevention of Delinquent Behavior*, Vol. 10, ed. John D. Burchard and Sara N. Burchard. Newbury Park, CA: Sage.

Werner, Emmy E., and Ruth S. Smith. 1977. *Kauai's Children Come of Age.* Honolulu: The University of Hawaii Press.

———. 1992. *Overcoming the Odds: High Risk Children from Birth to Adulthood.* London: Cornell University Press.

Whitaker, Agnes H., Ronon Van Rossem, Judith F. Feldman, Irvin Sam Schonfeld, Jennifer A. Pinto-Martin, Carolyn Torre, David Shaffer, and Nigel Penth. 1997. Psychiatric outcomes in low-birth-weight children at age 6 years: Relation to neonatal cranial ultrasound abnormalities. *Archives of General Psychiatry* 54:847–56.

Wilson, James Q., and Richard Herrnstein. 1985. *Crime and Human Nature.* New York: Simon and Schuster.

Wolfgang, Marvin. 1983. Delinquency in two birth cohorts. In *Prospective Studies of Crime and Delinquency*, ed. K. Van Dusen and Sarnoff A. Mednick. Boston: KIuwer-Nijhoff.

Wolfgang, Marvin, Terence Thornberry, and Robert Figlio. 1987. *From Boy to Man, From Delinquency to Crime.* Chicago: University of Chicago Press.

World Health Organization, Expert Committee on Maternal and Child Health. 1950. *Public Health Aspect of Low Birthweight* (WHO Technical Report Series, No. 27). Geneva, Switzerland: WHO.

REVIEW QUESTIONS

1. Which factor in this study represents Moffitt's high-risk factor for neuropsychological problems? Do you agree with the use of this measure as an indicator of this risk?

2. For which groups does this study seem to indicate Moffitt's theory is most applicable?

3. Knowing the results of this study, what would you recommend in terms of reducing chronic offenders in our society?

❖

READING

In this selection, Alex Piquero, David Farrington, and Alfred Blumstein provide a comprehensive review of the concepts, issues, and propositions presented by the criminal career framework, on which developmental theories are based. The developmental paradigm is, by definition, based on following the criminal activity of individuals over time and exploring the reasons why they offend when they do, as well as why these individuals don't offend at certain times. The authors explore the history of the criminal career perspective, as well as the key concepts of participation, offending frequency, duration, and co-offending patterns. Piquero et al. also explore the policy implications that can be gathered, given the current state of research on these various dimensions.

SOURCE: Adapted from Alex R. Piquero, David P. Farrington, and Alfred Blumstein, "Criminal Career Paradigm: Background, Recent Developments, and the Way Forward," *International Annals of Criminology* 41 (2003): 243–69. Copyright © 2003 Chicago University Press. Reprinted by permission.

Then the authors explore the issues related to chronic offenders, who are the small percentage of society (approximately 5% to 8% of the population) who commit the vast majority of violent and serious offenses. Piquero et al. also provide policy recommendations, especially regarding implications for incarceration policies. Finally, the authors review the extant research on career length and desistence (or the ceasing of offending in an individual's life), as well as prescribing issues that future research should examine.

While reading this selection, readers should consider the various concepts, such as frequency and duration, regarding their own offending careers, or the offending careers of people they know well. Furthermore, readers should examine the incarceration policies suggested by Piquero et al., and to consider whether they agree or disagree with their suggestions.

Criminal Career Paradigm

Background, Recent Developments, and the Way Forward

Alex R. Piquero, David P. Farrington, and Alfred Blumstein

Researchers have long been interested in the patterning of criminal activity throughout the course of criminal careers. Early on, Quetelet (1831) recognized that the age was closely related to the propensity for crime. Using data on crimes committed against persons and property in France from 1826 to 1829, Quetelet found, that crimes peaked in the late teens through the mid twenties. Since Quetelet's findings, a number of researchers have pursued the relationship between age and crime, across cultures and historical periods, and for a number of different crime types (Hirschi and Gottfredson 1983). Research on the relationship between age and crime has been one of the most studied issues within criminology (Farrington, 1986; Steffensmeier et al. 1989; Tittle and Grasmick 1993).

The relationship between age and crime raises the question of the degree to which the aggregate pattern displayed in the age/crime curve is similar to—or different from—the pattern of individual careers and whether conclusions about individuals can be validly drawn from aggregate data. For example, how

far does the observed peak of the aggregate age/crime curve reflect changes within individuals as opposed to changes in the composition of offenders? In other words, is the peak in the age/crime curve a function of active offenders committing more crime, or is it a function of more individuals actively offending during those peak years?

Within individuals, to what extent is the slowing past the peak age a function of deceleration in continued criminal activity or stopping by some of the individuals? Across individuals, how much of the age/crime curve can be attributed to the arrival/initiation and departure/termination of different individuals? How about the role of co-offending? How much of the continuation of offending by lone/solo offenders is attributable to identifying theirs as the key criminal careers of long duration, with their co-offenders serving merely as transients with shorter careers? How much of the age/crime curve for any particular crime type is a consequence of individuals persisting in offending, but switching from less serious crime types early in the career to more serious crime types as they get older? What about the relationship between past and future offending? Is it due to some causal factors or changes in causal factors (state dependence), unobserved individual differences (persistent heterogeneity), or some combination?

These questions are central to theory, as well as policy, especially those policies that are geared toward incapacitative effects of criminal sanctions, as well as to changes in the criminal career (e.g., rehabilitation or criminalization patterns as a result of actions by the criminal justice system). For example, if crime commission and arrest rates differ significantly among offenders and over the career, the effect of sentence length on overall crime will depend on who is incarcerated, and for how long (Petersilia 1980:325). Addressing these and related issues requires knowledge about individual criminal careers, their initiation, their termination and the dynamic changes between these end points (Blumstein et al. 1986).

In 1983, a Panel on Research on Criminal Careers was convened by the National Academy of Science at the request of the U. S. National Institute of Justice and was charged with evaluating the feasibility of predicting the future course of criminal careers, assessing the effects of prediction instruments in reducing crime through incapacitation, and reviewing the contribution of research on criminal careers to the development of fundamental knowledge about crime and criminals. This report outlined a novel approach of asking questions regarding the longitudinal patterning of criminal activity over the life course, i.e., the criminal career paradigm (Blumstein et al. 1986).

Since publication of the report, numerous theoretical, empirical, and policy issues have surfaced regarding the longitudinal patterning of criminal careers. One concerned the relevance (or lack thereof) of criminal career research for criminology generally, and public policy in particular. Gottfredson and Hirschi (1990) levied a series of critiques against the criminal career approach in which they claimed that attempts to identify career criminals and other types of offenders were doomed to failure. Perhaps the most important issue they raised concerns causality. Although the criminal career paradigm necessitates a longitudinal focus in order to study both the between- and within-individual patterning of criminal activity, Gottfredson and Hirschi questioned whether longitudinal research designs could actually resolve questions of causal order. They also argued that, since correlations with offending were relatively stable over the life-course, cross-sectional designs were suitable for studying the causes of crime.

This paper summarizes background and recent developments regarding the criminal career paradigm. Section I provides a brief review of the criminal career paradigm, as well as an overview of the empirical findings generated by criminal careers research, with a concentration on the dimensions of criminal careers. Section II presents a discussion of selected policy implications including the identification of career criminals and policies associated with sentence duration. Section III offers an agenda for future theoretical, empirical, and methodological research. The full conclusion of our report may be found in the essay "The Criminal Career Paradigm," published in *Crime and Justice: A Review of Research, Volume 30* (Piquero; Farrington, and Blumstein 2003).

⊠ I. The Criminal Career Paradigm

At its most basic level, a criminal career is the "characterization of the longitudinal sequence of crimes committed by an individual offender" (Blumstein et al. 1986:12). This definition helps to focus researchers' attention on entry into a career when or before the first crime is committed and dropout from the career when or after the last crime is committed. The criminal career paradigm recognizes that individuals start their criminal activity at some age, engage in crime at some individual crime rate, commit a mixture of crimes, and eventually stop. Hence, the criminal career approach emphasizes the need to investigate issues related to why and when people start offending (onset), why and how they continue offending (persistence), why and if offending becomes more frequent or serious (escalation) or specialized, and why and when people stop offending (desistance). The study of criminal careers does not imply that offenders necessarily derive their livelihood exclusively or even predominantly from crime; instead, the concept is intended only as a means of structuring the longitudinal sequence of criminal events associated with an individual in a systematic way (Blumstein et al. 1982:5). In sum, the criminal career approach focuses on both between—and within—individual changes in criminal activity over time.

A. Dimensions of a Criminal Career

1. Participation

The criminal career approach partitions the aggregate crime rate into two primary components: "participation," the distinction between those who commit crime and those who do not; and "frequency," the rate of offending among active offenders, commonly denoted by the Greek letter λ (Blumstein et al. 1986:12). Participation is measured by the fraction of a population ever committing at least one crime before some age or currently active during some particular observation period. In any period, active offenders include both new offenders whose first offense occurs during the observation period, and persisting offenders who began criminal activity in an earlier period and continue to be active during the observation period. Importantly, the longer the average duration of offending, the greater the contribution of persisters to measured participation in successive observation periods.

Estimates of ever-participation in criminal activity vary across reporting method (they tend to be much higher with self-report than with official records which are a filtered subset of self-reports), the crimes in which participation is being measured (there is more participation in less serious criminal activity), the level of threshold of involvement (police contact, arrest, conviction), and the characteristics and representativeness of the sample (high school students, college students, general population, offender-based, etc.). In general, ever-participation estimates are fairly common across data sets and consistent with most criminological findings.

There is a relatively high rate of participation among males in criminal activity (Elliott et al. 1987:502). Blumstein et al. (1986) reported that about 15 percent of urban males are arrested for an index offense by age eighteen, and about 25 to 45 percent of urban males are arrested for a non-traffic offense by age eighteen. Visher and Roth's (1986) overview of several longitudinal studies employing police and court records indicates a lifetime prevalence estimate of 40 to 50 percent, with slightly higher rates for blacks and much lower rates among females. Stattin et al.'s (1989) longitudinal study of Swedish males and females revealed that by age 30, 37.7% of Swedish males and 9% of Swedish females were registered for a criminal offense. The cumulative prevalence of self-reported offenses is even more striking. For example, in the Cambridge study, Farrington (2003) found that 96 percent of the males had reported committing at least one of ten specified offenses (including burglary, theft, assault, vandalism, and drug abuse) up to age thirty-two. Kelley et al. (1997) used self-reported data on serious violence from three longitudinal studies funded by the Office of Juvenile Justice and Delinquency Prevention, the Causes and Correlates studies, and found that 39 percent of Denver males, 41 percent of Pittsburgh males, 40 percent of Rochester males, 16 percent of Denver females, and 32 percent of Rochester females reported committing at least one serious violent act by age sixteen.

Regardless of whether official or self-report records are used to study prevalence, three main conclusions emerge. First, male participation rates are typically higher than those for females, and especially so for the more serious offenses. Second, black participation rates are typically higher than those for whites, especially when participation is examined via official records as opposed to self-reports (Hindelang et al. 1981). In self-reports, blacks have also been found to report continuing their violent offending at higher rates than whites (Elliott 1994). Third, there is a strong relationship between age and participation. In particular, the probability of initiating a criminal career at a given age is highest between thirteen and eighteen, on the lower end for self-report estimates and on the higher end for arrest and conviction records, with little to no gender difference (Moffitt et al. 2001). Also, evidence on the probability of committing an offense at a given age is mixed, with some research indicating a consistent increase through the mid-teens to a peak at age nineteen and then subsequent decline, while other research indicates a decline in self-reported participation through the teens (Elliott et al. 1983; Thornberry 1989; Lauritsen 1998). Studying demographic differences in prevalence remains controversial. For example, Hindelang et al. (1981) argued that there is a race difference in the validity of self-reported delinquency measures, which leads to a serious underestimation of black males' prevalence rates.

2. Key Dimensions of Active Criminal Careers

The criminal career paradigm encompasses several dimensions of active criminal careers including offending frequency, duration, crime type mix and seriousness, and co-offending patterns.

a. Offending Frequency. The offending rate for individual offenders, λ, reflects the frequency of offending by individuals who are actively engaged in crime (Blumstein et al. 1986:55).

Much criminal career research has been concerned with estimating the individual offending frequency of active offenders during their criminal careers (Blumstein and Cohen 1979; Cohen 1986; Loeber and Snyder 1990).

Blumstein et al. (1986) summarized variation in λ for active offenders by gender, age, and race. Regarding gender, they found little variation in frequency across males and females (i.e., the ratios are generally 2:1 or less) for most crimes (Blumstein at al. 1986:67–68). Thus, if active in a crime type, females commit crimes at rates similar to those of males (for an exception see Wikström 1985). Regarding age, Blumstein et al. reported little change with age in offense-specific frequency rates for active offenders, but when all offense types are combined, there tended to be an increase during the juvenile years and a decrease during the adult years. In the Rand Inmate surveys, there appeared to be some evidence of general stability of λ over age (Chaiken and Chaiken 1982). The number of active crime types declined with age in the Rand survey, but crime-specific frequencies tended to be stable (Peterson and Braiker 1980). Finally, although research based on official records tends to indicate that there is not a strong relationship between offending frequency and demographic characteristics, some recent self-report data on serious violence tends to indicate otherwise (Elliott 1994).

Spelman (1994) summarized current knowledge on offending frequencies. First, there are different values for the average offense frequencies across studies because researchers provide different definitions and operationalizations of the offense rate. Second, most of the variation in offense rates can be attributed to differences in the populations sampled and especially where in the criminal justice system they are sampled. Third, the average offender commits around eight crimes per year, while offenders who are incarcerated at some point in their lives commit thirty to fifty crimes per year, and the average member of an incoming prison cohort commits between sixty and 100 crimes per year. Fourth, criminals do not commit crimes all the time; in other words, there is evidence that many offenders spend long periods of time in which they commit no crimes. Fifth, the distribution of offending frequencies is highly skewed, with a few offenders committing crimes at much higher than average rates.

b. Duration, the Interval between Initiation and Termination. One aspect of the criminal career paradigm that has received a great deal of research attention is initiation, or the onset of antisocial and criminal activity (Farrington et al. 1990). Several studies have reported higher recidivism rates among offenders with records of early criminal activity as juveniles (Blumstein et al. 1986). Although many researchers argue that individuals who begin offending early will desist later, and thus have lengthy careers (Hamparian et al. 1978; Krohn et al. 2001), there has been much less research on the duration of criminal careers, or an individual's criminal career (Piquero, Brame, and Lynam 2003). It is more tenable, however, to measure a rate of desistance for an identified group of offenders (Bushway et al. 2001). Research on desistance, or the termination of a criminal career, has received even less attention because of difficulties in measurement and operationalization (Laub and Sampson 2001).

The two most common approaches for studying career termination have been through providing estimates of termination probabilities after each arrest, and estimating the time between the first and last crimes committed. Regarding termination probabilities, Blumstein et al. (1986) calculated persistence probabilities for six different data sets and found that after each subsequent event (i.e., police contact, arrest, conviction, etc.), the persistence probability increases, reaching a plateau of .7 to .9 by the fourth event across all data sets. Farrington, Lambert, and West (1998) used conviction data to calculate recidivism probabilities for

Cambridge study males through age thirty-two and found that after the third offense, the recidivism probability ranged from .79 to .91 through the tenth offense.

A number of studies have attempted to derive estimates of career duration, typically measured as career length in years. Three major studies conducted in the 1970s estimated career lengths to be between five and fifteen years (Greenberg 1975; Shinnar and Shinnar 1975; Greene 1977). In 1982, Blumstein, Cohen, and Hsieh conducted the most detailed study of criminal career duration and used data on arrests rather than on arrestees to estimate career lengths, and concluded that criminal careers are relatively short, averaging about five years for offenders who are active in index offenses as young adults. Residual careers, or the length of time still remaining in careers, increase to an expected ten years for index offenders still active in their thirties. Persistent offenders who begin their adult careers at age eighteen or earlier and who are still active in their thirties are most likely to be persistent offenders and are likely to continue to commit crimes for about another ten years (Visher 2000).

Spelman (1994) studied career lengths with data from the three-state Rand Inmate Survey, and developed estimates of total career lengths of about six or seven years (Spelman 1994). Spelman showed that young and inexperienced offenders, those in the first five years of their career, were more likely than older offenders to drop out each year, but after five years the rate of dropout leveled off, rising only after the twentieth year as an active offender. Farrington (2003) examined the duration of criminal careers in the Cambridge study using conviction data to age forty and found that the average duration of criminal careers was 7.1 years. Excluding one-time offenders whose duration was zero, the average duration of criminal careers was 10.4 years. Piquero, Brame, and Lynam (2003) studied the length of criminal careers using data from a sample of serious offenders paroled from California Youth Authority institutions in the 1970s and found that the average career length was 17.27 years, with little difference between white (16.7 years) and non-white parolees (17.7 years).

c. Crime Type Mix and Seriousness. The mix of different offense types among active offenders is another important criminal career dimension. The study of crime-type mix involves studying seriousness (the tendency to commit serious crimes throughout one's criminal career), escalations (the tendency to move toward more serious crimes as one's career progresses), specialization (the tendency to repeat the same offense type on successive crimes), and crime-type switching (the tendency to switch types of crimes and/or crime categories on successive crimes).

Diverse methodological techniques have been employed to investigate specialization, or the tendency to repeat the same offense type on successive crimes. Using official records, some research provides evidence in favor of some small degree of specialization (Bursik 1980; Rojek and Erickson 1982; Smith and Smith 1984), but most find that generality is the norm throughout offending careers (Farrington et al. 1988; Wolfgang et al. 1972; Nevares et al. 1990; Tracy et al. 1990). At the same time, important differences in specialization are observed between adults and juveniles such that specialization appears to be stronger in magnitude for adult rather than for juvenile offenders (Le Blanc and Fréchette 1989; Piquero et al. 1999). On the other hand, self-report data from the Rand studies suggest that, although there is some evidence of property specialization (Spelman 1994), incarcerated offenders tend to report much more generality than speciality (Petersilia et al. 1978; Peterson and Braiker 1980; Chaiken and Chaiken 1982).

Some scholars have investigated specialization in violence. Using official records, Farrington (1989) and Piquero (2000) reported

little evidence of specialization in violence in the Cambridge study or the Philadelphia perinatal cohort, and that the commission of a violent offense in a criminal career is a function of offending frequency: frequent offenders are more likely to accumulate a violent offense in their career. Similar results have been obtained by Capaldi and Patterson (1996) with self-report data from the Oregon Youth Study.

Directly related to the specialization issue is the switching that occurs across clusters of crime types. Clusters represent natural groupings of offense types (violence, property, other), and research indicates that adult offenders display a stronger tendency to switch among offense types within a cluster and a weaker tendency to switch to offense types outside a cluster, but the strong partitioning is not as sharp among juveniles (Blumstein et al. 1986; Cohen 1986). Adult offenders and incarcerated juveniles are more likely to commit offenses within a cluster than to switch to offenses outside a cluster (Rojek and Erickson 1982; Blumstein et al. 1988). Drug offenders, however, do not tend to switch to either violent or property offenses.

d. Co-offending Patterns. Another important criminal career feature is whether a person commits on offense alone or with others (Reiss 1986). Little empirical work has been completed on co-offending, and even less information exists regarding the group criminal behavior of youths in transition to adult status or of adult offenders at different ages. In the Cambridge Study, Reiss and Farrington (1991) report that the incidence of co-offending is greatest for burglary and robbery, and that juvenile offenders primarily commit their crimes with others, whereas adult offenders primarily commit their crimes alone. Although the decline in co-offending may, at first glance, be attributed to co-offenders dropping out, it seems to occur because males change from co-offending in their teenage years to lone offending in their twenties. In the Swedish

Borlänge study, Sarnecki (1990) found that 45 percent of all youths suspected of offense at some stage during the six-year study period could be linked together in a single large network that accounted for most offenses. Recently, Sarnecki (2001) used data from all individuals aged twenty or less who were suspected of one or more offenses in Stockholm during 1991–1995 to study the extent and role of co-offending and uncovered that 60 percent of the individuals had a co-offender at some point. Interestingly, he also found that males tended to co-offend primarily with other males, but among females, the proportion of girls choosing other females was lower than the proportion of boys choosing other males as co-offenders. Conway and McCord (2002) conducted the first co-offending study designed to track patterns of violent criminal behavior over an eighteen-year period (1976–1994) among a random sample of 400 urban offenders and their accomplices in Philadelphia. Using crime data collected from court records and "rap sheets," they found that nonviolent offenders who committed their first co-offense with a violent accomplice were at increased risk for subsequent serious violent crime, independent of the effects of age and gender.

B. Policy Issues

The criminal career paradigm suggests three general orientations for crime control strategies: prevention, career modification, and incapacitation. Knowledge concerning the patterning of criminal careers is intimately related to these policy issues. Prevention strategies, including general deterrence, are intended to reduce the number of nonoffenders who become offenders. Career modification strategies, including individual deterrence and rehabilitation, are focused on persons already known to be criminals and seek to reduce the frequency or seriousness of their crimes. In addition, these strategies encourage the termination of ongoing criminal careers

through mechanisms such as job training and drug treatment. Incapacitative strategies focus on the crimes reduced as a result of removing offenders from society during their criminal careers. Two types of incapacitation are general, or collective and selective which focuses on the highest frequency offenders. These three crime control strategies are intimately related to specific laws, including habitual offender statutes, truth-in-sentencing laws, three-strikes laws, and mandatory minimum sentences laws.

1. Crime Control Strategies

Of all crime control strategies, the criminal career paradigm has focused extensive attention on incapacitation. General or collective incapacitation strategies aim to reduce criminal activity as a consequence of increasing the total level of incarceration while selective incapacitation policies focus primarily on offenders who represent the greatest risk of future offending. The former approach is consistent with the equal treatment concerns of a just-deserts sentencing policy while the latter focuses as much on the offenders as the offense. Importantly, the degree to which selective incapacitation policies are effective depends on the ability to distinguish high- and low-risk offenders, and to identify them early enough before they are about to terminate criminal activity. Three related issues arise: the ability to classify individual offenders in terms of their projected criminal activity; the quality of the classification rules; and the legitimacy of basing punishment of an individual on the possibility of future crimes rather than only on the crimes already committed (and the consequent level of disparity that is considered acceptable).

Regarding collective incapacitation, Blumstein et al. (1986) suggest that achieving a 10 percent reduction in crime may require more than doubling the existing inmate population. However, under selective incapacitation policies, long prison terms would be reserved primarily for offenders identified as most likely to continue committing serious crimes at high rates. Blumstein et al. conclude that selective incapacitation policies could achieve 5 to 10 percent reductions in robbery with 10 to 20 percent increases in the population of robbers in prison, while much larger increases in prison populations are required for collective incapacitation policies.

2. Relationship to Laws

Both collective and selective incapacitation policies are directly influenced by laws and policies that govern criminal justice decisions regarding the punishment of offenders. For example, habitual offender statutes give special options to prosecutors for dealing with repeat offenders. Truth-in-sentencing laws are intended to increase incapacitation by requiring offenders, particularly violent offenders, to serve a substantial portion of their prison sentence, and parole eligibility and good-time credits are restricted or eliminated. Three-strikes laws provide that any person convicted of three, typically violent, felony offenses, must serve a lengthy prison term, usually a minimum term of twenty-five-years-to-life. Mandatory-minimum sentence laws require a specified sentence and prohibit offenders convicted of certain crimes from being placed on probation, while other statutes prohibit certain offenders from being considered for parole. Mandatory-minimum sentence laws can also serve as sentencing enhancement measures, requiring that offenders spend additional time in prison if they commit particular crimes in a particular manner (e.g., committing a felony with a gun). The net effect of these laws is to increase prison populations by incarcerating certain kinds of offenders or increasing the sentence length of those offenders convicted for certain types of crimes.

C. "Chronic" Offenders

Criminologists have long recognized that a small group of individuals is responsible for a majority of criminal activity. Wolfgang et al. (1972) focused attention on the chronic offender by applying that label to the small group of 627 delinquents in the 1945 Philadelphia birth cohort who committed five or more offenses by age seventeen (hereafter five-plus). This group constituted just 6 percent of the full cohort of 9,945 males and 18 percent of the delinquent subset of 3,475, but was responsible for 5,305 offenses, or 52 percent of all delinquency in the cohort through age seventeen. The chronic offenders were responsible for an even larger percentage of the more serious, violent offenses. The finding that a small subset of sample members are responsible for a majority of criminal activity is supported by data from other longitudinal data sets, including the second 1958 Philadelphia birth cohort (Tracy et al., 1990), the Puerto Rico Birth Cohort Study (Nevares et al., 1990), the Dunedin New Zealand Multidisciplinary Health Study (Moffitt et al. 2001), the Philadelphia (Piquero 2000) and Providence (Piquero and Buka 2002) perinatal projects, the Racine, WI birth cohorts (Shannon 1982), the Cambridge study (Farrington 2003), and also by cohort studies in Sweden (Wikström 1985), Finland (Pulkkinen 1988), and Denmark (Guttridge et al. 1983). The finding is also replicated across gender and race (Moffitt et al. 2001; Piquero and Buka 2002), and emerges from both official and self-report data. Research also indicates that chronic offenders exhibit an early onset, a longer career duration, and involvement in serious offenses—including person/violent-oriented offenses—than other offenders (Loeber and Farrington 1998).

The five-plus cutoff advanced by Wolfgang et al. has been employed in several studies; however, since theoretical and empirical definitions of chronicity have yet to be established, questions have been raised about the extent to which similar definitions of chronicity should be used across gender (Farrington and Loeber 1998; Piquero 2000), as well as the relatively arbitrary designation of five-plus offenses as characteristic of chronicity (Blumstein et al. 1985). Blumstein et al. (1985) raised other concerns with the use of five-plus as the chronicity cut point. They argued that the chronic offender calculation, which was based on the full cohort, overestimates the chronic offender effect because many cohort members will never be arrested. Instead, they urge that the ever-arrested subjects should be the base used to calculate the chronic offender effect. With the base, the 627 chronics with five-plus arrests represented 18 percent of those arrested, as opposed to 6 percent of the cohort. Blumstein and colleagues also argued that the proportion of chronic offenders observed by Wolfgang et al. could have resulted from a homogenous population of persisters. Blumstein et al. (1985) tested the hypothesis that all persisters (those with more than three arrests) could be viewed as having the same rearrest probability. Such an assumption could not be rejected. Although those with five-plus arrests accounted for the majority of arrests among the persisters, such a result could have occurred even if all subjects with three or more arrests had identical recidivism probabilities. Thus, the chronic offenders who were identified retrospectively as those with five or more arrests could not have been distinguished prospectively from nonchronics with three or four arrests.

▧ II. Policy Implications

Research on criminal careers has direct import for decision-making in the criminal justice system. In this section, we address four implications of criminal career research: the role of criminal career research in policy and individual decision-making, individual prediction of offending frequencies (λ), sentence duration, and research on career length and desistance and its relation to intelligent sentencing policy.

A. Role of Criminal Career Research in Policy and Individual Decision-Making

A principal example of the importance of criminal career research for criminal justice policy is criminal career length. Three-strikes and selective incapacitation philosophies assume that high-rate offenders will continue to offend at high rates and for long periods of time if they are not incarcerated. From an incapacitative perspective, incarceration is only effective in averting crimes when it is applied during an active criminal career. Thus, incarceration after the career ends or when a career is abating, is wasted for incapacitation purposes (Blumstein et al. 1982:70). By identifying career lengths, especially residual career lengths, policymakers can better target incarceration on offenders whose expected remaining careers are longest. Incarceration policies should be based on career duration distribution information. The more hardcore committed offenders with the longest remaining careers are identifiable only after an offender has remained active for several years (Blumstein et al. 1982). Earlier and later in criminal careers, sanctions will be applied to many offenders who are likely to drop out shortly anyway (Blumstein et al. 1982:71). The benefits derived from incapacitation will vary depending on an individual's crime rate and the length of his or her remaining criminal career. Continuing to incarcerate an offender after his/her career ends limits the usefulness of incarceration.

B. Individual Prediction of λ

Rand's second inmate survey highlighted the extreme skewness of the distribution of λ for a sample of serious criminals (Chaiken and Chaiken 1982; Visher 1986). Naturally, the identification of a small number of inmates who reported committing several hundred crimes per year led to the search for a method to identify these offenders in advance. If high-rate offenders cannot be identified prospectively, then crime control efforts will be hampered (Visher 1987). In this section, we highlight two related issues: the difficulty in identifying high-λ individuals, and the alleviation of the concern over prediction by "stochastic selectivity."

1. Difficulty in Identifying High- λ Individuals

Although high-λ individuals emerge in the aggregate, it has been difficult to identify specific individuals. Greenwood and Turner (1987) used data consisting of follow-up criminal history information on the California inmates who were included in the original Rand survey and who had been out of prison for two years to examine the extent to which Greenwood's seven-item prediction scale succeeded in predicting recidivism. The scale was not very effective in predicting post-release criminal activity when the recidivism measure was arrest. The majority of released inmates, regardless of whether they were predicted to be low- or high-rate offenders, were rearrested within two years. Greenwood and Turner also created a measure of the offender's annual arrest rate (i.e., the number of arrests per year of street time) for the follow-up sample and defined high-rate offenders as those inmates who had an actual arrest rate greater than 0.78. They found that the seven-item scale was less accurate in predicting annual arrest rates than it was in predicting re-incarceration.

There are also concerns related to the false positive prediction problem in identifying high-λ individuals. For example, Visher (1986:204-5) reanalyzed the Rand second inmate survey and found that not only were the estimates of λ for robbery and burglary sensitive to choices in computation (i.e., handling missing data, street time, etc.), but also that some inmates reported annual rates of 1,000 or more robberies or burglaries, thus strongly affecting the distribution of λ, and especially its mean. Finally, Visher's analysis of the Greenwood scale for identifying high-rate offenders indicated that 55 percent of the classified high-rate group (27 percent of the total sample) were false positives who did not commit crimes at high rates. In fact, the prediction scale worked better in identifying low-rate offenders. Recently, Auerhahn (1999) replicated Greenwood and Abrahams's (1982) selective incapacitation study with a representative sample of California state prison inmates, and found that the scale's overall predictive accuracy was 60 percent, indicating a great deal of error in identifying serious, high-rate offenders.

2. Concern and Need for Prediction Alleviated by "Stochastic Selectivity"

Many analyses of the crime control potential of increasing incarceration rely on a single estimate of mean λ derived from prison inmates and applying it indiscriminately to all related populations of offenders (Canela-Cacho et al. 1997). This assumes that all offenders engage in the same amount (λ) of criminal behavior—regardless of whether they are in prison or jail, or free in the community—and that the probability of their detection and incarceration is equal. However, measures of λ derived from arrestee/convictee populations display a strong selection bias because individuals who have gone through the criminal justice process are unlikely to be representative of the total offender population. This selection bias could be because samples of arrestees have a higher propensity for arrest or different offending frequencies. A highly heterogeneous distribution of offending frequency in the total population of offenders combines with relatively low imprisonment levels to lead to substantial selectivity of high-λ offenders among resident inmates and a correspondingly low mean value of λ among those offenders who remain free (Canela-Cacho et al. 1997). "Stochastic selectivity," then, draws into prison new inmates disproportionately from the high end of the λ distribution of free offenders. Further, the higher the incarceration probability following a crime, the deeper into the offender pool incarceration will reach, and the lower will be the incapacitation effect associated with the incoming cohorts (Canela-Cacho et al. 1997).

Using data from the second Rand inmate survey, Canela-Cacho et al. studied the issue of stochastic selectivity and found that the proportion of low-λ burglars and robbers among free offenders was much larger than among resident inmates, while at the high end of the offending frequency distribution there was a larger proportion of high-λ burglars and robbers among resident inmates than among free offenders. Thus, selectivity occurred naturally as high-λ offenders experienced greater opportunities for incarceration through the greater number of crimes they committed (Canela-Cacho el al. 1997:142), thereby obviating the need for efforts to explicitly identify individual high-λ offenders.

C. Sentence Duration

Information about crime rates and career lengths is particularly useful for incapacitation and incarceration decisions and policies, and such knowledge can also provide useful information regarding the intelligent use of incapacitation and may even provide powerful arguments against lengthy incapacitation policies. Principal among these is the decision regarding sentence length. Many current sentencing policies are based on the assumption that high-rate offenders will continue committing crimes at high rates and for lengthy periods, and thus prescribe lengthy incarceration stints. The extent to which this policy is effective however, is contingent on the duration of a criminal career.

Much debate regarding sentence length has centered on three-strikes policies. These policies severely limit judges' discretion because they prescribe a mandatory prison sentence of (typically) twenty-five-years-to-life. The incapacitation effectiveness of three-strikes laws, however, depends on the duration of criminal careers. To

the extent that sentencing decisions incarcerate individuals with short residual career lengths, a three-strikes law will waste incarceration resources (Stolzenberg and D'Alessio 1997:466).

Stolzenberg and D'Alessio (1997) used aggregate data drawn from the ten largest cities in California to examine the impact of California's three-strikes law on serious crime rates and found that the three-strikes law did not decrease serious crime or petty theft rates below the level expected on the basis of preexisting trends. Zimring et al. (1999) obtained a sample of felony arrests (and relevant criminal records) in Los Angeles, San Francisco, and San Diego, both before and after the California law went into effect to study the three-strikes issue. Two key findings emerged from their study. First, the mean age at arrest for two strikes and above was 34.6 years. This is particularly important because: "[O]n average the two or more strikes defendant has an almost 40 percent longer criminal adult career behind him (estimated at 16.6 years) than does the no-strikes felony defendant. All other things being equal, this means that the twenty-five-years-to-life mandatory prison sentence will prevent fewer crimes among the third-strike group than it would in the general population of felons because the group eligible for it is somewhat older" (Zimring et al. 1999:34). Second, when comparing crime trends in the three cities before and after the law, Zimring et al. found that there was no decline in the crimes committed by those targeted by the new law. In particular, the lower crime rates in 1994 and 1995 (just immediately after the three-strikes law went into effect) were evenly spread among targeted and non-targeted populations, suggesting that the decline in crime observed after the law went into effect was not due to the law.

D. Research on Career Length and Desistance

Sentencing practices involving lengthy sentence durations assume that affected offenders will continue to commit crime at a high rate and for a long period. To the extent that this is the case, incapacitation policies will avert crimes and thwart continued careers. However, to the extent that offenders retire before the expiration of a lengthy sentence, shorter career durations will reduce the effects of lengthy sentences. Using data from Florida, Schmertmann et al. (1998) concluded that the aging of prison populations under three-strikes policies in that state will undermine their long-run effectiveness. In particular, they noted that the policies will cause increases in prison populations due to the addition of large numbers of older inmates who are unlikely to commit future offenses.

The key to the sentence duration issue, and why estimates of criminal career duration are so important, rests on the characteristics of the person years—not the people—that are removed from free society as a result of such policies (Schmertmann et al. 1998:458). Such policies will be effective only to the extent that they incarcerate offenders during the early stages of their criminal careers when they are committing crimes at a high-rate.

Unfortunately, research on career duration and desistance is in its infancy. Knowledge on this subject will be important for furthering criminal justice policy and the cost-effective use of criminal justice resources.

III. Directions for Future Criminal Career Research

Evidence on criminal career issues cuts to the heart of theory and policy. On the theoretical side, knowledge on the correlates of criminal career dimensions is relevant to the necessity for general versus typological models. If research indicates that the correlates of one offending dimension are similar to another offending dimension, then more general and non-dimension-specific theories are warranted. If the correlates of one offending dimension are different from another offending dimension, then the causal process(es) underlying these two particular dimensions are probably different and different explanations and theories are required.

Better knowledge on various criminal career dimensions would aid policy initiatives designed to prevent initial involvement, curtail current offending, and accelerate the desistance process. If research suggests that poor parental socialization is related to early initiation, then prevention efforts should include parent-training efforts. Similarly, if drug use is associated with continued involvement in delinquent and criminal behavior, then intervention efforts should include drug treatment. Finally, if some set of correlates is associated with desistance, then policy efforts may wish to provide for specific prevention and intervention efforts.

Knowledge on career length and residual career length could best inform criminal justice policies because it deals directly with sentencing and incapacitation policies that are now driven more by ideology than by empirical knowledge. For example, if residual criminal career lengths average around five years, criminal justice policies advocating multi-decade sentences waste scarce resources. Similarly, if offenders are incarcerated in late adulthood when their residual career lengths have diminished, incarceration space will be wasted, and health care costs will increase, thereby further straining scarce resources.

Empirical study of criminal careers requires data collection for large samples of individuals beginning early in life and continuing for a lengthy period into adulthood. Such data are needed if questions surrounding initiation, continuation, and desistance are to be adequately addressed, and this is especially the case among serious offenders for which little longitudinal data exists (Laub and Sampson 2001; Piquero et al. 2002). The use of such longitudinal data, of course, brings with it several potential problems that need to be considered including methodological issues such as street time (Piquero et al. 2001), mortality (Eggleston, Laub, and Sampson 2003) and sample attrition (Brame and Piquero 2003), as well as statistical issues that deal with various modeling strategies and assumptions (Bushway et al. 1999; Nagin 1999; Raudenbush 2001). Nevertheless, continued data collection and research are important to identify and study unaddressed and unresolved criminal career issues, and to update thirty-year-old estimates.

Information derived from criminal career research is important to advance fundamental knowledge about offending and to assist criminal justice decision-makers in dealing with offenders. Much more important criminal career research lies on the horizon nationally and internationally, and we look forward to seeing this research emerge.

◪ References

Auerhahn, Kathleen. 1999. "Selective Incapacitation and the Problem of Prediction." *Criminology* 37: 703–34.

Blumstein, Alfred, and Jacqueline Cohen. 1979. "Estimation of Individual Crime Rates from Arrest Records." *Journal of Criminal Law and Criminology* 70: 561–85.

Blumstein, Alfred, Jacqueline Cohn, Somnath Das, and Soumyo D. Moitra. 1988. "Specialization and Seriousness during Adult Criminal Careers." *Journal of Quantitative Criminology* 4: 303–45.

Blumstein, Alfred, Jacqueline Cohen, and Paul Hsieh. 1982. *The Duration of Adult Criminal Careers.* Final report submitted to National Institute of Justice, August 1982. Pittsburgh, PA: School of Urban and Public Affairs, Carnegie-Mellon University.

Blumstein, Alfred, Jacqueline Cohen, Jeffrey A. Roth, and Christy A. Visher, eds. 1986. *Criminal Careers and "Career Criminal."* 2 vols. Panel on Research on Criminal Careers, Committee on Research on Law Enforcement and the Administration of Justice. Commission on Behavioral and Social Sciences and Education, National Research Council. Washington, D.C.: National Academy Press.

Blumstein, Alfred, David P. Farrington, and Soumyo Moitra. 1985. "Delinquency Careers: Innocents, Desisters, and Persisters." In *Crime and Justice: An Annual Review of Research,* Vol. 6, ed. Michael Tonry and Norval Morris. Chicago: University of Chicago Press.

Brame. Robert, and Alex R. Piquero. 2003. "The Role of Sample Attrition in Studying the Longitudinal Relationship between Age and Crime." *Journal of Quantitative Criminology.*

Bursik, Robert J., Jr. 1980. "The Dynamics of Specialization in Juvenile Offenses." *Social Forces* 58: 851–64.

Bushway, Shawn, Robert Brame, and Raymond Paternoster. 1999. "Assessing Stability and Change in Criminal Offending: A Comparison of Random Effects, Semiparametric, and Fixed Effects Modeling Strategies." *Journal of Quantitative Criminology* 15: 23–64.

Bushway, Shawn D., Alex R. Piquero, Lisa M. Broidy, Elizabeth Cauffman, and Paul Mazerolle. 2001. "An Empirical Framework for Studying Desistance as a Process." *Criminology* 39: 491–515.

Canela-Cacho. José E., Alfred Blumstein, and Jacqueline Cohen. 1997. "Relationship between the Offending Frequency (λ) of Imprisoned and Free Offenders." *Criminology* 35: 133–76.

Capaldi, Deborah N., and Gerald R. Patterson. 1996. "Can Violent Offenders Be Distinguished from Frequent Offenders: Prediction from Childhood to Adolescence." *Journal of Research in Crime and Delinquency* 33: 206–31.

Chaiken, Jan M., and Marcia R. Chaiken. 1982. *Varieties of Criminal Behavior.* Rand Report R-2814-NIJ. Santa Monica, CA: Rand Corporation.

Cohen, Jacqueline. 1986. "Research on Criminal Careers: Individual Frequency Rates and Offense Seriousness." In *Criminal Careers and "Career Criminals,* Vol. 1, ed. Alfred Blumstein, Jacqueline Cohen, Jeffrey A. Roth, and Christy A. Visher. Washington, DC: National Academy Press.

Conway, Kevin P., and Joan McCord. 2002. "A Longitudinal Examination of the Relation between Co-offending with Violent Accomplices and Violent Crime." *Aggressive Behavior* 28: 97–108.

Eggleston, Eliane, John H. Laub, and Robert J. Sampson. 2003. "Examining Long-term Trajectories of Criminal Offending: The Glueck Delinquents from Age 7 to 80." *Journal of Quantitative Criminology,* forthcoming.

Elliott, Delbert S. 1994. "1993 Presidential Address—Serious Violent Offenders: Onset, Developmental Course, and Termination." *Criminology* 32: 1–22.

Elliott, Delbert S., Suzanne S. Ageton, David Huizinga, Barbara Knowles, and R. Canter. 1983. *The Prevalence and Incidence of Delinquent Behavior: 1976–1980.* National Youth Survey, Report No. 26. Boulder, CO: Behavioral Research Institute.

Elliott, Delbert S., David Huizinga, and Barbara Morse. 1987. "Self-Reported Violent Offending: A Descriptive Analysis of Juvenile Violent Offenders and Their Offending Careers." *Journal of Interpersonal Violence* 1: 472–514.

Farrington, David P. 1986. "Age and Crime." In *Crime and Justice: An Annual Review of Research,* Vol. 7, ed. Michael Tonry and Norval Morris. Chicago: University of Chicago Press.

———. 1989. "Self-Reported and Official Offending from Adolescence to Adulthood." In *Cross-National Research in Self-Reported Crime and Delinquency,* ed. Malcolm W. Klein. Dordrecht: Kluwrer.

———. 2003. "Key Results from the First Forty Years of the Cambridge Study in Delinquent Development." In *Taking Stock of Delinquency: An Overview of Findings from Contemporary Longitudinal Studies,* ed. Terence P. Thornberry and Marvin D. Krohn. New York: Kluwer/Plenum.

Farrington, David P., Sandra Lambert, and Donald J. West. 1998. "Criminal Careers of Two Generations of Family Members in the Cambridge Study in Delinquent Development." *Studies on Crime and Crime Prevention* 7: 85–106.

Farrington, David P., and Rolf Loeber. 1998. "Major Aims of This Book." In *Serious & Violent Juvenile Offenders: Risk Factors and Successful Interventions,* ed. Rolf Loeber and David P. Farrington. Thousand Oaks, CA: Sage.

Farrington, David P., Rolf Loeber, Delbert S. Elliott, J. David Hawkins, Denise Kandel, Malcolm Klein, Joan McCord, David Rowe, and Richard Tremblay. 1990. "Advancing Knowledge about the Onset of Delinquency and Crime." In *Advances in Clinical and Child Psychology,* ed. Bernard Lahey and A. Kazdin. New York: Plenum.

Farrington, David P., Howard N. Snyder, and Terrence A. Finnegan. 1988. "Specialization in Juvenile Court Careers." *Criminology* 26: 461–87.

Gottfredson, Michael R., and Travis Hirschi. 1990. *A General Theory of Crime.* Stanford, CA: Stanford University Press.

Greenberg, David F. 1975. "The Incapacitative Effect of Imprisonment: Some Estimates." *Law and Society Review* 9: 541–80.

Greene, M. A. 1977. The Incapacitative Effect of Imprisonment on Policies of Crime. Unpublished PhD thesis, School of Urban and Public Affairs, Carnegie-Mellon University, Pittsburgh, PA. (University Microfilms, Ann Arbor, MI).

Greenwood, Peter W., and Allan Abrahams. 1982. *Selective Incapacitation.* Rand Report R-2815-NIJ. Santa Monica, CA: Rand.

Greenwood, Peter W., and Susan Turner. 1987. *Selective Incapacitation Revisited: Why the High-Rate Offenders Are Hard to Predict.* Rand Report R-3397-NIJ. Santa Monica, CA: Rand.

Guttridge, P., W. F. Gabrielli, Jr., Sarnoff A. Mednick, and Kathertne T. Van Dusen. 1983. "Criminal Violence in a Birth Cohort." In *Prospective Studies of Crime and Delinquency,,* ed. Katherine T. Van Dusen and Sarnoff A. Mednick. Boston: Kluwer-Nijhoff.

Hamparian, D. M., R. Schuster, S. Dinitz, and J. Conrad. 1978. *The Violent Few: A Study of Dangerous Juvenile Offenders.* Lexington, MA: Lexington Books.

Hindelang , Michael, Travis Hirschi, and Joseph Weis. 1981. *Measuring Delinquency.* Beverly Hills, CA: Sage.

Hirschi, Travis, and Michael G. Gottfredson. 1983. "Age and the Explanation of Crime." *American Journal of Sociology* 89: 552–84

Kelley, Barbara Tatem, David Huizinga, Terence P. Thornberry, and Rolf Loeber. 1997. *Epidemiology of Serious Violence.* Office of Juvenile Justice Bulletin. Washington, DC: U.S. Department of Justice, Office of Juvenile Justice and Delinquency Prevention.

Krohn, Marvin D., Terence P. Thornberry, Craig Rivera, and Marc Le Blanc. 2001. "Later Delinquency Careers." In *Child Deinquents: Development, Intervention, and Service Needs,* ed. Rolf Loeber and David P. Farrington. Thousand Oaks, CA: Sage.

Laub, John H., and Robert J. Sampson. 2001. "Understanding Desistance from Crime." In *Crime and Justice: A Review of Research,* Vol. 28, ed. Michael Tonry. Chicago: University of Chicago Press.

Lauritsen, Janet. 1998. "The Age-Crime Debate: Assessing the Limits of Longitudinal Self-Report Data." *Social Forces* 77: 127–55.

Le Blanc, Marc, and Marcel Fréchette. 1989. *Male Criminal Activity from Childhood through Youth: Multilevel and Developmental Perspectives.* New York: Springer-Verlag.

Loeber, Rolf, and David P. Farrington, eds. 1998. *Serious & Violent Juvenile Offenders: Risk Factors and Successful Interventions.* Thousand Oaks, CA: Sage.

Loeber, Rolf, and Howard N. Snyder. 1990. "Rate of Offending in Juvenile Careers: Findings of Constancy and Change in Lambda." *Criminology* 28: 97–110.

Moffitt, Terrie E., Avshalom Caspi, Michael Rutter, and Phil A. Silva. 2001. *Sex Differences in Antisocial Behaviour: Conduct Disorder, Delinquency, and Violence in the Dunedin Longitudinal Study.* Cambridge, UK: Cambridge University Press.

Nagin, Daniel S. 1999. "Analyzing Developmental Trajectories: A Semi-Parametric, Group-Based Approach." *Psychological Methods* 4: 139–77.

Nevares, Dora, Marvin E. Wolfgang, and Paul E. Tracy. 1990. *Delinquency in Puerto Rico: The 1970 Birth Cohort Study.* New York: Greenwood Press.

Petersilia, Joan. 1980. "Criminal Career Research: A Review of Recent Evidence." In *Crime and Justice: An Annual Review of Research,* Vol. 2, ed. Norval Morris and Michael Tonry. Chicago: University of Chicago Press.

Petersilia, Joan, Peter W. Greenwood, and Marvin Lavin. 1978. *Criminal Careers of Habitual Felons.* Washington, DC: National Institute of Law Enforcement and Criminal Justice, Law Enforcement Assistance Administration, U.S. Government Printing Office.

Peterson, Mark A., and Harriet B. Braiker. 1980. *Doing Crime: A Survey of California Prison Inmates.* Report R-2200-DOJ. Santa Monica, CA: Rand.

Piquero, Alex R. 2000. "Assessing the Relationships between Gender, Chronicity, Seriousness, and Offense Skewness in Criminal Offending." *Journal of Criminal Justice* 28:103–16.

Piquero, Alex R., Alfred Blumstein, Robert Brame, Rudy Haapanen, Edward P. Mulvey, and Daniel S. Nagin. 2001. "Assessing the Impact of Exposure Time and Incapacitation on Longitudinal Trajectories of Criminal Offending." *Journal of Adolescent Research* 16:54–74.

Piquero, Alex R., Robert Brame, and Donald Lynam. 2003. "Do the Factors Associated with Life-Course-Persistent Offending Relate to Career Length?" *Crime and Delinquency,* forthcoming.

Piquero, Alex R., David P. Farrington, and Alfred Blumstein. 2003. "The Criminal Career Paradigm." In *Crime and Justice: A Review of Research,* Vol. 30, ed. Michael Tonry. Chicago: University of Chicago Press.

Piquero, Alex R., Robert Brame, Paul Mazerolle, and Rudy Haapanen. 2002. "Crime in Emerging Adulthood." *Criminology* 40: 137–70.

Piquero, Alex R., and Stephen L. Buka. 2002. "Linking Juvenile and Adult Patterns of Criminal Activity in the Providence Cohort of the National Collaborative Perinatal Project." *Journal of Criminal Justice* 30: 1–14.

Piquero, Alex R., Raymond Paternoster, Paul Mazerolle, Robert Brame, and Charles W. Dean. 1999. "Onset Age and Offense Specialization." *Journal of Research in Crime and Delinquency* 36: 275–99.

Pulkkinen, L. 1988. "Delinquent Development: Theoretical and Empirical Considerations." In *Studies of Psychosocial Risk: The Power of Longitudinal Data,* ed. Michael Rutter. Cambridge, UK: Cambridge University Press.

Quetelet, Adolphe. 1831. *Research on the Propensity for Crime at Different Ages.* 1984 edition translated by Sawyer F. Sylvester. Cincinnati, OH: Anderson Publishing Company.

Raudenbush, Stephen W. 2001. "Toward a Coherent Framework for Comparing Trajectories of Individual Change." In *New Methods for the Analysis of Change,* ed. Linda M. Collins and Aline G. Sayers. Washington, DC: American Psychological Association.

Reiss, Albert J., Jr. 1986. "Co-Offender Influences on Criminal Careers." In *Criminal Careers and "Career Criminals,"* ed. Alfred Blumstein, Jacqueline Cohen, Jeffrey A. Roth, and Christy A. Visher. Washington, DC: National Academy Press.

Reiss, Albert J., Jr., and David P. Farrington. 1991. "Advancing Knowledge about Co-Offending: Results from a Prospective Longitudinal Survey of London Males." *Journal of Criminal Law and Criminology* 82: 360–95.

Rojek, D. G., and M. L. Erickson. 1982. "Delinquent Careers: A Test of the Career Escalation Model." *Criminology* 20: 5–28.

Sarnecki, Jerzy. 1990. "Delinquent Networks in Sweden." *Journal of Quantitative Criminology* 6: 31–51.

———. 2001. *Delinquent Networks . Youth Co-Offending in Stockholm.* Cambridge, UK: Cambridge University Press.

Schmertmann, Carl P., Adansi A. Amankwaa, and Robert D. Long. 1998. "Three Strikes and You're Out: Demographic Analysis of Mandatory Prison Sentencing." *Demography* 35: 445–63.

Shannon, Lyle. 1982. *Assessing the Relationship of Adult Criminal Careers to Juvenile Careers.* Washington, DC: U.S. Department of Justice, Office of Juvenile Justice and Delinquency Prevention.

Shinnar, Shlomo, and Reuel Shinnar. 1975. "The Effects of the Criminal Justice System on the Control of Crime: A Quantitative Approach." *Law and Society Review* 9: 581–611.

Smith, D. Randall, and William R. Smith. 1984. "Patterns of Delinquent Careers: An Assessment of Three Perspectives." *Social Science Research* 13: 129–58.

Spelman, William. 1994. *Criminal Incapacitation.* New York: Plenum.

Stattin, Håkan, David Magnusson, and Howard Reichel. 1989. "Criminal Activity at Different Ages: A Study Based on a Swedish Longitudinal Research Population." *British Journal of Criminology* 29: 368–85.

Steffensmeier, Darrell J., Emilie Andersen Allan, Miles D. Harer, and Cathy Streifel. 1989. "Age and the Distribution of Crime." *American Journal of Sociology* 94: 803–31.

Stolzknberg, Lisa, and Stewart J. D'Alessio. 1997. "'Three Strikes and You're Out': The Impact of California's New Mandatory Sentencing Law on Serious Crime Rates." *Crime and Delinquency* 43: 457–69.

Thornberry. Terence P. 1989. "Panel Effects and the Use of Self-Reported Measures of Delinquency in Longitudinal Studies." In *Cross-National Research in Self-Reported Crime and Delinquency,* ed. Malcolm W. Klein. Dordrecht: Kluwer Academic.

Tittle, Charles R., and Harold G. Grasmick. 1993. "Criminal Behavior and Age: A Test of Three Provocative Hypotheses." *Journal of Criminal Law and Criminology* 88: 309–42.

Tracy, Paul E., Marvin E. Wolfgang, and Robert M. Figlio. 1990. *Delinquency Careers in Two Birth Cohorts.* New York: Plenum.

Visher, Christy A. 1986. "The Rand Inmate Survey: A Re-Analysis." In *Criminal Careers and " Career Criminals,"* Vol. 2, ed. Alfred Blumstein, Jacqueline Cohen, Jeffrey A. Roth, and Christy A. Visher. Washington, DC: National Academy Press.

———. 1987. "Incapacitation and Crime Control: Does a 'Lock 'em Up' Strategy Reduce Crime?" *Justice Quarterly* 4: 513–43.

———. 2000. "Career Criminals and Crime Control." In *Criminology: A Contemporary Handbook,* ed. Joseph F. Sheley. 3rd ed. Belmont, CA: Wadsworth.

Visher, Christy A., and Jeffrey A Roth. 1986. "Participation in Criminal Careers." In *Criminal Careers and "Career Criminals,"* Vol. 1, ed. Alfred Blumstein, Jacqueline Cohen, Jeffrey A. Roth, and Christy A. Visher. Washington, DC: National Academy Press.

Wikström, Per-Olof H. 1985. *Everyday Violence in Contemporary Sweden: Situational and Ecological Aspects.* Stockholm: National Council for Crime Prevention, Sweden, Research Division.

Wolfgang, Marvin E., Robert M. Figlio, and Thorsten Sellin. 1972. *Delinquency in a Birth Cohort.* Chicago: University of Chicago Press.

Zimring, Franklin E., Sam Kamin, and Gordon Hawkins. 1999. *Crime and Punishment in California: The Impact of Three Strikes and You're Out.* Berkeley, CA: Institute of Governmental Studies Press, University of California.

REVIEW QUESTIONS

1. Explain in your own words some of the concepts that Piquero et al. point out as key concepts of the career criminal paradigm, such as offending frequency, duration, and so on. Which of the concepts do you feel are most important in determining which individuals pose the greatest danger to society?

2. Which policy implications suggested by Piquero et al. do you feel make the most sense? Which make the least sense to you, and why?

3. Do you know any individuals in your life whom you would classify as chronic offenders (if not, consider stories in the media/press)? What do you think caused them to become such habitual offenders? What do you think are the primary causes for their chronic offending, and what types of policies would you implement to prevent others from becoming such persistent offenders?

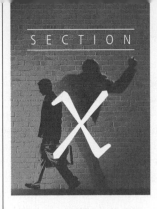

INTEGRATED THEORETICAL MODELS AND NEW PERSPECTIVES OF CRIME

The introduction of this section will introduce integrated theories, those in which two or more traditional theories are merged into one cohesive model. We will then discuss the pros and cons of integrating theories and explain the various ways theories can be integrated. A review of various integrated theories will demonstrate the many ways different theories have been merged to form a more empirically valid explanation for criminal behavior.

◪ Introduction

This introduction examines relatively recent developments in criminological development, with virtually all of these advances taking place in the last couple of decades. Specifically, we discuss the types of models that modern explanatory formulations seem to have taken, namely integrated and theories. Of course, other unique theoretical frameworks have been presented in the last 20 years, but most of the dominant models presented during this time fall into the category of integrated theories.

Integrated theories attempt to integrate two or more traditionally separate models of offending to form one unified explanatory theory. Given the empirical validity of most of the theories discussed in previous sections, as well as the failure of all of these previously examined theories to explain all variation in offending behavior, it makes sense that theorists would try to combine the best of several theories, while disregarding or de-emphasizing the concepts and propositions that don't seem to be as scientifically valid in each of the frameworks. Furthermore, some forms of theoretical integration deal with only concepts or propositions, while others vary by level of analysis (micro vs. macro, or both). Although such integrated formulations sound attractive and appear to be sure-fire ways to develop sounder explanatory models of behavior, they have a number of weaknesses and criticisms. In this introduction, we explore these issues, as well as the best known and most accepted integrated theories that are currently being examined and tested in the extant criminological literature.

◩ Integrated Theories

About 30 years ago at a conference at the State University of New York at Albany, leading scholars in criminological theory development and research came together to discuss the most important issues in the growing area of theoretical integration.[1] Some integrated theories go well beyond formulating relationships between two or more traditionally separate explanatory models; they actually fuse the theories into one, all-encompassing framework. The following sections will examine why integrated theories became popular over the last several decades while discussing different types of integrated theories, the strengths and weaknesses of theoretical integration, and several of the better known and respected integrated theories.

The Need for Integrated Theories in Criminology

The emphasis on theoretical integration is a relatively recent development, which has evolved due to the need to improve the empirical validity of our traditional theories, which suffer from lack of input from various disciplines.[2] This was the result of the history of criminological theory, which we discussed in prior sections of this book. Specifically, most 19th-century theories of criminal behavior are best described as based on a single-factor (e.g., IQ) or limited-factor (e.g., stigmata) reductionism. Later, in the early 1900s, a second stage of theoretical development involved the examination of various social, biological, and psychological factors, which became known as the multiple-factor approach and is most commonly linked with the work of Sheldon and Eleanor Glueck.

Finally, in the latter half of the 20th century, a third stage of theoretical development and research in criminology became dominant, which represented a backlash against the multiple-factor approach. This stage has been called *systemic reductionism*, which refers to

---

[1]For a compilation of the papers presented at this conference, see Steven F. Messner, Marvin D. Krohn, and Allen E. Liska, eds., *Theoretical Integration in the Study of Deviance and Crime: Problems and Prospects* (Albany: State University of New York Press, 1989).

[2]This discussion is largely drawn from Charles F. Wellford, "Towards an Integrated Theory of Criminal Behavior," in Messner et al., *Theoretical Integration,* 119–27.

the pervasive attempts to explain criminal behavior in terms of a particular system of knowledge.[3] For example, a biologist who examines only individuals' genotype will likely not be able to explain much criminal behavior because he is missing a lot of information about the environment in which the person lives, such as level of poverty or unemployment. For the last 50 years, the criminological discipline in the United States has been dominated by sociologists, which is largely due to the efforts and influence of Edwin Sutherland (see Section VII). Thus, as one expert has observed:

> It is not surprising to find that most current explanations of criminal behavior are sociologically based and are attempts to explain variations in the rates of criminal behavior as opposed to individual instances of that behavior . . . [e]ven when the effort is to explain individual behavior, the attempt is to use exposure to or belief in cultural or social factors to explain individual instances or patterns of criminal behavior. . . . We find ourselves at the stage of development in criminology where a variety of sociological systemic reductionistic explanations dominate, all of which have proven to be relatively inadequate (to the standard of total explanation, prediction, or control) in explaining the individual occurrence or the distribution of crime through time or space.[4]

Other criminologists have also noted this sociological dominance, with some going as far to claim that:

> Sutherland and the sociologists were intent on turning the study of crime into an exclusively sociological enterprise and that they overreacted to the efforts of potential intruders to capture some of what they regarded as their intellectual turf.[5]

This dominance of sociology over the discipline is manifested in many obvious ways. For example, virtually all professors of criminology have a Doctorate of Philosophy (Ph.D.) in sociology or criminal justice, and virtually no professors have a degree in biology, neuropsychology or other fields that obviously have important influence on human behavior. Furthermore, virtually no undergraduate (or even graduate) programs in criminal justice and criminology currently require students to take a course that covers principles in biology, psychology, anthropology, or economics; rather, virtually all the training is sociology based. In fact, most criminal justice programs in the United States do not even offer a course in biopsychological approaches toward understanding criminal behavior, despite the obvious need for and relevance of such perspectives.

Many modern criminologists now acknowledge the limitations of this state of systemic reductionism, which limits theories to only one system of knowledge, in this case

---

[3]Ibid., 120.

[4]Ibid., 120–21.

[5]Quote from Stephen Brown, Finn-Aage Esbensen, and Gilbert Geis, *Criminology: Explaining Crime and Its Context*, 4th ed. (Cincinnati, OH: Anderson, 2001), 251, regarding claims made by Robert Sampson and John Laub, "Crime and Deviance Over the Life Course: The Salience of Adult Social Bonds," *American Sociological Review* 55 (1990): 609–27. Also, see discussion in Richard A. Wright, "Edwin H. Sutherland," *Encyclopedia of Criminology*, Vol. 1, ed. Richard A. Wright and J. Mitchell Miller (New York: Routledge, 2005).

sociology.[6] What resulted is a period of relative stagnation in theoretical development, with experts regarding the 1970s as "one of the least creative periods in criminological history."[7] In addition, the mainstream theories that were introduced for most of the 1900s (such as differential association, strain, social bonding, and labeling, which we reviewed in previous sections) received limited empirical support, which should not be too much of a surprise given that they were based on principles of only one discipline, namely sociology. Thus, it has been proposed that integrated theories evolved as a response to such limitations and the need to revitalize progress in the area of criminological theory building.[8] After reviewing some of the many integrated theories that have been proposed in the last two decades, we think readers will agree that the approach of combining explanatory models has indeed helped stimulate much growth in the area of criminological theory development. But before examining these theories, it is important to discuss the varying forms they take.

Different Forms of Integrated Theories

There are several different types of integrated theories, which are typically categorized by the way that the creator of the integrated model proposes the theories fuse together in a useful way. The three most common forms of propositional theoretical integration, meaning that they synthesize theories based on their postulates, are known as end-to-end, side-by-side, and up-and-down integration.[9] First, we will explore each of these types before discussing a few more variations of combining theoretical concepts and propositions.

End-to-End Integration. The first type, **end-to-end theoretical integration,** typically is used when theorists expect that one theory will come before or after another in terms of temporal ordering of causal factors. This type of integration is more developmental in the sense that it tends to propose a certain ordering of the component theories that are being merged. For example, an integrated theory may claim that most paths toward delinquency and crime have their early roots in the breakdown of social attachments and controls (i.e., social bonding theory), but later, the influence of negative peers (i.e., differential association) becomes more emphasized. Thus, such a model would look like this:

Weak Social Bond → Negative Peer Associations → Crime

Such an integrated model is referred to as *end-to end* (or "sequential") integration, which is appropriately named because it conveys the linkage of the theories based on the temporal ordering of two or more theories in their causal timing.[10] Specifically, the

---

[6]Wellford, "Toward an Integrated Theory"; Sampson and Laub, "Crime and Deviance."

[7]Wellford, "Toward an Integrated Theory," 120, citing Frank P. Williams III, "The Demise of Criminological Imagination: A Critique of Recent Criminology," *Justice* 1 (1984): 91–106.

[8]Wellford, "Toward an Integrated Theory."

[9]This typology is largely based on Travis Hirschi, "Separate and Unequal Is Better," *Journal of Research in Crime and Delinquency* 16 (1979): 34–37.

[10]Allen E. Liska, Marvin D. Krohn, and Steven F. Messner, "Strategies and Requisites for Theoretical Integration in the Study of Crime and Deviance," in Messner et al., *Theoretical Integration,* 1–19.

breakdown of social bonds comes first, followed by the negative peer relations. Another way of saying this is that the breakdown of social bonds is expected to have a more "remote" or indirect influence on crime, which is mediated by differential peer influences; on the other hand, according to this model, peer influences are expected to have a more "immediate," or direct effect on crime, with no mediating influences of other variables. This model is hypothetical and presented to illustrate the end-to-end form of integration, but we will see that some established frameworks have incorporated similar propositions regarding social bonding and differential association/reinforcement theories.

Many of the traditionally separate theories that we have examined in previous introductions tend to differ in their focus on either remote or immediate causal factors.[11] For instance, one of the assumptions of differential association theory is that psychological learning of crime is based on day-to-day (or even moment-to-moment) learning, so the emphasis is on more immediate causes of crime, namely interactions with peers and other significant others. On the other hand, other theories tend to focus more on remote causes of crime, such as social disorganization and strain theory, which place the emphasis on social structure factors, such as relative deprivation or industrialization, that most experts would agree are typically not directly implicated in situational decisions to engage in an actual criminal act but are extremely important nonetheless.

This seems to be conducive to using end-to-end integration, and often, the theories seem to complement each other quite well, as in our hypothetical example in which social control theory proposes the remote cause (i.e., weakened bonds) and differential association theory contributes the more direct, proximate cause (i.e., negative peer influence). On the other hand, two or more theories that focus only on more immediate causes of crime would be harder to combine because they both claim they are working at the same time, in a sense competing against the other for being the primary direct cause of criminal activity; thus, they would be unlikely to fuse together as nicely and would not complement one another. Also, some theorists have argued that end-to-end integration is simply a form of *theoretical elaboration,* which we will discuss later. Another major limitation of end-to-end integration is the issue of whether the basic assumptions of the included theories are consistent with one another. We will discuss this in the following section, which discusses criticism of integration. But first we must examine the other forms of theoretical integration.

Side-by-Side Integration. Another type of integrated theory is called **side-by-side (or "horizontal") integration.** In the most common form of side-by-side integration, cases are classified by a certain criterion (e.g., impulsive versus planned), and two or more theories are considered parallel explanations depending on what type of case is being considered. So when the assumptions or target offenses of two or more theories are different, a side-by-side integration is often the most natural way to integrate them. For example, low self-control theory may be used to explain impulsive criminal activity, whereas rational choice theory may be used to explain criminal behavior that involves planning, such as white-collar crime. Traditionally, many theorists would likely argue that low self-control and rational choice theory are quite different, almost inherently opposing

---

[11]Much of our discussion of these forms of integration is taken from Liska et al., "Strategies and Requisites."

perspectives of crime. However, contemporary studies and theorizing have shown that the two models complement and fill gaps in the other.[12] Specifically, the rational choice/deterrence framework has always had a rather hard time explaining and predicting why individuals often do very stupid things for which there is little payoff and a high likelihood of getting in serious trouble, so the low self-control perspective helps fill in this gap by explaining that some individuals are far more impulsive and are more concerned with the immediate payoff (albeit often small) rather than any long-term consequences.

An illustration of this sort of side-by-side integration of these two theories might look something like this:

For most typical individuals:

High Self-Control → Consideration of Potential Negative Consequences →
Deterred From Committing Crime

For more impulsive individuals/activities:

Low Self-Control → Desire for Immediate Payoff → Failure to Consider
Consequences → Decision to Commit Criminal Act

This side-by-side integration shows how two different theories can each be accurate, depending on what type of individual or criminal activity is being considered. As some scholars have concluded, rational choice theory is likely not a good explanation for homicides between intimates (which tend to be spontaneous acts), but it may be very applicable to corporate crime.[13]

Up-and-Down Integration. Another way of combining two or more theories is referred to as **up-and-down (or "deductive") integration**, and is generally considered the classic form of theoretical integration because it has been done relatively often in the history of criminological theory development, even before it was considered integration. This often involves increasing the level of abstraction of a single theory so that postulates seem to follow from a conceptually broader theory. Up-and-down integration can take two prevalent forms: theoretical reduction and theoretical synthesis.

---

[12]Daniel Nagin and Raymond Paternoster, "Enduring Individual Differences and Rational Choice Theories of Crime," *Law and Society Review* 27 (1993): 467–96; Alex Piquero and Stephen Tibbetts, "Specifying the Direct and Indirect Effects of Low Self-Control and Situational Factors in Offenders' Decision Making: Toward a More Complete Model of Rational Offending," *Justice Quarterly* 13 (1996): 481–510; Stephen Tibbetts and Denise Herz, "Gender Differences in Factors of Social Control and Rational Choice," *Deviant Behavior* 17 (1996): 183–208; Stephen Tibbetts and David Myers, "Low Self-Control, Rational Choice, and Student Test Cheating," *American Journal of Criminal Justice* 23 (1999): 179–200; Bradley Wright, Avshalom Caspi, and Terrie Moffitt, "Does the Perceived Risk of Punishment Deter Criminally Prone Individuals? Rational Choice, Self-Control, and Crime," *Journal of Research in Crime and Delinquency* 41 (2004): 180–213; for a review, see Stephen Tibbetts, "Individual Propensities and Rational Decision-Making: Recent Findings and Promising Approaches," in *Rational Choice and Criminal Behavior: Recent Research and Future Challenges*, ed. Alex Piquero and Stephen Tibbetts (New York: Routledge, 2002), 3–24.

[13]Liska et al., "Strategies and Requisites"; for a review, see Brown et al., *Explaining Crime*; for a recent study of rational choice being applied to corporate crime, see Nicole Leeper Piquero, M. Lyn Exum, and Sally S. Simpson, "Integrating the Desire-for-Control and Rational Choice in a Corporate Crime Context," *Justice Quarterly* 22 (2005): 252–81.

Theoretical reduction is typically done when it becomes evident "that theory A contains more abstract or general assumptions than theory B and, therefore, that key parts of theory B can be accommodated within the structure of theory A."[14] Regarding theoretical reduction, we have discussed this form of integration previously in this book, without actually calling it by this name. For example, in the introduction to Section VII, we discussed how differential reinforcement theory subsumed Sutherland's differential association theory. By equating concepts contained in both theories, the authors of differential reinforcement argued somewhat effectively that the learning that takes place through interactions with primary groups is one type of conditioning.[15] As you will recall, the main point is that the concepts and propositions of differential reinforcement theory are more general than, but entirely consistent with, those of differential association theory, such that the latter is typically a specific form of the former model. In other words, differential reinforcement is a much more broad, general theory, which accounts for not only differential association (i.e., classical conditioning), but also many other theoretical concepts and propositions (operant conditioning, modeling/imitation).

This same type of theoretical reduction was also discussed in the introduction to Section V when we noted that general strain theory subsumed Merton's traditional strain theory. Specifically, traditional strain theory focused on only one type of strain: failure to achieve positively valued goals. In comparison, while general strain theory also places an emphasis on failure to achieve positively valued goals, it is more general and broad because it also focuses on two other forms of strain: loss of positively valued stimuli and exposure to noxious stimuli. Therefore, it seems to make sense that general strain theory would subsume traditional strain theory because the concepts and principles of traditional strain appear to simply represent a specific type of general strain and can be fully accounted for by the more general version of the theory.[16]

Despite the obvious efficiency in theoretical reduction, many theorists have criticized such subsuming of theories by another. Specifically, many experts in the social sciences view such deduction as a form of theoretical imperialism because the theory being deduced essentially loses its unique identity.[17] Therefore, the very phrase *theoretical reduction* generally has a negative connotation among scholars. In fact, one of the most accepted reductions in the 20th century criminological literature, the differential association-differential reinforcement synthesis we referred to above, has been condemned as a "revisionist takeover . . . a travesty of Sutherland's position."[18] So even the most widely known and cited forms of theoretical reduction have been harshly criticized, which gives readers some idea of how frowned upon such subsuming of theories by another is considered by social scientists.

---

[14]Liska et al., "Strategies and Requisites," 10.

[15]Liska et al., "Strategies and Requisites," 13; Robert L. Burgess and Ronald Akers, "Differential Association-Reinforcement Theory of Criminal Behavior," *Social Problems* 14 (1966): 128–46; Ronald Akers, *Deviant Behavior: A Social Learning Approach*, 3rd ed., (Belmont, CA: Wadsworth, 1985).

[16]Robert Agnew, "Foundation for a General Strain Theory of Crime and Delinquency," *Criminology* 30 (1992): 47–87; Robert Agnew and Helen R. White, "An Empirical Test of General Strain Theory," *Criminology* 30 (1992): 475–99.

[17]Liska et al., "Strategies and Requisites."

[18]Quote from Ian Taylor, Paul Walton, and Jock Young, *The New Criminology*, (New York: Harper & Row, 1973), 131–132; see discussion in Liska et al., "Strategies and Requisites," 13.

The other form of up-and-down integration is referred to as *theoretical synthesis,* which is "done by abstracting more general assumptions from theories A and B, allowing parts of both theories to be incorporated in a new theory C."[19] This form of up-and-down integration is more uncommon in social science than theoretical reduction, perhaps because it is more difficult to achieve successfully because it, by definition, requires a new theory that essentially necessitates new concepts and/or hypotheses that are not already found in the original component models. Furthermore, if the constituent theories are competing explanations, then it is quite likely that new terminology will have to be created or incorporated to resolve these differences.[20] However, if theoretical synthesis can be done correctly, it is perhaps the type of integration that provides the most advancement of theory development because, in addition to bringing together previously independent models, it also results in a new theory with predictions and propositions that go beyond the original frameworks.

One of the best known and most accepted (despite its critics) examples of theoretical synthesis is Elliott's integrated model, which we review in detail later in this introduction.[21] For our purposes here, it is important only to understand why Elliott's model is considered theoretical synthesis. This is because Elliott's model integrates the concepts and propositions of various theories (e.g., social control and differential association) and contributes additional propositions that did not exist in the component theories.

Levels of Analysis of Integrated Theories

Beyond the variation of types of propositional theoretical integration discussed above, such synthesis of explanatory models also differs in terms of the level of analysis that is being considered. Specifically, the integrated models can include component theories of micro-micro, macro-macro, or even micro-macro (called "cross-level") combinations. For example, Elliott's integrated theory is a micro-micro level theory, which means that all of the component theories that make up the synthesized model refer to the individual as the unit of analysis. Although these models can provide sound understanding for why certain individuals behave a certain way, they typically do not account for differences in criminality across groups (e.g., gender, socioeconomic groups, etc.).

On the other hand, some integrated models include theories from only macro levels. A good example of this type of integration is seen in Bursik's synthesis of conflict theory and the social disorganization framework.[22] Both of these component theories focus only on the macro (group) level of analysis, thus neglecting the individual level, such as explaining why some individuals do or do not commit crime even in the same structural environment.

---

[19]Liska et al., "Strategies and Requisites," 10.

[20]David Wagner and Joseph Berger, "Do Sociological Theories Grow?" *American Journal of Sociology* 90 (1985): 697–728.

[21]Delbert Elliott, Suzanne S. Ageton, and Rachelle J. Canter, "An Integrated Theoretical Perspective on Delinquent Behavior," *Journal of Research in Crime and Delinquency* 16 (1979): 3–27; also see discussion in Delbert Elliott, "The Assumption that Theories can be Combined with Increased Explanatory Power: Theoretical Integrations," in *Theoretical Methods in Criminology,* ed. Robert F. Meier (Beverly Hills, CA: Sage, 1985), 123–49; discussion of Elliott's model as an example of theoretical synthesis can be found in Charles R. Tittle, "Prospects for Synthetic Theory: A Consideration of Macro-Level Criminological Activity," in Messner et al., *Theoretical Integration,* 161–78.

[22]Robert J. Bursik, "Political Decisionmaking and Ecological Models of Delinquency, Conflict and Consensus," in Messner et al., *Theoretical Integration,* 105–18.

The most complicated integrated theories, at least in terms of levels of analysis, are those that include both micro and macro models. Such models are likely the most difficult to synthesize successfully because it involves bringing together rather unnatural relationships between individual-based propositions and group-level postulates. However, when done effectively, such a model can be rather profound in terms of explanation of crime. After all, one of the primary objectives of a good theory is to explain behavior across as many circumstances as possible. Thus, a theory that can effectively explain why crime occurs in certain individuals as well as certain groups would be better than a theory that explains crime across only individuals or groups.

An example of this is Braithwaite's theory of reintegrative shaming, which begins with an individual (or micro) level theory of social control/bonding—which he refers to as **interdependency**—and then relates levels of this concept to a group or community (or macro) level theory of bonding—which he refers to as **communitarianism**.[23] This theory will be discussed in far more detail later, but it is important here to acknowledge the advantages of explaining both the micro (individual) and macro (group) level of analysis in explanations of criminal behavior, such as in the theory of reintegrative shaming.

Ultimately, the levels of analysis of any component theories are an important consideration in the creation of integrated models of crime. It is particularly important to ensure that the merging of certain theories from within or across particular levels makes rational sense and, most important, advances our knowledge and understanding of causes of crime.

Additional Considerations Regarding Types of Integration

Beyond the basic forms of propositional integration models and levels of analysis, there are two additional types of integration—conceptual integration and theoretical elaboration—that are quite common, perhaps even more common than the traditional forms discussed above. **Conceptual integration** involves the synthesis of models in which "the theorist equates concepts in different theories, arguing that while the words and terms are different, the theoretical meanings and operations of measurement are similar."[24] Essentially, the goal in such a formulation is to take the primary constructs of two or more theories and to merge them into a more general framework that aids understanding in explaining behavior by unifying terms that fundamentally represent similar phenomena or issues. Such formulations are considered less intrusive on the component theories than the propositional integrations we discussed previously.

One of the first and most cited examples of conceptual integration was provided by Pearson and Weiner in 1985.[25] As shown in Table 10.1, Pearson and Weiner attempted to map the various concepts of numerous criminological theories. This is done by creating a category that numerous concepts from various theories appear to fit into through their inclusion in

---

[23]John Braithwaite, *Crime, Shame and Reintegration* (Cambridge, UK: Cambridge University, 1989).

[24] Liska et al., "Strategies and Requisites," 15; definition adapted from Frank S. Pearson and Neil Alan Weiner, "Toward an Integration of Criminological Theories," *Journal of Criminal Law and Criminology* 76 (1985): 116–50.

[25]Pearson and Weiner, "Toward an Integration"; see review in Thomas J. Bernard and Jeffrey B. Snipes, "Theoretical Integration in Criminology," in *Crime and Justice Review*, Vol. 20, ed. M. Tonry (Chicago, IL: University of Chicago Press, 1996), 301–48.

Table 10.1 Mapping of the Selected Criminological Theories Into the Integrative Structure

INTEGRATIVE CONSTRUCTS

Selected Criminological Theories	MICRO-LEVEL								MACRO-LEVEL			
	ANTECEDENT						CONSEQUENCES OR FEEDBACK		SOCIAL STRUCTURAL PRODUCTION AND DISTRIBUTION OF:			
	INTERNAL				EXTERNAL							
	Utility Demand (Deprivation)	Behavioral Skill	Rules of Expedience	Rules of Morality	Signs of Favorable Opportunities (Descriminative Stimuli)	Behavioral Resources	Utility Reception	Information Acquisition	Utilities	Opportunities	Rules of Expedience and Morality	Belief About Sanctioning Practices
Differential Association	✓	✓	✓	✓			✓					
Negative Labeling	✓	✓	✓	✓			✓					
Social Control:												
1. attachment	✓		✓				✓					
2. commitment	✓		✓		✓			✓				
3. involvement					✓			✓				
4. belief				✓								
Occurrence	✓		✓		✓		✓	✓				✓
Economic	✓	✓	✓		✓		✓	✓	✓		✓	
Routine Activities	✓	✓	✓		✓	✓			✓	✓		
Neutralization		✓	✓	✓							✓	
Relative Deprivation	✓	✓	✓				✓		✓	✓		
Strain	✓	✓	✓	✓	✓		✓	✓	✓	✓		
Normative (Culture) Conflict				✓							✓	
Generalized Strain and Normative (Culture) Conflict	✓	✓	✓	✓	✓		✓	✓	✓	✓	✓	
Marxist-Critical/ Group Conflict	✓		✓	✓			✓		✓	✓	✓	

explanatory models of criminal behavior and across level of explanation. Although based on a social learning/differential reinforcement perspective, Pearson and Weiner's model attempts to include concepts of virtually all existing theories of crime and delinquency models of behavior. A particular strength is that this model includes feedback or behavioral consequence elements, as well as classifying each model by unit of measurement and analysis.

For example, as illustrated in Table 10.1, the conceptual model shows that differential association theory has concepts that apply to all internal aspects and one of the two concepts at the consequences at the micro level, but none of the aspects at the external-micro level or any of the four macro-level concepts. On the other hand, Marxist critical theory is shown to apply to three of the internal-micro and one of the consequences-level concepts and to most of the macro-level concepts as well. Although some of the theories that we discuss in this book are included, largely because Pearson and Weiner's conceptual integration was done in the mid-1980s, this model includes most of the dominant theoretical frameworks that were prevalent in the criminological literature at the time of their formulation. Thus, this conceptual model remains as a sort of prototype for future attempts to conceptually integrate established theories explaining criminal conduct.

One notable strength of their conceptual integration is the inclusion of most mainstream theories—differential association, labeling, social control, deterrence/economic, routine activities, neutralization, strain, cultural, conflict—at the time in which they created their integrated framework. Another strength of their integrated framework is the fact that they account for concepts at different levels of analysis, namely micro (individual) and macro (group) levels, as shown in Table 10.1. So the framework clearly does a good job at creating links between the most prominent theories in terms of the concepts that they propose as the primary causes of crime.

However, Pearson and Weiner's conceptually integrated framework does have several notable weaknesses. One of the most prominent criticisms of the conceptual model is that it is based on a single theory, specifically social learning theory (tied to Akers's differential reinforcement/social learning theory discussed in Section VII), and the authors never really provide a strong argument for why they chose this particular base framework for their model. Another major weakness of this conceptual framework is that it completely neglects many biological and biosocial factors—hormones, neurotransmitters, toxins—that have been consistently shown to have profound effects on human behavior. Still, despite the criticisms of this integrated framework, respected theorists have noted that Pearson and Weiner's "integration of these concepts into a consistent, coherent framework is impressive," but the model has never received much attention in the criminological literature.[26]

However, despite the obvious tendency to simplify formulations through conceptual integration, many critics have claimed that such attempts are simply a means toward propositional integration. Thus, despite its categorization as conceptual, this type of integration is not necessarily seen as such, and it may be seen by many as a form of deductive integration. However, many experts have noted that conceptual integration is actually not a form of side-by-side integration nor is it end-to-end integration; rather, it is what it says it is: conceptual integration. It is nothing more and nothing less.[27] Therefore, it appears that conceptual integration is a distinct derivative form of integration and that it can and should occur

---

[26]Ronald Akers and Christine Sellers, *Criminological Theories: Introduction, Evaluation, and Application*, 4th ed. (Los Angeles, CA: Roxbury, 2004), 288.

[27]Liska et al., "Strategies and Requisites," 15.

independently if it helps to advance understanding of why people commit crime, particularly for integrated models of offending. As the most knowledgeable experts recently concluded, "establishing some conceptual equivalence is necessary for deductive integration."[28]

An additional variation of integration is **theoretical elaboration**, which is a strategy that takes an existing theory that is arguably underdeveloped and then further develops it by adding concepts and/or propositions to include other components from other theories. Many critics of traditional theoretical integration have argued that theoretical elaboration is a more attractive strategy because existing theories of offending are not developed enough to fully integrate them.[29] An example of theoretical integration can be seen in the expansion of rational choice theory, which had traditionally focused on ratios of perceived costs and benefits, to include deontological constructs such as moral beliefs. Specifically, studies have consistently shown that ethical constraints condition the influence of expected consequences on criminal behavior, ranging from violence to academic dishonesty to white-collar crime.[30] Although they are not without critics, most rational choice scholars appear to be in agreement that such elaboration advanced the theoretical framework and made it a more accurate explanation of human behavior. In sum, theoretical integration, like conceptual integration, is an option for merging and improving theoretical models without completely synthesizing two or more entire paradigms.

Criticisms and Weaknesses of Integrated Theories

A number of criticisms have been leveled at theoretical integration. Perhaps one of the most obvious and prevalent, not to mention extremely valid, criticisms is the argument that caution should be taken in attempting to integrate theories that have apparent contradictions or inconsistencies in their postulates.[31] As we have seen, different theories are based on varying, often opposite, perspectives of human nature. While most versions of strain theory (e.g., Merton's theory) assume that human beings are born with a natural tendency toward being good, most variations of control theory (e.g., Hirschi's social bonding and self-control theories) assume that humans are innately selfish and hedonistic. At the same time, most versions of learning theory (e.g., Sutherland's theory of differential association and Akers's differential reinforcement theory) assume that humans are born with a blank slate, in other words, people are born with neither good nor bad tendencies but rather learn their morality moment-to-moment from social interaction.

---

[28]Ibid., 16.

[29]Liska et al., "Strategies and Requisites," 16; Travis Hirschi, "Exploring Alternatives to Integrated Theory," in Messner et al., *Theoretical Integration*, 37–50; Terence P. Thornberry, "Reflections on the Advantages and Disadvantages of Theoretical Integration," in Messner et al., *Theoretical Integration*, 51–60; Robert F. Meier, "Deviance and Differentiation," Messner et al., *Theoretical Integration*, 199–212.

[30]A. Etzioni, *The Moral Dimension: Toward a New Economics* (New York: Free Press, 1988); Ronet Bachman, Raymond Paternoster, and Sally Ward, "The Rationality of Sexual Offending: Testing a Deterrence/Rational Choice Conception of Sexual Assault," *Law and Society Review* 26 (1992): 401–19; Raymond Paternoster and Sally Simpson, "Sanction Threats and Appeals to Morality: Testing a Rational Choice Model of Corporate Crime," *Law and Society Review* 30 (1996): 378–99; Stephen G. Tibbetts, "College Student Perceptions of Utility and Intentions of Test Cheating," PhD dissertation, University of Maryland-College Park (Ann Arbor, MI:UMI, 1997); for a review and discussion, see Sally Simpson, Nicole Leeper Piquero, and Raymond Paternoster, "Rationality and Corporate Offending Decisions," in Piquero and Tibbetts, *Rational Choice*, 25–40.

[31]Much of this discussion is taken from Brown et al., *Explaining Crime*, and Messner et al., *Theoretical Integration*.

Obviously, attempts to integrate such theories face the problematic hurdles of dealing with such obvious contradictions, and many formulations that do merge some of these perspectives simply do not deal with this issue. The failure to acknowledge, let alone explain, such inconsistencies is likely to result in regression of theoretical development instead of leading to progress in understanding, which is the primary goal of integrating theories in the first place.

Other experts have argued that any attempts to integrate theories must unite three different levels of analysis, which include individual (micro), group (macro), and microsituational, which merges the micro level with spontaneous context.[32] While we agree that this would be the ideal for theory formulation, we know of no theory that does so. Therefore, this proposition is more of an ideal that has not yet been attempted. So we are inclined to go with the best that has been offered by expert theorists. Still, in the future, it is hoped that such an integrated model will be offered that addresses all of these aspects in an explanatory model.

Another argument against theoretical integration is the stance that explanatory perspectives of crime are meant to stand alone. This is the position taken by Travis Hirschi, who is one of the most cited and respected scholars in criminology over the last 40 years.[33] As others have noted, Hirschi informed his position on this debate unequivocally when he stated that "separate and unequal is better" than integrating traditionally independent theoretical models.[34]

This type of perspective is also called *oppositional tradition* or *theoretical competition*, because the separate theories are essentially pitted against one another, in a form of battle or opposition. Although scientists are trained always to be skeptical of their own beliefs and open to other possibilities, especially the desire to refine theoretical models that are shown to be invalid, it is surprising that such a position exists. Despite the rather unscientific nature of this stance, such a position of oppositional tradition has many supporters. Specifically, one of the most respected and cited criminologists, Travis Hirschi claimed:

> The first purpose of oppositional theory construction is to make the world safe for a theory contrary to currently accepted views. Unless this task is accomplished, there will be little hope for the survival of the theory and less hope for its development. Therefore, oppositional theorists should not make life easy for those interested in preserving the status quo. They should instead remain at all times blind to the weaknesses of their own position and stubborn in its defense. Finally, they should never smile.[35]

Unfortunately, this position against theoretical integration is presented in a very unscientific tone. After all, scientists should always be critical of their own views and theory. By stating that theorists should be "blind" to opposing viewpoints and "stubborn" in their own perspective's defense, Hirschi advocates a position that is absolutely against science. Still, this statement, albeit flawed, shows the extreme position against theoretical integration and is favorable toward having each independent theory standing opposed to

---

[32]James Short, "The Level of Explanation Problem in Criminology," in *Theoretical Methods in Criminology*, ed. R. Meier (Beverly Hills, CA: Sage, 1985), 42–71; see also James Short, "On the Etiology of Delinquent Behavior," *Journal of Research in Crime and Delinquency* 16 (1979): 28–33.

[33]Bernard and Snipes, "Theoretical Integration"; Hirschi, "Exploring Alternatives."

[34]Hirschi, "Separate but Unequal."

[35]Hirschi, "Exploring Alternatives," 45.

others. It is our position that this stand does not have much defense, which is shown by Hirschi's lack of rational argument. Furthermore, it is generally acceptable to smile when presenting any theoretical or empirical conclusions, even when they involve opposition to or acceptance of integrated theoretical models of criminal behavior.

Perhaps one of the most important criticisms against Hirschi's and others' criticisms of integrated models is that most traditional models alone only explain a limited amount of variation in criminal activity. Elliott and colleagues have claimed:

> Stated simply, the level of explained variance attributable to separate theories is embarrassingly low, and, if sociological explanations for crime and delinquency are to have any significant impact upon future planning and policy, they must be able to demonstrate greater predictive power.[36]

While some put this estimate at 10 to 20 percent of the variance in illegal activities, this is simply an average across different theories and various forms of deviant behavior.[37] However, this range is an overestimate regarding many studies on the accuracy of separate theories, which tend to show weak support (well under 10 percent explained variation in offending), particularly for some tests of social bonding and strain models, as well as others.[38]

On the other hand, this estimated range of explained variance underestimates the empirical validity of some theoretical frameworks that consistently show a high level of explained variation in certain criminal behaviors. For example, a large number of studies examining Akers's differential reinforcement/social learning theory (discussed in Section VII), which examined not only a wide range of samples (in terms of age, nationality, and other demographic characteristics) but also a large range of deviant activities (e.g., cigarette smoking to drug usage to violent sexual crimes) consistently account for more than 20 percent of variation in such behaviors.[39] Specifically, most of these studies estimate that

---

[36]Delbert Elliott, David Huizinga, and S. Ageton, *Explaining Delinquency and Drug Use* (Beverly Hills, CA: Sage, 1985), 125, as quoted in Bernard and Snipes, "Theoretical Integration," 306.

[37]This estimate can be found in Bernard and Snipes, "Theoretical Integration," 306, but is based on the estimates of others, as discussed in this work.

[38]See review in Akers and Sellers, *Criminological Theories*, particularly p. 97.

[39]Ronald Akers, Marvin Krohn, Lonn Lanza-Kaduce, and Maria Radosevich, "Social Learning and Deviant Behavior: A Specific Test of a General Theory," *American Sociological Review* 44 (1979): 635–55; Marvin Krohn and Lonn Lanza-Kaduce, "Community Context and Theories of Deviant Behavior: An Examination of Social Learning and Social Bonding Theories," *Sociological Quarterly* 25 (1984): 353–71; Lonn Lanza-Kaduce, Ronald Akers, Marvin Krohn, and Marcia Radosevich, "Cessation of Alcohol and Drug Use Among Adolescents: A Social Learning Model," *Deviant Behavior* 5 (1984): 79–96; Ronald Akers and John Cochran, "Adolescent Marijuana Use: A Test of Three Theories of Deviant Behavior," *Deviant Behavior* 6 (1985): 323–46; Ronald Akers and Gang Lee, "Age, Social Learning, and Social Bonding in Adolescent Substance Use," *Deviant Behavior* 19 (1999): 1–25; Marvin Krohn, William Skinner, James Massey, and Ronald Akers, "Social Learning Theory and Adolescent Cigarette Smoking: A Longitudinal Study," *Social Problems* 32 (1985): 455–473; Ronald Akers and Gang Lee, "A Longitudinal Test of Social Learning Theory: Adolescent Smoking," *Journal of Drug Issues* 26 (1996): 317–43; Ronald Akers, Anthony La Greca, John Cochran, and Christine Sellers, "Social Learning Theory and Alcohol Behavior Among the Elderly," *Sociological Quarterly* 30 (1989): 625–38; Scot Boeringer, Constance Shehan, and Ronald Akers, "Social Contexts and Social Learning in Sexual Coercion and Aggression: Assessing the Contribution of Fraternity Membership," *Family Relations* 40 (1991): 558–564; Sunghyun Hwang and Ronald Akers, "Adolescent Substance Use in South Korea: A Cross-Cultural Test of Three Theories," in *Social Learning Theory and the Explanation of Crime: A Guide for the New Century,* ed. Ronald Akers and Gary Jensen (New Brunswick, NJ: Transaction, 2003), 39–64.

Akers's social learning model explains up to 68 percent or more of the variation in certain deviant behaviors, with the lowest estimate around 30 percent.

Obviously, not all independent theories of crime lack empirical validity, so this does not support critics' claims that traditionally separate theories of crime do not do a good job in explaining criminal behavior. However, it is also true that many of the theories that do the best job in empirical tests for validity are those that are somewhat integrated, in the sense that they often have been formed by merging traditional theories with other constructs and propositions, much like Akers's differential reinforcement theory, which added more modern psychological concepts and principles (e.g., operant conditioning and modeling) to Sutherland's traditional theory of differential association (see Section VII). So in a sense, an argument can be made that theoretical integration (or at least theoretical elaboration) had already occurred, which made this theory far more empirically valid than the earlier model.

Another example of the high level of empirical validity of existing models of offending can be found in some models of rational choice, which have been revised through theoretical elaboration and have explained more than 60 percent of the explained variation in deviant behavior.[40] However, much of the explanatory power of such frameworks relies on incorporating the constructs and principles of other theoretical models, which is what science is based on; specifically, they revise and improve theory based on what is evident from empirical testing. After all, even some of the harshest critics of theoretical integration admit that traditional theories do not "own" variables or constructs.[41] For example, Hirschi claimed that:

> Integrationists somehow conclude that variables appear . . . with opposition theory labels attached to them. This allows them to list variables by the theory that owns them. Social disorganization theory . . . might own economic status, cultural heterogeneity, and mobility. . . . Each of the many variables is measured and . . . the theories are ranked in terms of the success of their variables in explaining variation in delinquency . . . such that integration is in effect *required* by the evidence and surprisingly easily accomplished.[42]

This is the way that science and theoretical development and revision are supposed to work, so in our opinion, this is exactly as it should be. All scientists and theoreticians should constantly be seeking to improve their explanatory models and be open to ways to do so, as opposed to being staunch supporters of one position and blind to existing evidence.

Despite the criticisms against theoretical integration, a strong argument has been made that theoretical competition and oppositional tradition is "generally pointless."[43] A big reason for this belief of proponents of theoretical integration is that various theories tend to explain different types of crime and varying portions of the causal processes for

---

[40]For example, see Tibbetts, "College Student Perceptions," which showed that an elaborated rational choice model explained more than 60 percent of variation in test cheating among college students.

[41]See discussion in Bernard and Snipes, "Theoretical Integration," 306–7.

[42]Hirschi, "Exploring Alternatives," 41, as cited (revised) in Bernard and Snipes, "Theoretical Integration," 307.

[43]Quote from Bernard and Snipes, "Theoretical Integration," 306, based on the rationale provided by Elliott et al., *Explaining Delinquency*.

behavior. For example, some theories focus more on property crimes while others place emphasis on violent crimes, and some theories emphasize the antecedent or root causes (e.g., genetics, poverty) of crime while others focus on more immediate causes (e.g., current social context at the scene). Given that there are multiple factors that contribute to crime and that different factors are more important for various types of crime, it only makes sense that a synthesis of traditionally separate theories must come together to explain the wide range of criminal activity that occurs in the real world.

Ultimately, there are both pros and cons regarding whether or not to integrate theories. It is our belief that theoretical integration is generally a good thing, as long as there is caution and attention given to merging models that have opposing assumptions, such as the natural state of human beings (e.g., good vs. bad vs. blank slate). But only after considerable empirical research will the true validity of integrated models be tested, and many have already been put to the test. We will now examine a handful of integrated theories that have been proposed over the last couple of decades, as well as the studies that have examined their empirical validity. Not surprisingly, some integrated and elaborated theories appear to be more valid than others, with most adding considerably to our understanding of human behavior and contributing to explaining the reasons why certain individuals or groups commit criminal behavior more than others.

Examples of Integrated Criminological Theory

We have already discussed the advantages and disadvantages of theoretical integration, as well as the ways in which traditionally separate explanatory models are combined to form a new synthesized framework. We will now review a number of the most prominent examples of theoretical integration that have been proposed in the last 30 years, which is largely the time period when most attempts at integration have been presented. We hope that readers will critique each theory based on the criteria that we have already discussed, particularly noting the empirical validity of each model as based on scientific observation and logical consistency of its propositions.

Elliott's Integrated Model. Perhaps the first and certainly the most prominent integrated model is that proposed by Delbert Elliott and his colleagues in 1979, which has become known as **Elliott's integrated model.**[44] In fact, this model "opened the current round of debate on integration," because it was essentially the first major perspective proposed that clearly attempted to merge various traditionally separate theories of crime.[45] Elliott's integrated framework attempts to merge strain, social disorganization, control, and social learning/differential association-reinforcement perspectives for the purpose of explaining delinquency, particularly in terms of drug use, as well as other forms of deviant behavior (see Figure 10.1).

As can be seen in Figure 10.1, the concepts and propositions of strain and social disorganization, as well as inadequate socialization, are considered antecedent (or root)

---

[44]Delbert Elliott, S. Ageton, and R. Cantor, "An Integrated Theoretical Perspective on Delinquent Behavior," *Journal of Research in Crime and Delinquency* 16 (1979): 3–27; for further elaboration and refinement of this theory, see Elliott et al., *Explaining Delinquency.*

[45]Bernard and Snipes, "Theoretical Integration," 310.

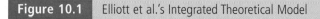

Figure 10.1   Elliott et al.'s Integrated Theoretical Model

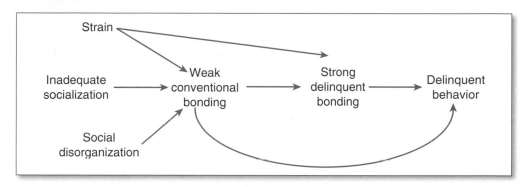

causes of delinquency. In other words, failing to achieve one's goals (i.e., strain theory) or coming from disadvantaged neighborhoods (i.e., social disorganization) are key causes for predisposing people to criminal behavior. Furthermore, the fact that many low-income households tend to lack adequate socialization, such as when a single parent has to work two or three jobs to make ends meet, is also a major root cause of delinquency.

Because this model clearly shows some models/constructs coming first (e.g., strain, social disorganization) and others coming later (e.g., weak bonding and then affiliations with delinquents), which lead to criminality, this is a good example of an "end-to-end" form of theoretical integration. This is an "end-to-end" form of integration because some models/concepts, such as strain, occur chronologically first, which then lead chronologically to other models/concepts, such as weak conventional bonding or strong delinquent bonding, which then leads to crime.

The notable ways in which this perspective becomes a true integrated model is seen in the mediating or intervening variables. Although some antecedent variables (such as strain) can lead directly toward delinquent activity, most of the criminal activity is theoretically predicted through a process that would include a breakdown of conventional bonding (i.e., social control/bonding theory), which occurs in many individuals who experience strain or social disorganization in their neighborhoods and inadequate socialization. Furthermore, individuals who have such a breakdown in conventional bonding tend to be more highly influenced by the associations that they make in the streets among their peers (i.e., differential association-reinforcement/social learning theory). According to Elliott's integrated theory, this factor—strong delinquent bonding—most directly results in delinquent behavior among most juvenile offenders.

One of the notable features of this theoretical model is that it allows for various types of individuals to become criminal. In other words, unlike traditionally separate frameworks that assume offenders expect not to achieve their goals (e.g., strain theory) or come from bad neighborhoods (e.g., Chicago/social disorganization theory), Elliott's integrated theory allows for a variety of possibilities when it comes to causal paths that explain how crime and delinquency develop in certain people and groups. This is what makes integrated models so much more powerful; namely, they bring several valid explanations together and allow for various possibilities to occur, all of which explain some criminality.

Whereas traditional theories largely only provide for one causal process, Elliott and his colleagues showed right from the first major integrated theory that different types of trajectories, or paths to crime, are possible.

One of the major criticisms of merging such theories, particularly strain (which assumes individuals are born relatively good) and control/social bonding (which clearly assumes that people are born relatively bad [i.e., selfish, greedy, aggressive, etc.]), is that they tend to have extremely different, even opposite, assumptions of human behavior. As a recent review of theoretical integration noted, Elliott and his colleagues attempted to circumvent this obvious contrast in basic assumptions by claiming that the model allows for variation in individual motivations for why people engage in delinquency and crime.[46] For example, they claim that failure to achieve one's goals (i.e., strain theory) is not always the motivation for crime; rather, crime can result from inadequate socialization or coming from a disadvantaged neighborhood.

Furthermore, Elliott's integrated model also allows for different forms of control or social bonding, with not all of the delinquents being required to have weak social bonds with conventional society. For instance, as can be seen in Figure 10.1, a person who has experienced strain (or failure to achieve one's goals) can move directly to the social learning/differential association-reinforcement variables of strong delinquent bonding, so that the weak conventional bonding construct is not a required causal process in this theoretical model. So while some critics claim that Elliott and his colleagues have combined theories that simply cannot be synthesized due to contrasting assumptions, we believe that the integrated theory did, in fact, find a logical and consistent way of showing how both models can be merged in a way that makes a lot of sense in the real world. On the other hand, the model does indeed claim that strain directly causes weak conventional bonding, and the theoretical framework implies that the probability of delinquency is highest when an individual experiences both strain and weak conventional bonding, so to some extent the critics make an important point regarding the logical consistency of the full model, which Elliott and his colleagues have not adequately addressed.[47]

Despite the presence of elements of the four traditionally separate theories (strain, social disorganization, social control/bonding, and social learning/differential association-reinforcement), Elliott and his colleagues identify this integrated theory with social control as opposed to any of the other perspectives, which they argue is a more general theory and can explain crime and delinquency across different levels of explanation. They also note that the social control perspective is more sociological, in the sense that it places more importance on the role of institutional structures in controlling criminal behavior.[48] Perhaps another reason why they identified their model with social control/bonding theory was because both intervening constructs represent types of bonds formed or not formed with others: weak conventional bonding and strong delinquent bonding. However, it is important to keep in mind that these constructs actually represent two traditionally separate theories, namely social bonding theory and differential association-reinforcement/social learning, respectively.

The authors of this textbook believe that when an integrated theory (which claims to be such an integrated model) chooses a primary or dominant theory as its basis, it does

---

[46]Ibid.

[47]Ibid.

[48]Ibid., 311–12.

not help in selling it to the scientific community, let alone others. A similar problem was seen in Pearson and Weiner's conceptual integration in their identifying a single traditionally separate theory—social learning/differential association-reinforcement—as the foundation or basis for their framework. Obviously, the first reaction by many theorists, even those that are not inherently against theoretical integration, would be somewhat cautious or even resistant. After all, why would a theory that claims to be integrated outright claim a single explanatory framework as its basis.

Rather, such models may be seen more so as examples of theoretical elaboration, which tends to start with the assumptions and concepts of one theory and draw from other models to improve the base model. Still, despite the criticisms toward identifying a single model as a basis for developing an integrated framework for explaining criminality, it is apparent that latitude has been shown for this flaw. It is important to note that Elliott's model is considered the first "true" attempt at theoretical integration and is still widely respected as the prototype and example of what an integrated model could and should be.

Regarding the empirical evidence for and against Elliott's integrated model, much of the evidence for it has been provided by Elliott and his colleagues through their testing of the theory. Specifically, much of the testing they have done has been via the National Youth Survey (NYS), a national survey collected and synthesized by criminologists at the University of Colorado, Boulder, which is where Elliott and most of the colleagues with whom he works are professors. This longitudinal measure of delinquency has been administered and analyzed for several decades, and it represents perhaps the most systematic collection of information from youth regarding key developmental variables and delinquency rates that has ever existed.

Most of the evidence from the National Youth Survey shows general strong support for Elliott's model.[49] However, some evidence showed few direct effects of strain and social control concepts, which is surprising given that the basis for the model was the social control/bonding elements of the theory.[50] In fact, the original hypothesis in the model—that strain and social control/bonding theories would have a direct effect on delinquency—were not observed.[51] Rather, only bonding to delinquent peers had a significant and strong effect on future criminality, which supports social learning/differential association-reinforcement theory. This strongly supports the social learning variables in the model and diminishes the claim by Elliott and his colleagues about the fundamental theoretical perspective of social control/bonding.

Furthermore, a critical review of Elliott's framework, as presented in his 1985 book (with Huizinga and Ageton), notes that a major problem with the integrated framework is that:

> The most puzzling feature of the theory is the inclusion of social disorganization in it as a causal factor, in the absence of any attempt to measure or test the importance of this factor. Presumably, the authors wished to claim that their theory was "more sociological than psychological."[52]

---

[49]See review in Akers and Sellers, *Criminological Theories*, 273–76.

[50]Elliott, Huizinga, and Ageton, *Explaining Delinquency*.

[51]Akers and Sellers, *Criminological Theories*, 275.

[52]David Farrington, book review of Elliott et al., *Explaining Delinquency*, in *British Journal of Addiction* 81 (1986): 433; embedded quote taken from p. 67 in Elliott et al.'s book.

This point is particularly important, given that virtually none of the tests of Elliott's integrated model have included social disorganization factors in their studies, even after this critical review was published more than 20 years ago. So Elliott and his colleagues should either drop this portion of his model from consideration in the model or provide an adequate test of it in relation to the rest of the framework. Although we would opt for the latter, one of these two alternatives must be chosen.

However, from the presentation that Elliott gave as his presidential address to the American Society of Criminology in 1993, it appears that the former alternative was chosen. Specifically, in his model of the onset of serious violent offending, all of the antecedent variables in his model could be explained by the other traditionally separate theoretical models in his framework. He included two bonding constructs (family and school), parental sanctions, stressful life events (i.e., strain), and early exposure to crime and victimization. Perhaps the last factor could be construed as related to social disorganization theory, but it is probably regarded more as a social learning/differential association-reinforcement variable, as is suggested from the original model.

In addition, this critique noted that the existing evidence shows that most delinquent activity is committed among groups of juveniles, and thus, youth who tend to commit illegal acts naturally associate with delinquents. Yet, the affiliation with delinquent peers is not considered the basic foundation of the theory—as tests of this model demonstrated—but rather an intervening variable in the model. The author of this critique claims that Elliott's model does not emphasize this point strongly enough and thereby does not provide "convincing evidence in this book that delinquent peers cause or facilitate offending."[53]

In light of these findings, even Elliott and his colleagues acknowledged that their integrated model best fit the data as a social learning/differential association-reinforcement framework of delinquency.[54] However, they chose to retain social control/bonding theory as the primary foundation of their integrated model, stating that "it is not clear that a social learning model would have predicted a conditional relationship between conventional bonding . . . and deviant bonding."[55] Elliott and colleagues went on to say that they did not attribute most of the explained variation in the model on social learning/differential reinforcement as being the most important model or construct because they claimed that such variables did not play a strong enough part in the indirect effects that were seen in the estimated models.

This is actually a logical position, given the fact that their integrated framework would, in fact, predict the social learning/differential association-reinforcement variables were the primary direct effects on delinquency, and the social control, strain, and social disorganization variables were considered primarily indirect all along, according to their model. However, some critics, especially the proponents of social learning/differential reinforcement theory, have claimed that:

---

[53]Farrington, book review, 433.

[54]Ibid.

[55]Ibid., based on quote from Elliott et al. *Explaining Delinquency*, 137.

In our opinion, even with the addition of the interactive effects of conventional bonding, the final model reported by Elliott et al. is more of a variation on social learning theory (with bonding modifications) than it is a variation on social bonding theory (with learning modifications).[56]

After all, the measures of delinquent peer bonds used by Elliott and his colleagues are essentially measures that have been used by theorists and researchers of social learning theorists over the last few decades, regarding differential associations, reinforcements, and modeling.[57]

Additional analyses using the National Youth Survey have tended to agree with the critics regarding the importance of social learning/differential reinforcement variables in predicting crime and delinquency. Specifically, one relatively recent reanalysis of the NYS data showed that social learning variables appeared to predict more variation in deviance than did the other models or constructs in the model.[58] Some have noted that the social bonding assumption—that strong attachments to others, regardless of who they are—is not supported by empirical research (see Section VII). This is true, and all future integrated models must address this issue if they include social control/bonding propositions or constructs in their models.

Despite these criticisms and empirical observations, it appears that Elliott's integrated model has contributed to our understanding of the development of delinquent behavior. In the least, it inspired other theoretical frameworks, some of which we review here. However, it should also be obvious that there are some valid criticisms of this model, such as claiming that the model is based on social control/bonding theory while depending heavily on the strong delinquent bonding that takes place in most cases of criminality, clearly implicating social learning/differential reinforcement theory. Again, we want to stress that a true integrated model should not place any emphasis on a particular theory, otherwise the critics will have a sound argument when findings show that one theory is more influential than another.

Thornberry's Interactional Theory. After Elliott's integrated model, presented in 1979, the next major integrated framework was that of Terrence Thornberry in 1987.[59] This model incorporated empirical evidence drawn since Elliott's presentation to create a unique and insightful model of criminality, which addressed an extremely important aspect that had never been addressed previously in criminological theory. Specifically, Thornberry's **interactional theory** was the first to emphasize reciprocal, or feedback effects, in the causal modeling of the theoretical framework.

---

[56]Akers and Sellers, *Criminological Theories*, 275.

[57]Ibid., 276.

[58]Robert Agnew, "Why Do They Do It? An Examination of the Intervening Mechanisms Between 'Social Control' Variables and Delinquency," *Journal of Research in Crime and Delinquency* 30 (1993): 245–66.

[59]Terrence Thornberry, "Toward an Interactional Theory of Delinquency," *Criminology* 25 (1987): 863–87; see also Thornberry, "Reflections"; Terrence Thornberry, A. Lizotte, M. Krohn, M. Farnworth, and S. Jang, "Testing Interactional Theory: An Examination of Reciprocal Causal Relationships Among Family, School and Delinquency," *Journal of Criminal Law and Criminology* 82 (1991): 3–35; Terrance Thornberry, Alan Lizotte, Marvin Krohn, Margaret Farnworth, and Sung Joon Jang, "Delinquent Peers, Beliefs, and Delinquent Behavior: A Longitudinal Test of Interactional Theory," *Criminology* 32 (1994): 47–83.

As a basis for his model, Thornberry combined social control and social learning models. According to Thornberry, both of these theories try to explain criminality in a straightforward, causal process and are largely targeted toward a certain age population.[60] Thornberry uniquely claims that the processes of both social control and social learning theory affect each other in a type of feedback process.

Thornberry's integrated model incorporates five primary theoretical constructs, which are synthesized in a comprehensive framework to explain criminal behavior. These five concepts are: commitment to school; attachment to parents; belief in conventional values (these first three are taken from social control/bonding theory); adoption of delinquent values; and association with delinquent peers (these last two are drawn from social learning/differential association-reinforcement theory). These five constructs, which most criminologists would agree are important in the development of criminality, are obviously important in a rational model of crime, so at first it does not appear that Thornberry has added much to the understanding of criminal behavior. Furthermore, Thornberry's model clearly points out that different variables will have greater effects at certain times; for example, he claims that association with delinquent peers will have more effect in mid-teenage years than at other ages.

What Thornberry adds beyond other theories is the idea of reciprocity or feedback loops, which no previous theory had mentioned, much less emphasized. In fact, much of the previous criminological literature had spent much time debating whether individuals become delinquent and then start hanging out with similar peers or whether individuals start hanging out with delinquent peers and then begin engaging in criminal activity. This has been the traditional "chicken or egg" question in criminology for most of the 20th century; namely, which came first, delinquency or bad friends. It has often been referred to as the "self-selection" versus "social learning" debate; in other words, do certain individuals select themselves to hang out with delinquents by their previous behavior, or do they learn criminality from delinquents with whom they start associating. One of the major contributions of Thornberry's interactional model is that he directly answered this question.

Specifically, Thornberry noted that most, if not all, contributors to delinquency (and criminal behavior itself) are related reciprocally. Thus, Thornberry postulated that engaging in crime leads to hanging out with other delinquents *and* that hanging out with delinquents leads to committing crimes. It is quite common for individuals to commit crime and then start hanging out with other peers who are doing the same, and it is also quite common for people to start hanging out with delinquent peers and then start committing offenses. Furthermore, it is perhaps the most likely scenario for a person to be offending and to be dealing with both the influences of past experiences and with peer effects as well.

As mentioned previously, Thornberry considers the social control/bonding constructs, such as attachments to parents and commitment to school, the most essential predictors of delinquency. Like previous theoretical models of social bonding/control, Thornberry's model puts the level of attachments and commitment to conventional society ahead of the degree of moral beliefs that individuals have toward criminal offending. However, lack of such moral beliefs leads to delinquent behavior, which in turn negatively

---

[60]Much of this discussion is taken from Bernard and Snipes, "Theoretical Integration," 314–16; and also Akers and Sellers, *Criminological Theories*, 278.

affects the level of commitment or attachments an individual may have built in his/her development. As Thornberry claimed:

> While the weakening of the bond to conventional society may be an initial cause of delinquency, delinquency eventually becomes its own indirect cause precisely because of its ability to weaken further the person's bonds to family, school, and conventional beliefs.[61]

Thus, the implications of this model are that variables relating to social control/bonding and other sources cause delinquency, which then becomes, in itself, a predictor and cause for the breakdown of other important causes of delinquency and crime.

For example, consider a person we shall call Johnny, who has an absent father and a mother who uses inconsistent discipline and sometimes harsh physical abuse of her son. He sees his mother's state of constant neglect and abuse as proof that belief in conventional values is wrong, and he becomes indifferent toward governmental laws; after all, his main goal is to survive and be successful. Because of his mother's psychological and physical neglect, Johnny pays no attention to school and rather turns to his older peers for guidance and support. These peers guide him toward behavior that gives him both financial reward (selling what they steal) and status in their group (respect from performing well in illegal acts). At some point, Johnny gets caught, and this makes the peers who taught him how to engage in crime very proud, while alienating him from the previous bonds he had with his school, where he may be suspended or expelled, and with his mother, who further distances herself from him. This creates a reciprocal effect or feedback loop to the previous factors, which was lack of attachment to his mother and commitment to school. The lowered level of social bonding/control with conventional institutions/factors (mother, school), and increased influence by the delinquent peers then leads Johnny to commit more frequent and more serious crimes.

Such a model, although complex and hard to measure, is logically consistent, and the postulates are sound. However, the value of any theory has to be determined by the empirical evidence that is found regarding its validity. Much of the scientific evidence regarding Thornberry's empirical model has been contributed by Thornberry and his colleagues.

Although the full model has yet to be tested, the researchers "have found general support for the reciprocal relationships between both control concepts and learning concepts with delinquent behavior."[62] One test of Thornberry's model used the longitudinal Rochester Youth Development Study to test its postulates.[63] This study found that the estimates of previous unidirectional models (nonreciprocal models) did not adequately explain the variation in the data. Rather, the results supported the interactional model, with delinquent associations leading to increases in delinquency while delinquency led to reinforcing peer networks and both directional processes working through the social environment. In fact, this longitudinal study demonstrated that once the participants had acquired

---

[61]Thornberry, "An Interactional Theory," 876, as quoted by Bernard and Snipes, "Theoretical Integration," 315.

[62]Bernard and Snipes, "Theoretical Integration," 316.

[63]Thornberry et al., "Delinquent Peers."

delinquent beliefs from their peers, the effects of these beliefs had further effects on their future behavior and associations, which is exactly what Thornberry's theory predicts.[64]

Perhaps the most recent test of Thornberry's interactional model examined the age-varying effects of the theory.[65] This study incorporated hierarchical linear modeling in investigating a sample of the National Youth Survey, a national sample of youth. The results showed that while the effects of delinquent peers were relatively close to predictions, peaking in the mid-teenage years, the predictions regarding the effects of family on delinquency were not found to be significant in the periods that were expected, although family was important during adolescence.

Unlike other authors of integrated theories, Thornberry specifically noted that he prefers to see his approach as "theoretical elaboration," and not full theoretical integration. While we commend Thornberry for addressing this concern, virtually all criminologists still consider his framework a fully integrated model. And in many ways, Thornberry's interactional model is far more integrated than others discussed here because it gives equal weight to the traditionally separate theoretical frameworks that are combined into his model, in the sense that both are considered antecedent and reciprocal in their effects on criminal behavior.

Braithwaite's Theory of Reintegrative Shaming. A unique integrated model that proposed the synthesis of several traditionally separate theories was presented in 1989, in a book entitled *Crime, Shame, and Reintegration.*[66] The theory of **reintegrative shaming** merges constructs and principles from several theories, primarily social control/bonding theory and labeling theory, with elements of strain theory, subculture/gang theory, and differential association theory. All of these theories are synthesized in a clear and coherent framework that is presented in both descriptive and graphic form. We will spend extra time discussing Braithwaite's theory because it discusses not only U.S. culture but also cultural and justice tendencies in Japanese culture.

Braithwaite's idea for the theory was obviously inspired by the cultural differences he observed in Eastern (particularly Japanese) culture, in terms of both socialization practices and the justice system, as compared to the Western world. (Note that Braithwaite is from Australia, which uses the same Western practices as England and the United States.) Specifically, he emphasizes the Japanese focus on the aggregate, such as family, school, or business, to which the individual belongs. In contrast, in many Western cultures, epitomized by U.S. culture, the emphasis is clearly placed on the individual.

This contrast has often been referred to as the "we" culture versus the "me" culture (Eastern versus Western emphases, respectively). Although this is seen in virtually all aspects of culture and policy practices, it is quite evident in people's names. In most Eastern cultures, people are known by their family name, which is placed first in the ordering. This shows the importance that is placed on the group to which they belong. In contrast, Western societies listing individual names first, implying a focus on the individual him/herself. These naming practices are a manifestation of a virtually all-encompassing

---

[64]Ibid.

[65]Sung Joon Jang, "Age-Varying Effects of Family, School, and Peers on Delinquency: A Multilevel Modeling Test of Interactional Theory," *Criminology* 37 (1999): 643–85.

[66]John Braithwaite, *Crime, Shame, and Reintegration* (New York: Cambridge University Press, 1989).

cultural difference regarding group dynamics and social expectations across societies, especially their justice systems. For example, it is quite common in Japan to receive a sentence of apologizing in public, even for the most serious violent crimes.[67] In his book, Braithwaite points out that after World War II, Japan was the only highly industrializing society that showed a dramatic decrease in its crime rate.

Criminological theory would predict Japan to have an increasing crime rate, given the extremely high density of urban areas due to rapid industrialization, especially on such a small amount of land, Japan being a large island. As Braithwaite describes it, the Japanese suffered from anomie after the war in the sense of a general breakdown of cultural norms, but the nation was able to deal with this anomic state despite the odds. Japan definitively decided not to follow the Western model of justice after the war; rather, they rejected the Western system of stigmatizing convicted felons. Instead, the Japanese implemented a system in which they reintroduce (hence "reintegration") the offenders to society via a formal ceremony in which citizens accept the offenders back into conventional society. In contrast, in the United States, we typically give our "ex-cons" about $75 on average and make them promise always to identify themselves as felons on legal documents.

In other cultural contrasts, Japan is extremely lenient in sentencing offenders to prison. In contrast, by rate of incarceration the United States is the most punitive developed nation. In Japan, Braithwaite notes that:

> Prosecution only proceeds in major cases . . . where the normal process of apology, compensation, and forgiveness by the victim breaks down. Fewer than 10 percent of those offenders who are convicted receive prison sentences, and for two-thirds of these, prison sentences are suspended. Whereas 45 percent of those convicted of a crime serve a jail sentence in the U.S., in Japan the percentage is under two.[68]

Public apology is the most common punishment among the Japanese, which strongly reflects the nature of honor in Japanese society, as well as pointing out the fundamental differences between how we deal with offenders. Braithwaite claims that this cultural and political difference has a huge impact on why crime rates in both nations have experienced such differential trends.

Most developing Western nations, including the United States, experienced a rising crime rate in the 1950s, 1960s, and 1970s. Braithwaite argues that this was likely due to culture and the differential treatment of its offenders, namely "it might be argued that this [the downward trend in crime rates after World War II in Japan] was a result of the re-establishment of cultural traditions of shaming wrongdoers, including the effective coupling of shame and punishment."[69] In contrast, Braithwaite claims that "one contention . . . is that the uncoupling of shame and punishment manifested in a wide variety of ways in many Western countries is an important factor in explaining the rising crime rates in those countries."[70]

---

[67] Ibid., 62.

[68] Ibid., citing findings from J. Haley, "Sheathing the Sword of Justice in Japan: An Essay on Law Without Sanctions," *Journal of Japanese Studies* 8 (1982): 265–81.

[69] Braithwaite, *Crime Shame*, 61, brackets are authors' paraphrasing.

[70] Ibid, 61.

Furthermore, in contrast to American society, the Japanese are typically less confrontational with authority. For example, some scholars have noted that the Japanese accept the authority of law, and police officers are considered similar to "elder brothers" who rely "on positive rather than negative reinforcement," when it comes to crime control.[71] This difference in the way officers and other authority figures are considered is likely due to the way that the Japanese view society, in terms of neighborhood, community, school, and work—the informal institutions of control that we have mentioned earlier in this book as having more effect on the crime rate than the formal institutions (i.e., police, courts, prisons).

Beyond discussing the cultural differences, Braithwaite's integrated theory addresses some of the most notable scientific observations regarding what types of individuals and groups are likely to commit crime. Specifically, Braithwaite states that most crime is committed by young, poor, single males who live in urban communities that have high levels of residential mobility; they are not likely attached to or do poorly in school, have low educational/occupational aspirations, are not strongly attached to their parents, associate with delinquent peers, and do not have moral beliefs against violating law.[72]

It is obvious from this list that Braithwaite incorporated some of the major theories and corresponding variables into his theoretical perspective. The emphasis on poor people who do not have high educational/occupational aspirations obviously supports strain theory, whereas the inclusion of urban individuals who live in communities with high residential mobility reflects the Chicago School/social disorganization theory. At the same time, Braithwaite clearly highlights the predisposition of people who have limited moral beliefs and weak attachments, which conjures images of social bonding/control theory; individuals having delinquent peers obviously supports differential association/reinforcement theory. Thus, a handful of theories are important in the construction of Braithwaite's integrated theory.

Braithwaite notes that much of the effectiveness of the Japanese system of crime control depends on two constructs: interdependency and communitarianism. Interdependency is the level of bonds that an individual has to conventional society, such as the degree to which they are involved or attached to conventional groups in society, which would include employment, family, church, and organizations. According to Braithwaite, interdependency is the building block of the other major theoretical construct in his integrated model: communitarianism. Communitarianism is the macro (group) level of interdependency, meaning that it is the cumulative degree of the bonds that individuals have with the conventional groups and institutions (e.g., family, employment, organizations, church, etc.) in a given society. Obviously, these theoretical constructs are highly based on the theory of social bonding/control, in the sense that they are based on attachments, commitment, and involvement in conventional society.

Braithwaite's model has a causal ordering that starts with largely demographic variables, such as age, gender, marital status, employment, and aspirations on the individual

---

[71]D. H. Bayley, *Forces of Order: Police Behavior in Japan and the United States* (Berkeley, CA: University of California Press, 1976).

[72]Braithwaite, *Crime, Shame*, 44–53.

level and urbanization and residential mobility on the macro (group) level. All of these factors are predicted to influence the level of interdependency and communitarianism in the model, which as we previously discussed is largely based on social control/bonding theory. Depending on which type of culture is considered, or what forms of shaming are used in a given jurisdiction, the various types of shaming that are administered are key in this integrated model.

According to Braithwaite, societies that emphasize reintegrative shaming, such as Japan, will reduce the rates of crime in their society. When an offender in Japan completes his or her sentence or punishment for committing a crime, the government will often sponsor a formal ceremony in which the offender is reintroduced or "reintegrated" back into conventional society. According to Braithwaite, we do not reintegrate offenders into our society after shaming them but rather stigmatize them, which leads them to associate only with people from criminal subcultures (e.g., drug users, gang members). Braithwaite claims that this leads to the formation of criminal groups and the grasping of illegitimate opportunities offered in the local community. Ultimately, this means that people who are not reintegrated into conventional society will inevitably be stigmatized and labeled as offenders, preventing them from becoming productive members of the community, even if their intentions are to do such.

Most empirical studies of Braithwaite's theory show mixed results, with most in favor of this theory, especially regarding its implementation for policy. Some tests have shown that reintegrative ideology regarding violations of law can have a positive impact on future compliance with the law; others have found that high levels of shaming in parental practices do not increase offending.[73] Studies in other countries, such as China and Iceland, show partial support for Braithwaite's theory.[74] While some studies have found encouraging results,[75] other recent studies have also found weak or no support for the theory.[76]

The most recent reports regarding the effects of shame show that outcomes largely depend on how shame is measured, which Braithwaite's theory largely ignores. There are, for example, episodic or situational shame states as well as long-term shame traits or propensities. Recent reviews of the literature and studies of how different types of shame are measured show that certain forms of shame are positively correlated with offending, whereas other forms of shame tend to inhibit criminal behavior.[77] If an individual persistently feels shame, they are more likely to commit criminal activity, but if persons who are

---

[73]Toni Makkai and John Braithwaite, "Reintegrative Shaming and Compliance with Regulatory Standards," *Criminology* 32 (1994): 361–386; Carter Hay, "An Exploratory Test of Braithwaite's Reintegrative Shaming Theory," *Journal of Research in Crime and Delinquency* 38 (2001): 132–53.

[74]Lu Hong, Zhang Lening, and Terance Miethe, "Interdependency, Communitarianism, and Reintegrative Shaming in China," *Social Science Journal* 39 (2002): 189–202; Eric Baumer, Richard Wright, Kristrun Kristinsdottir, Helgi Gunnlaugsson, "Crime, Shame, and Recidivism: The Case of Iceland," *British Journal of Criminology* 42 (2002): 40–60.

[75]Nathan Harris, Lode Walgrave, and John Braithwaite, "Emotional Dynamics in Restorative Conferences," *Theoretical Criminology* 8 (2004): 191–210; Eliza Ahmed and Valerie Braithwaite, "'What, Me Ashamed'? Shame Management and School Bullying," *Journal of Research in Crime and Delinquency* 41 (2004): 269–94.

[76]Bas Van Stokkom, "Moral Emotions in Restorative Justice Conferences: Managing Shame, Designing Empathy," *Theoretical Criminology* 6 (2002): 339–61; Charles Tittle, Jason Bratton, and Marc Gertz, "A Test of a Micro-Level Application of Shaming Theory," *Social Problems* 50 (2003): 592–617; Lening Zhang and Sheldon Zhang, "Reintegrative Shaming and Predatory Delinquency," *Journal of Research in Crime and Delinquency* 41 (2004): 433–53.

not predisposed to feel shame perceive that they would feel shame for doing a given illegal activity, then they would be strongly inhibited from engaging in such activity.[78] This is consistent with findings from another recent study that demonstrated that the effect of reintegrative shaming had an interactive effect on delinquency.[79]

Furthermore, Braithwaite's theory does not take into account other important self-conscious emotions, such as guilt, pride, embarrassment, and empathy, that are important when individuals are deciding whether or not to commit criminal behavior. Although some theorists claim that shame is the key social emotion, studies show that they are clearly wrong. Rather, many emotions, such as guilt and embarrassment, are based on social interaction and self-consciousness; in many ways, they are just as inhibitory or rehabilitating as shame.[80] The literature examining rational choice theory is also an indication of the effects of other emotions than shame in influencing decisions on offending.[81]

Still, Braithwaite's reintegrative theory provides an important step ahead in theoretical development, particularly in terms of combining explanatory models to address crime rates, as well as correctional policies/philosophies, across various cultures. Specifically, Braithwaite's theory makes a strong argument that Eastern (particularly Japanese) policies of reintegration and apology for convicted offenders are beneficial for their culture. Would this system work in Western culture, especially the United States? The answer is definitely unknown, but it is unlikely that such a model would work well in the United States. After all, most of the chronic, serious offenders in the United States would not be highly deterred by having to apologize to the public, and many might consider such an apology or reintegration ceremony as an honor, or a way to show that they weren't really punished. This may not be true, but we would not know this until such policies were implemented fully in our system. However, programs along these lines have been implemented in our country, and some of these programs will be reviewed later in this book.

Tittle's Control-Balance Theory. One of the more recently proposed models of theoretical integration is that of Charles Tittle's control balance theory. Presented by Tittle in 1995, control-balance integrated theory proposes that (1) the amount of control to which one is *subjected* and (2) the amount of control that one can *exercise* determine the probability of deviance occurring. In other words, the balance between these two types of control, he argued, can predict the *type* of behavior that is likely to be committed.[82]

In this integrated theoretical framework, Tittle claimed that a person is least likely to offend when he or she has a balance of controlling and being controlled. On the other

---

[77]Stephen Tibbetts, "Shame and Rational Choice in Offending Decisions," *Criminal Justice and Behavior* 24 (1996): 234–55.

[78]Stephen Tibbetts, "Self-Conscious Emotions and Criminal Offending," *Psychological Reports* 93 (2003): 201–31.

[79]Hay, "Exploratory Test."

[80]For an excellent review, see June Price Tangney and K. Fischer, *Self-Conscious Emotions: The Psychology of Shame, Guilt, Embarrassment, and Pride* (New York: Guilford Press, 1995).

[81]See Harold Grasmick, Brenda Sims Blackwell, and Robert Bursik, "Changes over Time in Gender Differences in Perceived Risk of Sanctions," *Law and Society Review* 27 (1993): 679–705; Harold Grasmick, Robert Bursik, and Bruce Arneklev, "Reduction in Drunk Driving as a Response to Increased Threats of Shame, Embarrassment, and Legal Sanctions," *Criminology* 31 (1993): 41–67; Tibbetts, "College Student Perceptions."

[82]Charles Tittle, *Control Balance: Toward a General Theory of Deviance* (Boulder, CO: Westview, 1995).

hand, the likelihood of offending will increase when these become unbalanced. If individuals are more controlled by external forces, which Tittle calls "control deficit," then the theory predicts they will commit predatory or defiant criminal behavior. In contrast, if an individual possesses an excessive level of control by external forces, which Tittle refers to as "control surplus," then that individual will be more likely to commit acts of exploitation or decadence. It is important to realize that this excessive control is not the same as excessive self-control, which would be covered by other theories examined in this introduction. Rather, Tittle argues that people who are controlling, that is, who have excessive control over others, will be predisposed toward inappropriate activities.

Early empirical tests of control-balance theory have reported mixed results, with both surpluses and deficits predicting the same types of deviance.[83] Furthermore, researchers have uncovered differing effects of the control balance ratio on two types of deviance that are contingent on gender. This finding is consistent with the gender-specific support found for Reckless's containment theory and with the gender differences found in other theoretical models.[84] Despite the mixed findings for Tittle's control balance theory of crime, this model of criminal offending still gets a lot of attention, and most of the empirical evidence has been in favor of this theory.

Hagan's Power-Control Theory. The final integrated theory that we will cover in this introduction deals with the influence of familial control and how that relates to criminality across gender. Power-control theory is another integrated theory that was proposed by John Hagan and his colleagues.[85] The primary focus of this theory is on the level of patriarchal attitudes and structure in the household, which are influenced by parental positions in the workforce.

Power-control theory assumes that in households where the mother and father have relatively similar levels of power at work (i.e., balanced households), mothers will be less likely to exert control on their daughters. These balanced households will be less likely to experience gender differences in the criminal offending of the children. However, households in which mothers and fathers have dissimilar levels of power in the workplace (i.e., unbalanced households) are more likely to suppress criminal activity in daughters. In such families, it is assumed that daughters will be taught to be less risk-taking than the males in the family. In addition, assertiveness and risky activity among the males in the house will be encouraged. This assertiveness and risky activity may be a precursor to crime, which is highly consistent with the empirical evidence regarding trends in crime related to gender. Thus, Hagan's integrated theory seems to have a considerable amount of face validity.

---

[83]Alex Piquero and Matthew Hickman, "An Empirical Test of Tittle's Control Balance Theory," *Criminology* 37 (1999): 319–42; Matthew Hickman and Alex Piquero, "Exploring the Relationships Between Gender, Control Balance, and Deviance," *Deviant Behavior* 22 (2001): 323–51.

[84]Matthew Hickman and Alex Piquero, "Exploring the Relationships"; for contrast and comparison, see Grasmick et al., "Changes Over Time"; Grasmick et al., "Reduction in Drunk Driving"; Tibbetts, "College Student Perceptions."

[85]John Hagan, *Structural Criminology* (Newark, NJ: Rutgers University Press, 1989); John Hagan, A. Gillis, and J. Simpson, "The Class Structure of Gender and Delinquency: Toward a Power-Control Theory of Common Delinquent Behavior," *American Journal of Sociology* 90 (1985): 1151–78; John Hagan, A. Gillis, and J. Simpson, "Clarifying and Extending Power-Control Theory, *American Journal of Sociology* 95 (1990): 1024–37; John Hagan, J. Simpson, and A. Gillis, "Class in the Household: A Power-Control Theory of Gender and Delinquency," *American Journal of Sociology* 92 (1987): 788–816.

Most empirical tests of power-control theory have provided moderate support, while more recent studies have further specified the validity of the theory in different contexts.[86] For example, one recent study reported that the influence of mothers, not fathers, on sons had the greatest impact on reducing the delinquency of young males.[87] Another researcher has found that differences in perceived threats of embarrassment and formal sanctions varied between more patriarchal and less patriarchal households.[88] Finally, studies have also started measuring the effect of patriarchal attitudes on crime and delinquency.[89] However, most of the empirical studies that have shown support for the theory have been done by Hagan or his colleagues. Still, power-control theory is a good example of a social control theory in that it is consistent with the idea that individuals must be socialized, and that the gender differences in such socialization make a difference in how people will act throughout life.

◪ Policy Implications

Many policy implications can be drawn from the various integrated theoretical models that we have discussed in this section introduction. We will focus on the concepts that were most prominent in this section, which are parenting and peer influences. Regarding the former, numerous empirical studies have examined programs for improving the ability of parents and expecting parents to be effective.[90] Such programs typically involve training high school students—or individuals/couples who are already parents—on how to be better parents. Additional programs include Head Start, and other preschool programs that attempt to prepare high-risk youth for starting school; these have been found to be effective in reducing disciplinary problems.[91]

Regarding peer influences, numerous programs and evaluations have examined reducing negative peer influences regarding crime.[92] Programs that emphasize prosocial peer groups are often successful, whereas others show little or no success.[93] The conclusion from most of these studies is that the most successful programs of this type are those that

---

[86]John Hagan, J. Simpson, and A. Gillis, "Class in the Household"; B. McCarthy and John Hagan, "Gender, Delinquency, and the Great Depression: A Test of Power-Control Theory," *Canadian Review of Sociology and Anthropology* 24 (1987): 153–77; Merry Morash and Meda Chesney-Lind, "A Reformulation and Partial Test of the Power-Control Theory of Delinquency," *Justice Quarterly* 8 (1991): 347–77; S. Singer and M. Levine, "Power-Control Theory, Gender, and Delinquency: A Partial Replication with Additional Evidence on the Effects of Peers," *Criminology* 26 (1988): 627–47.

[87]B. McCarthy, John Hagan, and T. Woodward, "In the Company of Women: Structure and Agency in a Revised Power-Control Theory of Gender and Delinquency," *Criminology* 37 (1999): 761–88.

[88]Brenda Sims Blackwell, "Perceived Sanction Threats, Gender, and Crime: A Test and Elaboration of Power-Control Theory," *Criminology* 38 (2000): 439–88.

[89]Brenda Sims Blackwell, "Perceived Sanction Threats"; Kristin Bates, Chris Bader, and F. Carson Mencken, "Family Structure, Power-Control Theory, and Deviance: Extending Power-Control Theory to Include Alternate Family Forms," *Western Criminology Review* 4 (2003): 170–90.

[90]See review in Brown et al., *Explaining Crime*, 425.

[91]Ibid.

[92]For a review, see Akers and Sellers, *Criminological Theories*, 102–8.

[93]Ibid.

focus on learning life skills and prosocial skills and use a curriculum based on cognitive-behavioral approaches.[94] This approach includes reinforcing positive behavior, clarifying rules of behavior in social settings, teaching life and thinking skills, and perhaps most important, thinking about the consequences of a given behavior before acting (hence cognitive-behavioral approach). Studies consistently show that programs using a cognitive-behavioral approach (i.e., think before you act) are far more successful than programs that emphasize interactions among peers or use psychoanalysis or other forms of therapy.[95]

Many other policy implications can be derived from integrated theories explaining criminal behavior, but parenting practices and peer influences are the primary constructs in most integrated models. Thus, these are the two areas that should be targeted for policy interventions, but they must be done correctly. For a start, policymakers could review the findings from empirical studies and evaluations and see that the earlier parenting programs start, particularly for high-risk children, the more effective they can be. Regarding the peer-influence programs, the emphasis on a cognitive-behavior therapy/training and life skills appears to be more effective than other approaches.

▧ Conclusion

In this section, we have reviewed what determines the types of integrated theories and what criteria make an integrated theory a good explanation of human behavior. We have also examined some examples of integrated theories that have been proposed in the criminological literature in the last 20 years. All of the examples represent the most researched and discussed integrated theories, and they demonstrate both the advantages and disadvantages of theoretical integration/elaboration. We hope that readers will be able to determine for themselves which of these integrated theories are the best in explaining criminal activity.

In this introduction, we have examined the various ways in which theoretical integration can be done, including forms of conceptual integration and theoretical elaboration. Furthermore, the criticisms of the different variations of integration and elaboration have been discussed. In addition, numerous examples of theoretical integration have been presented, along with the empirical studies that have been performed to examine their validity.

Finally, we discussed the policy implications that can be recommended from such integrated theories. Specifically, we concluded that the influence of early parenting and peer-influences are the two most important constructs across these theoretical models. Furthermore, we concluded that "the earlier, the better" regarding the parenting programs. We also concluded that a cognitive-behavioral approach, which includes life skills, is most effective for peer-influence programs.

▧ Section Summary

- ◆ Theoretical integration is one of the more contemporary developments in criminological theorizing. This approach brings with it many criticisms, yet arguably many advantages.

---

[94]Ibid., 108.

[95]Ibid.

- Types of theoretical integration include end-to-end, side-by-side, and up-and-down types.
- Conceptual integration appears to be a useful, albeit rarely explored, form of theoretical integration.
- Theoretical elaboration is another form of integration, which involves using one theory as the base or primary model and then incorporating concepts and propositions from other theories to make the primary model stronger.
- A number of seminal integrated theories have been examined using empirical evidence and appear to have enhanced our understanding of criminal behavior.
- Some of the theoretical models that have been proposed (such as Elliott et al.'s model) have been supported by empirical research.
- Theoretical integration models have many critics, who claim that the assumptions, concepts, and propositions of the mixed theories are counterintuitive.

KEY TERMS

Braithwaite's reintegrative shaming theory

Communitarianism

Conceptual integration

Elliott's integrated model

End-to-end theoretical integration

Integrated theories

Interdependency

Side-by-side (or "horizontal") integration

Theoretical elaboration

Thornberry's interactional model

Up-and-down integration

DISCUSSION QUESTIONS

1. What is the definition of theoretical integration, and why can such theories be beneficial?

2. Describe what end-to-end theoretical integration is and provide an example of such integration.

3. Describe what side-by-side theoretical integration is and provide an example of such integration.

4. Describe what up-and-down theoretical integration is and provide an example of such integration.

5. Discuss the difference between theoretical elaboration, theoretical integration, and conceptual integration.

6. What are the major strengths and weaknesses of theoretical integration?

7. In your opinion, what is the best of the integrated models? Why do you believe this is the best integrated model?

8. What do you believe is the weakest integrated model? Why?

Integrated Theories of Crime

  http://www.indiana.edu/~theory/Kip/IC.htm

John Braithwaite/Reintegrative Theory

  http://www.realjustice.org/library/braithwaite06.html

Various Integrated Theories of Crime

  http://law.jrank.org/pages/821/Crime-Causation-Sociological-Theories-Integrated-theories.html

READING

In this selection, Travis Hirschi argues that theoretical integration is against the best interests of the development of criminological theory. Hirschi's idea of integrated models is that they "divide the child in two," "giving the larger half" to other competing theories, which is consistent with what was seen in the introduction to this section.

While reading this selection, readers should take Hirschi's arguments into consideration in deciding whether or not integrated models are beneficial in understanding why individuals engage in criminal offending. However, readers should also feel free to question the rationale that Hirschi presents for dismissing such integrated frameworks.

Separate and Unequal Is Better

Travis Hirschi

Some argue that the assumptions of strain, control, and differential association theories of delinquency are fundamentally incompatible (Hirschi, 1969; Kornhauser, 1978), while others deny that they are necessarily incompatible and suggest the possibility of a compromise theory satisfactory to all concerned. For sheer reasonableness, the integrationist approach would seem to have much to commend it: Why should we continue to squabble over petty

SOURCE: Travis Hirschi, "Separate and Unequal Is Better," *Journal of Research in Crime and Delinquency* 16 (1979): 34–37. Copyright © 1979 Sage Publications, Inc. Used by permission of Sage Publications, Inc.

differences when, with a little concession here and a minor modification there, the larger truth we all love so well would be better served?[1] As a matter of fact, however, integration turns out to be more difficult than this question suggests. When Elliott, Ageton, and Canter (1979) actually face the Solomonic task of resolving the conflicting claims of the three perspectives, they find themselves compelled to agree that it cannot be done. Their solution is to use the terms and ignore the claims of control theory. This allows them to divide the child in two, giving the larger half to differential association and the remainder to strain theory.

Since it seems late to be inquiring after the child's health, it may be more useful here to look briefly at the mediating devices available to those who would resolve the conflicts among these traditional perspectives. For all their popular appeal, the repeated failure of integrationist approaches suggests that there are inherent difficulties that preclude attainment of their avowed goal.

⊠ End to End

One procedure open to the integrationist is to put partial theories end to end so that they describe a developmental sequence. The dependent variables of prior theories become the independent variables of subsequent theories. An overall increase in the ability to account for the final dependent variable is not expected because the last theory in the sequence presumably absorbs the predictive power of those preceding it. In fact, given the reasonable notion that the power and complexity of a theory should somehow be proportioned to what it explains, sequential integrations will normally explain less than the sum of things explained by their constituent theories. In other words, sequential integrations presuppose important outcomes (e.g., sustained patterns of delinquent behavior) and are

inappropriate as explanations of trivial events (e.g., crossing a neighbor's yard without permission). In sequential integrations that cross disciplinary lines, questions about which partial theory is most important will usually seem to be off the point. In others, where each of the constituent theories was originally advanced as a direct explanation of the outcome variables, integrationists face a formidable task: They must argue in effect that all theories but the last one in the sequence are wrong. Unless based on evidence not previously available, this argument will undercut the integrationists' claims to impartiality, since they will have become parties to the dispute they proposed to mediate.

This abstract description of the end-to-end strategy is of course partly based on the discussion by Elliott et al. (But not entirely. For roughly parallel integrations of the same perspectives—and some of the same features—see Lofland and Stark, 1965.) To my mind, it illustrates that Elliott et al. have on the whole been true to the logic of their procedure. It also illustrates that this procedure cannot resolve differences among the theories in question. Those whose minds are closed to the idea that, for example, "access to and involvement in delinquent learning and performance structures is a necessary . . . variable in the etiology of delinquent behavior" are unlikely to be persuaded by the argument that this "postulate" is a necessary consequence of an open-minded procedure.

⊠ Side by Side

Another procedure is to put partial theories side by side and segregate the cases to which they are considered applicable (Warren and Hindelang, 1979). This increases explained variance, avoids questions of incompatibility, and answers the question of relative importance. (Other things being equal, a theory that applies to 40 percent of the cases is "better" than one applying to 30 percent.) Since the side-by-side approach avoids the theory overload problem, it is useful in explaining delinquent acts differing widely in significance.

Being all virtue and no apparent defect, the side-by-side "integration" would seem to be the way to go. The reluctance of Elliott et al. to exploit this strategy fully seems to result from the fact that in the side-by-side approach the definition of delinquency is unrestricted. This approach does not allow a single definition tailored to the needs of a particular theory. Instead, it leaves each sub-theory free to define delinquency in its own terms. Having constructed their definition with a causal sequence and differential association in mind, Elliott et al. cannot reopen the definitional question without reopening the question of the adequacy of their integrated theory. (A pure side-by-side approach would require consideration of control theory as a possible explanation of at least some forms of delinquent behavior.)

The major difficulty inherent in the side-by-side approach, as far as I can determine, is that no one has as yet come up with a way of segregating cases that produces the results suggested. (The notoriously unsuccessful social-class-specific approach to theory is a good example of the side-by-side procedure in operation.) Elliott et al. segregate cases on the basis of the strength of initial "bonds to the conventional order." Those with formerly strong bonds are said to follow a path to delinquency different from the path followed by those who have never developed such bonds.[2] As is usually true with side-by-side integrations, procedures for identifying the two groups are not provided. Furthermore, evidence for the existence of the strain theory path is exceedingly vague. Put another way, I interpret the Elliott et al. side-by-side effort as a bow in the direction of "the most influential and widely used contemporary formulation in the sociology of delinquent behavior," a formulation whose continued influence and wide use are something of a mystery, to say the least.[3]

⊠ Up and Down

Integrationists may also raise the level of abstraction to the point where these partial theories become specific applications of a general theory of deviance—or law (e.g., Black, 1976). Such efforts should lead to greater explained variance because what were once considered unrelated processes may now be seen to bear on the same outcome. This procedure too has the defects of its virtues: a marked tendency on the part of integrationists to accept without question the truth of any partial theory their general theory subsumes. (The greater the number of partial theories accounted for, the greater and more powerful the general theory.) What appears to be unusual reasonableness on the part of the integrationist may then turn out to be nothing more than failure to invoke required scientific bases of discrimination.[4]

To some extent, all of these integrative procedures are employed by Elliott et al. As can be seen from their general properties, the life of the integrationist is not easy. He or she is forced to make theoretical decisions (and to abide by their consequences) unguided by a consistent theoretical perspective. If, in desperation, one or another theoretical perspective is adopted, claims to even-handed treatment of all perspectives are no longer tenable.

I think we should be pleased to find that attainment of the integrationist's goal is so difficult. A "successful" integration would destroy the healthy competition among ideas that has made the field of delinquency one of the most interesting and exciting fields in sociology for some time. To their credit, Elliott et al. eventually abandon the integrationist perspective in favor of the theory of differential association. To my mind, differential association is at least of some historical interest.[5] I would not be able to speak so highly of an integrated theory of delinquency.

⊠ Notes

1. Students in schools of criminal justice, where the boundaries between disciplines are often obscure, are easily irritated by a focus on differences among theories within discipline. If all disciplines are "really saying the same thing," within-discipline differences are obviously too petty to merit attention.

2. Elliott et al. occasionally mention that strain theory requires a repeated-measures design, while research from within the control perspective has been cross-sectional or "static." I do not find their equation of *superior research design* and *superior theory* convincing. A repeated-measures design may indeed allow the researcher to examine the effects of prior states, but it does not guarantee that such effects exist. In fact, given the empirical record, I would expect the strain theorist to be extremely leery of the hypothesis that prior and subsequent states interact in their effects on delinquency.

3. "We fail to find any support for Cloward and Ohlin's disjunction hypothesis" (Elliott and Voss, 1974:170). "There is little or no support for the hypothesis that a tendency to attribute blame externally . . . increases the likelihood of delinquency. . . ." (Elliott and Voss (1974:170). "Strain theories have a decided defect: they are not consistent with the evidence" (Hirschi, 1969:228). "On the basis of available evidence, Cloward and Oblin are wrong" (Hirschi and Hindelang, 1977:578). "There are no delinquent gangs of the types described by Cloward and Ohlin" (Kornhauser, 1978:159).

4. In the empirical ("down") equivalent of this procedure, the integrationist simply combines variables from various theories on the grounds that some may explain cases left unexplained by the others. This is the "multiple regression" approach, where predictability is increased by inclusion of variables unrelated to variables already considered. In most applications of this procedure, unfortunately, the presumption of orthogonality among independent variables owes much of its plausibility to the fact that it has not yet been tested.

5. In his 1939 statement of differential association, Edwin Sutherland limited his dependent-variables to systematic crimes in order "to postpone consideration of the very trivial criminal acts" (Cohen et al., 1956:21). In their 1979 statement of essentially the same theory, Elliott et al. restrict their dependent variable to "sustained patterns of illegal behavior" because they "are not concerned . . . with the isolated delinquent act."

By 1947 Sutherland had abandoned "systematic" because a psychiatrist could find no more than 2 systematic criminals among the 2,000 inmates of Indiana State Prison and because his own students found it "most difficult . . . to determine objectively whether a prisoner was a criminal systematically or adventitiously" (Cohen et al., 1956:21). I suspect that in the next go around Elliott and his colleagues will reach similar conclusions about "sustained patterns of illegal behavior." They will see that it eliminates too much of interest and is too difficult (impossible?) to operationalize to justify its retention in a theory of delinquency. (I suspect too that they will see that this definition of delinquency reopens the question of the theoretical relevance of self-report results, since it has not been the definition employed in self-report research.)

⊠ References

Black, D. 1976 *The Behavior of Law*. New York: Academic Press,

Cohen, A., et al. 1956. *The Sutherland Papers*. Bloomington: Indiana University Press.

Elliott, D. S., S. S. Ageton, and R. J. Canter. 1979. "An Integrated Theoretical Perspective-off Delinquent Behavior." *Journal of Research in Crime and Delinquency* 16(1): 3–27.

Elliott, D. S., and H. Voss. 1974. *Delinquency and Dropout*. Lexington, MA: D. C. Heath.

Farrell, R. A., and V. L. Swigert. 1978. *Social Deviance*. Philadelphia: Lippincott.

Hirschi, T. 1969. *Causes of dlinquency*. Berkeley: University of California Press.

Hirschi, T., and M. J. Hindelang. 1977. "Intelligence and Delinquency: A Revisionist Review." *American Sociological Review* 42(4): 571–87.

Kornhauser, R. R. 1978. *Social Sources of Delinquency: An Appraisal of Analytic Models*. Chicago: University of Chicago Press.

Lofland, J., and R. Stark, R. 1965. "Becoming a World Saver: A Theory of Conversion to a Deviant Perspective." *American Sociological Review* 30(6): 862–75.

Warren, M. Q., and M. J. Hindelang. 1979. "Current Explanations of Offender Behavior." Pp. 162–82 H. Toch (Ed.), *Psychology of crime and criminal justice*, ed. H. Toch. New York: Holt, Rinehart and Winston.

REVIEW QUESTIONS

1. What does Hirschi have to say about integrated models generally? Do you agree or disagree?

2. What does Hirschi have to say about "end-to-end" integrated models versus "side-by-side" integrated models? Do you agree or disagree? Can you provide an example for both types of theoretical integration?

3. Finally, what does Hirschi have to say about "up-and-down" integrated models? Do you agree or disagree with his stance on this form of integrated frameworks? What does Hirschi conclude, and do you agree?

❖

READING

In this selection, Delbert Elliott, Suzanne Ageton, and Rachelle Canter provide one of the earliest, and still one of the best-known and respected examples of theoretical integration. We shall see that Elliott et al. merge various concepts and propositions from at least three traditional theoretical perspectives, namely strain theory, social control theory, and various social learning theories.

One of the beneficial aspects of Elliott et al.'s integrated theory is that it attempts to merge various versions of each of these perspectives into a cohesive, unitary explanation of delinquency/criminal behavior. For example, the authors go beyond the original strain theory proposed by Merton in 1938 and also take into consideration the subsequent versions of strain theory that appeared in later decades (such as Cloward and Ohlin's theory of differential opportunities, as well as Cohen's ideas regarding the formation of gang subcultures). Also, when explaining how control theories were integrated into Elliott et al.'s model, the authors do not simply rely on Hirschi's version of social bonding but also consider some of the other versions of control theory (such as control theories by Reiss, Nye, Reckless as well as Matza's theory of drift). Finally, the authors don't simply merge Sutherland's theory of differential association but also take into account the more modern versions of social learning theory, such as Akers's model of differential reinforcement and Bandura's model of imitation/modeling. Although even more modern versions of each of these theoretical perspectives have been introduced (e.g., general strain theory) since Elliott et al. proposed this model in 1979, it is obvious from this selection that the authors did their best to integrate these three major perspectives using the most recent scientific evidence they had in the late-1970s.

SOURCE: Delbert Elliott, Suzanne S. Ageton, and Rachelle J. Canter, "An Integrated Theoretical Perspective on Delinquent Behavior," *Journal of Research in Crime and Delinquency* 16 (1979): 3–27. Copyright © 1979 Sage Publications, Inc. Used by permission of Sage Publications, Inc.

While reading this selection, readers are encouraged to consider the validity of the model by considering themselves or others they grew up with and examining whether the development of the various stages proposed by Elliott et al. seems to fit with the delinquents or criminals they have known. Furthermore, if you were to create such an integrated model, would you have emphasized any concepts or propositions from these three theoretical perspectives (or those from other theories) that the authors do not seem to focus on or merge into their "comprehensive" explanation of why individuals become delinquents? Finally, readers should consider whether it is rational to merge theoretical perspectives that have opposing basic assumptions, such as merging strain theory, which claims that individuals are born good, with perspectives such as control theory, which claims that people are born bad and must be taught or controlled to be good.

An Integrated Theoretical Perspective on Delinquent Behavior

Delbert S. Elliott, Suzanne S. Ageton, and Rachelle J. Canter

Previous Theories: Strain and Control

Anomie/Strain Perspective

... Strain theory has become the most influential and widely used contemporary formulation in the sociology of delinquent behavior. A specific application of strain theory to delinquency has been proposed by Cloward and Ohlin (1960) and, more recently, by Elliott and Voss (1974). Cloward and Ohlin's work is of particular interest to us because their formulation, like that proposed here, represents an attempt to integrate and extend current theoretical positions. Although their theory has been viewed primarily as an extension of the earlier work of Durkheim and Merton, it is equally an extension of the differential association perspective and the prior work of Sutherland (1947). Indeed, much of its significance lies in the fact that it successfully integrated these two traditional perspectives on the etiology of delinquent behavior.

Cloward and Ohlin maintain that limited opportunity for achieving conventional goals is the motivational stimulus for delinquent behavior. The specific form and pattern of delinquent behavior are acquired through normal learning processes within delinquent groups. Experiences of limited or blocked opportunities (a result of structural limitations on success) thus lead to alienation (perceived anomie) and an active seeking out of alternative groups and settings in which particular patterns of delinquent behavior are acquired and reinforced (social learning).

Merton, Cloward and Ohlin have conceptualized the condition leading to anomie in terms of differential opportunities for achieving socially valued goals. Differential access to opportunity creates strain; this is postulated to occur primarily among disadvantaged, low-SES [socioeconomic status] youths, resulting in the concentration of delinquent subcultures in low-SES neighborhoods. It is important to note, however, that Cloward and Ohlin have changed the level of explanation from the macrosociological level which characterized Durkheim's work to an individual level. It is the *perception* of limited access to conventional goals that motivates the *individual* to explore deviant means. This change in level of explanation was essential for the integration of strain and learning perspectives.

Elliott and Voss's more recent work (1974) has attempted to deal with the class-bound assumptions inherent in strain theory. Their formulation extends Cloward and Ohlin's classic statement in the following three ways: (1) The focus on limited opportunities was extended to a wider range of conventional goals. (2) The goal-means disjunction was modified to be logically independent of social class. (3) The role of social learning in the development of delinquent behavior was further emphasized. Elliott and Voss have proposed a sequential, or developmental, model of delinquency: (1) Limited opportunities or failure to achieve conventional goals serves to (2) attenuate one's initial commitment to the normative order and (3) results in a particular form of alienation (normlessness), which serves as a "permitter" for delinquency, and (4) exposure to delinquent groups, which provides learning and rewards for delinquent behavior for those whose bonds have undergone the attenuation process.

From this perspective, aspiration-opportunity disjunctions provide motivation for delinquent behavior. As compared with Merton and Cloward and Ohlin, Elliott and Voss view *both* goals and opportunities as variables. They postulate that middle-class youths are just as likely to aspire beyond their means as are low-SES youths. While the absolute levels of aspirations and opportunities may vary by class, the discrepancies between personal goals and opportunities for realizing these goals need not vary systematically by class. Given Durkheim's (1897/1951, p. 254) view that poverty restrains aspirations, Elliott and Voss have postulated that aspiration-opportunity disjunctions would be at least as great, if not

greater, among middle-class youths. In any case, the motivational stimulus for delinquent behavior in the form of aspiration-opportunity discrepancies or goal failure is viewed as logically independent of social class.

Normlessness, the expectation that one must employ illegitimate means to achieve socially valued goals (Seeman, 1959), is postulated to result from perceived aspiration-opportunity disjunctions. When a person cannot reach his or her goals by conventional means, deviant or illegitimate means become rational and functional alternatives. When the source of failure or blockage is perceived as external—resulting from institutional practices and policies—the individual has some justification for withdrawing his or her moral commitment to these conventional norms. In this manner, a sense of injustice mitigates ties to conventional norms and generates normlessness.

Once at this point in the developmental sequence, the relative presence or absence of specific delinquent learning and performance structures accounts for the likelihood of one's behavior. The time-ordering of the exposure to delinquency variable is not explicit. It may predate failure or it may be the result of seeking a social context in which one can achieve some success. While the exposure may result in the acquisition of delinquent behavior patterns, actual delinquent behavior (performance) will not result until one's attachment to the social order is neutralized through real or anticipated failure, and the delinquent behavior has been reinforced. The results of research relative to this set of propositions have been generally encouraging. . . .

While considerable empirical support for an integrated strain-learning approach to delinquency has been amassed, most of the variance in delinquency remains unexplained. If the power of this theoretical formulation is to be improved, some basic modification is required. One avenue is suggested by the weak predictive power of the aspiration-opportunity discrepancy variables. . . . [In some studies], limited academic success at school and failure in one's relationship with parents were predictive, but only weakly. To some extent, the low strength of these predictors might be anticipated, since they are the initial variables in the causal sequence and are tied to delinquency only through a set of other conditional variables. On the other hand, the strong emphasis placed on these specific variables in strain theories seems questionable, given the available data. It might be argued that the difficulty lies in the operationalization or measurement of the relevant goal-opportunity disjunctions. However, we are inclined to reject this position because previous findings as to this postulated relationship have been generally weak and inconclusive (Spergel, 1967; Short, 1964; Elliott, 1962; Short, Rivera, and Tennyson, 1965; Jessor et al., 1968; Hirschi, 1969; Liska, 1971; and Brennan, 1974). Furthermore, there is substantial evidence in the above-mentioned studies that many adolescents engaging in significant amounts of delinquent behavior experience no discrepancies between aspirations and perceived opportunities. The lack of consistent support for this relationship suggests that failure or anticipated failure constitutes only one possible path to an involvement in delinquency.

The Control Perspective

The different assumptions of strain and control theories are significant. Strain formulations assume a positively socialized individual who violates conventional norms only when his or her attachment and commitment are attenuated. Norm violation occurs only after the individual perceives that opportunities for socially valued goals are blocked. Strain theory focuses on this attenuation process. Control theories, on the other hand, treat the socialization process and commitment to conventional norms and values as problematic. Persons differ with respect to their commitment

to and integration into the conventional social order. . . .

From a control perspective, delinquency is viewed as a consequence of (1) lack of internalized normative controls, (2) breakdown in previously established controls, and/or (3) conflict or inconsistency in rules or social controls. Strain formulations of delinquency appear to be focusing on those variables and processes which account for the second condition identified by Reiss (1951): attenuation or breakdown in previously established controls. On the other hand, most control theorists direct their attention to the first and third conditions, exploring such variables as inadequate socialization (failure to internalize conventional norms) and integration into conventional groups or institutions which provide strong external or social controls on behavior. From our perspective, these need not be viewed as contradictory explanations. On the contrary, they may be viewed as alternative processes, depending on the outcome of one's early socialization experience.

For example, Hirschi (1969) has argued that high aspirations involve a commitment to conventional lines of action that functions as a positive control or bond to the social order. Strain theories, on the other hand, view high aspirations (in the face of limited opportunities) as a source of attenuation of attachment to the conventional order. Recognizing this difference, Hirschi suggested that the examination of this relationship would constitute a crucial test of the two theories. Empirically, the evidence is inconsistent and far from conclusive. One possible interpretation is that both hypotheses are correct and are part of different etiological sequences leading to delinquent behavior.

Empirical studies using the control perspective have focused almost exclusively on the static relation of weak internal and external controls to delinquency without considering the longer developmental processes. These processes may involve an initially strong commitment to and integration into society which becomes attenuated over time, with the attenuation eventually resulting in delinquency. The source of this difficulty may lie in the infrequent use of longitudinal designs. Without a repeated-measure design, youths with strong bonds which subsequently become attenuated may be indistinguishable from those who never developed strong bonds.

⬛ An Integrated Strain-Control Perspective

Our proposed integrated theoretical paradigm begins with the assumption that different youths have different early socialization experiences, which result in variable degrees of commitment to and integration into conventional social groups. The effect of failure to achieve conventional goals on subsequent delinquency is related to the strength of one's initial bonds. Limited opportunities to achieve conventional goals constitute a source of strain and thus a motivational stimulus for delinquency only if one is committed to these goals. In contrast, limited opportunities to achieve such goals should have little or no impact on those with weak ties and commitments to the conventional social order.

Limited opportunities to achieve conventional goals are not the only experiences which weaken or break initially strong ties to the social order. Labeling theorists have argued that the experience of being apprehended and publicly labeled delinquent initiates social processes which limit one's access to conventional social roles and statuses, isolating one from participation in these activities and relationships and forcing one to assume a delinquent role (Becker, 1963; Schur, 1971, Kitsuse, 1962; Rubington & Weinberg, 1968; Ageton & Elliott, 1974; and Goldman, 1963). It has also been argued that the effects of social disorganization or crisis in the home (divorce, parental strife and discord, death of a parent) and/or community (high rates of mobility, economic depression, unemployment) attenuate or break one's ties to society (Thomas & Znaniecki, 1927; Shaw, 1931; Savitz, 1970; Monahan, 1957; Toby, 1957; Glueck & Glueck, 1970; Andry, 1962; and Rosen, 1970).

In sum, we postulate that limited opportunities, failure to achieve valued goals, negative labeling experiences, and social disorganization at home and in the community are all experiences which may attenuate one's ties to the conventional social order and may thus be causal factors in the developmental sequence leading to delinquent behavior for those whose early socialization experiences produced strong bonds to society. For those whose attachments to the conventional social order are already weak, such factors may further weaken ties to society but are not necessary factors in the etiological sequence leading to delinquency.

Our basic conceptual framework comes from control theory, with a slightly different emphasis placed on participation in and commitment to delinquent groups. Further, it identifies a set of attenuating/bonding experiences which weaken or strengthen ties to the conventional social order over time. Our focus is on experiences and social contexts which are relevant to adolescents. A diagram of our proposed theoretical scheme is shown in Figure 1. The rows in Figure 1 indicate the direction and sequence of the hypothesized relationships. While the time order designated in Figure 1 is unidirectional, the actual relationships between initial socialization, bonding/attenuation processes, normative orientations of groups, and behavior are often reciprocal and reinforcing. We have also presented the variables in dichotomized form to simplify the model and the discussion of its major elements.

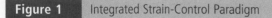

| **Figure 1** | Integrated Strain-Control Paradigm |

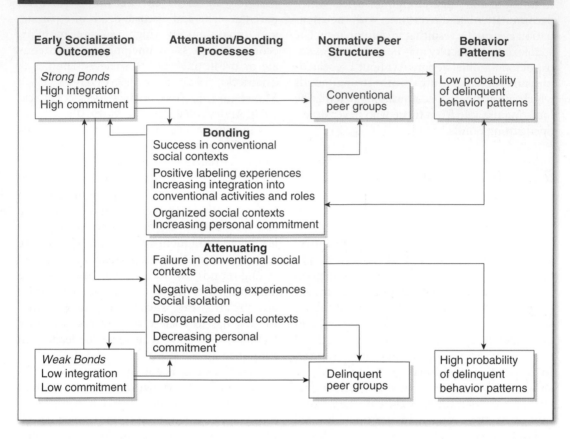

Bonds

Control theorists disagree about sources of control, but they all accept the central proposition that delinquent behavior is a direct result of weak ties to the conventional normative order. In operationalizing control theory, major emphasis has been placed on the bond(s) which tie a person to society. Hirschi (1969) conceptualized four elements of this bond. First, attachment implies a moral link to other people and encompasses such concepts as conscience, superego, and internalization of norms. Commitment, the second factor, is the rational element in the bond. Hirschi views commitment to conformity as an investment in conventional lines of action, such as an educational or occupational career. Other theorists have tied the concept of commitment to such notions as "stake in conformity" (Goode, 1960) and "side bets" (Becker, 1960). Involvement is the time and energy dimension of the bond for Hirschi. Given the limits of time and energy, involvement in conventional activities acts as a social constraint on delinquent behavior. The final bond, *belief,* refers to one's acceptance of the moral validity of social rules and norms. According to Hirschi, this psychological element of the bond is effective as long as a person accepts the validity of the rules. If one denies or depreciates the validity of the rules, one source of control is neutralized.

Other control theorists, such as Reiss (1951), Nye (1958), and Reckless (1967) use a more general classification of bonds as internal (personal) and external (social) controls. Hirschi's dimensions are not easily placed into these two general categories, although Hirschi identifies attachment as an internal and involvement as an external element of the bond (1969, p. 19). We believe that distinguishing internal controls, whose locus is within the person (beliefs, commitment, attitudes, perceptions), from external controls, whose locus is in the surrounding social and physical milieu, poses fewer difficulties and produces greater conceptual clarity than is found in Hirschi's four concepts.

The external, or social, bond we have defined as *integration*. By this, we refer to involvement in and attachment to conventional groups and institutions, such as the family, school, peer networks, and so on. Those persons who occupy and are actively involved in conventional social roles are, by this definition, highly integrated. Group controls exist in the form of sanctioning networks (the formal and informal rules and regulations by which the behavior of social role occupants or group members is regulated). This conceptualization of integration is akin to Hirschi's concepts of involvement and commitment.

The internal, or personal, bond is defined as *commitment*. Commitment involves personal attachment to conventional roles, groups, and institutions. At another level, it reflects the extent to which one feels morally bound by the social norms and rules and the degree to which one internalizes or adopts those norms as directives for action. Our notion of commitment is akin to Hirschi's concepts of attachment and belief. Integration and commitment together constitute the bonds which tie an individual to the prevailing social order. High levels of integration and commitment imply strong bonds and general insulation from delinquent behavior. Conversely, low social integration and commitment presuppose weak bonds and a susceptibility to delinquent behavior. All gradations of integration and commitment are possible.

Building Social Control: The Bonding/Attenuation Processes

The inclusion of the bonding/attenuation process in the model suggests that, throughout adolescence, youths are involved in experiences and processes which attenuate or reinforce their childhood bonds to the conventional social order. Adolescence is a critical life period, both psychologically and socially. As youths make the transition from childhood to adulthood, the level of involvement in the immediate family declines and they move into new and more complex social settings at school and in the community. For one who developed strong childhood bonds, such factors as (1) success experiences at school and in the larger community, (2) positive labeling in these new settings, and (3) a continuous, stable, harmonious home life constitute positive reinforcements of initially strong bonds and continuing insulation from delinquency. For some, the transition is not as smooth, and failure, negative labeling, isolation, and rejection occur in these new social settings; these, in turn, may create difficulties in the youth's relationship with his family. The net effect of these new experiences may be a weakening of one's integration into . . . these social groups and institutions and an increasing likelihood of involvement in delinquent behavior. Finally, for those who never developed strong bonds during childhood, bonding/attenuation experiences will either strengthen the weak bonds, thus reducing the likelihood of delinquency, or further attenuate them, thus maintaining or increasing the probability of delinquent behavior.

We do not propose that this specific set of variables exhausts the possible experiences or conditions which might attenuate or reinforce one's bonds to society during adolescence. Rather, we have purposely selected those conditions and experiences which prior theory and research have suggested as critical variables to illustrate the major dimensions of the paradigm.

Delinquent Learning and Performance Structures

A major criticism of control theory has been that weak bonds and the implied absence of restraints cannot alone account for the specific form or content of the behavior which results. They may account for a state of "drift," as described by Matza (1964), but they do not explain why some youths in this state turn to delinquency, drug use, and various unconventional subcultures, while others maintain an essentially conforming pattern of behavior; nor can they account for emerging patterns of delinquency which may be unique to particular ages or birth cohorts. We therefore postulate that access to and involvement in delinquent learning and performance structures is a necessary (but not sufficient) variable in the etiology of delinquent behavior. Following Sutherland (1947), we maintain that delinquent behavior, like conforming behavior, presupposes a pattern of social relationships through which motives, rationalizations, techniques, and rewards can be learned and maintained (Burgess & Akers, 1966a and 1966b; Akers, 1977; Bandura, 1969, 1973; and Mischel, 1968). Delinquent behavior is thus viewed as behavior which has social meaning and must be supported and rewarded by social groups if it is to persist.

By the time children enter adolescence, virtually all have been sufficiently exposed to criminal forms of behavior to have "learned" or acquired some potential for such acts. The more critical issue for any theory of delinquency is why and how this universal potential is transformed into delinquent acts for some youths and not others. For most learning theorists, a distinction is made between learning and performance and the latter is directly tied to reinforcements (Rotter, 1954; Bandura & Walters, 1963; Mischel, 1968; and Bandura, 1969). . . .

According to the present social learning formulation, learning or acquisition of novel responses is regulated by sensory and cognitive processes; learning may be facilitated by reinforcement but does not depend on it (e.g., Bandura & Walters, 1963; Hebb, 1966). Direct and vicarious reinforcements are, however, important determinants of response selection in performance.

The delinquent peer group thus provides a positive social setting that is essential for the performance and maintenance of delinquent patterns of behavior over time. Those committed to conventional goals, although they may have been exposed to and learned some delinquent behaviors, should not establish patterns of such behavior unless (1) their ties to the conventional social order are neutralized through some attenuating experiences and (2) they are participating in a social context in which delinquent behavior is rewarded. In social learning terms, they may have acquired or learned delinquent behavior patterns, but the actual performance and maintenance of such behavior are contingent on attenuation of their commitment to conventional norms and their participation in a social context supportive of delinquent acts. Alternatively, for those with weak ties and commitments to the conventional social order, there is no reason for a delay between acquisition and performance of delinquent acts.

In the causal sequence described by strain theory, the individual holds conventional goals but is unable to attain them by conventional means. If attachment to the goals is strong enough, it may support delinquent behavior without participation in delinquent groups, for attaining these goals may provide sufficient reinforcement to maintain the behavior. Therefore, our model shows one direct route to delinquent behavior from attenuating experiences, without mediating group support for delinquency. We view this as the atypical case, however, and postulate that it is difficult to sustain this causal sequence for extended periods of time.

Involvement in a delinquent group is a necessary condition for sustained patterns of delinquency among persons who do not subscribe to conventional goals (the weakly socialized person described by control theory). Individual patterns of delinquency (without group support) are more viable for those committed to conventional goals because there are generally shared expectations and social supports for achievement of those goals. For youths with weak bonds, involvement in a delinquent peer group serves this support function. Cohen (1966) has observed that delinquency often involves a desire for recognition and social acceptance, and, therefore, requires group visibility and support. Maintenance of delinquent behavior patterns should require some exposure to and participation in groups supporting delinquent activities. Though not a necessary condition for delinquent behavior among those with initially strong bonds, contact with delinquent groups should, nevertheless, increase the likelihood of sustained delinquent behavior.

Delineation of the delinquent peer group as a necessary condition for maintenance of delinquent behavior patterns represents an extension of previous statements of control theory. . . . It is one thing to be a social isolate with weak bonds to conventional peer groups and another to be highly committed to and integrated into a delinquent peer group. Both persons may be characterized as having weak bonds to the social order, with few conventional restraints on their behavior; but those committed to and participating in delinquent peer groups have some incentive and social support for specifically delinquent forms of behavior. We agree with Hirschi's (1969) and Hepburn's (1976) argument that those with a large stake in conformity (strong bonds) are relatively immune to delinquent peer group influence. However, we postulate that, in addition to weak bonding and an absence of restraints, some positive motivation is necessary for sustained involvement in delinquent

behavior. In the absence of positive motivation, we would not predict significant involvement in delinquency across time even for those with weak bonds, for there is no apparent mechanism for maintaining such behavior (Brennan, Huizinga, & Elliott, 1978). It may be that some exploratory, "primary" forms of delinquency (Lemert, 1951) may occur without group support, or that this constitutes a pathological path to delinquency, but the maintenance of delinquent behavior patterns usually requires some exposure to and participation in groups supporting delinquent activity.

In sum, we postulate that bonding to conventional groups and institutions insulates one from involvement in delinquent patterns of behavior and that bonding to deviant groups or subcultures facilitates and sustains delinquent behavior. When examining the influence of social bonds, it is critical that the normative orientation of particular groups be taken into account. This focus on the normative orientations of groups is the central theme in subcultural theories of delinquency (Cohen, 1955; Cloward and Ohlin, 1960; and Miller, 1958) and constitutes an important qualification to a simple interpretation of the relationship between social bonds and delinquency. This position has an empirical as well as a theoretical base. . . .

Delinquent Behavior

Delinquent behavior is viewed as a special subclass of deviant behavior. While deviance includes all violations of all prevailing norms, delinquent behavior includes only violations of statutory proscriptive norms, or, as they are usually called, laws. Thus, delinquent behavior takes on special meaning because (1) there is generally broad community consensus for these norms, (2) virtually all persons are aware that these specific proscriptions are enforced by official sanctions, and (3) the risk of detection and punishment influences the performance of delinquent acts.

We are not concerned here with the isolated delinquent act. Our focus is on sustained patterns of delinquent behavior, whether the person involved is socially or self-defined as a delinquent or nondelinquent person. Although our definition of delinquency subsumes one characteristic of a delinquent role (sustained patterns of delinquent behavior), it is our view that continuing involvement in delinquency may not necessarily involve the enactment of a delinquent role (Becker, 1963). There is empirical evidence that many embezzlers, auto thieves, check forgers, shoplifters, and persons involved in violent assaults against persons (including rape) do not view themselves as criminal or delinquent (Gibbons, 1977; Lemert, 1951, 1953; Cameron, 1964; Robin, 1974; Gauthier, 1959; and Gebhard et al., 1965). Furthermore, many adolescents involved in sustained patterns of delinquent behavior are never apprehended and publicly labeled as delinquent persons, and have neither a public nor a self-definition as a delinquent or criminal person (Sykes and Matza, 1957; Reiss, 1962; Cameron, 1964; Hirschi, 1969; Kelly, 1977; and Jensen, 1972). Thus, our conceptualization of delinquency focuses on sustained patterns of illegal behavior and is logically independent of the concept of delinquent role.

Etiological Paths to Delinquency

There are two dominant etiological paths to delinquency in the paradigm shown in Figure 1. The first involves an integration of traditional control theory and social-learning theory. Weak integration into and commitment to the social order, absence of conventional restraints on behavior, and high vulnerability to the influence of delinquent peer groups during adolescence characterize the socialization experiences related to the first path. Depending on the presence and accessibility of conventional and delinquent peer groups, some weakly bonded youths turn to delinquency while others maintain an essentially conforming pattern of behavior or a legal, but unconventional, lifestyle. The crucial element in this path is the delinquent peer group. Weakly bonded youths may not hold conventional aspirations (as for academic success), but they do share in more general aspirations for friendship and acceptance, as well as status and material rewards, which may be offered through participation in a group. Given an absence of conventional restraints and access to delinquent groups, the reasons for involvement are not unlike those for involvement in more conventional peer groups during adolescence.

The second path represents an integration of traditional strain and social-learning perspectives. Youths who follow this path develop strong bonds to the conventional social order through their socialization experiences. The crucial element in this sequence is the attenuation, or weakening, of these bonds. Attenuating experiences during adolescence involve personal failure to achieve conventional goals and/or threats to the stability and cohesion of one's conventional social groups. Once one's bonds are effectively weakened, like those who never developed strong bonds, one is free to explore alternative means for goal achievement and to participate in delinquent or unconventional groups.

In most instances, this path also involves participation in peer groups which tolerate or encourage delinquent forms of behavior. It is our view that truly individual adaptations to this situation are unlikely to survive long enough to generate detectable patterns of delinquent behavior. However, two possible subtypes deserve mention. The diagram of this integrated paradigm shows a direct causal path from initially strong bonds and subsequent attenuation experiences to delinquent behavior patterns. Under some circumstances, participation in groups providing reinforcements for delinquent acts is unnecessary. Attenuating experiences are sufficient to motivate repeated acts of delinquency, which are attempts to regain conventional rewards through unconventional means. This pattern involves the classic strain model, in which the person retains a strong commitment to conventional goals and values and uses illegal means as a temporary expedient. The attenuation process is only partial, and these youths retain some commitment to and integration into conventional groups. We anticipate such patterns to be of relatively short duration and to involve highly instrumental forms of delinquent behavior. Patterns of theft may characterize this etiological path.

A second subtype corresponds to that described generally by Simon and Gagnon (1976) in their article on the anomie of affluence. This path involves those whose commitments to conventional goals are attenuated by a decreasing gratification derived from goal achievement. Unlike the previously described subtype, which involved failure to achieve conventional success goals because of limited means or abilities, this type has ability and a ready access to legitimate means and is successful by conventional standards. The failure to derive personal gratification from "success" results in an attenuation of the commitment to these success goals and sets in motion a search for alternative goals whose attainment will provide a greater measure of personal gratification. This path to delinquency clearly requires participation in social groups in which delinquent behavior patterns can be learned and reinforced. This pattern of delinquency is characterized by a search for new experiences, which frequently involves illegal forms of behavior, such as illicit drug use and sex-related offenses.

At a more tentative level, we postulate that the two major paths (1) typically involve different forms of personal alienation and (2) result in different self-images and social labels. Conceptually, alienation plays a slightly different role within strain and control perspectives. From a control perspective, alienation, in the form of powerlessness, societal estrangement, and social isolation, directly reflects a weak personal commitment to conventional groups and norms. For strain theory, however, alienation represents a crucial intervening variable linking failure to delinquency. It is evidence of the attenuation of one's commitment bond or, in Hirschi's (1969) terms, the neutralization of "moral obstacles" to delinquency. In the form of alienation described by Cloward and Ohlin (1960), the neutralization is achieved through a blaming process in which failure is attributed to others or to general societal injustice. These same elements are present in Sykes and Matza's (1957) techniques of neutralization. Cartwright et al. (1966) and Cartwright (1971) identify four types of alienation which provide this direct encouragement, justification, or permission for delinquency: normlessness, futility, lack of trust, and perceived indifference. If we assume some relationship between the two causal paths and social class, there is some indirect empirical support for the hypothesis that the form of alienation is tied to the strength of one's initial commitment bond. . . .

We also hypothesize that those with initially strong bonds are less likely to view themselves as delinquent, even when they are involved in sustained patterns of delinquent behavior. Such persons are more likely to come from advantaged backgrounds and to have prosocial self-images. Consequently, they are likely to view their delinquent acts as temporary expedients, retaining at least a partial commitment to conventional goals. The probability of apprehension and public labeling by the police and courts is also much lower for such youths. In contrast, those who never developed strong bonds to the social order are more vulnerable to labeling processes and thus more likely to be viewed as delinquents by themselves and by others (Jensen, 1972). This may account, in part, for the persistent view among law enforcement officials and the general public that most delinquents are poor and/or nonwhite, in spite of the compelling evidence that the incidence of delinquent behavior is unrelated to these variables.

Summary and Discussion

. . . . We believe the synthesis of traditional strain, social control, and social-learning perspectives into a single paradigm has several advantages over a conceptualization which treats each theory as separate and independent. First, the provision for multiple etiological paths to delinquency in a single paradigm presents a more comprehensive view. The integration of strain and control perspectives assumes that these two paths are independent and additive and that their integration will account for more variance in sustained patterns of delinquent behavior than either can explain independently. Independent tests of these traditional perspectives in the past have often failed to include the variables necessary to test alternative explanations, and even when such variables were available, the alternative explanations were assumed to be competitive and were thus evaluated with respect to the relative strengths of the two competing hypotheses (Hirschi, 1969; and Eve, 1977). Such an approach misses the possibility that both hypotheses are correct and are accounting for different portions of the variance in delinquency. We have also suggested that different patterns of delinquency may be tied to alternative etiological paths; for example, we postulated that one of the strain paths (limited means/goal failure) should produce forms of delinquency which are considered very instrumental by conventional values. The alternative strain path (attenuated commitment to conventional goals) should result in less instrumental forms of delinquency,

since it characteristically involves a search for new experiences (e.g., drug use) rather than attempts to achieve conventional goals.

Second, we believe that our integrated paradigm is consistent with previous empirical findings and offers some insight into contradictory findings. Previous research using the social control perspective has established a relationship between the strength of one's bonds and social class, with low-SES and minority youths characterized by weaker bonds (Nye, 1958; Gold, 1963; McKinley, 1964; and Hirschi, 1969). In contrast, the attenuated commitment strain path has been associated with affluence, and the limited means-strain path seems most relevant to working-class youths. The combined effect seems consistent with the observed class distribution of self-reported delinquent behavior. Our assumption that weakly bonded youths run the greatest risk of official processing (because of greater surveillance in their neighborhoods, more traditional forms of delinquent behavior, and limited resources with which to avoid processing in the justice system) would account for the observed class distribution of official measures of delinquency. . . .

References

Ageton, S., & Elliott, D. S. (1974). The effects of legal processing on delinquent orientations. *Social Problems, 22,* 87–100.

Akers, R. (1977). *Deviant behavior: A social learning perspective.* Belmont, CA: Wadsworth.

Andry, R. G. (1962). Parental affection and delinquency. In M. E. Wolfgang, L. Savitz, & N. Johnston (Eds.), *The sociology of crime and delinquency* (pp. 342–352). New York: Wiley.

Bandura, A. (1969). *Principles of behavior modification.* New York: Holt, Rinehart & Winston.

Bandura, A. (1973). *Aggression: A social learning analysis.* Englewood Cliffs, NJ: Prentice Hall.

Bandura, A., & Walters, R. H. (1963). *Social learning and personality development.* New York: Holt, Rinehart & Winston.

Becker, H. S. (1960). Notes on the concept of commitment. *American Journal of Sociology, 66,* 32–40.

Becker, H. S. (1963). *Outsiders.* New York: Free Press.

Brennan, T. (1974). *Evaluation and validation regarding the National Strategy for Youth Development: A review of findings* (Report submitted to the Office of Youth Development). Boulder, CO: Behavioral Research and Evaluation Corporation.

Brennan, T., Huizinga, D., & Elliott, D. S. (1978). *The social psychology of runaways.* Lexington, MA: D. C. Heath.

Burgess, R. L., & Akers, R. L. (1966a). Are operant principles tautological? *Psychological Record, 16,* 305–312.

Burgess, R. L., & Akers, R. L. (1966b). A different association-reinforcement theory of criminal behavior. *Social Problems, 14,* 128–147.

Cameron, M. O. (1964). *The booster and the snitch.* New York: Free Press.

Cartwright, D. S. (1971). *Summary of conceptual issues in the National Strategy for Delinquency Prevention* (Document No. 34 in Center for Action Research). Boulder, CO: University of Colorado, Bureau of Sociological Research.

Cartwright, D. S., Reuterman, N. A., & Vandiver, R. I. (1966). *Multiple-factor approach to delinquency.* Boulder: Department of Psychology, University of Colorado.

Cloward, R. A., & Ohlin, L. E. (1960). *Delinquency and opportunity—A theory of delinquent gangs.* New York: Free Press.

Cohen, A. K. (1955). *Delinquent boys: The culture of the gang.* Glencoe, IL: Free Press.

Cohen, A. (1966). *Deviance and control.* Englewood Cliffs, NJ: Prentice Hall.

Durkheim, E. (1951). *Suicide: A study of sociology.* Glencoe, IL: Free Press. (Original work published 1897)

Gauthier, M. (1959). The psychology of the compulsive forger. *Canadian Journal of Corrections, 1,* 62–69.

Gebhard, P. H., et al. (1965). *Sex offenders.* New York: Harper & Row.

Gibbons, D. C. (1977). *Society, crime and criminal careers* (3rd ed.). Englewood Cliffs, NJ: Prentice Hall.

Goldman, N. (1963). *The differential selection of juvenile offenders for court appearance.* Washington, DC: National Council on Crime and Delinquency.

Goode, W. J. (1960). Norm commitment and conformity to role status obligation. *American Journal of Sociology, 64,* 246–258.

Elliott, D. S. (1962). Delinquency and perceived opportunity. *Sociological Inquiry, 32,* 216–227.

Elliott, D. S., & Voss, H. (1974). *Delinquency and dropout.* Lexington, MA: D. C. Heath.

Eve, R. (1977). *The efficacy of strain, culture conflict and social control theories for explaining rebelliousness among high school students.* Unpublished manuscript, University of Texas at Arlington.

Glueck, S., & Glueck, E. (1950). *Unraveling juvenile delinquency.* Cambridge, MA: Harvard University Press.

Gold, M. (1963). *Status forces in delinquent boys.* Ann Arbor: University of Michigan, Institute for Social Research.

Hebb, D. O. (1966). *Psychology.* Philadelphia: Saunders.

Hepburn, J. R. (1976). Testing alternative models of delinquency causation. *Journal of Criminal Law and Criminology, 67,* 450–460.

Hirschi, T. (1969). *Causes of delinquency.* Berkeley: University of California Press.

Jensen, G. F. (1972). Delinquency and adolescent self-conceptions: A study of the personal relevance of infraction. *Social Problems, 20,* 84–103.

Jessor, R., et al. (1968). *Society, personality and deviant behavior: A study of a tri-ethnic community.* New York: Holt, Rinehart & Winston.

Kelly, D. H. (1977). The effects of legal processing upon a delinquent's public identity: An analytical and empirical critique. *Education, 97,* 280–289.

Kitsuse, J. I. (1962). Societal reaction to deviant behavior: Problems of theory and method. *Social Problems, 9*(Winter), 247–256.

Lemert, E. M. (1951). *Social pathology.* New York: McGraw-Hill.

Lemert, E. M. (1953). An isolation and closure theory of naive check forgery. *Journal of Criminal Law, Criminology and Police Science, 44,* 296–307.

Liska, A. E. (1971). Aspirations, expectations and delinquency: Stress and additive models. *Sociological Quarterly, 12,* 99–107.

Matza, D. (1964). *Delinquency and drift.* New York: Wiley.

McKinley, D. G. (1964). *Social class and family life.* New York: Free Press.

Miller, W. B. (1958). Lower class culture as a generating milieu of gang delinquency. *Journal of Social Issues, 14*(3), 5–19.

Mischel, W. (1968). *Personality and assessment.* New York: Wiley.

Monahan, T. P. (1957). Family status and the delinquent child: A reappraisal and some new findings. *Social Forces, 35,* 250–258.

Nye, F. I. (1958). *Family relationships and delinquent behavior.* New York: Wiley.

Reiss, A. J., Jr. (1951). Delinquency as the failure of personal and social controls. *American Sociological Review, 16,* 196–207.

Reiss, A. J., Jr. (1961). The social integration of queers and peers. *Social Problems, 9,* 102–120.

Robin, G. (1974). The American customer: Shopper or shoplifter? *Police, 8,* 6–14.

Rosen, L. (1970). The broken home and male delinquency. In M. E. Wolfgang, L. Savitz, & N. Johnston (Eds.), *The sociology of crime and delinquency* (2nd ed., pp. 484–495). New York: Wiley.

Rotter, J. B. (1954). *Social learning and clinical psychology.* Englewood Cliffs, NJ: Prentice Hall.

Rubington, E. R., & Weinberg, M. S. (Eds.). (1968). *Deviance: The interactionist perspective.* New York: Macmillan.

Savitz, L. (1970). Delinquency and migration. In M. E. Wolfgang, L. Savitz, & N. Johnston (Eds.), *The sociology of crime and delinquency* (2nd ed., pp. 473–480). New York: Wiley.

Schur, E. M. (1971). *Labeling deviant behavior.* New York: Harper & Row.

Seeman, M. (1959). On the meaning of alienation. *American Sociological Review, 24,* 783–791.

Shaw, G. (1931). *Delinquency areas.* Chicago: University of Chicago Press.

Short, J. F., Jr. (1964). Gang delinquency and anomie. In M. B. Clinard (Ed.), *Anomie and deviant behavior* (pp. 98–127). New York: Free Press.

Short, J. F., Jr., Rivera, R., & Tennyson, R. A. (1965). Perceived opportunities, gang membership and delinquency. *American Sociological Review, 30,* 56–67.

Simon, W., & Gagnon, J. H. (1976). The anomie of affluence: A post Mertonian conception. *American Journal of Sociology, 82,* 356–378.

Spergel, I. (1967). Deviant patterns and opportunities of pre-adolescent Negro boys in three Chicago neighborhoods. In M. W. Klein (Ed.), *Juvenile gangs in context: Theory, research and action* (pp. 38–54). Englewood Cliffs, NJ: Prentice Hall.

Sutherland, E. H. (1947). *Criminology.* Philadelphia: J. B. Lippincott.

Sykes, G. M., & Matza, D. (1957). Techniques of neutralization: A theory of delinquency. *American Sociological Review, 22,* 664–670.

Thomas, W. I., & Znaniecki, F. (1927). *The Polish peasant in Europe and America.* New York: Knopf.

Toby, J. (1957). The differential impact of family disorganization. *American Sociological Review, 22,* 505–512.

REVIEW QUESTIONS

1. In the integrated model proposed by Elliott et al., which theoretical perspective is considered to be antecedent, or most important in the early stages, for developing delinquent or criminal tendencies? Do you agree with this portion of the model?

2. Do you think it is rational to merge theoretical perspectives that have opposing basic assumptions, such as merging strain theory, which claims that individuals are born good, with perspectives such as control theory, which claims that people are born bad and must be taught or controlled to be good? Or with learning theories, such as differential association/reinforcement, that assume individuals are born neither good nor bad, but rather as a "blank slate"? If you agree with such integration, explain the rationale for your reasoning. If not, why?

3. Ultimately, what do you think of Elliott et al.'s model? Do you think it was a good early attempt for an integrated theory, or do you think it is rather weak given the information they had at that time (1979)?

READING

In this selection, Carter Hay examines the empirical validity of a more modern and well-respected integrated theoretical model, namely John Braithwaite's theory of reintegrative shaming. We examined Braithwaite's integrated model in the introduction to this section, and Hay provides one of the first empirical tests of the theory's propositions regarding individuals. The author does this with an American sample, which consists of 197 adolescents from one region in the United States.

While reading this selection, readers should consider the extent to which the primary concepts or variables in the model are represented, such as the way youths are shamed negatively or in a more positive, reintegrative form. Also, readers should consider whether the effects of reintegrative shaming reduce delinquency among youth.

An Exploratory Test of Braithwaite's Reintegrative Shaming Theory

Carter Hay

Dynamic may be the best term to describe the current state of criminological theory. The past two decades have seen unprecedented growth in both the number and variety of theories of crime causation. Bernard and Snipes (1996:301–2) point to 16 theories of crime that emerged just between 1985 and 1994. If we consider theories that they omitted as well as those put forth since 1994, at least 25 theories of crime have been put forth since 1985.[1]

The diversity of these theories is as striking as their sheer number. Prominent recent theories run the gamut of intellectual perspectives, including those that are Marxist (Colvin and Pauly 1983; Currie 1997), neoclassical (Cornish and Clarke 1986; Gottfredson and Hirschi 1990; Wilson and Herrnstein 1985), feminist (Messerschmidt 1993), symbolic-interactionist (Matsueda 1992), macro-evolutionary (Messner and Rosenfeld 1994), life course developmental (Moffitt 1993; Sampson and Laub 1993), biosocial (Cohen and Machalek 1988; Moffitt 1993), and integrative (Braithwaite 1989; Cullen 1994; Tittle 1995; Thornberry 1987; Vila 1994).

This proliferation of theory is in some sense welcomed, given the limited explanatory power of traditional theories of crime. Elliott (1985:124) notes that the dominant crime theories—social control, social learning, and strain—typically account for no more than 20 percent of the variation in crime and delinquency. New theories therefore present the potential for improved explanation of crime.

It is also true, however, that the recent growth in theory comes at some risk. If new theories are not (1) empirically tested and (2) discarded when support is lacking, the wave of new theory may prove overwhelming. Theoretical criminology could become fragmented as a result of an immense and confusing number of theories that have either limited or unknown levels of explanatory power. Moreover, the sheer number of theories would require considerable specialization among criminologists, thereby hindering communication across theoretical lines. To the extent that such fragmentation does in fact emerge, the question "What causes crime?" might become answerable in even less lucid terms than is presently the case.

Bernard and Snipes (1996) already see the field as fragmented, and Liska, Krohn, and Messner (1989:1) would seem to agree when they refer to the "seemingly chaotic state of affairs" in theoretical criminology. Bernard and Snipes's central argument is that criminology has too many theories, and the resulting fragmentation impedes scientific progress. They warn that cynicism about criminological theorizing may already be rampant and may increase in the future. Bernard and Snipes cite theoretical integration as one solution to this problem, arguing that integration can reduce the number of crime theories and at the same time improve their overall quality.

Although the importance of integration should not be dismissed, an equally important solution is simply to ensure that recent theories receive sufficient empirical scrutiny. If new theories are to illuminate rather than obscure our understanding of the causes of crime, explicit tests of them are needed. With few exceptions (e.g., Gottfredson and Hirschi's 1990 self-control theory and Agnew's 1992 general strain theory), such research has not been done. This article addresses that void by

reporting an exploratory test of one prominent new theory of crime: John Braithwaite's (1989) reintegrative shaming theory.

◤ Reintegrative Shaming Theory

John Braithwaite first put forth reintegrative shaming theory (RST) in the book *Crime, Shame and Reintegration* (Braithwaite 1989). The theory's essential argument is that the precise ways in which societies, communities, and families sanction deviance affect the extent to which their members engage in predatory criminal behavior. The key explanatory variable in the theory is shaming, which Braithwaite defines as any social process that expresses disapproval of a sanctioned act such that there is the intent or effect of invoking moral regret in the person being shamed (p. 100). Braithwaite is explicit about how shaming contrasts with a more classical view of sanctioning: "Shaming, unlike purely deterrent punishment, sets out to *moralize* [emphasis added] with the offender to communicate reasons for the evil of her actions" (p. 100).

Shaming is not, however, a uniform type of sanctioning—it can be done in different ways and in different contexts. Braithwaite (1989) distinguishes between two types of shaming. First, shaming is reintegrative when it reinforces an offender's membership in the community of law-abiding citizens. This prevents the shamed individual from adopting a deviant master status and is accomplished when shaming (1) maintains bonds of love or respect between the person being shamed and the person doing the shaming, (2) is directed at the evil of the act rather than the evil of the person, (3) is delivered in a context of general social approval, and (4) is terminated with gestures or ceremonies of forgiveness (pp. 100–1).

Reintegrative shaming is contrasted with stigmatization, which is disintegrative shaming in which little or no effort is made to forgive offenders or affirm the basic goodness of their character and thus reinforce their membership in the community of law-abiding citizens. Stigmatization can be seen essentially as shaming in the absence of reintegration—it is the converse of each of the four aspects of reintegration mentioned earlier. The primary importance of stigmatization is that it treats offenders as outcasts and provokes a rebellious and criminal reaction from them: "Shaming that is stigmatizing . . . makes criminal subcultures more attractive because these are in some sense subcultures which reject the rejectors" (Braithwaite 1989:102).

RST's basic prediction therefore is quite simple: There should be a negative relationship between the use of reintegrative shaming and the extent of criminal behavior. This prediction is applicable to both micro- and macro-level units of analysis. That is, just as individuals who are exposed to reintegrative shaming should commit fewer crimes, communities or societies with high levels of reintegrative shaming should have low aggregate rates of crime.

The focus of this study is the micro-level portion of the theory, and three further points about the theory should be emphasized. First, Braithwaite (1989) specifies the antecedent of reintegrative shaming: Whether individuals are exposed to reintegrative shaming to begin with should be a function of their involvement in interdependent relationships. Braithwaite describes interdependency in the following way:

> Interdependency is a condition of individuals. It means the extent to which individuals participate in networks wherein they are dependent on others to achieve valued ends and others are dependent on them. . . . Interdependency is approximately equivalent to the social bonding, attachment and commitment of control theory. (pp. 99–100)

Braithwaite (1989) argues that several individual characteristics generally found to reduce crime—age (being younger than 15 and older than 25), being married, being female, and having a job—do so because they increase involvement in interdependent relationships and therefore increase the chances that rule-violating behavior will be met with reintegrative shaming rather than stigmatization.

Second, key to the theory is the interaction between reintegration and shaming: The combination of reintegration and shaming should have an effect on offending that exceeds the sum of each variable's independent effects. In fact, strictly speaking, neither variable should have an independent effect on offending (Braithwaite 1989:99; Makkai and Braithwaite 1994:371–72). Instead, the effects of shaming should be conditional on the level of reintegration. When reintegration is high, shaming should be negatively related to offending. When reintegration is low, shaming has a stigmatizing effect and should be positively related to offending.

A third point to emphasize about RST is that it limits its scope to the explanation of predatory offenses against persons and property. The reason for this restricted focus is that reintegrative shaming is thought to be relevant only for offenses in which there is consensus regarding their moral wrongfulness. When such consensus is lacking, shaming should not be influential.

▨ The Current Study

This study reports an exploratory test of the micro-level predictions of RST. As has been noted elsewhere (Braithwaite 1989:120-21; Hay 1998:424), there are substantial barriers to testing the macro-level portions of RST. Using survey research to measure reintegrative shaming across a sample of communities or societies would be prohibitively expensive. Moreover, researchers have yet to identify proxies for community- or societal-wide levels of reintegrative shaming.

This study instead will examine the more testable micro-level arguments of RST. Specifically, adolescents are taken as the units of analysis, and the focus is on the relationship between their perceptions of their parents' sanctioning methods and their reports of predatory delinquency. This strategy would seem acceptable to Braithwaite (1989), who noted that "the best place to see reintegrative shaming at work is in loving families" (p. 56). The analysis will address three principal research questions: (1) What variables predict the extent to which parents respond to adolescent rule-violating behavior with reintegration and shaming? (2) Do reintegration and shaming statistically interact to affect predatory delinquency, or are their effects additive? (3) Are any observed effects of reintegration and shaming on delinquency merely a result of spuriousness?

Data and Method

Sample. The data used to examine the research questions mentioned above come from anonymous, self-administered questionnaires completed by a sample of 197 adolescents taken from a single urban area in a southwestern state of the United States. Respondents were located through their attendance in a single high school located in the central part of the urban area. This school was chosen because of its diversity in race/ethnicity and socioeconomic status. All students enrolled in physical education classes during the fall of 1998 were invited to participate in the study.[2] Roughly 60 percent of students completed and returned the necessary consent forms and took part in the study.

Table 1 shows the demographic characteristics of the sample. Most respondents are between the ages of 14 and 17, and the sex ratio of the sample is roughly 1 to 1. The sample is diverse in terms of race/ethnicity, with Whites,

Table 1	Characteristics of the Sample (N = 197)	
Variable	**Category**	**Percentage of Sample**
Age	14	25.8
	15	30.8
	16	24.7
	17	13.2
	Other	5.5
Sex	Male	47.8
	Female	52.2
Race	White	40.9
	Hispanic	32.0
	African American	19.9
	Other	7.2
Family structure	Mother and father	44.8
	Two-parent other	16.0
	Single parent	39.2
Mother's education	High school diploma or less	44.8
Father's education	High school diploma or less	43.2

Hispanics, and Africans Americans all highly represented. Family disruption is common, with only 45 percent of respondents living in a household with their biological mother and father. Also, close to 45 percent of respondents' parents have a high school diploma or less.

Measurement of key variables. As more quantitative tests of RST emerge, knowledge of how reintegration and shaming should be measured will emerge as well. For now, however, this study makes essentially the same qualifying statement that has appeared in two prior quantitative analyses of RST (Makkai and Braithwaite 1994:368; Zhang 1995:251): Measuring reintegration and shaming is no easy task, and the measures used in this study should be improved on as more research is conducted.

Consistent with Braithwaite's (1989:100) discussion of shaming, the key measurement concern was to identify sanctions that involved "moralizing." Three survey items were selected and scaled. With these items, respondents were asked to indicate how much importance their mother and father place on three goals when reacting to a violation of any rule considered important by the parent: convincing the adolescent that what he or she did was immoral or unfair, making him or her feel guilty or ashamed for what was done, and having him or her "make up" for the actions by apologizing or helping to erase any harm that was done."[3]

Parents' use of reintegration was measured with a scale comprising four items that correspond to the four aspects of reintegration discussed earlier. The first two items asked respondents to indicate how much they agreed or disagreed that their parents see them as good people even when upset with them and treat them with respect when they are disciplining them. For the last two items, respondents assessed how likely it is that in response to their rule violations, parents would tell them that they are "bad kids" and eventually express their forgiveness to the adolescents.

All of the shaming and reintegration items asked respondents to assess their mother and father separately. Preliminary analyses indicated that mothers and fathers did not differ in terms of their use of shaming and reintegration or the effects of those variables on predatory offending. Scores for mothers and fathers were therefore combined to create an overall score for parents. Responses to the shaming and reintegration scales were coded so that high scorers perceive high levels of the two variables.

Delinquency was measured with two different variables. The first is a nine-item scale of early childhood antisocial behavior that will be used to protect against spuriousness and examine whether childhood behavioral problems predict parents' use of reintegrative shaming. For these items, respondents were asked to indicate the earliest age at which they

committed nine offenses: breaking into a building or house, stealing things worth $50 or less, stealing things worth more than $50, purposely damaging or destroying property, taking a car for a drive without the owner's permission, getting in a fight with someone with the idea of seriously hurting him or her, carrying a hidden weapon such as a knife or gun, drinking alcohol, and using marijuana or some other drug. For each item, responses were coded 1 if the respondent reported committing the act by age 11 or earlier and 0 if he or she had not. The nine items were summed to produce a measure of childhood antisocial behavior, with possible scores ranging from 0 to 9 and a mean of 1.24.

The second delinquency variable is a measure of adolescents' projected involvement in the first seven offenses listed above. Importantly, each of the seven acts is an illegal violent or property offense and therefore fits within RST's emphasis on predatory offenses. These survey items asked respondents the following question: "If you found yourself in a situation where you had the chance to do the following things [each of the seven offenses], how likely is it that you would do each one?" Respondents chose an answer from a scale of 0 to 10, where 0 means there is no chance that they would commit the act, 5 means there is a "50-50 chance," and 10 means that they definitely would commit the act. The seven items were averaged to create a projected predatory delinquency scale, with scores ranging from 0 to 10 and a mean of 2.55. The Cronbach's alpha for the scale is .90.

This measure of projected delinquency—rather than self-reported past delinquency—was used as a dependent variable to avoid the causal order problems that often arise in cross-sectional delinquency research. Cross-sectional studies that use self-reported past offending as the dependent variable suffer inherent causal order problems if independent variables are measured with items about respondents' current attitudes and perceptions—any assertion of appropriate causal order is questionable if current attitudes and perceptions are used to explain past behavior. Thus, the clear strength of a projected offending variable is that it allows for clear conceptualization of causal ordering with cross-sectional data (Tittle 1977:586).

Measures of projected offending have been used effectively by a number of other researchers (e.g., Bachman, Paternoster, and Ward 1992; Grasmick, Bursik, and Arneklev 1993; Grasmick and Bursik 1990; Jensen and Stitt 1982; Tittle 1980). A few things attest to their validity. First, variables that are known to predict self-reported offending in longitudinal designs have also been found to predict projected offending (Jensen and Stitt 1982; Tittle 1980). Second, using a two-wave panel design, Green (1989) found that projections of future deviance were highly correlated with actual subsequent deviance ($r = .80$). The current evidence therefore suggests that measures of projected offending are useful for avoiding the causal order problems associated with cross-sectional data.

⊠ Analysis

The Prediction of Reintegration and Shaming

The first issue to consider involves the antecedents of parental reintegration and shaming— that is, what variables predict that adolescents' parents will respond to their rule-violating behavior with reintegration and shaming? Braithwaite (1989) argues that involvement in interdependent relationships should be the most important factor; moreover, many demographic variables known to predict offending may do so because they increase involvement in interdependent relationships.

To consider that possibility, three regression equations were estimated for both dependent variable, reintegration and shaming. Equation

(1) includes five exogenous variables: four demographic variables (age, sex, family structure, and race/ethnicity) and a measure of childhood antisocial behavior.[4] Although Braithwaite (1989) makes no reference to how childhood behavior problems affect parental reintegration and shaming during adolescence, much research suggests that parent-child interactions are highly reciprocal (e.g., Cohen and Brook 1995; Kandel and Wu 1995; Sampson and Laub 1993).

Equations (2) and (3) consider the role of parent-child interdependency. Equation (2) adds individual measures of three parent-child interdependency variables—adolescent attachment to parents, adolescent perceptions of parents' feelings of attachment for them, and adolescent reports of instrumental and intimate communication with parents.[5] If Braithwaite's (1989) predictions are correct, the parent-child interdependency variables should have significant effects on reintegration and shaming, and their inclusion in the model should substantially weaken any effects of the exogenous variables that are observed in equation (1). Equation (3) examines the same issue, but rather than considering interdependency as three separate variables, a combined measure is used. This combined measure is simply an average of its three component parts.

Table 3 shows the results of the analysis and reveals moderate support for RST's predictions. Looking first at the results of equation (1) for reintegration, only race/ethnicity and childhood antisocial behavior significantly affect parents' use of reintegration. Consistent with Braithwaite's (1989) predictions, both effects are reduced (childhood antisocial behavior is rendered insignificant) when the interdependency variables are introduced into equation (2). The effects of both attachment to parents and perceived parental attachment to the child are statistically and substantively significant. The effect of parent-child communication is less impressive but is nevertheless in the predicted direction and higher than those

| Table 2 | Standardized Coefficients and Fit for Ordinary Least Squares Regressions of Reintegration and Shaming on Exogenous Variables and Parent-Child Interdependency |

Independent Variable	1	2	3
Reintegration			
Age	−.04	−.03	−.04
Sex	.02	−.04	−.03
Race/ethnicity	−.13*	−.09*	−.08
Family structure	−.11	−.08	−.09
Childhood antisocial behavior	−.23*	−.05	−.04
Attachment to parents	—	.40*	—
Perceived parental attachment to child	—	.30*	—
Parent-child communication	—	.12	—
Parent-child interdependency	—	—	.72*
R^2	.09	.57	.56
Shaming			
Age	−.01	.00	−.01
Sex	−.05	−.07	−.06
Race/ethnicity	−.05	−.02	−.04
Family structure	−.11	−.10	−.10
Childhood antisocial behavior	−.06	−.01	−.01
Attachment to parents	—	−.01	—
Perceived parental attachment to child	−.17	—	—
Parent-child communication	—	.08	—
Parent-child interdependency	—	—	.20*
R^2	.02	.07	.06

NOTE: Age was measured as a continuous variable in years. Sex, race/ethnicity, and family structure were measured with dummy variables: Sex was coded 1 for males and 0 for females, race/ethnicity was coded 1 for non-Whites and 0 for Whites, and family structure was coded 1 for non-intact and 0 for intact homes.

*$p < .05$.

observed for the exogenous variables. Equation (3) shows a similar pattern, revealing that the combined interdependency variable has a substantial effect ($\beta = .72$) on reintegration.

Perhaps most notable about these results is the substantial increase in explained variance that occurs when the interdependency variables are included in the model. In equation (1), the explained variance for reintegration is less than 10 percent, but in equations

(2) and (3), explained variance is nearly 60 percent.

The results for shaming are generally less supportive of RST's predictions. The results for equation (1) reveal that none of the exogenous variables have an effect on shaming, and they combine to explain just 2 percent of the variation in parental shaming. When the three interdependency variables are included in equation (2), explained variance improves to

7 percent, although none of the three have a statistically significant effect on shaming. This is due in part to the shared variance between the three measures of interdependency—equation (3) reveals that when they are combined into a single measure, the effect of interdependency on shaming (β = .20) is significant. Nevertheless, the explained variance (6 percent) in equation (3) remains low. By and large, the effect of parent-child interdependency on parental shaming is relatively modest but apparent nonetheless.

Testing for a Reintegration-Shaming Interaction

The next issue to consider involves the way in which reintegration and shaming affect predatory delinquency. RST predicts that the two interact, but it may be that their effects are additive rather than interactive—that is, the combination of reintegration and shaming may not produce an effect that exceeds the sum of each variable's independent effects on delinquency (Hay 1998:430–31).

The typical strategy for detecting an interaction between two variables is to enter a multiplicative term into an ordinary least squares (OLS) equation that includes parameters for the two main effects. This is not the preferred strategy, however, for assessing the interaction in question here. Jointly considering reintegration and shaming yields four parental sanctioning combinations, including sanctioning that is (1) high in both reintegration and shaming, (2) high in reintegration but low in shaming, (3) low in reintegration but high in shaming, and (4) low in both reintegration and shaming. Using a multiplicative term to express this interaction would obscure how each combination is distinct from the others.

To capture these distinctions, respondents' reintegration and shaming scores were collapsed into two categories—high or low—on the basis of whether they were above or below the median. This created the four combinations

listed above: high reintegration/high shaming, high reintegration/low shaming, low reintegration/high shaming, and low reintegration/low shaming. Dummy variables were then constructed to represent the four categories.

Equation (1) of Table 4 regresses projected predatory offending on three of the dummy variables (low reintegration/low shaming is the reference category) as well as control variables for sex, age, family structure, race/ethnicity, and childhood antisocial behavior. This model does not directly bear on the issue of interaction but rather reveals how projected offending is distributed across the four categories of reintegration and shaming. The relatively large, negative coefficient for each dummy variable indicates that each category is associated with lower projected offending than the low-reintegration/low-shaming category omitted from the model. Consistent with RST, this strongly suggests the delinquency-generating effects of being exposed to neither reintegration nor shaming.

Table 3	Standardized Coefficients and Fit for Ordinary Least Squares Regression of Predatory Crime on Reintegration and Shaming Dummy Variables and Main Effects

Independent Variable	1	2
High reintegration/ high shaming	−.27*	.07
High reintegration/ low shaming	−.25*	−.02
Low reintegration/ high shaming	−.24*	−.08
Reintegration	—	−.29*
Shaming	—	−.16**
R^2	.33	.36

NOTE: Low reintegration/low shaming is the omitted category. Both equations include controls for age, sex, race/ethnicity, and childhood antisocial behavior.

*$p < .05$. **$p < .08$.

In contrast to the RST's predictions, however, the negative effect of the high-reintegration/high-shaming category—the reintegrative shaming category—is not statistically different from the effects of the other two categories included in the model. In fact, that the low-reintegration/high-shaming category—what Braithwaite (1989) refers to as stigmatization—has a negative rather than positive coefficient is itself contradictory to RST. Braithwaite clearly predicts that stigmatization should produce a higher rate of projected offending than any other category.

Although informative, equation (1) does not address the issue of interaction because main effects for reintegration and shaming were not included in the model and thus not partialled out from the effects of the dummy variables. Equation (2) includes the two main effects. The results for equation (2) suggest that reintegration and shaming affect offending additively rather than interactively. The main effects of reintegration ($\beta = -.29$) and shaming ($\beta = -.16$) on predatory delinquency are significant (at $p = .08$ for shaming) and of at least moderate strength. Of principal importance is that none of the three dummy variables has an effect on delinquency that is significantly different from the omitted category.[6]

Two things should be noted about this finding of no interaction. First, it runs counter to the findings of Makkai and Braithwaite (1994), who observed that reintegration and shaming from health inspectors interact to affect nursing homes' compliance with regulatory standards. Second, although a finding of no interaction runs counter to RST, it does not suggest that reintegration and shaming are without importance. Quite the opposite, both variables are significantly and negatively related to projected offending, even after controlling for one another; dummy variables representing the different categories of reintegration and shaming; and theoretically relevant control variables, including childhood antisocial behavior. A tentative conclusion that can therefore be made is that reintegration and shaming appear to affect delinquency additively in these data. Each variable—but especially reintegration—exerts an independent effect on delinquency, but the combination of the two produces no effect marginal to those independent effects.

Testing for Spuriousness

Before concluding that reintegration and shaming are significant causes of offending, the possibility of spuriousness must be ruled out. The apparent effect of reintegration and shaming on delinquency may simply reflect that all three result from the same prior cause. The analyses shown in Table 4 partially protected against this by including controls for basic demographic variables and childhood antisocial behavior. Perhaps the most important source of spuriousness, however, may be parent-child interdependency—that is, reintegration and shaming may appear to be causes of delinquency only because all three are by-products of parent-child interdependency.

To consider that possibility, two OLS equations were estimated, and the results are shown in Table 5. Equation (1) regresses projected predatory delinquency on reintegration, shaming, and controls for age, sex, family structure, race/ethnicity, and childhood antisocial behavior. Consistent with the earlier findings, moderately large effects of reintegration and shaming are revealed. Both effects are statistically significant, but the standardized effect of reintegration ($\beta = -.24$) is greatest. Equation (2) of Table 5 tests for spuriousness by adding the composite parent-child interdependency variable that combines the three separate interdependency variables used earlier. If the effects of reintegration and shaming are causal effects, they should persist even when controlling for interdependency, given RST's clear prediction that interdependency is a cause of reintegration and shaming rather than a mediator of their effects on offending (Braithwaite 1989:99).

Table 4	Standardized Coefficient and Fit for Ordinary Least Squares Regressions of Predatory Crime on Reintegration, Shaming, and Parent-Child Interdependency	
Independent Variable	1	2
Reintegration	−.24*	.05
Shaming	−.17*	−.12*
Parent-child interdependency	—	−.41*
R^2	.35	.42

NOTE: Both equations include controls for age, sex, race/ethnicity, family structure, and childhood antisocial behavior.

*$p < .05$.

The results for equation (2) suggest that reintegration's effect is spurious but that the effect of shaming is not. When parent-child interdependency is included in the equation, the effect of reintegration is entirely eliminated (the beta goes from −.24 to .05), whereas the negative effect of shaming is reduced less dramatically (the beta goes from −.17 to −.12) and remains statistically significant. Including the interdependency variable resulted in only a moderate increase (from .35 to .42) in R^2, pointing to the overlap between reintegration and interdependency.[7]

Admittedly, the failure of reintegration to retain its significance when interdependency was controlled may result from inadequate measurement of reintegration. Recall that reintegration was measured with just 4 items, whereas interdependency was measured with 12. Moreover, the newness of RST and the lack of prior quantitative tests precluded the use of empirically validated measures of either reintegration or shaming. Most of the interdependency items, on the other hand, have been used effectively in delinquency research over the past several decades. Given the obvious conceptual overlap between parental reintegration and parent-child interdependency, this analysis may have favored the better measured variable.

The enduring effect of shaming, however, is worth emphasizing. Despite the same measurement limitations faced with reintegration, equation (2) of Table 5 reveals that shaming significantly reduces predatory delinquency even after controlling for reintegration, parent-child interdependency, childhood antisocial behavior, and important demographic variables such as age, sex, race/ethnicity, and family structure. Shaming is arguably the most novel aspect of RST, and this analysis suggests its potential importance to the explanation of predatory delinquency.

Discussion and Conclusion

The past two decades have seen an explosion in theories of crime and delinquency causation, but these theories have received limited empirical attention. This article sought to address that void by testing the micro-level portion of reintegrative shaming theory as it applies to the sanctioning methods used by parents with adolescents. Because this is one of the first explicit tests of RST, firm conclusions about the theory's validity and defects are not warranted. Nevertheless, three key findings are worth reflecting on: (1) the strong relationship between parent-child interdependency and reintegration, (2) the apparent spuriousness of the reintegration-delinquency relationship, and (3) the durable independent effect of shaming on delinquency.

First, consistent with RST's predictions, this study revealed a strong relationship between the level of parent-child interdependency and parents' use of reintegration. Interdependency also had a significant effect on shaming, but its effect on reintegration was notably stronger—interdependency and reintegration share about 50 percent of their variation. If nothing else, this strong effect clearly supports the idea that parents who have a close relationship with their children are likely to sanction in such away that reinforces that close relationship.

But this strong statistical relationship between parent-child interdependency and reintegration contributes to a second key finding: An initial effect of reintegration on delinquency disappeared when parent-child interdependency was held constant. This was interpreted as evidence that the initial effect of reintegration was spurious—reintegration was related to delinquency only because each was the result of interdependency. This interpretation is consistent with RST's specification of causal order: Interdependency affects the level of reintegration, which in turn affects delinquency. Because reintegration is supposed to be the variable most proximate to delinquency, its effect should have remained.

An alternative possibility with these two variables is that their actual causal order is opposite what RST predicts—it may be that reintegrative sanctioning of children leads to high parent-child interdependency rather than vice versa. If that were the case, the findings presented here would be just as expected. Future research might be usefully directed toward sorting out the causal order of these variables, paying special attention to the possibility that parental reintegration and parent-child interdependency are reciprocally related to one another in the same way that many other delinquency-causing variables are (Thornberry et al. 1994).

A third key finding of this study involves the durable, independent effect of shaming on delinquency. Shaming—arguably the most novel aspect of RST (see Hay 1998:423)—was significantly and negatively related to projected delinquency even when controlling for age, sex, family structure, race/ethnicity, childhood antisocial behavior, and parent-child interdependency. Moreover the negative effect of shaming was not dependent on the level of reintegration—the effect was strong when reintegration was at both high and low levels

(see equation [1] of Table 4). Although this clearly suggests the potential importance of shaming, strictly speaking, it is not entirely supportive of RST, which argues that shaming should be negatively related to delinquency only when reintegration is high.

One possibility to consider in future research is that contrary to RST, moral-based sanctions such as shaming may rarely be stigmatizing, even when used in the absence of high reintegration or interdependency. Research in developmental psychology has consistently shown that moral- and reason-based sanctions are generally effective at controlling adolescents, especially relative to sanctions based on coercion and intimidation (see, e.g., Bandura and Walters 1959; Baumrind 1991; Patterson, Reid, and Dishion 1992; Sears, Maccoby, and Levin 1957). In short, RST's emphasis on the harmful effects of stigmatization may be appropriate, but shaming as defined by RST may not be a source of such stigmatization, even in the absence of reintegration. Future research may consider that stigmatization is likely to result not from moral-based sanctions such as shaming but rather, from such things as physical punishment, harsh verbal attacks, and similar forms of intensely antagonistic sanctioning.

In concluding, it can be noted that like most empirical tests of theory, the findings in this study are neither uniformly supportive nor unsupportive. The findings here do suggest, however, that RST and its central concept of shaming are worthy of greater empirical attention than they have received thus far. Future empirical tests may involve micro-level tests of the theory such as the one reported here or, conversely, macro-level analyses of community or societal rates of crime. Or better yet, future tests should consider the micro/macro linkages that can be derived from RST. At any rate, increased attention to RST and other recent theories will be necessary if those theories are to contribute to rather than complicate our understanding of the causes of crime.

◩ Notes

1. The theories listed by Bernard and Snipes (1996:301-2) are forth by Schwendinger and Schwendinger (1985), Wilson and Herrnstein (1985), Cornish and Clarke (1986), Mawson (1987), Thornberry (1987), Cohen and Machalek (1988), Hagan (1988), Katz (1988), Bernard (1989), Braithwaite (1989), Jeffery (1989), Gottfredson and Hirschi (1990), Agnew (1992), Walters (1992), Sampson and Laub (1993), and Messner and Rosenfeld (1994). Additional theories have been put forth by Pearson and Weiner (1985), Matsueda (1992), Messerschmidt (1993), Moffitt (1993), Sherman (1993), Cullen (1994), Vila (1994), Tittle (1995), Currie (1997), Wood et al. (1997), and Heimer and De Coster (1999).

2. Physical education is a required course for all students except those on athletic teams. Rather than taking normal physical education classes, athletes enroll in a class specific to their sport. Most students at this school fulfill the physical education requirement during their first two years of high school.

3. There is some question as to whether RST is a theory of crime causation or recidivism (Hay 1998:432). From a measurement standpoint, this is an important question. If RST is a theory of crime causation, then measurement should focus on reintegration and shaming in response to *any* rule-violating behavior, including noncriminal acts. On the other hand, if RST is solely a theory of recidivism, reintegration and shaming should be relevant only in response to criminal acts.

Braithwaite (1989) is not clear on the issue, so different interpretations are possible. The position taken here is that RST is a theory of both causation and recidivism. Whether it is treated as one or the other in a given study will depend largely on which sanctioning institution is the focus. For example, if the micro-level theory is tested by examining the extent of reintegrative shaming found in juvenile court dispositions, this would obviously be examining RST as a theory of recidivism because individuals would not be exposed to the sanctioning had they not already committed a crime. However, when parental sanctioning is the focus, it makes more sense to examine RST as a theory of crime causation because the vast majority (perhaps all) of parental sanctions will be directed at noncriminal rule violations. Moreover, parental sanctioning begins at an age prior to when crime is possible. So the question here becomes. "How does parental reintegrative shaming in response to *all* rule-violations, including noncriminal violations, prevent involvement in serious criminal violations?"

4. Admittedly, RST makes no prediction that age differences within this range (roughly 14–17) will explain offending or exposure to reintegration and shaming. Rather, it predicts that age differences should exist between those inside this range and those outside of it. Thus, including this demographic variable does not explicitly test RST's predictions about the effects of age on reintegration and shaming.

5. The interdependency variables were measured in a way consistent with Braithwaite's (1989) argument that interdependency approximates Hirschi's (1969) notion of the social bond. Because parental sanctioning is the focus of this study, interdependency between the parent and child in particular was emphasized. The following scales were used, with adolescents indicating how much they agree or disagree with each statement. Each item was answered separately for mothers and fathers, and the scores were combined to create an overall score for parents. All items were coded so that high values correspond to high interdependency.

Adolescent attachment to parents:

> I'm closer to my mother/father than a lot of kids are to theirs.

> Having a good relationship with her/him is important to me.

> I would like to be the kind of person she/he is.

Adolescent perceptions of parents' feelings of attachment:

> She/he is interested in what I do.

> She/he encourages me to discuss my problems with her/him.

> I think she/he shows more interest in my brothers and sisters than in me.

> Other mothers/fathers seem to show more interest in their children than mine does in me.

> She/he tries to understand my problems and worries.

Parent-child communication:

> I often share my thoughts and feelings with my mother/father.

> I enjoy letting her/him in on my "big" moments.

I often talk to her/him about problems that I am facing.

I often talk to her/him about my plans for the future.

6. Multicollinearity is a concern with regression equations that include interaction terms and the variables used to construct them. In this case, examination of variance inflation factors (VIFs) indicated no cause for concern. VIFs for the three interaction dummy variables were 3.35, 3.23, and 4.49—well below the point at which multicollinearity typically is seen as problematic (see Myers 1990:369; Stevens 1992:77). An additional equation was estimated without the control variables. This lowered the VIFs for the dummy variables slightly to 3.17, 3.10, and 4.23. This equation once again indicated that the interaction dummy variables were not significantly related to crime, but the two main effects were.

7. The F test $(F = 19.9, df = 1,166)$ for this difference is, however, significant at $p < .01$. Other analyses indicate that the R^2 for the model in equation (2) of Table 5 is also marginally higher than that found for an equation that only includes interdependency $(R^2 = 40)$, but this difference $(F = 2.00, df = 2.168)$ is not significant.

⊠ References

Agnew, Robert. 1992. "Foundation for a General Strain Theory of Crime and Delinquency." *Criminology* 30: 47–87.

Bachman, Ronet, Raymond Paternoster, and Sally Ward. 1992. "The Rationality of Sexual Offending: Testing a Deterrence/Rational Choice Conception of Sexual Assault." *Law and Society Review* 26: 343–72.

Bandura, Albert, and Richard H. Walters, 1959. *Adolescent Aggression.* New York: Ronald Press.

Baumrind, Diana. 1991. "The Influence of Parenting Style on Adolescent Competence and Substance Use." *Journal of Early Adolescence* 11: 56–95.

Bernard, Thomas J. 1989. "A Theoretical Approach to Integration." Pp.137–59 in *Theoretical Integration in the Study of Crime and Deviance: Problems and Prospects,* ed. Steven F. Messner, Marvin D. Krohn, and Allen E. Liska. Albany: State University of New York Press.

Bernard, Thomas J., and Jeffrey B. Snipes. 1996. "Theoretical Integration in Criminology." Pp. 301–48

in *Crime and Justice: A Review of Research,* ed. Michael Tonry. Chicago: University of Chicago Press.

Braithwaite, John. 1989. *Crime, Shame and Reintegration.* New York: Cambridge University Press.

———. 1999. "Restorative Justice: Assessing Optimistic and Pessimistic Accounts." *Crime and Justice: A Review of Research* 25: 1–127.

Cohen, Lawrence E., and Richard Machalek. 1988. "A General Theory of Expropriative Crime." *American Journal of Sociology* 94: 465–501.

Cohen, Patricia and Judith Brook. 1995. "The Reciprocal Influence of Punishment and Child Behavior Disorder." In *Coercion and Punishment in Long-Term Perspectives,* ed. Joan McCord. New York: Cambridge University Press.

Colvin, Mark, and John Pauly. 1983. "A Critique of Criminology: Toward an Integrated Structural-Marxist Theory of Delinquency Production." *American Journal of Sociology* 89: 513–51.

Cornish, Derek B., and Ronald V. Clarke, eds. 1986. *The Reasoning Criminal: Rational Choice Perspectives on Offending.* New York: Springer-Verlag.

Cullen, Francis T. 1994. "Social Support as an Organizing Concept for Criminology: Presidential Address to the Academy of Criminal Justice Sciences." *Justice Quarterly* 11: 527–59.

Currie, Elliott. 1997. "Market, Crime, and Community: Toward a Mid-Range Theory of Post-Industrial Violence." *Theoretical Criminology* 2: 147–72.

Elliott, Delbert. 1985. "The Assumption That Theories Can Be Combined with Increased Explanatory Power." Pp.123–49 in *Theoretical Methods in Criminology,* ed. Robert F. Meier. Beverly Hills, CA: Sage.

Gottfredson, Michael R., and Travis Hirschi. 1990. *A General Theory of Crime.* Stanford, CA. Stanford University Press.

Grasmick, Harold G., and Robert J. Bursik, Jr. 1990. "Conscience, Significant Others, and Rational Choice: Extending the Deterrence Model." *Law and Society Review* 24: 837–61.

Grasmick, Harold G., Robert J. Bursik, Jr., and Bruce J. Arneklev. 1993. "Reduction in Drunk Driving as a Response to Increased Threats of Shame, Embarrassment, and Legal Sanctions." *Criminology* 31: 41–67.

Green, Donald E. 1989. "Measures of Illegal Behavior in Individual Level Deterrence Research." *Journal of Research in Crime and Delinquency* 26: 253–75.

Hagan, John. 1988. *Structural Criminology.* New Brunswick. NJ: Rutgers University Press.

Hay, Carter. 1998. "Parental Sanctions and Delinquent Behavior: Toward Clarification of Braithwaite's Theory of Reintegrative Shaming." *Theoretical Criminology* 2: 419–43.

Heimer, Karen, and Stacy De Coster. 1999. "The Gendering of Violent Delinquency." *Criminology* 37: 277–317.

Hirschi, Travis. 1969. *Causes of Delinquency.* Berkeley: University of California Press.

Jeffery, C. R. 1989. *Criminology.* Englewood Cliffs. NJ: Prentice Hall.

Jensen, Gary, and B. Grant Stitt. 1982. "Words and Misdeeds: Hypothetical Choices versus Past Behavior as Measures of Deviance." Pp. 33–54 in *Deterrence Reconsidered,* ed. John Hagan. Beverly Hills, CA: Sage.

Kandel, Denise, and Ping Wu. 1995. "Disentangling Mother-Child Effects." In *Coercion and Punishment in Long-Term Perspectives,* ed. Joan McCord. Cambridge, UK: Cambridge University Press.

Katz, Jack. 1988. *The Seductions of Crime.* New York: Basic Books.

Liska, Allen E., Marvin D. Krohn, and Steven F. Messner. 1989. "Strategies and Requisites for Theoretical Integration in the Study of Crime and Deviance." Pp. 1–19 in *Theoretical Integration in the Study of Deviance and Crime: Problem and Prospects,* ed. Steven F. Messner, Marvin D. Krohn, and Allen E. Liska. Albany: State University of New York Press.

Makkai, Toni, and John Braithwaite. 1994. "Reintegrative Shaming and Compliance with Regulatory Standards." *Criminology* 32: 361–83.

Matsueda, Ross L. 1992. "Reflected Appraisals, Parental Labeling, and Delinquency: Specifying a Symbolic Interactionist Theory." *American Journal of Sociology* 97: 1577–1611.

Mawson, A. R. 1987. *Transient Criminality* New York: Praeger.

Messerschmidt, James M. 1993.*Masculinities and Crime: Critique and Reconceptualization of Theory.* Lanham, MD: Rowman & Littlefield.

Messner, Steven F., and Richard Rosenfeld. 1994. *Crime and the American Dream.* Belmont. CA: Wadsworth.

Moffitt, Terrie E. 1993. "Adolescence-Limited and Life-Course-Persistent Antisocial Behavior: A Developmental Taxonomy." *Psychological Review* 100: 674–701.

Myers, R. 1990. Classical and Modern Regression with Applications. 2nd ed. Boston: Duxbury.

Patterson, Gerald R., John B. Reid, and Thomas J. Dishion. 1992. *Antisocial Boys.* Eugene, OR: Castalia.

Pearson, Frank, and Neil Alan Weiner. 1985. "Toward an Integration of Criminological Theories." *Journal of Criminal Law and Criminology* 76: 116–50.

Sampson, Robert, and John Laub. 1993. *Crime in the Making: Pathways and Turning Points through Life.* Cambridge, MA: Harvard University Press.

Schwendinger, H., and J. S. Schwendinger. 1985. *Adolescent Subcultures and Delinquency.* New York: Praeger.

Sears, R., E. Maccoby, and H. Levin. 1957. *Patterns of Child Rearing.* Evanston. IL: Row, Peterson.

Sherman, Lawrence W. 1993. "Defiance, Deterrence, and Irrelevance: A Theory of the Criminal Sanction." *Journal of Research in Crime and Delinquency* 30: 445–73.

Stevens, James. 1992. *Applied Multivariate Statistics for the Social Sciences.* 2nd ed. Hillsdale, NJ: Lawrence Erlbaum.

Tittle, Charles. 1977. "Sanction Fear and the Maintenance of Social Order." *Social Forces* 55: 579–96.

———. 1980. *Sanctions and Social Deviance: The Question of Deterrence.* New York: Praeger.

———. 1995. *Control Balance: Toward a General Theory of Deviance.* Boulder, CO: Westview.

Thornberry, Terence P. 1987. "Toward an Interactional Theory of Delinquency." *Criminology* 25: 863–91.

Thornberry, Terence P., Alan J. Lizotte, Marvin D. Krohn, Margaret Farnsworth, and Sung Joon Jang. 1994. "Delinquent Peers, Beliefs, and Delinquent Behavior: A Longitudinal Test of Interactional Theory." *Criminology* 32: 47–84.

Tyler, Tom R. 1990. *Why People Obey the Law.* New Haven. CT: Yale University Press.

Vila, Bryan J. 1994. "A General Paradigm for Understanding Criminal Behavior: Extending Evolutionary Ecological Theory." *Criminology* 32: 311–60.

Walters, G. 1992. *Foundations of Criminal Science.* New York: Praeger.

Wilson, James Q., and Richard J. Herrnstein. 1985. *Crime and Human Nature.* New York: Simon & Schuster.

Wood, Peter B., Walter R. Gove, James A. Wilson, and John K. Cochran. 1997. "Nonsocial Reinforcement and Habitual Criminal Conduct: An Extension of Learning Theory." *Criminology* 35: 335–66.

Zhang, Sheldon X. 1995. "Measuring Shaming in an Ethnic Context." *British Journal of Criminology* 35: 248–62.

REVIEW QUESTIONS

1. According to Hay, what types of factors seem to be important in determining whether parents use reintegrative disciplinary strategies, as opposed to more stigmatizing techniques? If you were (or are) a parent, do these strategies make sense to you in the disciplining of your own child(ren)?

2. From the results of this study, do the effects of reintegration and shaming tend to interact to predict offending? If so, how? If not, give your reasons why they do not seem to interact.

3. What other factors may account for the findings of this study, specifically reintegration and shaming? Did Hay find evidence that other such factors seemed to account for such findings?

4. Do you think Braithwaite's model of reintegrative shaming could work in the United States as a whole? Explain your reasoning for your answer.

❖

READING

In this selection, Terence Thornberry presents an integrated model that primarily merges two theoretical perspectives, specifically control theory and differential reinforcement/ social learning theory. However, the primary distinction between previous integrated frameworks and Thornberry's model is that Thornberry takes into account the reciprocal, or feedback, effects that certain variables may have on the increase of delinquent/criminal behavior.

For example, Thornberry claims that weak bonds to parents (a control concept) may lead to association with delinquent peers (a differential association/social learning concept), which is likely to lead to parents having even weaker bonds with the individual/youth. So it is proposed that the fact that a youth gets in trouble is likely to lead to more alienation or weaker bonds with the youth's parents, which is quite likely in reality. Such an effect is a good example of a reciprocal effect or feedback effect, and this is just one of the many reciprocal effects that are proposed in this model. Thornberry's integrated model of offending is full of such reciprocal (or feedback) effects, and this is what distinguishes his model from previous integrated frameworks, or any of the traditional theoretical models that attempted to explain offending.

While reading this selection, readers should consider their own experience or those of others they know who were caught once (and maybe arrested) for a relatively minor charge. The social "fallout" from this initial arrest may have resulted in more adverse reactions from parents (or others), and that made the other factors, such as delinquent peers, more prominent. Readers may notice that this type of reciprocal or feedback loop resembles labeling theory,

SOURCE: Terence P. Thornberry, "Toward an Interactional Theory of Delinquency," *Criminology* 25 (1987): 863–91. Reprinted by permission of the American Society of Criminology.

in which a person commits an initial (often minor) offense and is labeled as an offender, leading to worse and more frequent offending. Although Thornberry does not specifically invoke labeling theory in his theory, there is an obvious comparison to be made between the reciprocal effects he discusses and the effects of stigmatization that labeling theorists make in their explanatory models. Still, Thornberry's integrated model was one of the first integrated frameworks that actually incorporated such reciprocal or feedback effects into the theory of why individuals engage in offending behavior.

Toward an Interactional Theory of Delinquency ——

Terence P. Thornberry

⚑ Origins and Assumptions

The basic premise of the model proposed here is that human behavior occurs in social interaction and can therefore best be explained by models that focus on interactive processes. Rather than viewing adolescents as propelled along a unidirectional pathway to one or another outcome—that is, delinquency or conformity—it argues that adolescents interact with other people and institutions and that behavioral outcomes are formed by that interactive process. For example, the delinquent behavior of an adolescent is formed in part by how he and his parents *interact* over time, not simply by the child's perceived, and presumably invariant, *level* of attachment to parents. Moreover, since it is an interactive system, the behaviors of others—for example, parents and school officials—are influenced both by each other and by the adolescent, including his or her delinquent behavior. If this view is correct, then interactional effects have to be modeled explicitly if we are to understand the social and psychological processes involved with initiation into delinquency, the maintenance of such behavior, and its eventual reduction.

Interactional theory develops from the same intellectual tradition as the theories mentioned above, especially the Durkheimian tradition of social control. It asserts that the fundamental cause of delinquency lies in the weakening of social constraints over the conduct of the individual. Unlike classical control theory, however, it does not assume that the attenuation of controls leads directly to delinquency. The weakening of controls simply allows for a much wider array of behavior, including continued conventional action, failure as indicated by school dropout and sporadic employment histories, alcoholism, mental illness, delinquent and criminal careers, or some combination of these outcomes. For the freedom resulting from weakened bonds to be channeled into delinquency, especially serious prolonged delinquency, requires an interactive setting in which delinquency is learned, performed, and reinforced. This view is similar to Cullen's structuring perspective which draws attention to the [indeterminacy] of deviant behavior. "It can thus be argued that there is an *indeterminate* and not a determinate or etiologically specific relationship between motivational variables on the one hand and any particular form of deviant behavior on the other hand" (Cullen, 1984, p. 5).

Although heavily influenced by control and learning theories, and to a lesser extent by strain and culture conflict theories, this is not an effort at theoretical integration as that term is usually used (Elliott, 1985). Rather, this paper is guided by what we have elsewhere called theoretical elaboration (Thornberry, 1987). In this instance, a basic control theory is extended, or elaborated upon, using available theoretical perspectives and empirical findings to provide a more accurate model of the causes of delinquency. In the process of elaboration, there is no requirement to resolve disputes among other theories—for example, their different assumptions about the origins of deviance (Thornberry, 1987, pp. 15–18); all that is required is that the propositions of the model developed here be consistent with one another and with the assumptions about deviance stated above.

⊠ Organization

The presentation of the interactional model begins by identifying the central concepts to be included in the model. Next, the underlying theoretical structure of the proposed model is examined and the rationale for moving from unidirectional to reciprocal causal models is developed. The reciprocal model is then extended to include a developmental perspective, examining the theoretical saliency of different variables at different developmental stages. Finally, the influence of the person's position in the social structure is explored. Although in some senses the last issue is logically prior to the others, since it is concerned with sources of initial variation in the causal variables, it is discussed last so that the reciprocal relationships among the concepts—the heart of an interactional perspective—can be more fully developed.

⊠ Theoretical Concepts

Given these basic premises, an interactional model must respond to two overriding issues. First, how are traditional social constraints over behavior weakened and, second, once weakened, how is the resulting freedom

channeled into delinquent patterns? To address these issues, the present paper presents an initial version of an interactional model, focusing on the interrelationships among six concepts: attachment to parents, commitment to school, belief in conventional values, associations with delinquent peers, adopting delinquent values, and engaging in delinquent behavior. These concepts form the core of the theoretical model since they are central to social psychological theories of delinquency and since they have been shown in numerous studies to be strongly related to subsequent delinquent behavior (see Elliott et al., 1985, Chs. 1-3, for an excellent review of this literature).

The first three derive from Hirschi's version of control theory (1969) and represent the primary mechanisms by which adolescents are bonded to conventional middle-class society. When those elements of the bond are weakened, behavioral freedom increases considerably. For that freedom to lead to delinquent behavior, however, interactive settings that reinforce delinquency are required. In the model, those settings are represented by two concepts: associations with delinquent peers and the formation of delinquent values which derive primarily from social learning theory. For the purpose of explicating the overall theoretical perspective, each of these concepts is defined quite broadly. Attachment to parents includes the affective relationship between parent and child, communication patterns, parenting skills such as monitoring and discipline, parent-child conflict, and the like. Commitment to school refers to the stake in conformity the adolescent has developed and includes such factors as success in school, perceived importance of education, attachment to teachers, and involvement in school activities. Belief in conventional values represents the granting of legitimacy to such middle-class values as education, personal industry, financial success, deferral of gratification, and the like.

Three delinquency variables are included in the model. Association with delinquent peers includes the level of attachment to peers, the delinquent behavior and values of peers, and their reinforcing reactions to the adolescent's own delinquent or conforming behavior. It is a continuous measure that can vary from groups that are heavily delinquent to those that are almost entirely nondelinquent. Delinquent values refer to the granting of legitimacy to delinquent activities as acceptable modes of behavior as well as a general willingness to violate the law to achieve other ends. Delinquent behavior, the primary outcome variable, refers to acts that place the youth at risk for adjudication; it ranges from status offenses to serious violent activities. Since the present model is an interactional one, interested not only in explaining delinquency but in explaining the effects of delinquency on other variables, particular attention is paid to prolonged involvement in serious delinquency. . . .

⊠ Model Specification

A causal model allowing for reciprocal relationships among the six concepts of interest—attachment to parents, commitment to school, belief in conventional values, association with

delinquent peers, delinquent values, and delinquent behavior—is presented in Figure 1. This model refers to the period of early adolescence, from about ages 11 to 13, when delinquent careers are beginning, but prior to the period at which delinquency reaches its apex in terms of seriousness and frequency. In the following sections the model is extended to later ages.

The specification of causal effects begins by examining the three concepts that form the heart of social learning theories of delinquency—delinquent peers, delinquent values, and delinquent behavior. For now we focus on the reciprocal nature of the relationships, ignoring until later variations in the strength of the relationships. Traditional social learning theory specifies a causal order among these variables in which delinquent associations affect delinquent values and, in turn, both produce delinquent behavior (Akers, Krohn, Lanza-Kaduce, & Radosevich, 1979; Matsueda, 1982). Yet, for each of the dyadic relationships involving these variables, other theoretical perspectives and much empirical evidence suggest the appropriateness of reversing this causal order. For example, social learning theory proposes that associating with delinquents, or more precisely, people who hold and reinforce delinquent values, increases the chances of

Figure 1   A Reciprocal Model of Delinquent Involvement at Early Adolescence

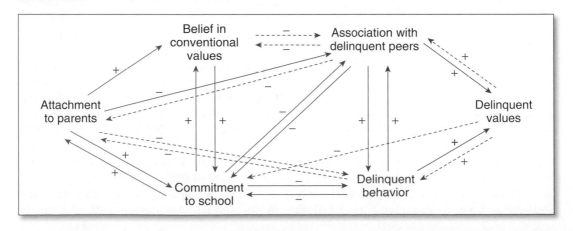

delinquent behavior (Akers, 1977). Yet, as far back as the work of the Gluecks (1950) this specification has been challenged. Arguing that "birds of a feather flock together," the Gluecks propose that youths who are delinquent seek out and associate with others who share those tendencies. From this perspective, rather than being a cause of delinquency, associations are the result of delinquents seeking out and associating with like-minded peers.

An attempt to resolve the somewhat tedious argument over the temporal priority of associations and behavior is less productive theoretically than capitalizing on the interactive nature of human behavior and treating the relationship as it probably is: a reciprocal one. People often take on the behavioral repertoire of their associates but, at the same time, they often seek out associates who share their behavioral interests. Individuals clearly behave this way in conventional settings, and there is no reason to assume that deviant activities, such as delinquency, are substantially different in this regard.

Similar arguments can be made for the other two relationships among the delinquency variables. Most recent theories of delinquency, following the lead of social learning theory, posit that delinquent associations lead to the formation of delinquent values. Subcultural theories, however, especially those that derive from a cultural deviance perspective (Miller, 1958) suggest that values precede the formation of peer groups. Indeed, it is the socialization of adolescents into the "lower-class culture" and its particular value system that leads them to associate with delinquent peers in the first place. This specification can also be derived from a social control perspective as demonstrated in Weis and Sederstrom's social development model (1981) and Burkett and Warren's social selection model (1987).

Finally, the link between delinquent values and delinquent behavior restates, in many ways, the basic social psychological question of the relationship between attitudes and behavior.

Do attitudes form behavior patterns or does behavior lead to attitude formation? Social psychological research, especially in cognitive psychology and balance models (for example, Festinger, 1957; Brehm and Cohen, 1962) points to the reciprocal nature of this relationship. It suggests that people indeed behave in a manner consistent with their attitudes, but also that behavior is one of the most persuasive forces in the formation and maintenance of attitudes.

Such a view of the relationship between delinquent values and behavior is consistent with Hindelang's findings: this general pattern of results indicates that one can "predict" a respondent's self approval [of illegal behaviors] from knowledge of that respondent's involvement/non-involvement [in delinquency] with fewer errors than vice-versa (1974, p. 382). It is also consistent with recent deterrence research which demonstrates that the "experiential effect," in which behavior affects attitudes, is much stronger than the deterrent effect, in which attitudes affect behavior (Paternoster, Saltzman, Waldo, & Chiricos, 1982; Paternoster, Saltzman, Chiricos, & Waldo, 1983).

Although each of these relationships appears to be reciprocal, the predicted strengths of the associations are not of equal strength during the early adolescent period (see Figure 1). Beliefs that delinquent conduct is acceptable [and] positively valued may be emerging, but such beliefs are not fully articulated for 11- to 13-year-olds. Because of their emerging quality, they are viewed as more effect than cause, produced by delinquent behavior and associations with delinquent peers. As these values emerge, however, they have feedback effects, albeit relatively weak ones at these ages, on behavior and associations. That is, as the values become more fully articulated and delinquency becomes positively valued, it increases the likelihood of such behavior and further reinforces associations with like-minded peers.

Summary. When attention is focused on the interrelationships among associations with delinquent peers, delinquent values, and delinquent behavior, it appears that they are, in fact, reciprocally related. The world of human behavior is far more complex than a simple recursive one in which a temporal order can be imposed on interactional variables of this nature. Interactional theory sees these three concepts as embedded in a causal loop, each reinforcing the others over time. Regardless of where the individual enters the loop the following obtains: delinquency increases associations with delinquent peers and delinquent values; delinquent values increase delinquent behavior and associations with delinquent peers; and associations with delinquent peers increases delinquent behavior and delinquent values. The question now concerns the identification of factors that lead some youth, but not others into this spiral of increasing delinquency.

⊠ Social Control Effects

As indicated at the outset of this essay, the promise of interactional theory is that the fundamental cause of delinquency is the attenuation of social controls over the person's conduct. Whenever bonds to the conventional world are substantially weakened, the individual is freed from moral constraints and is at risk for a wide array of deviant activities, including delinquency. The primary mechanisms that bind adolescents to the conventional world are attachment to parents, commitment to school, and belief in conventional values, and their role in the model can now be examined.

During the early adolescent years, the family is the most salient arena for social interaction and involvement and, because of this, attachment to parents has a stronger influence on other aspects of the youth's life at this stage than it does at later stages of development. With this in mind, attachment to parents is predicted to affect four other variables. Since youths who are attached to their parents are sensitive to their wishes (Hirschi, 1969, pp. 16–19), and, since parents are almost universally supportive of the conventional world, these children are likely to be strongly committed to school and to espouse conventional values. In addition, youths who are attached to their parents, again because of their sensitivity to parental wishes, are unlikely to associate with delinquent peers or to engage in delinquent behavior.

In brief, parental influence is seen as central to controlling the behavior of youths at these relatively early ages. Parents who have a strong affective bond with their children, who communicate with them, who exercise appropriate parenting skills, and so forth, are likely to lead their children towards conventional actions and beliefs and away from delinquent friends and actions.

On the other hand, attachment to parents is not seen as an immutable trait, impervious to the effects of other variables. Indeed, associating with delinquent peers, not being committed to school, and engaging in delinquent behavior are so contradictory to parental expectations that they tend to diminish the level of attachment between parent and child. Adolescents who fail at school, who associate with delinquent peers, and who engage in delinquent conduct are, as a consequence, likely to jeopardize their affective bond with their parents, precisely because these behaviors suggest that the "person does not care about the wishes and expectations of other people" (Hirschi, 1969, p. 18), in this instance, his or her parents.

Turning next to belief in conventional values, this concept is involved in two different causal loops. First, it strongly affects commitment to school and in turn is affected by commitment to school. In essence, this loop posits a behavioral and attitudinal consistency in the conventional realm. Second, a weaker loop is posited between belief in conventional values and associations with delinquent peers. Youths who do not grant legitimacy to conventional values are more apt to associate with delinquent friends who share those views, and those friendships are likely to attenuate further their beliefs in conventional values. This reciprocal specification is supported by Burkett and Warren's findings concerning religious beliefs and peer associations (1987). Finally, youths who believe in conventional values are seen as somewhat less likely to engage in delinquent behavior.

Although belief in conventional values plays some role in the genesis of delinquency, its impact is not particularly strong. For example, it is not affected by delinquent behavior, nor is it related to delinquent values. This is primarily because belief in conventional values appears to be quite invariant; regardless of class of origin or delinquency status, for example, most people strongly assert conventional values (Short & Strodtbeck, 1965, Ch. 3). Nevertheless, these beliefs do exert some influence in the model, especially with respect to reinforcing commitment to school.

Finally, the impact of commitment to school is considered. This variable is involved in reciprocal loops with both of the other bonding variables. Youngsters who are attached to their parents are likely to be committed to and succeed in school, and that success is likely to reinforce the close ties to their parents. Similarly, youths who believe in conventional values are likely to be committed to school, the primary arena in which they can act in accordance with those values, and, in turn, success in that arena is likely to reinforce the beliefs.

In addition to its relationships with the other control variables, commitment to school also has direct effects on two of the delinquency variables. Students who are committed to succeeding in school are unlikely to associate with delinquents or to engage in substantial amounts of serious, repetitive delinquent behavior. These youths have built up a stake in conformity and should be unwilling to jeopardize that investment by either engaging in

delinquent behavior or by associating with those who do. Low commitment to school is not seen as leading directly to the formation of delinquent values, however. Its primary effect on delinquent values is indirect, via associations with delinquent peers and delinquent behavior (Conger, 1980, p. 137). While school failure may lead to a reduced commitment to conventional values, it does not follow that it directly increases the acceptance of values that support delinquency.

Commitment to school, on the other hand, is affected by each of the delinquency variables in the model. Youths who accept values that are consistent with delinquent behavior, who associate with other delinquents, and who engage in delinquent behavior are simply unlikely candidates to maintain an active commitment to school and the conventional world that school symbolizes.

Summary. Attachment to parents, commitment to school, and belief in conventional values reduce delinquency by cementing the person to conventional institutions and people. When these elements of the bond to conventional society are strong, delinquency is unlikely, but when they are weak the individual is placed at much greater risk for delinquency. When viewed from an interactional perspective, two additional qualities of these concepts become increasingly evident.

First, attachment to parents, commitment to school, and belief in conventional values are not static attributes of the person, invariant over time. These concepts interact with one another during the developmental process. For some youths the levels of attachment, commitment, and belief increase as these elements reinforce one another, while for other youths the interlocking nature of these relationships suggests a greater and greater attenuation of the bond will develop over time.

Second, the bonding variables appear to be reciprocally linked to delinquency, exerting a causal impact on associations with delinquent peers and delinquent behavior; they also are causally affected by these variables. As the youth engages in more and more delinquent conduct and increasingly associates with delinquent peers, the level of his bond to the conventional world is further weakened. Thus, while the weakening of the bond to conventional society may be an initial cause of delinquency, delinquency eventually becomes its own indirect cause precisely because of its ability to weaken further the person's bonds to family, school, and conventional beliefs. The implications of this amplifying causal structure [are] examined below. First, however, the available support for reciprocal models is reviewed and the basic model is extended to later developmental stages. . . .

Developmental Extensions

The previous section developed a strategy for addressing one of the three major limitations of delinquency theories mentioned in the introduction, namely, their unidirectional causal structure. A second limitation is the non-developmental posture of most theories which tend to provide a cross sectional picture of the factors associated with delinquency at one age, but which do not provide a rationale for understanding how delinquent behavior develops over time. The present section offers a developmental extension of the basic model.

Middle Adolescence

First, a model for middle adolescence, when the youths are approximately 15 or 16 years of age is presented (Figure 2). This period represents the highest rates of involvement in delinquency and is the reference period, either implicitly or explicitly, for most theories of delinquent involvement. Since the models for the early and middle adolescent periods have essentially the same structure and causal relationships (Figures 1 and 2), discussion focuses on the differences between them and does not repeat the rationale for individual causal effects.

Perhaps the most important difference concerns attachment to parents which is involved in relatively few strong relationships. By this point in the life cycle, the most salient variables involved in the production of delinquency are likely to be external to the home, associated with the youth's activities in school

| Figure 2 | A Reciprocal Model of Delinquent Involvement at Middle Adolescence |

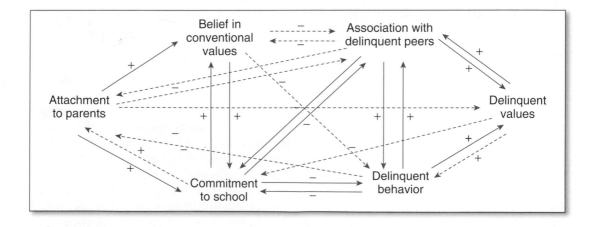

and peer networks. This specification is consistent with empirical results for subjects in this age range (Johnson, 1979, p. 105; and Schoenberg, 1975, quoted in Johnson. Indeed, Johnson concludes that "an adolescent's public life has as much or more to do with his or her deviance or conformity than do 'under-the-roof' experiences" (1979, p. 116).

This is not to say that attachment to parents is irrelevant; such attachments are involved in enhancing commitment to school and belief in conventional values, and in preventing associations with delinquent peers. It is just that the overall strength of parental effects [is] weaker than at earlier ages when the salience of the family as a locus of interaction and control was greater. The second major change concerns the increased importance of delinquent values as a causal factor. It is still embedded in the causal loop with the other two delinquency variables, but now it is as much cause as effect. Recall that at the younger ages delinquent values were seen as emerging, produced by associations with delinquent peers and delinquent behavior. Given their emergent nature, they were not seen as primary causes of other variables. At mid-adolescence, however, when delinquency is at its apex, these values are more fully articulated and have stronger effects on other variables. First, delinquent values are seen as major reinforcers of both delinquent associations and delinquent behavior. In general, espousing values supportive of delinquency tends to increase the potency of this causal loop. Second, since delinquent values are antithetical to the conventional settings of school and family, youths who espouse them are less likely to be committed to school and attached to parents. Consistent with the reduced saliency of family at these ages, the feedback effect to school is seen as stronger than the feedback effect to parents.

By and large, the other concepts in the model play the same role at these ages as they do at the earlier ones. Thus, the major change from early to middle adolescence concerns the changing saliency of some of the theoretical concepts. The family declines in relative importance while the adolescent's own world of school and peers takes on increasing significance. While these changes occur, the overall structure of the theory remains constant. These interactive variables are still seen as mutually reinforcing over time.

Later Adolescence

Finally, the causes of delinquency during the transition from adolescence to adulthood, about ages 18 to 20, can be examined (Figure 3). At these ages one should more properly speak of crime than delinquency, but for consistency we will continue to use the term delinquency in the causal diagrams and employ the terms delinquency and crime interchangeably in the text.

Two new variables are added to the model to reflect the changing life circumstances at this stage of development. The more important of these is commitment to conventional activities which includes employment, attending college, and military service. Along with the transition to the world of work, there is a parallel transition from the family of origin to one's own family. Although this transition does not peak until the early 20s, for many people its influence is beginning at this stage. Included in this concept are marriage, plans for marriage, and plans for childrearing. These new variables largely replace attachment to parents and commitment to school in the theoretical scheme; they represent the major sources of bonds to conventional society for young adults. Both attachment to parents and commitment to school remain in the model but take on the cast of exogenous variables. Attachment to parents has only a minor effect on commitment to school, and commitment

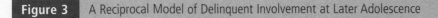

Figure 3	A Reciprocal Model of Delinquent Involvement at Later Adolescence

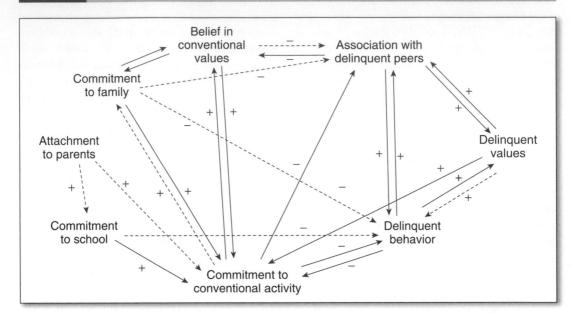

to school is proposed to affect only commitment to conventional activities and, more weakly, delinquent behavior.

The other three variables considered in the previous models—association with delinquent peers, delinquent values, and delinquent behavior—are still hypothesized to be embedded in an amplifying causal loop. As indicated above, this loop is most likely to occur among adolescents who, at earlier ages, were freed from the controlling influence of parents and school. Moreover, via the feedback paths delinquent peers, delinquent values, and delinquent behavior further alienate the youth from parents and diminish commitment to school. Once this spiral begins, the probability of sustained delinquency increases.

This situation, if it continued uninterrupted, would yield higher and higher rates of crime as the subjects matured. Such an outcome is inconsistent with the desistance that has been observed during this age period (Wolfgang, Thornberry, and Figlio, 1987).

Rates of delinquency and crime begin to subside by the late teenage years, a phenomenon often attributed to "maturational reform." Such an explanation, however, is tautological since it claims that crime stops when adolescents get older, because they get older. It is also uninformative since the concept of maturational reform is theoretically undefined. A developmental approach, however, offers an explanation for desistance. As the developmental process unfolds, life circumstances change, developmental milestones are met (or, for some, missed), new social roles are created, and new networks of attachments and commitments emerge. The effects of these changes enter the processual model to explain new and often dramatically different behavioral patterns. In the present model, these changes are represented by commitment to conventional activity and commitment to family.

Commitment to conventional activity is influenced by a number of variables, including earlier attachment to parents, commitment to school, and belief in conventional values. And once the transition to the world of work is made, tremendous opportunities are afforded for new and different effects in the delinquency model. Becoming committed to conventional activities, work, college, military service, and so on—reduces the likelihood of delinquent behavior and associations with delinquent peers because it builds up a stake in conformity that is antithetical to delinquency. Moreover, since the delinquency variables are still embedded in a causal loop, the effect of commitment to conventional activities tends to resonate throughout the system. But, because of the increased saliency of a new variable, commitment to conventional activities, the reinforcing loop is now set in motion to *reduce* rather than increase delinquent and criminal involvement. The variable of commitment to family has similar, albeit weaker, effects since the transition to the family is only beginning at these ages. Nevertheless, commitment to family is proposed to reduce both delinquent associations and delinquent values and to increase commitment to conventional activity. In general, as the individual takes on the responsibilities of family, the bond to conventional society increases, placing additional constraints on behavior and precluding further delinquency.

These changes do not occur in all cases, however, nor should they be expected to since many delinquents continue on to careers in adult crime. In the Philadelphia cohort of 1945, 51% of the juvenile delinquents were also adult offenders, and the more serious and prolonged the delinquent careers were, the greater the odds of an adult career (Wolfgang et al., 1987, Ch. 4). The continuation of criminal careers can also be explained by the nature of the reciprocal effects included in this model. In general, extensive involvement in delinquency at earlier ages feeds back upon and weakens attachment to parents and commitment to school (see Figures 1 and 2). These variables, as well as involvement in delinquency

itself, weaken later commitment to family and to conventional activities (Figure 3). Thus, these new variables, commitment to conventional activities and to family, are affected by the person's situation at earlier stages and do not "automatically" alter the probability of continued criminal involvement. If the initial bonds are extremely weak, the chances of new bonding variables being established to break the cycle towards criminal careers are low and it is likely that criminal behavior will continue. . . .

▧ Structural Effects

Structural variables, including race, class, sex, and community of residence, refer to the person's location in the structure of social roles and statuses. The manner in which they are incorporated in the interactional model is illustrated here by examining only one of them, social class of origin. Although social class is often measured continuously, a categorical approach is more consistent with the present model and with most theories of delinquency that incorporate class as a major explanatory variable—for example, strain and social disorganization theories. For our purposes, the most important categories are the lower class, the working lower class, and the middle class.

The lower class is composed of those who are chronically or sporadically unemployed, receive welfare, and subsist at or below the poverty level. They are similar to Johnson's "underclass" (1979). The working lower class is composed of those with more stable work patterns, training for semiskilled jobs, and incomes that allow for some economic stability. For these families, however, the hold on even a marginal level of occupational and economic security is always tenuous. Finally, the middle class refers to all families above these lower levels. Middle-class families have achieved some degree of economic success and stability and can reasonably expect to remain at that level or improve their standing over time.

The manner in which the social class of origin affects the interactional variables and the behavioral trajectories can be demonstrated by comparing the life expectancies of children from lower- and middle-class families. As compared to children from a middle-class background, children from a lower class background are more apt to have (1) disrupted family processes and environments (Conger, McCarty, Wang, Lahey, & Kroop, 1984; Wahler, 1980); (2) poorer preparation for school (Cloward and Ohlin, 1960); (3) belief structures influenced by the traditions of the American lower class (Miller, 1958); and (4) greater exposure to neighborhoods with high rates of crime (Shaw & McKay, 1942; Braithwaite, 1981). The direction of all these effects is such that we would expect children from lower-class families to be initially less bonded to conventional society and more exposed to delinquent values, friends, and behaviors.

As one moves towards the working lower class, both the likelihood and the potency of the factors just listed decrease. As a result, the initial values of the interactional variables improve but, because of the tenuous nature of economic and social stability for these families, both the bonding variables and the delinquency variables are still apt to lead to considerable amounts of delinquent conduct. Finally, youths from middle-class families, given their greater stability and economic security, are likely to start with a stronger family structure, greater stakes in conformity, and higher chances of success, and all of these factors are likely to reduce the likelihood of initial delinquent involvement.

In brief, the initial values of the interactional variables are systematically related to the social class of origin. Moreover, since these variables are reciprocally related, it follows logically that social class is systematically related to the behavioral trajectories described above. Youngsters from the lowest classes have the highest probability of moving forward on a trajectory of increasing delinquency. Starting from a position of low bonding to conventional institutions and a high delinquency environment, the reciprocal nature of the interrelationships leads inexorably towards extremely high rates of delinquent and criminal involvement. Such a view is consistent with prevalence data which show that by age 18, 50%, and by age 30, 70% of low SES minority males have an official police record (Wolfgang et al., 1987).

On the other hand, the expected trajectory of middle-class youths suggests that they will move toward an essentially conforming lifestyle, in which their stakes in conformity increase and more and more preclude serious and prolonged involvement in delinquency. Finally, because the initial values of the interactional variables are mixed and indecisive for children from lower working-class homes, their behavioral trajectories are much more volatile and the outcome much less certain.

Summary. Interactional theory asserts that both the initial values of the process variables and their development over time are systematically related to the social class of origin. Moreover, parallel arguments can be made for other structural variables, especially those associated with class, such as race, ethnicity, and the social disorganization of the neighborhood. Like class of origin, these variables are systematically related to variables such as commitment to school and involvement in delinquent behavior, and therefore, as a group, these structural variables set the stage on which the reciprocal effects develop across the life cycle. . . .

✂ References

Akers, R. (1977). *Deviant behavior: A social learning perspective.* Belmont, CA: Wadsworth.

Akers, R. L., Krohn, M. D., Lanza-Kaduce, L., & Radosevich, M. (1979). Social learning theory and deviant behavior. *American Sociological Review, 44,* 635–655.

Braithwaite, J. (1981). The myth of social class and criminality reconsidered. *American Sociological Review, 46,* 36–58.

Brehm, J. W., & Cohen, A. R. (1962). *Explorations in cognitive dissonance.* New York: Wiley.

Burkett, S. R., & Warren, B. O. (1987). Religiosity, peer influence, and adolescent marijuana use: A panel study of underlying causal structures. *Criminology, 25,* 109–131.

Cloward, R. A., & Ohlin, L. E. (1960). *Delinquency and opportunity—A theory of delinquent gangs.* New York: Free Press.

Conger, R. D. (1980). Juvenile delinquency: Behavior restraint or behavior facilitation? In T. Hirschi & M. Gottfredson (Eds.), *Understanding crime.* Beverly Hills, CA: Sage.

Conger, R. D., McCarty, J. A., Wang, R. K., Lahey, B. B., & Kroop, J. P. (1984). Perception of child, child-rearing values, and emotional distress as mediating links between environmental stressors and observed maternal behavior. *Child Development, 55,* 2234–2247.

Cullen, F. T. (1984). *Rethinking crime and deviance theory: The emergence of a structuring tradition.* Totowa, NJ: Rowman and Allanheld.

Elliott, D. S. (1985). The assumption that theories can be combined with increased explanatory power: Theoretical integrations. In R. F. Meier (Ed.), *Theoretical methods in criminology.* Beverly Hills, CA: Sage.

Elliott, D. S., Huizinga, D., & Ageton, S. S. (1985). *Explaining delinquency and drug use.* Beverly Hills, CA: Sage.

Festinger, L. (1957). *A theory of cognitive dissonance.* Stanford, CA: Stanford University Press.

Glueck, S., & Glueck, E. (1950). *Unraveling juvenile delinquency.* Cambridge, MA: Harvard University Press.

Hindelang, M. J. (1974). Moral evaluations of illegal behaviors. *Social Problems, 21,* 370–384.

Hirschi, T. (1969). Causes of delinquency. Berkeley: University of California Press.

Johnson, R. E. (1979). *Juvenile delinquency and its origins.* Cambridge, UK: Cambridge University Press.

Matsueda, R. (1982). Testing social control theory and differential association. *American Sociological Review, 47,* 489–504.

Miller, W. B. (1958). Lower class culture as a generating milieu of gang delinquency. *Journal of Social Issues, 14*(3), 5–19.

Paternoster, R., Saltzman, L. E., Chiricos, T. G., & Waldo, G. P. (1983). Perceived risk and social control: Do sanctions really deter? *Law and Society Review, 17,* 457–479.

Paternoster, R., Saltzman, L. E., Waldo, G. P., & Chiricos, T. G. (1982). Perceived risk and deterrence: Methodological artifacts in perceptual deterrence research. *Journal of Criminal Law and Criminology, 73,* 1238–1258.

Shaw, C., & McKay, H. D. (1942). *Juvenile delinquency and urban areas.* Chicago: University of Chicago Press.

Short, J. F., Jr., & Strodtbeck, F. L. (1965). *Group process and gang delinquency.* Chicago: University of Chicago Press.

Thornberry, T. P. (1987a). *Reflections on the advantages and disadvantages of theoretical integration.* Paper presented at the Albany Conference on Theoretical Integration in the Study of Crime and Deviance.

Thornberry, T. P. (1987b). Towards an interactional theory of delinquency. *Criminology, 25,* 863–891.

Wahler, R. (1980). The insular mother: Her problems in parent-child treatment. *Journal of Applied Behavior Analysis, 13,* 207–219.

Weis, J. G., & Sederstrom, J. (1981). *The prevention of serious delinquency: What to do?* Washington, DC: U.S. Department of Justice.

Wolfgang, M. E., Thornberry, T. P., & Figlio, R. M. (1987). *From boy to man—From delinquency to crime: Followup to the Philadelphia Birth Cohort of 1945.* Chicago: University of Chicago Press.

REVIEW QUESTIONS

1. Explain what is meant by reciprocal (or feedback) effects. Use a non-crime example to explain such effects.

2. Can you think of concepts or theoretical models other than labeling that Thornberry could or should have incorporated into his integrated model of crime? Explain your rationale for including such concepts or models.

3. Using your own life, or others you know, give an example of someone who follows the causal model that is proposed by Thornberry. If you can't come up with one, explore the media for someone who does.

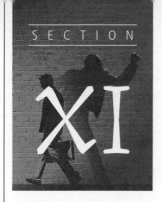

APPLYING CRIMINOLOGICAL THEORY TO POLICY

The introduction of this section will discuss how criminological theories that we have discussed throughout the book have been used to make current criminal justice policies more effective in preventing and reducing crime. Given that we have examined some of the policy implications derived from theoretical models throughout this book, we will provide only a brief review of the policy implications regarding each category of theories in this section. Furthermore, we will include only one published article here because many of the previous selections in this book have included discussions and recommendations regarding the policy implications that their specific studies suggest. Thus, the selection in this section is meant to summarize the way policy has been altered by criminological research in U.S. history, as well as suggestions for future policy changes based on the current state of research in criminology and criminal justice.

⬚ Introduction

Criminological theories have been used in informing and development of policy ever since the father of deterrence, Cesare Beccaria, developed the first rational theory of crime in the mid-18th century. Because many of these resulting policies have been discussed in the previous sections, this introduction will be rather brief, focusing on the policies that have received the most attention by researchers and policymakers. Readers should realize that this section is not meant to be a conclusive review of all policies that can be applied from a particular type of theory; on the contrary, this is only a "short list" of the numerous policies that can be derived from a given theoretical framework.

⬚ Policy Applications Derived From the Classical School: Deterrence Theory

Although most modern criminologists and theoreticians in the field do not consider deterrence theory to be one of the more robust models for understanding or reducing illegal activity,[1] there is no doubt that policy making in our criminal justice system is dominated by the ideas of the Classical School, especially deterrence theory.[2] This is seen in most policies used throughout the system by its three major components: law enforcement, courts, and corrections.

Law Enforcement

Perhaps the most important assumption in applying deterrence theory to policy making in the criminal justice system is that individuals are rational. Unfortunately, much research suggests that people often engage in behaviors that they know are irrational and that individuals' cognitive differences can limit rational decision making.[3] Criminologists often refer to this limitation as "bounded rationality."[4] Therefore, it is not surprising that many attempts by police and other criminal justice authorities to deter potential offenders do not seem to have much effect in preventing crime.

Of course, the simple existence of a policing authority tends to deter people to some extent. When policing breaks down in given jurisdictions, riots and looting often follow, and it can become a free-for-all among a large mob, which typically contains otherwise law-abiding people. This is one of the reasons why it is typically illegal for police unions to conduct a strike that would leave no police officers available for responding to calls. Looking at deterrence another way, studies show that the vast majority of drivers speed on highways. For example, one recent study by researchers at Purdue University revealed that 21 percent of motorists think it's perfectly safe to exceed the speed limit by 5 mph, another 43 percent saw no risk in going 10 mph over the limit, and another 36 percent saw no risk in driving 20 mph over the speed limit.[5] This finding is consistent with other studies on drivers' attitudes regarding speeding and limits.

---

[1]Lee Ellis and Anthony Walsh, "Criminologists' Opinions about Causes and Theories of Crime and Delinquency," *The Criminologist* 24 (1999): 1–4.

[2]George Vold, Thomas Bernard, and Jeffrey Snipes, *Theoretical Criminology,* 5th ed. (Oxford, UK: Oxford University Press, 2002).

[3]For a review of the extant research on this topic, see Alex Piquero and Stephen Tibbetts, *Rational Choice and Criminal Behavior* (New York: Routledge, 2002).

[4]For more recent discussion on the complexity of developing policies based on deterrence/rational choice models, see Travis Pratt, "Rational Choice Theory, Crime Control Policy, and Criminological Relevance," *Criminology and Public Policy* 7 (2008): 43–52.

[5]For an example, see a news report regarding a study by researchers at Purdue University in which the vast majority of drivers believed going over the speed limit was acceptable and safe: see link to news story at: http://www.themotorreport.com.au/12358/study-shows-motorists-ambivalent-to-speed-limits/

However, when police cruisers are seen, most people will slow down and be more careful, which is in a sense a form of deterrence. For example, most individuals' heart rate will go up a few notches when they see a police cruiser or highway patrol car, even if they are not doing anything wrong. This is a good example of how deterrence can result from simply seeing the police presence. It is likely we can all relate to this, especially if we are speeding or engaging in other illegal behavior (e.g., talking on the cell phone while driving, etc.).

On the other hand, research is less clear about whether increasing the number of police in a given area, often referred to as "saturated patrol," can significantly reduce crime. Some early studies, especially one key experiment involving increased random patrol in certain "beats" in Kansas City, found no deterrent effect from saturated patrol.[6] However, a recent review of the extant literature, hereafter referred to as the Maryland Report (because all key authors of the report were professors at University of Maryland), concluded that extra police who focus on specific locations (i.e., hot spots), as opposed to a larger "beat" area, can prevent crime.[7] This strategy of cracking down on specific locations applies to deterrence theory and also incorporates another type of theory that assumes a rational individual: routine activities theory. That theory assumes that motivated offenders will be less likely to commit an offense if there is a presence of guardianship, so adding police to a specific location reduces the likelihood that they will have the opportunity to commit criminal activity at that "hot spot."

This same Maryland Report also noted that proactive arrests for drunk driving have consistently been found to reduce such behavior.[8] The effectiveness of reducing drunk driving this way probably involves deterring individuals from continuing such activity; that effect is not surprising given the fact that drunk driving often is committed by middle- and upper-class people, who have a lot to lose if they don't alter their behavior. The Maryland Report also concludes that the same type of phenomenon is seen with arresting employed (but not unemployed) individuals for domestic violence; to clarify, the employed people have more to lose, so they tend to be deterred from recidivating for domestic violence.[9]

The Maryland Report also offers conclusions about which policing strategies do *not* seem to prevent or deter crime: more rapid response time to 911 calls, arrests of juveniles for minor crimes, arrest crackdowns at drug markets, and community policing programs with no clear focus on crime risk factors.[10] Finally, it is interesting to note that the Maryland Report acknowledges that the government invests far more federal funds in policing—$2 billion per year—than in any other crime prevention institution, despite the fact that there are still many questions about how effective this funding strategy is, given the wide variation in how each police agency uses this funding and the departments' lack of understanding about what types of programs may be effective versus those that

---

[6]George Kelling, *The Kansas City Preventive Patrol Experiment* (Washington, DC: Police Foundation, 1974).

[7]Lawrence Sherman, Denise Gottfredson, Doris MacKenzie, John Eck, Peter Reuter, and Shawn Bushway, *Preventing Crime: What Works, What Doesn't, What's Promising: A Report to the United States Congress* (Washington, DC: US Department of Justice, 1997).

[8]Ibid.

[9]Ibid.

[10]Ibid.

aren't.[11] Thus, it would be very beneficial if policymakers and policing authorities would examine the findings of recent empirical research to direct their resources to policies that actually work in reducing crime.

Courts/Diversion Programs/Corrections

Numerous policies implemented by courts and diversion strategies are based, at least to a large extent, on deterrence. One of the most notorious of the diversion programs is the various "scared straight" programs that became popular in the 1970s.[12] These programs involved having prisoners talk to youth about the realities of their lives, which was supposed to "scare" them into refraining from recidivating and to make them decide to go "straight." Unfortunately, virtually all of these programs were found to be ineffective, and some evaluations showed that young participants had significantly higher rates of recidivism.[13] Perhaps the intense experience of being scared by actual inmates had the effect of stigmatizing the youth who went through these programs. Regardless of the reasons, empirical research has consistently found that scared straight programs do not deter or prevent future offending, and the Maryland Report concluded this after reviewing the evaluation literature.

▲ **Photo 11.1**   "Scared Straight" programs were designed to expose delinquents to "heart-to-heart" talks with inmates with the aim of literally scaring them into becoming straight or nondelinquent.

---

[11]Ibid.

[12]For a review of these programs and evaluations of them, see Richard Lundman, *Prevention and Control of Juvenile Delinquency*, 2nd ed. (Oxford, UK: Oxford University Press, 1993).

[13]Lundman, *Prevention and Control*.

In quite a different strategy, some judges have moved toward using shaming strategies to deter offenders from recidivating.[14] For example, many cities have begun posting the names (and sometimes pictures) of men who are arrested for soliciting prostitutes. Also, some judges have started forcing individuals convicted of drunk driving to place glow-in-the-dark bumper stickers on their car, as a condition of their probation, which state "BEWARE: Convicted Drunk Driver!," or to walk in front of stores with a sign that says, "I am a thief, I stole from Wal-Mart!" These are just some of the examples of a growing trend that attempts to deter individuals by threatening them with shaming practices if they choose to engage in such illegal behavior. Such sentences apply elements of rational choice theory, particularly its ideas about the strong deterrent influence of informal factors, such as community, family, employment, and church. Unfortunately, there have been virtually no empirical evaluations of the effectiveness of such shaming penalties.

Another strategy courts have used for deterring individuals is to develop special sentencing enhancements for habitual offenders, such as California's "Three Strikes You're Out." Using the common term from baseball, the three-strikes-you're-out refers to sentencing schemes that mandate life imprisonment for three-time felons, although each state differs in how this formula works. Washington state was the first to implement a three-strikes law, and today, most states have some version of a three-strikes law.[15] As a specific deterrent, such a strategy clearly works because a third-time felon will be in prison for life; however, the three-strikes law was also intended to be a general deterrent and to prevent offenders from recidivating.

Empirical evaluations of the deterrent effect of this law provide no clear result. One study of California's three-strikes law suggests that it reduced crime,[16] whereas several other studies showed negligible effects. In some cases, three-strikes laws may have actually increased violence,[17] perhaps due to the desperation and "nothing-left-to-lose" mentality of offenders, who know they face a life sentence if apprehended.[18] Also, the vast majority of offenders prosecuted under the three-strikes penalty are nonviolent (i.e., drug offenders, property offenders, etc.). This is true in California, one of the states that most frequently uses this type of sentencing policy. For example, a man who had no violent crime on his record stole some DVDs from a store and was given a life sentence, which was upheld by the U.S. Supreme Court, because of two prior offenses. In some cases, these types of sentencing policies have actually forced the release of violent offenders to make room for habitual nonviolent drug and property offenders. This issue was also mentioned as a serious problem in the Maryland Report.

---

[14]For a variety of examples of judges using such shaming sanctions, see link to story at: http://www.msnbc.msn.com/id/19957110/.

[15]David Shichor and D. K. Sechrest, eds., *Three Strikes and You're Out: Vengeance as Social Policy* (Thousand Oaks, CA: Sage, 1996).

[16]J. Shepherd, "Fear of the First Strike: The Full Deterrent Effect of California's Two- and Three-Strikes Legislation," *Journal of Legal Studies* 31 (2002): 159–201.

[17]Lisa Stolzenberg and S.J. D'Alessio, "Three-Strikes-You're-Out: The Impact of California's New Mandatory Sentencing Law on Serious Crime Rates," *Crime and Delinquency* 43 (1997): 457–69; M. Males and D. Macallair, "Striking Out: The Failure of California's 'Three-Strikes and You're Out Law," *Stanford Law and Policy Review* 11 (1999): 65–72.

[18]T. Kovandzic, J. J. Sloan III, and L. M. Vieraitis, "Unintended Consequences of Politically Popular Sentencing Policy: The Homicide-Promoting Effects of 'Three Strikes' in U.S. Cities (1980-1999)," *Criminology and Public Policy* 1 (2002): 399–424.

Thus far, successful deterrent policies in the courts and corrections components of the criminal justice system seem rare. According to the Maryland Report, one of the only court-mandated policies that seems promising in deterring offending behavior is orders of protection for battered women.[19] Still, far more brain-storming and evaluations are needed of deterrence-based policies.

✄ Policy Applications Derived From Biosocial Theories

Although a multitude of policy strategies are related to the biosocial perspective (see Section IV), there has been a reluctance to claim that these policies derive from such a theoretical framework due to the stigma of any reference to biology in curbing crime. However, the Maryland Report suggests the importance of identifying and treating early head and bodily trauma, concluding that some of the most consistently supported early-intervention programs for such physiological problems are those that involve weekly infant home visitation.[20] Experts have recently suggested many more strategies, such as mandatory health insurance for pregnant mothers and their children, which is likely one of the most efficient ways to reduce crime in the long run.[21] Furthermore, all youth should be screened for levels of toxins (e.g., lead), as well as abnormal levels of neuro-transmitters (e.g., serotonin) and hormones (e.g., testosterone).[22] Furthermore, greater efforts should be made to ensure that any type of abuse or other form of head trauma gets special attention.

Ultimately, despite the neglect that biosocial models of crime receive in terms of both recognition and policy implications, there is no doubt that this area is crucial if we hope to advance our understanding and create more efficient policies regarding criminal behavior. It is time that all criminologists recognize the degree to which human behavior results from physiological disorders. We all have a brain, about three pounds in weight, that determines the choices we make. Criminologists must acknowledge the influence of biological or physiological factors that influence this vital organ, or the discipline will be behind the curve in terms of understanding why people commit (or do not commit) criminal offenses.

✄ Policy Applications Derived From Social Structure Theories

Since the introduction of Shaw and McKay's theory of social disorganization (see Section VI), there have been numerous attempts to prevent crime in lower-class, urban areas where crime is concentrated. One of the most straightforward strategies involves forming a neighborhood watch in high-risk neighborhoods.[23] After reviewing the extant evaluations

---

[19]Sherman et al., *Preventing Crime.*

[20]Ibid.

[21]John Wright, Stephen Tibbetts, and Leah Daigle, *Criminals in the Making: Criminality Across the Life Course* (Thousand Oaks, CA: Sage, 2008).

[22]Ibid.

[23]Sherman et al., *Preventing Crime.*

of such neighborhood watch programs, the Maryland Report concluded that they do not seem to work in reducing crime in neighborhoods. Perhaps this is because the cohesive communities that form such watches already have low crime rates; the neighborhoods that have the highest crime rates tend to be those that are the least cohesive or are in the worst areas regarding crime. Furthermore, the report explicitly concludes that community mobilization against crime in high-crime, inner-city poverty areas does not work, probably for the same reasons. Specifically, the most crime-ridden areas tend to have a high turnover of poor residents, who have virtually no investment in the neighborhood; they plan on moving out of the area as soon as they can.

Also related to Shaw and McKay's theory of social disorganization, the Maryland Report concluded that some promising strategies (although there was not enough information to make a strong conclusion) included gang violence prevention programs that focus on reducing gang cohesion, as well as volunteer mentoring programs, such as the Big Brothers/Big Sisters program, which appears to be especially good for reducing substance abuse among youth.[24] Thus, it appears that focusing on high-risk youth may be one way of effectively targeting resources and strategies for reducing crime, as opposed to trying to address the crime issues of an entire neighborhood or area of a city. This is consistent with meta-analyses of evaluation studies, which show that the most effective programs are those that focus on the individuals that have the highest risk for offending or recidivism.[25] It is also consistent with studies showing that diversion and rehabilitation programs that target the offenders who are most at risk tend to have more effect than those targeting medium- or low-risk offenders, probably due to the fact that the lower risk offenders likely won't offend again anyway.[26]

In regard to Merton's strain theory (see Section V), there have been significant attempts to address the more global issues of deprivation that result from poverty in U.S. society. Perhaps the best example is the War on Poverty, a federal program introduced by President Lyndon B. Johnson during the 1960s. Although an enormous amount of money and other resources were spent in trying to aid the poor in order to reduce crime rates in America, crime rates soared during this period. Many experts noted that the federal investment was not efficiently handed out, but the rise in crime during this period still casts doubt on the effectiveness of such a strategy.

Rather, the Maryland Report concludes that vocational training for adult male offenders has consistently lowered offending rates, as have intensive residential training programs for at-risk youth (such as Job Corps).[27] This conclusion appears consistent with our previous finding that a focus on certain high-risk individuals is more effective than programs addressing an entire group, such as all types of offenders. Perhaps this is because not all offenders have the target characteristic, but such programs tend to be effective for those who do. For example, some offenders are thrust into drug programs even though they have never used drugs or alcohol. Also, some offenders are coerced into

---

[24]Ibid.

[25]Patricia Van Voorhis, Michael Braswell, and David Lester, *Correctional Counseling and Rehabilitation* (Cincinnati, OH: Anderson, 1999).

[26]Ibid.

[27]Sherman et al., *Preventing Crime.*

engaging in vocational programs when they already have skills in a certain trade. Thus, it makes sense to focus resources (e.g., vocational programs, drug programs, etc.) on the people who are most at risk for the given risk factor.

▧ Policy Applications Derived From Social Process Theories

Related to Section VII, ever since the introduction of Sutherland's differential association theory in the 1930s, numerous programs to reduce crime and delinquency have been based on the assumption that youth must be provided with prosocial interactions with their parents and peers. Many of these programs have shown poor results, probably due to the fact that mixing offenders with other offenders only leads to more offending, ironically due to associations with other offenders. However, many of these programs have shown success or at least promise.

According to the Maryland Report, evidence consistently supports programs that involve long-term frequent home visitation combined with preschool, as well as family therapy by clinical staff for at-risk youth.[28] Furthermore, the Maryland Report stresses the success of school programs that aim at clarifying and communicating norms about various behaviors by establishing school rules and consistently enforcing them through positive reinforcement and school-wide campaigns.[29] Other effective school programs focus on various competency skills, such as developing self-control, stress-management, responsible decision making, problem solving, and communication skills.[30] On the other hand, the Maryland Report noted some school programs that do *not* reduce delinquent behavior, such as counseling students in a peer-group context (see reasoning above), as well as substance use programs that simply rely on information dissemination or fear arousal (such as the traditional DARE program).[31] Ultimately, programs that teach youths to develop cognitive or thinking-skills are beneficial, as opposed to those that simply focus on information about drugs or criminal offending. The bottom line for successful programs appears to be that they must instruct youths to deal with everyday influences in a healthy way (according to a cognitive-behavioral model), as opposed to just giving them information about crime and drugs.

▧ Policy Applications Derived From Social Reaction and Conflict Theories

Most of the policy implications that were derived from labeling and other forms of social reaction theories (see Section VIII) are often referred to as "the Ds" because they include diversion, decriminalization, and deinstitutionalization, which all became very popular in

---

[28]Ibid.

[29]Ibid.

[30]Ibid.

[31]Ibid.

the late 1960s and early 1970s, when the popularity of labeling theory peaked. Consistent with labeling theory, all three emphasize the "hands-off" doctrine, which seeks to reduce the stigmatization that may occur from being processed or incarcerated by the formal justice system. Diversion includes a variety of programs, such as drug courts or community service programs, which attempt to remove an offender from the traditional prosecutorial and trial process as soon as possible, often with the promise of the charge being expunged or removed from their permanent records. Despite empirical research showing mixed results, some of these diversion programs, such as drug courts, have shown promise.[32]

Decriminalization is the lessening of punishments, which is typically recommended for nonviolent crimes to lessen the stigma and penalty related to an offense. This can be distinguished from legalization, which makes an act completely legal; rather, when a certain behavior is decriminalized, it remains illegal, but the penalties are significantly reduced. For example, the possession of small amounts of marijuana is decriminalized in some states (e.g., California); instead of being arrested and possibly jailed, a person found holding only a small amount is given a ticket, similar to a speeding ticket, which requires payment of a fine. Similarly, deinstitutionalization refers to the avoidance of incarcerating minor, nonviolent offenders, especially if they are juveniles committing status offenses (e.g., running away, truancy, etc.). Some have argued that this policy would lead to higher rates of recidivism, but this has not become a problem according to statistics from police and FBI reports.

The bulk of policies that have been derived from conflict theories have emphasized attempting to make laws, enforcement, and processing through the justice system more equitable in terms of social class. For example, the federal sentencing guidelines were recently reviewed for drug possession because it became clear that they were unfair in how they punished drug possession offenders. Specifically, according to the 1980s guidelines, possession of an ounce of crack cocaine earned a defendant a sentence 100 times longer than an offender who possessed an ounce of powder cocaine. Not surprisingly, crack cocaine is typically used and sold by lower class minorities, whereas cocaine powder is typically used and sold by upper- or middle-class Whites. The disparity in this form of sentencing, which clearly penalizes a certain group of people more than another, has recently become a target for reform by the Obama administration, so there are current efforts to address this issue. This is just one example of how laws and processing are significantly disproportionate in the sense that certain groups (typically lower class minorities) are arrested, punished, and locked up far more than well-to-do offenders. Studies showing higher conviction rates and longer sentences for offenders who must rely on public defenders to represent them consistently support the high degree of disparities in the criminal justice system, which policies derived from conflict theory attempt to address.

Recently, these hands-off policies have been less attractive to authorities and politicians who want to be seen as hard on criminal offenders so they will be elected or retained in office. Still, due to recent economic crises in many state and local communities, it is likely that many jurisdictions will reconsider these approaches and find alternative approaches for dealing with nonviolent offenders via alternative or diversion strategies for these individuals.

---

[32]Ibid.

☒ Policy Applications Derived From Integrated and Developmental Theories

Few policy strategies have been derived from integrated or developmental theories, probably because most policymakers are either unaware of such theoretical frameworks or because the theories are so complicated that it is difficult for practitioners to apply them. However, one exception to this rule is Braithwaite's theory of shaming and reintegration (see Section X). Braithwaite has gone to great lengths to apply his theory in Australia, and some evaluations of this approach have shown promise, even if not in the United States. Interventions stress the restorative concepts of his theory, particularly the idea of reintegrating offenders back into society, as opposed to stigmatizing them.[33] This strategy emphasizes holding offenders accountable, yet, it tries to bring them back into conventional society as quickly and efficiently as possible. In contrast, the U.S. system typically gives prisoners a nominal amount of funds (in most jurisdictions, about $200), and they must report themselves as convicted felons for the rest of their lives on all job applications. Would Braithwaite's model work here in the United States? We don't know, but it seems that there could be a better way to incorporate offenders back into society in such a way that they are not automatically set up to fail.

Another area of policy implications from developmental theory is that involving early childhood. Recent research has shown that the earlier resources are provided—for maternal care during pregnancy, as well as infant care, especially for high-risk infants—the less likely that such infants will be disposed toward engaging in criminal offending.[34] Thus, far more resources and funding should be given to early development of youth, especially for high-risk mothers who are pregnant.

☒ Conclusion

This introduction briefly examined some of the policy implications that can be derived from the theories discussed throughout this text. Other policy strategies have been tried, but we have focused on those that have received the most attention. It is up to each reader to consider what other types of programs can or should be derived from the various theories explored in this book.

☒ Section Summary

- ◆ Ever since the beginning of rational criminological theorizing in the 18th century, policy implications have always been derived from the popular theories of a given time period.
- ◆ Policies derived from Classical School/deterrence theory have seen mixed results, largely due to the limited rationality of people due to psychological and physiological propensities. However, some of these policy implications appear to have some success, such as proactive policing and targeting high-crime locations, known as

---

[33]John Braithwaite, "Restorative Justice: Assessing Optimistic and Pessimistic Accounts," in *Crime and Justice: A Review of Research*, Vol. 25, ed. Michael Tonry (Chicago: University of Chicago Press, 1999), 1–27.

[34]Wright et al., *Criminals in the Making*.

"hot spots." On the other hand, many of the court-implemented policies based on deterrence principles, such as "three strikes you're out," have not shown much promise as they are currently administered.

◆ Policies derived from biosocial theories emphasize the need for more medical care, especially during the pre- and perinatal stages of life. The earlier biological and social risk factors can be addressed, the better.

◆ Many policies can be derived from social structure theories, but these have not always been found successful: for example, the War on Poverty in the 1960s and various forms of neighborhood watch programs in the last few decades. However, some programs do appear to work, such as those that provide vocational training so people can pull themselves out of poverty.

◆ Social process theories also have numerous policy applications, many of them successful, such as family visitations by professionals to make sure the family environment is not problematic, as well as programs that teach individuals (typically youth) cognitive-behavioral skills on how to handle peer pressure and other social situations.

◆ Social reaction theories have inspired such policies as diversion and decriminalization. Conflict theories have inspired such policies as revising sentencing guidelines to make them more equitable according to the seriousness of a given act.

◆ Developmental and integrative theories have not inspired too many policies yet, due to the complexity of such models. However, it is quite likely that in the near future, such theoretical frameworks will be able to inform policymakers.

DISCUSSION QUESTIONS

1. What types of policies have been derived from the Classical School/deterrence theory? Which have been found to be successful and which have not?

2. What types of policies have been derived from biosocial theories?

3. What types of policies have been derived from social structure policies?

4. What types of policies have been derived from social process theories?

5. What types of policies have been derived from social reaction and conflict theories?

6. What types of policies have been derived from integrated/developmental theories?

7. What pattern can be seen from the successful programs across all the theories discussed in this book? What patterns do you see in the failure of programs across all theories in this book?

WEB RESOURCES

The University of Maryland Report of What Works and What Doesn't

http://www.ncjrs.org/works/

Policy Implications Derived from Criminological Theories

http://www.cjpf.org/crime/crime.html

http://www.ncpa.org/iss/cri/

http://www.crimepolicy.org/

READING

▲ **Photo 11.2** James Q. Wilson (1931–)

In this selection, James Q. Wilson gives a contemporary review of the various policies that are recommended by the most recent research that exists in the scientific criminological literature. However, keep in mind while reading this selection that Wilson does have some bias toward certain types of theories and assumptions, as does every researcher (whether or not they admit to it).

Wilson provides a variety of explanations for the recent reduction in crime, such as his claim that when crimes decreased in the mid-1990s, one "of the reasons may have been a drop in crack dealing." Wilson explains that another reason for the large decrease is that "crime is also reduced when a nation increases its use of prisons." These two factors are the primary criteria that Wilson acknowledges as important as determinants in why crime/delinquency reduced so much in the last decade.

While reading this selection, readers should consider whether they agree with Wilson's policy recommendations. Furthermore, they should consider the fact that the data are on Wilson's side; specifically, as incarceration rates have soared over the last 15 years, the rate of serious street crimes (e.g., homicide) has fallen drastically. Still, other crimes have increased substantially, and it is up to readers to decide what types of policy recommendations make the most sense for dealing with crimes in the future.

Source: James Q. Wilson, "Crime and Public Policy," in *Crime: Public Policies for Crime Control,* ed. J. Q. Wilson and J. Petersilia (Oakland, CA: ICS Press, 2002), Chapter 19, 537–57. Reprinted with permission of the editor, Joan Petersilia.

Crime and Public Policy

James Q. Wilson

There are two great questions about crime: why do people differ in the rate at which they commit crimes, and why do crime rates in a society rise and fall? You might suppose that the answer to these two questions would be the same. If we can explain why people differ in their criminality, then all we need to do is add up these individual differences in order to know how much crime there is in their society. But in fact, the answers to these questions are not at all the same. The forces that put people at risk for committing crimes overlap only in part with the factors that influence how much crime a nation will experience. For example, we have come to understand that young men are much more likely to break laws than are young women (or older men), but most societies have roughly the same share of young men. Despite that, many nations will differ dramatically in their crime rates.

Over the last 40 years, social scientists have made great gains in explaining why some people are more likely than others to commit crimes but far smaller gains in understanding a nation's crime rate. We know, for example, much more than we once did about the influence of biological, familial, and neighborhood factors that put people at risk for—or protect them from—criminality. But though we also know much more than we once did about the effect on crime of criminal sentences and the opportunities for reducing crime rates by prevention and treatment methods, we cannot easily extract from this knowledge a good explanation of why some nations have much higher crime rates than others or useful lessons for driving down the crime rate in our nation.

Consider prisons. We are more confident than we once were that the rate at which offenders are sent to prison, other things being equal, will affect the crime rate: The higher the probability of punishment, the lower the crime rate. But that knowledge does not make it easy to explain why nations differ in their crime rates, in part because the international data we would need to make such judgments are often lacking and in part because there seem to be so many other factors—youth, culture, opportunities, the availability of guns—that also affect crime rates. No one, so far as I am aware, has ever explained a nation's crime rate by taking systematically into account all of the variables that might affect it. (Unfortunately, this inability has not prevented some scholars from confidently explaining to journalists why the crime rate has gone up or gone down.) Moreover, it is no simple matter to drive down crime rates by relying more on imprisonment. Doing that requires changing the effectiveness of policing, altering the behavior of prosecutors and judges, and coping with the costs tangible and intangible, of having a large prison population.

The same problems confront efforts to reduce the crime rate by means of prevention, employment, and rehabilitation programs. We are getting better at evaluating such efforts and drawing lessons from them, but we have no idea what would be the cumulative effect of putting in place all of the best programs on a large scale. One reason is that the best ones are often hard to make larger. They begin as small, carefully managed efforts that draw on the talents of highly talented, strongly motivated people. Scaling up such programs requires one to believe that what a few skilled people can do can also be done by a large number of less skilled ones and that managing a big program is no different from managing a small one. For example, putting some at-risk young people into imaginative day-care programs, creating job skills for a roomful of juveniles, and making probation a meaningful experience for a few dozen offenders may well reduce the risk of criminality, but doing the same thing for a million youngsters, a million juveniles, or a million probationers is no easy matter.

Moreover, one of the most frequently suggested remedies for crime—keeping unemployment low—turns out to have only a modest effect on the crime rate. There is evidence that higher unemployment does increase the rate of property crime, but that relationship for the United States is quite modest. Property crime rates fall by about 2 percent for every 1 percent cut in the unemployment rate, but this means that if the unemployment rate is cut dramatically—say, from 8 percent to 4 percent—the rate of property crime will only decrease by 8 percent. Now, if the United States had a serious depression and unemployment soared to very high levels, cutting it down to 4 percent would probably make a big difference in property crime, but that has not happened in this country since the 1930s, and that was a time when we had no reliable crime measures with which to test the effect of reducing unemployment. Moreover, the fact that high unemployment rates are found in certain neighborhoods does not explain why the crime rate in those areas is so high. But another aspect of the labor problem—the existence of many people who neither have nor are looking for a job—may well contribute to high crime rates. Understanding the connection between low labor force participation and high crime is complicated, however, because we are not certain in which direction causality runs (does low labor force participation cause crime, or does a lot of crime make illegal activities more rewarding than work?), and we are not certain what underlying factors (such as weak families, poor schooling, strong gangs, or frequent drug use) may simultaneously cause both low labor force participation and high levels of crime.

And dealing with crime means not only preventing it but judging it. Suppose we were confident that either imprisonment or rehabilitation would reduce the crime rate sharply. We are not free to simply put these ideas in place; the reason is that ordinary citizens view crime in moral as well as practical terms. When a person is convicted of a crime, people expect that he or she will be punished in a way that reflects the gravity of the offense and the prior record of the offender as well as in a way that will reduce the crime rate. Suppose we learned that a $5,000 fine, if routinely imposed, would cut the rate at which women are raped. I doubt that many people would agree that the fine was enough punishment, whatever its effect on the frequency of rapes. To most of us, the crime is sufficiently appalling that we would demand that the offender spend a long time in prison whether or not the length of imprisonment had any effect on the rate at which rapes occur. Or to take a different example, suppose that rapists could be cured of their habit by putting them in a treatment program outside of prison, perhaps one located in a clinic. No matter how successful the treatment, people would expect more severe measures to be employed as a way of making clear how gravely society views the offense.

And the same argument can be made in the opposite direction. If a sentence of five years in prison would reduce the robbery rate and a sentence of ten years would add no further deterrent value, would we be justified in increasing the penalty to ten years just because people detest robbers? Perhaps, but a good argument can be made that by doing this we are not only making no difference in the robbery rate, we are also consuming scarce prison space that would be put to better use by confining another robber for five years. Morality and utility sometimes lead to the same result, but just as often they may be in conflict.

When we debate public policy toward crime, we are arguing about issues that necessarily combine moral and utilitarian considerations. Social scientists can help people understand the utility of different strategies, but they are not very good at helping people cope with the morality of these efforts. This is not to say that social scientists should avoid making moral judgments; inevitably they will make them, and that is fine so long as those judgments are made explicitly and not left as the hidden subtext of a purportedly scientific claim.

In the remainder of this chapter, I wish to address three questions that concern many readers: Why does America have so much crime? Why did the crime rate go down in the late 1990s? What, if anything, can be done to keep it down?

⌧ Crime in America

Americans should stop thinking of their country as more crime-ridden than is the case in Europe. In general, America has less property crime and more violent crime than do other industrialized nations. There are two ways of calculating crime rates—police reports and victim reports. In general, the two measures lead to roughly comparable results. Using police data, the burglary rate in Australia, Austria, Canada, England, Germany, and the Netherlands is much higher than it is in the United States; only in France and Sweden is it lower. Using victim reports, the differences in burglary rates look much the same except that Germany now has a lower rate than does America. Motor vehicle theft is probably an even better measure of property crime because automobiles are insured, and thus their loss is more likely to be reported to the police. Your car is more likely to be stolen in Australia, Austria, Canada, England, and Sweden than it is in the United States.

But America indisputably has a higher murder rate than does any other industrialized nation. The American rate is three to five times higher than it is in most of Europe. But for another violent crime, robbery, the differences as reported by victims are not very large. The American rate is about the same as it is in Australia, Canada, England, France, and Italy.

Of course, robbery rates are harder to measure accurately than are homicide rates: the latter produce dead bodies, while the former may range from minor lunch-yard thefts to armed attacks on banks, making the victims of these widely various crimes differ in their inclination to call the authorities.

We sometimes suppose that our high murder rate is the result of violent television, dangerous drugs, or the widespread ownership of guns. These factors may have some effect (guns almost surely do), but bear in mind that the homicide rate in New York City has been 10 to 15 times higher than that rate in London for at least 200 years (Monkkonen 2001). We were killing each other at a high rate long before movies, television, heroin, and cocaine were invented. And though the availability of guns probably affects the murder rate, please know that the rate at which Americans kill each other without using guns—that is, with clubs, knives, and fists—is three times greater than the nongun homicide rate in England. Robberies in this country are more likely to be fatal than robberies in England, in part because American robbers are more likely to be armed, but the death rate from robberies committed by unarmed robbers is also much higher in the United States than in England (Zimring and Hawkins 1997). Even without a gun in their hands, American robbers are more violent than English ones.

We are a violent nation. Why? If you are asked this question on a final examination, you are in luck, because there is a long list of possible answers and precious little scientific evidence as to which one, or which combination of them, is correct. Write down any of the following: We are more violent because the country was settled by immigrants who created farms and villages before there was any government to protect them, and so they defended themselves against each other and against Native Americans. We are more violent because America has never had a landed aristocracy that tended to monopolize violent power; since wealthy knights could not own all the swords and guns, their ownership was spread among everybody, some of whom liked to settle quarrels with their use. We are more violent because for three centuries many people lived on a frontier where they hunted for food and fought against Native Americans. We are more violent because the country was filled by people from many different nations and ethnic groups who struggled, sometimes violently, to define against one another an appropriate standard of behavior. We are a more violent nation because whites used force to subdue black slaves and then slowly put an end to slavery by a war between the North and the South that killed 140,000 soldiers, injured another quarter million, and left a lot of people armed and angry. We are a more violent nation because a culture of honor emerged in the South that placed a high value on maintaining status by means of duels and killings. We are a more violent nation because the government had, for most of its history, only weak powers over people and for many centuries took little notice of crimes committed by African Americans on other African Americans. We are a more violent nation because Prohibition created large criminal gangs that relied on force to maintain their illegal control of the liquor trade. And so on.

⊠ The Recent Decline in Crime

Crime rates rose dramatically from the early 1960s until around 1980, leveled off a bit until the mid-1980s, then increased sharply in the late 1980s, and finally, in the late 1990s, dropped just as sharply. Some of this change was the result of changes in the age distribution of the population: in the 1960s we had a lot of young people, in the 1990s we had relatively fewer. Crime went up when there were more young men around and down when there were not so many. But the age explanation only

accounts for a small fraction of either the increase or the decrease in crime; by the 1990s, there may have been proportionately fewer young people, but the ones who existed had per capita, a higher rate of criminality.

One of the puzzles of crime in the 1980s can be solved by disaggregating crime along age lines. Murder and robbery rates hit a high around 1980, then started to decline, and then shot up again in the late 1980s and early 1990s. The reason in part is that in the late 1980s and early 1990s young people became more criminal, probably because many were part of the crack cocaine trade. The crime rate was falling for a variety of reasons—the population was getting older, more offenders were in prison—but when the crack drug trade opened up after 1985, many young men got involved in it, acquired guns to defend their illegal activities against competitors, and increased the rate at which they killed people. By the mid-1990s, however, murder and robbery rates were dropping sharply. One of the reasons may have been a drop in crack dealing, as many would-be users turned against a drug that was sending so many addicts to hospitals, prisons, and morgues.

But the age and drug factors are probably only one part of the story. At least two other things happened, and a third may have. The third factor can be mentioned immediately because we have no facts that bear on the matter. Perhaps the crime rate has dropped because the culture has changed. If the public has become more concerned about crime it may, through families, neighborhoods, and schools, invest more heavily in persuading young people to stay out of trouble and in using shame to penalize those who do get into trouble. A good case can be made that the emancipation of the individual from traditional restraints imposed by families and neighbors contributed to the steep rise in crime that occurred in virtually every industrialized nation between the early 1960s and the 1990s despite high and rising levels of economic achievement (Wilson 1983). It is possible that this emancipation was reduced a bit by the popular aversion to crime and a desire to reimpose some older restrictions. We may have moved from an era that let us "do our own thing" to one that urged us to "do what society expects." But this speculation is merely that; we have no way of evaluating it.

A different explanation has more facts to support it. As Americans worried more and more about crime, they invested more heavily in self-defense. They moved to the suburbs or to gated communities, bought condos in buildings with armed guards, equipped their cars with alarms and special locks, and avoided dangerous neighborhoods. No one knows how much these changes affected crime rates, but they surely had some influence, since crime requires not only an offender but an opportunity to offend. Most ordinary street crime occurs near where offenders live; if the neighborhood becomes harder to attack, there will be less crime, with rather few would-be offenders moving to easier targets some distance away.

But this investment in self-defense also imposes a cost. When the middle class moves out of an urban neighborhood in search of a safer life in the suburbs, it leaves behind poorer people who cannot move, and these may become more vulnerable to crime. Crime is more common in unstable neighborhoods where people move in and out frequently, because it is hard to maintain in these places an adequate sense of mutual trust and a shared willingness to intervene in local affairs in order to reduce crime. What Sampson calls the "collective efficacy" of a neighborhood is reduced when people with a stake in and a talent for sustaining that efficacy move away.

Crime is also reduced when a nation increases its use of prisons. This has been a deeply controversial topic among criminologists, with some praising it as a way of deterring crimes and locking up offenders (Wilson 1983) and others denouncing it as an excessive reaction. But one can estimate the extent to which imprisonment makes a difference.

William Spelman suggests, on the basis of his calculations, that the increased use of prison can explain about 25 percent of the reduction in crime (Spelman 2000). This estimate, though carefully done, is still only an estimate; the real impact of prison could be much higher or somewhat lower. It cannot be zero, however, because we know from many studies that the median nondrug-dealing offender when free on the street commits about 12 crimes a year (Chaiken and Chaiken 1982; DiIulio and Piehl 1991; Spelman 1994). When 1,000 more offenders are taken off the street, there are potentially 12,000 fewer crimes. The actual decrease, of course, is much harder to calculate, since it depends on how many crimes are committed by newly imprisoned (instead of already imprisoned) offenders. That number will vary greatly by state, depending on how tough each state is about the use of prison. A state such as Texas, which sends a high proportion of its offenders to prison, will gain less from sending a few more to prison (because the additional inmates will have less serious crime rates) than will a state that sends only a small fraction of its convicts to prison.

Moreover, national differences in the willingness to use prison may help us understand why nations differ in their crime rates. Many of the European nations where the rates of burglary and auto theft are higher than they are in the United States display less willingness to use prison. We must beware, however, of such gross simplifications as that America is "the most punitive nation." Much depends on the kind of crime we analyze and whether we look at the stock of people in prison or the rate (or flow) at which people enter prison, and whether we distinguish between the length of sentence and the probability of getting a sentence.

Allowing for all of this, it becomes clear that the United States and most European nations are equally likely to use prison for serious offenders such as those who commit homicide and robbery, that America is more likely to send burglars to prison than are these other nations, and that until recently the time spent in prison was about the same for America, Canada, and England. But this last factor—how long each offender spends in prison—has gone up in the United States more than in other nations. For both drug crimes and property crimes such as burglary, America is more punitive.

Some scholars (Farrington and Langan 1992; Farrington and Wikstrom 1993) have suggested that America's greater use of prison for burglary and other property crimes may help explain why there are more such crimes committed in nations that do not rely so heavily on prison. The facts are consistent with this view though of course it is impossible to know whether they prove it. England, for example, had a lower burglary rate than did the United States during the early 1980s. But by the mid-1980s England's rate had caught up with that of the U.S., and by the early 1990s the former was much higher than the latter. This change in burglary rates between the two nations coincided with an increase in the probability of going to prison here and a decrease in that probability in England. Whether this coincidence amounts to a cause is, of course, impossible to say with any confidence.

Since no nation wants a high crime rate, one may wonder why some nations are more willing to use prison to reduce that rate than are others. The answer, I think, is politics. In the United States, district attorneys and many judges are elected, and in a lot of states the citizens, by voting on initiative issues, can alter the laws affecting punishment. These facts make American officials highly sensitive to what voters want. In England, by contrast, neither prosecutors nor judges are elected, there is no way for the voters to change sentencing laws by means of an initiative measure, and much of the policy toward offenders is determined by appointed officials in the Home Office, who have little interest in how voters think. As a result, America uses prison more than does England.

Some people think that Americans let politics influence sentencing policies too much. The voters, the argument goes, are unduly influenced by horrific crimes and get carried away with a desire for retribution. No doubt this sometimes happens. Voters can insist on increasing the maximum sentence for, say, burglary without realizing that many burglars will get no prison sentence at all; they will be punished by being placed on probation, and so the longer maximum sentence will affect hardly any offenders. Moreover, a long sentence for burglary may add little to the deterrent effect of prison if (as is often the case) the certainty of punishment is more important than its severity. As a result, we may end up with a few burglars growing old in prison while a lot more enjoy probation.

But just as often the voters may get matters right. They may insist on mandatory minimum sentences that reduce the chances of probation for certain offenders (at the cost, of course, of big bills for prison construction), or they may support a way of increasing the deterrent effect of prison. One controversial example is California's "Three Strikes" law, which requires most offenders who have already been convicted of two crimes to spend at least 25 years in prison when they are convicted of a third. The controversy over this law involves several matters: How serious should the crimes be before they count as a strike? How much discretion should prosecutors and judges have in deciding that a prior conviction is not a strike? Will this law actually affect crime, and if so, will it do it at a cost the state can afford to pay?

The last issue has been hotly debated among scholars. One early study (Greenwood et al. 1994) predicted that the law in California would produce a small reduction in crime at the cost of a very expensive prison-building program. This claim was based on the effect of Three Strikes on incapacitating offenders—that is, on reducing crime simply by keeping criminals off the street for longer periods of time. But the law might also have a deterrent effect: it might discourage offenders who have two strikes from taking the risk of getting a third one. The early study had no way of estimating deterrence because the law had yet to be enforced. But since its enforcement, research has suggested that it does indeed have a deterrent value, such that the crime rate in California fell after the law was implemented by more that would have been predicted in the absence of the law (Chen 1998, 2000).

The value of prison will vary depending on the type of crime being committed. For property crimes such as robbery, burglary, and larceny, a higher probability of prison may reduce the number of such crimes. But for drug offenses, there may be no deterrent effect at all. An incarcerated robber is not immediately replaced on the street by a would-be robber who notices that a vacancy has occurred in the robbing business. But an incarcerated drug dealer may well be replaced on the street by a drug-dealing organization that realizes one of its members is locked up and so recruits a new dealer to replace him. It is also possible, of course, that high penalties for drug dealing will discourage any would-be dealers from volunteering for the job. We do not know which process operates, but we certainly have no reason at present for thinking that imprisoning drug dealers will have the same effect on the narcotics traffic as imprisoning robbers has on the robbery business.

In short, some unknown combination of age, culture, prisons, and urban conditions have shaped the American crime rate in ways that are hard to predict.

▨ Can More Be Done to Reduce Crime?

There is no silver bullet that will reduce crime, much less eliminate it, Indeed, the "crime rate" itself is a misleading number. What people care about is the risk they face in the neighborhoods where they live, and this risk varies dramatically across the country. The murder rate in South Dakota is only one-twelfth of what it is in Louisiana, and in Louisiana the rate in New Orleans is almost ten times higher than what it is in Lafayette. Someone living in Sioux Falls, South Dakota, is 70 times safer from murder than someone living in New Orleans, and within New Orleans some neighborhoods are much safer than others. In 1996, 81 American cities accounted for half of all the murders in the nation. In that same year, one-third of all the robberies in the country occurred in just two states, California and New York. The risks people confront vary so greatly that any responsible criminal justice policy has to be attuned to the needs of particular places. What works in New York City may not work in a New York City suburb.

We have heard a great deal about the effectiveness of the New York City Police Department in sharply reducing the crime rate in that city during the late 1990s. I am persuaded that new management policies and street strategies at the NYPD did have a lot to do with that reduction, but we must bear in mind that the crime rate has also fallen, though not by nearly as much, in cities where no special police efforts were made.

While no one can satisfactorily explain changes in crime rates, there are at least three things that push that rate up—drugs; guns; and personal, familial, and neighborhood factors. The first two forces are objects of much controversy, with some people saying that we ought to legalize drugs and criminalize guns while others argue that we ought to criminalize drugs and legalize guns. The first group believes that legalizing drugs would reduce thefts motivated by a desire to buy drugs, eliminate the need for violence to protect the drug trade, and get rid of the corruption that accompanies illegal trafficking. Accompanying drug legalization, many supporters believe, there should be a sharp reduction in the public's access to guns, because guns, unlike drugs, are inherently dangerous. Their opponents made exactly the opposite arguments: drugs are too dangerous for free consumption to be allowed, while guns are a valuable form of personal self-protection. Guns serve legal as well as illegal purposes, and drugs differ greatly in the effects they have on people. Picking a set of useful policies requires one to make some careful distinctions.

The third factor that affects the crime rate—personal familial, and neighborhood forces—suffers not so much from controversy as from despair. Hardly anyone doubts that our personality, how we were raised, and where we live affect our risk of becoming an offender, but for most people it is impossible to think of any way to make a significant difference in these risk factors. Let me offer my own suggestions on these topics.

Drugs

Illegal drugs contribute to crime by causing some people to steal in order to buy them and other people to use force or bribery to maintain their control over the supply.

There is no chance that heroin, cocaine, PCP, or methamphetamines will be legalized, nor do I think there is any good reason why they should be. Since this change will not happen, I shall spare you my arguments about why it should not happen, and ask you to focus on the real strategic options that the country faces. There are essentially two strategic factors: reducing demand and reducing supply. It is fair to say that supply reduction—destroying drug supplies abroad, attacking smugglers who bring drugs into this country, and arresting drug dealers on the street—has been the chief strategy that this country has pursued. No doubt there have been some gains from this effort, but any fair assessment of the drug trade will lead you to understand, I think, that as long as there is a demand for drugs a supply of them will be generated. If opium (with which

to make heroin) is no longer supplied by Turkey, it will be supplied by Myanmar or Afghanistan; if coca leaves (with which to make cocaine) are no longer produced in Colombia, they will be produced in Peru; if one drug-dealing cartel is broken up, another will rise to take its place; if trucks can no longer smuggle drugs, airplanes will. It is impossible to think of any product for which there is a strong human demand that will not be made available by somebody, somewhere.

Reducing the demand for drugs is difficult but not impossible, provided we focus our energies on people who use drugs at high rates and are vulnerable to social control. Those who meet these two tests are convicted offenders—those in prison, on parole, and on probation. A majority of them report being involved with drugs at the time of their arrest, and they number in excess of 5 million persons. They fall under the coercive power of the state because as inmates they are bound by prison rules, and as probationers or parolees they live in the community subject to certain state controls that can be exercised without violating constitutional bans on unreasonable search and seizure.

These facts suggest the need to expand prison-based drug treatment programs so they are available to and required of every inmate who had a pre-arrest drug problem. It is expensive but essential to do what we can to reduce the dependence on drugs of people under our control. A second implication, one already embraced by more than 400 American cities, are drug courts. These are staffed by judges dedicated to handling offenders on probation who have a record of drug abuse and who have not committed a serious crime. The judges require that these convicts enter and stay in drug treatment programs as a condition of remaining on the street. We do not yet know the long-term effect of drug courts, but they do seem to provide at least short-term benefits to the offenders subject to them.

However, drug courts reach only a small fraction of drug-abusing offenders, because they serve only those who are on probation for having committed minor (as opposed to major) crimes. But many probationers and most parolees have committed major offenses. Therefore, it is important to direct probation and parole officers to keep these people under close watch. A "coerced abstinence" program calls for frequent drug tests (perhaps twice a week) of probationers and parolees, with those who fail the tests quickly punished by (initially) a brief return to jail, with further violations followed by longer returns. The goal of this coercion is not to punish drug users but to make staying in a community-based drug treatment program more attractive than repeated trips to jail. We know from other studies that drug treatment programs are valuable in more or less direct proportion to how long addicts remain in them, and that coercion as a way of inducing participation does not weaken the program's effects and may, in fact, strengthen them (Satel 1999; Anglin and Hser 1990).

For this program to work, the penalties for failing a street drug test must be swift and certain, though not especially severe. But swiftness and certainty cannot be achieved by referring probationers and parolees back to the criminal courts for a new hearing; that takes much too long and has an uncertain outcome. Rather, the (temporary) revocation of probation or parole must be an administrative matter—revoking probation or parole—done on the spot by probation and parole officers. This can be accomplished by a statute that makes clear that no probationer or parolee can refuse a drug test or can remain on the street if he or she fails one.

To create jail or prison space for those breaking the rules governing probationers and parolees on the street, it will be necessary to release many inmates who are now confined solely because they were guilty of possessing illegal drugs. This is no small number; perhaps one-third of all state prison inmates are there on drug charges, and a significant fraction of these have committed no other offense. Since we have no evidence that imprisoning them reduces the drug trade, they are occupying scarce prison space that ought to be used for offenders who have committed property or violent crimes or who have failed to live by the rules governing street life for probationers and parolees.

More can probably be done to reduce drug demand among employees. We know from experience with the military that drug use was cut sharply by a policy of frequent tests accompanied by zero tolerance for drug use. The prevalence of drug use in the military fell from about 27 percent in 1980 to about 3 percent in 1994 (Mehay and Pacula 2000). It is hard, but not impossible, to extend this policy to private employers.

▧ Guns

Much the same logic ought to govern our effort to reduce the availability of guns to people who are at risk of using them illegally. The vast majority of all guns are owned by people who never commit a crime and for whom gun ownership represents an opportunity to hunt, to shoot at targets, and to enhance their self-defense. Though controversial, the study by John Lott (2000) suggests that states in which people without a criminal or mental health record are allowed to carry concealed firearms on the street reduce the rate at which street crimes occur. Lott's argument is that a would-be offender must take into account the ability of a would-be victim to resist an attack. The chances of effective resistance go up if some unknown but not trivial percentage of would-be victims is armed. Gun carrying thus deters some crimes without, so far as we can tell from the evidence, increasing the rate at which persons authorized to carry weapons harm other people. Even guns kept at home have some deterrent value as is suggested by the fact that "hot burglaries"—that is, burglaries at a time when the owners are at home—are much less common in the United States, where millions of people have guns at home, than they are in England, where very few people have weapons in their residences. The Lott study has been challenged by other scholars (Duggan 2000; Ayres and Donohue 1999), and so the ultimate effect of private gun carrying on crime remains in doubt. But the doubt is important to remember because so many advocates of tougher gun control have no doubts at all.

Many people carry weapons illegally. They may be known criminals, or young men looking to protect themselves in a fight, or gang members protecting their turf or looking for a chance to get even. Guns in the wrong hands create problems, as is evidenced by drive-by shootings in which gang members shoot at rivals and miss, killing instead an innocent child or adult standing nearby. Given this, it would be difficult to say that illegally carried weapons cause no harm.

The task of the police is to get such guns off the streets. Doing so would, in my view, have a greater impact on violent crime than further tightening of the rules governing how guns are sold. The vast majority of gun sales are arranged by law-abiding citizens; making the rules tougher for such sales will chiefly affect law-abiding citizens. A significant fraction of all guns used in crimes are stolen or borrowed (Wright and Rossi 1986). If point-of-sale restrictions become much tougher, an even larger fraction of such guns will be acquired by theft or in the black market.

The law governing police searches of ordinary persons is vague. In general, it suggests that the police need only a reasonable suspicion that the citizen has committed or is about to commit a crime in order for them to have the authority to stop and pat down an individual to see if a gun is in his or her clothing. But the police can readily stop and search parolees and probationers for guns, and should do so at every opportunity. Since most persons released from prison commit new crimes, keeping guns out of their hands reduces the risk that those

weapons will we used in those crimes. But stopping probationers and parolees is not enough; they must be penalized if they are illegally carrying a weapon. Unfortunately, some research suggests that a large fraction of people from whom illegally carried weapons have been taken are given no meaningful penalty. For them, carrying a gun has no cost.

One new way of dealing with these people has been pioneered in such cities as Baltimore, Birmingham, and Providence. There gun courts, modeled on drug courts, impose sentences on people caught illegally carrying weapons. Some of these courts deal with all offenders, some only with juveniles; some require their defendants and their families to take special firearms safety courses, others send the defendants to boot camps or impose strict probation rules, and yet others do all of these things. We have as yet little published evidence on their effect, but one study suggests than the recidivism rate of juveniles going through the Birmingham gun court is about half that of young people processed in the ordinary way (Braun 2001).

But for citizens who are neither probationers nor parolees, there is at present no obvious way of both seizing illegal weapons and observing their constitutional right to be free of an improper search. That can change if the government proceeds with its efforts to develop a portable device that will enable the police to identify a gun under the clothing of a person from a distance of 20 or 30 feet. Such devices exist in the prototype stage, but are still too large to be of much practical value. With advanced versions of this technology, it will be possible for the police to see not simply a lump of iron in a person's pocket but the precise outline of the weapon. With that information, the police could then stop and question the person. If he or she has a license, they are free to move on; if not, they can be arrested and the gun seized.

The evidence we have so far is that street gun searches make a difference. In one neighborhood that tried this, the number of guns seized almost tripled. Gun-related crimes in that area dropped by 58 percent, compared with only a 29 percent crime decrease in a control neighborhood. Increasing gun seizures seems to cut gun crimes.

There is, of course, a risk in greater police efforts at preventing crime on the streets. Today that risk is called racial profiling, by which is meant an effort to stop and question people on the basis of their racial or ethnic identity. Suppose we ask the police to patrol intensively criminal "hot spots," to focus on known high-rate offenders who are free on the street, and to stop people who are suspected of carrying weapons. This means that a disproportionate share of such stops will involve African Americans and Hispanics. This is *not* racial profiling if the police have good grounds apart from race for making their interventions. It *would* be racial profiling if automobile drivers were stopped simply because they were black or Hispanic. In between these two extremes are many tough cases.

▧ Prevention

Everyone would like to prevent people from committing crimes in the first place, but almost everyone also knows that this is not an easy task. *Almost* everyone, for a few people have devoted themselves to designing and running crime-prevention programs that they are convinced will stop young people from becoming serious offenders. The two central questions that such efforts raise are these: Do these programs really work? If they work, can they be made larger? We have learned a great deal about how to answer the first question but not much about how to answer the second.

There are hundreds, perhaps thousands, of crime-prevention programs under way. Many may work (and, of course, all their leaders think they work). But which actually work can only be determined by a rigorous evaluation. Not many have been evaluated in this way. A rigorous

evaluation requires four things to be done: First, people must be assigned randomly to either the prevention program or a control group. Random assignment virtually eliminates the chance that those in the program will differ in some unknown way from those not in it. Random assignment is better than trying to match people in the two because we probably will not know (or even be able to observe) all the ways by which they should be matched. Second, the prevention must actually be applied. Sometimes people are enrolled in a program but do not in fact get the planned treatment. Third, the positive benefit, if any, of the program must last for at least one year after the program ends. It is not hard to change people while they are in a program; what is difficult is to make the change last afterward. Fourth, if the program produces a positive effect (that is, people in it are less likely to commit crimes than similar ones not in it), that program should be evaluated again in a different location. Some programs will work once because they are run by exceptional people or in a community that facilitates its success; the critical test is to see if they will run when tried elsewhere using different people.

The Center for the Study and Prevention of Violence at the University of Colorado has been working for many years to identify programs that meet these four tests. They have (so far) found about a dozen prevention programs that, on the basis of rigorous evaluations, seem to work. Among the successful programs are these: Big Brothers/Big Sisters, a national effort to match adults with young people from single-parent homes so as to give, on a part-time basis, serious adult care to youngsters at risk for delinquency or drug abuse; nurse home visitations, a program in several cities that sends nurses to the homes of poor, often unmarried, pregnant women to advise them on child care and parenting; and various school-based and family-oriented programs that teach anger management, ways of resisting peer pressure, and strategies for improving motivation and the acceptance of personal responsibility. Those that seem to work contain few great surprises: they are the ones that make the school as a whole an effective, well-managed institution with firm but fair discipline and high expectations about student behavior. Small add-on programs—individual counseling, recreational opportunities, many courses designed to "teach" good behavior—often have little effect.

Cities and states typically do not invest in what works but only in what is popular. A few crime-prevention and criminal rehabilitation programs have passed rigorous evaluation, but these are not widely copied; instead, other programs that have never been evaluated or have failed an evaluation are funded. For example, two programs aimed at preventing drug abuse and reducing delinquency, DARE (Drug Abuse Resistance Education) and "boot camps," are immensely popular despite the dearth of credible evidence that they make a lasting difference. To their credit, the leaders of DARE have recently announced that they plan to make important changes in their program.

Imagine what would happen if we sold commercial products the way we prevent crime. For a business firm to operate the way crime-prevention programs function, it would have to market a new product without testing it with a small number of consumers, mount advertising for it without finding out whether the ads reach customers, and spend money on the product without wondering whether it ever returned a profit. If you run a corporation that way, the firm will soon be bankrupt and you will be out of a job.

Indeed, the problem is even more profound when we understand how criminal justice money is spent. The police and the prisons get most of it. Criminologists know full well that I am no opponent of either policing or prisons; I have argued for decades in favor of better police strategies and an expansion of prison use. I still believe that they have made a big difference in the crime reduction that has occurred in this country. But if we are to do

more, we have to spend more money and design better strategies for the neglected part of the system—probation and parole officers and crime-prevention programs.

There are about 5 million people today under the control of the criminal justice system, but only one-fifth of them are in prison. The rest are on the streets as probationers and parolees, even though most of these offenders have committed serious crimes—one or more felonies. Managing people in the community is even more important than managing them in prisons, because while they are on the streets they can harm all of us. And in time most people on parole will return to prison unless we become more skillful at changing their behavior.

But politically improving community-based corrections cannot be done by asking the public for funds to pay for community-based corrections (that sounds much too flabby) or to hire more probation and parole officers (they sound like people who will be too easy on criminals). There is no way, I think, to improve what we do on the street without making crystal clear what specific goals we hope to achieve. We want to reduce the demand for drugs among criminal offenders and get illegally carried weapons out of the pockets of dangerous people. That should be the message that politicians take to the public. As it turns out, the best way to do these things is to hire more probation and parole officers and give them some tough marching orders.

Much the same message needs to be broadcast with respect to crime prevention and rehabilitation. More people support these efforts than believe in community-based corrections, but the political emphasis must be on achieving here, in this city or county, the same gains that we know were achieved elsewhere by a scientifically validated effort. That is hard to do, not because ordinary voters will fight these efforts, but because the people who run programs that do not work will fight hard to keep their share of public money. The chief problems facing crime-prevention programs are three: We have to worry about making a successful small program into an equally successful large one, we have to convince citizens that there are programs that really work, and we have to win an interest-group war with those who run programs that do not work. None of this is easy.

⬚ Thinking About Crime

But there is another, more ideological war that also must be fought. People who write about crime often think that their political preferences should dictate their policy recommendations. If you are a liberal you will tend to blame crime on racism and unemployment and argue in favor of more civil rights and better jobs—whatever the facts may be. If you are a conservative, you will tend to blame crime on personal irresponsibility and argue in favor of capital punishment and longer terms in prison—whatever the facts may be. Picking your way carefully through this ideological morass is difficult, and no one, including the authors of this book, is entirely successful at doing it. But trying to think clearly and factually about crime is a goal all of us ought to share.

That goal ought to lie at the heart of the federal government's efforts to control crime. Obviously, federal authorities have important laws to enforce and are ideally situated to deal with crimes that cross state lines or involve international conspiracies. But ultimately, law enforcement remains a local matter. Police, sheriffs, prosecutors, public defenders, prisons, and treatment and prevention programs are overwhelmingly local responsibilities. Cities and counties investigate the vast majority of crimes, arrest the vast majority of offenders, and try and imprison the great majority of inmates.

But there is one thing that local authorities cannot do very well: find out what works. In part this is because city and state leaders will be criticized if they spend scarce money on "research" when the public wants them to attack crime. And in part it is because even if local leaders had the money and the political opportunity to spend on research, those studies would benefit all cities and states and not just the one that spent the money. Moreover, money used to find out if a popular program really cuts crime may well reveal that it does not, and so the authorities that supported the program will be embarrassed by their effort to evaluate it. For all these reasons, state and local expenditures on evaluating crime control projects are costly, risky, and rare.

Federal authorities do not face quite the same constraints. They do not run most crime-prevention programs; cities, counties, and states do. As a result, Washington can evaluate what other people are doing. And since there is only one national government that collects tax money from everybody, whatever lessons about crime control it learns can be shared with every taxpayer (which is to say, with every city, county, and state).

Over the years, Washington has conducted many evaluations, largely through the National Institute of Justice (a part of the Justice Department) and in part through other research agencies concerned with science and public health. But the money it spends on evaluations is typically but a small fraction of what it spends on aiding cities and states to fight crime. And when Congress, following public opinion, becomes worried about crime, it tends to react by telling the FBI to help investigate what the police and sheriffs are already investigating and by creating new, tougher federal penalties for that offense.

Now there is nothing intrinsically wrong with that reaction, though sometimes one worries that Washington is moving in the direction of nationalizing our law-enforcement system. But there is something wrong with not trying to find out what works. Yet when someone proposes such an attempt, some political leader dismisses the idea as "just another piece of research." (Of course, sometimes they are right; crime control, like every piece of social science, is littered with bad research projects that either tell us nothing or have little effect on how law enforcement operates.) What I mean by "research" is chiefly evaluations of ideas about how to reduce crime.

Evaluations, as we have already seen in this and other chapters, are not easily done well, but it is vital that good ones be attempted. The chief federal role in domestic law enforcement should be to encourage and fund such research. No one else will do it.

◪ References

Anglin, M. Douglas, and Yih-Ing Hser. 1990. "Treatment of Drug Abuse." Pp. 393–460 in *Drugs and Crime*, ed. Michael Tonry and James Q. Wilson, 393–460. Chicago: University of Chicago Press.

Ayres, Ian, and John Donohue. 1999. "Nondiscretionary Concealed Weapons Laws." *American Law and Economics Review* 1:436–70.

Braun, Stephen. (2001). "'Gun Court' Mixes Hard Justice." *Los Angeles Times,* February 15.

Chaiken, Jan, and Marcia Chaiken. 1982. *Varieties of Criminal Behavior.* Santa Monica, CA: RAND.

Chen, Elsa Y. 1998. "Estimating the Impacts of Three Strikes and You're Out and Truth in Sentencing on Crime and Corrections." Paper presented to Midwestern Political Science Association.

———. 2000. "Three Strikes and Truth in Sentencing." Ph.D. diss., Department of Political Science, UCLA.

DiIulio, John J., and Anne Piehl. 1991. "Does Prison Pay?" *Brookings Review* 28–35.

Duggan, Mark. 2000. "More Guns, More Crime." NBER Working Paper 7967. Cambridge, MA: National Bureau of Economic Research.

Farrington, David P., and Patrick A. Langan. 1992. "Changes in Crime and Punishment in England and America in the 1980s." *Justice Quarterly* 9:5–46.

Farrington, David P., and Per-Olof Wikstrom. 1993. "Changes in Crime and Punishment in England and Sweden in the 1980s." In *Studies in Crime and Crime Prevention* 2: 142–70.

Greenwood, Peter, C. Peter Rydell, Allan F. Abrahamse, Nathan P. Caulkins, James Chiesa, Karyn E. Model, and S. P. Klein. 1994. "Three Strikes and You're Out: Estimated Benefits and Costs of California's New Mandatory Sentencing Law." Santa Monica, CA: RAND.

Lott, John R. 2000. *More Guns, Less Crime,* 2d ed. Chicago: University of Chicago Press.

Mehay, Stephen, and Rosalie Liccardo Pacula. 2000. "The Effectiveness of Workplace Drug Prevention Programs: Does 'Zero Tolerance' Work?" NBER Working Paper 7383. Cambridge, MA: National Bureau of Economic Research.

Monkkonen, Eric. 2001. *Murder in New York City.* Berkeley: University of California Press.

Satel, Sally. 1999. *Drug Treatment: The Case for Coercion.* Washington, DC: AEI Press.

Spelman, William. 1994. *Criminal Incapacitation.* New York: Plenum.

———. 2000. "The Limited Importance of Prison Expansion." In *The Crime Drop in America,* ed. Alfred Blumstein and Joel Wallman. New York: Cambridge University Press.

Wilson, James Q. 1983. *Thinking About Crime.* Rev. ed. New York: Basic Books.

Wright, James, and Peter H. Rossi. 1986. *Armed and Dangerous: A Survey of Felons.* New York: Aldine de Gruyter.

Zimring, Franklin E., and Gordon Hawkins. 1997. *Crime Is Not the Problem: Lethal Violence in America.* New York: Oxford University Press.

DISCUSSION QUESTIONS

1. What do you think of Wilson's comment that the level of dealing crack cocaine played a large part in the crime rates in the United States? What types of policy implications do you think are beneficial based on this phenomenon, if you agree with this assessment? If you don't agree with assessment, what variables do you think contributed most to the crime surge at this time?

2. Analyze Wilson's estimation of the use of prison incarceration rates as the way to reduce crime. Do you agree or disagree, and why?

3. What does Wilson have to say about community corrections (e.g., parole or probation), and do you agree with his assessment? Explain your answer and expand on how you would improve such community correctional models of justice.

Glossary

Actus reus: In legal terms, whether the offender actually engaged in a given criminal act. This concept can be contrasted with *mens rea,* which is a concept regarding whether the offender had the intent to commit a given act. This concept is important, especially in situations in which juveniles or mentally disabled individuals engage in offending.

Adaptations to strain: As proposed by Merton, the five ways that individuals deal with feelings of strain; see Conformity, Innovation, Rebellion, Retreatism, and Ritualism.

Adolescence-limited offenders: A type of offender labeled in Moffitt's developmental theory; such offenders commit crimes only during adolescence and desist from offending once they reach their twenties or adulthood.

Adoption studies: Studies that examine the criminality of adoptees as compared to the criminality of their biological and adoptive parents; such studies consistently show that biological parents have more influence on children's criminal behavior than the adoptive parents who raised them.

Age of Enlightenment: A period of the late 17th century to 18th century in which philosophers and scholars began to emphasize the rights of individuals in society. This movement emphasized the rights of individuals to have a voice in their government and to exercise free choice; it also included the idea of the social contract and other important assumptions that influenced our current government and criminal justice system.

Anomie: A concept originally proposed by Durkheim, which meant "normlessness" or the chaos that takes place when a society (e.g., economic structure) changes very rapidly. This concept was later used by Merton in his strain theory of crime, where he redefined it as a disjunction between the emphasis placed on conventional goals and the conventional means used to achieve such goals.

Atavistic/Atavism: The belief that certain characteristics/behaviors of a person are a throwback to an earlier stage of evolutionary development.

Autonomic nervous system/ANS: The portion of the nervous system that consists of our anxiety levels, such as the "fight or flight" response, as well as our involuntary motor activities (e.g., heart rate). Studies consistently show that lower levels of ANS functioning are linked to criminality.

Bourgeoisie: A class/status that Karl Marx assigned to the dominant, oppressing owners of production, who are considered the elite class due to ownership of companies, factories, and so on. Marx proposed that this group created and implemented laws that helped retain the dominance of this class over the proletariat or working class.

Braithwaite's reintegrative shaming theory: This integrated theoretical model merges constructs and principles from several theories, primarily social control/bonding theory and labeling theory, with elements of strain theory, subculture/gang theory, and differential association theory.

Brutalization effect: The predicted tendency of homicides to increase after an execution, particularly after high-profile executions.

Central nervous system/CNS: The portion of the nervous system that largely consists of the brain and spinal column and is responsible for our voluntary motor activities. Studies consistently show that low functioning of the CNS (e.g., slower brain wave patterns) is linked to criminal behavior.

Cerebrotonic: The type of temperament or personality associated with an ectomorphic (thin) body type; these people tend to be introverted and shy.

Certainty of punishment: Certainty of punishment is one of the key elements of deterrence; the assumption is that when people commit a crime, they will perceive a high likelihood of being caught and punished.

Chicago School of criminology: A theoretical framework of criminal behavior that is often referred to as the Ecological School or the theory of social disorganization; it emphasizes the environmental impact of living in a high-crime neighborhood and asserts that this increases criminal activity. This model applies ecological principles of invasion, domination, and succession in explaining how cities grow and the implications this has on crime rates. Also, this model emphasizes the level of organization (or lack thereof) in explaining crime rates in a given neighborhood.

Classical conditioning: A learning model that assumes that animals, as well as people, learn through associations between stimuli and responses; this model was primarily promoted by Pavlov.

Classical School: The Classical School of criminological theory is a perspective that is considered the first rational model of crime, one that was based on logic rather than supernatural/demonic factors; it assumes that crime occurs after a rational individual mentally weighs the potential good and bad consequences of crime and then makes a decision about whether or not to engage in a given behavior; this model is directly tied to the formation of deterrence theory and assumes that people have free will to control their behavior.

Clearance rate: The percentage of crimes reported to police that result in an arrest, or an identification of a suspect who cannot be apprehended (due to death of suspect, or fleeing, etc.) In other words, the authorities essentially have a very good idea of who committed the crime, so it is considered "solved." The clearance rate can be seen as rough estimate of the rate at which crimes are solved.

Collective conscience: According to Durkheim, the extent of similarities or likeness that people in a society share. The theory assumes that the stronger the collective conscience, the less crime in that community.

College boy: A type of lower-class male youth identified by Cohen who has experienced the same strains and status frustration as his peers but responds to his disadvantaged situation by dedicating himself to overcoming the odds and competing in the middle-class schools despite the unlikely chances for success.

Communitarianism: A concept in Braithwaite's theory of reintegration, which is a macro-level measure of the degree to which the individuals are connected or interdependent on mainstream society (via organizations or groups).

Concentric circles theory: A model proposed by Chicago School theorists; assumes that all cities grow in a natural way that universally has the same five zones (or circles/areas). For example, all cities have a central Zone I, which contains basic government buildings, as well as a Zone II, which was once residential but is being "invaded" by factories. The outer three zones involve various forms of residential areas.

Conceptual integration: A type of theoretical integration in which a theoretical perspective consumes or uses concepts from many other theoretical models.

Concordance rates: Rates at which twin pairs either share a trait (e.g., criminality) or the lack of the trait; for example, either both twins are criminal or neither is criminal; discordant would be if one of the pair is criminal, and the other is not.

Conflict gangs: A type of gang identified by Cloward and Ohlin that tends to develop in neighborhoods with weak stability and little or no organization; gangs are typically relatively disorganized and lack the skills and knowledge to make a profit through criminal activity. Thus, their primary illegal activity is violence, which is used to gain prominence and respect among themselves and the neighborhood.

Conflict theory: Theories of criminal behavior which assume that most people disagree on what the law should be and/or that law is used as a tool by those in power to keep down other groups.

Conformity: In strain theory, an adaptation to strain in which an individual buys into the conventional means of success and also buys into the conventional goals.

Consensual perspective: Theories that assume virtually everyone is in agreement on the laws and therefore assume no conflict in attitudes regarding the laws/rules of society.

Containment theory: A control theory proposed by Reckless in the 1960s, which presented a model that emphasized internal and social pressures to commit crime, which range from personality predispositions to peer influences, as well as internal and external constraints, ranging from personal self-control to parental control, that determine whether an individual will engage in criminal activity. This theory is often criticized as being too vague or general, but it advanced criminological theory by providing a framework in which many internal and external factors were emphasized.

Control-balance theory: An integrated theory originally presented by Tittle, which assumes that the amount of control to which one is subjected, as compared to the amount of control that one can exercise, determines the probability of deviant behavior and the types of deviance that are committed by that individual. In other words, the balance or imbalance between these two types of control can predict the amount and type of behavior likely to be committed.

Control theories: A group of theories of criminal behavior that emphasize the assumption that humans are born selfish and have tendencies to be aggressive and offend and that individuals must be controlled, typically by socialization and discipline, or from internalized self-control that has been developed in their upbringing.

Corner boy: A type of lower-class male youth identified by Cohen who has experienced the same strains and status frustration as others but responds to his disadvantaged situation by accepting his place in society as someone who will somewhat passively make the best of life at the bottom of the social order. As the label describes, they often "hang out" on corners.

Correlation or covariation: A criterion of causality that requires a change in a predictor variable (X) to be consistently associated with some change (either positive or negative) in the explanatory variable (Y). An example would be unemployment (X) being related to criminal activity (Y).

Craniometry: The field of study that emphasized the belief that the size of the brain or skull reflected superiority or inferiority, with larger brains/skulls being considered superior.

Criminal gangs: A type of gang identified by Cloward and Ohlin that forms in lower-class neighborhoods with an organized structure of adult criminal behavior. Such neighborhoods are so organized and stable that their criminal networks are often known and accepted by the conventional citizens. In these neighborhoods, the adult gangsters mentor neighborhood youth and take them under their wing. Such gangs tend to be highly organized and stable.

Criminology: The scientific study of crime and the reasons why people engage (or don't engage) in criminal behavior, as well as the study of why certain trends occur or groups of people seem to engage in criminal behavior more than others.

Critical/radical feminism: A perspective of feminist theory that emphasizes the idea that many societies (such as the United States) are based on a structure of patriarchy, in which males dominate virtually every aspect of society, such as law, politics, family structure, and the economy.

Cross-sectional studies: A form of research design modeling in which a collection of data is taken at one point in time (often in survey format).

Cultural/subcultural theories: A perspective of criminal offending which assumes that many offenders believe in a normative system that is distinctly different than, and often at odds with, the norms accepted by conventional society.

Cytogenetic studies: Studies of crime that focus on the genetic makeup of individuals, with a specific focus on abnormalities in their chromosomal makeup. An example is XYY instead of the normal XX (females) and normal XY (males).

Dark figure: The vast majority of major crime incidents that never get reported to police due to the failure of victims to file a police report; covers most criminal offending, with the exception of homicide and motor vehicle theft, which are almost always reported to police.

Decriminalization: A policy related to labeling theory, which proposes less harsh punishments for some minor offenses, such as the possession of small amounts of marijuana; for example, in California, an offender gets a ticket/fine for this offense rather than being officially charged and prosecuted.

Deinstitutionalization: A policy related to labeling theory; proposes that juveniles or those accused of relatively minor offenses should not be locked up in jail or prison.

Delinquent boy: A type of lower-class male youth identified by Cohen who responds to strains and status frustration by joining with similar others in a group to commit a crime.

Determinism: The assumption that human behavior is caused by factors outside of free will and rational decision making (e.g., biology, peer influence, poverty, bad parenting); it is the distinctive, primary assumption of positivism (as opposed to the Classical School of criminological theory, which assumes free will/free choice).

Deterrence theory: The theory of crime associated with the Classical School, which proposes that individuals will make rational decisions regarding their behavior. This theory focuses on three concepts: the individual's perception of (1) certainty of punishment, (2) severity of punishment, and (3) the swiftness of punishment.

Developmental theories: Perspectives of criminal behavior, which are also to some extent integrated but are distinguished by their emphasis on the evolution of individuals' criminality over time. Specifically, developmental theories tend to look at the individual as the unit of analysis, and such models focus on the various aspects of the onset, frequency, intensity, duration, desistence, and other aspects of the individuals' criminal career.

Deviance: Behaviors that are not normal; includes many illegal acts, as well as activities that are not necessarily criminal but are unusual and often violate social norms, such as burping loudly at a formal dinner or wearing inappropriate clothing.

Differential association theory: A theory of criminal behavior that emphasizes the association with significant others (peers, parents, etc.) in learning criminal behavior. This theory was originally presented by Sutherland.

Differential identification theory: A theory of criminal behavior that is very similar to differential association theory; the major difference is that differential identification theory takes into account associations with persons/images that are presented in the media (e.g., movies, TV, sports, etc.). This model was originally proposed by Glaser.

Differential reinforcement theory: A theory of criminal behavior that emphasizes various types of social learning, specifically classical conditioning, operant conditioning, and imitation/modeling. This theory was originally presented by Burgess and Akers and is one of the most supported theories according to empirical studies.

Diversion: A set of policies related to labeling theory that attempt to get an offender out of the formal justice system as quickly as possible; an offender might perform a service or enter a rehabilitation program instead of serving time in jail or prison. Often, if an offender successfully completes the contract, the official charge/conviction for the crime is expunged or eliminated from the official record.

Dizygotic twins: Also referred to as fraternal/nonidentical twins, these are twin pairs that come from two separate eggs (zygotes) and thus share only 50% of the genetic makeup that can vary.

Dopamine: A neurotransmitter that is largely responsible for "good feelings" in the brain; it is increased by many illicit drugs (e.g., cocaine).

Dramatization of evil: A concept proposed by Tannenbaum in relation to labeling theory, which states that often when relatively minor laws are broken, the community tends to overreact and make a rather large deal out of it ("dramatizing" it). A good example is when a very young offender sprays graffiti on a street sign, and the neighborhood ostracizes that youth.

Drift theory: A theory of criminal behavior in which the lack of social controls in the teenage years allows for individuals to experiment in various criminal offending, often due to peer influence, without the individuals buying into a criminal lifestyle; this theory was introduced by David Matza.

Ecological School/perspective: See Chicago School of criminology.

Ectoderm: A medical term for the outer layer of tissue in our bodies (e.g., skin, nervous system).

Ectomorphic: The type of body shape associated with an emphasis on the outer layer of tissue (ectoderm) during development; these people are disposed to be thin.

Ego: The only conscious domain of the psyche, according to Freud, it functions to mediate the battle between id and superego.

Elliott's integrated model: Perhaps the first major integrated perspective proposed that clearly attempted to merge various traditionally separate theories of crime. Elliott's integrated framework attempts to merge strain, social disorganization, control, and social learning/differential association-reinforcement perspectives for the purpose of explaining delinquency, particularly in terms of drug use, as well as other forms of deviant behavior.

Empirical validity: Refers to the extent to which a theoretical model is supported by scientific research. In criminology, empirical research has consistently supported a number of theories and consistently refuted others.

Endoderm: A medical term for the inner layer of tissue in our bodies (e.g., digestive organs).

Endomorphic: The type of body shape associated with an emphasis on the inner layer of tissue (endoderm) during development; these people are disposed to be obese.

End-to-end theoretical integration: A type of theoretical integration that conveys the linkage of the theories based on the temporal ordering of two or more theories in their causal timing. This means that one theory (or concepts from one theory) precedes another theory (or concepts from another theory) in terms of causal ordering or timing.

Equivalency hypothesis: A "mirror image" tendency, which is the observed phenomenon that virtually all studies have shown; the characteristics, such as young, male, urban, poor, or minority, tend to have the highest rates of criminal offending *and* the highest rates of victimization. This hypothesis has one important exception, namely lower class offenders tend to have higher rates of theft against middle- to upper-class households/individuals).

Eugenics: The study of and policies related to the improvement of the human race via discriminatory control over reproduction.

Experiential effect: The extent to which individuals' previous experience has an effect on their perceptions of how certain or severe criminal punishment will be when they are deciding whether or not to offend again.

Family studies: Studies that examine the clustering of criminality in a given family.

Feeblemindedness: A technical, scientific term in the early 1900s meaning those who had significantly below average levels of intelligence.

Focal concerns: The primary concept of Walter Miller's theory, which asserts that all members of the lower class focus on a number of concepts they deem important: fate, autonomy, trouble, toughness, excitement, and smartness.

Formal controls: Factors that involve the official aspects of criminal justice, such as police, courts, and corrections (e.g., prisons, parole, probation).

Frontal lobes: A region of the brain that is, as its name suggests, located in the frontal portion of the brain; most of the "executive functions" of the brain, such as problem solving, take place here, so it is perhaps the most vital portion of the brain and what makes us human.

General deterrence: Punishments given to an individual are meant to prevent or deter other potential offenders from engaging in such criminal activity in the future.

General strain theory: Although derived from traditional strain theory, this theoretical framework assumes that people of all social classes and economic positions deal with frustrations in routine daily life, to which virtually everyone can relate; includes more sources of strain than traditional strain theory.

Hot spots: Specific locations, such as businesses, residences, or parks, that experience a high concentration of crime incidents; a key concept in routine activities theory.

Hypotheses: Specific predictions that are based on a scientific theoretical framework and tested via observation.

Id: A subconscious domain of the psyche, according to Freud, with which we are all born; it is responsible for our innate desires and drives (such as libido [sex drive]); it battles the moral conscience of the superego.

Imitation and modeling: A learning model that emphasizes that humans (and other animals) learn behavior simply by observing others; this model was most notably proposed by Bandura.

Index Offenses: Also known as Part I offenses, according to the FBI Uniform Crime Report, these are eight common offenses: murder, forcible rape, aggravated assault, robbery, burglary, motor vehicle theft, larceny, and arson. All reports of these crimes, even when they do not result in an arrest, are recorded to estimate crime in the nation and various states/regions.

Informal controls: Factors like family, church, or friends that do not involve official aspects of criminal justice, such as police, courts, and corrections (e.g., prisons).

Innovation: In strain theory, an adaptation to strain in which an individual buys into the conventional goals of success but does not buy into the conventional means for getting to the goals.

Integrated theories: Theories that combine two or more traditional theories into one combined model.

Interdependency: A concept in Braithwaite's theory of reintegration, which is a micro-level measure of the degree to which the individuals are connected or interdependent on mainstream society (via organizations or groups).

Interracial: When an occurrence (such as a crime event) involves people of different race/ethnicity, such as a White person committing crime against a Black person.

Intraracial: When an occurrence (such as a crime event) involves people of the same race/ethnicity, such as a White person committing crime against another White person.

Labeling theory: A theoretical perspective that assumes that criminal behavior increases because certain individuals are caught and labeled as offenders; their offending increases because they have been stigmatized as "offenders." Most versions of this perspective also assume that certain people (e.g., lower class or minorities) are more likely to be caught and punished. Another assumption of most versions is that if such labeling did not occur, the behavior would stop; this assumption led to numerous policy implications, such as diversion, decriminalization, and deinstitutionalization.

Learning theories: Theoretical models that assume criminal behavior of individuals is due to a process of learning from others the motivations and techniques for engaging in such behavior. Virtually all of the variations of the learning perspective propose that the processes involved in a person learning how and why to commit crimes are the same as those involved in learning to engage in conventional activities (e.g., riding a bike, playing basketball).

Legalistic approach: A way of defining behaviors as crime; includes only acts that are specifically against the legal codes of a given jurisdiction. The problem with such a definition is that what is a crime in one jurisdiction is not necessarily a crime in other jurisdictions.

Liberal feminism: One of the areas of feminist theories of crime that emphasizes the assumption that differences between males and females in offending were due to the lack of female opportunities in education, employment, etc., as compared to males.

Life-course persistent offenders: A type of offender, as labeled by Moffitt's developmental theory; such people start offending early and persist in offending through adulthood.

Logical consistency: Refers to the extent to which concepts and propositions of a theoretical model makes sense, in terms of both face value and regarding the extent to which the model is consistent with what is readily known about crime rates/trends.

Low self-control theory: A theory that proposes that individuals either develop self-control by the time they are about age 10 or do not. Those who do not develop self-control will manifest criminal/deviant behaviors throughout life. This perspective was originally proposed by Gottfredson and Hirschi.

Macro-level of analysis: Theories that focus on group or aggregated scores and measures as the unit of analysis, as opposed to individual rates.

Mala in se: Acts that are considered inherently evil and that virtually all societies consider to be serious crimes; an example is murder.

Mala prohibita: The many acts that are considered crimes primarily because they have been declared bad by the legal codes in that jurisdiction; in other places and times, they are not illegal; examples are gambling, prostitution, and drug usage.

Marxist feminism: A perspective of crime that emphasizes men's ownership and control of the means of economic production; similar to critical/radical feminism but distinguished by its reliance on the sole concept of economic structure, in accordance with Marx's theory.

Mechanical societies: In Durkheim's theory, these societies were rather primitive with a simple distribution of labor (e.g., hunters & gatherers) and thus a high level of agreement regarding social norms and rules because nearly everyone is engaged in the same roles.

Mens rea: In legal terms, this means "guilty mind" or intent. This concept involves whether or not offenders actually knew what they were doing and meant to do it.

Mesoderm: A medical term for the middle layer of tissue in our bodies (e.g., muscles, tendons, bone structure).

Mesomorphic: The type of body shape associated with an emphasis on the middle layer of tissue (mesoderm) during development; these people are disposed to be athletic or muscular.

Micro-level of analysis: Theories that focus on individual scores and measures as the unit of analysis, as opposed to group or aggregate rates.

Minor physical anomalies (MPAs): Physical features, such as asymmetrical or low-seated ears, which are believed to indicate developmental problems, typically problems in the prenatal stage in the womb.

Modeling/Imitation: A major factor in differential reinforcement theory, which proposes that much social learning takes place via imitation or modeling of behavior; for example, when adults/parents say "bad" words, their children begin using those words.

Moffitt's developmental theory/taxonomy: A theoretical perspective proposed by Moffitt, in which criminal behavior is believed to be caused by two different causal paths: (1) adolescence-limited offenders commit their crimes during teenage years due to peer pressure (2) life-course persistent offenders commit antisocial behavior throughout life, starting very early and continuing on throughout their lives, because of the interaction between their neuropsychological deficits and criminogenic environments in their upbringing.

Monozygotic twins: Also referred to as identical twins, these are twin pairs that come from a single egg (zygote) and thus share 100% of their genetic makeup.

National Crime Victimization Survey (NCVS): One of the primary measures of crime in the United States; collected by the Department of Justice and the Census Bureau, based on interviews with victims of crime. This measure started in the early 1970s.

Natural areas: The Chicago School's idea that all cities naturally contain identifiable clusters, such as a Chinatown or Little Italy, and neighborhoods that have low or high crime rates.

Negative punishment: A concept in social learning in which people are given a punishment by removing something that they enjoy/like (e.g., taking away driving privileges for a teenager).

Negative reinforcement: A concept in social learning in which people are given a reward by removing something that they dislike (e.g., not being on curfew or not having to do their "chores").

Neoclassical School: Neoclassical School of criminological theory is virtually identical to the Classical School (both assume free will, rationality, social contract, deterrence, etc.), except that it assumes that aggravating and mitigating circumstances should be taken into account for purposes of sentencing/punishing an offender.

Neurotransmitters: Nervous system chemicals in the brain and body that help transmit electric signals from one neuron to another, thus allowing healthy communication in the brain and to the body.

Neutralization theory: A theory of criminal behavior that emphasizes the excuses or neutralization techniques that are used by offenders to alleviate the guilt (or to excuse) their behavior, when they know that their behavior is immoral; this theory was originally presented by Gresham Sykes and David Matza. In their theory, they presented five key "techniques," ways that offenders alleviate their guilt or excuse their behavior, which they know is wrong; since they presented this idea in the 1960s, other techniques have been added, especially regarding white-collar crime.

Non-index offenses: Also known as Part II offenses, more than two dozen crimes that are considered relatively less serious than Index crimes and must result in an arrest to be recorded by the FBI; therefore, the data on such results are far less reliable because the vast majority of reports of these crimes do not result in an arrest and are not in the annual FBI report.

Operant conditioning: The learning model that takes place in organisms (such as humans), based on association between an action and feedback that occurs after it has taken place; for example, a rat running a maze can be trained to run the maze faster based on rewards (reinforcement), such as cheese, as well as punishments, such as electric shocks; introduced and promoted by B. F. Skinner.

Organic societies: In the Durkheimian model, those societies that have a high division of labor and thus a low level of agreement about societal norms, largely because everyone has such different roles in society, leading to very different attitudes about the rules and norms of behavior.

Paradigm: A unique perspective of a phenomenon; has an essential set of assumptions that significantly oppose those of other existing paradigms or explanations of the same phenomenon.

Parsimony: Essentially means "simple"; a characteristic of a good theory, meaning that it explains a certain phenomenon, in our case criminal behavior, with the fewest possible propositions/concepts.

Phenotype: An observed manifestation of the interaction of genotypical traits with the environment, such as height (which depends largely on genetic disposition and diet, as exhibited by Asians or Mexicans who are raised in the United States and thus grow taller).

Phrenology: The science of determining human dispositions based on distinctions (e.g., bumps) in the skull, which is believed to conform to the shape of the brain.

Physiognomy: The study of facial and other bodily aspects to identify developmental problems, such as criminality.

Pluralist (conflict) perspective: A theoretical assumption that instead of one dominant and other inferior groups in society, there are a variety of groups that lobby and compete to influence changes in law; most linked to Vold's theoretical model.

Policy implications: The extent to which a theory can be used to inform authorities about how to address a given phenomenon; in this case, ways to help law enforcement, court, and prison officials reduce crime/recidivism.

Positive punishment: A concept in social learning in which an individual is given a punishment by doing something they dislike (e.g., spanking, time-out, grounding, etc.).

Positive reinforcement: A concept in social learning in which an individual is given a reward by providing something they like (e.g., money, extending curfew, etc.).

Positive School: The Positive School of criminological theory is a perspective that assumes individuals have no free will to control their behavior. Rather, the theories of the Positive School assume that criminal behavior is "determined" by factors outside of free choices made by the individual, such as peers, bad parenting, poverty, or biology.

Postmodern feminism: A perspective that says women as a group cannot be understood, even by other women, because every person's experience is unique; therefore, there is no need to measure or research such experiences. This perspective has been criticized as being "anti-science" and thus, irrelevant; from this perspective, there is no need to investigate any cases because the findings cannot be generalized beyond the specific individual.

Power-control theory: An integrated theory of crime that assumes that in households where the mother and father have relatively similar levels of power at work (i.e., balanced households), mothers will be less likely to exert control on their daughters.

These balanced households will be less likely to experience gender differences in the criminal offending of the children. However, households in which mothers and fathers have dissimilar levels of power in the workplace (i.e., unbalanced households) are more likely to suppress criminal activity in daughters, but more criminal activity is likely in the boys of the household.

Primary deviance: A concept in labeling theory originally presented by Lemert; the type of minor, infrequent offending that people commit before they are caught and labeled as offenders. Most normal individuals commit this type of offending due to peer pressure and normal social behavior in their teenage years.

Proletariat: In Marx's conflict theory, the proletariat is the oppressed group of workers who are exploited by the bourgeoisie, an elite class that owns the means of production; according to Marx, the proletariat will never truly profit from their efforts because the upper class owns and controls the means of production.

Radical (or critical) feminism: see Critical/radical feminism.

Rational choice theory: A modern, Classical School-based framework for explaining crime that includes the traditional formal deterrence aspects, such as police, courts, and corrections and adds other informal factors that studies show consistently and strongly influence behavior, specifically informal deterrence factors (such as friends, family, community, etc.) and also the benefits of offending, whether they be monetary, peer status, or physiological (the "rush" of engaging in deviance).

Reaction formation: A Freudian defense mechanism applied to Cohen's theory of youth offending, which involves adopting attitudes or committing behaviors that are opposite of what is expected, for example, by engaging in malicious behavior as a form of defiance; youth buy into this antinormative belief system so that they will feel less guilt for not living up to the standards they are failing to achieve and so they can achieve status among their delinquent peers.

Rebellion: In strain theory, an adaptation to strain in which an individual buys into the idea of conventional means and goals of success but does not buy into the current conventional means or goals.

Reintegrative Shaming theory: See Braithwaite's Reintegrative shaming theory.

Relative deprivation: The perception that results when relatively poor people live in close proximity to relatively wealthy people. This concept is distinct from poverty in the sense that a poor area could mean that nearly everyone is poor, but relative deprivation inherently suggests that there is a notable amount of wealth and poor in a given area.

Retreatism: In strain theory, an adaptation to strain in which an individual does not buy into the conventional goals of success and also does not buy into the conventional means.

Retreatist gangs: A type of gang identified by Cloward and Ohlin that tends to attract individuals who have failed to succeed in both the conventional world and the criminal or conflict gangs of their neighborhoods. Members of retreatist gangs are no good at making a profit from crime, nor are they good at using violence to achieve status, so the primary form of offending in retreatist gangs is usually drug usage.

Ritualism: In strain theory, an adaptation to strain in which an individual buys into the conventional means of success (e.g., work, school, etc.) but does not buy into the conventional goals.

Routine activities theory: An explanation of crime that assumes that most crimes are committed during the normal daily activities of people's lives; it assumes that crime/victimization is highest in places where three factors come together in time and place: motivated offenders, suitable/attractive targets, and absence of guardian; this perspective assumes a rational offender who picks targets due to opportunity.

Scenario (vignette) research: Studies that involve providing participants with specific hypothetical scenarios and then asking them what they would do in that situation; typically, they are also asked about their perceptions of punishment and other factors related to that particular situation.

Scientific Method: The method used in all scientific fields to determine the most objective results and conclusions regarding empirical observations.

This method involves testing hypotheses via observation/data collection and then making conclusions based on the findings.

Scope: Refers to the range of criminal behavior that a theory attempts to explain, which in our case can be seen as the amount of criminal activity a theory can account for, such as only violent crime or only property crime, or only drug usage; if a theory has a very large scope, it would attempt to explain all types of offending.

Secondary deviance: A concept in labeling theory originally presented by Lemert; the type of more serious, frequent offending that people commit after they get caught and are labeled offenders. Individuals commit this type of offending because they have internalized their status as offenders and often have resorted to hanging out with other offenders.

Selective placement: A criticism of adoption studies, arguing that adoptees tend to be placed in households that resemble that of their biological parents; thus, adoptees from rich biological parents are placed in rich adoptive households.

Self-report data: One of the primary ways that crime data are collected, typically by asking offenders about their own offending; the most useful for examining key causal factors in explaining crime (e.g., personality, attitudes, etc.).

Serotonin: A neurotransmitter that is key in information processing; low levels are linked to depression and other mental illnesses; the neurotransmitter most consistently linked to criminal behavior in its deficiency.

Severity of punishment: One of the key elements of deterrence; the assumption is that a given punishment must be serious enough to outweigh any potential benefits gained from a crime (but not too severe, so that it causes people to commit far more severe offenses to avoid apprehension); in other words, this theoretical concept advises graded penalties that increase as the offender recidivates/reoffends.

Side-by-side (or "horizontal") integration: A type of theoretical integration in which cases are classified by a certain criteria (e.g., impulsive versus planned) and two or more theories are considered

parallel explanations based on what type of case is being considered. Thus, there are two different paths in which a case is predicted to go, typically based on an initial variable (such as low or high self-control).

Social bonding theory: A control theory proposed by Hirschi in 1969 which assumes that individuals are predisposed to commit crime and that the conventional bond that is formed with the individual prevents or reduces their offending. This bond is made up of four constructs: attachments, commitment, involvement, and moral beliefs regarding committing crime.

Social contract: An Enlightenment ideal/assumption that stipulates there is an unspecified arrangement among citizens of a society in which they promise the state/government not to commit offenses against other citizens (to follow the rules of a society), and in turn they gain protection from being violated by other citizens; violators will be punished.

Social Darwinism: The belief that only the beneficial (or "fittest") societal institutions or groups of people survive or thrive in society.

Social disorganization. *See* Chicago School of criminology.

Social dynamics: A concept proposed by Comte that describes aspects of social life that alter how societies are structured and that pattern the development of societal institutions.

Socialist feminism: Feminist theories that moved away from economic structure (e.g., Marxism) as the primary detriment for females and placed a focus on control of reproductive systems. This model believes that women should take control of their own bodies and their reproductive functions via contraceptives.

Social sciences: A category of scientific disciplines or fields of study that focus on various aspects of human behavior, such as criminology, psychology, economics, sociology, or anthropology; typically use the scientific method for gaining knowledge.

Social statics: A concept proposed by Comte to describe aspects of society that relate to stability and social order, which allow societies to continue and endure.

Soft Determinism: The assumption that both determinism (the fundamental assumption of the Positive School of criminology) and free will/free choice (the fundamental assumption of the Classical School) play a role in offenders' decisions to engage in criminal behavior. This perspective can be seen as a type of compromise or "middle-road" concept.

Somotonic: The type of temperament or personality associated with a mesomorphic (muscular) body type; these people tend to be risk-taking and aggressive.

Somatotyping: The area of study, primarily linked to William Sheldon, that links body type to risk for delinquent/criminal behavior. Also, as a methodology, it is a way of ranking body types based on three categories: endomorphy, mesomorphy, and ectomorphy (see other entries).

Specific deterrence: Punishments given to an individual are meant to prevent or deter that particular individual from committing crime in the future.

Spuriousness: When other factors (often referred to as Z factors) are actually causing two variables (X and Y) to occur at the same time; it may appear as if X causes Y, when in fact they are both being caused by other Z factor(s). To account for spuriousness, which is required for determining causality, researchers must ensure that no other factors are causing the observed correlation between X and Y. An example is when ice cream sales (X) are related to crime rates (Y); the Z variable is warm weather, which increases the opportunity for crime because more people and offenders are interacting.

Stake in conformity: A significant portion of Toby's control theory, which applies to virtually all control theories, which refers to the extent to which individuals have investments in conventional society. It is believed, and supported by empirical studies, that the higher the stake in conformity an individual has, the less likely he or she will engage in criminal offending.

Stigmata: The physical manifestations of atavism (biological inferiority), according to Lombroso; he claimed that if a person had more than 5, he/she was a born criminal, meaning that a person or feature of an individual is a throwback to an earlier stage of evolutionary development and inevitably would be a chronic offender. An example is very large ears, or very small ears.

Strain theory: A category of theories of criminal behavior in which the emphasis is placed on a sense of frustration (e.g., economy) in crime causation; hence, the name "strain" theories.

Structural-functional theory: The theoretical framework proposing that society is similar to a living organism, in that all portions of society function together to form a functioning system. Specifically, like a human body, which depends on numerous interrelated organs to survive, a society depends on a variety of vital parts (i.e., societal institutions such as education, economy, family) that work together as a well-functioning system.

Subterranean values: Those norms that individuals have been socialized to accept (e.g., violence) in certain contexts in a given society; an example would be the popularity of boxing or Ultimate Fighting Championship events in American society, even though violence is generally viewed negatively. Another example is the romanticized nature and popularity of crime movies, such as *The Godfather* and *Pulp Fiction*.

Superego: A subconscious domain of the psyche, according to Freud; it is not part of our nature but must be developed through early social attachments; it is responsible for our morality and conscience; it battles the subconscious drives of the id.

Swiftness of punishment: Swiftness of punishment is one of the key elements of deterrence; the assumption is that the faster punishment occurs after a crime is committed, the more an individual will be deterred in the future.

Symbolic interaction: The theoretical perspective proposed by Mead and related to the labeling perspective; proposes that many social interactions involve symbolism, which occurs when individuals interpret each other's words or gestures and then act based on the meaning of those interactions.

Tabula rasa: The assumption that when people are born, they have a "blank slate" regarding morality and that every portion of their ethical/moral beliefs is determined by the interactions that occur in the way they are raised and socialized. This is a key assumption of virtually all learning theories.

Techniques of neutralization: See Neutralization theory.

Telescoping: The human tendency in which events are perceived to occur much more recently in the past than they actually did, causing estimates of crime events to be overreported; in the NCVS measure asking respondents to estimate their victimization over the last six months, respondents often report victimization that happened before the six month cut-off date.

Temporal lobes: The region of the brain located above our ears, which is responsible for a variety of functions; located right above many primary limbic structures, which govern our emotional and memory functions.

Temporal ordering: The criterion for determining causality; requires that the predictor variable (X) precedes the explanatory variable (Y) in time.

Testability: Refers to the extent that a theoretical model can be empirically or scientifically tested through observation and empirical research.

Theoretical elaboration: A form of theoretical integration that uses a traditional theory as the framework for the theoretical model but also adds concepts/propositions from other theories.

Theoretical reduction: *See* Up-and-down integration.

Theory: A set of concepts linked together by a series of propositions in an organized way to explain a phenomenon.

Thornberry's interactional model: This integrated model of crime was the first major perspective to emphasize reciprocal, or feedback effects, in the causal modeling of the theoretical framework.

Trajectory: A path that someone takes in life, often due to the transitions (see Transitions).

Transitions: Events that are important in altering trajectories toward or against crime, such as marriage, employment.

Twins separated at birth studies: Studies that examine the similarities between identical twins who are separated in infancy; research indicates that such twins are often extremely similar even though they grew up in completely different environments.

Twin studies: Studies that examine the relative concordance rates for monozygotic versus dizygotic twins, with virtually every study showing that identical twins (monozygotic) tend to be far more concordant for criminality than fraternal (dizygotic) twins.

Uniform Crime Report (UCR): An annual report published by the FBI in the Department of Justice, which is meant to estimate most of the major street crimes in the United States. It is based on police reports/arrests throughout the nation, and started in the 1930s.

Up-and-down integration: A type of theoretical integration that is generally considered the classic form of theoretical integration because it has been done relatively often in the history of criminological theory development. This often involves increasing the level of abstraction of a single theory so that postulates seem to follow from a conceptually broader theory, such as differential reinforcement theory assuming virtually all of the concepts and assumptions of differential association theory.

Utilitarianism: A philosophical concept that is often applied to social policies of the Classical School of criminology, which relates to the "greatest good for the greatest number."

Viscerotonic: The type of temperament or personality associated with an endomorphic (obese) body type; these people tend to be jolly, lazy, and happy-go-lucky.

Zone in transition: In the Chicago School/social disorganization theory, this zone (labeled Zone II) was once residential but is becoming more industrial because it is being invaded by the factories; this area of a city tends to have the highest crime rates due to the chaotic effect that the invasion of factories has on the area.

Credits and Sources

Introduction to the Book

Photo I.1: © 2009 Jupiterimages Corporation.

Photo I.2: © 2009 Jupiterimages Corporation.

Photo I.3: © Getty Images.

Photo I.4: © Getty Images.

Figure I.1: National Center for Health Statistics. (2002). *Vital statistics.* Washington, DC: U.S. Department of Health and Human Services.

Section I

Photo 1.1: © Getty Images.

Photo 1.2: © Getty Images.

Photo 1.3: © 2009 Jupiterimages Corporation.

Section III

Photo 3.1: © 2009 Jupiterimages Corporation.
Photo 3.2: © Getty Images.
Photo 3.3: © 2009 Jupiterimages Corporation.
Figure 3.1a, b, c: Created by Sandy Sauvajot.

Section IV

Photo 4.1: © 2009 Jupiterimages Corporation.

Photo 4.2: © AP/Worldwide Photos.

Section VI

Figure 6.2: Clifford Shaw and Henry McKay, *Juvenile Delinquency in Urban Areas* (Chicago: University of Chicago Press, 1972), 69. Copyright © The University of Chicago Press.

Section VII

Photo 7.1: Used with permission of the Estate of Donald Cressey.

Section VIII

Photo 8.1: © 2009 Jupiterimages Corporation.

Photo 8.2: © 2009 Jupiterimages Corporation.

Section IX

Photo 9.1: © 2009 Jupiterimages Corporation.

Section X

Figure 10.1: Delbert S. Elliott, David Huizinga, and Suzanne S. Ageton, *Explaining Delinquency and Drug Use* (Beverly Hills, CA: Sage, 1985), p. 66.

Table 10.1: Frank S. Pearson and Neil Alan Weiner, "Toward an Integration of Criminological Theories," *Journal of Criminal Law and Criminology* 76 (1985): 116–50 (Table p. 130). Copyright © 2004–2008 *The Journal of Criminal Law and Criminology*—Northwestern University School of Law.

Section XI

Photo 11.1: © AP/Worldwide Photos.

Photo 11.2: © Associated Press.

Index

About the Authors

Stephen G. Tibbetts is a professor in the Department of Criminal Justice at California State University, San Bernardino. He earned his undergraduate degree in criminology and law at University of Florida and his masters and doctorate degrees at University of Maryland, College Park. For more than a decade, he worked as an officer of the court (juvenile) in both Washington County, Tennessee, and San Bernardino County, California, providing recommendations for disposing numerous juvenile court cases. He has published more than 30 scholarly publications in scientific journals (including *Criminology* and *Justice Quarterly*), as well as several books, all examining various topics regarding criminal offending or policies to reduce such behavior. He was awarded the Outstanding Professional Development Award from the College of Social and Behavioral Sciences at California State University, San Bernardino, in 2009. His recent research interests include developmental and biosocial factors in predicting offending, particularly factors that affect brain function, as well as testing traditional theoretical models of criminal offending and gang intervention/prevention strategies. His most recent book is a co-edited anthology, *American Youth Gangs at the Millennium*, which was given a Choice award by the American Library Association as an Outstanding Academic Title.

Craig Hemmens holds a J.D. from North Carolina Central University School of Law and a PhD in criminal justice from Sam Houston State University. He is the director of the Honors College and a professor in the Department of Criminal Justice at Boise State University, where he has taught since 1996. He has previously served as academic director of the Paralegal Studies Program and chair of the Department of Criminal Justice. He has published 10 books and more than 100 articles on a variety of criminal justice-related topics. His primary research interests are criminal law and procedure and corrections. He has served as the editor of the *Journal of Criminal Justice Education*. His publications have appeared in *Justice Quarterly*, the *Journal of Criminal Justice, Crime and Delinquency*, the *Criminal Law Bulletin*, and the *Prison Journal*.

Supporting researchers
for more than 40 years

Research methods have always been at the core of SAGE's publishing program. Founder Sara Miller McCune published SAGE's first methods book, *Public Policy Evaluation*, in 1970. Soon after, she launched the *Quantitative Applications in the Social Sciences* series—affectionately known as the "little green books."

Always at the forefront of developing and supporting new approaches in methods, SAGE published early groundbreaking texts and journals in the fields of qualitative methods and evaluation.

Today, more than 40 years and two million little green books later, SAGE continues to push the boundaries with a growing list of more than 1,200 research methods books, journals, and reference works across the social, behavioral, and health sciences. Its imprints—Pine Forge Press, home of innovative textbooks in sociology, and Corwin, publisher of PreK–12 resources for teachers and administrators—broaden SAGE's range of offerings in methods. SAGE further extended its impact in 2008 when it acquired CQ Press and its best-selling and highly respected political science research methods list.

From qualitative, quantitative, and mixed methods to evaluation, SAGE is the essential resource for academics and practitioners looking for the latest methods by leading scholars.

For more information, visit **www.sagepub.com**.

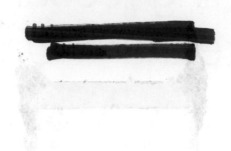